Indoor Environmental Quality and Occupant Comfort

Indoor Environmental Quality and Occupant Comfort

Editors

Yue Wu
Zheming Liu
Zhe Kong

MDPI • Basel • Beijing • Wuhan • Barcelona • Belgrade • Manchester • Tokyo • Cluj • Tianjin

Editors

Yue Wu	Zheming Liu	Zhe Kong
Harbin Institute of Technology	Northeastern University	Southeast University
Harbin, China	Shenyang, China	Nanjing, China

Editorial Office
MDPI
St. Alban-Anlage 66
4052 Basel, Switzerland

This is a reprint of articles from the Special Issue published online in the open access journal *Buildings* (ISSN 2075-5309) (available at: https://www.mdpi.com/journal/buildings/special_issues/Environmental_Comfort).

For citation purposes, cite each article independently as indicated on the article page online and as indicated below:

LastName, A.A.; LastName, B.B.; LastName, C.C. Article Title. *Journal Name* **Year**, *Volume Number*, Page Range.

ISBN 978-3-0365-8184-2 (Hbk)
ISBN 978-3-0365-8185-9 (PDF)

Contents

About the Editors

Yue Wu

Yue Wu (1988-1) is an associate professor at Harbin Institute of Technology (HIT) and a supervisor of doctoral students. Wu has published 2 pieces of scholarly work and 42 academic papers. She also serves as a reviewer for more than ten international and national journals.

Zheming Liu

Zheming Liu (1988-1) is a lecturer and deputy head of the architecture department at Northeastern University, China, and a postdoctoral researcher at the Architectural Design and Research Institute of HIT Co., Ltd. His research interests include urban microclimate, indoor environmental quality, and building energy efficiency.

Zhe Kong

Zhe Kong (1985-4) is a lecturer at Southeast University. She obtained a doctoral degree at the University of Wisconsin-Milwaukee and finished her post-doctoral program at Singapore University of Technology and Design. Her research interests focus on lighting environments of urban and building scales.

Editorial

Indoor Environmental Quality and Occupant Comfort

Yue Wu [1,*], Zheming Liu [2] and Zhe Kong [3]

[1] Key Laboratory of Cold Region Urban and Rural Human Settlement Environment Science and Technology, School of Architecture, Harbin Institute of Technology, Harbin 150001, China

[2] Jangho Architecture College, Northeastern University, Shenyang 110819, China; liuzheming@mail.neu.edu.cn

[3] School of Architecture, Southeast University, Nanjing 210096, China; kongzhe@seu.edu.cn

[*] Correspondence: wuyue@hit.edu.cn

Citation: Wu, Y.; Liu, Z.; Kong, Z. Indoor Environmental Quality and Occupant Comfort. *Buildings* **2023**, *13*, 1400. https://doi.org/10.3390/buildings13061400

Received: 8 May 2023
Accepted: 12 May 2023
Published: 29 May 2023

1. Introduction

Positive indoor environments can improve occupant comfort and well-being by inducing positive perceptual outcomes. Uncomfortable environments, including noise, improper temperature, humidity, dim lighting [1,2], poor air quality, and unpleasant smells [3,4], may impede the quality of life and negatively affect occupants' experiences [5,6]. Furthermore, prolonged exposure to suboptimal indoor environments may lead to adverse changes in individual health conditions. In order to ensure positive indoor environments for occupants, perceptual quality assessment has been introduced and extensively studied in recent years. Influenced by indoor environmental quality (IEQ), it is necessary and beneficial to explore how humans perceive and what effects the environment brings. While there is still a limited understanding of the intrinsic neurological and biological mechanisms of human perception, it is still worthwhile to investigate indoor environmental quality from this perspective.

Despite ongoing research efforts in this area, the underlying mechanisms linking environmental factors and their perceptual effects on users still need to be fully understood. Furthermore, as researchers explore these relationships, additional challenges emerge in terms of psychological and sociological methodologies. It is necessary for researchers and practitioners in built environments to address these issues. Therefore, this Special Issue aims to gather articles that discuss indoor environmental quality and occupant comfort. The articles in this Special Issue encompass different research categories, ranging from conceptual analyses and reviews to research papers. The studies presented here investigate the characterization and perception of both individual indoor environments as well as complex environmental interactions, along with their management and design implications. The focuses of these investigations include both theoretical aspects (including the relationships between environmental quality and psychological or physiological effects) and methodological aspects (including protocols and procedures for gathering objective and subjective data).

2. Research Themes

Considering the broad scope of this Special Issue's call, the topics and research questions addressed by the submissions are diverse. We have identified common themes and clustered all published papers under three categories: (1) design-based optimization of indoor environment performance, (2) comfort evaluation of IEQ, and (3) the impact of IEQ on psychophysiology. These contributions help to advance the scientific conversation concerning these important issues.

2.1. Design-Based Optimization of Indoor Environment Performance

Improving indoor environmental performance and reducing building energy consumption through active and passive design has become an important research field. The articles published in this Special Issue are representative of this research field. Scholars focus on how to effectively improve indoor thermal comfort [7], air quality [8], lighting performance [9–12], and sound insulation performance [13], and reduce building energy consumption by optimizing building equipment [7–9], interior spaces [10–12], and building structures [13]. In addition, the functional types and geographical locations of the research objects are different. Zhang et al. [8] studied a new ventilation system for rural houses in severely cold regions of China in winter. Piraei et al. [10], Ma et al. [11], and Jia et al. [12] explored the optimal design of daylighting performance of heritage buildings in high-latitude areas and classrooms in mid-latitude areas. Qu et al. [13] conducted a study on the sound insulation performance of interior partition structures for a hotel. These studies have provided a reference and evaluation basis for architectural design to improve indoor environmental performance.

2.2. Comfort Evaluations of IEQ

The articles published on this topic are representative of the exploration of comfort evaluations, and researchers have found that the positive design of the environment can significantly improve comfort. They cover a diverse range of occupant types, including studies on the elderly [14], visually impaired individuals [15], and infants [16]. The examined architectural types and geographic locations also vary, including historical residents in Zanzibar [17], school buildings under both severely cold areas in China [18] and mild climate areas in Japan [16], and office buildings in North China and America [19]. These articles have investigated the impact of acoustic [15,20], lighting [14], and thermal [16–19] indoor environmental quality on comfort.

2.3. Impact of IEQ on Psychophysiology

IEQ has significant impacts on occupants, both psychologically and physiologically. Researchers have conducted studies on public buildings, such as offices [21], schools [22,23], shopping malls [24], hotels [25,26], and elderly facilities [27], by using a variety of methods, including field measurements, simulations, behavioral observations, questionnaires, and interviews, as well as tools such as virtual reality (VR) and electroencephalography (EEG). Some interesting findings deserve attention: interior natural lightscapes can influence physiological indicators [21], whereas colors can impact emotional indicators [22,25]. These findings potentially provide supporting data for further developments in the therapeutic effects of indoor environments on health.

3. Concluding Remarks

Although the three themes discussed above do not comprehensively cover all the aspects of IEQ and occupant comfort, they highlight some "hot topics" relevant to researchers. It is worth noting that as a research area examining the impacts of the environment on occupants, this field is still evolving and expanding to a certain extent, particularly with the integration of research methods from fields such as neuroscience, computer science, psychology, and sociology into architectural research. Examples of such methods include virtual reality techniques, Heart Rate Variability (HRV) and EEG measurement, on-site investigation, and behavioral observation. The interdisciplinary nature of these approaches helps researchers to better understand how the built environment affects human perception, as well as how architectural design can support better human experiences. In future research, it is important to engage a wider range of stakeholders in the discussion, including the general public, government, investors, building professionals, designers, and artists.

Conflicts of Interest: The authors declare no conflict of interest.

References

1. Kong, Z.; Utzinger, D.M.; Freihoefer, K.; Steege, T. The impact of interior design on visual discomfort reduction: A field study integrating lighting environments with POE survey. *Build. Environ.* **2018**, *138*, 135–148. [CrossRef]

2.   Kong, Z.; Jakubiec, J.A. Evaluations of long-term lighting qualities for computer labs in Singapore. *Build. Environ.* **2021**, *194*, 107689. [CrossRef]
3.   Wu, Y.; Kang, J.; Mu, J. Assessment and simulation of evacuation in large railway stations. *Build. Simul.* **2021**, *14*, 1553. [CrossRef]
4.   Zhang, S.; Zheng, J.; Wu, Y. Field study of air environment perceptions and influencing factors in waiting spaces of general hospitals in winter cities. *Build. Environ.* **2020**, *183*, 107203. [CrossRef]
5.   Zhang, H.; Liu, Z. Influence of winter natural ventilation on thermal environment of university dormitories under central heating mode in severe cold regions of China. *Archit. Sci. Rev.* **2023**, *66*, 226–241. [CrossRef]
6.   Kong, Z.; Liu, Q.; Li, X.; Hou, K.; Xing, Q. Indoor lighting effects on subjective impressions and mood states: A critical review. *Build. Environ.* **2022**, *224*, 109591. [CrossRef]
7.   Zhu, X.; Liu, J.; Zhu, X.; Wang, X.; Du, Y.; Miao, J. Experimental Study on Operating Characteristic of a Combined Radiant Floor and Fan Coil Cooling System in a High Humidity Environment. *Buildings* **2022**, *12*, 499. [CrossRef]
8.   Zhang, B.; Cai, X.; Liu, M. Study on a New Type of Ventilation System for Rural Houses in Winter in the Severe Cold Regions of China. *Buildings* **2022**, *12*, 1010. [CrossRef]
9.   Budhiyanto, A.; Chiou, Y.-S. Prototyping a Lighting Control System Using LabVIEW with Real-Time High Dynamic Range Images (HDRis) as the Luminance Sensor. *Buildings* **2022**, *12*, 650. [CrossRef]
10. Piraei, F.; Matusiak, B.; Lo Verso, V.R.M. Evaluation and Optimization of Daylighting in Heritage Buildings: A Case-Study at High Latitudes. *Buildings* **2022**, *12*, 2045. [CrossRef]
11. Ma, J.; Yang, Q. Optimizing Annual Daylighting Performance for Atrium-Based Classrooms of Primary and Secondary Schools in Nanjing, China. *Buildings* **2023**, *13*, 11. [CrossRef]
12. Jia, Y.; Liu, Z.; Fang, Y.; Zhang, H.; Zhao, C.; Cai, X. Effect of Interior Space and Window Geometry on Daylighting Performance for Terrace Classrooms of Universities in Severe Cold Regions: A Case Study of Shenyang, China. *Buildings* **2023**, *13*, 603. [CrossRef]
13. Qu, T.; Wang, B.; Min, H. Lightweight Composite Partitions with High Sound Insulation in Hotel Interior Spaces: Design and Application. *Buildings* **2022**, *12*, 2184. [CrossRef]
14. Fu, Y.; Wu, Y.; Gao, W.; Hui, R. The Effect of Daylight Illumination in Nursing Buildings on Reading Comfort of Elderly Persons. *Buildings* **2022**, *12*, 214. [CrossRef]
15. Wu, Y.; Huo, S.; Mu, J.; Kang, J. Sound Perception of Blind Older Adults in Nursing Homes. *Buildings* **2022**, *12*, 1838. [CrossRef]
16. Genjo, K. Assessment of Indoor Climate for Infants in Nursery School Classrooms in Mild Climatic Areas in Japan. *Buildings* **2022**, *12*, 1054. [CrossRef]
17. Liu, C.; Xie, H.; Ali, H.M.; Liu, J. Evaluation of Passive Cooling and Thermal Comfort in Historical Residential Buildings in Zanzibar. *Buildings* **2022**, *12*, 2149. [CrossRef]
18. Zhang, J.; Li, P.; Ma, M. Thermal Environment and Thermal Comfort in University Classrooms during the Heating Season. *Buildings* **2022**, *12*, 912. [CrossRef]
19. Wang, X.; Mu, T.; Zhang, L.; Zhang, W.; Zhang, L. A Simplified Thermal Comfort Calculation Method of Radiant Floor Cooling Technology for Office Buildings in Northern China. *Buildings* **2022**, *12*, 483. [CrossRef]
20. Glean, A.A.; Gatland, S.D., II; Elzeyadi, I. Visualization of Acoustic Comfort in an Open-Plan, High-Performance Glass Building. *Buildings* **2022**, *12*, 338. [CrossRef]
21. Kong, Z.; Hou, K.; Wang, Z.; Chen, F.; Li, Y.; Liu, X.; Liu, C. Subjective and Physiological Responses towards Interior Natural Lightscape: Influences of Aperture Design, Window Size and Sky Condition. *Buildings* **2022**, *12*, 1612. [CrossRef]
22. Tao, W.; Wu, Y.; Li, W.; Liu, F. Influence of Classroom Colour Environment on College Students' Emotions during Campus Lockdown in the COVID-19 Post-Pandemic Era—A Case Study in Harbin, China. *Buildings* **2022**, *12*, 1873. [CrossRef]
23. Hamida, A.; Eijkelenboom, A.; Bluyssen, P.M. Profiling Students Based on the Overlap between IEQ and Psychosocial Preferences of Study Places. *Buildings* **2023**, *13*, 231. [CrossRef]
24. Deng, H.; Xu, Y.; Deng, Y. Is the Shortest Path Always the Best? Analysis of General Demands of Indoor Navigation System for Shopping Malls. *Buildings* **2022**, *12*, 1574. [CrossRef]
25. Xu, J.; Li, M.; Cao, K.; Zhou, F.; Lv, B.; Lu, Z.; Cui, Z.; Zhang, K. A VR Experimental Study on the Influence of Chinese Hotel Interior Color Design on Customers' Emotional Experience. *Buildings* **2022**, *12*, 984. [CrossRef]
26. Xu, J.; Li, M.; Huang, D.; Wei, Y.; Zhong, S. A Comparative Study on the Influence of Different Decoration Styles on Subjective Evaluation of Hotel Indoor Environment. *Buildings* **2022**, *12*, 1777. [CrossRef]
27. Wang, H.; Hou, K.; Kong, Z.; Guan, X.; Hu, S.; Lu, M.; Piao, X.; Qian, Y. "In-Between Area" Design Method: An Optimization Design Method for Indoor Public Spaces for Elderly Facilities Evaluated by STAI, HRV and EEG. *Buildings* **2022**, *12*, 1274. [CrossRef]

 *buildings*

Article

Sound Perception of Blind Older Adults in Nursing Homes

Yue Wu [1,2], Sijia Huo [1,2], Jingyi Mu [1,2,*] and Jian Kang [3,*]

1   School of Architecture, Harbin Institute of Technology, Harbin 150001, China
2   Key Laboratory of Cold Region Urban and Rural Human Settlement Environment Science and Technology, Ministry of Industry and Information Technology, Harbin 150001, China
3   UCL Institute for Environmental Design and Engineering, The Bartlett, University College London (UCL), Central House, 14 Upper Woburn Place, London WC1H 0NN, UK
*   Correspondence: mujingyi@hit.edu.cn (J.M.); j.kang@ucl.ac.uk (J.K.)

Abstract: The number of blind older adults is gradually increasing with the aging of world's population, and their needs and perception of sound are specific. This study investigated the behavioral activities of blind older adults and the dominant sound sources through on-site observation of an all-blind nursing home in China, and it used semi-structured interviews to obtain the sound perceptions of blind older adults. The findings showed that the daily behavioral activities can be categorized into basic living activity, leisure activity, social activity; and physical activity. The dominant sound sources included human, equipment, informational, and environmental sounds. This study developed a sound perception model of blind older adults in nursing homes, which takes three levels: sound requirements, acoustic environment, and sound cognition. Firstly, the blind older adults have a basic understanding of sound from the perspective of their living needs, then they feel the sound environment from the perspective of the living environment, and finally, they perceive the acoustic environment from the dimension of sound cognition in conjunction with contextual memory. This study sheds a light on the aural diversity of older adults, which is expected to support the inclusive design of nursing homes for older adults with visual impairments.

Keywords: blind older adults; nursing homes; acoustic environment; sound perception; aural diversity; inclusive design

Citation: Wu, Y.; Huo, S.; Mu, J.; Kang, J. Sound Perception of Blind Older Adults in Nursing Homes. *Buildings* **2022**, *12*, 1838. https:// doi.org/10.3390/buildings12111838

Academic Editor: Cinzia Buratti

Received: 30 August 2022
Accepted: 20 October 2022
Published: 1 November 2022

Publisher's Note: MDPI stays neutral with regard to jurisdictional claims in published maps and institutional affiliations.

1. Introduction

With at least 2.2 billion people suffering from vision impairment [1], an estimated 1.8 billion suffering from presbyopia [2], and 196 million having age-related macular degeneration, which is expected to increase to 288 million by 2040 [3], vision impairment along with aging has become a global public-health concern. In addition to congenital anomalies and accidents, the cataract, glaucoma, age-related macular degeneration, diabetic retinopathy, and presbyopia are common causes of vision impairment [4]. The number of people with vision impairment is gradually increasing with the increase of aging population, with 43.3 million people worldwide expected to be blind in 2020 [5].

To address the difficulties that visually impaired or blind people have in mobility [6–10], non-verbal sounds and/or speech, such as musical cues, are often used to convey shapes and figures [11,12], because visually impaired people acquire spatial information through hearing to deal with various challenges, especially in unfamiliar environments [13–15]. Miura et al. examined the mobility situation and mobility needs of visually impaired people and found that they could deal with impairment through acquiring auditory information with various strategies, such as rotating the head to hear environmental sound more clearly, or tapping the floor hard with a cane or foot to enhance reflective or reverberant sounds [16]. However, sound has both positive and negative effects on visually impaired people: footsteps and stick echoes can complement acoustic signals, and certain continuous sounds make orientation easier, while noise can interfere with sounds that provide directional guidance [17]. Rychtarikova suggested that the known indoor sound sources are preferred by blind people and argued that the acoustic environment could help people to extract unfamiliar indoor space information, while excessive noise affects positioning [18]. In conclusion, the architectural acoustic environment is significant for people with vision impairment.

A poor indoor acoustic environment can not only harm the health of older adults [19–21], but may also hinder their everyday life in nursing homes [22]. Zeng studied the living environment of older adults in different functional rooms in 11 nursing homes in Guangzhou, China, and found that the acoustic setting had the most significant impact on the subjective evaluations of the older adults among all the physical settings of nursing homes [23]. Some studies, however, found that the acoustic quality of the building environment designed for the general population was largely inappropriate for the disabled and the elderly with hearing loss [24]. Other studies also found that the acoustic environment was the second crucial environmental parameter, just after the light, affecting the behavior and health of dementia patients [25], and a good acoustic environment was vital in helping to delay the onset and progression of Alzheimer's disease [26,27].

Soundscape first proposed in 1929 [28] was later defined as the study of the effects of the acoustic environment on the physical responses or behavioral characteristics of creatures living within it [29]. The research on soundscape was gradually applied to the urban and architectural design, and formally introduced in the 16th International Congress on Acoustics [30]. And International standard ISO 12913-1 (2014) defines soundscape as "[the] acoustic environment as perceived or experienced and/or understood by a person or people, in context" [31]. Due to the important progress of soundscape research in both the natural and social sciences in recent years, many studies have explored the understanding and perception of soundscape in urban construction [32–35] and building type [36–40], and investigated the subjective perception of soundscape in specific populations such as children [41,42], the elderly [43,44], and people with disabilities [45,46]. In terms of the soundscape perception of blind people, Rychtarikova used in-depth interviewing with blind adults to understand their experience of the built environment and discussed the issue of the inclusiveness of soundscape [47], and reviewed the studies on topics related to sound and soundscape perception of blind people, and concluded that blind people perceive the reality also in a multisensory way [18]. Mediastika et al. found that the dimension of eventfulness of park soundscape and the dimensions of pleasantness and space of mall soundscape were the most prominent factors and suggested that the visually impaired used hearing to perceive the danger and direction of the soundscape [48,49]. Hearing differences lead to auditory diversity, and researchers have found differences in hearing levels among infants, adolescents, adults and older adults. We do not yet know how blind older adults perceive sound differently from others?

Generally, blind older adults can only choose to stay at home or live with other older adults in nursing homes [50,51], which brings up the needs to study the inconveniences of the environment for blind older adults. This study examined the only nursing home in China that provides the environmental and service support for blind older adults. As shown in Figure 1, this study first investigated the architectural environment and space of the nursing home, then observed the daily behavioral activities of the blind residents in the nursing home, and finally conducted semi-structured interviews with the blind residents on their sound perceptions, aiming to answer two main questions: 1. How do the blind older adults behave and what are the sound sources in nursing homes for them? 2. How do the blind older adults in nursing homes perceive sounds? This study expects to help nursing homes provide the better environmental support for blind older adults.

Figure 1. Research framework.

2. Materials and Methods

2.1. Field Survey

The all-blind nursing home where the study was conducted is located in Shenyang, Liaoning Province, Northeast China. It has five floors with a total of 96 care units. As shown in Figure 2, the first floor of the building composes of a multifunctional hall, activity rooms, office rooms, and rooms for other purposes, mainly for the daily activities of the blind older adults and the staff, the second to fifth floors of the building are for care units where the blind older adults live. In addition to the rooms for their daily activities, there are other care units for them to engage in social interaction and the corridors for them to stroll during inclement weather. There were about 30 to 50 blind older residents and 12 staff members at the time of study.

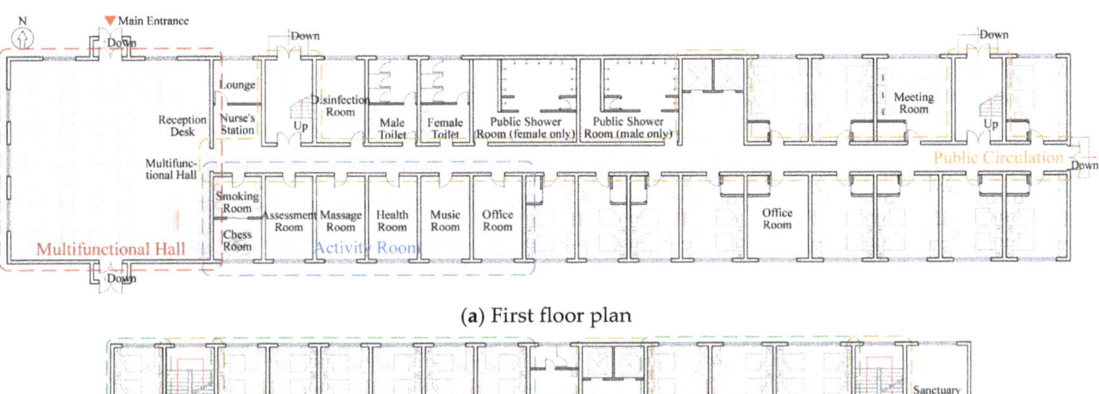

(a) First floor plan

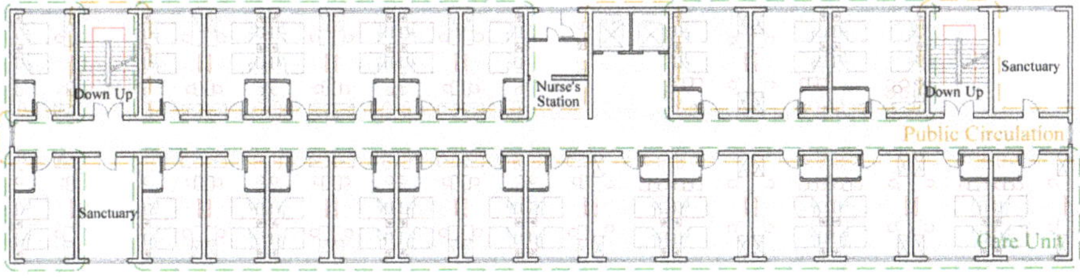

(b) Third floor plan

Figure 2. Floor plan of the nursing home. (Note: ⸺ Multifunctional Hall ⸺ Activity Room ⸺ Public Circulation ⸺ Care Unit).

Image acquisition is an important research method for observing the behavior of the observed [52–56]. To avoid the collision between the camera and the blind older adults, this study was conducted during the period from 5:00 to 20:00 when the residents were awake [21], on 20 July 2021 and 27 December 2021, with typical summer and winter weathers, respectively. The number and type of activities performed by the blind older adults in the six types of spaces—multifunctional hall, activity room, outdoor space, public circulation, and personal and other care units—were monitored and recorded in the nursing home. In addition, during the two days of the study, the researcher circled the nursing home every hour and recorded the sound sources that could be perceived indoors.

2.2. Field Interviews

In this study, the semi-structured interviews were conducted with the blind older adults and the staff members living in the nursing home in terms of the sound perception of the nursing home. The interview questions were:

Q1. What do you think of the acoustic environment of the nursing home?

Q2. What are your sound perceptions in the nursing home in your daily life?

Q3. What sounds do you think need to be added to, or subtracted from, the nursing home?

The researchers took down in shorthand in the memos the participants' answers to the questions, and their own findings and reflections; afterwards the memos were transcribed; finally the transcripts were coded and entered into the Nvivo 11 software for analysis.

The definition of blindness was based on the criteria in Chinese National Standard on Disability Classification and Classification of Persons with Disabilities [57]. Two field interviews were conducted on 21 July and 28 December 2021, followed by the two behavior observations on the next days, respectively. And the data from 37 interviews and the basic information of the participants were collected, as shown in Table 1 and Figure 3. In order to not cause psychological stress to the blind older adults and to reduce their vigilance, the researchers were introduced by the nursing workers into the care units, to explain the purpose of the study and obtain the consent of the participants before conducting the interviews. The interviews were relayed and recorded. In accordance with local legislation and institutional requirements, this study was ethically reviewed by the Harbin Institute of Technology. Afterwards, the interview data were coded, induced, and clustered based on classical grounded theory [58–63] to construct a sound perception model of the blind older adults in nursing homes.

Table 1. Respondent profiles.

Measures	Items	The Blind Older Adults	Staff	All Respondents
Gender	Male	14	3	17
	Female	17	3	20
Age	50–	3	3	6
	50–59	12	3	15
	60–69	7	0	7
	70+	9	0	9

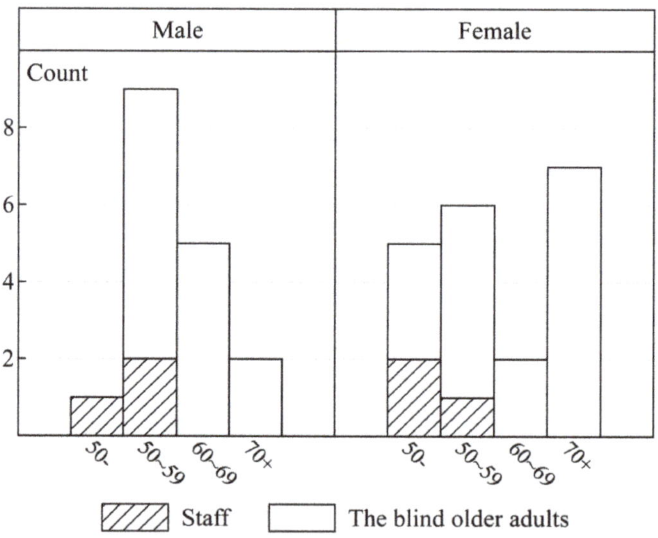

Figure 3. Age and sex distribution of the respondents.

3. Results

3.1. Behavioral activities and Sound Sources

3.1.1. The Behavioral activities of the Blind Older Adults

As shown in Figure 4, the blind older adults living in the nursing home have a consistent schedule for their activities. They usually wake up one after another at 5:00 a.m. Breakfast is served from 7:30 a.m. to 8:00 a.m. Before breakfast, the blind older adults rest in their rooms or stroll in the courtyard. In the morning, they may spend time in activity rooms doing activities, such as singing, dancing, massaging, and playing cards, etc., or in their living rooms for entertainments, such as surfing the Internet, listening to songs, news, playing musical instruments, etc., and may also move to others' rooms to gossip together. After lunch, some blind older adults take a walk outside to digest food, while the others rest in their rooms. After a while, the residents nap in their own care units. From 14:30 to 16:30, most of the blind older adults gather in the multifunctional hall for some group activities, such as performing, playing games, and making dumplings, while the others entertain in their own rooms. After dinner, most of the blind older adults choose to walk, work out, or talk outside, while some remain in their rooms. After 19:00, all the residents retreat to their rooms to rest, and most of the blind older adults take their shower in the public shower room. After 20:00, the nursing home is quiet, for the residents fall sleep one after another. In winter, the cold weather reduces the residents' willingness to go outside, so most of them choose to walk back and forth in the corridors for exercise, wake up later in the morning and spend more time in the care units.

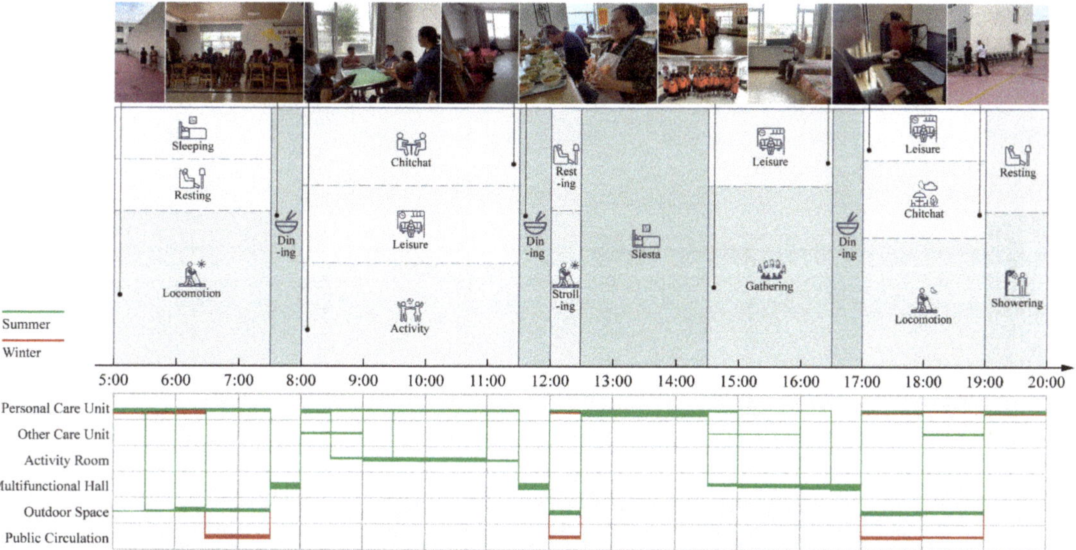

Figure 4. The daily behavior and space utilization of the blind older adults.

3.1.2. Sound Sources in the Nursing Home

Based on the data about the building layout, site conditions, user behavior, and sound sources in the nursing home, the study categorized four main types of sound sources: human sound, equipment sound, informational sound, and environmental sound. Firstly, the human sound refers to the sound generated by the behavioral activities of the blind older adults, such as talking, walking, and doing activities. Secondly, the equipment sound refers to the sound generated by the equipment and machinery in supporting the normal operation of the nursing home and the recreation of the blind residents. Thirdly, the informational sound refers to the sound generated to signal the movement and activities of the blind older adults, for example, a variety of audible messages set up at key points both indoors and outdoors in the nursing home. Finally, environmental sound refers to the sound in the surroundings of the nursing home, which can be heard inside the nursing home, such as traffic and natural sound. The types of sound sources in the nursing home are shown in Figure 5.

As shown in Figure 5a, the indoor sound sources in the nursing home are:

- Talking sound: It is generated in transport, activity and living spaces to meet the living, recreational, and social needs of the blind older adults.
- Walking sound: It is caused by walking with whistles, bells, claps, and other actions to avoid a collision; or moving with crutches and in wheelchairs.
- Activity sound: It is caused by the acts of singing, working out, playing musical instruments, and washing clothes, etc., in addition to the sounds of entertainment in the activity room.
- Prompt sound: It is equipped at the multi-function hall and corridors of the nursing home to deliver notices of meals and gatherings, in addition to the voice in the lifts that indicate the floor.
- Equipment sound: It is the mechanical sound generated by machinery that supports the operation of the nursing home and small appliances for the residents, such as the screen-reading sounds from electronic devices for the blind older adults, and the voice sounds from smart speakers.

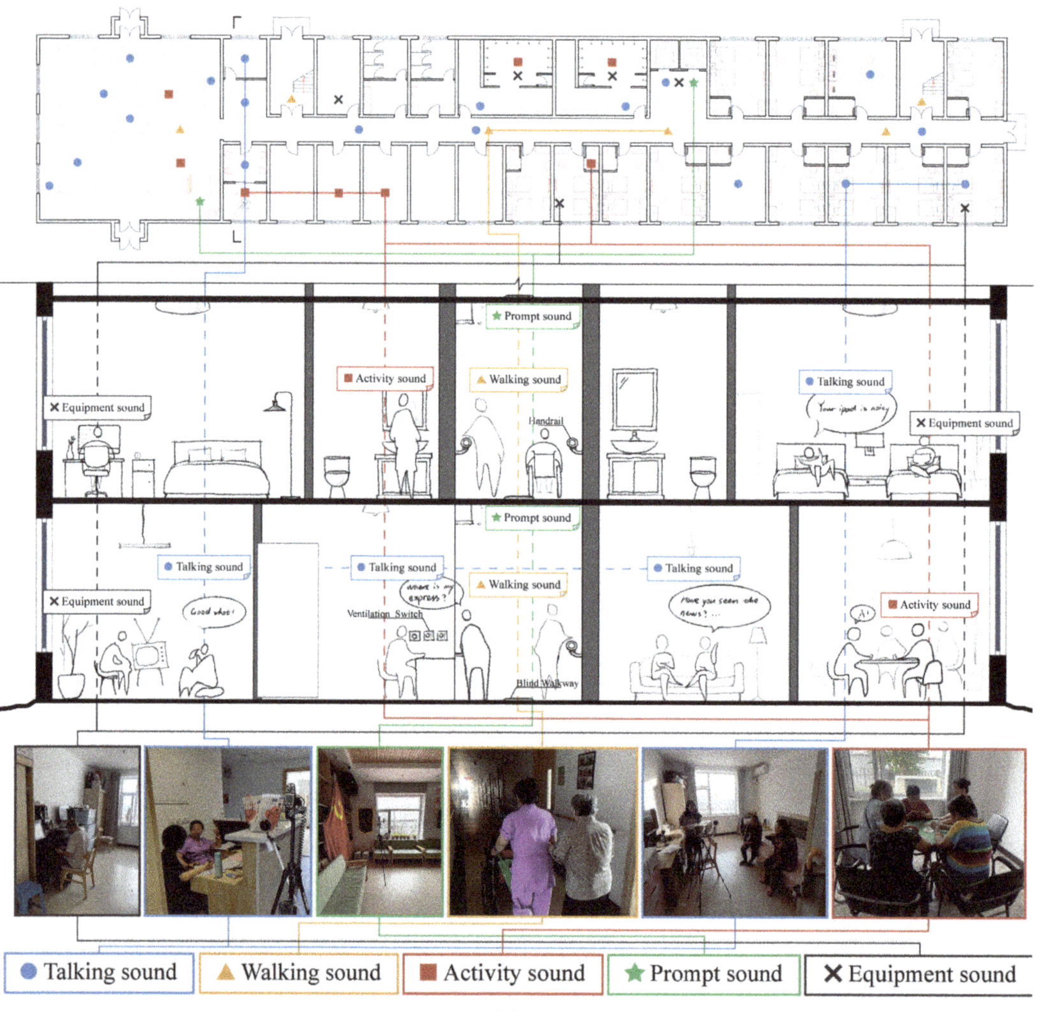

(**a**)

Figure 5. *Cont.*

Figure 5. (**a**) Indoor sound sources in the nursing home; (**b**) outdoor sound sources in the nursing home.

Moreover, as shown in Figure 5b, the outdoor sound sources that can be heard indoors in the nursing home are:

- Talking sound: It refers to the sound of talking when the blind older adults gather together, before breakfast or after dinner.
- Activity sound: it is the sound generated by the walking or working out of the blind older adults outside before and/or after their three meals, when the weather is good.
- Broadcast sound: It is a music broadcast set up next to the waterscape in the northwest of the nursing home; to help the blind older adults orient themselves and count laps.
- Traffic sound: It is generated by moving vehicles and alternating signals, from the western and northern sides of the nursing home which are adjacent to traffic roads.
- Natural sound: It refers to fountain sound on the waterscape platform in the northwest of the nursing home, and the rustling or birdsong from the greenery and trees around the nursing home.

3.2. Sound Perception of the Blind Older Adults

3.2.1. Sound Perception Model Development

1. Open coding

This study did the open coding, which is the process of coding data line by line and conceptualizing and categorizing them, and breaking, crushing, and re-integrating them through continual comparison [64]. The 25 sets of raw data from the first interview were summarized to the level of concepts and categories, as shown in Table 2. The initial extraction of memos was carried out, resulting in 74 labels (aa); then, the 74 labels generated 22 initial concepts (a) through the process of grouping the labels with the same content into one concept; finally the results obtained from the conceptualization were re-refined, resulting in nine initial categories (A).During the iterative process of comparing the categories, three core categories emerged, in terms of the sound perception of the blind older adults in nursing homes: "sound requirements" (A2, A5, A6), "acoustic environment" (A3, A7, A8), and "sound cognition" (A1, A4, A9).

Table 2. Open coding scheme of the sound perception.

Labeling (aa)	Conceptualizing Data (a)	Categorizing Data (A)
Equipment noise	Noisy feeling (aa1, aa19, aa20, aa41, aa56, aa68)	Sound perception (a1, a2)
Outdoor sounds can be heard when windows are opened	Sounds heard (aa2, aa51)	Noise controls (a3, a4, a17)
No outdoor sound can be heard with windows closed	Noise prevention (aa3, aa7, aa8, aa42)	Room assignment (a5, a16)
Affected by talking sound	Noise annoyance (aa4, aa5, aa6, aa38, aa45, aa46, aa52, aa54, aa60, aa62, aa69)	Soundscape memories (a6, a9, a14)
Affected by equipment sound	Acoustic layout (aa9, aa24, aa49)	Lifestyle needs (a7, a13, a15)
Annoyed by the sound of traffic signals	Emotional trigger (aa10, aa59)	Sound functions (a8, a11, a19, a21)
Closing doors can block sound	Sound insulation needs (aa11, aa67)	Loudness design (a10, a18)
Playing sounds to mask other sounds	Recreational means (aa12, aa13, aa16, aa31, aa36, aa37, aa44, aa53, aa55, aa57, aa64)	Sound source settings (a12, a20)
North room noisy	Sound preference (aa14, aa15, aa32, aa30)	Sound context (a22)
Traffic sound could cause fear	Moderate loudness (aa17)	
Good sound insulation may delay calling for help	Wayfinding role (aa18, aa21, aa34, aa35, aa43, aa65)	
Listening to news for entertainment	Sound source needs (aa22, aa52, aa66, aa71)	
Singing for pleasure	Quiet demand (aa23)	
Enjoying birdsong	Psychological dependence (aa25)	
Enjoying the sound of flowing water	Space requirements (aa26)	
Listening to novels for entertainment	Zoning methods (aa27)	
Broadcast sound should be designed to be just audible	Rule constraints (aa28, aa48, aa72)	
Prompt sound for residents who do not know Braille when finding their way	Excessive loudness (aa29, aa41, aa58)	
Others talking loudly	Information transfer (aa39, aa40)	
Others playing loudly	Sound source impact (aa33, aa47)	
Orientation by sound	Avoiding danger (aa50, aa63, aa70)	
. . . . *	Lively atmosphere (aa61)	

* The rest of the labels are detailed in Appendix A, Table A1.

2.   Selective coding

After the core categories were preliminarily determined, the data obtained for the second time underwent selective coding. In the process, only the content related to the core categories of the blind older adults' sound perception in nursing homes was encoded [65]. As shown in Table 3, three new initial concepts were extracted and grouped into the categories of "sound perception" (a1, a2, a23), "sound source setting" (a12, a20, a24), and "sound context" (a22, a25); no new core categories were created.

3.   Theoretical coding

This study also did the theoretical coding, to organize the implicit relationships between the categories formed in the process of substantive coding to build a theory [66]. In this study, three core categories of sound requirements, acoustic environment, and sound cognition were identified in the substantive coding process, and these core categories were found in a recursive relationship: first, the blind older adults obtain a basic understanding of sound from the starting point of their living needs; then they feel the acoustic environment of nursing homes from the perspective of the living environment, and finally; they perceive the acoustic environment from the dimension of sound cognition in combination with contextual memory. The sound perception model of the blind older adults in nursing homes is shown in Figure 6.

Table 3. Selective coding scheme of the sound perception.

Labeling (aa)	Conceptualizing Data (a)	Categorizing Data (A)
Sharp sound will stimulate the heart of old adults	a12. Sound source needs (aa22, aa33, aa52, aa66, aa71, **aa75**, **aa79**, **aa86**) *	A1. Sound perception (a1, a2, **a23**)
Old residents are light sleepers and need quiet at night	a13. Quiet demand (aa23, **aa76**)	A8. Sound source settings (a12, a20, **a24**)
Broadcast sound for positioning and counting laps	a11. Wayfinding role (aa18, aa21, aa34, aa35, aa43, aa65, **aa77**, **aa82**)	A9. Sound context (a22, **a25**)
Staff greet each other with as little noise as possible	a17. Rule constraints (aa28, aa48, aa72, **aa78**)	
Prompt sound should be added to public shower	**a23. Becoming habits (aa80)**	
Getting used to loud noises	a20. Sound source impact (aa47, **aa81**)	
People who do not need a prompt sound will feel noisy	**a24. Seasonal influences (aa83)**	
Positioning wind chimes in canteens for orientation	a2. Sounds heard (aa2, aa51, **aa84**)	
No fountain broadcast in winter	a9. Sound preference (aa14, aa15, aa32, aa30, **aa85**)	
No nature sound can be heard	a6. Emotional trigger (aa10, aa59, **aa87**)	
Enjoying natural sound	a4. Noise annoyance (aa4, aa5, aa6, aa38, aa45, aa46, aa52, aa54, aa60, aa62, aa69, **aa88**)	
Equipment noise cannot be avoided	a22. Lively atmosphere (aa61, **aa89**)	
Annoyed by equipment noise	**a25. Social environment (aa90)**	
Affected by talking sound while resting		
Lively with sound		
The sound of cars means that people are nearby		

* Bolded parts are newly-emerging codes.

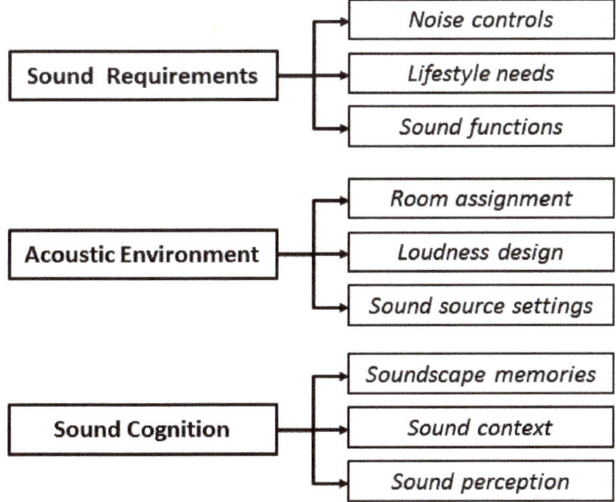

Figure 6. Sound perception model of the blind older adults in nursing homes.

3.2.2. Sound Perception of the Blind Older Adults in Nursing Homes

1. Sound requirements

Participants in the interviews indicated that sound was a vital ingredient in their daily lives, helping them to access information about the language and the environment, to avoid danger, and to be a constant companion and a major means of entertainment. However, sound, for example noise, may interfere with the life and rest of blind older adults. In addition, some participants stated that "I need a quiet environment when I speculate (aa23)", "Blind people make more noise than normal people, so we need a separate space (aa26)", and "We need to hear the outside world to make sure we are in a social environment (aa73)". They also emphasized that "The sound insulation of nursing homes should not be too good, otherwise it will block old residents' calls for help (aa11)". It can be seen that the sound requirements of the blind older adults vary with the demands of life.

2. Acoustic environment

It was found that he acoustic environment influenced the sound perception of blind residents through the room allocation, sound source, and loudness setting in nursing homes. Since the blind older adults relied on the acoustic environment to understand and remember their current living environment [13], nursing homes created sound sources at particular points to convey key messages. For example, some staff members stated, "The sound of music broadcast makes it easy for the residents to determine their position relative to the building as they walk around the outdoor circular walkway, and can help them remember roughly how many laps they have taken (aa77). The music broadcast is not played in the winter when older adults will go out less (aa83)". However, one resident stated that "Broadcast sound is too loud and affects the judgement of orientation (aa47)". Therefore, the sound source setting and loudness design need to complement each other to achieve the best possible delivery. In addition, some participants would like to have "more audible warnings in public showers (aa79), in places difficult to walk (aa22) and when others pass by (aa50)", and some stated that "Some places have prompt sound but I don't know how to turn (aa66)". These statements revealed the problems existing in the set-up of sound sources used to suggest the routes and warn of dangers in the acoustic environment of the nursing home which needs to be optimized.

The allocation of rooms by the managers was also found impact the acoustic environment of the nursing homes. For example, some participants felt that "the eastern part of the home is too noisy". It may be because the east side of the nursing home is close to the traffic road, or may also be due to the managers placing in the east the senior residents who tend to make more noise because of their hearing loss and reduced mobility. However, a noisier environment is not necessarily bad, for some blind residents expressed their preference to a livelier environment. Therefore, conducting relevant investigations would benefit optimizing the layout of the acoustic environment and reasonably allocating rooms for the residents of nursing homes.

3. Sound cognition

The analysis revealed that sound cognition in the nursing homes was related to sound perception, soundscape memory, and sound context in minds of blind older adults. Wang investigated the acoustic demands in facilities for the elderly, and found that the most common unwanted sounds in general among older adults were people talking and other noises [67]. In the current study, most of residents interviewed expressed their understanding and acceptance, even though they often sensed the talking sound of others and felt the noise in the nursing home. And some residents interviewed even considered the sound of people talking a blessing stating that "We chat together in a good atmosphere (aa61)", and "There are sounds that will feel lively (aa89)". Since talking can be used to avoid loneliness [68], talking sound can make blind older adults feel that they are in a social environment.

The subjective preferences, personal history, and life experiences of the blind older adults produce different subjective perceptions of sound, and some perceptions were combined with the scene at the time to form a soundscape memory [35]. However, there are individual differences in soundscape memory. Taking traffic sounds as an example, some residents felt "fearful of traffic sound (aa10)" and "annoyed by the sound of traffic signals (aa6)". However, some reported that "hearing traffic sound makes us feel safe (aa59)", "the sound of cars means that people are nearby (aa90)", or "traffic sound makes me feel that we are living in a busy urban area (aa74)". Therefore, in designing the soundscape to optimize the acoustic environment, the blind residents' sound perception and soundscape memory and the context in which the sound is represented, need to be seriously considered. The reasonable control in the soundscape dimension and the enhanced soundscape design would enhance the health and well-being of the blind older adults living in nursing homes.

4. Discussion

4.1. The Relationship between Behavior, Sound, and Space

According to the behavioral findings in Section 3.1, the daily behavioral activities of blind older adults in nursing homes can be categorized into basic living activity, leisure activity, social activity, and physical activity [56]. As shown in Figure 7, firstly, the basic needs of the blind older adults were met in the care unit (living) and the multifunctional hall (food) in the nursing homes; secondly, their leisure and recreation generally occurred in their personal care units or activity rooms, while social activities mostly occurred in the multifunctional hall, outdoor space, and others' care units; finally, the physical activities occurred in the outdoor space and public circulation of the nursing homes.

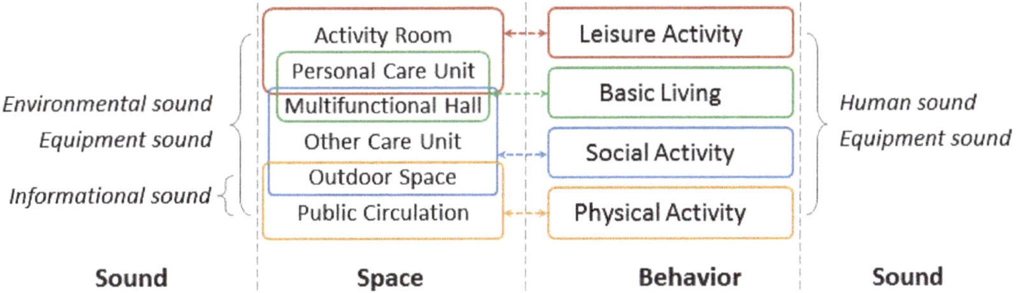

Figure 7. The relationship between behavior, sound, and space in nursing homes.

In this study, human, equipment, and environmental sounds were revealed as common sound sources in nursing homes [52,69–73]. Some of these studies found that staff sound was a major sound source in nursing homes, accounting for, for example, 26% [69] or 34% [72] of the total sound. However, the current study revealed that the staff members at the nursing home had already tried their best to avoid making noise around blind older adults so as to not confuse the information, as one staff member stated that "staff greet each other with as little noise as possible (aa78)". In addition, the nursing homes set up the informational sound at the spatial nodes to provide assistance for the blind older adults, and blind older adults were more dependent on electronic devices for leisure activities. The nursing home that we investigated had more informational and equipment sounds, and fewer staff sounds, than ordinary nursing homes. In conclusion, the daily behavioral activities of the blind older adults generated the massive human sound; the building site environment brought the environmental sound to nursing homes; as such, the normal operation of nursing homes and the equipment and machinery supporting the residents produced equipment sound. And the architectural spaces of the nursing homes helped the blind older adults accomplish the daily life behavioral activities, and these sounds were both indispensable and unavoidable. Meanwhile, exploring the sound perception of the blind older adults in nursing homes helped to understand the impact of these sounds on blind residents.

4.2. Comparison with the Soundscape Framework from ISO 12913-1

Firstly, Figure 8 shows the differences between the sound perception framework of the blind older adults developed in this study and the ISO soundscape framework. On the one hand, sound not only conveys verbal information and carries recreational activities, but also helps blind people orient, remember, and recognize things. It is a complement to the lack of vision [74,75] and a crucial tool for blind people to perceive the external world. On the other hand, sound may also interfere with the daily activities and rest of the blind older adults, such as noise. Noise can cause not only mental health problems, such as irritability, insomnia, and depression, but also physical health problems such as tinnitus, dizziness, heart disease, and cognitive impairment [76]. Miller systematically reviewed the effects of noise on people and suggested that older adults need a less noisy environment for both communication and sleep [19]. Harris found that adding noise into reverberant conditions resulted in significantly poorer speech recognition in older subjects than in younger subjects [77]. Maschke et al. examined the effected of noise on people of different ages and found that the background noise affects the comprehension of older people about conversation [78]. Moreover, the residents may feel annoyed, when exposed to noise, thereby causing various negative emotions. Noise annoyance, as a major effect of noise, was revealed as a multifaceted psychological concept, including behavioral and evaluative aspects [79]. The blind older adults, therefore, need some sound sources to meet their demands, and avoid the interference caused by unwanted sound sources. In contrast to others, the sound requirements of the blind older adults have to be fulfilled. As for the undesirable sounds, from the perspective of the building users, they presented their response to noise, suggesting the importance of noise prevention and rule constraints. The nursing homes could meet the resident needs for noise prevention and control by reducing noise at the sources (e.g., aa42. purchasing silencing devices), blocking means in transmission (e.g., aa7. closing doors and windows), using sound masking at reception (e.g., aa8. playing sounds to mask other sounds), or adding human intervention (e.g., aa28. establishing management rules and aa48. exercising self-regulation). Moreover, from the perspective of designers, they may consider slope construction, plant greening, water features, and noise barriers around the building site, to block environmental noise from the outside and utilize the internal partition walls indoors to isolate equipment noise and materials such as acoustic cotton, acoustic panels, and felt to retrofit care units with poor sound insulation.

Secondly, some of sound requirements of blind older adults can be fulfilled with the acoustic environment of nursing homes. Since different building materials and reverberation times can indicate changes in the surroundings or space for blind residents, designers can help blind older adults understand their location by varying the reverberation time and footstep echoes caused by changes in the space volume, height or floor material, which managers of nursing homes may add to the interior nodes of nursing homes the wall plants or small ornaments such as wind chimes and birdcages, to create beautiful sounds for blind residents and help blind older adults make orientation judgments. In addition, as for acoustic environments, some designers are more focused on controlling the physical parameters of sound, while others intend to use ideal sounds to create a good acoustic environment [61]. The blind older adults put forward their needs for sound and the acoustic environment from the perspective of users, providing a new design direction for designers. As for the acoustic design of the living environment of the elderly or the disabled, especially the visually impaired, the should reasonable room layout, the sound insulation measures and the special needs of users should also be considered.

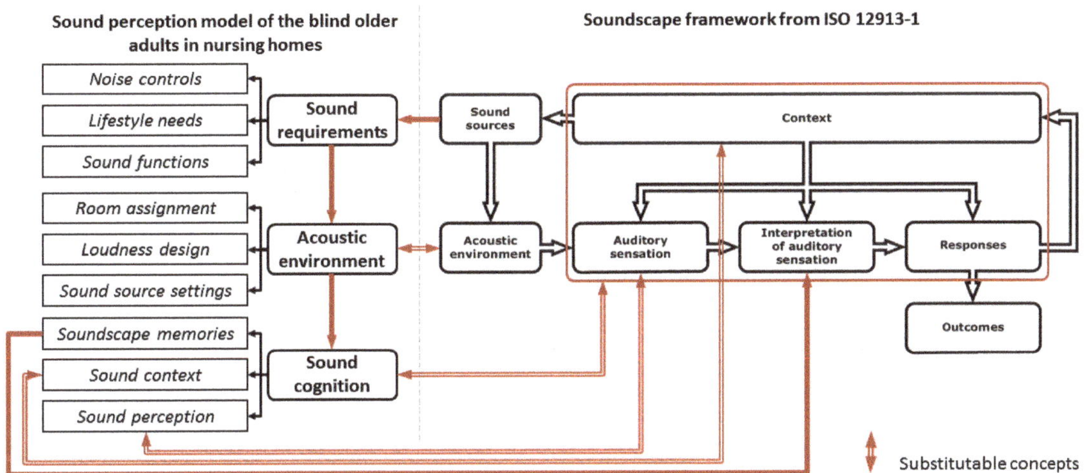

Figure 8. Differences between the sound perception framework of blind older adults and the ISO soundscape framework [31].

Thirdly, the acoustic environment describes the sound and defines the environment from the level of energy, while the soundscape mainly discusses people's perception of sound energy at the psychological level [80]. People's perception of sound depends on the acoustic environment, their attention, current activities, expectations, and prior knowledge [81]. Soundscape represents the relationship between people and the acoustic environment [82]. Similarly, sound perception refers to auditory sensation in ISO's soundscape framework in this study. Since blind people use known information to form mental maps [83] to support action [13,84,85] and hearing is involved in the perception, processing, and evaluation of information along with mental processes [86]; the soundscape memories forming during this mental process become part of the interpretation of auditory sensation. Furthermore, they are influential factors in the soundscape perception of blind older adults along with the sound context. Therefore, designers can add musical fountains, running water and plants to create a soundscape suitable for the site conditions, for the following reasons: the fountain has strong plasticity and can adapt to different sizes, and the running water can produce a signature sound of linear space. Moreover, plants can not only block the noise, but also create a good auditory feeling with wind or rain, and additional bird feed trees can be planted at the peripheral green belt of nursing homes attracts birds, for the songs of bird can increase the layers of natural sound. In sum, these strategies can be applied to enrich the soundscape layout of nursing homes.

4.3. Limitations

First, only one all-blind nursing home was investigated in this study, and thus the study based on this one case may not be comprehensive enough to explore all aspects of the issue. Second, this study focused on the subjective perception of blind older adults, without considering the physical environment, building materials of nursing homes, and the influence of the building interior decoration on the acoustic environment. Future studies may explore the complex relationship between the physical environment and sound perception of blind older adults living in nursing homes.

5. Conclusions

This study explored the relationship between space, behavior, and sound through on-site observations and semi-structured interviews in the only all-blind nursing home in

China, investigating the sound perception of the blind older adults in the nursing home, and developing a model of sound perception.

The daily behavioral activities of blind older adults in the nursing home can be categorized into basic living activity, leisure activity, social activity, and physical activity. The architectural spaces of the nursing homes help the blind older adults accomplish the daily life behavioral activities the sounds of nursing homes are both indispensable and unavoidable. The dominant sound sources that can be perceived inside the nursing home included human sound, equipment sound, informational sound, and environmental sound. The daily behavioral activities of the blind older adults generated massive human sound; the nursing home produced the informational sound to support the mobility of the blind older adults; the normal operations of nursing homes and the equipment and machinery produced equipment sound; the building site environment brought environmental sound into nursing homes. And the blind home had more equipment sound and less staff sound than other types of nursing homes, for it needs informational sound to provide location information for blind residents.

The blind older adults' perception of nursing home acoustics consists of three levels: sound requirements, acoustic environment, and sound cognition. Firstly, the blind older adults highlighted their needs for sound and their opinions about noise from the perspective of their daily lives. Second, the blind older adults evaluated the current situation of the acoustic environment in the nursing home from the perspective of the residential environment and suggest the improvement. Finally, the sound perception experience of the blind older adults was reflected on the dimension of sound cognition, combining the sound context with the residents' experience. The blind older adults, therefore, need sound sources to meet their requirements, and avoid the interference caused by other sound sources. The sound requirements of the blind older adults need to be fulfilled, especially the need for the setting of the informational sound source. In addition, soundscape memory, as part of the interpretation of auditory sensation, was found affect the sound cognition of blind older adults along with the sound context.

This study shows that older adults having aural- diversity experience and auditory impairments/hearing loss affected their perception of the acoustic environment, and confirms that the visual impairment brought about the differences in sound perception and was related to the differences in the daily behavioral activity and sound requirements of the blind older adults. Since sound is a necessary element for them to perceive direction, activity and other people, silence is not necessary for them during the daytime. Therefore, the aural diversity of the older adults places a higher demand on designers and institution managers for inclusion and refinement to better serve the needs of blind older adults with different visual impairments.

Author Contributions: Conceptualization, Y.W., J.M. and J.K.; methodology, Y.W. and J.K.; software, S.H.; validation, Y.W. and J.M.; investigation, S.H. and Y.W.; data curation, S.H.; writing—original draft preparation, S.H.; writing—review and editing, Y.W. and S.H.; supervision, J.K.; funding acquisition, J.M. and J.K. All authors have read and agreed to the published version of the manuscript.

Funding: This work was supported by the Youth Program of National Natural Science Foundation of China (grant number: 52208017), and Ministry of Science and Technology of China (grant number: G2021179030L).

Institutional Review Board Statement: The study was conducted in accordance with the Declaration of Helsinki and was approved by the Ethics Committee of School of Architecture, Harbin Institute of Technology.

Informed Consent Statement: Informed consent was obtained from all subjects involved in the study.

Data Availability Statement: The original data from the study are included in the text and Appendix A; further inquiries can be directed to the respective author.

Acknowledgments: We thank the blind older adults who participated in this study, the staff who helped us survey the nursing home and reach out to participants, and the Harbin Institute of Technology's Visually Impaired Aging Test Team for their contributions to data collection.

Conflicts of Interest: The authors declare no conflict of interest.

Appendix A

Table A1. The rest of the labels in the open coding.

aa22. Add sound in places difficult to walk	aa23. Needing a quiet environment	aa24. Noise in public areas
aa25. Dependence on sound	aa26. Needing a separate space	aa27. Partitioning by demand
aa28. Establishing management rules	aa29. Broadcast sound is too loud	aa30. Do not like the sound too loud
aa31. Chatting in the computer channel for entertainment	aa32. like to listen to sounds	aa33. The external acoustic environment is important
aa34. Sound for direction finding	aa35. Listen to the sound to find the way more convenient	aa36. Like musical instruments
aa37. Likes to play computers	aa38. Human sound has little effect	aa39. Prompt sound does not work after familiarity
aa40. Floor prompt sound is necessary	aa41. The sound of opening and closing the door is loud	aa42. Purchasing silencing devices
aa43. Put the radio to confirm the position	aa44. Listen to book machine entertainment	aa45. Affected by mechanical sound
aa46. Influenced by the sound of others' activities	aa47. Broadcast sound is too loud and affects the judgement of orientation	aa48. Exercising self-regulation
aa49. The eastern part of the home is too noisy	aa50. There should be a prompt sound when people pass by	aa51. There is a talking sound in the morning when the window is opened
aa52. Hope to block the mechanical sound	aa53. Smart speakers' function has defects	aa54. The equipment with the speakers on when opening the door will affect others
aa55. Sound is the main means of entertainment	aa56. Computer screen reading noise	aa57. Daily chat entertainment
aa58. The music broadcast sound is loud	aa59. Hearing traffic sound makes us feel safe	aa60. Traffic sound affects sleep
aa61. Chat together in a good atmosphere	aa62. Wake up at night with sound	aa63. Clap and whistle when I walk
aa64. Chatting with smart speakers	aa65. Sound should be added at the corners	aa66. Some places have prompt sounds but I do not know how to turn
aa67. Poor sound insulation of the door	aa68. The sound of the door opening in the stairwell is particularly noisy	aa69. Affected by walking sound
aa70. Need to add the sound of emergency assembly	aa71. Want more natural sounds	aa72. The equipment with the speakers on will affect others
aa73. Hearing outside sounds to identify myself in the social environment	aa74. Traffic sound makes us feel that we are living in a busy urban area	

References

1. World Health Organization. *World Report on Vision*; World Health Organization: Geneva, Switzerland, 2019.
2. Fricke, T.R.; Tahhan, N.; Resnikoff, S.; Papas, E.; Burnett, A.; Ho, S.M.; Naduvilath, T.; Naidoo, K.S. Global prevalence of presbyopia and vision impairment from uncorrected presbyopia: Systematic review, meta-analysis, and modelling. *Ophthalmology* **2018**, *125*, 1492–1499. [CrossRef] [PubMed]
3. Wong, W.L.; Su, X.; Li, X.; Cheung, C.M.G.; Klein, R.; Cheng, C.Y.; Wong, T.Y. Global prevalence of age-related macular degeneration and disease burden projection for 2020 and 2040: A systematic review and meta-analysis. *Lancet Glob. Health* **2014**, *2*, e106–e116. [CrossRef]
4. Swenor, B.K.; Ehrlich, J.R. Ageing and vision loss: Looking to the future. *Lancet Glob. Health* **2021**, *9*, e385–e386. [CrossRef]

5. Bourne, R.; Steinmetz, J.D.; Flaxman, S.; Briant, P.S.; Taylor, H.R.; Resnikoff, S.; Casson, R.J.; Abdoli, A.; Abu-Gharbieh, E.; Afshin, A.; et al. Trends in prevalence of blindness and distance and near vision impairment over 30 years: An analysis for the Global Burden of Disease Study. *Lancet Glob. Health* **2021**, *9*, e130–e143. [CrossRef]

6. Swenor, B.K.; Lee, M.J.; Varadaraj, V.; Whitson, H.E.; Ramulu, P.Y. Aging with vision loss: A framework for assessing the impact of visual impairment on older adults. *Gerontologist* **2020**, *60*, 989–995. [CrossRef]

7. Biegel, D.E.; Petchers, M.K.; Snyder, A.; Beisgen, B. Unmet needs and barriers to service delivery for the blind and visually impaired elderly. *Gerontologist* **1989**, *29*, 86–92. [CrossRef]

8. Haegele, J.A.; Famelia, R.; Lee, J. Health-related quality of life, physical activity, and sedentary behavior of adults with visual impairments. *Disabil. Rehabil.* **2017**, *39*, 2269–2276. [CrossRef]

9. Blasch, B.B.; Stuckey, K.A. Accessibility and mobility of persons who are visually impaired: A historical analysis. *J. Vis. Impair. Blind.* **1995**, *89*, 417–422. [CrossRef]

10. Laliberte Rudman, D.; Egan, M.Y.; McGrath, C.E.; Kessler, D.; Gardner, P.; King, J.; Ceci, C. Low vision rehabilitation, age-related vision loss, and risk: A critical interpretive synthesis. *Gerontologist* **2016**, *56*, e32–e45. [CrossRef]

11. Alty, J.L.; Rigas, D.I. Communicating graphical information to blind users using music: The role of context. In Proceedings of the SIGCHI Conference on Human Factors in Computing Systems, Los Angeles, CA, USA, 18–23 April 1998; pp. 574–581.

12. Rigas, D.I.; Alty, J.L. The use of music in a graphical interface for the visually impaired. In *Human-Computer Interaction INTERACT'97*; Springer: Boston, MA, USA, 1997; pp. 228–235.

13. Hersh, M. Mental maps and the use of sensory information by blind and partially sighted people. *ACM Trans. Access. Comput. (TACCESS)* **2020**, *13*, 1–32. [CrossRef]

14. Façanha, A.R.; Darin, T.; Viana, W.; Sánchez, J. O&M Indoor Virtual Environments for People Who Are Blind: A Systematic Literature Review. *ACM Trans. Access. Comput. (TACCESS)* **2020**, *13*, 1–42.

15. Basyazici-Kulac, B.; Ito-Alpturer, M. A phenomenological study of spatial experiences without sight and critique of visual dominance in architecture. In Proceedings of the Envisioning Architecture: Design, Evaluation, Communication Conference, Politecnico di Milano, Milan, Italy, 25–28 September 2013; pp. 167–174.

16. Miura, T.; Ebihara, Y.; Sakajiri, M.; Ifukube, T. Utilization of auditory perceptions of sounds and silent objects for orientation and mobility by visually-impaired people. In *2011 IEEE International Conference on Systems, Man, and Cybernetics*; IEEE: Piscataway, NJ, USA, 2011; pp. 1080–1087.

17. Braf, P.G. *The Physical Environment and the Visually Impaired*; ICTA Information Centre: Bromma, Sweden, 1974.

18. Rychtarikova, M. How do blind people perceive sound and soundscape? *Akustika* **2015**, *23*, 1–4.

19. Miller, J.D. Effects of noise on people. *J. Acoust. Soc. Am.* **1974**, *56*, 729–764. [CrossRef] [PubMed]

20. Peris, E.; Fenech, B. Associations and effect modification between transportation noise, self-reported response to noise and the wider determinants of health: A narrative synthesis of the literature. *Sci. Total Environ.* **2020**, *748*, 141040. [CrossRef] [PubMed]

21. Xie, H.; Zhong, B.; Liu, C. Sound environment quality in nursing units in Chinese nursing homes: A pilot study. *Build. Acoust.* **2020**, *27*, 283–298. [CrossRef]

22. *A Report of the 1954 Wisconsin State Fair Hearing Survey by the Research Center of the Subcommittee on Noise in Industry of the Committee on Conservation of Hearing of the American Academy of Ophthalmology and Otolaryngology*; Los Angeles, Research Center: Los Angeles, CA, USA, 1957.

23. Zeng, Y. *The Effect of Living Acoustical Environment on the Speech Communication of the Elderly People*; South China University of Technology: Guangzhou, China, 2018. (In Chinese)

24. Davies, W.J.; Cox, T.J.; Kearon, A.T.; Longhurst, B.J.; Webb, C.L. Hearing loss in the built environment: The experience of elderly people. *Acta Acust. United ACUSTICA* **2001**, *87*, 610–616.

25. Van Hoof, J.H.S.M.; Kort, H.S.M.; Duijnstee, M.S.H.; Rutten, P.G.S.; Hensen, J.L.M. The indoor environment and the integrated design of homes for older people with dementia. *Build. Environ.* **2010**, *45*, 1244–1261. [CrossRef]

26. Wong, J.K.W.; Skitmore, M.; Buys, L.; Wang, K. The effects of the indoor environment of residential care homes on dementia suffers in Hong Kong: A critical incident technique approach. *Build. Environ.* **2014**, *73*, 32–39. [CrossRef]

27. Devos, P.; Aletta, F.; Thomas, P.; Petrovic, M.; Vander Mynsbrugge, T.; Van de Velde, D.; De Vriendt, P.; Botteldooren, D. Designing supportive soundscapes for nursing home residents with dementia. *Int. J. Environ. Res. Public Health* **2019**, *16*, 4904. [CrossRef]

28. Grąno, J.G. Reine geographie. *Acta Geogra.* **1929**, *2*, 1–202.

29. Schafer, R.M. *The Soundscape: Our Sonic Environment and the Tuning of the World*; Simon and Schuster: New York, NY, USA, 1993.

30. Axelsson, Ö. Soundscape revisited. *J. Urban Des.* **2020**, *25*, 551–555. [CrossRef]

31. *ISO 12913-1:2014*; Acoustics—Soundscape—Part 1: Definition and Conceptual Framework. International Organization for Standardization: Geneva, Switzerland, 2014.

32. Arras, F.; Massacci, G.; Pittaluga, P. Soundscape perception in cagliari, Italy. In Proceedings of the Euronoise, Naples, Italy, 19–21 May 2003.

33. Dubois, D.; Guastavino, C.; Raimbault, M. A cognitive approach to urban soundscapes: Using verbal data to access everyday life auditory categories. *Acta Acust. United Acust.* **2006**, *92*, 865–874.

34. Guastavino, C. The ideal urban soundscape: Investigating the sound quality of French cities. *Acta Acust. United Acust.* **2006**, *92*, 945–951.

35. Liu, F.; Kang, J. A grounded theory approach to the subjective understanding of urban soundscape in Sheffield. *Cities* **2016**, *50*, 28–39. [CrossRef]
36. Yu, B.; Kang, J.; Ma, H. Development of indicators for the soundscape in urban shopping streets. *Acta Acust. United Acust.* **2016**, *102*, 462–473. [CrossRef]
37. Skånberg, A.; Öhrström, E. Adverse health effects in relation to urban residential soundscapes. *J. Sound Vib.* **2002**, *250*, 151–155. [CrossRef]
38. Tardieu, J.; Susini, P.; Poisson, F.; Lazareff, P.; McAdams, S. Perceptual study of soundscapes in train stations. *Appl. Acoust.* **2008**, *69*, 1224–1239. [CrossRef]
39. Kang, J.; Zhang, M. Semantic differential analysis of the soundscape in urban open public spaces. *Build. Environ.* **2010**, *45*, 150–157. [CrossRef]
40. Berglund, B.; Nilsson, M.E. On a tool for measuring soundscape quality in urban residential areas. *Acta Acust. United Acust.* **2006**, *92*, 938–944.
41. Nilsson, M.E.; Axelsson, Ö.; Berglund, B. Children's and adults' perception of soundscapes at school. In Proceedings of the EuroNoise, Naples, Italy, 19–21 May 2003.
42. Ma, H.; Su, H.; Cui, J. Characterization of soundscape perception of preschool children. *Build. Environ.* **2022**, *214*, 108921. [CrossRef]
43. Wang, L.; Chen, F.; Li, J.; Wang, J. Perception of Soundscape for the Elderly in Urban Communities: Field Study Based on Three Communities in Changsha, China. In *International Conference on Applied Human Factors and Ergonomics*; Springer: Cham, Switzerland, 2021; pp. 303–312.
44. Wang, R.H. *Strategy for Acoustic Environment Design of Care Facilities for the Elderly Based on Soundscape Memory*; Harbin Institute of Technology: Harbin, China, 2021. (In Chinese)
45. Van den Bosch, K.A.; Andringa, T.C.; Peterson, W.; Ruijssenaars, W.A.; Vlaskamp, C. A comparison of natural and non-natural soundscapes on people with severe or profound intellectual and multiple disabilities. *J. Intellect. Dev. Disabil.* **2017**, *42*, 301–307. [CrossRef]
46. Van den Bosch, K.A.; Andringa, T.C.; Başkent, D.; Vlaskamp, C. The role of sound in residential facilities for people with profound intellectual and multiple disabilities. *J. Policy Pract. Intellect. Disabil.* **2016**, *13*, 61–68. [CrossRef]
47. Rychtarikova, M.; Herssens, J.; Heylighen, A. Towards more inclusive approaches in soundscape research. In *INTER-NOISE 2012. Proceedings-National Conference on Noise Control Engineering, 19–22 August 2012, New York, NY, USA*; Institute of Noise Control Engineering: New York, NY, USA, 2012; Volume 6, pp. 4572–4579.
48. Mediastika, C.E.; Sudarsono, A.S.; Kristanto, L.; Tanuwidjaja, G.; Sunaryo, R.G.; Damayanti, R. Appraising the sonic environment of urban parks using the soundscape dimension of visually impaired people. *Int. J. Urban Sci.* **2020**, *24*, 216–241. [CrossRef]
49. Mediastika, C.E.; Sudarsono, A.S.; Kristanto, L. Indonesian shopping malls: A soundscape appraisal by sighted and visually impaired people. *Archit. Eng. Des. Manag.* **2022**, *18*, 184–203. [CrossRef]
50. Wahl, H.W.; Oswald, F.; Zimprich, D. Everyday competence in visually impaired older adults: A case for person-environment perspectives. *Gerontologist* **1999**, *39*, 140–149. [CrossRef]
51. Horowitz, A. Vision impairment and functional disability among nursing home residents. *Gerontologist* **1994**, *34*, 316–323. [CrossRef]
52. Mu, J.; Kang, J.; Wu, Y. Acoustic environment of comprehensive activity spaces in nursing homes: A case study in Harbin, China. *Appl. Acoust.* **2021**, *177*, 107932. [CrossRef]
53. Meng, Q.; Liu, S.; Kang, J. Effect of children on the sound environment in fast-food restaurants. *Appl. Acoust.* **2020**, *162*, 107201. [CrossRef]
54. Jo, H.I.; Jeon, J.Y. The influence of human behavioral characteristics on soundscape perception in urban parks: Subjective and observational approaches. *Landsc. Urban Plan.* **2020**, *203*, 103890. [CrossRef]
55. Ma, X.; Tian, Y.; Du, M.; Hong, B.; Lin, B. How to design comfortable open spaces for the elderly? Implications of their thermal perceptions in an urban park. *Sci. Total Environ.* **2021**, *768*, 144985. [CrossRef]
56. Han, S.; Song, D.; Xu, L.; Ye, Y.; Yan, S.; Shi, F.; Zhang, Y.; Liu, X.; Du, H. Behaviour in public open spaces: A systematic review of studies with quantitative research methods. *Build. Environ.* **2022**, *223*, 109444. [CrossRef]
57. Standardization Administration of the PRC, Ministry of Civil Affairs of the PRC, CDPF. *GB/26341-2010*; National Standard on Disability Classification and Classification of Persons with Disabilities (implemented in May, 2011). China Standard Publishing House: Beijing, China, 2011.
58. Bollo, C.; Collins, T. The power of words: Grounded theory research methods in architecture & design. *Archit. Complex. Des. Syst. Soc. Environ. J. Proc.* **2017**, 87–94.
59. Lianto, F. Grounded Theory Methodology in Architectural Research. In *Journal of Physics: Conference Series*; IOP Publishing: Bristol, UK, 2019; Volume 1179, p. 012102.
60. Burke, R.L.; Veliz-Reyes, A. Socio-spatial relationships in design of residential care homes for people living with dementia diagnoses: A grounded theory approach. *Archit. Sci. Rev.* **2021**, 1–15. [CrossRef]
61. Liu, X.; Kang, J.; Ma, H.; Wang, C. Comparison between architects and non-architects on perceptions of architectural acoustic environments. *Appl. Acoust.* **2021**, *184*, 108313. [CrossRef]

62. Fiebig, A.; Schulte-Fortkamp, B. Soundscapes and their influence on inhabitants—New findings with the help of a grounded theory approach. *J. Acoust. Soc. Am.* **2004**, *115*, 2496. [CrossRef]
63. Glaser, B.G.; Strauss, A.L. *The Discovery of Grounded Theory: Strategies for Qualitative Research*; Routledge: Chicago, IL, USA, 1967.
64. Glaser, B. *Basics of Grounded Theory Analysis*; Sociology Press: Mill Valley, CA, USA, 1992.
65. Glaser, B.G. *Theoretical Sensitivity*; Sociology Press: Mill Valley, CA, USA, 1978.
66. Glaser, B.G. *Doing Grounded Theory: Issues and Discussions*; Sociology Press: Mill Valley, CA, USA, 1998.
67. Wang, L.; Kang, J. Acoustic demands and influencing factors in facilities for the elderly. *Appl. Acoust.* **2020**, *170*, 107470. [CrossRef]
68. Neves, B.B.; Sanders, A.; Kokanović, R. "It's the worst bloody feeling in the world": Experiences of loneliness and social isolation among older people living in care homes. *J. Aging Stud.* **2019**, *49*, 74–84. [CrossRef]
69. Joosse, L.L. Sound levels in nursing homes. *J. Gerontol. Nurs.* **2011**, *37*, 30–35. [CrossRef]
70. McClaugherty, L.; Valibhai, F.; Womack, S.; Desai, P. Physiological and psychological effects of noise on healthcare professionals and residents in long-term care facilities and enhancing quality of life. *Dir. (Cincinnati Ohio)* **2000**, *8*, 98–100.
71. Aletta, F.; Vander Mynsbrugge, T.; Van de Velde, D.; De Vriendt, P.; Thomas, P.; Filipan, K.; Botteldooren, D.; Devos, P. Awareness of 'sound' in nursing homes: A large-scale soundscape survey in Flanders (Belgium). *Build. Acoust.* **2018**, *25*, 43–59. [CrossRef]
72. Webber, A.; Martin, J.; Alessi, C.; Josephson, K.; Harker, J. Nighttime noise in the post-acute rehabilitation nursing home setting. In *Gerontologist*; Gerontological Society AMER: Washington, DC, USA, 2004; Volume 44, p. 38.
73. Aletta, F.; Botteldooren, D.; Thomas, P.; Vander Mynsbrugge, T.; De Vriendt, P.; Van de Velde, D.; Devos, P. Monitoring sound levels and soundscape quality in the living rooms of nursing homes: A case study in Flanders (Belgium). *Appl. Sci.* **2017**, *7*, 874. [CrossRef]
74. Xu, L.L.; Zhao, X.Y. Auditory strengths, neural mechanisms and educational insights for the blind. *J. Suihua Coll.* **2018**, *38*, 84–89. (In Chinese)
75. Chen, H.Y. Non-visual depth perception—A design study for the blind. *Art J.* **2008**, 62–66. (In Chinese)
76. Berglund, B.; Lindvall, T.; Schwela, D.H.; World Health Organization. *Guidelines for Community Noise*; World Health Organization: Geneva, Switzerland, 1999.
77. Harris, R.W.; Reitz, M.L. Effects of room reverberation and noise on speech discrimination by the elderly. *Audiology* **1985**, *24*, 319–324. [CrossRef]
78. Maschke, C.; Rupp, T.; Hecht, K. The influence of stressors on biochemical reactions-A review of present scientific findings with noise. *Int. J. Hyg. Environ. Health* **2000**, *203*, 45–53. [CrossRef]
79. Guski, R.; Felscher-Suhr, U.; Schuemer, R. The concept of noise annoyance: How international experts see it. *J. Sound Vib.* **1999**, *223*, 513–527. [CrossRef]
80. Baquero Larriva, M.T.; Higueras García, E. Differences in Perceptions of the Urban Acoustic Environment in Older Adults: A Systematic Review. *J. Popul. Ageing* **2021**, 1–33. [CrossRef]
81. Kang, J.; Aletta, F.; Gjestland, T.T.; Brown, L.A.; Botteldooren, D.; Schulte-Fortkamp, B.; Lercher, P.; van Kamp, I.; Genuit, K.; Fiebig, A.; et al. Ten questions on the soundscapes of the built environment. *Build. Environ.* **2016**, *108*, 284–294. [CrossRef]
82. Zhang, M.; Kang, J. Towards the evaluation, description, and creation of soundscapes in urban open spaces. *Environ. Plan. B Plan. Des.* **2007**, *34*, 68–86. [CrossRef]
83. Takamiya, S.; Hamada, S. Information used by visually impaired people while walking. *Transp. Res. Rec.* **1998**, *1636*, 104–109. [CrossRef]
84. Quinones, P.A.; Greene, T.; Yang, R.; Newman, M. Supporting visually impaired navigation: A needs-finding study. In *CHI'11 Extended Abstracts on Human Factors in Computing Systems*; Association for Computing Machinery: New York, NY, USA, 2011; pp. 1645–1650.
85. Watabe, T.; Niitsuma, M. Mental map generation assistance tool using relative pitch difference and angular information for visually impaired people. In *2013 IEEE 4th International Conference on Cognitive Infocommunications (CogInfoCom)*; IEEE: Piscataway, NJ, USA, 2013; pp. 255–260.
86. Majerova, H. The aspects of spatial cognitive mapping in persons with visual impairment. *Procedia-Soc. Behav. Sci.* **2015**, *174*, 3278–3284. [CrossRef]

Article

Subjective and Physiological Responses towards Interior Natural Lightscape: Influences of Aperture Design, Window Size and Sky Condition

Zhe Kong [1,*], Keming Hou [2], Zhongyu Wang [3], Feifei Chen [3], Yunhao Li [2], Xinyue Liu [2] and Chengyu Liu [3]

[1] School of Architecture, Southeast University, Nanjing 210096, China
[2] College of Architecture and Urban Planning, Qingdao University of Technology, Qingdao 266033, China
[3] School of Instrument Science and Engineering, Southeast University, Nanjing 210096, China
* Correspondence: kongzhe@seu.edu.cn

Abstract: Indoor daylighting quality has impacts on occupants' physical and psychological aspects. Although daylighting design metrics have strictly restricted the amount of sunlight penetration, studies have shown occupants' preference towards an appropriate amount of sunlight and distributions. Currently, insufficient studies have focused on the composition of interior daylighting distributions. Therefore, this paper presents a laboratory experiment exploring the psychological influences of sunlight patterns under immersive virtual reality scenes. The sunlight patterns are created by a combination of nine aperture designs, two window sizes and two sky types. The experiment collects 41 valid architecture students' assessments and their physiological responses. Degrees of eight adjectives, including pleasantness, calmness, interest, excitement, complexity, spaciousness, satisfaction with exterior view amount and brightness, are rated by the participants. Physiological data of heart rates and electroencephalogram are collected. According to the analysis, both the aperture designs and sky types have influences upon subjective responses. The large window enhances beta oscillations and beta power on the right prefrontal lobe area, and the clear sky attenuates the theta rhythm on the pre frontal lobe areas. These findings indicate the important influence of natural lightscape compositions created by aperture designs and sky types upon occupants' psychological processes.

Keywords: sunlight patterns; aperture design; immersive virtual reality; EEG; ECG; sky type

Citation: Kong, Z.; Hou, K.; Wang, Z.; Chen, F.; Li, Y.; Liu, X.; Liu, C. Subjective and Physiological Responses towards Interior Natural Lightscape: Influences of Aperture Design, Window Size and Sky Condition. *Buildings* **2022**, *12*, 1612. https://doi.org/10.3390/buildings12101612

Academic Editors: Eusébio Z.E. Conceiçã and Mark Bomber

Received: 10 August 2022
Accepted: 29 September 2022
Published: 5 October 2022

Publisher's Note: MDPI stays neutral with regard to jurisdictional claims in published maps and institutional affiliations.

1. Introduction

Lighting quality, especially daylighting quality, plays an important role within offices [1]. Daylighting quality influences occupants from both biopsychological and psychological processes [2,3]. To create sufficient and comfortable daylighting environments, many studies have explored the performance of daylighting design metrics from the biopsychological processes [4–8]. However, there are still insufficient studies concerning the psychological processes of daylight. With the development of building forms, as well as façade and aperture designs, many buildings provide diverse sunlight patterns and daylighting compositions to individuals [9], which demonstrates the necessity of exploring daylighting psychological influences.

Different interior sunlight patterns and daylighting compositions are influenced by many factors. According to previous studies, spatial properties, window properties and sky types are three primary influencing factors. Spatial properties include spatial dimension and spatial context [10]. Window properties include window dimension [10], window number [11], window shape [12] and aperture design (or façade design) [13]. Sky types include overcast skies and clear skies [10,14–18], varying in sunlight existence and luminance distributions. Interior sunlight patterns and daylighting compositions are composed by a combination of multiple design parameters, defined as natural a lightscape by Wu [19].

Different interior sunlight patterns and daylighting compositions trigger various subjective lighting impressions and mood states [10,14–17,20], which requires further exploration.

Due to the difficulty of controlling sky conditions in the real world, simulation-based methods have commonly been employed. Researchers have utilized simulation software to generate various daylighting distributions under different sky conditions and presented these daylighting scenes by using monitors or projection screens [21,22]. This method solves the problem of controlling sky conditions with the lack of providing 3D spatial perceptions. Another improved method of using virtual reality (VR) scenes to both control sky conditions and provide subjective 3D perception has been rapidly developed and used in the last five years. Although VR headsets have limited ranges of luminance values, valid studies have demonstrated no statistically significant differences of subjective visual assessments of the lighting scenes between VR headsets and in the real world [23,24]. In other words, immersive VR is an effective tool for representing interior lighting scenes with spatial perceptions.

Additionally, given that some previous studies concluded insignificant effects of lighting environments on subjective assessments [25–27], physiological data are recommended to be employed for reflecting participants' physiological reactions at a high level of sensitivity. Therefore, this study utilizes immersive VR to create various natural lightscapes composed of a combination of aperture designs, window sizes and sky types. Subjective assessments and physiological data of both electroencephalogram (EEG) and electrocardiogram (ECG) are collected. The primary research question is to find out whether different daylighting compositions affect subjective responses.

2. Materials and Methods

2.1. Experimental Conditions and Materials

This study aims to explore subjective evaluations and psychological data according to various patterns of sunlight and shadows resulting from a combination of aperture design, window size and sky type. Nine aperture designs derived from architecture projects were selected due to their results of various sunlight patterns. Figure 1 presents the aperture design drawn from the associated architecture projects. As Figure 1 shows, representative aperture designs of five types (Design 1 to Design 8), along with a regular design of façade—curtain walls (Design 6)—were selected for comparison in this study.

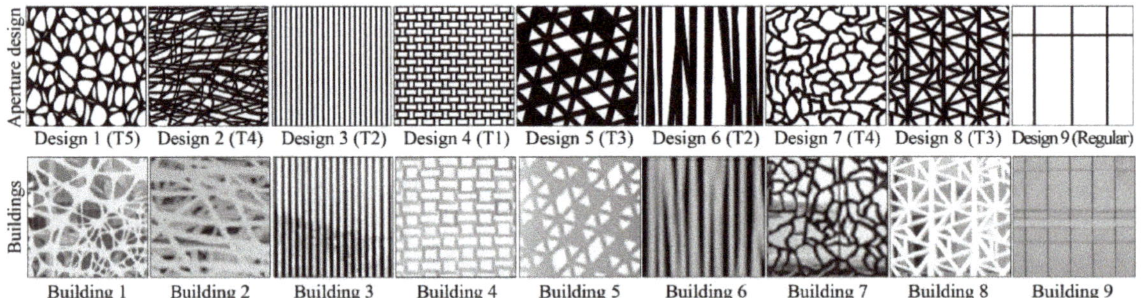

Figure 1. Nine aperture designs (the upper row) and the original buildings (the lower row). Building 1: Tokyo Airspace, Japan; Building 2: Art Wall, Qatar; Building 3: Freshwater House, Australia; Building 4: MAC headquarters, Thailand; Building 5: Panama Diamond Exchange, Panama; Building 6: Selcuk Ecza Headquarters, Turkey; Building 7: MuCEM, France; Building 8: Nakara Residential Hotel, France; Building 9: BMW Experience Center, Chengdu, China.

All nine aperture designs were employed in the same open-plan office. Figure 2 illustrates the office layout and section, the spatial dimensions and aperture dimensions for the large window. A mezzanine floor containing 12 workstations was designed to increase spatial richness. A hallway with the exact window design as the office was added to prolong the participants' walk. The red dot at the left end of the hallway on the layout demonstrates the starting point where a participant stood in each scene. The 3D models were generated in SketchUp [28] and rendered in Enscape, which integrates building information modelling (BIM) and visualization and converts 3D models to Virtual Reality (VR) environments [29]. Interior materials were determined based on the material properties of regular classrooms on the Sipai Lou campus at Southeast University, which architecture undergraduate and graduate students are familiar with. The open-plan office was embedded into an urban environment to enhance the spatial realism.

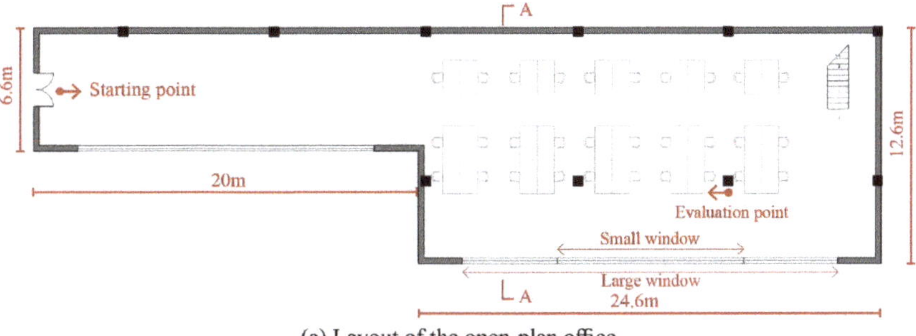

(a) Layout of the open-plan office

(b) A-A section perspective

Figure 2. (**a**) Layout and (**b**) Section perspective of the open-plan office.

2.2. Experimental Design

This study followed a mixed $9 \times 2 \times 2$ full factorial design. The design consisted of the within-subjects and between-subjects factor of aperture design (seven aperture design variations for each participant), as well as the between-subjects factors of window size (a small window of 10 m in width and 3.5 m in height and a large window of 20 m in width and 3.5 m in height) and sky type (overcast sky and clear sky with low sun angle). The size and location of both small and large windows are illustrated by the red lines on the layout of the open-plan office, Figure 2. A combination of aperture designs, window sizes and sky types produced 36 unique scenes. All four scenes of aperture design one (D1) are shown in Figure 3. All nine aperture designs with the large window under the clear sky are shown in Figure 4. An HTC Vive Pro headset was used to present the stimuli. Due to current technical limitations, the maximum luminance of the device measured at the level of the lens is 92 cd/m^2. The headset used an AMOLED display with a resolution of 1440×1600 pixels per eye, a refresh rate of 90 Hz and a 110° diagonal field of view [30].

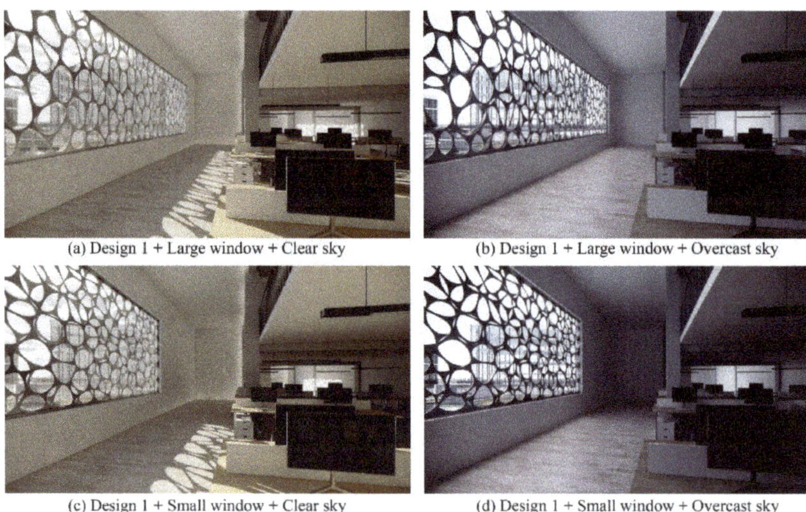

(a) Design 1 + Large window + Clear sky

(b) Design 1 + Large window + Overcast sky

(c) Design 1 + Small window + Clear sky

(d) Design 1 + Small window + Overcast sky

Figure 3. Four scenes for Design One with the combination of window size and sky type.

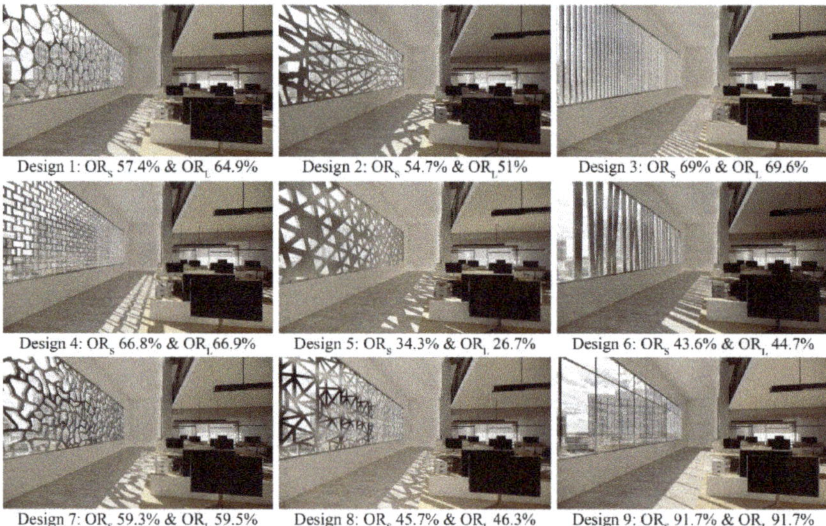

Figure 4. Nine aperture designs with the large window under the clear sky (OR$_S$ represents the openness ratio for the small window, and OR$_L$ represents the openness ratio for the large window).

2.3. Measures

In this study, self-reported assessments were collected by digital surveys. Psychological data, including electroencephalogram (EEG) and electrocardiogram (ECG), were collected.

2.3.1. Subjective Questionnaires

Each participant needed to complete three questionnaires: a start-up questionnaire, an evaluation questionnaire and an ending questionnaire. Table 1 lists all the questions for the three questionnaires. The start-up questionnaire contained the demographic information of gender, age, eyesight, grade and major, as well as the subjective current state of fatigue and alertness (Questions 4 and 5 in the Ending questionnaire of Table 1). The evaluation questionnaire contained eight questions followed Moscoso and Chamilothori's studies, which explored subjective assessments of daylighting patterns through VR scenes [10,14–17]. The ending questionnaire asked participants to report their current state of eye soreness, vision clearness, mind freshness, fatigue and alertness after the experiment. All the questions were designed on an 11-scale from 0 (Not at all) to 10 (Very) except for Karolinska Sleepiness Scale (KSS), which used a 9-point scale [31]. Given that each participant evaluated seven scenes and answered the evaluation questionnaire seven times, the eight questions were asked in a random order.

Table 1. The evaluation questionnaire and ending questionnaire.

Evaluation questionnaire		
No.	Theme	Question
1	Pleasantness	How pleasant is this space?
2	Calmness	How calming is this space?
3	Interest	How interesting is this space?
4	Excitement	How exciting is this space?
5	Complexity	How complex is this space?
6	Spaciousness	How spacious is this space?
7	Satisfaction with views	How satisfied are you with the amount of outside view?
8	Brightness	How bright is this space?
Ending questionnaire		
No.	Theme	Question
1	Eye soreness	How sore do your eyes feel?
2	Vision clearness	How clear is your vision?
3	Mind freshness	How fresh does your head feel?
4 *	Fatigue *	How fatigued do you feel?
5 *	KSS *	The Karolinska Sleepiness Scale
6	Open-ended question	Do you have any other symptoms or suggestions?

* Questions included in both start-up and ending questionnaires.

2.3.2. Physiological Measures

A heart rate sensor developed and validated by Liu et al. [32] was set at 500 Hz for collecting heart rate variability (HRV). Before the experiment, the sensor was stuck on the chest. The raw data stream was preprocessed with the BioSPPy Python software [33], and the parameters of time–domain mean and median of beat interval, denoted as $Mean_{NN}$ and $Median_{NN}$, were extracted for each participant evaluating each scene. Moreover, an EEG device, developed and validated by Liu et al. [34], was set at 256 Hz and positioned FP1 and FP2 channels on pre frontal lobe areas for collecting EEG data. Following the data processing steps in [35], EEG data were processed using the EEGLAB plugin [36] for MATLAB 2022 [37]. To exclude the influences of participant movements, thinking and speaking, only the EEG data of 40 s while participants standing still within each scene were extracted for further analysis. Apparent spikes or oscillations outside of the normal range were discarded. Since the AC working frequency caused the EEG spectrum to sag at 50 Hz, signals between 1 Hz and 48 Hz were removed. EEG signal noises such as eye movement artifacts, ECG artifacts and skin artifacts were also moved. The time domain signal was converted to the frequency domain by Fast Fourier Transform (FFT), which were separated into four bands: δ (0.5 Hz~4 Hz), θ (4 Hz~8 Hz), α (8 Hz~12 Hz) and β (12 Hz~30 Hz). Representative values of the time domain means ($Mean_{\delta}$, $Mean_{\theta}$, $Mean_{\alpha}$ and $Mean_{\beta}$), time domain standard deviations (SD_{δ}, SD_{θ}, SD_{α} and SD_{β}) and the average power of the EEG signal ($PowerMean_{\delta}$, $PowerMean_{\theta}$, $PowerMean_{\alpha}$ and $PowerMean_{\beta}$) of each participant's left and right FP channels of each scene were extracted.

2.4. Experimental Protocol

The experiment was conducted between the end of April and the end of May, lasting around a month in 2022. The experiment was conducted in the lighting lab, which is located on the second floor, at the northeast corner of Zhongda Yuan. Indoor air temperature and relative humidity were monitored throughout the entire data collection. Throughout the entire experiment, the indoor air temperature varied between 22.3 °C and 26.6 °C, with a mean of 23.7 °C and a standard deviation (SD) of 1.3 °C. The relative humidity varied between 41.7% and 61.8%, with a mean of 50.5% and a SD of 7%. The variations of indoor air temperature and relative humidity satisfied occupants' basic requirement [38].

Figure 5 illustrates the experimental protocol for each participant. After entering the lab, the participant was introduced to the whole experimental procedure. Once the participant agreed to contributing to this study, a consent form was signed. The participant first filled out the start-up questionnaire and then wore the psychological instruments with an investigator's assistance. Once the investigator ensured the stability of the EEG and ECG, the participant was assisted to wear the HTC Vive pro headset. After wearing the headset, the participant stayed still for 30 s to ensure that the psychological data was stable.

-8	-5	-1	0	Scene 1	4	16	20	32	34
Welcome introduction & Consent form & Start–up questionnaire	Wear physiological instrument & VR headset	Chromatic adaptation	Practice in VR	30-second Exploration, 40–second Standing still & 80–second Verbal questionnaire	Scene 2, 3, 4		Rest	Scene 5, 6, 7	Ending questionnaire & Gift card
Preparation		**Exercises**		**Seven VR Scenes**					**Ending**

Figure 5. VR experiment protocol.

Before presenting the first scene, a single color was displayed in VR for 15 s to ensure chromatic adaptation. When presenting the first scene, each participant was told to imagine that they were working in this open-plan office. Participants entered at the starting point (the starting point on the layout of Figure 2). They were asked to freely explore each scene for 30 s, including walking within the office or going upstairs to the mezzanine floor. Participants used controllers to move forward in each scene and slowly turned around if necessary. After a 30 s walk, participants were instructed to stand on the predetermined evaluation spot (the evaluation point on the layout of Figure 2) and stand still to look towards the entrance to ensure that their field of view was composed of half window view and half interior office. Figures 3 and 4 present the associated field of view when a participant stood at the evaluation point. After standing still for 40 s, they verbally responded to the eight questions of the evaluation questionnaire while remaining immersed in the VR environment. Once the evaluation questionnaire was completed, the participant would wait for the second scene. The participant repeated the same procedure in the second scene as in the first scene. Each scene took around 2.5 min. Figure 6 shows the experimental scenes with six of the participants.

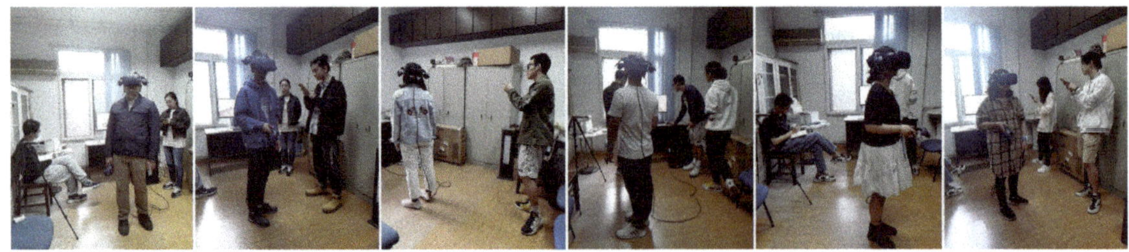

Figure 6. VR experimental scenes with six random participants.

In order to control the experimental duration to minimize subject fatigue, each participant only evaluated seven scenes consisting of a combination of seven aperture designs, two window sizes and two sky types in random order. After completing four scenes, participants were asked to either take a 2 min break or carry on to complete the remaining three scenes. The entire experiment for each participant lasted between 43 and 55 min. The investigator observed the psychological data throughout each participant's entire experiment.

2.5. Participants

Since participants with or without design background evaluated aperture designs differently [39], this study merely concentrated on participants with at least four years of architecture design experience. In other words, first year to third year undergraduates majoring in architecture, as well as undergraduates and graduates majoring in urban planning and landscape, were not recruited.

From an initial sample size of 44 participants, two were excluded as they demonstrated Virtual Reality (VR) sickness during the experiment, and one was excluded due to the controllers being out of power. The resulting sample size thus corresponded to 41 participants, with 21 males and 20 females. Participant age ranged from 21 to 33 years with a mean of 25.2 years and a SD of 3.2 years. All 41 participants majoring in architecture were right-handed. Table 2 lists the average eyesight of each participants' left and right eyes, as well as their grades.

Table 2. Eyesight and grade information of 41 participants.

Eyesight		Grade	
Diopters	Count	Grade	Count
Normal vision	3	Fourth- or fifth-year undergraduates	10
Mild myopia < −3.00 diopters	17	Master students	25
Moderate myopia −3.00 to −6.00 diopters	17	Ph.D. students	6
High myopia −6.00 to −9.00	4		

Given that the aperture design nine (D9) represented a regular –wall design, it was assessed by all participants. Therefore, each participant evaluated six aperture designs plus D9. In the end, D1 to D4, along with D6 and D7 were evaluated by 31 participants; D5 and D8 were evaluated by 30 participants; D9 was evaluated by 41 participants. Large windows and small windows were assessed 143 and 144 times, respectively. The overcast sky and the clear sky were assessed 145 and 142 times, respectively. In order words, all levels of each independent variable were evaluated by a relatively equal number of participants unless it was designed in particular (D9).

2.6. Statistical Analysis

Since this experiment mixed within–between design with repeated measures for the factor of aperture design, a Linear Mixed Model (LMM) was used, which has been employed by previous studies [14,40]. The identification number of each participant was specified as a random factor, and the aperture design, window size and sky type were specified as fixed effects. The interactive effects between two factors were also explored: aperture design × window size, aperture design × sky type and window size × aperture design. Potential confounding factors, including individual differences (gender, age and eyesight) and experimental procedure (scene order and the first aperture design) were added as covariates. All the factors were analyzed as ordinal variables. A factor was removed from the model whenever it was not significant. In this paper, aperture design, window size and sky type were the three fixed factors. The lightscape scene presentation order, gender and age were the three covariates. Additionally, the three studied interactions were aperture design × window size, aperture design × sky type and window size × sky type. Therefore, a Bonferroni-corrected significance level $\alpha' = 0.05/((3 + 3 + 3) \times 8) = 0.0007$ was used to account for the multiple analyses, since there were eight questions. In other words, any *p*-value greater than 0.0007 failed to reject the null hypothesis.

3. Results

3.1. Effects of Aperture Design, Window Size and Sky Type on Subjective Responses

The interaction of design × window size failed to meet the adjusted significance threshold (all p-values > 0.0045). The interaction of design × sky was only significant for impressions of brightness (F(8, 247.26) = 3.64, p = 0.00051). Additionally, gender was insignificant for all eight attributes, and age was only significant for impressions of calmness (F(8, 39.582) = 10.733, p = 0.002194). Therefore, all interaction terms and gender were excluded from the final models.

Table 3 presents an overview of the LMM analyses for the main factor of interest in the final models. Grey cells indicate that p-values are greater than 0.0045, and the correlations between the associated predictors and attributes are insignificant. Significant effects of aperture design are found for six attributes except for pleasantness. Significant effects of sky type are also found for seven attributes except for satisfaction with views. Detailed LMM results concerning each factor will be discussed below.

Table 3. Results of the LMM analysis for the main factors of aperture design, window size and sky type on all subjective responses.

Predictor	Attribute	df	F	p-Value
	Pleasantness	229.43	2.21	0.028
	Calmness	232.15	4.82	0.0000165 **
	Interest	229.41	6.90	<0.00001 ***
Aperture design	Excitement	228.81	4.98	<0.00001 ***
	Complexity	229.19	16.52	<0.00001 ***
	Spaciousness	223.56	4.89	<0.000014 **
	Satisfaction with view amount	223.11	15.35	<0.00001 ***
	Brightness	228.97	2.99	0.0033
	Pleasantness	224.55	0.20	0.66
	Calmness	224.94	0.01	0.92
	Interest	223.80	1.00	0.318
Window size	Excitement	224.16	5.03	0.026
	Complexity	224.20	3.80	0.053
	Spaciousness	221.54	6.25	0.013
	Satisfaction with view amount	219.87	1.16	0.283
	Brightness	224.43	4.24	0.041
	Pleasantness	227.26	104.57	<0.00001 ***
	Calmness	229.11	55.43	<0.00001 ***
	Interest	226.96	184.00	<0.00001 ***
Sky type	Excitement	226.72	190.72	<0.00001 ***
	Complexity	226.98	171.55	<0.00001 **
	Spaciousness	222.36	5.66	0.018
	Satisfaction with view amount	221.47	17.89	0.0000342 **
	Brightness	226.92	165.86	<0.00001 ***

Note: * p < 0.0007, ** p < 0.00014, *** p < 0.00001.

Table 4 lists the marginal and conditional R^2 for the LMM of each dependent variable. Compared to R^2 marginal, the proportion of explained variance in the model is greatly increased for R^2 conditional. According to Ferguson's suggestions [41], one R^2 marginal and four R^2 marginal in bold were moderate effects (R^2 > 0.5), and the remaining R^2 were minimum practical effects (R^2 > 0.2) with the exception of impressions of spaciousness (R^2 marginal = 0.149).

Table 4. Results of the LMM analysis for the main factors of aperture design, window size and sky type on all dependent variables.

Attribute	R^2 Marginal	R^2 Conditional
Pleasantness	0.322	0.427
Calmness	0.304	0.352
Interest	0.456	0.525
Excitement	0.454	0.546
Complexity	0.506	0.581
Spaciousness	0.149	0.488
Satisfaction with view amount	0.294	0.495
Brightness	0.417	0.518

3.2. Perceptual Differences between Aperture Design

To further investigate the effect of aperture design on subjective responses, post-hoc pairwise analyses were conducted for all combinations of aperture design for each dependent variable. As the pairwise comparisons are too numerous to be depicted in a table, Table 5 lists the estimated marginal means (EMM), SE and 95% confidence intervals (CI) for all attributes. Grey cells indicate that there is no statistically significant difference between two designs. White cells indicate that there is at least one other aperture design statistically different from this aperture design in terms of associated attributes. The maximum and minimum EMMs of nine aperture designs associated with each attribute are in bold. The following paragraphs describe differences between aperture designs in details.

A Spearman test was conducted between the eight subjective questions, the result shows that subjective impressions of interest, excitement and complexity are mutually strongly correlated with the p-values lower than 0.001 and the correlation coefficient varies between 0.725 and 0.82. Therefore, these three attributes will be discussed in one section. Subjective impressions of calmness will be individually discussed due to its negative correlations with interest ($R^2 = -0.284$), excitement ($R^2 = -0.37$), complexity ($R^2 = -0.425$) and brightness ($R^2 = -0.144$). Subjective impressions of spaciousness and satisfaction with exterior view amount will be discussed in one section, due to their moderate correlation ($R^2 = 0.439$). Finally, subjective impressions of pleasantness and brightness will be discussed in one section due to their insignificant difference.

3.2.1. Impressions of Interest, Excitement and Complexity

For impressions of interest, D3 is different from D1 (B = 1.566, $p = 0.0076$), D2 (B = 1.630, $p = 0.0046$) and D7 (B = -1.7264, $p = 0.0019$), respectively. Compared to these three aperture designs (D1, D2 and D7), D3 induced decreases in interest ratings varying between 14.2% and 15.6%. D9 is different from six aperture designs, including D1 (B = 1.991, $p < 0.0001$), D2 (B = 2.0544, $p < 0.0001$), D4 (B = 1.421, $p = 0.0113$), D5 (B = 1.5264, $p = 0.0049$), D7 (B = 2.1511, $p < 0.0001$) and D8 (B = 0.5879, $p = 0.0044$). Given that D9 was rated the least interesting aperture design with an EMM of 5.05, those six aperture designs increased subjective ratings of interest varying between 12.9% and 19.5%. Neither D3 nor D6 is different from D9 in terms of subjective ratings of interest.

For impressions of excitement, D6 is different from D1 (B = 1.4507, $p = 0.0189$) and D7 (B = -1.5896, $p = 0.0067$) with a decrease in subjective ratings of excitement of 13.2% and 14.5%, respectively. As shown in Table 5, D9 is different from five aperture designs (D1, D2, D5, D7 and D8) but similar to three aperture designs (D3, D4 and D6). Nonetheless, D9 was rated the least exciting aperture design with an EMM of 4.86.

For impressions of complexity, D3 differs from D1 (B = 1.6495, $p = 0.0044$), D5 (B = -1.7222, $p = 0.0022$) and D7 (B = -2.2452, $p < 0.0001$) with a decrease in subjective ratings of complexity varying between 15% and 20.4%. Besides D3, D7 is also different from D4 (B = -1.5892, $p = 0.0076$) and D6 (B = -1.7540, $p = 0.0022$) with an increase in subjective ratings of complexity of 14.5% and 15.9%, respectively. Due to the lowest EMM of 3.43, D9 is different from all remaining eight aperture designs.

Table 5. Estimated marginal means (EMM), standard errors (SE) and 95% confidence intervals (CI) per attribute and aperture design (without interaction).

		D1	D2	D3	D4	D5	D6	D7	D8	D9
Pleasantness	EMM	7.58	7.42	6.67	7.52	6.38	6.44	6.67	6.94	6.96
	SE	0.334	0.337	0.33	0.33	0.334	0.336	0.336	0.352	0.297
	95% CI	[6.65, 8.52]	[6.48, 8.36]	[5.75, 7.6]	[6.6, 8.45]	[5.45, 7.32]	[5.5, 7.38]	[5.73, 7.61]	[5.96, 7.92]	[6.13, 7.79]
Calmness	EMM	6.98	7.4	7.85	7.63	6.64	7.64	5.43	6.94	7.56
	SE	0.362	0.365	0.357	0.357	0.362	0.363	0.363	0.381	0.319
	95% CI	[5.97, 5.97]	[6.38, 6.38]	[6.85, 6.85]	[6.64, 6.64]	[5.63, 5.63]	[6.62, 6.62]	[4.42, 4.42]	[5.88, 5.88]	[6.67, 6.67]
Interest	EMM	7.04	7.1	5.48	6.47	6.58	5.91	7.2	6.64	5.05
	SE	0.314	0.316	0.31	0.31	0.314	0.315	0.315	0.331	0.278
	95% CI	[6.16, 7.92]	[6.22, 7.99]	[4.61, 6.34]	[5.61, 7.34]	[5.7, 7.45]	[5.03, 6.79]	[6.32, 8.08]	[5.71, 7.56]	[4.27, 5.83]
Excitement	EMM	6.63	6.2	5.51	5.87	6.3	5.18	6.77	6.2	4.86
	SE	0.314	0.317	0.31	0.31	0.314	0.316	0.316	0.331	0.279
	95% CI	[5.75, 7.51]	[5.31, 7.08]	[4.64, 6.37]	[5, 6.74]	[5.42, 7.18]	[4.3, 6.06]	[5.88, 7.65]	[5.28, 7.13]	[4.08, 5.64]
Complexity	EMM	6.67	6.1	5.02	5.67	6.74	5.51	7.26	6.31	3.43
	SE	0.32	0.323	0.317	0.316	0.32	0.322	0.322	0.338	0.284
	95% CI	[5.77, 7.56]	[5.2, 7.01]	[4.13, 5.9]	[4.79, 6.56]	[5.84, 7.63]	[4.61, 6.41]	[6.36, 8.16]	[5.36, 7.25]	[2.64, 4.23]
Spaciousness	EMM	7.69	7.95	6.92	7.85	6.38	6.69	6.94	7.42	7.85
	SE	0.325	0.328	0.323	0.323	0.325	0.327	0.327	0.339	0.298
	95% CI	[6.77, 8.6]	[7.02, 8.87]	[6.01, 7.82]	[6.94, 8.75]	[5.46, 7.29]	[5.77, 7.61]	[6.02, 7.86]	[6.47, 8.37]	[7.02, 8.69]
Satisfaction with view amount	EMM	6.88	6.09	5.87	6.57	4.66	6.01	5.77	6.01	8.58
	SE	0.343	0.346	0.34	0.34	0.343	0.345	0.345	0.359	0.31
	95% CI	[5.92, 7.85]	[5.12, 7.06]	[4.92, 6.82]	[5.62, 7.52]	[3.7, 5.62]	[5.05, 6.98]	[4.81, 6.74]	[5, 7.02]	[7.71, 9.45]
Brightness	EMM	7.3	6.81	7.16	7.6	6.26	6.33	7.13	6.52	7.6
	SE	0.337	0.34	0.333	0.333	0.337	0.339	0.339	0.354	0.299
	95% CI	[6.36, 8.24]	[5.86, 7.76]	[6.23, 8.09]	[6.67, 8.53]	[5.32, 7.2]	[5.39, 7.28]	[6.19, 8.08]	[5.53, 7.51]	[6.77, 8.44]

3.2.2. Impressions of Calmness

For impressions of calmness, D7 statistically differs from D2 (B = 1.9708, $p < 0.0033$), D3 (B = 2.41889, $p = 0.0001$), D4 (B = 2.20156, $p = 0.0004$), D6 (B = 2.20605, $p = 0.0005$) and D9 (B = -2.13048, $p = 0.0003$). Since D7 was rated as the least calm scene with an EMM of 5.43, it decreased the calmness of the participants between 17.9% and 22%. Besides D7, there are no statistically significant differences between other aperture designs.

3.2.3. Impressions of Spaciousness and Satisfaction with View Amount

For impressions of spaciousness, with the lowest EMM of 6.38, D5 statistically significant differs from D1 (B = 1.30958, $p = 0.0158$), D2 (B = 1.56863, $p = 0.0013$), D4 (B = 1.46963, $p = 0.0035$) and D9 (B = -1.47755, $p = 0.0012$), with a decrease in spaciousness impressions varying between 11.9% and 14.3%. Additionally, with a slightly greater EMM of 6.669, D6 is statistically significantly different from D2 (B = 1.25904, $p = 0.0292$) and D9 (B = -1.16796, $p = 0.0286$) with a decrease in spaciousness impressions of 11.5% and 10.5%, respectively.

For satisfaction with exterior view amount, D5 was also rated as the least satisfying aperture design with the lowest EMM of 4.66. D5 is statistically significantly different from D1 (B = 2.22535, $p < 0.0001$), D2 (B = 1.42785, $p = 0.0246$), D4 (B = 1.91267, $p = 0.0003$), D6 (B = -1.35500, $p = 0.0431$) and D9 (B = -3.92385, $p < 0.0001$) with the satisfaction rating decreasing between 13% and 35.6%. However, D5 presents no difference from D3, D7 or D8 in terms of satisfaction with exterior view amount. Given the greatest satisfaction level of EMM of 8.58, D9 is also different from the other seven aperture designs, in addition to D5, with a decrease in satisfaction rating percentage varying between 15.5% and 25.5%.

The openness ratios (OR) for D1 to D9, as marked in Figure 4, demonstrate that D9 (91.7%) and D5 (34.3% and 26.7%) are the top and bottom designs in terms of OR, respectively. In other words, subjective responses to impressions of spaciousness and satisfaction with exterior view amount could be ascribed to OR, especially to D9 and D1.

3.2.4. Impressions of Pleasantness and Brightness

Finally, none of the nine aperture designs presents statistically significant differences in terms of subjective assessments of pleasantness or brightness. For pleasantness impressions, the greatest rated aperture design is D1 with an EMM of 7.58, while the lowest rated aperture design is D5 with an EMM of 6.38. All nine aperture designs presented EMMs of pleasantness greater than 6, meaning that none of the aperture designs was rated as unpleasant. For brightness impressions, the greatest rated aperture designs are D4 and D9 with the same EMM of 7.6, while the lowest rated aperture design is D5 with an EMM of 6.26. Similarly, all nine aperture designs present EMMs of brightness above 6.

3.3. Perceptual Differences between Sky Type

Table 6 lists the EMM and SE for all eight dependent variables, along with the estimated B and SE for the pairwise comparisons between the overcast sky and the clear sky. Only the spaciousness impressions had no statistically significant differences between the overcast and clear skies, although the EMM of 7.11 under the overcast sky was slightly lower than that of 7.49 under the clear sky. Participants rated the six attributes of pleasantness, interest, excitement, complexity, satisfaction with exterior view amount and brightness as greater under the clear sky than under the overcast sky. Compared to the overcast sky, the clear sky increases subjective impressions of pleasantness by 18.5%, interest by 23.9%, excitement by 24.2%, complexity by 22.9%, satisfaction with exterior view amount by 7.1% and brightness by 23.1%. On the contrary, However, compared to the overcast sky, the spatial calmness is decreased by 15.8% under the clear sky.

3.4. Effects of Aperture Design, Window Size and Sky Type on EEG

Table 7 presents the LMM analyses for the main factors of aperture design, window size and sky type on EEG representative predictors. Given the 12 EEG predictors, only the ones with p-values within 0.05 are shown in Table 7. However, since only three main factors were tested herein, the adjusted *p*-value of the initial 0.05 was adjusted to 0.0021 $(0.005/(3 \times 8))$. Although both the mean of beta power and S.D. of beta rhythm of the right prefrontal lobe area are influenced by aperture designs, the *p*-values are greater than 0.0021. Neither the S.D. of alpha rhythm are influenced by the window size. However, mean power of beta (F(1, 203.41) = 5.8359, *p* = 0.00166) and the S.D. of beta rhythm (F(1, 203.41) = 5.9674, *p* = 0.00154) on the right prefrontal lobe area are influenced by window size. Table 8 lists the estimated marginal means (EMM), SE and 95% confidence intervals (CI) of sky types for $\log_{10}$ of $PowerMean_\beta$ (B = 0.0337, *p* = 0.0166) and $\log_{10}$ of SD_β (B = 0.0337, *p* = 0.0166). The large window resulted in higher oscillations and stronger power of the beta rhythm on the right prefrontal lobe area than the small window did.

Table 6. Estimated marginal means (EMM), standard errors (SE), 95% confidence intervals (CI) for all eight dependent variables, along with the estimates B and standard errors (SE) for pairwise comparisons.

	Overcast Sky			Clear Sky		
	EMM	SE	95% CI	EMM	SE	95% CI
Pleasantness	5.94	0.184	[5.52, 6.36]	7.97	0.185	[7.55, 8.39]
Calmness	7.99	0.184	[7.57, 8.40]	6.25	0.186	[5.83, 6.68]
Interest	5.07	0.167	[4.69, 5.45]	7.70	0.168	[7.31, 8.08]
Excitement	4.62	0.172	[4.22, 5.01]	7.28	0.174	[6.88, 7.67]
Complexity	4.60	0.174	[4.20, 4.99]	7.12	0.175	[6.71, 7.52]
Spaciousness	7.11	0.225	[6.59, 7.62]	7.49	0.226	[6.97, 8.01]
Satisfaction with view amount	5.88	0.217	[5.38, 6.38]	6.66	0.219	[6.16, 7.17]
Brightness	5.70	0.189	[5.27, 6.13]	8.24	0.190	[7.80, 8.67]

Table 7. Results of the LMM analysis for the main factors of aperture design, window size and sky type on EEG predictors.

EEG Predictors	Independent Variables	df	F	*p*-Value
$\log_{10}(SD_{\alpha\text{-right}})$	Window size	203.26	4.58	0.03354
$\log_{10}(PowerMean_{\beta\text{-right}})$	Design	203.77	2.4094	0.01664
	Window size	203.41	5.8359	0.00166 *
$\log_{10}(SD_{\beta\text{-right}})$	Design	203.77	2.3918	0.01743
	Window size	203.41	5.9674	0.00154 *
$Mean_{\theta\text{-left}}$	Sky type	88.074	8.2583	0.00208 *
$Mean_{\theta\text{-right}}$	Sky type	4.7805	11.0845	0.00123 *

Note: * *p* < 0.0021, ** *p* < 0.0004, *** *p* < 0. 0001.

Table 8. Estimated marginal means (EMM), standard errors (SE) and 95% confidence intervals (CI) per EEG attribute and window size (without interaction).

	Large Window			Small Window		
	EMM	SE	95% CI	EMM	SE	95% CI
$\log_{10}(PowerMean_{\beta\text{-right}})$	−3.70	0.0459	[−3.80, −3.61]	−3.74	0.0459	[−3.83, −3.65]
$\log_{10}(SD_{\beta\text{-right}})$	−3.70	0.0459	[−3.80, −3.61]	−3.74	0.0459	[−3.83, −3.65]

Regarding the sky type, $\log_{10}$ of $Mean_\theta$ on both the left (F(1, 88.074) = 8.2583, *p* = 0.00208) and right (F(1, 4.7805) = 11.0845, *p* = 0.00123) prefrontal lobe areas are influenced by the sky types. Table 9 lists the estimated marginal means (EMM), SE and 95% confidence intervals (CI) of the sky types for $\log_{10}$ of $Mean_{\theta\text{-left}}$ (B = -3.56×10^{-6}, *p* = 0.0045) and $\log_{10}$ of $Mean_{\theta\text{-right}}$ (B = -4.27×10^{-6}, *p* = 0.0010). The mean theta values on both the left and right prefrontal lobe areas under the overcast sky were three times greater than the mean theta values under the clear sky. In other words, the clear sky attenuates the theta rhythm on the prefrontal lobe areas.

Table 9. Estimated marginal means (EMM), standard errors (SE) and 95% confidence intervals (CI) per EEG attribute and sky type (without interaction).

		Overcast Sky			Clear Sky	
	EMM	SE	95% CI	EMM	SE	95% CI
$\text{Mean}_{\theta\text{-left}}$	7.74×10^{-6}	1.94×10^{-6}	$[3.84 \times 10^{-6}, 1.16 \times 10^{-5}]$	4.18×10^{-6}	1.93×10^{-6}	$[2.98 \times 10^{-7}, 8.07 \times 10^{-6}]$
$\text{Mean}_{\theta\text{-right}}$	7.96×10^{-6}	1.73×10^{-6}	$[4.49 \times 10^{-6}, 1.14 \times 10^{-5}]$	3.69×10^{-6}	1.72×10^{-6}	$[2.37 \times 10^{-7}, 7.14 \times 10^{-6}]$

3.5. Effects of Aperture Design, Window Size and Sky Type on ECG

No significant effects are found for aperture design, window size or sky type on ECG data (all p values > 0.0021). In other words, neither Mean_{NN} nor Median_{NN} was influenced by aperture design, window size or sky type herein.

4. Discussion

4.1. Influences of Aperture Design

Subjective assessments of spatial interest, excitement and complexity were mutually positively correlated. The aperture designs that were rated more interesting were also rated more exciting and complex. D7, D1 and D5 were reported more interesting, exciting and complex than the remaining designs, which agrees with Chamilothori's conclusion that participants responded more positively towards irregular than regular aperture designs [13]. Additionally, D9, representing regular curtain walls, D3, representing vertical shades, and D6, representing vertical shades with a low degree of variations, were reported less interesting, exciting and complex. D9 was rated the highest levels in calmness, satisfaction with exterior view amounts and brightness. Subjective greatest satisfaction with exterior view amounts was in alignment with Abboushi and colleagues' conclusion in [42]. It is understandable that no aperture design was significantly different in terms of brightness, given that the VR headset provides a limited range of luminance. All nine aperture designs were rated as EMM greater than 6, meaning that participants were pleased with all aperture designs on average. Nonetheless, subjective VR experience might have contributed to their positive responses.

Finally, physiological data of neither EEG nor ECG was influenced by aperture design. The insignificant results of ECG is in line with Chamilothori's conclusion in [17] but different from Chamilothori's conclusion that participants showed a larger decrease in heart rate while exposed to the irregular condition in [13]. In other words, influences of aperture designs upon physiological measures are still inconsistent and cannot be drawn a general conclusion.

4.2. Influences of Window Size

Concerning window size, no statistically significant difference was found among subjective assessments. However, beta variations and powers on the prefrontal right lobe areas were enhanced more by the large window than the small window. According to the theory of left and right brain segmentation proposed by Roger Sperry [43], the left hemisphere is mainly responsible for functions such as logical understanding, memory, time, judgment, classification, analysis and writing, whereas the right hemisphere is mainly responsible for memory, intuition, emotion, vision, imagination, inspiration and thinking. Greater power means and variance of beta band under the large window indicate a strong intensity of EEG signal oscillations and higher levels of neuronal activity. Although no significant subjective assessments concerning the window size was observed, EEG data might be more sensitive to reflect the different influences of window size.

4.3. Influences of Sky Type

The sky types influence both subjective assessments and psychological data. Compared to the overcast sky, the clear sky results in higher levels in subjective impressions of pleasantness, interest, excitement, complexity, brightness and satisfaction with exterior view amount, but lower levels in subjective impressions of calmness. Moreover, the overcast sky results in greater levels of mean theta rhythm on both left and right prefrontal lobe areas. Unlike Moscoso et al. and Chamilothori et al.'s conclusion that sky types have no significant influence on subjective assessments [10,40], the contradictory conclusion reported in this study might be caused by the different simulation method. In both Moscoso et al. and Chamilothori et al.'s studies, they projected simulated luminance maps to create 3D VR scenes with a fixed evaluation spot for the participants. However, this study uses Enscape to create interior simulation scenes, where the participants could freely explore each scene. However, the rendered scenes exaggerated the color temperature difference between the overcast and clear skies. As shown in Figure 3, in addition to the existence of sunlight patterns, the warm color appearance under the clear sky and the cool color appearance under the overcast sky might affect subjective assessments and their theta rhythm in high likelihood. Further studies are required to calibrate and validate the method of using Enscape for creating VR scenes in terms of daylighting appearance.

4.4. Research Limitations

In addition to the exaggerated color temperature differences between the overcast and clear skies, this study contains the following limitations. First, the participants in the studies are experts in architecture designs. Assessments of non-experts were not included in this study. Second, this study only measured the EEG on the left and right prefrontal lobe areas, other channels require further exploration. Third, this study only included 41 participants' valid responses. More participants with diverse distributions of age ranges could provide more concrete conclusions.

5. Conclusions

This paper presents a laboratory experiment using VR scenes. Nine aperture designs, two window sizes and two sky types were explored. Forty-one participants with four years or longer architecture background evaluated eight aspects of interior natural lightscape. Physiological data of both ECG and EEG of the prefrontal lobe areas were collected during the experiment. According to the data analysis, the aperture designs have impacts on subjective assessments of calmness, interest, excitement, complexity, spaciousness and satisfaction with exterior views. The sky types influence subjective assessments of pleasantness, calmness, interest, excitement, complexity, spaciousness, brightness and satisfaction with exterior views. Compared to the small window, the large window enhances beta oscillations and beta power on the right prefrontal lobe area. Compared to the overcast sky, the clear sky attenuates the theta rhythm on the prefrontal lobe areas.

Based on the subjective responses, the design recommendations are proposed as below:

1. Regular windows (Design 9) rated as the least interesting, exciting and complex design are appropriate in spaces where a calm and stable atmosphere is required. In other words, classrooms and offices, where occupants need to focus on their work rather than to be distracted by sunlight patterns, are suitable environments in which to employ regular window design.
2. Regular-shape aperture designs (Designs 3, 4, and 6) were also rated with low levels of interest, excitement and complexity, as well as high level of calmness. Similar to regular windows, these designs (Designs 3, 4, and 6) are also suitable for spaces with the requirement of a relatively calm and stable atmosphere.
3. Irregular shape aperture designs (Designs 1, 2, 7 and 8) were rated with high levels of interest, excitement and complexity, as well as low level of calmness. These aperture designs, as well as the sunlight patterns, are more likely to trigger participants' positive feelings and responses. Spaces that require a lively and dynamic atmosphere, such as hotel lobbies, shopping malls and art museums, could employ these irregular aperture designs.

4   Finally, openness ratios of aperture designs, which had great a impact on subjective impressions of spatial spaciousness and satisfaction with exterior view amounts, need be considered during the aperture and façade design.

Author Contributions: Conceptualization, Z.K., K.H. and C.L.; methodology, Z.K., K.H. and C.L.; software, Z.W. and F.C.; validation, Z.K., K.H. and C.L.; formal analysis, Z.W., F.C., Y.L. and X.L.; investigation, Z.W. and F.C.; resources, Z.K., K.H. and C.L.; data curation, Z.K.; writing—original draft preparation, Z.K. and K.H.; writing—review and editing, Z.K., K.H. and C.L.; visualization, Z.K., Y.L. and X.L.; supervision, C.L.; project administration, Z.K.; funding acquisition, K.H. All authors have read and agreed to the published version of the manuscript.

Funding: This work was funded by National Natural Science Foundation of China (Grant No. 52208012), Natural Science Foundation of China Youth Science Foundation (Grant No. 51908301), National Natural Science Foundation of China Youth Science Foundation (Grant No. 51908111), General project of Shandong social science planning and research project (Grant No. 21CSHJ06), Shandong Natural Science Foundation Training Fund (Grant No. ZR2019PEE034) and Qingdao social science planning and research project (Grant No. QDSKL2101178).

Acknowledgments: The authors would like to thank these six students for their help with data collection: Ruyan Zhang, Xingqi Kong, Pengxuan Ning, Jinghao Ni, Jiayue Wang and Qilin Liu. The authors also would like to thank the 44 participants, who contributed their time, enthusiasm and valuable feedback to this experiment.

Conflicts of Interest: The authors declare no conflict of interest.

References

1. Veitch, J.A.; Charles, K.E.; Farley, K.M.J.; Newsham, G.R. A model of satisfaction with open-plan office conditions: COPE field findings. *J. Environ. Psychol.* **2007**, *27*, 177–189.
2. Veitch, J. Psychological Processes Influencing Lighting Quality. *J. Illum. Eng. Soc.* **2001**, *30*, 124–140.
3. Veitch, J.A. Lighting Quality Contributions from Biopsychological Processes. *J. Illum. Eng. Soc.* **2001**, *30*, 3–16.
4. Kong, Z.; Zhang, R.; Ni, J.; Ning, P.; Kong, X.; Wang, J. Towards an integration of visual comfort and lighting impression: A field study within higher educational buildings. *Build. Environ.* **2022**, *216*, 108989.
5. Kong, Z.; Jakubiec, J. Instantaneous lighting quality within higher educational classrooms in Singapore. *Front. Arch. Res.* **2021**, *10*, 787–802.
6. Kong, Z.; Jakubiec, J.A. Evaluations of long-term lighting qualities for computer labs in Singapore. *Build. Environ.* **2021**, *194*, 107689.
7. Hirning, M.B.; Isoardi, G.L.; Garcia-Hansen, V.R. Prediction of discomfort glare from windows under tropical skies. *Build. Environ.* **2017**, *113*, 107–120.
8. Van Den Wymelenberg, K.G. Visual Comfort, Discomfort Glare, and Occupant Fenestration Control: Developing a Research Agenda. *Leukos* **2014**, *10*, 207–221.
9. Chamilothori, K. Perceptual effects of daylight patterns in architecture in School of Architecture. *Civ. Environ. Eng.* **2019**.
10. Moscoso, C.; Chamilothori, K.; Wienold, J.; Andersen, M.; Matusiak, B. Window Size Effects on Subjective Impressions of Daylit Spaces: Indoor Studies at High Latitudes Using Virtual Reality. *Leukos* **2020**, *17*, 242–264.
11. Lam Lo, W.; Steemers, K. The art of brightness and darkness a critical investigation on daylighting quality. In Proceedings of the 26th Conference on Passive and Low Energy Architecture, Quebec, QC, Canada, 22–24 June 2009.
12. Dogrusoy, I.T.; Tureyen, M. A field study on determination of preferences for windows in office environments. *Build. Environ.* **2007**, *42*, 3660–3668.
13. Chamilothori, K.; Chinazzo, G.; Rodrigues, J.; Dan-Glauser, E.S.; Wienold, J.; Andersen, M. Subjective and physiological responses to façade and sunlight pattern geometry in virtual reality. *Build. Environ.* **2019**, *150*, 144–155.
14. Moscoso, C.; Chamilothori, K.; Wienold, J.; Andersen, M.; Matusiak, B. Regional differences in the perception of daylit scenes across Europe using virtual reality. Part I: Effects of window size. *Leukos* **2021**, *18*, 294–315.
15. Rockcastle, S.; Amundadottir, M.; Andersen, M. Contrast measures for predicting perceptual effects of daylight in architectural renderings. *Light. Res. Technol.* **2016**, *49*, 882–903.
16. Rockcastle, S.F.; Chamilothori, K.; Andersen, M. An Experiment in Virtual Reality to Measure Daylight-Driven Interest in Rendered Architectural Scenes. In *Building Simulation 2017*; Pubilsher: San Francisco, CA, USA, 2017.
17. Chamilothori, K.; Wienold, J.; Moscoso, C.; Matusiak, B.; Andersen, M. Subjective and physiological responses towards daylit spaces with contemporary façade patterns in virtual reality: Influence of sky type, space function, and latitude. *J. Environ. Psychol.* **2022**, *82*, 101839.
18. Kong, Z.; Utzinger, D.M.; Freihoefer, K.; Steege, T. The impact of interior design on visual discomfort reduction: A field study integrating lighting environments with POE survey. *Build. Environ.* **2018**, *138*, 135–148.

19. Wu, S. Main Points of Lightscape. *South Archit.* **2017**, *3*, 4–6.
20. Kong, Z.; Liu, Q.; Li, X.; Hou, K.; Xing, Q. Indoor lighting effects on subjective impressions and mood states: A critical review. *Build. Environ.* **2022**, *224*, 109591.
21. Abboushi, B.; Elzeyadi, I.; Taylor, R.; Sereno, M. Fractals in architecture: The visual interest, preference, and mood response to projected fractal light patterns in interior spaces. *J. Environ. Psychol.* **2018**, *61*, 57–70.
22. Moscoso, C.; Matusiak, B.; Svensson, U.P. Impact of window size and room reflectance on the perceived quality of a room. *J. Archit. Plan. Res.* **2015**, *3*, 294–306.
23. Chamilothori, K.; Wienold, J.; Andersen, M. Adequacy of Immersive Virtual Reality for the Perception of Daylit Spaces: Comparison of Real and Virtual Environments. *Leukos* **2018**, *15*, 203–226.
24. Abd-Alhamid, F.; Kent, M.; Bennett, C.; Calautit, J.; Wu, Y. Developing an Innovative Method for Visual Perception Evaluation in a Physical-Based Virtual Environment. *Build. Environ.* **2019**, *162*, 106278.
25. Veitch, J. Revisiting the Performance and Mood Effects of Information about Lighting and Fluorescent Lamp Type. *J. Environ. Psychol.* **1997**, *17*, 253–262.
26. Küller, R.; Wetterberg, L. Melatonin, cortisol, EEG, ECG and subjective comfort in healthy humans: Impact of two fluorescent lamp types at two light intensities. *Light. Res. Technol.* **1993**, *25*, 71–81.
27. Boray, P.F.; Gifford, R.; Rosenblood, L. Effects of warm white, cool white and full-spectrum fluorescent lighting on simple cognitive performance, mood and ratings of others. *J. Environ. Psychol.* **1989**, *9*, 297–307.
28. Trimble Inc. *SketchUp*; Trimble Inc.: Munich, Germany, 2022.
29. Enscape GmbH. *ENSCAPE*; Enscape GmbH: Karlsruhe, Germany, 2022.
30. Pro, V. VIVE United States, (n.d.). 2022. Available online: https://www.vive.com/us/product/vive-pro/ (accessed on 30 July 2022).
31. Åkerstedt, T.; Gillberg, M. Subjective and objective sleepiness in the active individual. *Int. J. Neurosci.* **1990**, *52*, 29–37.
32. Liu, Y.; Li, L.; Li, Y.; Li, J. A wearable device to collect and display pulse and ECG data. In *Southeast University*; Yu, J., Ed.; Nanjing Zhonglian Patent Agency Co.: Nanjing, China, 2022.
33. Carreiras, C.; Alves, A.P.; Lourenço, A.; Canento, F.; Silva, H.; Fred, A. *BioSPPy: Biosignal Processing in Python*, Python package version 0.6.1. 2015.
34. Xing, Y.; Xiao, Z.; Wang, Z.; Li, R.; Li, J.; Liu, C. *A Device for Testing the Performance of ECG Electrodes*; Xu, X., Ed.; Nanjing Zhonglian Patent Agency, Co.: Nanjing, China, 2022.
35. Chen, M.F.; Zhao, L.; Li, B.; Yang, L. Depression evaluation based on prefrontal EEG signals in resting state using fuzzy measure entropy. *Physiol. Meas.* **2020**, *41*, 095007.
36. GitHub. *EEGLAB*; GitHub: San Francisco, CA, USA, 2022.
37. The MathWorks. *MATLAB*; The MathWorks, Inc.: Natick, MA, USA, 2022.
38. ASHRAE. *Thermal Environmental Conditions for Human Occupancy*; ASHRAE: Atlanta, GA, USA, 2020.
39. Sawyer, A.O.; Niermann, M.; Groat, L.N. The use of environmental aesthetics in subjective evaluation of daylight quality in office buildings. In Proceedings of the IES Annual Conference, Indianapolis, IN, USA, 8 November 2015.
40. Chamilothori, K.; Wienold, J.; Moscoso, C.; Matusiak, B.; Andersen, M. Regional Differences in the Perception of Daylit Scenes across Europe Using Virtual Reality. Part II: Effects of Façade and Daylight Pattern Geometry. *Leukos* **2022**, *18*, 316–340.
41. Ferguson, C.J. An effect size primer: A guide for clinicians and researchers. *Prof. Psychol. Res. Pract.* **2009**, *40*, 532–538.
42. Abboushi, B.; Elzeyadi, I.; Van Den Wymelenberg, K.; Taylor, R.; Sereno, M.; Jacobsen, G. Assessing the Visual Comfort, Visual Interest of Sunlight Patterns, and View Quality under Different Window Conditions in an Open-Plan Office. *Leukos* **2020**, *17*, 321–337.
43. Pearce, J. The "split brain" and Roger Wolcott Sperry (1913–1994). *Rev. Neurol.* **2019**, *175*, 217–220.

Article

Influence of Classroom Colour Environment on College Students' Emotions during Campus Lockdown in the COVID-19 Post-Pandemic Era—A Case Study in Harbin, China

Weiyi Tao, Yue Wu *, Weifeng Li and Fangfang Liu *

Key Laboratory of Cold Region Urban and Rural Human Settlement Environment Science and Technology, Ministry of Industry and Information Technology, School of Architecture, Harbin Institute of Technology, Harbin 150001, China

* Correspondence: wuyuehit@hit.edu.cn (Y.W.); liufangfang@hit.edu.cn (F.L.)

Abstract: Campus lockdown during COVID-19 and the post-pandemic era has had a huge negative effect on college students. As a vital part of interior teaching spaces, colour deeply influences college students' mental health and can be used for healing. Nevertheless, research on this topic has been limited. Based on colour psychology and colour therapy, this paper discusses the relationship between interior teaching space colours (hue and brightness) and emotions among college students. The HAD scale and questionnaire survey method were used. It was concluded that: (1) Anxiety and depression were prominent among the college student population during the quarantine of the university due to the epidemic. (2) Warm colours have an advantage over both cold and neutral colours in creating pleasure, relaxation, and mental attention, with the second in line being the cold and the last being the neutral. Warm colours make it pleasant for individuals while cold colours boost attention. (3) When subjects have higher values of anxiety and depression, they are less satisfied with the colour of the teaching space. (4) In most cases, there is no significant difference in the colour preference of teaching spaces across the gender, grade, and major groups, with females having a higher preference for warm high-brightness classrooms than males. These findings provide crucial ideas for future interior teaching space design and enrich the theories in colour psychology.

Keywords: COVID-19; classroom colour environment; college students' mental health; HAD scale

Citation: Tao, W.; Wu, Y.; Li, W.; Liu, F. Influence of Classroom Colour Environment on College Students' Emotions during Campus Lockdown in the COVID-19 Post-Pandemic Era—A Case Study in Harbin, China. *Buildings* **2022**, *12*, 1873. https://doi.org/10.3390/buildings12111873

Academic Editor: Diego Pablo Ruiz Padillo

Received: 29 August 2022
Accepted: 26 October 2022
Published: 3 November 2022

Publisher's Note: MDPI stays neutral with regard to jurisdictional claims in published maps and institutional affiliations.

1. Introduction

Since its emergence, the new coronavirus has had various adverse effects on the mental health of the general public. In the first year of COVID-19, the prevalence of depression and anxiety disorders is estimated to have increased by 25% [1,2]. Studies [3] show that people have had varied degrees of psychological problems throughout the outbreak. Stress, anxiety, and depression values remained high after two weeks and did not decrease over time. This was exacerbated by the isolation and confinement caused by the epidemic, with symptoms such as mood disorders, depression, stress, poor mood, irritability, insomnia, and post-traumatic stress disorder [4]. Of all populations, students are among the most prone to have severe psychological problems [5]. A comparative study found that children who experienced isolation had post-traumatic stress values four times higher than those who did not [6]. In an Italian study [7], late bedtimes and late wakeups were particularly common among the student population, and sleep quality was also affected during the period of isolation, with 27.8% reporting depression symptoms and 34.3% showing anxiety symptoms. In a pre-and post-closure survey of Chinese university students, the mean PANAS-NA (negative affect) scale score fell from 2.38 (0.79) to 2.24 (0.80), and the mean anxiety-depression score on the PHQ-4 scale changed from 0.95 (0.65) to 0.76 (0.61), with significant reductions in both values. This indicates that there is a significant increase in anxiety and depression symptoms among students, and the negative effects of school closure are becoming more prominent [8]. The epidemic school closure greatly harms the mental health of college students and requires urgent interventions.

China has been implementing the "Dynamic Zero-COVID" approach since the pandemic outbreak in 2020. In the context of that, universities in China have also been undergoing a dynamic lockdown for nearly three years. Because of numerous elements like Chinese universities offering practically every student housing and the Chinese population figure being relatively significant, Chinese universities also have a higher volume of students living on campus than foreign universities. Therefore, if no protection measure is conducted, the virus is more likely to transmit on the Chinese campus and cause a disastrous outcome. Based on all the facts and according to policies in districts, the university operates in closure, resulting in students being forced to stay in campus walls. During the closure, students reside in dormitories and can wander around inside the university, conducting ordinary tasks including studying indoors and playing sports outside, etc. If needed, medical resources are supplied. This action is temporary, and the campus will be reopened once the external epidemic is under control. However, we can observe complaints from students on social media like Weibo, writing about anxiety from long-time closure. Some mention that their psychological condition gets worse because of the feeling of being restricted on campus. College students' mental health is affected by campus lockdown. However, there is a lack of post-pandemic closure studies with Chinese characteristics. The current studies globally on epidemic closure mostly focus on two contexts—home isolation and confinement [4]—whereas there is a lack of discussion on campus lockdown when students are required to stay in university. In addition, most of the studies focus on the early stage of the new corona outbreak [3,5], and there is a lack of studies on the normalization of closure in the post-epidemic period. Furthermore, studies focusing on college students' mental health during post-pandemic college lockdown are also relatively few; the existing literature mostly analyses the changes in the psychological state of adolescents [9,10] during school closure or isolation. Some studies have focused on the psychological changes of college students [8], but also failed to suggest creative strategies of mitigation.

Existing literature indicates that colour can relieve pressure to some extent. Colour psychology is a branch of psychological science that believes that colour has various psychological and behavioural effects [11]. Some psychologists [12] have shown that 83% of the information humans obtain is from visual sources, and that colour predominates in this visual information. Extensive research discusses the psychological, cognitive, physiological, and behavioural effects of colour [13,14]. Some studies indicate the psychological impact of colour in terms of dynamism, size and quantity, and warmth and coolness [15,16]. Since each colour has its own wavelength and frequency, when the body absorbs its specific energy, it might change the original energy in the body. Therefore, colour can be used for the treatment of physical illnesses and psychological problems [17], which has led to the concept of colour healing. Colour therapy, derived from colour psychology, is a technique of psychotherapy that promotes recovery by allowing the patient to see and feel a colourful environment that causes stimulation of the brain and emotions [18]. Colour has been found to have features that affect the patients' physiological activities, emotions in daily life, cognitive processing, and other changes in mental activity [19]. In addition to the use of colour in the medical environment, the use of colour as an adjuvant to therapy to relieve the psychological distress of patients has been increasingly recognized and applied in existing research [20]. It has been shown [17] that colour therapy can have the same or a similar effect on any group of people, regardless of the cultural background. It is known that colour preference could be influenced by differences in age, sex, and geographical region. Additionally, factor analysis and cluster analysis indicated some relation between colour preference and the subjects' lifestyles [21]. For instance, Great Britain has a strong preference for G categories and a warm-greyish colour image is preferred. Italy has a preference for R and Y categories and a warm-clear image is preferred [22]. However, Chinese people have specific colour preferences. For example, black on red signifies happiness to Chinese people, and therefore the colour combination is commonly used for wedding invitations [23]. Red is not only consistently associated with "active," "hot", and "vibrant", but it also conveys additional meaning ("pleasant") in China [23]. There are also some scholars who point out that the "red preference" phenomenon is observed in Chinese adults. Light colours are preferred the most in terms of chroma-lightness level [24]. Based on all the facts, colour therapy has great potential in reality. However, the current application of colour healing in China is limited. Although there are now discussions on the application of colour psychology for campus space design [25], they are failing to incorporate the current state of college student's mental health and failing to apply various theories of colour healing. Therefore, we must consider the possibility of applying colour therapy in the teaching space to alleviate the anxiety and depression of college students. If the environmental colours of interior teaching spaces can be used to reduce anxiety and depression values, it will considerably enhance the mental health of college students.

To assess students' psychological condition, the HAD scale is used as the measurement tool. The Hospital Anxiety and Depression Scale (HAD) was created by Zigmond and Snaith in 1983 to screen for anxiety and depression in general hospital patients. The scale consists of two separate scales. One is the Hospital Anxiety Scale (HADA) and the other is the Hospital Depression Scale (HADD). It has been translated into several national versions and is widely used in medical assessment. There are studies of national versions of the scale, such as in Spanish [26], Chinese [27], Norwegian [28], and Arabic [29], as well as studies of applicable populations, such as patients with fibromyalgia syndrome [30], tinnitus [31], office workers [32], the elderly [27], and patients with oral burning syndrome [33]. In a Spanish study [30], the subscales "anxiety" and "depression" were evaluated separately, and both scales were found to be highly reliable and accurate (HADA = 0.80, HADD = 0.85). Some studies have found that the HAD scale has 80% sensitivity and 90% specificity, considering it a good screening tool for anxiety and depression in older adults in Cantonese-speaking areas [27]. It has also been suggested that the HAD scale is more useful in the assessment of depression [31]. A Norwegian study [28] revealed the high internal consistency of the scale with a substantial sample size (65,648 participants). In summary, most of the studies corroborate the scientific validity, reliability, and validity of the HAD scale and therefore support the application of the scale in the assessment of anxiety and depression in various domains. Yet there are few cases of applying the HAD scale to research in China, more focus being on particular patients [34,35] and application in the field of education being neglected. There are no examples of assessing students' mental health. Therefore, it is practical and feasible in this paper to apply the HAD scale to assess anxiety and depression among college students.

The research specifically focuses on college students who must stay inside campus because of the pandemic prevention policy. The group's features are quite different from those who can only stay in the dormitory or those home commuting subjects. Therefore, this study has its speciality in geography, timing, groups, and so on. This study will supplement the gap of existing research. Based on the mental health problems of college students in the post-epidemic school closure normalization, using the HAD scale to assess the relevant indicators, we lead to conclusions of colour healing to provide a reference for subsequent space design and psychotherapy. In conclusion, this study aims to provide references and suggestions for the development of campus teaching space environment design in the post-epidemic era, and it is also an innovative attempt to intervene in the mental health of college students from the perspective of colour healing. The following hypotheses are proposed and tested:

H1. *During the closure of colleges and universities due to the epidemic, there is a high prevalence of anxiety and depression in the college student population.*

H2. *Neutral, warm, and cold teaching spaces and teaching spaces with adjusted lightness shifts have different effects on college students' emotions.*

H3. *There is a significant difference between different anxious and depressed groups in judging the effect of teaching space on mood.*

H4. *There are significant differences between demographic characteristics in judging the effect of instructional space on mood.*

2. Methods

In this study, a questionnaire was distributed and filled out through the "Questionnaire Star" platform to collect the subjects' emotional evaluation of the teaching space with different colour characteristics. The research idea is shown in Figure 1.

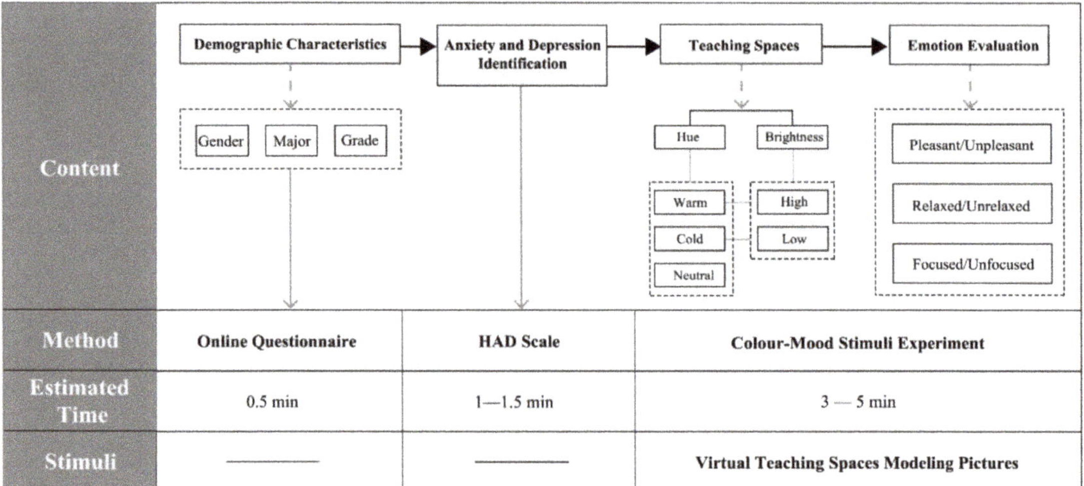

Figure 1. Research scheme.

2.1. Questionnaire Setting

The questionnaire (Appendix A) for this investigation consisted of three parts. Part I: The subjects were asked about their demographic information, including gender, education, and major. Part II: The subjects were tested on the Hospital Anxiety and Depression Scale (HAD scale). The test contains 14 questions (Table 1), and subjects make choices based on their past week. From the outcomes we obtained the subject's level of anxiety and depression. The HAD scale consists of two subscales, anxiety and depression, for anxiety (A) and depression (D), each with 7 questions. Each item is assessed on a 4-point scale, with single-sign ratings summing to anxiety ratings and double-sign ratings summing to depression ratings. A single scale score of 0–7 indicates no depression or anxiety, a total score of 8–10 indicates possible or "borderline" anxiety and depression, and a total score of 11–20 indicates possible significant anxiety or depression.

Part III: Conducting the observation of virtual teaching spaces with different colour differences was carried out. In this experiment, two different teaching spaces were used as prototypes. The initial model was built with Revit 2021 then rendered and post-adjusted with Lumion 11. According to the variation of hue and lightness, 14 different virtual spaces are constructed (Figure 2). Immediately after the observation of each set of virtual teaching spaces, the subjects filled out a questionnaire on the level of pleasure, relaxation, and mental attention for the scene to obtain their subjective feelings about the pictures. The chromaticity analysis was conducted to explore the effect of neutral, warm, and cold classrooms on human emotions. The brightness analysis was conducted to investigate the effect of warm and cold classrooms on human emotions at both high and low brightness levels. The reference data is the mean and standard deviation of the questionnaire scores. In the colour and brightness selection section, three pairs of two-level adjectives, "pleasant/unpleasant", "relaxed/unrelaxed" and "focused/unfocused", were used to evaluate different colour teaching spaces. The standard is a 5-point semantic difference, using a scoring system from 1 to 5. The lower the score, the more negative is the emotion.

Table 1. HAD scale.

Title	Options			
1. *I feel nervous (or painful)*	Not at all	Sometimes	Most of the time	Almost all the time
2. *I am still interested in the things I used to be interested in*	Definitely the same	Not as much as before	Only a little	Basically no more
3. *I felt some fear as if I had a feeling that something terrible was going to happen*	Not at all	A little, but it doesn't bother me	Yes, but not too serious	Very sure and very serious
4. *I can laugh and see the funny side of things*	I do this a lot.	I am not so much anymore.	Definitely not too much now	Not at all
5. *My heart is full of worries*	Occasionally so	From time to time, but not often	Often	Most of the time
6. *I feel happy*	Most of the time	Sometimes	Not often.	Not at all
7. *I can sit at ease and relax*	Affirmation	Frequently	Not often	Not at all
8. *I lose interest in my appearance (dressing)*	I still care as much as ever	I may not care very much	Not as caring as I should be	Affirmation
9. *I was a little fidgety as if I felt compelled to move*	Not at all	Not much	Quite a bit	A bit too much indeed
10. *I look forward to the future with a happy heart*	Almost like this	It doesn't quite work that way	Rarely do you do this	Almost never do this
11. *I suddenly have a sense of panic*	Not at all	Not often	from time to time	Very often indeed
12. *I seem to feel that people have become dull*	Not at all	Sometimes	Very often	Almost all the time
13. *I feel a shivering fear*	Not at all	Sometimes	Very often	Very often
14. *I can enjoy a good book or a good radio or TV program*	Often	Sometimes	Not often	Rarely

2.2. Participants

One hundred and ten participants were recruited to fill out the questionnaire through the Questionnaire Star platform in April–May 2022. During the period from 11 April to 15 May, most of the subjects were quarantined on campus due to the epidemic closure and were unable to enter or leave the campus freely.

A total of 110 valid questionnaires were returned in this study, with an effective rate of 100%. The numerical characteristics of the demographic variables can be seen according to the analysis results in Table 2, which reflect the distribution of the respondents in this survey and where the mean value represents the trend among them and the standard deviation represents the fluctuation. According to the results of the frequency analysis of each variable, it can be seen that the distribution meets the requirements of the sample survey. For example, among the gender survey results, the proportion of males is 65.5%, and the proportion of females is 34.5%. This shows that the results of this survey focus on male colour preference. In terms of academic distribution, the largest category is undergraduates, including the highest proportion of junior students. In terms of professional distribution, the highest proportion is engineering students, up to 73.6%, indicating that the subjects are mainly science and technology students.

(a) Neutral Colour Classrooms

(b) Warm Colour Classrooms

(c) Cold Colour Classrooms

Figure 2. Virtual environment modelling of teaching spaces. Note: This computer model is built based on actual classrooms in a university in Harbin, China. The size, furniture, and material are in accord with reality. There are two types of classrooms. One is small and could contain no more than 20 students. The other is large and could hold no more than 200 students. The pictures above are taken from two angles which is of both eyes' perspective, aiming to offer subjects a more immersive experience. Since most Chinese classrooms are decorated with coating materials/paint (Figure 3), we built models without considering the materials' influence. Because wall colour takes up the most percentage of classroom colour, this time we only picked wall colour as the variation. The wall is changed from neutral colour to warm colour and cold colour. Then the lightness of different colours is altered.

Figure 3. Teaching spaces in China (online) [36–42].

Table 2. Frequency analysis of demographic variables.

Variables	Options	Frequency	Percentage	Crowd	Average Value	Standard Deviation
Gender	Male	72	65.5%	1	1.35	0.48
	Female	38	34.5%			
Grade	Freshman year	5	4.5%	3	4.11	1.89
	Sophomore	14	12.7%			
	Junior	39	35.5%			
	Senior Year	15	13.6%			
	Master	18	16.4%			
	PhD	18	16.4%			
	Other	1	0.9%			
Category	Humanities and Social Sciences	16	14.5%	3	2.76	0.95
	Science	7	6.4%			
	Engineering	81	73.6%			
	Medicine	2	1.8%			
	Art Studies	1	0.9%			
	Other	3	2.7%			
	Total	110	100.0%			

2.3. Data Analysis

The data analysis software used for the study was SPSS 27. The reliability validity of the dependent variables was first examined, and correlation tests were used to assess whether there was a relationship between the dependent variables, after which the mean and standard deviation of the data were calculated. A repeated measures ANOVA was used to assess the effect of differences in the colour of the teaching space environment on participants' emotions in that context. A one-way ANOVA was used to test whether there were significant differences in the emotional perceptions of the teaching environment space between different anxious and depressed groups. One-way ANOVA was performed afterwards to test whether there were significant differences in the effects of education and major on the participants' emotions; independent samples *t*-test was used to test whether there were significant differences in the effects of gender on the participants' emotions in the difference of colour in the teaching space. The data obtained were presented in graphical or tabular form.

2.4. Reliability Validity Test

SPSS 27 was used to implement the process of reliability and validity analysis. First, we conducted reliability statistics on 14 HAD scale items and 21 questionnaire items respectively. It was found that their standard reliability coefficients were 0.907 and 0.925, which were very close to 1, meaning that the reliability was very high. Then, we conducted reliability statistics on all the questions in the questionnaire. According to the results of the reliability analysis of the overall scale, the Cronbach α coefficient based on standardized items was 0.837. It shows that the analysis results have high reliability. The validity analysis of the questionnaire was carried out by the test process through the approach of exploratory factor analysis in SPSS 27. According to the results of the exploratory factor analysis, the coefficient of the KMO test was 0.823, and the range of the coefficient of the KMO test was between 0 and 1. The closer to 1, the better is the validity of the questionnaire. According to the significance of the sphericity test, it can also be concluded that the significance of this test is infinitely close to 0. The significance is significantly less than 0.005, and the original hypothesis is rejected, indicating the questionnaire has good validity.

3. Results

3.1. Anxiety-Depression Evaluation

This paper analyses and discusses the results of the "Anxiety" and "Depression" sub-scales. For the "Anxiety" scale, a score of 7 or below was defined as healthy and asymptomatic, a score of 8–10 was defined as critical, and a score of 11–20 was defined as significantly anxious. Of the 110 subjects, 64.5% were healthy and asymptomatic, 20% were critical, and 15.5% showed serious anxiety symptoms. The "Depression" scale results were assessed the same way as the "Anxiety" scale. Of the 110 subjects, 68% were in a healthy state, 20% were in a critical state, and 12% had significant depressive symptoms. From the data, it can be concluded that more than 30% of the subjects suffered from mild or significant anxiety or depression, indicating that the phenomenon of anxiety and depression is prominent in this group which supports the validity of the opening H1.

3.2. The Effect of Colour on Mood

3.2.1. Correlation Analysis of Colour and Mood

According to the results of the correlation analysis presented in Figure 4, it can be observed that there are significant correlations among all variables. The correlation coefficients of the scores of all variables were more than 0, except for the negative correlation coefficient of the depression score. So, the anxiety-depression score was negatively correlated with the rest of the hue and lightness scores and positively correlated with the scores of all the remaining variables. For example, the correlation coefficient between the anxiety-depression score and the warm colour score is -0.318 **, which means that they are significantly correlated at the 99% significance level and are negatively correlated. By analogy, this can explain the correlation between all other variables. The higher the anxiety-depression value, the lower is the colour score. The higher the hue score, the higher is the lightness score.

3.2.2. Analysis of the Effect of Colour on Mood

The colours were divided into two parts: hue contrast (neutral, warm, and cold colours) and brightness contrast. The data in Tables 3 and 4 were scored according to three emotional criteria: pleasure, relaxation, and focus, as well as the box plots in Figure 5. According to the sphericity test results, p-value is less than 0.05, and the data does not fulfil the sphericity hypothesis. Combined with the results obtained from the multivariate test ($p = 0.000$), it can be found that the results demonstrate a statistically significant difference ($p < 0.001$), indicating that there is a significant difference in the effect of different colour classrooms on mood. This corroborates the validity of H2 at the beginning of this paper.

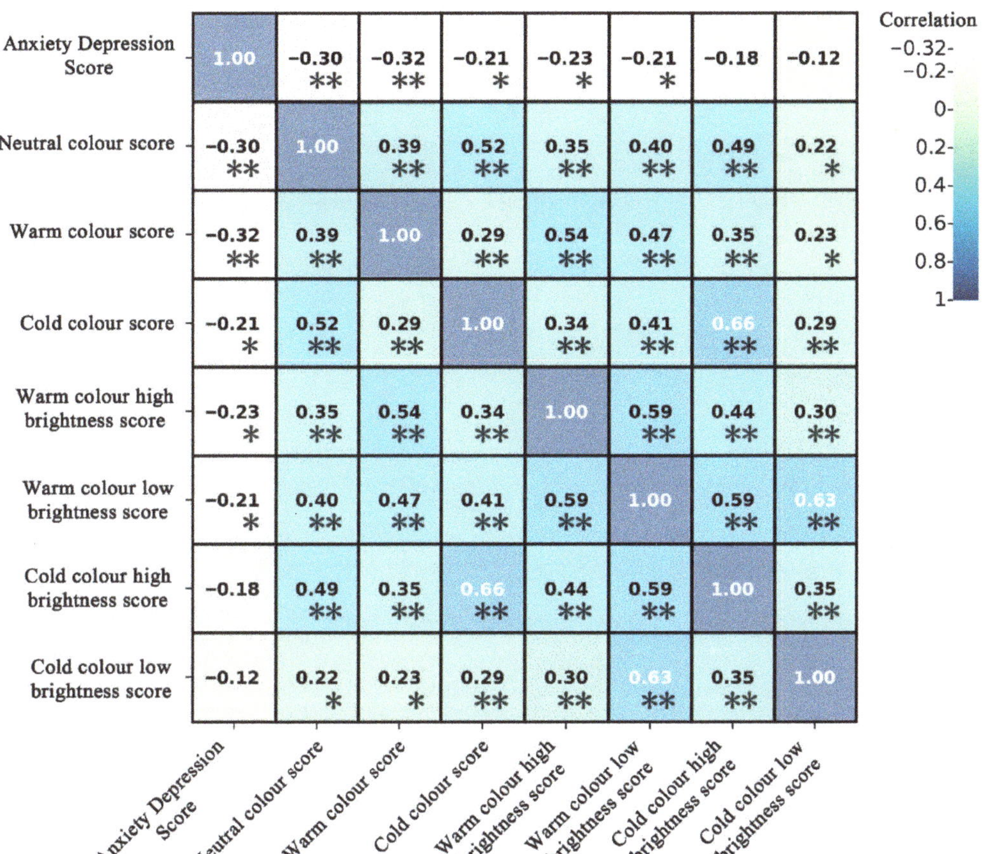

Figure 4. Correlation test among dimensions. Note: More asterisk "*" imply a stronger correlation.

Table 3. Colour hue mean and standard deviation.

	Hue					
	Neutral		Warm		Cold	
	M	SD	M	SD	M	SD
Pleasant/ Unpleasant	2.945	0.887	3.381	0.846	2.954	0.892
Relaxed/ Unrelaxed	2.81	0.869	3.427	0.795	3.009	0.914
Focused/ Unfocused	3.063	0.793	3.327	0.779	3.109	0.922

Note: Options are scored on a 5-point scale, with scores 1–5 corresponding to negative to positive emotions.

As can be observed from Table 3, for the three emotional criteria of pleasure, relaxation, and attention, the subjects' scores all showed with the mean values: warm classroom (WC) > cold classroom (CC) > neutral classroom (NC). This indicated that the healing effect of warm colours is greater than that of cold colours, and the healing effect of cold colours is greater than that of neutral colours. Combined with Table 4, it can be seen that the mean values of subjects' scores showed a trend of warm colour high brightness (WC-H) > cold colour high brightness (CC-H) > warm colour low brightness (WC-L) > cold colour low brightness (CC-L), indicating that the healing effect of high brightness is greater than that of low brightness based on colour hue.

Table 4. Mean and standard deviation of brightness.

	Brightness							
	Warm High Brightness		Warm Low Brightness		Cold High Brightness		Cold Low Brightness	
	M	SD	M	SD	M	SD	M	SD
Pleasant/ Unpleasant	3.518	0.993	2.981	0.878	3.173	0.876	2.764	1.013
Relaxed/ UnRelaxed	3.554	0.915	3.063	0.827	3.218	0.860	2.754	0.969
Focus/ Unfocused	3.427	0.893	3.082	0.920	3.255	0.818	2.836	0.934

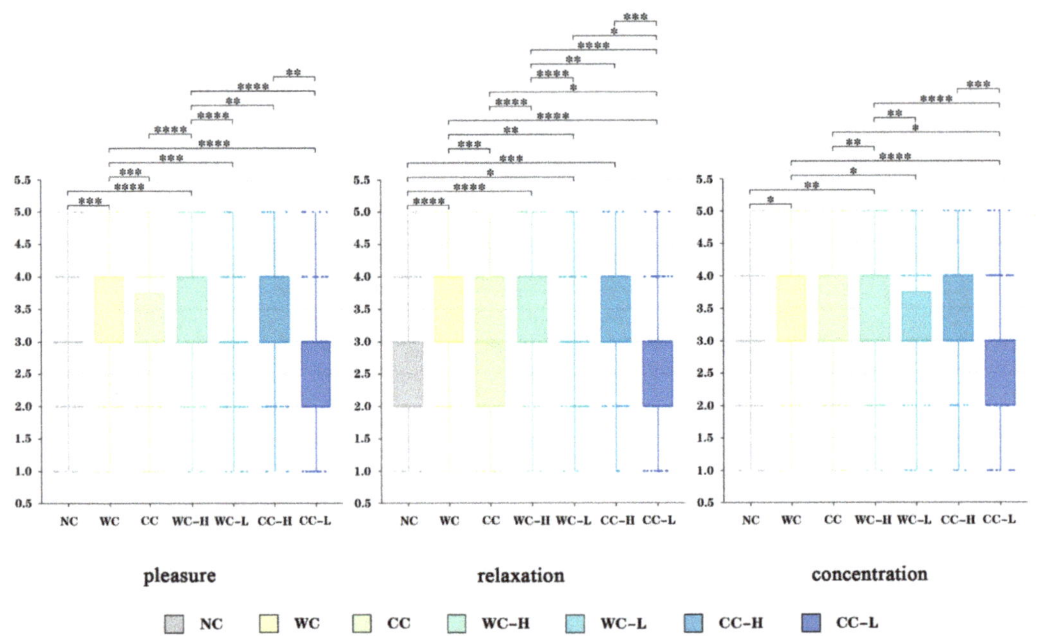

Figure 5. Box plot of emotion scores. Note: More asterisks "*" imply a stronger correlation.

Warm classrooms have the highest mean value of 3.381 in the "pleasant/unpleasant" category while cold classrooms do not differ significantly from neutral classrooms. The difference between the mean scores of warm and cold colours after adjusting the brightness is large (almost 0.1–0.2 points), and the mean value of 3.518 in warm high-brightness classrooms even exceeds that of warm classrooms itself. It is clear that warm colours play a pleasurable role in the emotional state, and higher brightness colours also make the mood more pleasant.

In the "relaxed/unrelaxed" category, warm colours have a significant relaxation effect, with a mean value of 3.427. Meanwhile, cold colours have a mean value of almost 0.2 points higher than neutral colours, which is a substantial difference. In terms of brightness, although the difference was still significant, the difference between the mean value of high brightness for cold colours and low brightness for warm colours decreased, while the mean value of low brightness for cool colours dropped to a minimum of 2.754.

In the "focused/unfocused" category, although the highest mean value was still for warm colours, the score decreased compared to the previous two moods (3.327). The difference between the cold and neutral colour classrooms was again not significant. The mean score for warm high luminosity also declined in this item. Relatively speaking, cold high luminosity scored 3.255 and it is the highest score among the three moods. The same is true for cold colours with low luminance, indicating that cool colours are easier to focus on mentally.

3.3. Different Effectiveness under Demographic Factors

3.3.1. Degree of Anxiety and Depression

Based on the results of the one-way ANOVA in Table 5, it can be seen that among the seven score dimensions, scores were significantly different across the anxiety population, as the significance tests were 0.02, 0.013, 0.016, 0.044, and 0.044. However, there was no significant difference in the cold-coloured low brightness classroom ($p = 0.476 > 0.05$). This corroborates the validity of H2 at the beginning of this paper.

Table 5. Results of the differences in the scores of each classroom on the three anxiety populations.

Variables	Options	N	Average Value	Standard Deviation	F	Significance	Multiple Comparisons
Neutral colour score	1	71	9.04	2.12	4.04	**0.02**	1 > 3, 2 > 3
	2	22	9.18	2.11			
	3	17	7.47	2.32			
Warm colour score	1	71	10.46	2.2	4.519	**0.013**	1 > 3, 2 > 3
	2	22	10.14	1.39			
	3	17	8.76	2.36			
Cold colour score	1	71	9.28	2.44	4.328	**0.016**	1 > 3, 2 > 3
	2	22	9.59	1.76			
	3	17	7.53	2.85			
Warm colour high brightness score	1	71	10.87	2.37	2.952	**0.057**	/
	2	22	10.18	2.24			
	3	17	9.35	2.83			
Warm colour low brightness score	1	71	9.28	2.36	3.214	**0.044**	1 > 3, 2 > 3
	2	22	9.59	1.89			
	3	17	7.88	2.23			
Cold colour high brightness score	1	71	9.65	2.42	3.216	**0.044**	2 > 3
	2	22	10.45	1.79			
	3	17	8.59	2.24			
Cold colour, low brightness score	1	71	8.35	2.7	0.747	**0.476**	/
	2	22	8.82	2.24			
	3	17	7.76	3.03			

Note: Where 1 represents people with no anxiety symptoms, 2 represents people with "critical" anxiety symptoms, and 3 represents people with significant anxiety symptoms. Value in bold means significant.

From the results of the multiple comparisons, it can be seen that, for different groups, the scores of "people without anxiety symptoms" are higher than those of "people with severe anxiety symptoms" and the scores of "people with possible anxiety symptoms" were also higher than those of "people with significant anxiety symptoms". Therefore, it can be concluded that the emotional satisfaction of "people with significant anxiety symptoms" with different colour spaces is significantly lower than that of the other two categories of anxious people, which is probably due to their high anxiety values. In the cold high brightness classroom, the scores of those with possible anxiety symptoms were greater than those with significant anxiety symptoms. Based on this result, it can be seen that "people with 'borderline' anxiety" feel pleasanter with the cold high brightness colour space than "people with severe anxiety".

According to the results of the one-way ANOVA in Table 6, it can be seen that among the seven score dimensions, the three categories of colour classroom scores, neutral colour score, warm colour score and warm high brightness, also differed significantly across depressed populations with significance tests of 0.041, 0.001, and 0.013, respectively, all significantly smaller than 0.05.

Table 6. Results of the differences in the scores of each classroom on the three depressed populations.

Variables	Options	N	Average Value	Standard Deviation	F	Significance	Multiple Comparisons
Neutral colour score	1	75	9.04	2.19	3.29	**0.041**	1 > 3, 2 > 3
	2	22	8.95	1.91			
	3	13	7.38	2.4			
Warm colour score	1	75	10.57	2.05	7.35	**0.001**	1 > 3, 2 > 3
	2	22	9.73	2.07			
	3	13	8.31	1.97			
Cold colour score	1	75	9.33	2.37	3.053	0.051	/
	2	22	9.09	2.29			
	3	13	7.54	2.9			
Warm colour high brightness score	1	75	10.93	2.42	4.485	**0.013**	1 > 3
	2	22	9.91	2.11			
	3	13	9	2.58			
Warm colour low brightness score	1	75	9.43	2.41	2.257	0.11	/
	2	22	8.68	1.52			
	3	13	8.15	2.51			
Cold colour high brightness score	1	75	9.88	2.48	1.931	0.15	/
	2	22	9.5	1.57			
	3	13	8.54	2.26			
Cold colour, low brightness score	1	75	8.37	2.74	1.272	0.284	/
	2	22	8.86	2.15			
	3	13	7.38	2.96			

Note: 1 represents people with no depressive symptoms, 2 represents people with "borderline" depressive symptoms, and 3 represents people with significant depressive symptoms. Value in bold means significant.

Based on the results of the multiple comparisons, it can be seen that for both the neutral and warm colour scores, the scores of "people without depression symptoms" and "people with 'borderline' depression symptoms scores were higher than the "people with significant depression symptoms" scores. It shows that "people with significant depression symptoms" are significantly less happy with neutral and warm colour spaces than the other two anxious groups, probably due to their high depression values. The "no depression symptoms" group scored higher than the "with significant depression symptoms" group for the high brightness classroom scores of warm colours. Based on this result, it can be seen that the "non-depressed" group felt pleasanter with the warm, high-light colour space than the "significantly anxious" group.

3.3.2. Gender

According to the results of the independent samples *t*-test in Table 7, it can be seen that there is no significant difference in most of the different colour and brightness classroom scores by gender, but only in the warm colour high brightness classrooms. The significance test for the difference between the scores of warm colour high brightness classrooms by gender is 0.033, which is less than 0.05, indicating that there is a difference in the degree of preference for warm colour high brightness classrooms among students of different genders. Based on the mean values, it can be seen that females rated slightly higher than males, thus females have a higher preference for warm-coloured high-brightness classrooms than males. The remaining variables are not statistically significantly different in terms of gender because the significance is greater than the standard 0.05, so the original hypothesis cannot be rejected.

Table 7. Analysis of the differences between dimensions in terms of gender.

Variables	Gender	Number of Cases	Average Value	Standard Deviation	t	Significance
Anxiety and Depression score	Male	72	13.22	7.575	1.248	0.215
	Female	38	11.32	7.697		
Neutral colour score (18–20)	Male	72	8.92	2.336	0.583	0.561
	Female	38	8.66	1.963		
Warm colour score (21–23)	Male	72	10.03	2.195	−0.724	0.471
	Female	38	10.34	2.109		
Cold colour score (24–26)	Male	72	9.25	2.281	1.039	0.301
	Female	38	8.74	2.777		
Warm colour high brightness score (27–29)	Male	72	10.14	2.44	−2.155	**0.033**
	Female	38	11.18	2.381		
Warm colour low brightness score (30–32)	Male	72	9.11	2.243	−0.101	0.92
	Female	38	9.16	2.444		
Cold colour high brightness score (33–35)	Male	72	9.69	2.329	0.303	0.763
	Female	38	9.55	2.345		
Cold colour low brightness score (36–38)	Male	72	8.6	2.51	1.32	0.19
	Female	38	7.89	2.911		

Note: Bold means significant.

3.3.3. Education Background

We divided the education into undergraduate and master's degrees for comparison and the majors into science and non-science disciplines for analysis. According to the results of the one-way ANOVA in Tables 8 and 9, it can be seen that because the significance is greater than the standard 0.05 for all, there is no significant difference in each dimension score in both education and major, so the original hypothesis cannot be rejected. That means, there is no significant difference in judging the influence of teaching space on emotion among different academic majors, and the opening H4 is overturned.

Table 8. Results of the variance analysis of each dimension in terms of education.

Variables	Options	N	Average Value	Standard Deviation	F	Significance	Multiple Comparisons
Anxiety and Depression score	Undergraduate	73	12.79	7.636	0.197	0.658	/
	Master's degree	37	12.11	7.724			
Neutral colour score (18–20)	Undergraduate	73	9.03	2.134	1.795	0.183	/
	Master's degree	37	8.43	2.328			
Warm colour score (21–23)	Undergraduate	73	10.07	2.03	0.213	0.646	/
	Master's degree	37	10.27	2.423			
Cold colour score (24–26)	Undergraduate	73	9.03	2.374	0.073	0.788	/
	Master's degree	37	9.16	2.662			
Warm colour high brightness score (27–29)	Undergraduate	73	10.77	2.378	2.597	0.11	/
	Master's degree	37	9.97	2.566			
Warm colour low brightness score (30–32)	Undergraduate	73	9.08	2.139	0.082	0.775	/
	Master's degree	37	9.22	2.626			
Cold colour high brightness score (33–35)	Undergraduate	73	9.64	2.33	0	0.992	/
	Master's degree	37	9.65	2.348			
Cold colour low brightness score (36–38)	Undergraduate	73	8.4	2.454	0.055	0.814	/
	Master's degree	37	8.27	3.07			

Table 9. Results of the analysis of the differences between the dimensions in terms of profession.

Variables	Options	N	Average Value	Standard Deviation	F	Significance	Multiple Comparisons
Anxiety and Depression score	Non-Scientific	22	11.91	8.28	0.201	0.655	/
	Science and Engineering	88	12.73	7.509			
Neutral colour score (18–20)	Non-Scientific	22	9.14	2.315	0.536	0.466	/
	Science and Engineering	88	8.75	2.188			
Warm colour score (21–23)	Non-Scientific	22	10.73	2.492	2.077	0.152	/
	Science and Engineering	88	9.99	2.059			
Cold colour score (24–26)	Non-Scientific	22	9.05	2.968	0.003	0.954	/
	Science and Engineering	88	9.08	2.34			
Warm colour high brightness score (27–29)	Non-Scientific	22	11.09	2.348	1.596	0.209	/
	Science and Engineering	88	10.35	2.478			
Warm colour low brightness score (30–32)	Non-Scientific	22	9.36	2.498	0.288	0.593	/
	Science and Engineering	88	9.07	2.263			

Table 9. *Cont.*

Variables	Options	N	Average Value	Standard Deviation	F	Significance	Multiple Comparisons
Cold colour high brightness score (33–35)	Non-Scientific	22	9.82	2.954	0.151	0.699	/
	Science and Engineering	88	9.6	2.158			
Cold colour low brightness score (36–38)	Non-Scientific	22	8.77	3.038	0.676	0.413	/
	Science and Engineering	88	8.25	2.57			

According to the analysis of the results, the different demographic variables do not differ significantly in judging the influence of teaching space on emotions, overturning the opening H4.

4. Discussion

4.1. Anxiety and Depression

There has been some investigations about the effects of quarantine on psychology. One study [43] compared psychological outcomes during quarantine with later outcomes and found that during quarantine, 7% (126 of 1656) showed anxiety symptoms. A study [44] of hospital staff who might have come into contact with SARS found that immediately after the quarantine period (9 days) ended, having been quarantined was the factor most predictive of symptoms of acute stress disorder. Some scholars did a review [4] of the psychological impact of quarantine using three electronic databases. Most reviewed studies reported negative psychological effects including post-traumatic stress symptoms, confusion, and anger. And conclusion 1, indicating prominent anxiety and depression among college students during campus lockdown, is in accord with existing findings. However, one study [45] compared undergraduates who had been quarantined with those not quarantined immediately after the quarantine period and found no significant difference between the groups in terms of post-traumatic stress symptoms or general mental health problems. Although this is inconsistent with our conclusion 1, it provides a new perspective for our future research, which can compare the students' psychological state before and after lockdown.

4.2. Colour and Mood

Though no research has revealed a one-to-one relationship between mood and colour [46], it is believed that different colours have corresponding emotional preferences and different degrees of health effects [47]. For example, warm colours stimulate the spirits and help relieve depression, while cool colours are more calming and relaxing for nerv-ousness [48, 49]. Conclusion 2 agrees with the basic theories of colour psychology. In addition, in the view of colour psychology, colours with higher brightness are more popular than those with lower brightness. Conclusion 2 verifies this theory and is consistent with prior studies [50]. The study of Costa Marco et al. [50] on the colour of college students' dorm rooms indicated that blue interior spaces facilitate various learning activities and make it easier for students to be calm and concentrated. Chong Gao et al. [51] found that patients with depression symptoms find it harder to recover when in blue interior spaces, compared with white and warm interior spaces. Yildirim et al. [52] in their study of living room colours also showed that warm colours were highly stimulating to evoke mood, while cool colours were more associated with "expanding space" and "resting". Bilal et al. [53] suggest that neutral colours, such as grey, can reduce the feeling of pleasure for guests in hotel rooms. Thus conclusion 3 correlates with existing studies.

In addition, existing studies related to colour psychology have indicated that there are significant differences in colour preferences between genders. For example, Costa Marco et al. [50] discovered substantial disparities between men and women in their preference for blue and purple dormitory spaces. Al-Rasheed [54] concluded that gender-specific preferences for colour exist in both Arabic and English cultural circles, with men preferring blue green. However, conclusion 4 is not fully consistent with the existing studies. In addition to gender, other studies focused on demographic elements such as age and income, like Cho [55] who identified substantial disparities in household income and age in terms of satisfaction with the interior colours of luxury stores.

4.3. Limitation

In addition, there are some limitations in this study.

This study is based on a relatively homogeneous geographical and cultural background, with subjects mostly coming from college students in Harbin, China, who are enrolled in universities with excellent academic reputations and good public images. In other regions, traits like language, lifestyle, weather, and ethnic background are all different. Comparison studies on different regions in China can be supplemented in the future. Additionally, this study mainly collected questionnaires during the school closure period. In future investigation, the range of subjects could be further expanded. A wide variety of students such as home commuting students, resident students, and even senior/junior students could also be considered.

This paper used the HAD scale to assess and classify the subjects' anxiety and depression symptoms. Future studies can increase the psychological assessment dimensions (e.g., the combination of multiple scales) to increase the credibility and accuracy of the evaluation. Apart from that, this study focused on the subjective feelings of the subjects, so the data obtained are subjective emotions. In the future, the physiological indicators of the subjects can be monitored and analysed in conjunction with real-life experiments. Moreover, only three emotional criteria, "pleasant/unpleasant", "relaxed/unrelaxed", and "focused/unfocused", were selected for evaluation, and there were few emotional indicators. Future studies can add emotional indicators to improve the evaluation.

In this study, for the sake of the controllability of the experiment and the accuracy of the results, other environmental components that affect indoor colour (e.g., light [56,57], furniture, material, etc.) were not discussed. Future studies may try to add relevant elements as variables to increase the exploration of more dimensions of colour in indoor teaching spaces. Furthermore, three hues and two kinds of lightness were selected for the study. The classification was simple and lacked specific colour values for support. In future, studies can take more colours and more colour dimensions (e.g., grey scale) into consideration, apply more detailed and specific classification methods, and combine colour parameters.

5. Conclusions

In this paper, a study was conducted on the emotional impact of environmental colour on college students in the indoor teaching space during the epidemic closure through a questionnaire survey method. The conclusions are as follows:

1. In the context of the "Dynamic Zero-Covid" policy, constant campus lockdown leads to prominent anxiety and depression among college students. More than 30% of the subject group suffered from mild or significant anxiety or depressive symptoms.
2. Chinese college students have colour preferences in teaching spaces. In the three teaching spaces of warm, cold, and neutral colours, warm colours have an advantage over both cold and neutral colours in creating pleasure, relaxation, and mental focus. Among the three types of teaching spaces, neutral colours provide the worst experience in terms of obtaining a positive mood. It is concluded that warm-coloured classrooms are more healing than cold-coloured classrooms, and cold-coloured classrooms are more healing than neutral-coloured classrooms. With the same colour hue, high brightness classrooms tend to have better healing effects than low brightness classrooms. The results might correlate with the Chinese colour preference for red and light colours.
3. There is a correlation between the teaching space colour score and the level of anxiety and depression of the subjects. When subjects have higher degrees of anxiety and depression, they are less satisfied with the colour of the teaching space. There are

some differences in the experiences of people with different anxiety and depression symptoms in different colours of teaching spaces.

4. In most cases, there is no significant difference in the colour preference of teaching spaces between the gender groups. However, there is a significant difference between males and females in warm high-brightness teaching spaces, with females having a higher preference for warm high-brightness classrooms than males. There is no significant difference in colour preference of teaching space among the different education groups.

Author Contributions: Conceptualization, Y.W. and W.T.; methodology, Y.W. and W.T.; software, W.L.; validation, W.T. and W.L.; formal analysis, W.T.; investigation, W.T. and W.L.; resources, W.T. and W.L.; data curation, W.L.; writing—original draft preparation, W.T. and W.L.; writing—review and editing, W.T., Y.W. and W.L.; visualization, W.L.; supervision, Y.W. and F.L.; project administration, W.T.; funding acquisition, F.L. All authors have read and agreed to the published version of the manuscript.

Funding: The research was funded by [the Ministry of Science and Technology of China] grant number [G2021179030L].

Conflicts of Interest: The authors declare no conflict of interest.

Appendix A

Table A1. The Colour Evaluation Questionnaire of Teaching Spaces.

1. Your gender is ( )	Male	Female			
2. Your current grade is ( )	Freshman	Sophomore	Junior year	Senior year	Grade five
	Master's degree	Doctor	Others		
3. Your major category is ( )	Humanities and Social Sciences	Natural Sciences	Engineering	Medicine	Arts
	Others				
4–17. HAD Scale (Table 1)					

Please observe photo group 1 carefully and answer questions 18–20 truthfully according to your feelings

18. This set of photos makes you feel ( )	very unpleasant	unpleasant	general	pleasure	very pleasant
19. This set of photos makes you feel ( )	very unrelaxed	unrelaxed	general	relaxed	very relaxed
20. This set of photos makes you feel ( )	very unfocused	unfocused	general	focused	very focused

Table A1. *Cont.*

Please observe photo group 2 carefully and answer questions 21–23 truthfully according to your feelings

21. This set of photos makes you feel ( )	very unpleasant	unpleasant	general	pleasure	very pleasant
22. This set of photos makes you feel ( )	very unrelaxed	unrelaxed	general	relaxed	very relaxed
23. This set of photos makes you feel ( )	very unfocused	unfocused	general	focused	very focused

Please observe photo group 3 carefully and answer questions 24–26 truthfully according to your feelings

24. This set of photos makes you feel ( )	very unpleasant	unpleasant	general	pleasure	very pleasant
25. This set of photos makes you feel ( )	very unrelaxed	unrelaxed	general	relaxed	very relaxed
26. This set of photos makes you feel ( )	very unfocused	unfocused	general	focused	very focused

Please observe photo group 4 carefully and answer questions 27–29 truthfully according to your feelings

27. This set of photos makes you feel ( )	very unpleasant	unpleasant	general	pleasure	very pleasant
28. This set of photos makes you feel ( )	very unrelaxed	unrelaxed	general	relaxed	very relaxed
29. This set of photos makes you feel ( )	very unfocused	unfocused	general	focused	very focused

Table A1. *Cont.*

Please observe photo group 5 carefully and answer questions 30–32 truthfully according to your feelings

30. This set of photos makes you feel ( )	very unpleasant	unpleasant	general	pleasure	very pleasant
31. This set of photos makes you feel ( )	very unrelaxed	unrelaxed	general	relaxed	very relaxed
32. This set of photos makes you feel ( )	very unfocused	unfocused	general	focused	very focused

Please observe photo group 1 carefully and answer questions 33–35 truthfully according to your feelings

33. This set of photos makes you feel ( )	very unpleasant	unpleasant	general	pleasure	very pleasant
34. This set of photos makes you feel ( )	very unrelaxed	unrelaxed	general	relaxed	very relaxed
35. This set of photos makes you feel ( )	very unfocused	unfocused	general	focused	very focused

Please observe photo group 1 carefully and answer questions 36–38 truthfully according to your feelings

36. This set of photos makes you feel ( )	very unpleasant	unpleasant	general	pleasure	very pleasant
37. This set of photos makes you feel ( )	very unrelaxed	unrelaxed	general	relaxed	very relaxed
38. This set of photos makes you feel ( )	very unfocused	unfocused	general	focused	very focused

References

1. UN. *The 2030 Agenda for Global Action and the Sustainable Development Goals*; United Nations: New York, NY, USA, 2015.
2. UN. *Convention on the Rights of Persons with Diabilities*; United Nations: New York, NY, USA, 2006.

3. Wang, C.; Pan, R.; Wan, X.; Tan, Y.; Xu, L.; McIntyre, R.S.; Choo, F.N.; Tran, B.; Ho, R.; Sharma, V.K.; et al. A longitudinal study on the mental health of the general population during the COVID-19 epidemic in China. *Brain Behav. Immun.* **2020**, *87*, 40–48. [CrossRef] [PubMed]
4. Brooks, S.K.; Webster, R.K.; Smith, L.E.; Woodland, L.; Wessely, S.; Greenberg, N.; Rubin, G.J. The psychological impact of quarantine and how to reduce it: Rapid review of the evidence. *Lancet* **2020**, *395*, 912–920. [CrossRef]
5. Wang, C.; Pan, R.; Wan, X.; Tan, Y.; Xu, L.; Ho, C.S.; Ho, R.C. Immediate psychological responses and associated factors during the initial stage of the 2019 coronavirus disease (COVID-19) epidemic among the general population in China. *Int. J. Environ. Res. Public Health* **2020**, *17*, 1729. [CrossRef] [PubMed]
6. Sprang, G.; Silman, M. Post-traumatic stress disorder in parents and youth after health-related disasters. *Disaster Med. Public Health Prep.* **2013**, *7*, 105–110. [CrossRef] [PubMed]
7. Marelli, S.; Castelnuovo, A.; Somma, A.; Castronovo, V.; Mombelli, S.; Bottoni, D.; Leitner, C.; Fossati, A.; Ferini-Stramb, L. Impact of COVID-19 lockdown on sleep quality in university students and administration staff. *J. Neurol.* **2021**, *268*, 8–15. [CrossRef] [PubMed]
8. Li, H.Y.; Cao, H.; Leung, D.Y.P.; Mak, Y.W. The Psychological Impacts of a COVID-19 Outbreak on College Students in China: A Longitudinal Study. *Int. J. Environ. Res. Public Health* **2020**, *17*, 3933. [CrossRef]
9. Panchal, U.; Salazar de Pablo, G.; Franco, M.; Moreno, C.; Parellada, M.; Arango, C.; Fusar-Poli, P. The impact of COVID-19 lockdown on child and adolescent mental health: A systematic review. *Eur. Child Adolesc. Psychiatry* **2021**, *8*, 1–27. [CrossRef]
10. Zhang, C.; Ye, M.; Fu, Y.; Yang, M.; Luo, F.; Yuan, J.; Tao, Q. The Psychological Impact of the COVID-19 Pandemic on Teenagers in China. *J. Adolesc. Health* **2020**, *67*, 747–755. [CrossRef]
11. Wan, J.; Zhou, Y.; Li, Y.; Su, Y.; Cao, Y.; Zhang, L.; Ying, L.; Deng, W. Research on Color Space Perceptions and Restorative Effects of Blue Space Based on Color Psychology: Examination of the Yijie District of Dujiangyan City as an Example. *Int. J. Environ. Res. Public Health* **2020**, *17*, 3137. [CrossRef]
12. Song, J.M. The theory of colour psychology, design careers and experiments. *Decoration* **2020**, *4*, 21–26. (In Chinese)
13. Elliot, A.J.; Maier, M.A. Color Psychology: Effects of Perceiving Color on Psychological Functioning in Humans. *Annu. Rev. Psychol.* **2014**, *65*, 95–120. [CrossRef]
14. Puntenney, I. Color Psychology and Color Therapy. *Am. J. Ophthalmol.* **1950**, *33*, 1619. [CrossRef]
15. Ou, L.C.; Luo, M.R.; Woodcock, A.; Wright, A. A study of colour emotion and colour preference. Part II: Colour emotions for two-colour combinations. *Colour Res. Appl.* **2004**, *29*, 292–298. [CrossRef]
16. Ou, L.C.; Luo, M.R.; Woodcock, A.; Wright, A. A study of colour emotion and colour preference. Part III: Colour preference modelling. *Colour Res. Appl.* **2004**, *29*, 381–389. [CrossRef]
17. Ahn, J.-H.; Kim, J.-D. Theoretical Consideration on Color Therapy. *J. Naturop.* **2013**, *2*, 75–76.
18. Glenn. Overview of Chinese and Western healing environments. *China Hosp. Archit. Equip.* **2013**, *14*, 25–28. (In Chinese)
19. Ai, M.; Liu, Y.H.; Qi, X.H.; Lei, L.; Chen, Y.S.; Li, P.; Yang, S. The effect of colour on human physiology and psychology. *Chin. J. Health Psychol.* **2015**, *2*, 317–320. (In Chinese)
20. Huang, S.-S.; Xu, A.; Zhou, A.-H. Research progress of colour psychology in the application of disease-assisted treatment. *Gen. Pract. Nurs.* **2020**, *18*, 288–290. (In Chinese)
21. Saito, M. Comparative Studies on Colour Preference in Japan and Other Asian Regions, with Special Emphasis on the Prefer-ence for White. *Colour Res. Appl.* **1996**, *21*, 35–49. [CrossRef]
22. Zhao, E.; Liu, T. A Study of the International Colour Sensibility through the Analysis of the Ethnic Colour Preference. *J. Korean Soc. Costume* **2012**, *62*, 38–52.
23. Madden, T.J.; Hewett, K.; Roth, M.S. Managing Images in Different Cultures: A Cross-National Study of Color Meanings and Preferences. *J. Int. Mark.* **2000**, *8*, 90–107. [CrossRef]
24. Zhang, Y.; Liu, P.; Han, B.; Xiang, Y.; Li, L. Hue, Chroma, and Preference in Chinese Adults: Age and Gender Differences. *Colour Res. Appl.* **2019**, *44*, 967–980. [CrossRef]
25. Yinan, S. The application of colour medicine in colleges and universities. *Art Sci. Technol.* **2014**, *27*, 42, 197. (In Chinese)
26. Terol-Cantero, M.C.; Cabrera-Perona, V.; Martín-Aragón, M. Hospital Anxiety and Depression Scale (HADS) review in Spanish Samples. *An. Psicol. Ann. Psychol.* **2015**, *31*, 494–503. [CrossRef]
27. Lam, C.L.K.; Pan, P.; Chan, A.W.T.; Chan, S.; Munro, C. Can the Hospital Anxiety and Depression (HAD) Scale be used on Chinese elderly in general practice? *Fam. Pract.* **1995**, *12*, 149–154. [CrossRef] [PubMed]
28. Mykletun, A.; Stordal, E.; Dahl, A. Hospital Anxiety and Depression (HAD) scale: Factor structure, item analyses and internal consistency in a large population. *Br. J. Psychiatry* **2001**, *179*, 540–544. [CrossRef]
29. El-Rufaie, O.E.F.; Absood, G.H. Retesting the validity of the Arabic version of the Hospital Anxiety and Depression (HAD) scale in primary health. *Soc. Psychiatry Psychiatr. Epidemiol.* **1995**, *30*, 26–31. [CrossRef]
30. Cabrera, V.; Martin-Aragon, M.; del Carmen Terol, M.; Nunez, R.; de los Angeles Pastor, M. Hospital Anxiety and Depression Scale (HADS) in fibromyalgia: Sensitivity and specificity analysis. *Ter. Psicol.* **2015**, *33*, 181–193. [CrossRef]
31. Melin, E.O.; Svensson, R.; Thulesius, H.O. Psychoeducation against depression, anxiety, alexithymia and fibromyalgia: A pilot study in primary care for patients on sick leave. *Scand. J. Prim. Health Care* **2018**, *36*, 123–133. [CrossRef]
32. Andrea, H.; Bültmann, U.; Beurskens AJ, H.M.; Swaen GM, H.; Van Schayck, C.P.; Kant, I.J. Anxiety and depression in the working population using the HAD Scale. *Soc. Psychiatry Psychiatr. Epidemiol.* **2004**, *39*, 637–646. [CrossRef]

33. Lamey, P.-J.; Lamb, A.B. The usefulness of the HAD scale in assessing anxiety and depression in patients with burning mouth syndrome. *Oral Surg. Oral Med. Oral Pathol.* **1989**, *67*, 390–392. [CrossRef]

34. Zhou, Y.; Zhang, H.; Zhang, X.Z.; Zhang, F. Evaluation of the psychological status of infertile women using the HAD scale. *J. Nurs.* **2005**, *20*, 65–66. (In Chinese)

35. Na, M.; Xiuqi, S.; Zhongjuan, Z.; Wandan, G. Clinical application significance of HAD scale in the treatment of elderly patients with arrhythmias of different complexity. *Harbin Med.* **2017**, *37*, 268–269. (In Chinese)

36. Available online: http://sxjy.chnu.edu.cn/Item/91321.aspx (accessed on 10 October 2022).

37. Available online: https://www.nipic.com/show/19457804.html (accessed on 10 October 2021).

38. Available online: https://www.nipic.com/show/18125100.html (accessed on 10 October 2022).

39. Available online: https://www.sohu.com/a/237068862_707429 (accessed on 10 October 2021).

40. Available online: https://www.meipian.cn/2cvb5cd3 (accessed on 10 October 2022).

41. Available online: https://baike.baidu.com/pic/%E6%95%99%E5%AE%A4/10996397/1/c75c10385343fbf2b2116b09e22ddd8065380cd721c7?fr=newalbum#aid=1528196564&pic=35a85edf8db1cb134954ff594d07414e9258d109dd67 (accessed on 10 October 2021).

42. Available online: https://baike.baidu.com/pic/%E6%95%99%E5%AE%A4/10996397/1/c75c10385343fbf2b2116b09e22ddd8065380cd721c7?fr=newalbum#aid=1528196564&pic=14ce36d3d539b6003af3b206531c222ac65c10381967 (accessed on 10 October 2021).

43. Jeong, H.; Yim, H.W.; Song, Y.J.; Ki, M.; Min, J.A.; Cho, J.; Chae, J.H. Mental health status of people isolated due to Middle East respiratory syndrome. *Epidemiol. Health* **2016**, *38*, e2016048. [CrossRef]

44. Bai, Y.; Lin, C.-C.; Lin, C.-Y.; Chen, J.-Y.; Chue, C.-M.; Chou, P. Survey of stress reactions among health care workers involved with the SARS outbreak. *Psychiatr. Serv.* **2004**, *55*, 1055–1057. [CrossRef]

45. Wang, Y.; Xu, B.; Zhao, G.; Cao, R.; He, X.; Fu, S. Is quarantine related to immediate negative psychological consequences during the 2009 H1N1 epidemic? *Gen. Hosp. Psychiatry* **2011**, *33*, 75–77. [CrossRef]

46. Fehrman, K.; Fehrman, C. *Color: The Secret Influence*; Prentice Hall: Upper Saddle River, NJ, USA, 2000.

47. O'Connor, Z. Colour psychology and colour therapy: Caveat emptor. *Color Res. Appl.* **2011**, *36*, 229–234. [CrossRef]

48. Wexner, L.B. The degree to which colours (hues) are associated with mood-tones. *Appl. Psychol.* **1954**, *38*, 432–435. [CrossRef]

49. Salonen, H.; Lahtinen, M.; Lappalainen, S.; Nevala, N.; Knibbs, L.D.; Morawska, L.; Reijula, K. Physical characteristics of the indoor environment that affect health and wellbeing in healthcare facilities: A review. *Intell. Build. Int.* **2013**, *5*, 3–25. [CrossRef]

50. Marco, C.; Sergio, F.; Mattia, N.; Iacopo, P. Interior Color and Psychological Functioning in a University Residence Hall. *Front. Psychol.* **2018**, *9*, 1580.

51. Gao, C.; Zhang, S. The restorative quality of patient ward environment: Tests of six dominant design characteristics. *Build. Environ.* **2020**, *180*, 107039. [CrossRef]

52. Yildirim, K.; Hidayetoglu, M.L.; Capanoglu, A. Effects of Interior Colors on Mood and Preference: Comparisons of Two Living Rooms. *Percept. Mot. Skills* **2011**, *112*, 509–524. [CrossRef] [PubMed]

53. Bilal, S.Y.; Aslanolu, R.; Olguntürk, N. Colour, emotion, and behavioural intentions in city hotel guestrooms. *Colour Res. Appl.* **2021**, *47*, 771–782. [CrossRef]

54. Al-Rasheed, A.S. An experimental study of gender and cultural differences in hue preference. *Front. Psychol.* **2015**, *6*, 30. [CrossRef]

55. Cho, J.Y.; Lee, E.-J. Impact of Interior Colors in Retail Store Atmosphere on Consumers' Perceived Store Luxury, Emotions, and Preference. *Cloth. Text. Res. J.* **2017**, *35*, 33–48. [CrossRef]

56. Kong, Z.; Zhang, R.; Ni, J.; Ning, P.; Kong, X.; Wang, J. Towards an integration of visual comfort and lighting impression: A field study within higher educational buildings. *Build. Environ.* **2022**, *216*, 108989. [CrossRef]

57. Kong, Z.; Liu, Q.; Li, X.; Hou, K.; Xing, Q. Indoor lighting effects on subjective impressions and mood states: A critical review. *Build. Environ.* **2022**, *224*, 109591. [CrossRef]

 *buildings*

Article

The Effect of Daylight Illumination in Nursing Buildings on Reading Comfort of Elderly Persons

Yao Fu [1,2], Yue Wu [1], Weijun Gao [2,3,*] and Rong Hui [4]

1   School of Architecture and Urban Planning, Shenyang Jianzhu University, Shenyang 110168, China; fuyao@sjzu.edu.cn (Y.F.); wuyue@stu.sjzu.edu.cn (Y.W.)
2   Faculty of Environmental Engineering, The University of Kitakyushu, Kitakyushu 808-0135, Japan
3   ISMART (Innovation Institute for Sustainable Maritime Architecture Research and Technology), Qingdao University of Technology, Qingdao 266011, China
4   Hunan Branch of CSCEC Southwest Design and Research Institute Co., Ltd., Changsha 410007, China; swqswk1992@163.com
*   Correspondence: gaoweijun@me.com

Abstract: Reading is one of the popular activities among elderly persons. A reasonable level of daylight illumination can ensure the visual comfort of reading for elderly persons. State arousal level and subjective comfort report are important parameters reflecting the effect of daylight illumination on visual comfort of reading in elderly persons. In this study, daylight illumination measurements were conducted in a nursing institution of Shenyang, China. Moreover, the methods of electrodermal activity (EDA) physiological index measurement and questionnaire scoring were used to compare and analyze the state arousal level and visual comfort of elderly persons under different illumination conditions. The results show that when elderly persons were involved in their daily reading activity, the acceptable daylight illumination range was between 300 and 1000 lx. When the daylight illumination was between 600 and 800 lx, the state arousal level and visual comfort was high; when it was 700 lx, the state arousal level and visual comfort was the highest. Although 500 and 900 lx both indicated neutral illumination, the evaluations were more consistent at 500 lx than at 900 lx. At 300, 400, and 1000 lx, visual comfort was poor and the state arousal level was low. At 300 lx, visual comfort was the worst and the state arousal level was the lowest. This study provides a reliable reference for architects to design the daylight conditions of the living spaces of the elderly.

Keywords: state arousal level; visual comfort; daylight illumination; elderly persons; reading behavior

Citation: Fu, Y.; Wu, Y.; Gao, W.; Hui, R. The Effect of Daylight Illumination in Nursing Buildings on Reading Comfort of Elderly Persons. *Buildings* **2022**, *12*, 214. https://doi.org/10.3390/buildings12020214

Academic Editor: Alessandro Cannavale

Received: 30 December 2021
Accepted: 8 February 2022
Published: 15 February 2022

Publisher's Note: MDPI stays neutral with regard to jurisdictional claims in published maps and institutional affiliations.

1. Introduction

Globally, an increasing number of people are aging; consequently, the quality of life of elderly persons has become the focus of the international community. It is crucial to provide a comfortable lighting environment for elderly persons. Due to visual decline [1], special consideration should be given to the lighting design of their living spaces. Reading is among the most popular leisure activities among elderly persons [2–5]. However, with age, the ciliary muscle loses its ability to contract and the pupil size decreases, resulting in presbyopia [6]. There is insufficient light intake for narrow pupils. Hence, elderly persons often need higher illumination to make up for the decline in visual ability. Studies have shown that illumination considerably influences the reading ability of elderly persons with low vision [7]. As the world's population ages, many countries and institutions have studied the illumination standard of reading for elderly persons and recommended specific values and ranges. However, lighting design standards in different countries do not have an agreed illumination value required for reading. The Architectural Lighting Design Standard of China [8] stipulates that, typically, reading lighting in the bedroom should be 300 lx. American National Standards Institute (ANSI) lighting standards [9] stipulate that reading lighting in the bedroom is 750 lx. The Lighting Handbook of Japan [10] stipulates that reading lighting in the bedroom is 600–1500 lx. Some scholars have conducted studies on the illumination preferred by seniors in terms of reading. Robert G. Davis conducted an experiment on the visual preference of elderly persons when reading under 1076, 107.6, and 10.76 illuminations. Results indicate that the favored illumination of elders was 1076 lx [11]. Zhang and Ma investigated the visual executive power and subjective comfort of elderly persons when reading under 50, 300, and 1000 lx. The results showed that elderly persons had the best visual executive power and relatively good subjective comfort when reading under 1000 lx [12]. The standards and research listed above were set or carried out under the assumption of artificial lighting. According to the living habits of the elderly persons, it is normal for them to read in daylight in front of windows. In the context of carbon peaking and carbon neutrality, it is also encouraged to make full use of daylight in daily activities. However, no detailed regulations apply to daylight illumination in the lighting design standards according to indoor activities of elderly persons in China, Japan, or the United States.

Illumination is a common parameter for evaluating the quality of daylight environment, and lighting evaluation methods such as DF, DA, and UDI all use illumination as the basic parameter. Illumination has an important supporting role for human visual function, and is not only the most important photometric indicator for elderly persons to complete reading tasks, but also one of the important indicators for evaluating the comfort and health associated with the lighting environment. Furthermore, elderly persons, as a special social group, have a certain dependence on daylight [13]. Physiologically, daylight can regulate the secretion of melatonin in their bodies and promote the absorption of calcium ions. Psychologically, daylight can reduce the risk of depression in elderly persons. Although they have different needs for daylight, they are highly satisfied with activities performed in a daylight environment [14]. As such, it is of great significance to focus on the impact of daylight illumination on the reading behavior of elderly persons in nursing buildings.

The main purpose of this study was to investigate daylight illumination of comfortable reading for elderly persons. "Comfort" is a state of relaxation and peace without physiological pressure, which is related to the overall state of a person. The full range of factors, both mental and physical, can be described as being involved in the state. Traditionally, subjective questionnaire surveys are commonly used in comfort-related research, such as human perceived comfort experience and comfort evaluation. This paper refers to the methods used in international research on emotions. Based on the view that "comfort" includes not only physical sensations, but also psychological factors [15], the combination of a physiological index measurement and a subjective questionnaire survey was used to analyze state arousal level and visual comfort in the reading of elderly persons. The effect of different levels of daylight illumination on reading by elderly persons was then explored.

2. Materials and Methods

2.1. Research Subjects

Elderly persons living in senior care institutions were the research subjects of this study. The enrolled subjects met the following requirements: self-care elderly people aged 60–89, graduated from primary school or above, with normal naked or corrected vision, no color blindness, cognitive impairment, or Alzheimer's disease, and reading behavior at least twice a week. A total of 30 seniors were enrolled, 15 men and 15 women.

2.2. Reading Materials

To reduce the impact of different reading materials, fonts, and other factors on the physiology and psychology of elderly persons, the length of each reading activity was controlled; also, the reading material was provided by our research group. The newspaper was selected as the experimental reading material (Figure 1). The background color of the newspaper was light gray and white. The font color was black, and font size was the 12 pt. for Han Chinese characters. Eight paragraphs from eight issues of the newspaper were randomly selected. The layout of each paragraph was the same with 100 words each.

The reading time of each paragraph was controlled at approximately 5 min, and each paragraph was in the same position of the newspaper. The content of the material was serial articles. Simple and understandable emotional stories were selected to eliminate personal preferences of elderly persons and reduce the impact of article content on their cognitive and emotional states.

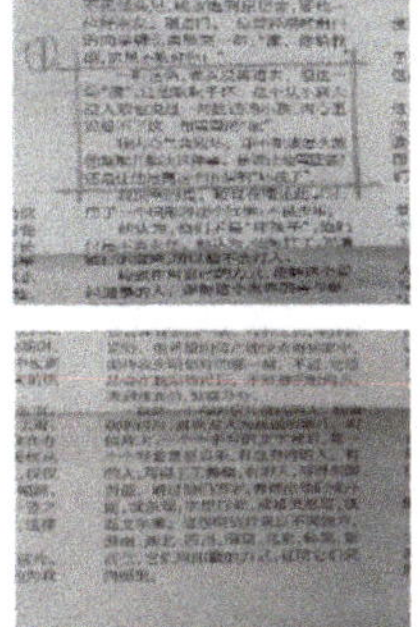
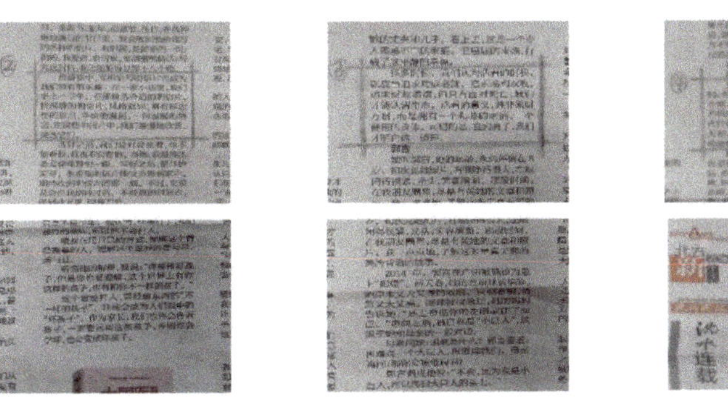

Figure 1. Reading materials.

2.3. Daylight Illumination Measurement

This study was based on the residential building of an elderly care institution in Dongling District, Shenyang, China. The building was established in 2014, with four floors and 12 households on each floor (Figure 2). All dormitories are single rooms, with a size of 3.6 × 6 m. These dorms include bedrooms, a leisure area, and independent bathrooms. The walls and ceilings in the room are painted white, and the floor is made of brown wood. The window in the room has a size of 2 × 1.8 m and is a height of 1 m from the ground, and faces south.

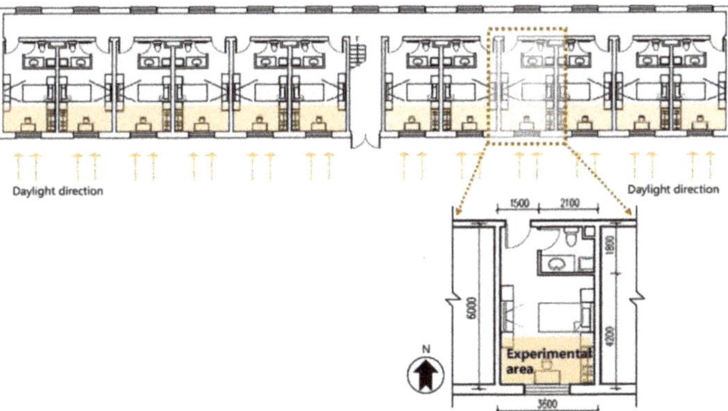

Figure 2. Floor orientation and room layout.

The elderly persons read experimental materials on the desk under the window. The desk is 0.75 m high, and the chair is 0.42 m high. After a survey of 30 elderly persons, their one-day activities were summarized in Table 1, and experimental time was finally determined. The illumination was measured between 8:00 and 10:00 every morning. Combined with the Architectural Lighting Design Standard of China, the illumination range of the experiment was set as 300–1000 lx. Prior to the start of the experiment, the elderly persons were free to change the illumination using the curtains. A TES-1330A illumination meter (TES Electrical Electronic Corp.) was used to measure the illumination before the elderly persons started reading. Electrodermal activity (EDA) data were received in real time using a laptop placed on the table at the rear left of the subject (Figure 3).

Table 1. The general situation of nursing homes.

Timetable	5:00–6:00	6:00–8:00	8:00–10:00	10:00–12:00	12:00–14:00	14:00–16:00	16:00–18:00	18:00–21:00	21:00–05:00
Bed room	B	I	A		K		L, E	D	K
Dining room		L		L					
Chess room			J			H			
Reading room									
Outdoor	F					G		N	
Foyer				M					

Types of activity: A. Reading books and newspapers; B. Getting up; C. Using computer; D. Watching TV; E. Doing housework; F. Fitness; G. Walking; H. Playing mahjong; I. Raising flowers; J. Play chess; K. Sleeping; L. Eating; M. Chatting; N. Dancing P. Others.

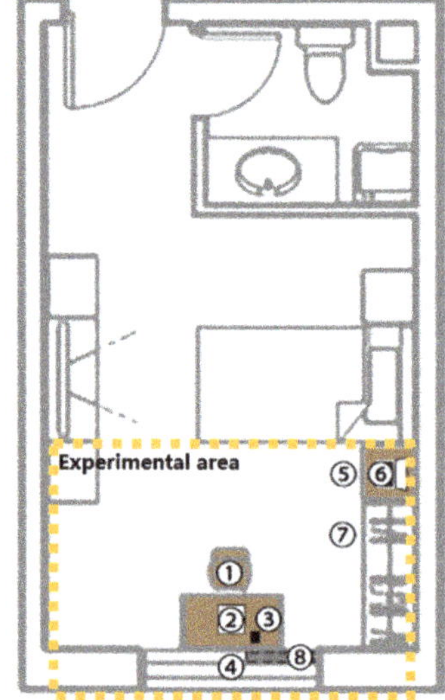

Figure 3. Experimental area plan.

The final illumination value was determined using the mean value of the illumination at the midpoint of each side of the reading material and the intersection of the diagonal (Figure 4). To reduce the fluctuation in daylight illumination, reading material with fewer words was first chosen in order to minimize the reading duration. Based on the experience during the pre-experiment, the elderly persons are not sensitive to changes in illumination within 50 lx. Thus, when the experiment was carried out, the illumination meter was placed close to the top of the reading material, and data that did not change beyond 50 lx throughout were adopted. A staff member stood at the left rear of the elderly persons and observed the illuminance meter's value on the table throughout the process to control the illumination change during the experiment (Figure 3) in order to avoid the problem of glare or uneven illumination during the experiment, which may have caused discomfort to the elderly persons. Before the experiment, the elderly persons could slightly adjust the position of the curtain, chair, and reading materials according to the actual situation (Figure 5). At the end of the experiment, elderly persons were asked if they encountered glare or excessive fluctuations in their visual field brightness. Data collected when this occurred were eliminated. The measurements were performed over 20 weeks from September 2017 to March 2018, and 217 valid data points were obtained.

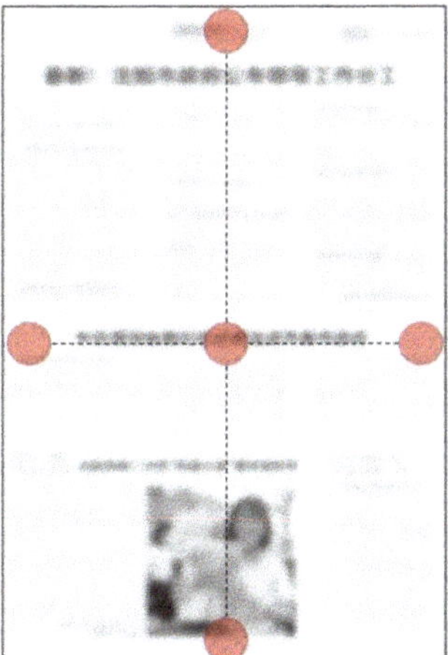

Figure 4. The selected points of the illumination value.

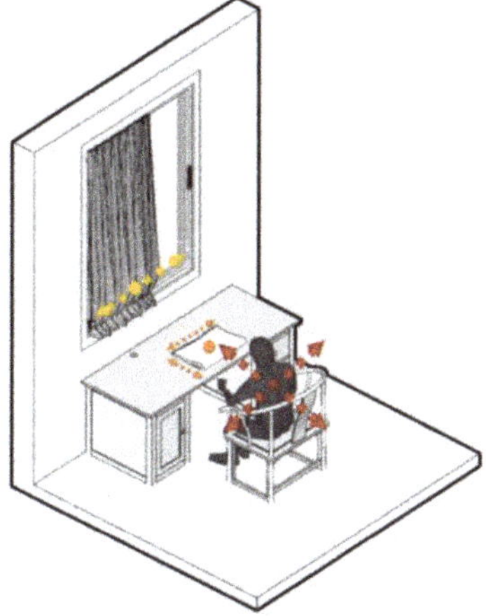

Figure 5. Adjustable position.

2.4. Visual Comfort Level Measurement

2.4.1. Measurement of Skin Conductance

When an individual engages in cognitive activities or is exposed to particular emotional stimuli, areas of the brain such as the anterior cingulate gyrus and the amygdala act, resulting in a sympathetic nervous system reaction [16]. The more excited the sympathetic nervous system becomes, the more stimulated the sweat glands become. In addition, the sweat glands secrete sweat to the skin surface through pores in the skin. When a balance between positive and negative ions in the secretion occurs, the skin conductance (SC) changes. Changes in SC thus indicate the activity level of specific areas of the brain during reading tasks, notably crucial cognitive and emotional activities. According to research, the neural systems of emotion and cognition are closely interwoven, and both positive and negative emotions play a role in learning and remembering [17]. Thus, it can be seen that the shift in SC indicates the state arousal level of the elderly persons during reading. However, it is possible that a well-lit environment facilitates the arousal or that the arousal compensates for the annoyance produced by a poor light environment. This was investigated further with the subjective questionnaire. The mean of the time-domain characteristics of the SC can reflect the average level of electrodermal activity in the statistical period [18]. The greater the absolute value of the mean change rate in different periods, the higher the state arousal degree.

An electrodermal activity (EDA) wireless physiological sensor (KingFar International Inc. Beijing, China) was used to collect and monitor the data of the SC indicators. Data storage and analysis were performed with a human–machine environment synchronization experimental platform (Ergo LAB). During reading, the elderly persons needed to wear the EDA wireless physiological sensor on the palm via an electrode so that the original physiological signal could be collected and transmitted to the Ergo LAB experimental platform in real time (Figure 6). The experimental platform has built-in filtering methods such as Smooth, Guass, and Hann, which can extract SC index data from the collected original signal and analyze the mean of SC time-domain characteristics in the corresponding time window.

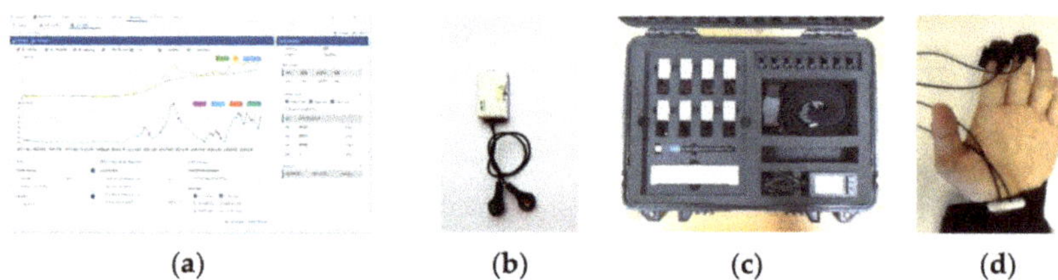

(a)   (b)   (c)   (d)

Figure 6. Instruments. (**a**) Human–machine environment synchronization experiment platform. (**b**) EDA modules. (**c**) Physiological sensing kit. (**d**) Wearing instruments.

2.4.2. Subjective Questionnaire

This research established two evaluation methods: a subjective comfort questionnaire and electrodermal activity measurement. The questionnaire was designed for the subjective visual comfort of elderly persons when reading at different illumination levels. The question was "What is your visual comfort level when reading at this illumination level?". The 5-point Likert scale method was adopted, in which five evaluation scales, from very uncomfortable to very comfortable, were used (Figure 7). There is no academic consensus on the concept of "visual comfort", and there are two widely used methods of evaluating visual comfort: One of them is the "no annoyance method", which states that no discomfort is considered comfortable. In other words, no physiological pain or irritation is considered to be visual comfort. Another is the "well-being method", which is based on subjective happiness and satisfaction [19]. In this study, "visual comfort" was defined as a state in which the reading task could be accomplished without physical or psychological stress by combining these two ways.

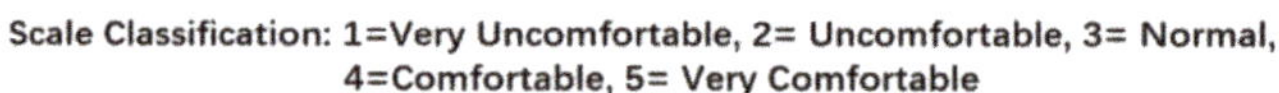

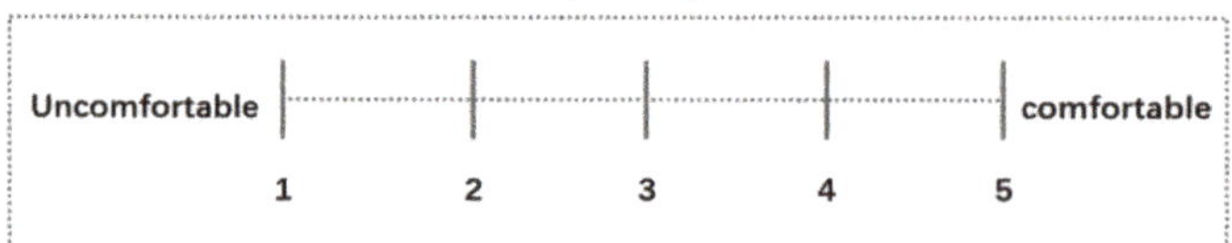

Figure 7. Subjective psychological evaluation scale.

At the end of reading, the persons enrolled in this study were asked whether they had fully seen and understood the reading materials. The data from the cases when they could not fully see and understand the reading materials were excluded. All respondents were informed of the purpose of the study and how the data would be used before filling out the questionnaire. According to the requirements of local legislation and institutions, this study did not require ethical review and approval. A total of 240 questionnaires were distributed, and 217 valid questionnaires were returned, with a recovery rate of 90%.

Q: What is your visual comfort level when reading at this illumination level?

2.5. Experimental Process

Before the experiment, the physiological signal baseline of the elderly persons was collected for 5 min in a daylight environment with approximately 300 lx illumination and uniform light. The mean value of the SC time-domain characteristics during this period was considered the baseline value. The collection site was on a chair next to a desk in the elderly person's own bedroom. The posture was a natural sitting posture, and the ambient illumination was adjusted using the curtains. The baseline only needed to be collected once, and there was no need to repeat it before every illumination.

After the baseline was determined, the participants could adjust their comfortable reading posture by themselves or adjust the illumination of the surrounding daylight by adjusting the curtains. After 3–5 min adjustment time, reading could start. The illumination value was randomly set in each experiment, in order to reduce the impact of the previous illumination value on the elderly persons; time was also left for them to make adjustments after each change in illumination. Throughout the reading process, the elderly persons wore physiological sensors. Ergo LAB human-machine experimental software marked the physiological data in real time at the beginning and end of reading. The recording stopped after the elderly persons finished their reading. Subsequently, the elderly persons began to complete the subjective questionnaire. The staff asked the elderly persons whether they experienced any discomfort such as glare during reading, and whether they fully saw and understood the content of the reading materials. The entire measurement process is illustrated in the figure below (Figure 8).

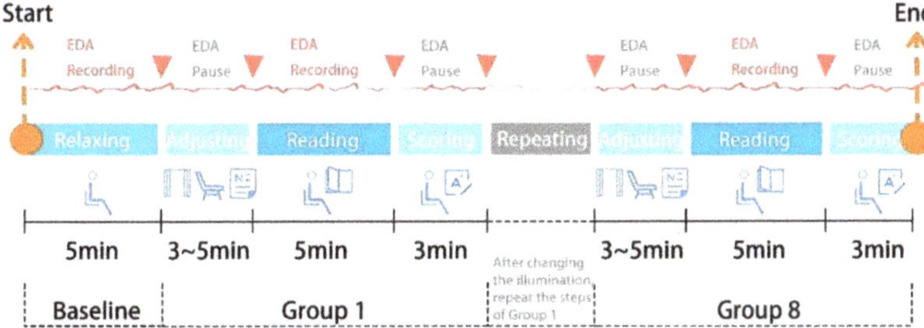

Figure 8. Experimental process.

3. Results

3.1. Daylight Illumination Measurement Results

A total of 217 illumination data points ranging from 300 to 1000 lx were used; this is the daylight illumination range that is acceptable for the elderly person to read. The questionnaire data and physiological data measured under different illumination levels were grouped according to the distribution of illumination data. The illumination difference corresponding to each group of data did not exceed 100 lx. Eight groups were recognized, i.e., "300 lx," "400 lx," "500 lx," "600 lx," "700 lx," "800 lx," "900 lx," and "1000 lx," with 25 to 30 valid data points under each group (Table 2).

Table 2. Illumination data summary.

Data Summary	
Illumination range	300–1000 lx
The total number of data	217
The number of data group	8
The number of valid data under each group	25–30

3.2. Influence of Daylight Illumination on Visual Comfort

3.2.1. Data Preprocessing

Analysis of Covariance

The level of basic electrodermal activity is related to personality characteristics. Individuals who have a higher basic electrodermal activity tend to be more introverted, nervous, and emotionally unstable. In contrast, individuals with a lower basic level are found to be more cheerful and outgoing, and have a more balanced mentality and better psychological adaptation. The time-domain mean of SC of each elderly person under different illumination was plotted on a scatterplot. According to this scatterplot (Figure 9), although the change in the electrodermal activity level of each elderly person was relatively small under different illumination conditions, a large difference in its values between different subjects was observed. This difference may originate from different physiques and cannot be artificially controlled. Covariance analysis was used to explore whether the individual's basic electrodermal activity level would affect the analysis of the SC time-domain characteristic mean under different illumination levels, and to clarify whether the SC time-domain mean values measured by different subjects could be directly compared.

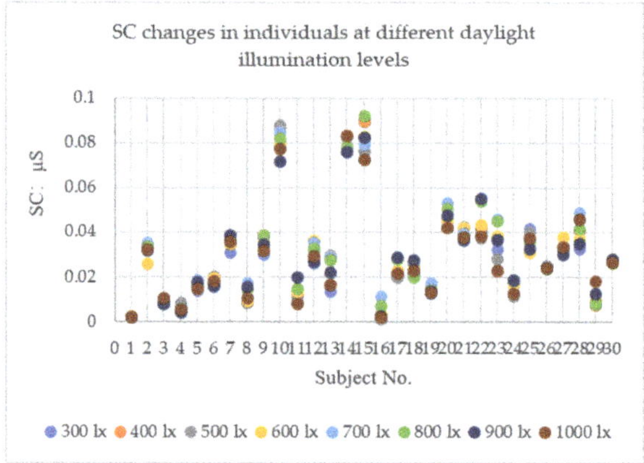

Figure 9. SC changes in individuals at different daylight illumination levels.

In this study, the SC time-domain characteristic mean values (hereinafter referred to as the baseline value) of the elderly persons during the baseline period was the covariate. The illumination was the independent variable, and the SC time-domain characteristic mean (hereinafter referred to as the SC mean value) under different illumination was the dependent variable. The statistical results of covariance are shown in the table below. The between-subjects effects test shows the baseline value of covariate p (Sig.) = 0.000 < 0.01 (Table 3). Hence, it can be considered that an interactive relationship existed between the SC mean value of the dependent variable and the baseline value of the covariate, indicating that a person's initial electrodermal activity level did affect the electrodermal activity level under different illumination conditions.

Table 3. Between-subjects effects test.

Dependent Variable: SC Time-Domain Mean Values					
Source	Type III Sum of Squares	Degrees of Freedom	Mean Square	F	*p* (Sig.)
Modified model	1033.506 [a]	8	129.188	525.045	0.000
Intercept Distance	9.049	1	9.049	36.775	0.000
Baseline	1028.447	1	1028.447	4179.796	0.000
Illumination	5.060	7	0.723	2.938	0.006
error	56.838	231	0.246		
Total	3256.463	240			
Total after correction	1090.344	239			

[a.] $R^2 = 0.948$ (After adjustment $R^2 = 0.946$).

After deducting the effect of covariate baseline on the experiment, *p* (Sig.) had a value of $0.006 < 0.05$ (Table 4), showing that the illumination of the independent variable still had a significant effect on the SC mean value of the dependent variable.

Table 4. Univariate tests.

Dependent Variable: SC Time-Domain Mean Values					
	Sum of Squares	Degrees Of Freedom	Mean Square	F	*p* (Sig.)
Contrast	5.060	7	0.723	2.938	0.006
error	56.838	231	0.246		

SC Time-Domain Mean Value Normalization

To eliminate the influence of each elderly person's initial electrodermal activity level, it was necessary to perform data normalization. The specific operation of normalization involves subtracting the SC mean value of an elderly person's reading under different illumination conditions from the baseline value, and dividing the result by the baseline value. The change rate Δk obtained is the normalized result. The Δk value indicates the state arousal level. A high Δk indicates high state arousal, whereas a low value indicates low state arousal. The equation below shows the normalization process, where Δk denotes the normalized change rate of the SC. $\overline{X}_{emotion}$ refers to the SC mean value under a certain illumination, and $\overline{x}_{calm}$ represents the baseline value. The formula is as follows [20]:

$$\Delta k = \frac{\overline{X}_{emotion} - \overline{x}_{calm}}{\overline{x}_{calm}}$$

3.2.2. State Arousal

Correlation Analysis

The variance homogeneity analysis of Δk data revealed a significance $p = 0.000 < 0.05$; thus, it was impossible to use the parametric test. Therefore, the nonparametric Kaplan–Meier (KM) analysis was performed to test the difference between each group of data. The obtained significance was $p = 0.006 < 0.05$, indicating a significant difference. As such, the data could be further analyzed. The illumination and Δk values did not satisfy the normal distribution; thus, Spearman correlation analysis was performed. The results show that the correlation coefficient was $0.108 > 0.01$ (Table 5), revealing a significant correlation.

Table 5. Spearman correlation coefficient of illumination and Δk.

			Illuminance	Δk
Spearman Rho	Illumination	Correlation coefficient	1.000	0.108
		Significance (2-tailed)	-	0.097
		Number of cases	217	240
	Δk	Correlation coefficient	0.108	1.000
		Significance (2-talied)	0.097	-
		Number of cases	217	240

Cumulative Analysis

First, analysis of variance (ANOVA) was performed on Δk under each group of illumination to test whether there was significant difference in state arousal under different illumination conditions. A variance homogeneity test revealed that p (Sig.) = 0.460 > 0.05; thus, ANOVA could be continued. The results are shown in the table below. p (Sig.) = 0 < 0.01 indicated a significant difference in state arousal under different illumination conditions (Table 6).

Table 6. ANOVA analysis results.

	Δk				
	Sum of Squares	**Degrees of Freedom**	**Mean Square**	**F**	***p* (Sig.)**
Between groups	51.523	7	7.360	6.203	0.000
Within group	242.047	204	1.187		
total	293.570	211			

The Δk value can indicate the arousal degree. The greater the value, the greater the arousal, whereas the smaller the value, the smaller the arousal degree. By accumulating the Δk value under each group of illuminations, the trend of arousal under different illumination conditions can be determined (Figure 10). For 600–900 lx, the state arousal degree was the largest. Under other illumination values, the state arousal degree was small (Table 7). However, the positive or negative state of each group remained unknown. Thus, further analysis in combination with the results of subjective questionnaire was required.

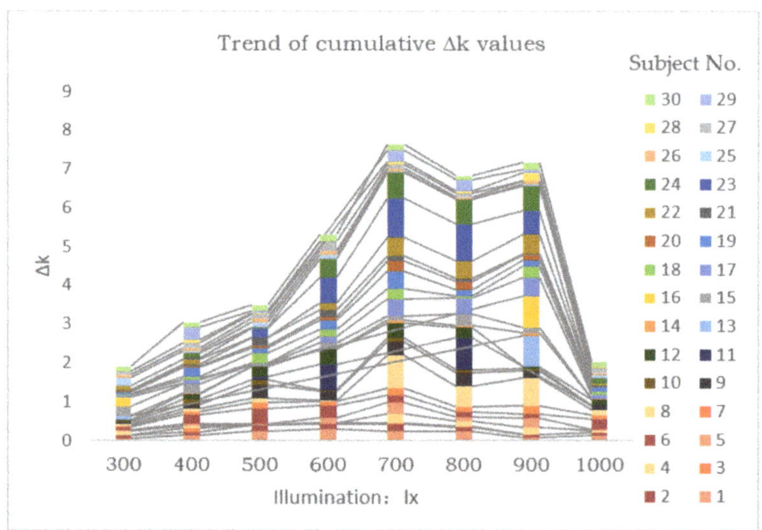

Figure 10. Trend of cumulative Δk value.

Table 7. Cumulative Δk value result.

Illumination (lx)	300	400	500	600	700	800	900	1000
Cumulative Δk	1.93	3.06	3.51	5.34	7.66	6.85	7.18	2.04

3.2.3. Visual Comfort

Correlation Analysis

The score data of visual comfort questionnaire were analyzed by variance homogeneity, and the significance was p = 0.387 > 0.05. Hence, one-way ANOVA was performed, and the significance determined was p = 0.000 < 0.05, indicating a significant difference; thus, in-depth analysis could be conducted. The illumination value and comfort score did not meet the normal distribution; therefore, the Spearman correlation analysis was performed. The results show that the correlation coefficient was 0.312 > 0.01, revealing a significant correlation (Table 8).

Table 8. Spearman correlation coefficient of illumination and visual comfort score.

			Questionnaire Score	Illuminance
Spearman Rho	Questionnaire score	Correlation coefficient	1.000	0.312
		Significance (2-tailed)	-	0.000
		Number of cases	240	240
	Illumination	Correlation coefficient	0.312	1.000
		Significance (2-tailed)	0.000	-
		Number of cases	240	240

Descriptive Analysis

The scores of visual comfort questionnaire of each group were counted, and the proportion of "Very Uncomfortable", "Uncomfortable", "Normal", "Comfortable", and "Very Comfortable" in each group was counted. From the questionnaire scores, it was found that (Figure 11):

1. When the illumination was 300 and 400 lx, the proportion of "Comfortable" and "Very Uncomfortable" below 3 points was very high, reaching 96.66% and 63.33%, respectively. At 500 lx, the proportion of "Normal" had the highest proportion, reaching 63.33%.
2. At 600 and 700 lx, the proportion of "Comfortable" and "Very Comfortable" more than 3 points was very high, accounting for 93.33% and 86.67%, respectively. At 800 lx, the proportion over 3 points still exceeded half, accounting for 56.67%. The proportion of "Normal" of 3 points also reached 40%, and the proportion of less than 3 points was only 3.33%.
3. At 900 lx, the proportion of "Uncomfortable" and "Very Uncomfortable" below 3 points was 30%. The largest proportion is 3 points for "Normal"., accounting for 46.67%. At 1000 lx, the proportion of "Uncomfortable" and "Very Uncomfortable" below 3 points was 40%, close to the proportion of "Normal" (46.67%).

Based on the above situation, it can be preliminarily concluded that 300 and 400 lx resulted in negative visual perception for most elderly persons, who were unable to read without physical or psychological stress. With the increase in illumination, the visual perception tended to be positive at 500 lx. At 600 and 700 lx, positive feedback was obtained from most elderly persons, and their feelings were biased towards comfort. When the illumination was 800 lx, the opinions began to diverge. Although the number of elderly persons who had normal feelings increased, more elderly persons felt comfort. At 900 lx, the number of elderly persons who considered they had normal feelings were in the majority, at nearly half. At 1000 lx, the number of the elderly persons who considered they had normal feelings was similar to that who were uncomfortable, but a small proportion of elderly persons felt comfortable.

By averaging the visual comfort scores under each group of illumination values, the general trend of visual comfort could be obtained (Table 9). The range of discomfort fell within 0–3 points. From the average value, 300, 400, and 1000 lx can be considered as negative discomfort in the range of less than 3 points. The scores under 900 and 500 lx were close to 3 points, which can be judged as normal. The scores under 600, 700, and 800 lx were in the range of more than 3 points, which can be considered as positive comfort.

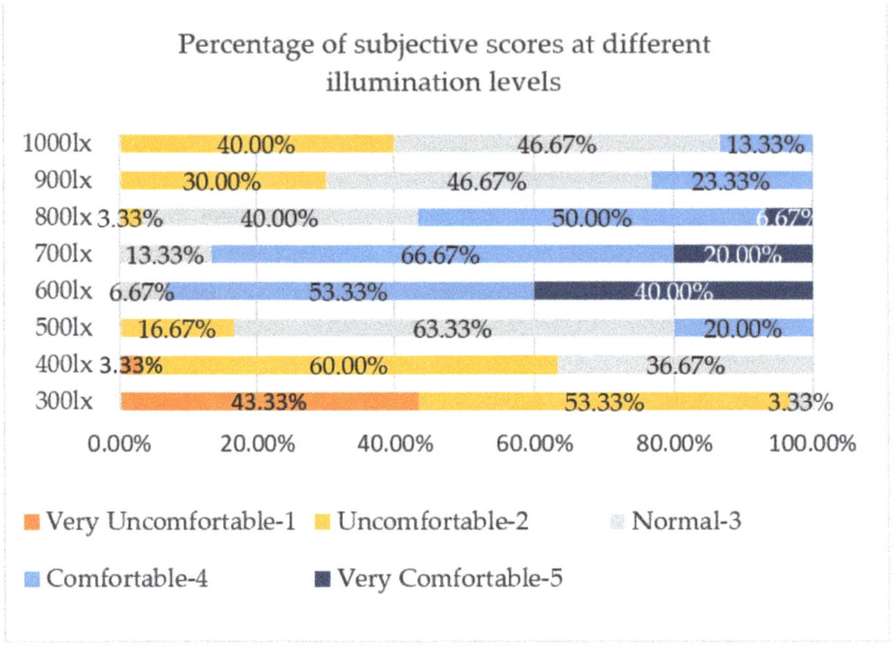

Figure 11. Percentage of subjective scores at different illumination levels.

Table 9. Average value of the visual comfort score.

Illuminatio (lx)	300	400	500	600	700	800	900	1000
Average score	1.6	2.36	3.04	4.36	4.07	3.6	2.9	2.69

3.3. Daylight Illumination Threshold Analysis

The average score of the visual comfort questionnaire of each group was compared with the accumulated Δk value. The comparison diagram is show in Figure 12.

Δk indicates state arousal level, and the visual comfort score indicates the trend of "Comfort" (>3 points), "Normal" (3.1> n > 2.9), and "Uncomfortable" (<3 points), which can be further divided into detail according to the level of 1–5 points (Table 10). According to the results of the subjective questionnaire, at 300–400 lx, the subjective visual comfort score was below 3, indicating discomfort. State arousal was also low, making it harder for the elderly persons to engage. When the illumination was increased to 500 lx, the situation improved, and most of the elderly persons were able to complete the reading task more easily. At the same time, the elderly persons' states of arousal rose during this procedure. At 600 lx, the subjective visual comfort score reached a maximum of 4.36 points, which was classified as "Very comfortable." This indicated that the elderly persons could read without feeling stressed, and their arousal level had improved slightly. When the light level was increased to 700 and 800 lx, the subjective comfort score decreased, but both scores were greater than 3 and still fell into the "comfort" category. At these illumination levels, the elderly persons were under greater pressure to complete the reading task than at 600 lx. Their state arousal was higher, which was a positive outcome. At 900 lx, the subjective visual comfort score remained at 3, despite the fact that state arousal remained high. Subjective visual comfort fell below a score of 3 at 1000 lx, and was classified as "uncomfortable", with a considerable decline in state arousal.

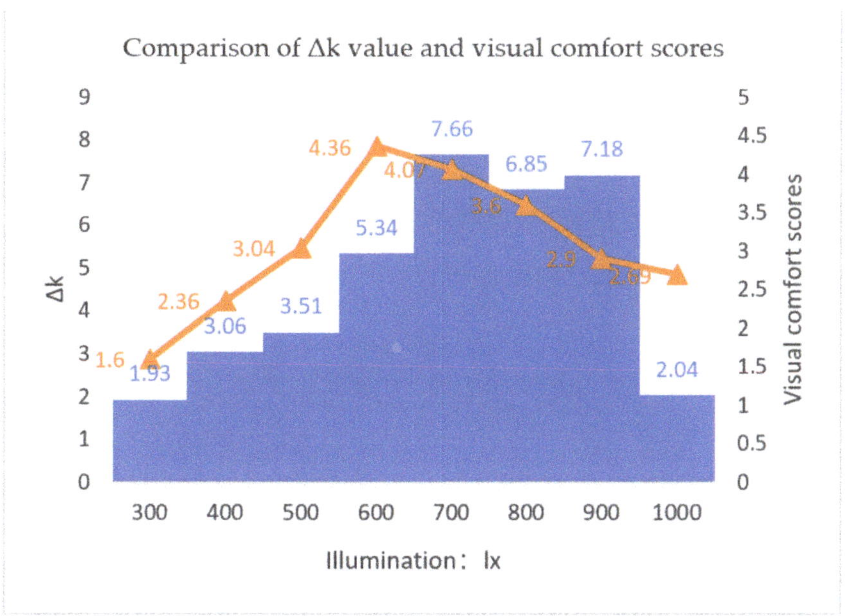

Figure 12. Comparison diagram of Δk and visual comfort scores.

Table 10. Ranking of scores of the visual comfort questionnaire.

Trend of Visual Feeling	Uncomfortable (n < 3 Points)			Normal (3.1 > n > 2.9)			Comfortable (n > 3 Points)	
Illumination (lx)	300	400	1000	900	500	800	600	700
Scores of visual comfort (n)	1.6	2.36	2.69	2.9	3.04	3.6	4.36	4.07
ArousalLevel (Δk)	1.93	3.06	2.04	7.18	3.51	6.85	5.34	7.66

4. Discussion

Based on the analysis of the experimental data, a trend can be noticed. When daylight illumination was set at 300–500 lx, subjective visual comfort scores rose in tandem with state arousal, showing that the elderly persons were gradually becoming more active and engaged. At an illumination level of 600–800 lx, the subjective visual comfort score dropped from its maximum point but remained in the "comfort" category, while state arousal levels continued to rise. Based on the elderly persons' subjective assessments, when the illumination was between 600 and 800 lx, the condition of state arousal was positive. It can be seen that physiological and psychological activity improved the elderly person's subjective perception. The levels of both 700 and the 600 lx had a subjective visual comfort score of 4 or more; 700 lx may be preferable since it resulted in a higher level of state arousal.

However, this did not necessarily apply to the 900 and 500 lx levels, which were both rated 3 out of 5 as "normal." According to the proportion of subjective visual comfort scores, more than half of the elderly persons assessed 500 lx as "normal." At 900 lx, several subjects experienced trouble adjusting the uniformity of light in their field of view, and more glare cases were reported. Some subjects also reported preferring the higher illumination of 900 lx, which helped them see more clearly. As a result, high state arousal can be both a positive and a negative outcome. In terms of consistency of evaluation, the 500 lx level is superior to that of 900 lx.

5. Conclusions

It can be seen from the experimental data that the peak of subjective visual comfort score appeared at 600 lx, and the peak of state arousal level appeared at 700 lx. The reason for this peak gap can be analyzed from the changes of the two indicators. When the daylight illumination was low (300 lx), as the illumination increased, the elderly persons invested more energy, and the visual comfort also improved. However, when the illumination was increased to 800 lx, the elderly persons felt a certain pressure when reading, and the increase in state arousal then offset a part of this pressure. When the illumination was higher than 1000 lx, the elderly persons appeared to have low state arousal, and the visual comfort was also greatly reduced.

Within the context of this experiment, the most comfortable illumination range for the elderly persons to read in daylight was 600–800 lx, and 700 lx was optimal. This level was higher than the illumination of 300 lx recommended for reading under artificial lighting in the Architectural Lighting Design Standard of China. This showed that, due to the dependence of the elderly persons on daylight, their demand for daylight illumination was higher than that under constant artificial lighting. This conclusion was also confirmed by a study in UK, in which individuals tolerated significantly higher levels of daylight illumination than CIBSE's typical artificial lighting recommendations unless there was glare or direct sunlight [21].

The main conclusions can be summarized as follows:

1. At daylight illumination of 300–500 lx, subjective visual comfort rose, in addition to state arousal, a process in which the elderly persons became progressively more engaged.
2. At daylight illumination of 600–800 lx, subjective comfort decreased from its highest point, but remained in the "comfort" category, while state arousal continued to increase.
3. At daylight illumination of 800 lx, the active state of the elderly compensated for some of the stress caused by the light environment.
4. At daylight illumination of 900 lx, although the subjective visual comfort rating was close to "normal", the comments were polarized and it was difficult to determine whether the increase in state arousal was due to positive or negative factors.
5. When the lighting was dim, the elderly persons had low state arousal and found it harder to engage in reading tasks. As the degree of illumination rose, so did the arousal level, and the body state of the elderly gradually became more active. When the illumination level exceeded a specific threshold, however, the arousal level dropped dramatically.
6. The most comfortable reading illumination level for the elderly persons was between 600 and 800 lx, with 700 lx providing the best performance. The ranking of the visual comfort levels of the daily illuminance values under reading behavior is shown in Figure 13.

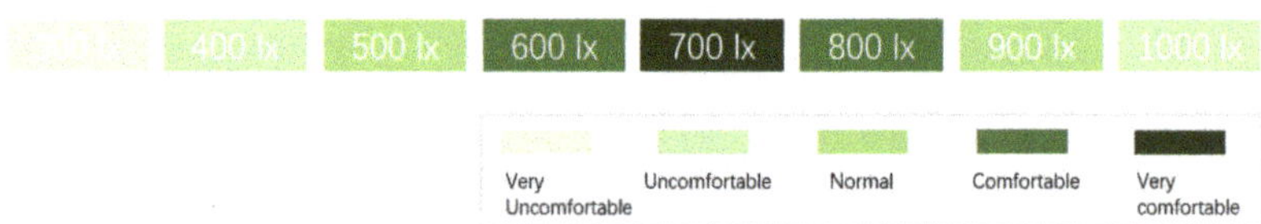

Figure 13. Schematic of illumination value and comfort level.

This experiment was chosen to take place in Shenyang, a city in China's harsh cold area, which is located at high latitudes. Long winters, low solar azimuths, and short daylight hours characterize cities in high latitudes. The elderly persons who live here are restricted to indoors and prefer the more daylight is brought indoors because they are psychologically closer to it. However, in cities located at low latitudes with high levels of daylight radiation (e.g., Guangdong), the issues to consider are quite different. On sunny days, low latitudes are characterized by intense daylight radiation, whereas cloudy situations result in insufficient daylight hours. Although it is critical to bring in as much daylight as possible, too much direct light may generate glare and diminish the comfort of the light environment. The elderly persons in various climatic zones may have different adaptations to daylight, and additional research in various regions is needed.

In this research, extraneous light environment indicators such as glare [22] and illumination uniformity [23] were controlled for, although these two indications are also crucial for evaluating the quality of the light environment. At the same time, the primary purpose of the light environment indicators is to inform the architects' lighting design strategy. The physiology, psychology, and behavior of the elderly persons are affected by design considerations such as building orientation, room sizes, window sizes and parameters, frames and position, types of glazing, transmission characteristics of glazing, cleanliness of glazing, and interior room surfaces [24,25]. The comfort of the daylight environment is a systematic "human-behavior-environment" problem, which is better suited to multifactorial research with the use of appropriate algorithms. Before that, however, experiments are needed to clarify the specific relationships between the different factors. The best design strategy based on daylight performance indicators may be discovered using enough experimental sample data and a multi-objective algorithm [26].

6. Limitation

First, the set illumination range was 300–1000 lx in the experiments. This was based on the recommended value of reading illumination under artificial lighting in the Architectural Lighting Design Standard of China and the pre-judgment of the actual situation in the pre-experiment. However, there were also cases where it exceeded 1000 lx or was less than 300 lx in the experiments. Since elderly persons with normal vision and reading habits were selected in this experiment, only in certain cases did elderly persons with low sensitivity to light believe that illumination higher than 1000 lx was more comfortable, or those with relatively better vision and high tolerance of daylight illumination considered that daylight below 300 lx was acceptable. It can be seen that the acceptance of daylight illumination was also different for the elderly persons with different physiological and psychological states. This research focused on a certain category of elderly persons and was not representative of a broad group of elderly persons. In addition, this experiment set the time from 8:00–10:00 in the morning according to the reading habits of the elderly persons. Table 1 summarizes the activities of the elderly persons in one day, among which watching TV, playing cards, playing chess, eating, and chatting were the activities enjoyed by the elderly in other periods; these activities were also not covered in this research. Finally, the site selected for this article was a rectangular room facing south, which was relatively simple. The daylight features of other rooms were different from those facing the south direction and would behave differently over time. In addition, there were complex geometries, which needed to be specifically analyzed according to the movement tracking of the elderly persons' activities. These questions all need to be further explored in the future.

Author Contributions: Conceptualization, Y.F. and R.H.; methodology, Y.F. and R.H.; software, R.H. and Y.W.; validation, Y.W.; formal analysis, Y.W.; investigation, R.H. and Y.W.; resources, Y.F.; data curation, R.H. and Y.W.; writing—original draft preparation, Y.W.; writing—review and editing, Y.F. and W.G.; visualization, Y.W.; supervision, Y.F. and W.G.; project administration, Y.F.; funding acquisition, Y.F. All authors have read and agreed to the published version of the manuscript.

Funding: This paper is supported and funded by two sponsors: "Xingliao talents plan" of the innovative leading talent project of Liaoning Province (tpjs2019001), Liaoning Provincial Natural Science Foundation Guiding Plan in 2019 (2019-zd-0656).

Institutional Review Board Statement: The study was conducted in accordance with the Declaration of Helsinki, and approved by the Ethics Committee of School of Architecture and Planning, Shenyang Jianzhu University.

Informed Consent Statement: Informed consent was obtained from all subjects involved in the study.

Data Availability Statement: Data are not publicly available due to restrictions regarding the privacy of the participants.

Acknowledgments: We are also grateful to the Cold Land Healthy City and Comfortable Building Research Center of the Shenyang Jianzhu University for providing the equipment support for this study, as well as Zitong Wang and Muye Wang for their experience and analysis regarding the data statistics.

References

1. van Bommel, W.J.M.; van den Beld, G.J. Lighting for work: A review of visual and biological effects. *Light. Res. Technol.* **2016**, *36*, 255–266. [CrossRef]
2. de Castro, V.C.; Carreira, L. Leisure activities and attitude of institutionalized elderly people: A basis for nursing practice. *Rev. Lat. Am. Enferm.* **2015**, *23*, 307–314. [CrossRef] [PubMed]
3. Zhang, W.; Feng, Q.; Lacanienta, J.; Zhen, Z. Leisure participation and subjective well-being: Exploring gender differences among elderly in Shanghai, China. *Arch. Gerontol. Geriatr.* **2017**, *69*, 45–54. [CrossRef] [PubMed]
4. Chauhan, M. Involvement of Elderly in Physical and Leisure Activities: Barriers and Challenges. In Proceedings of the Humanizing Work and Work Environment 2015, Mumbai, India, 7–9 December 2015.
5. Gümüş, H.; Erbaş, Ü. The relationship between leisure activity types selected by older adults and their income. *Stud. Perieget.* **2020**, *29*, 87–98. [CrossRef]
6. Lafosse, E.; Wolffsohn, J.; Talens-Estarelles, C.; García-Lázaro, S. Presbyopia and the aging eye: Existing refractive approaches and their potential impact on dry eye signs and symptoms. *Contact Lens Anterior Eye* **2020**, *43*, 103–114. [CrossRef] [PubMed]
7. Legge, G.E. *Psychophysics of Reading in Normal and Low Vision*; CRC Press: Boca Raton, FL, USA, 2006.
8. GB50034-2013. *Standard for Lighting Design of Buildings*; MOHURD and AQSIQ: Beijing, China, 2013.
9. Illuminating Engineering Society of North America. *Lighting for the Aged and Partially Sighted Committee. Recommended Practice for Lighting and the Visual Environment for Senior Living*; Illuminating Engineering Society of North America: New York, NY, USA, 1998.
10. The Illuminating Engineering Institute of Japan. *Lighting Handbook*; Ohm Corporation: Nagoya, Japan, 2003.
11. Davis, R.G.; Garza, A. Task Lighting for the Elderly. *J. Illum. Eng. Soc.* **2002**, *31*, 20–32. [CrossRef]
12. Zhang, Y.; Ma, J. Research of suitable reading lighting conditions for the elderly. *Chin. J. Gerontol.* **2007**, *27*, 2331–2333. [CrossRef]
13. Liu, M.; Shen, S.; Luo, X. The Application Strategy of Natural Light In Buildings For The Aged. *Chin. Hosp. Archit. Equip.* **2018**, *19*, 76–79. [CrossRef]
14. Kuboshima, Y.; McIntosh, J. Understanding the Spatial Requirements that Facilitate Personal Leisure Activities of the High-Needs Elderly. *J. Geriatr. Med.* **2020**, *2*. [CrossRef]
15. Kolcaba, K.Y.; Kolcaba, R.J. An analysis of the concept of comfort. *J. Adv. Nurs.* **1991**, *16*, 1301–1310. [CrossRef] [PubMed]
16. Critchley, H.D. Electrodermal responses: What happens in the brain. *Neuroscientist* **2002**, *8*, 132–142. [CrossRef] [PubMed]
17. Tyng, C.M.; Amin, H.U.; Saad, M.N.M.; Malik, A.S. The influences of emotion on learning and memory. *Front. Psychol.* **2017**, *8*, 1454. [CrossRef] [PubMed]
18. Storm, H. Skin conductance and the stress response from heel stick in preterm infants. *Arch. Dis. Child.-Fetal Neonatal Ed.* **2000**, *83*, F143–F147. [CrossRef] [PubMed]
19. Iacomussi, P.; Radis, M.; Rossi, G.; Rossi, L. Visual comfort with LED lighting. *Energy Procedia* **2015**, *78*, 729–734. [CrossRef]
20. Sugimine, S.; Saito, S.; Takazawa, T. Normalized skin conductance level could differentiate physical pain stimuli from other sympathetic stimuli. *Sci. Rep.* **2020**, *10*, 10950. [CrossRef] [PubMed]
21. Lee, E.S.; DiBartolomeo, D.L.; Selkowitz, S.E. The effect of venetian blinds on daylight photoelectric control performance. *J. Illum. Eng. Soc.* **1999**, *28*, 3–23. [CrossRef]
22. Quek, G.; Wienold, J.; Khanie, M.S.; Erell, E.; Kaftan, E.; Tzempelikos, A.; Konstantzos, I.; Christoffersen, J.; Kuhn, T.; Andersen, M. Comparing performance of discomfort glare metrics in high and low adaptation levels. *Build. Environ.* **2021**, *206*, 108335. [CrossRef]
23. Eltaweel, A.; Su, Y.; Mandour, M.A.; Elrawy, O.O. A novel automated louver with parametrically-angled reflective slats; design evaluation for better practicality and daylighting uniformity. *J. Build. Eng.* **2021**, *42*, 102438. [CrossRef]
24. Guide, C.L. *10: Daylighting and Window Design*; The Chartered Institution of Building Services Engineers: London, UK, 1999.
25. Tahbaz, M.; Djalilian, S.; Mousavi, F.; Kazemzade, M. The effect of architectural details on daylight distribution inside a room. In Proceedings of the International Conference CISBAT 2015 Future Buildings and Districts Sustainability from Nano to Urban Scale, Lausanne, Switzerland, 9–11 September 2015; pp. 295–300.
26. Ziaee, N.; Vakilinezhad, R. Multi-objective optimization of daylight performance and thermal comfort in classrooms with light-shelves: Case studies in Tehran and Sari, Iran. *Energy Build.* **2022**, *254*, 111590. [CrossRef]

Article

Visualization of Acoustic Comfort in an Open-Plan, High-Performance Glass Building

Aldo A. Glean [1,*,†], Stanley D. Gatland II [2,†] and Ihab Elzeyadi [3,†]

1   CertainTeed Corporation, Saint-Gobain Research North America, Northborough, MA 01532, USA
2   Saint-Gobain North America, Malvern, PA 19355, USA; stanley.d.gatland@saint-gobain.com
3   School of Architecture + Environment, University of Oregon, Eugene, OR 97403, USA; ihab@uoregon.edu
*   Correspondence: aldo.glean@saint-gobain.com
†   These authors contributed equally to this work.

Abstract: The aesthetic and functional appeal of high-performance, open-plan office buildings presents special challenges. Extensive use of glass at the building's perimeter to improve visual comfort and office communication can negatively impact acoustic comfort without proper design considerations. This study investigates the utility of a novel visualization approach to documenting the interactional impact of acoustical comfort on the health and well-being of occupants in an open-office environment. Room acoustic measurements of background noise and speech transmission index were conducted and distraction distances were calculated and visualized using a mapping technique. In addition, a comprehensive pre- and post-occupancy evaluation protocol was employed. The paper illustrates the reliability of the visualization approach to aid in the interpretation and comparison of various open-office acoustic solutions from a human-centric acoustic environment perspective.

Keywords: acoustic comfort; speech transmission index; speech intelligibility; open-plan office

Citation: Glean, A.A.; Gatland, S.D., II; Elzeyadi, I. Visualization of Acoustic Comfort in an Open-Plan, High-Performance Glass Building. *Buildings* **2022**, *12*, 338. https://doi.org/10.3390/buildings12030338

Academic Editors: Yue Wu, Zheming Liu and Zhe Kong

Received: 29 January 2022
Accepted: 27 February 2022
Published: 11 March 2022

Publisher's Note: MDPI stays neutral with regard to jurisdictional claims in published maps and institutional affiliations.

1. Introduction

Architectural acoustics is an important consideration in the overall comfort of open-plan offices. Researchers [1–7] have made considerable effort to investigate the acoustics of open-office spaces and present potential solutions for improving their indoor environmental quality. Examples of these solutions are (1) configuring the floor plan to effectively balance and acoustically separate areas; (2) designing collaborative and transition spaces with sound-absorbing surfaces; (3) employing high-performance, noise-reducing interior partitions, and exterior facades; and (4) adding background noise control and enhancement measures for selective workstation. While these measures have proven to help achieve a comfortable indoor work environment, the lack of approaches to visualize their effectiveness have resulted in limitations of their applications in practice. This paper reports on the applicability of a novel visualization approach to aid in the interpretation and comparison of various open-office acoustic solutions from a human-centric acoustic environment perspective.

Poor acoustic indoor environments have been shown [8–11] to negatively impact the cognitive performance and well-being of occupants, thus negatively affecting their productivity. The presence of irrelevant speech from single and multi-talkers in an open-plan office, along with noise from printers, phones, and HVAC systems can lead to a reduction in the quality of the acoustic environment. However, in some cases, HVAC systems can improve the acoustic comfort in open-plan offices by providing sound masking. Yadav et al. [12,13] have studied the impact of both single and multi-talkers on sound pressure level and their effect on the cognitive and task performance of occupants in an open-office environment. Their studies [13] investigated the potential benefit of the "babble effect", acting as a sound-masking source, thus improving the acoustic comfort of the space, but its impact was found to be limited. Sound masking, which is the addition of background noise to reduce the impact of distracting sound sources, is one of the common solutions [14–16] for improving the acoustic comfort within open-plan offices. In some instances, multiple voices in open-plan offices can be a potential source of sound masking [17], thus improving the acoustic environment.

Acoustic comfort within an open-plan office can be characterized by the calculated distraction distance, which is based on sound transmission index (STI) measurements. This relates to speech intelligibility as a function of distance from the source. Haapakangas et al. [18] conducted an analysis of 21 open-plan offices using surveys and room acoustic measurements of distraction distance, spatial decay of speech, speech level at 4 m from the speaker, and background noise level. Their findings showed that the perception of noise was mainly related to background speech and that an increase in distraction distance predicted an increase in disturbance by noise. The other measured quantities were not significant standalone indicators of noise disturbances. Distraction distance is an important metric in characterizing acoustic comfort. Despite the apparent benefits of visualizing distraction distance as a tool to understand and compare the acoustic environment, there is a current gap in the literature of examples suggesting visualization procedures and techniques for practical application. This limitation impacts the comparative analysis of acoustical treatment evaluations and presents barriers to employing innovative acoustic treatments to open-plan office environments.

This study investigates the utility of novel visualization and a building performance evaluation technique to document the interactional impact of acoustical comfort on the health and well-being of occupants in an open-office environment. In this study room acoustic measurements were conducted from which the distraction distance was calculated. The employee experience was captured using a comprehensive Pre-Post Occupancy Evaluation (PPOE) protocol. This paper illustrates a visualization strategy to aid in the interpretation and comparison of various open-office acoustic solutions, which is also correlated to survey responses by occupants from a human-centric acoustic environment perspective.

2. Materials and Methods

2.1. Building Design-Acoustic Comfort

2.1.1. Traditional Building

A corporate headquarters was relocated from a traditional cellular private office complex (pre-move) to a LEED platinum-certified open-concept plan office (post-move) in 2015. The traditional late 1960's corporate campus consisted of four low-rise, multistory buildings having perimeter offices and centrally located cubicle workstations. Transparent glass and gypsum wall partitions separated the perimeter and central spaces. The building envelope consisted of a brick facade and large single-glazed fixed casement windows. Workstations were divided using 1524 mm tall fabric covered partitions. Space lighting was created using ballasted tube florescent lights integrated into the suspended acoustical ceiling. Numerous hard sound reflective surfaces were created in the open plan space by lighting fixtures, furniture, metal ceiling panels, gypsum and glass partitions. The traditional building exterior is presented in Figure 1a along with a typical workstation in Figure 1b.

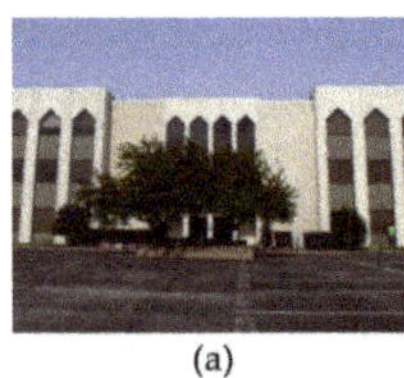

(a) (b) (c) (d)

Figure 1. (**a**) Traditional building complex. (**b**) Typical workstation of traditional building. (**c**) LEED-certified platinum building. (**d**) Workstations in new LEED certified building.

2.1.2. LEED-Certified Platinum Building

The renovated four-story, all-glass-facade LEED platinum-certified building consisted of large, centrally located workstations with perimeter hallways and collaborative areas. Transparent glass and gypsum wall partitions separated the perimeter collaborative and central spaces. Individual workstations were separated by fabric-covered partitions with colored glazing extensions having varying heights of 1047 mm, 1448 mm, and 1625 mm. Slim profile LED luminaries with sharp cut-off angles were integrated into the suspended acoustical ceiling to provide additional space lighting. Hard reflective surfaces were limited to the perimeter of the large open-plan office space. An active sound masking system was implemented in all open-plan spaces to increase acoustic comfort at the 800 plus workstations. The LEED certified platinum building exterior is presented in Figure 1c along with a typical workstation in Figure 1d.

Table 1 provides a comparison of the traditional building complex and the new LEED-certified building, mainly from an acoustic surface characteristic perspective. The main differences highlighted are the office layouts, addition of collaborative areas, and perimeter hallways as opposed to perimeter offices. With this new open-office concept the addition of sound masking and ceiling tiles with higher absorption were critical to the acoustic comfort within the space.

Table 1. Comparison of acoustic comfort spatial and surface characteristics between office buildings.

Acoustic Comfort Spatial and Surface Characteristics		
	Traditional Office Building (Valley Forge, PA/Pre-move)	LEED-Certified Office Building (Malvern, PA/Post-move)
office plan	perimeter offices central workstations	open-concept office space central workstations perimeter hallways perimeter collaborative areas
perimeter wall	glass-NRC 0.04, gypsum board-NRC 0.06	glass-NRC 0.04
partition wall	glass-NRC 0.04, gypsum board-NRC 0.06	glass-NRC 0.04, gypsum board-NRC 0.06
floor	carpet-NRC 0.15	carpet-NRC 0.20
workstation partition	fabric-NRC 0.40	fabric-NRC 0.40
ceiling	mineral board tile-NRC 0.70	fiberglass tile-NRC 0.95

2.2. Measurement Procedure

The acoustic comfort experienced in an open-plan office can be assessed through the measurement of acoustic parameters such as spatial decay rate of speech, speech transmission index (STI), distraction distance, privacy distance and background noise levels. STI and background noise levels were measured to evaluate the indoor environmental quality (IEQ) of the LEED platinum-certified building described in Section 2.1.

The speech transmission index represents the transmission quality of speech with respect to intelligibility. Background noise levels represent the measured sound pressure level in absence of any occupants within the evaluated space. Both distraction and privacy distances are determined from the STI measurements and represent the distance from the speaker at which the STI falls below 0.50 and 0.20 respectively. ISO 3382-3 [19] provides a detailed description of these acoustic parameters along with the standardized measurement procedure which was used to acquire the data presented.

Speech is essentially the modulation of a band of noise. Measurements of speech transmission index (STI) were conducted using the STIPA (STI for Public Address Systems) method [20]. This method is an alternative to the full STI measurement, which uses a Gaussian noise signal as the carrier and is divided into seven-octave bands from 125–8000 kHz, where each band is modulated by fourteen modulation frequencies in one-third octave bands, from 0.63–12.5 Hz. This results in 98 combinations, which creates a time-consuming measurement. In the STIPA method a total of 12 modulation indices are measured from carrier frequencies in one-third octave bands ranging from 0.63 to 12.5 Hz. The modulation represents the combination of modulation and carrier frequencies used to mimic speech excitation. A Nor140 (Norsonic sound level meter) was used to measure the STI using the STIPA method at specific distances from the excitation source within the space. The sound level meter was also used to measure the background noise level at various workstations throughout the open office, in accordance with ISO 3382-3.

Workstations within the LEED-certified open office building were evaluated for the speech transmission index and background noise levels with and without sound masking. Ideally, the STI values will drop below 0.50 between 2 and 4 m. The background levels are expected to fall below a noise criteria (NC) level of 40 dB. Noise criteria are single numbers used to define goals for maximum allowable noise in a given space.

Measurements at various workstations allowed for a comprehensive acoustic comfort mapping of distraction distance from the STI. The maps can be used for quick visualization and spatial assessment of the expected acoustic comfort of the open-plan space with regard to distraction and privacy distances.

2.3. Post-Occupancy Evaluation Survey

The Space Performance Evaluation Questionnaire (SPEQ^{TM}) is an online occupant's survey with developed categories and scales representative of the most important issues identified by occupants to impact their comfort, satisfaction, performance, and health, as perceived and experienced in their work environments. The survey semantics and linguistics have been designed based on proven language constructs that represent layperson descriptions of their physical environment. SPEQ^{TM} was cross-tested and calibrated in field and lab settings. The survey was peer-reviewed by an expert panel of 20 professional building scientists, psychologists, space planners, architects, and physicians.

The data collected by the questionnaire evaluates 30 pre- and post-occupancy issues in 76 questions classified into seven main categories. All questions contained a skip-logic approach to skip irrelevant information based on the occupant's responses. This made the instrument more effective and reduced respondents' fatigue. The average response time of the questionnaire is 12 min, with a minimum of 8 and a maximum of 20 min. All answers were recorded on a five-point Likert scale, semantic-differential scale, a numerical open scale, or a categorized aggregated scale. The scales allowed for continuous data that was easily analyzed using simple and more complex statistical modeling and regressions. In addition, open-ended responses were encouraged for all questionnaire categories to allow occupants to voice their opinions without restrictions. The questionnaire contained a forced response to a consent form and pre-set reminders for missed question responses. Respondents were allowed to skip questions on the second attempt for most questions, with the ability to skip demographic questions to maintain the respondent's privacy. Neither compensation nor a fee was administered for a respondent to respond to the questionnaire. To facilitate statistical analysis, questionnaire responses were recorded into a numerical scale of one to five, such that five was "strongly agree", three was "neutral", and one was "strongly disagree". Subjective responses from SPEQ^{TM} questions were paired with objective acoustical measurements identified in Section 2.2 using timestamps for cross-tabulation analyses.

3. Results

The level of background noise in a space directly impacts speech intelligibility. Background noise level is governed by noise criteria (NC). For an open-office plan with forced-air distribution systems, the ASA S.12.2-2008 [21] recommended noise criteria (NC) curve limit is NC-40 or less. At NC-40 the corresponding sound pressure level range is 46 dB$_A$ to 48 dB$_A$.

Background noise levels were measured in the LEED building with and without the sound masking system turned on during occupied hours. The dots in Figure 2 indicate locations on the third floor at which the background levels were measured. The average sound pressure levels in the north wing with and without the sound masking system turned on were 46.6 dB$_A$ and 41.1 dB$_A$, respectively. The sound masking system generated white noise in the space through an evenly distributed loudspeaker system located in the ceiling plenum. This increased the average background noise to the desired level to achieve good speech privacy between workstations. In Figure 2, the green dots indicate values that met the design criteria during occupied hours. Red dots indicate values that exceeded 48 dB$_A$ due to conversations (red unfilled ovals highlight locations) occurring near the measurement location. However, this value was only exceeded with the noise contribution from the sound masking system.

Figure 2. Background noise survey north wing of LEED-certified building with sound masking on (**left**) and off (**right**). The green dots denote values that met the design criteria during occupied hours whereas the red dots denote values exceeding 48 dB$_A$. Design criteria were exceeded in some instances due to nearby conversation, denoted by the red, unfilled oval shapes.

Workstations were evaluated for spatial sound distribution of the STI to assess the distraction distance from the speaker at which the speech transmission index fell below 0.50. A comprehensive acoustic comfort mapping strategy was applied in order to fully characterize the space and provide visuals of spatial performance. STI measurements were conducted in the northeast wing of the building with and without the activation of the sound masking system. Additionally, the impact of speaker height (standing or seated) on distraction distance was assessed.

STI data is often presented in a table format or an X–Y graphical plot. Figure 3 (WS3154) shows the location, and STI as a function of distance from the speaker at different angles. This dataset illustrates the scenario of a speaker standing versus sitting with sound masking and captures the impact on the STI. The distraction distance target (STI 0.50-0.30 at 2–4 m) was met for the 90-degree orientation with the standing speaker, and for all orientations except the direct (0-degree orientation) and a single 45-degree orientation when the speaker was seated. Given the layout of the workstations relative to the location of the speaker the differences observed with the orientation of the listener to the speaker were expected.

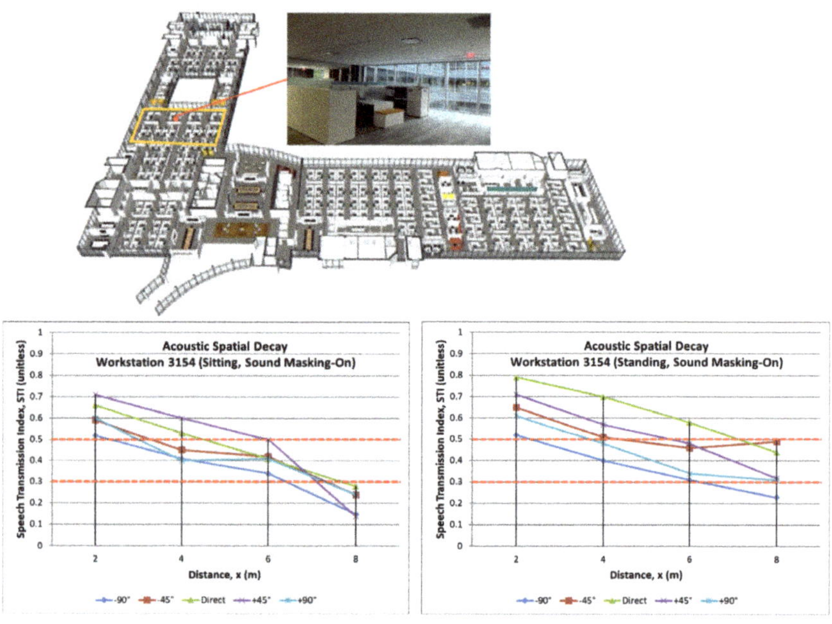

Figure 3. Office map depicting location on the third floor, north wing, at which measurements were conducted along with a snapshot of the interior space (**top**). Speech transmission index (STI) as a function of distance at different angles to a sitting speaker (**bottom left**) and a standing speaker (**bottom right**) with sound masking. The red dashed lines indicate a STI target value between 2–4 m.

Tables and line plots have been used extensively in the presentation of STI data. The plots are easy to interpret and present the data. However, this can at times be a challenge for architects, designers, owners, and developers, who are the solution decision-makers and would prefer the most user-friendly presentation of such data. A color-coded visualization of the STI acoustic comfort mapping can provide a quick and intuitive interpretation of the overall acoustic performance of the space. Overlaying the surface plot on a scaled drawing of the space can provide additional insights under various conditions.

To convey the space performance using the approach to STI acoustic comfort mapping, concentric semicircles were placed at 2-m intervals from the simulated speaker in a sitting position (Figure 4, left). The color map superimposed on the concentric circles convey the change in STI (Figure 4, right), and, by extension, speech intelligibility, at distances away from the speaker. Detailed scenarios are presented in Figure 5 which illustrate the STI measurements for standing and sitting with the sound masking system on and off. The plots, which can be interpreted relatively quickly, show the positive impact when the sound masking system is on and of the lower distraction distance (lower STI) in the seated versus standing positions.

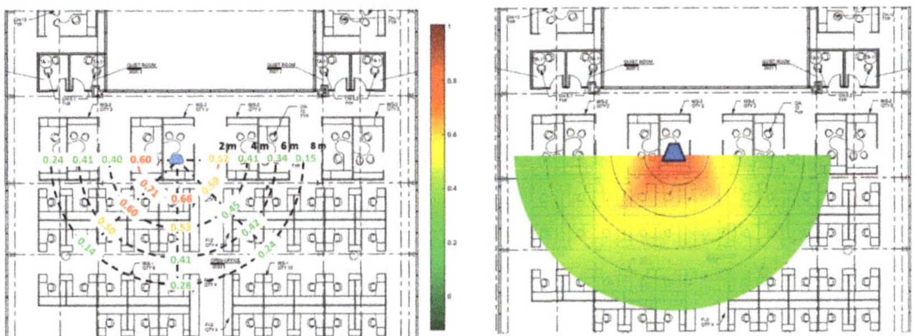

Figure 4. Acoustic comfort mapping with STI of north-facing workstation at the LEED building with the speaker in sitting position and sound masking on. (**Left**) Scaled drawing depicting STI measurements at various location and distances. (**Right**) Associated surface plot highlighting areas from good to bad STI rating.

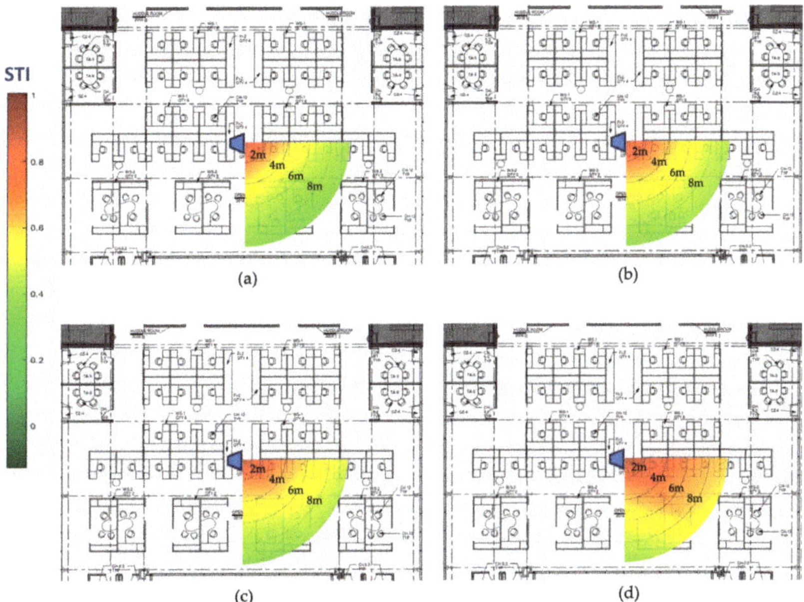

Figure 5. Acoustical comfort visualization through STI surface plots of measurements from the north wing workstation area on the third floor of the building. (**a**) Speaker in sitting position with sound masking on. (**b**) Speaker in sitting position with sound masking off. (**c**) Speaker in standing position with sound masking on. (**d**) Speaker in sitting position with sound masking off. The best- and worst-case scenarios for comfort shown in (**a,d**) respectively.

4. Discussion

4.1. Human-Centric Acoustic Comfort

The human-centric acoustic comfort model conceptualizes both the quality and quantity of the acoustical environment based on the occupant's perspective [22]. In this model, spatial and temporal impacts of comfort are evaluated physically and perceptually. To test the visualization approach proposed to acoustic comfort and its impacts on occupants' satisfaction, the SPEQ^{TM} survey was administered to the occupants of the traditional building prior to the move and one year after the move to the new LEED-certified building respectively. Focus groups were employed before the administration of the surveys to ensure that occupants were well habituated in the building and were even in their perception of both positive and negative attributes of the indoor environment. In addition, the after-move survey was administered one-year post-move to limit bias in perceptions between the new and old buildings.

Response to the survey was very high, with 42% of the 800 employees completing the survey prior to the move and 44% of the 800 plus employees completing the questionnaire one year after the move and relocation. Data tabulation and coding were performed on both the physical and human-response data sets. Physical measurements and survey responses were spatially tagged and statistically analyzed using the software Statistical Package for the Social Sciences (SPSS). This analysis ensured that spatial visualizations of physical acoustics metrics, such as STI, were correlated with occupants' perceptions of acoustical comfort and satisfaction responses to the SPEQ^{TM} questionnaire.

4.2. Occupants Perspective Analysis

To examine the correlation between acoustic comfort visualizations and occupants' perceptions, we calculated the mean difference in dissatisfaction and mean difference in satisfaction for the various aspects related to occupant perceptions of acoustical comfort (Table 2) in a traditional and a LEED-certified building. The reference in these calculations is the LEED building. For example, a negative sign (-) in dissatisfaction means a reduction in dissatisfaction responses for that item, hence more improvement. Similarly, a negative sign (-) in satisfaction means a reduction in satisfaction in this item, hence a lack of improvement and vice versa.

Table 2. Differences in perceived satisfaction of acoustic comfort attributes inside the LEED-certified building as compared with the traditional building.

Aspect	Mean Difference in Dissatisfaction (%)	Mean Difference in Satisfaction (%)	Total Perceived Difference (%)
overall acoustical satisfaction	−19.5	+22.7	42.2
noise from people talking in your area	−7.2	+12.0	19.2
noise from people talking in the hallway	−17.3	+16.9	34.2
noise from ventilation systems	−27.9	+31.2	59.1
noise from equipment	−25.3	+30.1	55.4
noise from lights	−10.6	+27.3	32.7
noise from outside the building	−10.4	+22.3	32.7

Employees responded to five different questions that collectively evaluated their acoustical comfort in the building pre- and post-occupancy between the traditional and LEED buildings. Across all questions, employees reported a higher level of acoustical comfort improvements of 42.2% in the LEED building as compared with the traditional building (Table 2). In addition, acoustical comfort improved across all parameters between 19% and 59%. The most improvement was the perceived reduction in noise of equipment and ventilation systems by 55.4% and 59.1%, respectively. Thanks to improvements in the building envelope and the provision of better daylighting and electric lighting systems, occupants reported an increased satisfaction of 37.9% for noise from lighting system and an increased satisfaction of 32.7% for outside building noise perception. The sound masking system and improvements in sound absorption material in the LEED building might be related to a perceived satisfaction with distractions of noise from people talking and walking in hallways despite being in an open-plan office. In general occupants perceived improvements in satisfaction of 19.2% for noise distractions due to talking and 34.2% improved satisfaction due to hallway noise as compared with the traditional building.

This improvement in satisfaction from the survey findings correlates well with the observation and interpretation of Figures 4 and 5. Within the workstation the STI falls within the range of 0.8–1.0 due to the close proximity to the speaker, and as a result, speech is highly intelligible. However, when sitting and the sound masking is on as illustrated in Figure 4, STI values, on average fall under 0.50 in the 4–6 m range. In this range the speech intelligibility has a fair rating, and a fair to poor rating at distances greater than 6 m. This means that occupants will be less distracted due to office conversations at those distances. Figure 5 clearly illustrates the change in distraction distance between standing vs sitting and sound masking on or off. The perceived satisfaction is expected to decrease to a minimum when the sound masking is off and the occupant is standing as the STI, on average, remains above 0.60 at a rage of 2–8 m. In general, the survey indicated occupants were satisfied with STI ranges of 0.2–0.4 outside of their immediate workstation.

5. Conclusions

Architectural acoustic designs for open-plan office spaces aim at reducing transmission between workstations while creating acoustical privacy within the individual workstation. This approach, however, doesn't usually result in a pleasant or comfortable work environment due to average numerical values that do not reflect a human-centric acoustical experience for the occupants, according to their spatial locations. Through the use of sound-absorbing surfaces, high-performance, noise-reducing interior partitions, and sound-masking systems, a more spatial acoustical comfort approach is feasible. Predicting the success of this approach in the field can be challenging. The use of STI mapping and visualizations is a tool that can help predict and estimate occupants' perceptions prior to occupancy. The procedure outlined in this paper offers a design and spatial visualization tool to help designers and building owners predict and estimate acoustical performance in spaces from the occupants' perspective.

The results show strong correlations between improved acoustic qualities of the retrofitted, high-performance, LEED-certified building in spaces visualized within the STI standards, and overall acoustical satisfaction (Table 2). A comparative analysis of the occupants' attitudes and perceptions of the impact of the building on their comfort levels reveal a significant perceptual change to the impact of acoustical design on the employee overall satisfaction. A comprehensive design and employment of spatial visualization tools to predict and implement metrics of indoor environmental quality (IEQ) might result in improved comfort, satisfaction, and the perception of well-being for the occupants. It is the hope that further integration of building systems concerning IEQ measures and comfort visualizations can better predict acoustical performance in the early design stages and lead to better work environments that are more responsive to occupant's needs and comfort expectations.

Author Contributions: All authorshave equally contributed to this manuscript. All authors have read and agreed to the published version of the manuscript.

Funding: This research received no external funding.

Institutional Review Board Statement: The study procedures involving human subjects, Protocol Number: 04252016.035, was approved by the University of Oregon Institutional Review Board as per 45 CFR Part 46.110. Upon further review, it was determined that this study qualifies for exemption as per the Common Rule regulations of Title 45 CFR 46.101(b)(2) on 28 April 2018.

Informed Consent Statement: Informed consent was obtained from all subjects involved in the study as a requirement to proceed with taking the SPEQ questionnaire.

Data Availability Statement: Data for this study is stored in an institutional depository protected by the University of Oregon IRB Protocol Number 04252016.035. Access to the data is only granted to key personnel involved in the study and who successfully passed the CITI training module.

Conflicts of Interest: The authors declare no conflict of interest.

References

1. Yadav, M.; Cabrera, D.; Love, J.; Kim, J.; Holmes, J.; Caldwell, H.; de Dear, R. Reliability and repeatability of ISO 3382-3 metrics based on repeated acoustic measurements in open-plan officies. *Appl. Acoust.* **2019**, *150*, 138–146. [CrossRef]
2. Lou, H.; Ou, D. A comparative field study of indoor environmental quality in two types of open-plan offices: Open-plan administrative offices and open-plan research offices. *Build. Environ.* **2019**, *148*, 394–404. [CrossRef]
3. Mikulski, W. Acoustic conditions in open plan office—Application of technical measures in a typical room. *Med. Pr.* **2018**, *69*, 153–165. [CrossRef] [PubMed]
4. Jahncke, H.; Hongisto, V.; Virjonen, P. Cognitive performance during irrelevant speech: Effects of speech intelligibility and office-task characteristics. *Appl. Acoust.* **2013**, *74*, 307–316. [CrossRef]
5. Jensen, K.L.; Arens, E.; Zagreus, L. Acoustical quality in office workstations, as assessed by occupant survey. In Proceedings of the 10th International Conference on Indoor Air Quality and Climate, Beijing, China, 4–9 September 2005; Volume 3, pp. 93–101.
6. Noble, M.R. Green buildings: Implications for acousticians. *J. Acoust. Soc* **2005**, *117*, 2378–2378. [CrossRef]
7. Hodson, M.; Khaleghi, A.; Richter, M.; Razavi, Z. Design and evaluation of noise-isolation measures for the natural-ventilation openings in a 'green' building. *Noise Control Eng. J.* **2009**, *57*, 493–506. [CrossRef]
8. Ebissou, A.; Parizet, E.; Chevret, P. Use of the Speech Transmission Index for the assessment of sound annoyance in open-plan offices. *Appl. Acoust.* **2015**, *88*, 90–95. [CrossRef]
9. Ebissou, A.; Parizet, E.; Chevret, P. Effect of masking noise on cognitive performance and annoyance in open plan offices. *Appl. Acoust.* **2016**, *88*, 44–55.
10. Haapakangas, A.; Hongisto, V.; Hyönä, J,; Kokko, J.; Keränen, J. Effect of masking noise on cognitive performance and annoyance in open plan offices. *Appl. Acoust.* **2014**, *86*, 1–16. [CrossRef]
11. Kang, S.; Ou, D.; Mak, C.M. The impact of indoor environmental quality on work productivity in university open-plan research offices. *Build. Environ.* **2017**, *124*, 78–79. [CrossRef]
12. Yadav, M.; Cabrera, D. Two simultaneous talkers distract more than one in simulated multi-talker environments, regardless of overall sound levels typical of open-plan offices. *Appl. Acoust.* **2019**, *148*, 46–54. [CrossRef]
13. Yadav, M.; Kim, J.; Cabrera, D.; de Dear, R. Auditory distraction in open-plan office environments: The effect of multi-talker acoustics. *Appl. Acoust.* **2019**, *126*, 68–80. [CrossRef]
14. Renz, T.; Leistner, P.; Liebl, A. Effects of the location of sound masking loudspeakers on cognitive performance in open-plan offices: Local sound masking is as efficient as conventional sound masking. *Appl. Acoust.* **2018**, *139*, 24–33. [CrossRef]
15. Renz, T.; Leistner, P.; Liebl, A. The Effect of Spatial Separation of Sound Masking and Distracting Speech Sounds on Working Memory Performance and Annoyance. *Acta Acust. United Acust.* **2018**, *104*, 611–622. [CrossRef]
16. Haapakangas, A.; Kankkunen E.; Hongisto, V.; Virjonen, P.; Oliva, D.; Keskinen, E. Effects of Five Speech Masking Sounds on Performance and Acoustic Satisfaction. Implications for Open-Plan Offices. *Acta Acust. United Acust.* **2011**, *97*, 614–655. [CrossRef]
17. Keus van de Poll, M.; Carlsson, J.; Marsh, J.E.; Ljung, R.; Odelius, J.; Schlittmeier, S.J.; Sundin, G.; Sörqvist, P. Unmasking the effects of masking on performance: The potential of multiple-voice masking in the office environment. *J. Acoust. Soc* **2015**, *138*, 807–816. [CrossRef] [PubMed]
18. Haapakangas, A.; Hongisto, V.; Eerola, M.; Kuusisto, T. Distraction distance and perceived disturbance by noise—An analysis of 21 open-plan offices. *J. Acoust. Soc* **2015**, *141*, 127–136. [CrossRef] [PubMed]
19. *ISO 3382-3; Acoustics—Measurement of Room Acoustic Parameters—Part 3: Open Plan Offices.* ISO: Genèva, Switzerland, 2012.
20. *Nor140 Sound Analyzer: Instruction Manual*; Norsonic: Lierskogen, Norway, 2021.
21. *ASA S12.2-2008; American National Standard Criteria for Evaluating Room Noise.* Acoustical Society of America: Melville, NY, USA, 2008.
22. Elzeyadi, I.M.K. Multi-Performance Spaces: Assessing the Impacts of an Innovative LEED™Platinum Educational Complex on the Triple Bottom Line. In *Adaptive Architecture: Changing Parameters and Practice*; Prieser W., Hardy A., Eds.; Taylor & Francis Group: Routledge, UK, 2017; pp. 34–43.

Article

A Simplified Thermal Comfort Calculation Method of Radiant Floor Cooling Technology for Office Buildings in Northern China

Xiaolong Wang [1,†], **Tian Mu** [2,†], **Lili Zhang** [1], **Wenke Zhang** [1,*] and **Linhua Zhang** [1,*]

[1] School of Thermal Engineering, Shandong Jianzhu University, Jinan 250101, China; wangxiaolong2017@sdjzu.edu.cn (X.W.); 2019030112@stu.sdjzu.edu.cn (L.Z.)
[2] School of Architecture and Urban Planning, Shandong Jianzhu University, Jinan 250101, China; 2019030114@stu.sdjzu.edu.cn
[*] Correspondence: zhangwenke@sdjzu.edu.cn (W.Z.); zhth0015@sdjzu.edu.cn (L.Z.)
[†] First two authors contributed equally to this paper and should be considered co-first authors.

Abstract: The increasing application of floor heating technology promotes the development of floor radiant cooling technology (abbreviated as FRC technology). Many office buildings in northern China try to use FRC technology to cool in summer, but thermal comfort is the key problem restricting the promotion of this technology. The thermal comfort problems of an office room with floor radiant cooling were studied in this paper by the methods of numerical simulation, control variable, and data fitting, and the experimental results were verified in multiple ways. It was found that, for an office room using floor radiant cooling, the effect of the floor surface temperature on thermal comfort was about 16%, while the effect of indoor air temperature was about 84%, and relative humidity had little effect on thermal comfort. A simplified thermal comfort calculation model was proposed, which could be used as an indicator to adjust the floor surface and indoor air temperature, or could be used to calculate the *PMV-PPD* value. The findings have guiding significance for the design and control of FRC technology.

Keywords: thermal comfort; *PMV-PPD*; radiant floor; cooling; office buildings

Citation: Wang, X.; Mu, T.; Zhang, L.; Zhang, W.; Zhang, L. A Simplified Thermal Comfort Calculation Method of Radiant Floor Cooling Technology for Office Buildings in Northern China. *Buildings* **2022**, *12*, 483. https://doi.org/10.3390/buildings12040483

Academic Editor: Cinzia Buratti

Received: 1 March 2022
Accepted: 11 April 2022
Published: 13 April 2022

Publisher's Note: MDPI stays neutral with regard to jurisdictional claims in published maps and institutional affiliations.

1. Introduction

Under the background of creating a low-carbon, healthy, and comfortable building environment [1,2], floor radiant cooling technology (following abbreviation as FRC technology) has attracted more and more attention all over the world. Compared with typical all-air conditioning, FRC technology is feasible for high-temperature cooling to achieve the purpose of energy saving [3], and it can also effectively use natural energy, such as surface water (groundwater/soil) cold, to realize "free cooling" or zero carbon emissions [4]. Many buildings have combined FRC technology with convective air conditioning, and proposed a new type of radiation–convection air conditioning, which creates nearly zero-carbon-emission buildings [5,6]. However, the thermal comfort of the indoor thermal environment is the key problem, which is also the main factor restricting the development of FRC technology [7].

The concept of floor radiant cooling comes from floor radiant heating technology (following abbreviation as FRH technology). People try to use the same coil in the floor structure of heating in the winter to pass high-temperature chilled water to cool the indoor thermal environment in the summer [8]. FRH has gradually replaced the traditional cast-iron radiator and has been widely used in urban residential buildings in the cold areas of China [9] and other regions [10,11]. More and more practical engineering experience is feeding back the floor heating technology and forming a virtuous circle, which makes the FRH technology very developed and mature [12], and also stimulates research interest in the thermal comfort of floor heating. Garcia [13] explored the possibilities of using radiant floor heating systems (RFHS) in removable glazed enclosed patios from the optimal range of the predicted mean vote (*PMV*) and predicted percentage of dissatisfaction (*PPD*). Zhang et al. [14] proposed an external wall radiant heating technology in passive buildings, that is, a wall implanted with heat pipes that can improve the utilization of absorbed solar energy, in this way increasing the *PMV* value in the transition season and reducing the dissatisfaction rate of people. Cho et al. [15] evaluated the response time of the intermittent operation of PB-Model (polybutylene pipe) floor heating and CT-Model (capillary tube) floor heating with initial water supply temperatures of 55 °C and 40 °C by measuring the time taken for the initial indoor air temperature to rise by 4 °C, which provided a reference basis for the start-up time of the indoor thermal environment with a low temperature, but this theory is not suitable for continuous central heating. Karacavus and Aydin [16] selected the thermal comfort of an office environment as the research object and evaluated the compliance of the office's general and local comfort levels with ISO 7730 [17]. The above thermal comfort analysis mainly focuses on the advantages of floor radiant heating, but when FRH technology is coupled with other heating methods, or when the floor receives solar radiation [18], it may also have the disadvantage of overheating [19], which can adversely affect thermal comfort. It can be seen that the indoor thermal comfort of rooms with FRH systems is good, and the advantages far outweigh the disadvantages. The indoor vertical temperature difference also meets the physiological functional requirements of "feet like heat and head like cool" [20,21]. This is also an important reason for the popularization and application of FRH technology.

Different from the heat-transfer process of floor heating, when the floor is used for cooling in the summer, the floor temperature is lower, and the human body has more exposed parts in the summer, so the traditional thermal comfort theory of floor heating is no longer suitable for floor cooling. Whether the vertical distribution of the indoor air temperature is contrary to the human body's thermal feeling, whether the asymmetric radiation temperature will cause discomfort in the exposed part of the human body, and how the human body can tolerate the FRC system cannot copy the findings related to "floor heating", which has stimulated extensive discussion and research. Zhang et al. [22] discussed the current application cases using FRC technology in China, and discussed the special characters and in relation to the topic of comfort and energy saving. Gu et al. [23] studied the thermal comfort of a conventional intermittently operating FRC system in hot and humid regions, and found that, when the system is subjected to sudden increases in the heat and moisture loads, a slow thermal response may occur. FRC systems are suitable for residential homes or offices with non-centralized air conditioning systems to handle transient heat and moisture load variations. The FRC system needs to be combined with other cooling or dehumidification systems to adapt to offices with non-centralized air conditioning systems to handle transient heat and moisture load variations. Du et al. [19] compared the thermal comfort of radiant cooling technology and convective air conditioning, and found that there is little practical difference in human subjective responses between the two systems in the continuous operating mode. Kwong et al. [24] subjectively and objectively evaluated the indoor thermal comfort of office buildings using a floor radiation and fresh air coupled cooling system, and found that the difference between the *PMV* and Actual Mean Vote (AMV) was greater than 0.5. Hu et al. [25] analyzed the intermittent operation of a radiant cooling system, and found that indoor comfort in the occupied period can be guaranteed with a reasonable pre-start time. Liu et al. [26] conducted a series of numerical simulations to evaluate the thermal comfort performance of a FRC system when combined with different ventilation systems, including mixed ventilation (MV), stratum ventilation (SV), displacement ventilation (DV), and ductless personalized ventilation (DPV). Most of the above analysis focuses on the deficiency of the thermal comfort of FRC systems and shows that the thermal comfort of a single FRC system is not up to the standard, which needs to be combined with other convection cooling ends to form a coupled cooling mode with complementary disadvantages.

To sum up, a large case of studies on floor heating have provided valuable practical experience and new methods for research on the thermal comfort problems of FRH and FRC systems, but still have several problems restricting the good thermal comfort of FRC systems to be solved. Existing researches on the thermal comfort of FRC systems were mainly based on the numerical simulation method [27], the human body sensation method (questionnaire survey method) [28], and the human body measurement method [29], and a few used the theoretical analysis method [30]. For the FRC system, the thermal comfort index is difficult to calculate and the contribution of the floor cooling end to thermal comfort is difficult to be evaluated for the following three reasons.

- The floor surface cools the indoor thermal environment in two ways coupling convection and radiation, and the heat transfer process is very complex [31];
- It is very complicated to calculate the radiation angular coefficient and average radiation temperature;
- Lack of practical application feedback and fitting empirical formulas.

How to quantitatively find the influence of indoor environmental factors on thermal comfort has become the primary problem to be solved. For this research goal, the following issues will be focused on for research and discussion.

- The effect of the changes of the main factors of the indoor thermal environment on the thermal comfort indicator in a room with floor cooling;
- Quantitative evaluation of the contribution of the floor cooling end to indoor thermal environment comfort;
- Finding a simplified calculation model of the thermal comfort indicator for FRC systems, which can be used to adjust the temperature value of the indoor thermal environment or calculate the value of the indoor thermal comfort indicator quickly.

The findings based on the thermal comfort evaluation indicator can provide a theoretical basis for the design of FRC systems and the control of the surface temperature of the floor cooling end.

2. Methods

2.1. Thermal Comfort Index and Influencing Factors

The Predicted Mean Vote (*PMV*) of human thermal comfort is a commonly used thermal comfort evaluation method in office buildings [32]. This method was first proposed by Fanger [33], and its calculation expression is:

$$PMV = [0.0303 \exp(-0.036M) + 0.0275] \cdot f(M,\ I_{cl},\ t_a,\ \gamma, v_a, \theta_{mrt}) \tag{1}$$

where M is the metabolic rate of the human body and I_{cl} is the thermal resistance of clothing. These two variables do not belong to the indoor thermal environment control variables. t_a is the indoor air temperature, γ is the indoor relative humidity, v_a is the air speed, and θ_{mrt} is the average radiation temperature, which represents the radiation heat transfer of each inner surface of envelopes to the human body and the asymmetrical radiation. The above four variables are the indoor thermal environment variables and also the key variables for the quantitative analysis of *PMV*.

The *PMV* index is divided into 7 levels that quantify the human thermal sensation, as shown in Table 1. The recommended value of *PMV* is $-0.5 \sim +0.5$ [34].

Table 1. *PMV* level and human thermal sensation.

Thermal Sensation	Hot	Warm	Slightly Warm	Moderate	Slightly Cool	Cool	Cold
PMV value	+3	+2	+1	0	−1	−2	−3

The *PMV* index also can be used to quantitatively evaluate the indoor thermal environment of office buildings in northern China, which use radiant floors for cooling in the summer. Considering the usage characteristics of office buildings, in the choice of *PMV* independent variable parameters, M takes the energy metabolism rate during sitting and I_{cl} uniformly takes the thermal resistance of summer clothing, as shown in Table 2.

Table 2. Parameter settings of office staff [35].

Personnel Status	Thermal Resistance of Summer Clothing (Shorts, Long Thin Pants, Short Sleeved Cardigan, Thin Socks, and Shoes)	Metabolic Rate
Sitting	0.080 m^2·K/W (0.5 CLO)	58.15 W/m^2 (1.0 met)

For the independent variable v_a, different from the traditional convection air conditioner, in a room with a FRC system, fan coil units or displacement ventilation are usually used to deal with fresh air, and people are in the return air area. At this time, the floor cooling end bears most of the cooling load [36], and the cooling load provided by the fresh air is small, so the wind speed around the human body is very small (<0.1 m/s) and changing the air velocity between 0 and 0.1 m/s has little impact on the indoor air distribution and *PMV* [37], so v_a can be set as a constant close to 0 m/s. For the average radiation temperature θ_{mrt}, this parameter is the key variable affecting the thermal comfort of a radiant air-conditioned room; it is also the calculation difficulty of using Equation (1) to calculate *PMV* [38], and it has a functional relationship with the internal surface temperature of the floor, ceiling, window, and four walls, which can be written in the form of the k function as Equation (2).

$$\theta_{mrt} = k(t_s, t_d, t_c, t_{w1\sim4}) = \sqrt[4]{t_s{}^4 \cdot F_{p-s} + t_d{}^4 \cdot F_{p-d} + t_c{}^4 \cdot F_{p-c} + t_{w1}{}^4 \cdot F_{p-w1} + \cdots + t_{w4}{}^4 \cdot F_{p-w4}} \tag{2}$$

where F_{p-i} is the angle factor between a person and surface i, t_s is the surface temperature of the floor, t_d is the inner surface temperature of the ceiling, t_c is the inner surface temperature of the inner shading of the window, and $t_{w1\sim4}$ is the inner surface temperature of four walls (the outer and the inner walls) of a room, respectively.

Different from the large vertical temperature difference in a floor-heated room, there is no significant vertical temperature gradient in a room using floor cooling [39], which is because the thermal pressure movement of indoor air and the floor surface with a low temperature have the opposite effect on the inner surface temperature of the ceiling; the rising of hot air causes t_d to increase, while the floor surface is facing the inner surface of the ceiling and is cooled by radiant heat transfer. These two aspects work together, resulting in uniform indoor air temperature; the vertical temperature gradient can be ignored. An office room is not a large-space building, the indoor space is limited, with the disturbance of personnel and air supply, and the inner surface temperature of each envelope is also restricted by the floor surface temperature t_s and indoor air temperature t_a, so the uncooled internal surface temperature has little difference. Variable t_i can be used to represent $t_d, t_c, t_{w1}, t_{w2}, t_{w3}$ and t_{w4} [5]. Equation (2) of variable θ_{mrt} can be expressed as follows:

$$\theta_{mrt} = \sqrt[4]{t_s{}^4 \cdot F_{p-s} + t_i{}^4 \cdot \left(F_{p-d} + F_{p-c} + F_{p-w1} + \cdots + F_{p-w4} \right)} \tag{3}$$

During the service time of office buildings from 9:00 to 17:00, t_i can also be expressed as a function q related only to the internal disturbance t_s and t_a [5], as follows:

$$t_i = q(t_s, t_a) \tag{4}$$

Combining Equations (3) and (4), θ_{mrt} can be considered as the dependent variable of indoor air temperature t_a and floor surface temperature t_s, which can be expressed in the form of the h function, as shown as Equation (5).

$$\theta_{mrt} = k(t_s, t_i) = k(t_s, q(t_s, t_a)) = h(t_a, t_s) \tag{5}$$

Through the above variable analysis and substituting Equation (5) into Equation (1), the functional expression of *PMV* can be written in the form of the following *g* function, as follows.

$$PMV = A \cdot f(t_a, \gamma, \theta_{mrt}) = A \cdot f(t_a, \gamma, h(t_a, t_s)) = g(t_a, t_s, \gamma) \qquad (6)$$

where *A* is a constant that can be calculated from $0.0303 \exp(-0.036M) + 0.0275$.

From Equation (6), *PMV* can be expressed as a dependent variable, with only the air temperature t_a, floor surface temperature t_s, and relative humidity γ as the independent variables. The expression of *PMV* can be obtained by solving Equation (6).

To solve the function with few independent variables, the controlled variable method is a commonly used method. This method can keep other parameters unchanged and study the relationship between a single independent variable and the dependent variable.

2.2. Numerical Simulation Methods

Numerical simulation is a good method for variable control analysis and finding the relationship between variables due to its flexibility, rapidity, and controllability. In this study, the TRNSYS (Transient System Simulation Program) [40] is used as the simulation software to discuss the relationships between three independent variables (t_s, t_a, and γ) and *PMV* in turn, and try to solve function *g*. The TRNSYS is an extremely flexible, visualized, and modularized transient process simulation software, which is mainly used to simulate the energy systems of buildings, especially radiant ceiling cooling systems and FRC technology, and can also be used to research the thermal comfort problems of *PMV-PPD* [3,40].

An original scale model of an office room in northern China was established in TRNBuild, and two cooling ends of the floor surface and fresh air were established in TRNSYS, as shown in Figure 1.

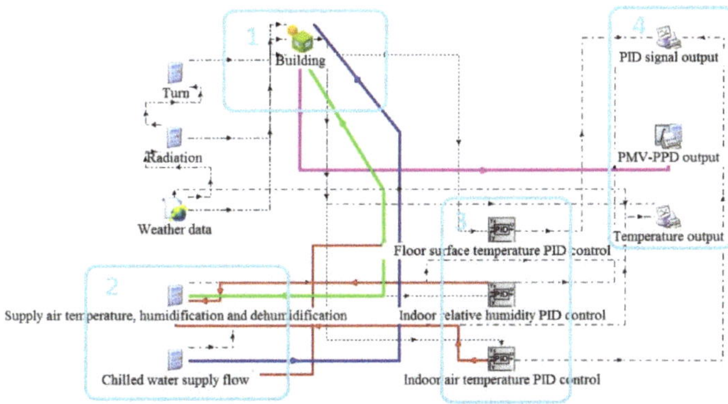

Figure 1. Numerical simulation diagram of the TRNSYS.

Figure 1 is mainly divided into four parts, and the key summary of numerical simulation is described as follows.

- Room model and two cooling ends (Part 1):

A typical office building in Jinan, China, was selected as the actual physical model, and the room size, buried pipe arrangement form, and floor structure are shown in Figure 2. Jinan is a cold region with the characteristics of a hot summer and cold winter. The office room belongs to the public buildings, its envelope structure parameters need to meet Table 3, and it faces south–north. The heat transfer coefficient of building envelopes is $0.45 \, \text{W}/(\text{m}^2 \cdot \text{K})$ [41], and they have great thermal inertia. Because the office buildings in northern areas have winter heating needs, the exterior walls are usually equipped with external insulation. The room has good insulation performance, good internal shading performance, and the ability to resist climate changes from the outdoor environment.

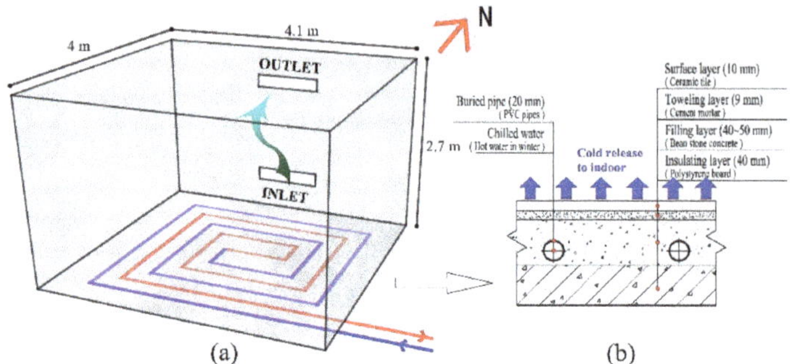

Figure 2. Schematic diagram of laying pipes and floor structure. (**a**) Layout form of floor laying pipe. (**b**) Floor structure and materials.

Table 3. Thermal performance limit of a public building's envelope in cold areas [41].

Envelopes of Public Buildings	Building Shape Coefficient $\leq$ 0.30	0.30 < Building Shape Coefficient $\leq$ 0.50
	Heat Transfer Coefficient	Heat Transfer Coefficient
Roof	$\leq$ 0.45 W/(m²·K)	$\leq$ 0.40 W/(m²·K)
Exterior wall (including non-transparent curtain wall)	$\leq$ 0.50 W/(m²·K)	$\leq$ 0.45 W/(m²·K)
Window to wall ratio	0.4	0.3

The setting of floor structure of floor heating or cooling does not use a certain module of the TRNSYS, although the TRNSYS is a modular simulation software [42]; the setting is completed in Part 1 (TRNBuild). Heating/cooling pipes are laid in the floor structure of a room, as shown in Figure 2. Chilled water is passed through the laying pipes for floor cooling, and the room also uses fresh air to cool the indoor environment together in the summer.

- Control of the indoor thermal environment (Part 2 and Part 3):

The cooling system has three PID controls, which respectively control the floor surface temperature t_s, indoor air temperature t_a, and indoor relative humidity γ. The floor surface temperature is controlled by chilled water variable flow regulation, and the indoor temperature is controlled by variable air supply temperature regulation. In practice, variable flow regulation can also be used to control the indoor air temperature. For the control of the relative humidity of indoor air, humidification and dehumidification are both allowed. The accuracy of the floor surface temperature and indoor air temperature can be controlled within ±0.05 °C, and the accuracy of the indoor relative humidity can be controlled within ±0.1%.

- Results output (Part 4):

Part 4 is the results output part, which shows the values of the *PMV-PPD* indicators of people in the room and the values of each temperature index of the indoor thermal environment. It is also necessary to output the PID control signal (0~1) because the PID control parameters need to be continuously adjusted according to the changes in the PID signal, and finally determine the PID control capability range in Part 2 and Part 3.

- Outdoor environmental meteorological parameters:

The typical daytime climate parameters of Jinan in the summer are selected as the outdoor climate changes.

- Work schedule:

  The working schedule of people in office buildings is generally from 9:00 to 17:00.

3. Results and Discussions

3.1. Solution of PMV

3.1.1. Effect of Floor Surface Temperature Changes on *PMV*

In rooms using radiant floors to cool, people usually set the indoor air temperature above 26 °C, but not higher than 29 °C, in the summer [41]. The surface temperature of the floor varies widely, and it can range from 18 °C to 25 °C (condensation is not considered in this paper). The indoor relative humidity is generally about 50% [43].

Using PID control to maintain the indoor relative humidity at 50% and to maintain the indoor air temperature at 26 °C, 27 °C, 28 °C, and 29 °C, the variation law of *PMV* with the floor surface temperature can be obtained by changing the floor surface temperature every 1 °C from 18 °C to 25 °C, as shown in Table 4.

Table 4. *PMV* values for different indoor air temperatures and floor surface temperatures.

Indoor Air Temperature/°C	Floor Surface Temperature Changes/°C							
	18	19	20	21	22	23	24	25
26	−0.59	−0.55	−0.50	−0.44	−0.38	−0.32	−0.26	−0.21
27	−0.34	−0.28	−0.23	−0.17	−0.12	−0.06	0.00	0.06
28	−0.07	0.00	0.04	0.09	0.15	0.20	0.26	0.32
29	0.21	0.25	0.32	0.36	0.41	0.47	0.53	0.59

From Table 4, it can be seen that the *PMV* increases monotonically with the increase in the floor surface temperature when the indoor air temperature remains constant, which indicates that people's warm feeling increases.

When the indoor air temperature was at 26 °C, all of the *PMVs* were negative values, no matter how the floor surface temperature changed, which meant that the indoor air temperature was too low. When the indoor air temperature was at 29 °C, all of the *PMVs* were positive values, no matter how the floor surface temperature changed, which meant that the indoor air temperature was too high and the single floor radiant cooling end gad the problem of insufficient cooling capacity; thus, it needs to cooperate with other cooling ways to reduce the indoor air temperature. When the indoor air temperatures were 27 °C and 28 °C, the *PMV* values changed from negative to positive with the temperature rise of the floor's surface.

As described above, the indoor air temperature of the room using FRC technology should be maintained at 27~28 °C. In the design of coupled cooling of floor radiation and fresh air, the indoor design temperature should be 1~2 °C higher than that of traditional convection air conditioning.

The finding is consistent with the results in [44,45], which proves the reliability and accuracy of the parameter setting of the numerical calculations.

3.1.2. Simplified Calculation Model of *PMV*

In order to find the functional relationship between t_s and *PMV*, the least-square method was used to fit the sample data in Table 4, as shown in Figure 3.

It can be seen from Figure 3 that t_s and *PMV* are linear, and the relationship of each function is as follows.

When the indoor air temperature is 29 °C, the relational formula is

$$PMV = 0.0543t_s - 0.7746 \tag{7}$$

When the indoor air temperature is 28 °C, the relational formula is

$$PMV = 0.0544t_s - 1.0460 \tag{8}$$

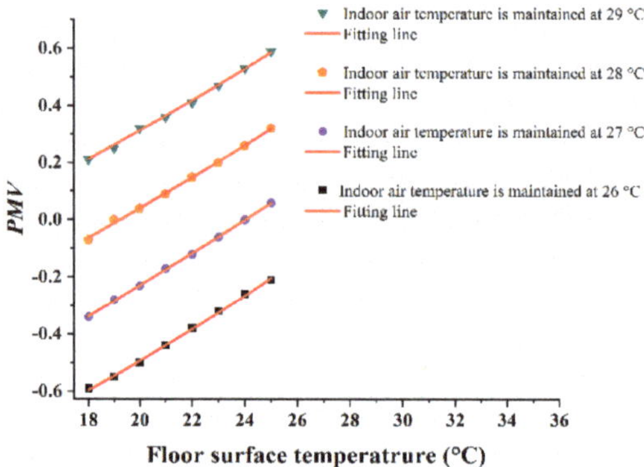

Figure 3. Fitting functional relationship between t_s and PMV.

When the indoor air temperature is 27 °C, the relational formula is

$$PMV = 0.0567t_s - 1.3608 \tag{9}$$

When the indoor air temperature at 26 °C, the relational formula is

$$PMV = 0.0561t_s - 1.6118 \tag{10}$$

The coefficient of determination R^2 is often used as an index to comprehensively measure the goodness of fit of a regression equation to sample observations, and its value range is 0 to 1—the closer to 1, the better the fit [46]. The determination coefficients R^2 of the above Equations (7)–(10) were all above 0.99, and the degree of fitting was very high. By observing Figure 3, it can be found that the slope of Equations (7)–(10) is almost equal, and the intercept is also linearly related to the indoor air temperature t_a. Based on this law, a function expression of PMV with the floor surface temperature t_s and indoor air temperature t_a can be obtained by combining the above four formulas:

$$PMV = g(t_s, t_a, 50\%) = 0.2826t_a + 5.5375 \times 10^{-2}t_s - 8.9709 \\ t_a \in [26, 29], \; t_s \in [18, 25] \tag{11}$$

As can be seen from Equation (11), PMV has a power relation with t_a and t_s. The coefficient before t_s is 5.5375×10^{-2}, while the coefficient before t_a is 0.2826. t_a and t_s are both the parameters characterizing temperature (the unit is °C), and both are positive. The former coefficients of t_a and t_s can be used for weighting the calculation to represent the effect on the PMV. The coefficient weighted calculation shows that the effect of t_s on the PMV is only 16%, while the influence of t_a on PMV is about 84%. The effect of t_a on PMV is much greater than that of t_s. The effect of the floor surface temperature on PMV is only about one-fifth of the indoor air temperature.

The results indicate that the indoor air temperature plays a leading role in PMV. For FRC technology, people are less sensitive to changes in the floor surface temperature, but more sensitive to indoor air temperature. The PMV changed by only about 0.055 for every 1 °C change in the floor surface temperature, and about 0.28 for every 1 °C change in the indoor air temperature.

Equation (11) can realize the following two functions:

(1). In the room using FRC technology, using Equation (11), the current *PMV* value can be calculated according to the current floor surface temperature and indoor air temperature so as to monitor the indoor thermal environment. The *PMV* value can also be obtained by referring to Figure 4.

(2). Equation (11) or Figure 4 can be used as a theoretical basis for regulating the floor surface temperature or indoor air temperature in a room using FRC technology, and can also be used to control the floor surface temperature and indoor air temperature to achieve the desired *PMV*.

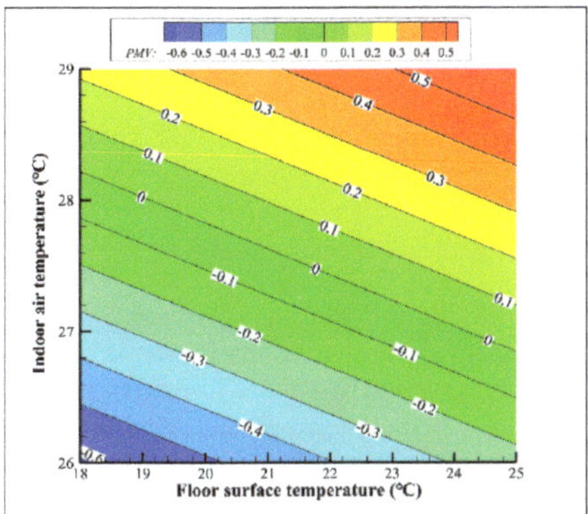

Figure 4. Temperature regulation diagram with *PMV* as the index.

3.2. Solution of PPD and Verification of Numerical Simulation

3.2.1. Solution of *PPD*

The *PMV* index represents the feelings of most people in the same environment, but there are physiological differences between people, so the *PMV* index does not necessarily represent the feelings of all people. For this, Fanger [47] proposed the predicted percent dissatisfied (*PPD*) index to indicate the dissatisfaction rate of the population with the thermal environment, and gave the quantitative relationship between *PPD* and *PMV* by the probability analysis method. *PPD* has the following relationship with *PMV* [48]:

$$PPD = 100 - 95 \cdot \exp\left[-0.2179PMV^2 - 0.03353PMV^4\right] \tag{12}$$

Bring Equation (11) into Equation (12). The visual relationship between the *PPD* and indoor air temperature t_a and floor surface temperature t_s is shown in Figure 5.

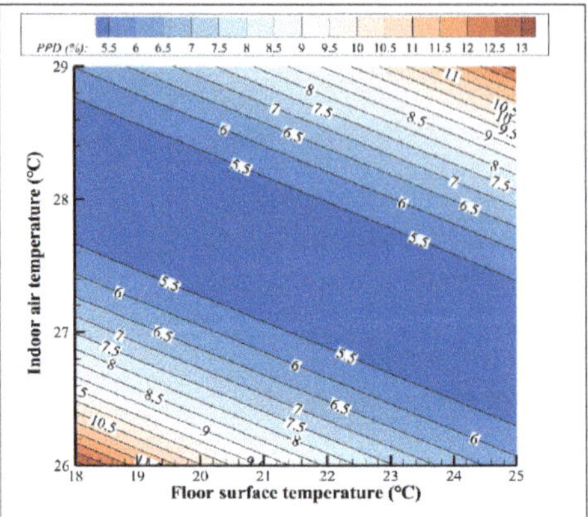

Figure 5. Relationship between *PPD* and two temperature indexes.

It can be seen from Figure 5 that the relationship between *PPD* and two temperature variables (t_a and t_s) is also a group of parallel lines. When the floor surface temperature and indoor air temperature are extremely distributed (both very high or low), the *PPD* value is large. When the floor surface temperature and indoor air temperature are extremely distributed (both high or low), the *PPD* value is large. Only by maintaining the dynamic balance between t_a and t_s can the *PPD* value be minimized; that is, when the indoor air temperature is very high, a lower floor surface temperature is required, and when the indoor air temperature is low, a higher floor surface temperature is required. For an ideal *PPD* value (5~5.5%), t_a and t_s should be distributed in the ideal range of Figure 5, which satisfies the following inequality Equation (13):

$$-0.1936 \cdot t_s + 31.143 < t_a < -0.1955 \cdot t_s + 32.282$$
$$t_s \in [18,\ 25\ °C]$$

(13)

3.2.2. Verification and Error Analysis

According to the usage habits of air conditioning, people generally set the indoor air temperature as an integer temperature. When the indoor air temperature is 26 °C, the *PPD* results of the numerical simulation of the present study are compared with a previous study [49]. The comparison is illustrated in Figure 6.

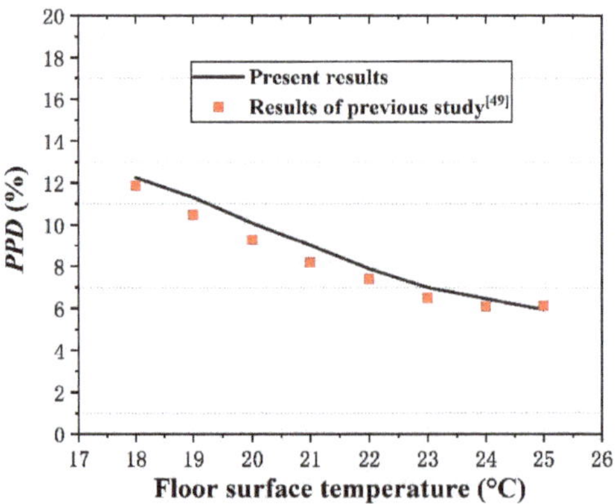

Figure 6. Results of the present study compared with a previous study [49].

A good agreement was achieved between the data of the previous study and present results, which showed the same trend of changes, and the maximum relative error was about 8.49%.

The mean absolute percentage error (*MAPE*) and the root mean square error (*RMSE*) [50] are employed to evaluate the accuracy of numerical simulation calculation models and boundary condition settings. The values of *MAPE* and *RMSE* were 6.64% and 0.60 respectively, which is acceptable.

This fully proves the rationality of the theoretical model and parameter settings used in the numerical simulations of the present study, and verifies the high reliability of the numerical results of this study. The numerical calculation model of the TRNSYS is verified.

3.3. Correction Equation of PMV

3.3.1. Effect of Relative Humidity on PMV

Changes in the relative humidity also have an impact on human thermal comfort. This section discusses the influence of changes in independent variable γ on the *PMV* and corrects the calculation expression of *PMV* to obtain a calculation model containing independent variable γ.

Using the control variable method, fix the values of the value of t_a and t_s, and select the working conditions of $t_a = 28\,°\text{C}$, $t_s = 23\,°\text{C}$ as condition I, $t_a = 28\,°\text{C}$, $t_s = 21\,°\text{C}$ as condition II, $t_a = 27\,°\text{C}$, $t_s = 24\,°\text{C}$ as condition III, $t_a = 27\,°\text{C}$, $t_s = 22\,°\text{C}$ as condition IV, and $t_a = 26\,°\text{C}$, $t_s = 25\,°\text{C}$ as condition V. For working conditions I~V, the indoor air relative humidity changes from 40% to 60% (in the summer, the relative indoor humidity of office buildings is generally between 40% and 60%), and a *PMV* value is output every 5% change in the relative humidity of the indoor air. The changes in the *PMV* are shown in Figure 7.

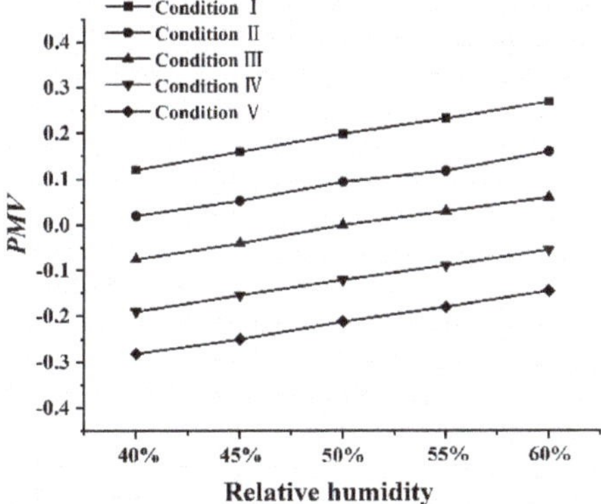

Figure 7. Influence of relative humidity changes on *PMV* under conditions I~V.

As can be seen from Figure 7, the *PMV* value increases with the increase in the relative humidity, which indicates that an increase in the relative humidity will increase people's warm feeling. This conclusion is consistent with the conclusion in [51], which verifies the reliability of numerical simulation from the side.

For each working condition in Figure 7, the changes in relative humidity had little impact on the *PMV*, and the maximum deviation (compared with the relative humidity at 50%) of the *PMV* under each working condition is less than 0.1.

It can be seen from Equation (12) that the *PMV* is the independent variable of *PPD*, and the change of the *PMV* is very small, so the change of *PPD* is also smaller.

3.3.2. Complete Calculation Expression of *PMV*

In Figure 7, it can be found that the sample points of each working condition are approximately connected in a straight line, and the five straight lines are almost parallel to each other. This shows that the effect of the change in the relative humidity on each working condition is almost the same. A straight line or a formula can be used to represent the deviation of the *PMV* caused by a change in the relative humidity. When the indoor relative humidity changes, the offset value Δ of the *PMV* under each working condition is shown in Table 5. When the relative humidity deviation is the same, the average offset $\overline{\Delta}$ of *PMV* is also given in Table 5.

Table 5. *PMV* deviation with relative humidity deviation (compared to 50%).

Indoor Relative Humidity	Relative Humidity Deviation	Δ_1	Δ_2	Δ_4	Δ_5	$\overline{\Delta}$
40%	−10%	−0.0800	−0.0700	−0.0750	−0.0700	−0.0720	−0.0734
45%	−5%	−0.0400	−0.0370	−0.0400	−0.0340	−0.0400	−0.0382
50%	0%	−0.0010	0.0040	0.0000	−0.0007	−0.0020	0.0001
55%	+5%	0.0320	0.0280	0.0300	0.0310	0.0300	0.0302
60%	+10%	0.0690	0.0700	0.0600	0.0660	0.0650	0.0660

It can be found from Table 5 that, when the indoor relative humidity deviation is the same, the *PMV* deviation of each working condition is almost the same, and the average deviation value can be used to replace the actual deviation value. The average offset $\overline{\Delta}$ can be used to correct the *PMV* when the relative humidity is not 50%. Through data fitting, the functional relationship between $\overline{\Delta}$ and γ is found:

$$\overline{\Delta} = 0.6944 \cdot (\gamma - 50\%) - 0.0031 \tag{14}$$

Equation (14) is the correction formula for *PMV*, and the complete calculation formula of *PMV* is as follows:

It can be found from Table 5 that, when the indoor relative humidity deviation is the same, the *PMV* deviation of each working condition is almost the same, and the average deviation value can be used to replace the actual deviation value. The average offset $\overline{\Delta}$ can be used to correct the *PMV* when the relative humidity is not 50%. Through data fitting, the functional relationship between $\overline{\Delta}$ and γ is found:

$$\overline{\Delta} = 0.6944 \cdot (\gamma - 50\%) - 0.0031 \tag{14}$$

Equation (14) is the correction formula for *PMV*, and the complete calculation formula of *PMV* is as follows:

$$
\begin{aligned}
PMV = g(t_s, t_a, \gamma) &= g(t_s, t_a, 50\%) + \overline{\Delta} \\
&= 0.2826 t_a + 5.5375 \times 10^{-2} t_s - 8.9709 + 0.6944 \\
&\quad \cdot (\gamma - 50\%) - 0.0031 \\
&= 0.2826 t_a + 5.5375 \times 10^{-2} t_s + 0.6944 \cdot \gamma - 9.3195
\end{aligned} \tag{15}
$$

$$t_a \in [26,\ 29\ °\mathrm{C}],\ t_s \in [18,\ 25\ °\mathrm{C}],\ \gamma \in [40\%, 60\%]$$

where t_a and t_s participate in the calculation with °C as the unit.

For Equation (15), because the relative humidity has no unit, the indoor air temperature and the floor surface temperature are in °C, so the coefficient before γ can no longer be used for weight calculation; however, according to this coefficient, we can know that, for every 5% change in relative humidity, PMV will change by about 0.0347.

Equation (15) is the complete expression form of $g(t_s, t_a, \gamma)$. PMV can be calculated according to the indoor air temperature t_a, floor surface temperature t_s, and indoor relative humidity γ. Among the three variables that affect PMV, air temperature is the main factor affecting PMV, floor surface temperature is a secondary factor affecting PMV, and relative humidity has little effect on PMV.

4. Conclusions

Some office buildings in northern China try to use radiant floors for cooling in the summer. Indoor thermal comfort is the key problem of a room using FRC technology, which was explored by the TRNSYS numerical simulation, control variable, and data fitting methods in this paper. The main conclusions are as follows.

(1). The indoor air temperature is the main factor affecting the *PMV* value. In rooms with floor radiant cooling, the indoor air temperature shall not be lower than 26 °C or higher than 29 °C. The floor surface temperature is the secondary factor affecting the *PMV*. The relative humidity of indoor air has little effect on the *PMV*;

(2). In the design of air conditioning engineering, the design temperature of the indoor air of a room using FRC technology should be 1~2 °C higher than the indoor design temperature of convective air conditioning.

A simplified calculation model of *PMV* for indoor thermal environment using floor radiant cooling was obtained. This simplified calculation model is suitable for changes in the indoor relative humidity, which contains three independent variables: the indoor air temperature t_a, floor surface temperature t_s, and indoor relative humidity γ. These three variables are the main factors affecting the indoor thermal environment. People can quickly calculate the *PMV* value based on this formula, or adjust one of the three variables with the *PMV* as an indicator to achieve the desired *PMV* value.

Additionally, the limitation of the adjusted temperature range of indoor air and floor surface caused by condensation on the floor surface has not been considered, which can be a research topic in the future.

Author Contributions: Conceptualization, L.Z. (Linhua Zhang); methodology, X.W.; software, X.W.; validation, T.M.; formal analysis, X.W.; investigation, T.M.; resources, L.Z. (Linhua Zhang); data curation, L.Z. (Lili Zhang); writing—original draft preparation, X.W.; writing—review and editing, W.Z.; visualization, X.W.; supervision, L.Z. (Linhua Zhang); project administration, W.Z.; funding acquisition, W.Z. All authors have read and agreed to the published version of the manuscript.

Funding: This research was funded by the Plan of Guidance and Cultivation for Young Innovative Talents of Shandong Provincial Colleges and Universities.

Institutional Review Board Statement: The study was conducted in accordance with the Declaration of Helsinki, and approved by the School of Thermal Engineering, Shandong Jianzhu University.

Informed Consent Statement: Informed consent was obtained from all subjects involved in the study.

Data Availability Statement: Data are not publicly available due to restrictions regarding the privacy of the participants.

Conflicts of Interest: The authors declare no conflict of interest.

Abbreviations

M	Metabolic rate W/m^2 (met)
I_{cl}	Thermal resistance of clothing m^2·K/W (CLO)
γ	Relative humidity
v_a	Air velocity m/s
θ_{mrt}	Mean radiant temperature °C
F_{p-i}	Angle factor between a person and surface i
t_a	Indoor air temperature °C
t_s	Floor surface temperature °C
t_d	Inner surface temperature of ceiling °C
t_c	Inner surface temperature of internal sunshade for the window °C
$t_{w1\sim4}$	Internal surface temperature of four walls °C
t_i	Average temperature of inner surfaces of other envelopes except floor °C
Δ	Offset
$\overline{\Delta}$	Average offset

FRC	Floor radiant cooling
FRH	Floor radiant heating
PMV	Predicted mean vote
PPD	Predicted percentage of dissatisfaction
MAPE	Mean absolute percentage error
RMSE	Root mean square error

References

1. Chen, M.; Ma, M.; Lin, Y.; Ma, Z.; Li, K. Carbon Kuznets Curve in China's Building Operations: Retrospective and Prospective Trajectories. *Sci. Total Environ.* **2022**, *803*, e150104. [CrossRef] [PubMed]
2. Yuan, F.; Yao, R.; Sadrizadeh, S.; Li, B.; Cao, G.; Zhang, S.; Zhou, S.; Liu, H.; Bogdan, A.; Croitoru, C.; et al. Thermal Comfort in Hospital Buildings—A Literature Review. *J. Build. Eng.* **2022**, *45*, e103463. [CrossRef]
3. Vangtook, P.; Chirarattananon, S. Application of Radiant Cooling as a Passive Cooling Option in Hot Humid Climate. *Build. Environ.* **2007**, *42*, 543–556. [CrossRef]
4. Wang, X.; Zhang, W.; Li, Q.; Wei, Z.; Lei, W.; Zhang, L. An Analytical Method to Estimate Temperature Distribution of Typical Radiant Floor Cooling Systems with Internal Heat Radiation. *Energy Explor. Exploit.* **2021**, *39*, 1283–1305. [CrossRef]
5. Wang, X.; Lei, W.; Zhang, W.; Zhang, Y.; Zhang, L. Simplified Calculation and Control Method of Floor Surface Temperature for Radiant Floor and Fresh Air Coupled Cooling Systems in Steady-State. *Case Stud. Therm. Eng.* **2021**, *27*, e101320. [CrossRef]
6. Dong, J.; Lan, H.; Liu, Y.; Wang, X.; Yu, C. Indoor Environment of Nearly Zero Energy Residential Buildings with Conventional Air Conditioning in Hot-Summer and Cold-Winter Zone. *Energy Built Environ.* **2021**, *3*, 129–138. [CrossRef]
7. Xing, D.; Li, N.; Zhang, C.; Heiselberg, P. A Critical Review of Passive Condensation Prevention for Radiant Cooling. *Build. Environ.* **2021**, *205*, e108230. [CrossRef]
8. Wang, X.; Man, Y.; Zhang, L.; Zhang, W.; Zhang, L. Dynamic Research on Cold Storage Performance of a Standard Radiant Floor Cooling System for Office Buildings in Northern China. *J. Therm. Sci.* **2022**, 1–11. [CrossRef]
9. Yan, H.; Yang, L.; Dong, M.; Hu, B.; Sun, Z.; Shi, F.; Yuan, G.; Bi, X. Thermal Comfort in Residential Buildings Using Bimetal Radiator Heating vs. Floor Heating Terminals. *J. Build. Eng.* **2022**, *45*, e103501. [CrossRef]
10. Shin, M.S.; Rhee, K.N.; Jung, G.J. Optimal Heating Start and Stop Control Based on the Inferred Occupancy Schedule in a Household with Radiant Floor Heating System. *Energy Build.* **2020**, *209*, e109737. [CrossRef]
11. Borinaga- Treviño, R.; Cuadrado, J.; Canales, J.; Rojí, E. Lime Mud Waste from the Paper Industry as a Partial Replacement of Cement in Mortars Used on Radiant Floor Heating Systems. *J. Build. Eng.* **2021**, *41*, e102408. [CrossRef]
12. Jiang, C.; Xie, G.; Wu, D.; Yan, T.; Chen, S.; Zhao, P.; Wu, Z.; Li, W. Experimental Investigation on an Energy-Efficient Floor Heating System with Intelligent Control: A Case Study in Chengdu, China. *Case Stud. Therm. Eng.* **2021**, *26*, e101094. [CrossRef]
13. Garcia, D.A. Can Radiant Floor Heating Systems Be Used in Removable Glazed Enclosed Patios Meeting Thermal Comfort Standards? *Build. Environ.* **2016**, *106*, 378–388. [CrossRef]
14. Zhang, Z.; Wu, M.; Yao, W. Performance of the Wall Implanted with Heat Pipes on Indoor Thermal Environment. *Indoor Built Environ.* **2021**, *31*, 878–894. [CrossRef]
15. Cho, J.; Park, B.; Lim, T. Experimental and Numerical Study on the Application of Low-Temperature Radiant Floor Heating System with Capillary Tube: Thermal Performance Analysis. *Appl. Therm. Eng.* **2019**, *163*, e114360. [CrossRef]
16. Karacavus, B.; Aydin, K. Numerical Investigation of General and Local Thermal Comfort of an Office Equipped with Radiant Panels. *Indoor Built Environ.* **2019**, *28*, 806–824. [CrossRef]
17. Sirhan, N.; Golan, S. Efficient PMV Computation for Public Environments with Transient Populations. *Energy Build.* **2021**, *231*, e110523. [CrossRef]
18. Li, T.; Merabtine, A.; Lachi, M.; Martaj, N.; Bennacer, R. Experimental Study on the Thermal Comfort in the Room Equipped with a Radiant Floor Heating System Exposed to Direct Solar Radiation. *Energy* **2021**, *230*, e120800. [CrossRef]
19. Du, H.; Lian, Z.; Lai, D.; Duanmu, L.; Zhai, Y.; Cao, B.; Zhang, Y.; Zhou, X.; Wang, Z.; Zhang, X.; et al. Comparison of Thermal Comfort between Radiant and Convective Systems Using Field Test Data from Chinese Thermal Comfort Database. *Build. Environ.* **2021**, *209*, 108685. [CrossRef]
20. Liu, J.; Kim, M.K.; Srebric, J. Numerical Analysis of Cooling Potential and Indoor Thermal Comfort with a Novel Hybrid Radiant Cooling System in Hot and Humid Climates. *Indoor Built Environ.* **2021**, *31*, 929–943. [CrossRef]
21. Tan, J.; Liu, J.; Liu, W.; Yu, B.; Zhang, J. Performance on Heating Human Body of an Optimised Radiant-Convective Combined Personal Electric Heater. *Build. Environ.* **2022**, *214*, e108882. [CrossRef]
22. Zhang, F.; Guo, H.A.; Liu, Z.; Zhang, G. A Critical Review of the Research about Radiant Cooling Systems in China. *Energy Build.* **2021**, *235*, e110756. [CrossRef]
23. Gu, X.; Cheng, M.; Zhang, X.; Qi, Z.; Liu, J.; Li, Z. Performance Analysis of a Hybrid Non-Centralized Radiant Floor Cooling System in Hot and Humid Regions. *Case Stud. Therm. Eng.* **2021**, *28*, e101645. [CrossRef]
24. Kwong, Q.J.; Arsad, M.A.; Adam, N.M. Evaluation of Indoor Thermal Environment in a Radiant-Cooled-Floor Office Building in Malaysia. *Appl. Mech. Mater.* **2014**, *564*, 228–233. [CrossRef]

25. Hu, Y.; Xia, X.; Wang, J. Research on Operation Strategy of Radiant Cooling System Based on Intermittent Operation Characteristics. *J. Build. Eng.* **2022**, *45*, e103483. [CrossRef]
26. Liu, J.; Li, Z.; Kim, M.K.; Zhu, S.; Zhang, L.; Srebric, J. A Comparison of the Thermal Comfort Performances of a Radiation Floor Cooling System When Combined with a Range of Ventilation Systems. *Indoor Built Environ.* **2020**, *29*, 527–542. [CrossRef]
27. Kastner, P.; Dogan, T. Eddy3D: A Toolkit for Decoupled Outdoor Thermal Comfort Simulations in Urban Areas. *Build. Environ.* **2022**, *212*, e108639. [CrossRef]
28. Wang, Z.; Cao, B.; Zhu, Y. Questionnaire Survey and Field Investigation on Sleep Thermal Comfort and Behavioral Adjustments in Bedrooms of Chinese Residents. *Energy Build.* **2021**, *253*, e111462. [CrossRef]
29. Zhou, X.; Liu, Y.; Zhang, J.; Ye, L.; Luo, M. Radiant Asymmetric Thermal Comfort Evaluation for Floor Cooling System—A Field Study in Office Building. *Energy Build.* **2022**, *260*, e111917. [CrossRef]
30. Ricciardi, P.; Ziletti, A.; Buratti, C. Evaluation of Thermal Comfort in an Historical Italian Opera Theatre by the Calculation of the Neutral Comfort Temperature. *Build. Environ.* **2016**, *102*, 116–127. [CrossRef]
31. Li, Q.; Zhang, Y.; Guo, T.; Fan, J. Development of a New Method to Estimate Thermal Performance of Multilayer Radiant Floor. *J. Build. Eng.* **2021**, *33*, e101562. [CrossRef]
32. Zheng, Z.; Zhang, Y.; Mao, Y.; Yang, Y.; Fu, C.; Fang, Z. Analysis of SET* and PMV to Evaluate Thermal Comfort in Prefab Construction Site Offices: Case Study in South China. *Case Stud. Therm. Eng.* **2021**, *26*, e101137. [CrossRef]
33. Park, J.; Choi, H.; Kim, D.; Kim, T. Development of Novel PMV-Based HVAC Control Strategies Using a Mean Radiant Temperature Prediction Model by Machine Learning in Kuwaiti Climate. *Build. Environ.* **2021**, *206*, e108357. [CrossRef]
34. Broday, E.E.; Ruivo, C.R.; Gameiro da Silva, M. The Use of Monte Carlo Method to Assess the Uncertainty of Thermal Comfort Indices PMV and PPD: Benefits of Using a Measuring Set with an Operative Temperature Probe. *J. Build. Eng.* **2021**, *35*, e101961. [CrossRef]
35. Omidvar, A.; Kim, J. Modification of Sweat Evaporative Heat Loss in the PMV/PPD Model to Improve Thermal Comfort Prediction in Warm Climates. *Build. Environ.* **2020**, *176*, e106868. [CrossRef]
36. Hesaraki, A.; Huda, N. A Comparative Review on the Application of Radiant Low-Temperature Heating and High-Temperature Cooling for Energy, Thermal Comfort, Indoor Air Quality, Design and Control. *Sustain. Energy Technol. Assess.* **2022**, *49*, e101661. [CrossRef]
37. Gao, S.; Ooka, R.; Oh, W. Experimental Investigation of the Effect of Clothing Insulation on Thermal Comfort Indices. *Build. Environ.* **2021**, *187*, e107393. [CrossRef]
38. Alfano, F.R.D.A.; Palella, B.I.; Riccio, G. Notes on the Calculation of the PMV Index by Means of Apps. *Energy Procedia* **2016**, *101*, 249–256. [CrossRef]
39. Zhang, D.L.; Wang, Z.J.; Zhang, X. Experimental Study on Dry Radiant Floor Cooling System Combined with Displacement Ventilation System. *Build. Sci.* **2009**, *25*, 38–42.
40. Brough, D.; Ramos, J.; Delpech, B.; Jouhara, H. Development and Validation of a TRNSYS Type to Simulate Heat Pipe Heat Exchangers in Transient Applications of Waste Heat Recovery. *Int. J. Thermofluids* **2021**, *9*, e100056. [CrossRef]
41. Ministry of Housing and Urban-Rural Development of the People's Republic of China. *Design Standard for Energy Efficiency of Public Buildings*; China Architecture & Building Press: Beijing, China, 2015.
42. Bordignon, S.; Emmi, G.; Zarrella, A.; De Carli, M. Energy Analysis of Different Configurations for a Reversible Ground Source Heat Pump Using a New Flexible TRNSYS Type. *Appl. Therm. Eng.* **2021**, *197*, e117413. [CrossRef]
43. Liu, C.; Zhang, Y.; Sun, L.; Gao, W.; Jing, X.; Ye, W. Influence of Indoor Air Temperature and Relative Humidity on Learning Performance of Undergraduates. *Case Stud. Therm. Eng.* **2021**, *28*, e101458. [CrossRef]
44. Qin, S.Y.; Cui, X.; Yang, C.; Jin, L.W. Thermal Comfort Analysis of Radiant Cooling Panels with Dedicated Fresh-Air System. *Indoor Built Environ.* **2021**, *30*, 1596–1608. [CrossRef]
45. Zhang, C.Y.; Wang, Z.J. Influence of Radiant Cooling Floor Surface on Inner Surface Temperature of Building Envelopes and Indoor Thermal Comfort. *Build. Sci.* **2008**, *24*, 79–84.
46. Dougherty, E.R.; Kim, S.; Chen, Y. Coefficient of Determination in Nonlinear Signal Processing. *Signal Processing* **2000**, *80*, 2219–2235. [CrossRef]
47. Fanger, P. Calculation of Thermal Comfort, Introduction of a Basic Comfort Equation. *ASHRAE Trans.* **1967**, *73*, III.4.1–III.4.20.
48. Cheung, T.; Schiavon, S.; Parkinson, T.; Li, P.; Brager, G. Analysis of the Accuracy on PMV—PPD Model Using the ASHRAE Global Thermal Comfort Database II. *Build. Environ.* **2019**, *153*, 205–217. [CrossRef]
49. Spielvogel, L. *Thermal Environmental Conditions for Human Occupancy*; ASHRAE: Atlanta, GA, USA, 2005.
50. Ahamed, M.S.; Guo, H.; Tanino, K. Evaluation of a Cloud Cover Based Model for Estimation of Hourly Global Solar Radiation in Western Canada. *Int. J. Sustain. Energy* **2019**, *38*, 64–73. [CrossRef]
51. d'Ambrosio Alfano, F.R.; Palella, B.I.; Riccio, G. The Role of Measurement Accuracy on the Thermal Environment Assessment by Means of PMV Index. *Build. Environ.* **2011**, *46*, 1361–1369. [CrossRef]

Article

Experimental Study on Operating Characteristic of a Combined Radiant Floor and Fan Coil Cooling System in a High Humidity Environment

Xuwei Zhu [1], Jiying Liu [1,2,*], Xiangyuan Zhu [1], Xiaole Wang [1], Yanqiu Du [1] and Jikui Miao [3]

[1] School of Thermal Engineering, Shandong Jianzhu University, Jinan 250101, China; dl_zxw9605@163.com (X.Z.); zhuxiangyuan20@sdjzu.edu.cn (X.Z.); wangxiaole20@sdjzu.edu.cn (X.W.); duyanqiu19@sdjzu.edu.cn (Y.D.)

[2] Built Environment Design and Research Institute, Shandong GRAD Group, Dezhou 253000, China

[3] School of Architecture and Urban Planning, Shandong Jianzhu University, Jinan 250101, China; mjikui2004@sdjzu.edu.cn

* Correspondence: jxl83@sdjzu.edu.cn

Citation: Zhu, X.; Liu, J.; Zhu, X.; Wang, X.; Du, Y.; Miao, J. Experimental Study on Operating Characteristic of a Combined Radiant Floor and Fan Coil Cooling System in a High Humidity Environment. *Buildings* **2022**, *12*, 499. https://doi.org/10.3390/buildings12040499

Academic Editor: Rafik Belarbi

Received: 11 March 2022
Accepted: 15 April 2022
Published: 17 April 2022

Publisher's Note: MDPI stays neutral with regard to jurisdictional claims in published maps and institutional affiliations.

Abstract: The combined radiant floor and fan coil cooling (RFCAFC) system is widely used due to its high comfort and large energy saving potential. In this study, as an example, the combined RFCAFC system was studied in a high humidity environment in Jinan, Shandong Province, China. The novelty of the combined RFCAFC system lies in its ability to automatically adjust the water supply temperature of the radiant floor and fan coil in real time according to outdoor meteorological conditions, achieving thermal comfort while eliminating the likelihood of condensation on the radiant floor surface. Days with similar outdoor meteorological conditions were grouped, and the comfort level and hourly cooling performance coefficient (EER_h) of different operating strategies for different outdoor meteorological conditions were monitored along with other evaluation indicators. The RFCAFC had good energy efficiency and comfort in a high humidity room environment. This study showed that the indoor vertical air temperature difference ranged from 1.6 to 1.8 °C, which met the ASHRAE55-2017 standard. The radiant floor surface temperature uniformity coefficient (S) fluctuated between 0.7 and 1.0, and the time it took the radiant floor surface temperature to reach 63.2% of the total variability range (τ_{63}) for different operation strategies based on different outdoor meteorological conditions ranged between 4.4 and 4.7 h, which was within the normal range. The proportion of the total cooling capacity contributed by fan coil cooling under low temperature and high humidity (LH), high temperature and low humidity (HL), and medium temperature and medium humidity (MM) were 68.0%, 73.8%, and 71.7%, respectively. Based on this study, the following recommendations for the combined cooling system can be made: (1) When the outdoor humidity is high, the radiant floor system should be turned on early to provide cooling capacity. When the outdoor temperature is high, the fan coil system should be turned on early to reduce the indoor temperature. (2) To reduce energy consumption and achieve efficient operation of the system, the radiant floor system should be continuously operated to maximize its contribution to the cooling capacity, while the fan coil can be operated intermittently. Natural cooling can be integrated to provide additional cooling capacity to the room in the hours preceding occupation (i.e., 7:00–9:00). (3) The operation strategy of the combined cooling system must be able to respond in real time to changes in outdoor meteorological conditions to prevent discomfort in times of extreme heat or humidity.

Keywords: radiant floor; fan coil; cooling; energy consumption; thermal comfort

1. Introduction

It is well known that energy consumed by buildings accounts for more than 40% all global energy consumption, and 60% of that goes to ensuring indoor thermal comfort for occupants [1,2]. Air conditioning systems, which play a crucial role in creating a thermally comfortable living environment, are used extensively since most people spend 80–90% of their time inside buildings. Because air conditioning systems make up such a large portion of total energy consumption, researchers are paying more attention to their optimization [3–6]. The goal is to reduce building energy consumption by optimizing air conditioning systems without reducing indoor thermal comfort. To achieve this, efficient operation strategies based on building envelope characteristics and loads and meteorological conditions are required. Over the years, many studies have investigated energy consumption by radiant cooling systems and found that, compared to conventional convection air conditioning systems, radiant cooling systems can reduce energy consumption by about 40% [7–11].

Radiant floor cooling (RFC) has attracted widespread attention because of its good thermal comfort, low energy consumption, building space saving, and low operating noise. Researchers have conducted many practical application studies and field measurement studies on radiant floor air conditioning systems [12–15]. Sourbron et al. compared the response times of RFC and convection air conditioning systems based on the thermal inertia of the envelope and proposed an operational strategy utilizing intermittent operation [14,16–18]. Romaní et al. proposed a water supply temperature operation strategy where the water supply temperature is continuously adjusted according to the outdoor temperature, but this strategy was more complicated than others. A water supply temperature operating strategy is well adapted to maintaining stable operation while accounting for uncertainties resulting from increased indoor heat loads or solar radiant heat gain throughout the day. However, RFCs are not able to immediately deal with changes in hot and cold loads due to the high thermal inertia of buildings, so other energy handling systems must be set up to buffer rapid change [19–23]. In light of this, Mikeska et al. demonstrated the advantages of using the RFC system at night or during off-peak hours, thus effectively utilizing low temperature air from natural ventilation and off-peak electricity prices. In this intermittent operation, when the system is stopped, heat accumulates in the air near the radiant floor surface, and when the system is started again, a strong temperature gradient is created near the radiant floor surface. This strategy increases the heat capacity of the radiant floor system and enables it to absorb heat for a short period of time while reducing energy consumption [19,24–26].

A summary of previous applications and studies on radiant cooling are listed in Table 1. Zarrella et al. studied radiant floor cooling with different ventilation systems using simulations, their results showed that RFC systems need to be combined with ventilation systems to achieve efficient and energy-saving operation [27,28]. Zhao et al. proposed a simple method for predicting the performance of RFC systems with solar radiation in steady states or large open spaces using a combination of experiments and simulations, which considered the role of solar radiation in RFC systems [29]. Jin et al. showed that the risk of condensation on a radiant surface is high during the start-up phase and that an air convector can effectively reduce the risk of condensation on the radiant surface [30,31]. Srivastava et al. studied a roof radiation system combined with different ventilation systems using both simulations and experiments, and their results showed that the radiant cooling system was able to maintain efficient operation in different climate zones [32,33]. Feng et al. showed that the cold transient rates of a radiant system are faster than those of an air system during heat gain by an experimental building [34]. Fernandez-H et al. studied a radiant floor with fan coil/displacement ventilation systems using experimental methods, and their results showed that radiant cooling systems not only improve energy efficiency, but also the system response time [35,36]. Joe et al. showed that different operating strategies can achieve different levels of energy savings while ensuring indoor thermal comfort [37,38].

Table 1. Summary of the existing research on radiant cooling.

Author	System	Building	Investigation Type	Anti-Condensation Method	Key Findings
Zarrella et al. [25]	Radiant floor + mechanical ventilation	Residential buildings	Simulation	Mechanical ventilation	RFC system must be integrated with ventilation systems for dehumidification purposes.
Seo et al. [28]	Radiant floor + natural ventilation/displacement ventilation		Simulation	Natural ventilation and displacement	In hot and humid climate, this hybrid system provides energy savings of approximately 19% compared to traditional RFC systems.
Zhao et al. [29]	Radiant floor + fan coil	Large open space	Simulation, experiment	Fan coil	A simple method for predicting the performance of radiant floor cooling systems with solar radiation in steady state is proposed.
Jin et al. [30]	Radiant roof		Simulation, experiment		During the start-up period, the temperature on the radiant surface drops rapidly. As the rate of radiation temperature drop increases, there is a risk of condensation on the radiant surface.
Keum-M and Leibundgut [31]	Radiant roof + air convector	Office buildings	Experiment	Air convectors	The use of air convectors helps reduce the risk of condensation on the radiant roof.
Schmelas et al. [32]	Radiant roof + mixed ventilation	Commercial buildings	Experiment	Mixed ventilation	The operation strategy of the combined system can maintain efficient operation for buildings with high thermal inertia.
Srivastava et al. [33]	Radiant roof and ventilation dehumidification system	Office buildings	Simulation, experiment	Ventilation/dehumidification	The integrated roof radiation and evaporative cooling system not only can be adapted to a variety of different climate zones but also have a large energy-saving potential.
Feng et al. [34]	Radiant roof + mixed air system	Experimental room	Experiment	Mixed air system	By studying the dynamic heat transfer in the room under the radiant system, the transient cooling rate of the radiant system is 18–21% higher than that of the air system during the heat gain.
Fernández-H et al. [35]	Radiant floor + air system		Simulation, experiment	Underfloor air supply	The combined system can reduce the response time to changes in the space thermal loads and increase the cooling capacity of the floor.

Table 1. *Cont.*

Author	System	Building	Investigation Type	Anti-Condensation Method	Key Findings
Gu et al. [36]	Radiant floor + fan coil	Office buildings	Experiment	Fan coil/outdoor air dehumidifiers	Outdoor air systems and fan coils improve response time of radiant system in addition to auxiliary dehumidification.
Joe and Karava [37]	Radiant floor + supper wall diffuser	Living lab	Experiment	Air handler	The MPC results show 34% cost savings compared to baseline feedback control during the cooling season.
Zhang et al. [38]	Radiant floor + displacement ventilation	Experimental room	Experiment	Displacement ventilation	By building a dynamic simplification model, the studied control strategy is approximately 17% more energy efficient than the conventional control strategy.

Convection air conditioning systems have short response times and are capable of rapid temperature adjustment compared to RFC systems. Here, to utilize the advantages of both RFC and traditional convection air conditioning systems, a combined radiant floor and fan coil cooling (RFCAFC) system was developed. As with conventional air conditioning systems, the adoption of suitable operation strategies is central to the energy consumption reduction strategy of the RFCAFC system [39]. Indeed, by studying operation strategies using combinations of radiant floors with different ventilation systems, Atienza et al. found that integrating an air conditioning system not only ensured indoor thermal comfort, but also resulted in a greater potential for energy savings. In addition, since weather conditions and indoor heat gain are dynamic and have a large impact on the performance of radiant cooling systems, an effective operation strategy is especially important [40,41].

Currently, RFC systems are not widely used in residential buildings and small offices without central air conditioning because of the high risk of condensation, slow response time, difficulty in controlling multiple zones, and insufficient floor cooling capacity due to intermittent use within high humidity environments [41–47]. Moreover, the existing field experiments on the operation strategies of combined radiant floor and fan coil cooling systems have been based on manual adjustments and have not implemented automated operation strategies. This means that RFC systems have been difficult to integrate into the automated systems of buildings to achieve efficient smart building functions. Based on the above problems, we built a RFCAFC automatic operation system and analyzed the impacts of different operating strategies on building energy consumption and thermal comfort in a high humidity environment. The RFCAFC automatic operation system was installed in Jinan, China. During the experiment, relevant parameters of the combined system were monitored and the operating characteristics and cooling capacity of the combined system under different outdoor meteorological conditions were analyzed. This study provides a strong reference and basis for further study of the operating characteristics of combined systems under high humidity environments and for further promotion of the combined system in small scale residential applications.

2. Experimental Description

2.1. Introduction of the Laboratory

The experimental laboratory was located in Jinan, China, which is in a hot summer and cold winter climate zone. The enclosed structure adopted aerated concrete integrated composite wall panels, which have low thermal conductivity and good thermal stability. The laboratory was a single building with length, width, and height of 4.00 m, 2.80 m, and 3.00 m, respectively, and a total construction area of 11.2 m^2. The exterior walls were all 250 mm aerated concrete integrated composite wall panels, and the heat transfer coefficient of the exterior walls was 0.536 W/(m^2·K). The south facing exterior wall had an aluminum framed double-glazed exterior window with an area of 2.4 m^2 and a heat transfer coefficient of 2.40 W/(m^2·K). The floor plan of the laboratory and building is shown in Figure 1.

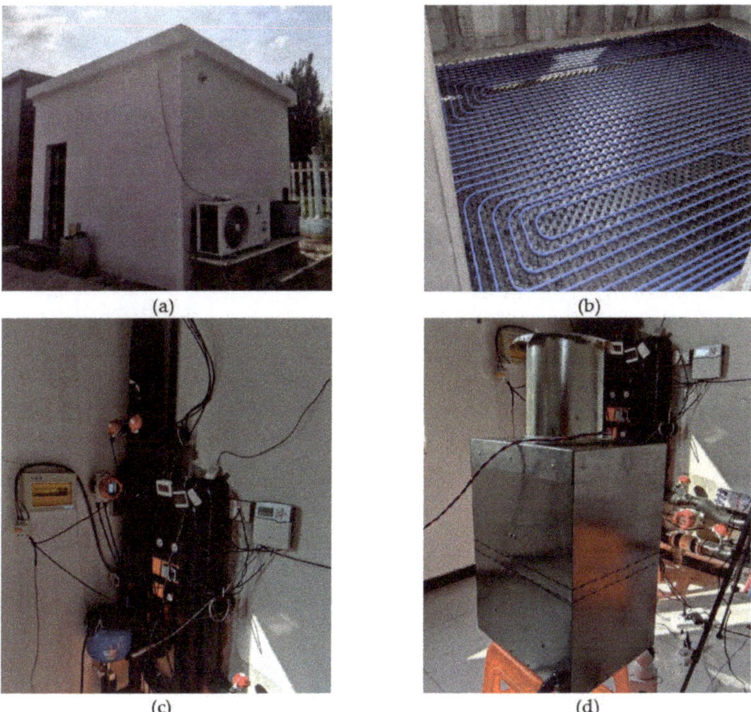

Figure 1. Photographs of the floor plan and experimental building. (**a**) Experimental site, (**b**) pipe with counter flow type laying, (**c**) control system components, and (**d**) indoor heat source.

The radiant floor coil adopted the double circuit dry buried pipe arrangement. The floor radiant coil pipe diameter was 12 mm, the pipe wall thickness was 2 mm, the coil spacing was 60 mm, and the pipe was PE-RT pipe. The radiant floor surface decoration material for the tile surface and radiant floor structure is shown in Figure 2. Its basic structure from the bottom to top was made of the following parts: floor layer, insulation layer, buried pipe layer, soil layer, screed layer and surface layer, where the insulation layer and buried pipe layer for integrated processing were combined into a radiation cooling module. The radiant floor thermal parameters are listed in Table 2. Indoor temperature sensors, humidity sensors, and anemometers were installed to measure indoor temperature, humidity, and vertical air temperature difference. The measurement points were arranged as shown in Figure 3. PT-100 thermal couplings were installed on the interior walls to measure the temperature of each wall's surface.

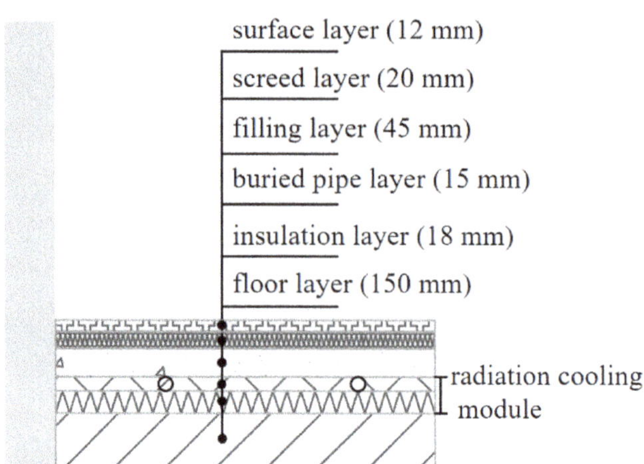

Figure 2. Radiant floor structure.

Table 2. Thermal parameters of the radiant floor.

Type	Thickness (mm)	Material	Density (kg/m^3)	Thermal Conductivity (W/m·k)
Floor layer	150	Aerated concrete	500	0.329
Insulation layer	18	Polystyrene	35	0.021
Buried pipe layer	15	Polythene	962	0.30
Filling layer	45	Cement mortar	1080	0.90
Screed layer	20	Cement	1800	0.93
Surface layer	12	Marble	2800	2.91

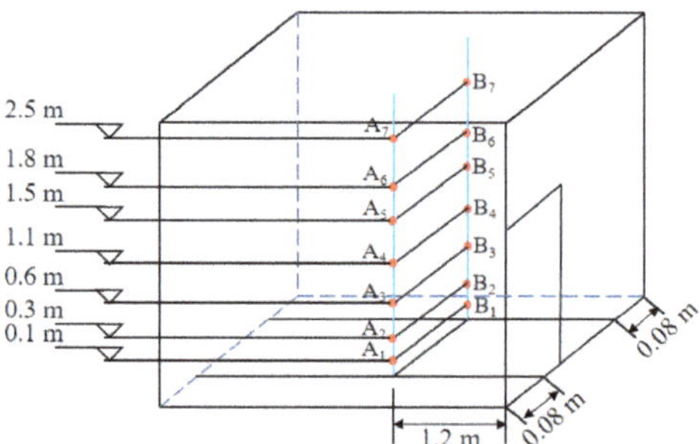

Figure 3. Location of the indoor measuring points.

2.2. Introduction of the Combined Cooling System

The operating principle of the RFCAFC system is shown in Figure 4. The RFCAFC system consisted of a cold source, a constant temperature water tank, a primary distribution system, a secondary distribution system (mixer pump), a safety component, and an air conditioning terminal.

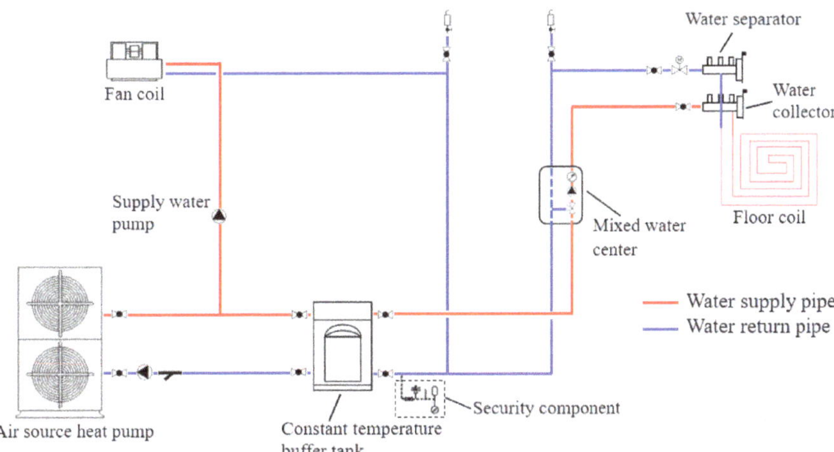

Figure 4. Schematic of the cooling system.

The combined cooling system used a single-phase inverter air source heat pump (KXD80/60EA-V) with a rated cooling capacity of 7.2 KW, an IPLV of 2.81, and a fan coil (FP-51) with a rated cooling capacity of 2.7 kW. To measure the cooling capacity of the combined cooling system, an electromagnetic flow meter (LDG-MIK DN25) and a water temperature sensor (MIKwrn-130) were installed on the supply and return pipes of the fan coil. To measure the energy consumption of the combined cooling system, an intelligent socket (KTBL03LM) was installed on the system power transmission and distribution system to record power consumption. The experimental equipment of the combined cooling system is shown in Figure 5 and the detailed parameters of the experimental equipment are shown in Table 3.

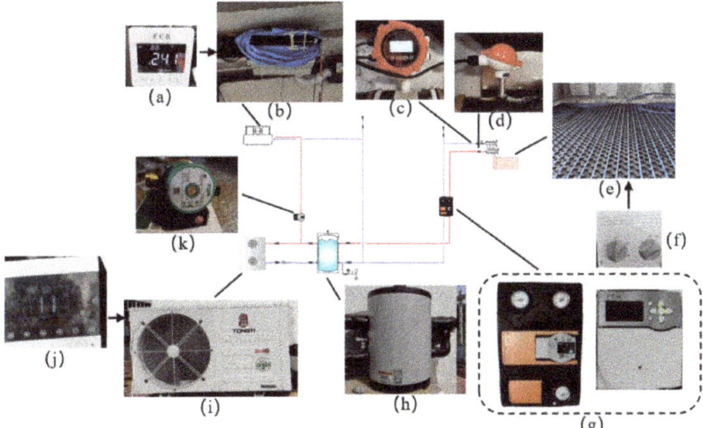

Figure 5. Illustrated schematic of the cooling system, (**a**) temperature controller of fan coil, (**b**) fan coil, (**c**) flow meter, (**d**) thermometer, (**e**) radiant floor coil, (**f**) temperature controller of radiant floor, (**g**) mixed water center, (**h**) constant temperature buffer tank, (**i**) air source heat pump, (**j**) controller of air source heat pump, (**k**) supply water pump.

Table 3. Summary of the experimental instrument operating parameters.

Parameters	Instruments	Range	Accuracy	Sampling Frequency
Room air temperature	Temperature and humidity meter	−20~80 °C	±0.5 °C	30 s
Room air relative humidity	Temperature and humidity meter	0~100%	±3%	30 s
Floor surface temperature	PT100-type thermometer	−100~300 °C	±0.5 °C	2 s
Ventilation air temperature	Air multi-parameter meter	−10~60 °C	±0.3 °C	30 s
Ventilation air relative humidity	Air multi-parameter meter	0~100%	±3%	30 s
Ventilation air velocity	Air multi-parameter meter	0~10 m/s	±0.015 m/s	30 s
Wall and ceiling temperature	PT100-type thermometer	−100~300 °C	±0.5 °C	2 s
Water temperature	PT100-type thermometer	−40~125 °C	±0.5 °C	2 s
Water flow rate	Flow meter	0~3.0 m³/h	±0.5%	2 s

In summer, chilled water produced by the air source heat pump first flows through the constant temperature buffer tank where it is brought to the required pressure by the circulating pump to overcome the resistance of the pipeline. The chilled water is sent to the air conditioner, passing through the temperature measuring instrument on the way, through the circulating pipeline made of polyvinyl chloride (PVC) plastic pipes.

As a radiant floor is more prone to condensation in high a humidity environment, the combined cooling system adopted the operation logic that the floor radiation water supply temperature should be higher than the indoor air dew temperature to prevent condensation. The specific implementation was as follows: The radiant floor water supply temperature always followed the indoor air dew temperature. An indoor dew thermostat collects the indoor air temperature and humidity and transmits the data to the central controller, which calculates the dew temperature based on the indoor temperature and humidity and the built-in enthalpy and humidity graph algorithm, and then adjusts the mixing valve at the mixing pump station to adjust the water temperature so it is higher than the dew point temperature by 2 °C, thus realizing the control logic flow, as shown in Figure 6.

The operation of the combined cooling system was realized primarily by setting the desired indoor temperature. In this system, when the indoor temperature is lower than the set value, the air source heat pump will run at a low frequency or be stopped, and when the indoor temperature is higher than the set value, the air source heat pump starts to run again. The low temperature chilled water produced during this operation was not only supplied to the radiant floor and fan coil air conditioning, but also stored in a thermostatic buffer tank. When the system needs low temperature chilled water again, it can be taken directly from the constant temperature buffer tank. This avoids frequently starting and stopping the air source heat pump, which prolongs the heat pump's service life and saves energy.

During the initial operating stage of the combined cooling system, due to the high humidity of the indoor environment, the radiant floor and fan coil are used together to provide cooling. When the fan coil stops running, the radiant floor provides all the cooling for the room.

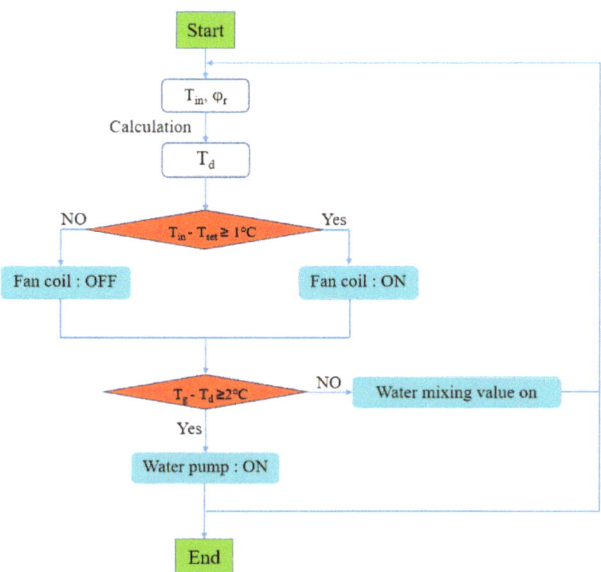

Figure 6. Control logic of the cooling system.

2.3. Uncertainty Analysis

In order to quantify the accuracy of measurements, an uncertainty analysis of the measured data was carried out based on the general law of uncertainty propagation [48]. The uncertainty was obtained using the following equations [49] and the results are listed in Table 4.

$$u_i = \left(\frac{w_i^2}{3}\right)^{0.5} \tag{1}$$

$$U_R = \left(\sum_{i=1}^{n} u_i^2\right)^{0.5} = \left(\frac{1}{3}\sum_{i}^{n} w_i^2\right)^{0.5} \tag{2}$$

where u_i is the standard uncertainty of the measured parameter; w_i is the accuracy of the device; and U_R is the standard uncertainty of the parameter determined by other measured parameters.

Table 4. Accuracy and uncertainty of measured parameters.

Parameters	Instruments	Uncertainty
Room air temperature	Temperature and humidity meter	0.29 °C
Room air relative humidity	Temperature and humidity meter	1.7%
Floor surface temperature	PT100-type thermometer	0.29 °C
Ventilation air temperature	Air multi-parameter meter	0.17 °C
Ventilation air relative humidity	Air multi-parameter meter	1.7%
Ventilation air velocity	Air multi-parameter meter	0.009 m/s
Wall and ceiling temperature	PT100-type thermometer	0.29 °C
Water temperature	PT100-type thermometer	0.29 °C
Water flow rate	Flow meter	0.29%

3. Experimental Cases and Evaluation Indices

3.1. Experimental Cases

These experiments were conducted in August and September, which are representative of the summer cooling season in Jinan, China. The outdoor temperature varied from 26 °C to 40 °C and the outdoor humidity varied from 35% to 94% during the test period, as shown in Figure 7. The outdoor weather was classified as overcast, sunny, or cloudy, which could be directly related to changes in outdoor total horizontal solar radiation. These climate characteristics provide conditions in which intermittent operation of the combined cooling system was required. During the experiment, the room temperature was set at 25 °C and the fan coil was set at the lowest operating air speed. During the operating time, two heated dummies (123.6 W and 137.5 W) were used to simulate indoor heat sources that approximated human bodies.

According to the temperature, humidity, and outdoor meteorological conditions, the environmental conditions were classified into three groups, as shown in Table 5. When the outdoor temperature and humidity ranges were 21–29 °C and 74–94%, respectively, the conditions were referred to as low temperature and high humidity (LH). When the outdoor temperature and humidity ranges were 24–38 °C and 42–90%, respectively, the weather conditions were referred to as high temperature and low humidity (HL). When the outdoor temperature and humidity ranges were 22–33 °C and 50–88%, respectively, the weather conditions were referred to as medium temperature and medium humidity (MM).

Table 5. Outdoor weather conditions classification table.

Condition No.	Experiment Time	Temperature and Humidity Characteristics	Temperature Range (°C)	Humidity Range (%)	Meteorological Conditions
1	20 August 2021 20 August 2021	LH	21~29	74~94	Overcast
2	11 September 2021 12 September 2021	HL	24~38	42~90	Sunny
3	19 August 2021 26 August 2021	MM	22~33	50~88	Cloudy

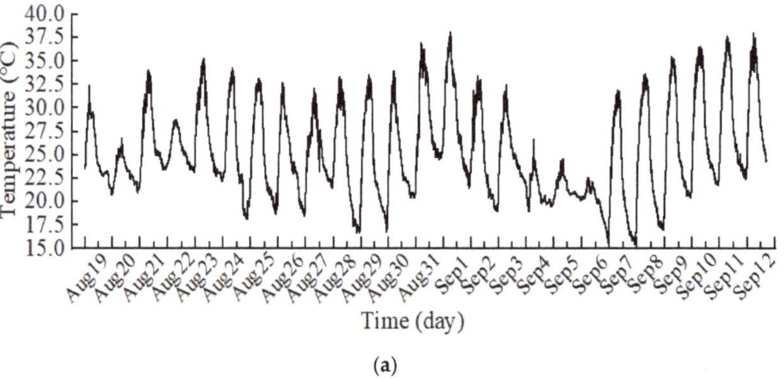

(a)

Figure 7. *Cont.*

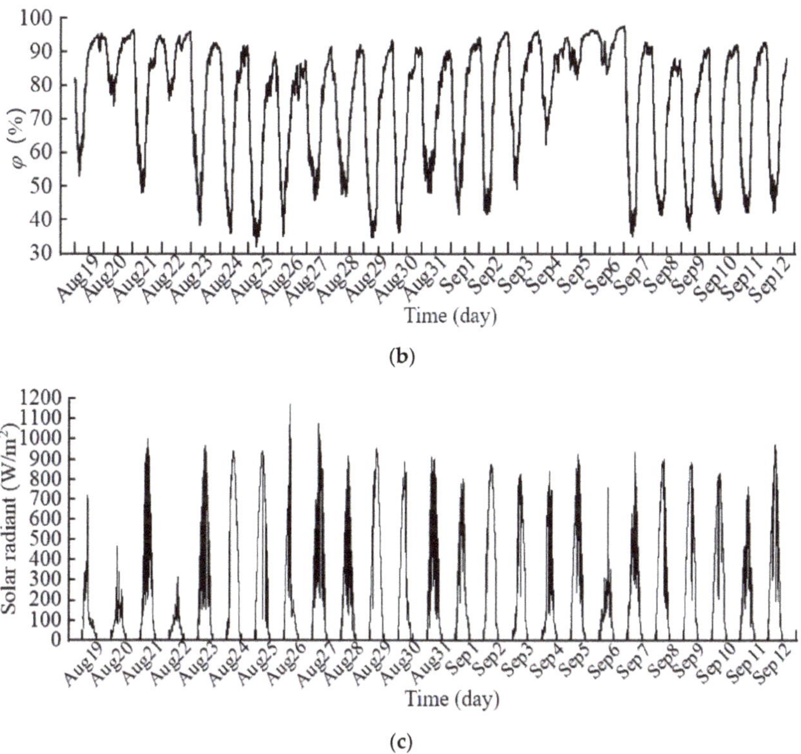

(b)

(c)

Figure 7. Meteorological conditions during the experiment, (**a**) outdoor air temperature, (**b**) outdoor air relative humidity, and (**c**) solar radiation.

Different outdoor weather conditions required different start-stop strategies in the RFCAFC system. The start-stop conditions of the combined cooling system during the experiment are shown in Table 6.

Table 6. Summary of the combined cooling system start/stop strategy under different outdoor weather conditions.

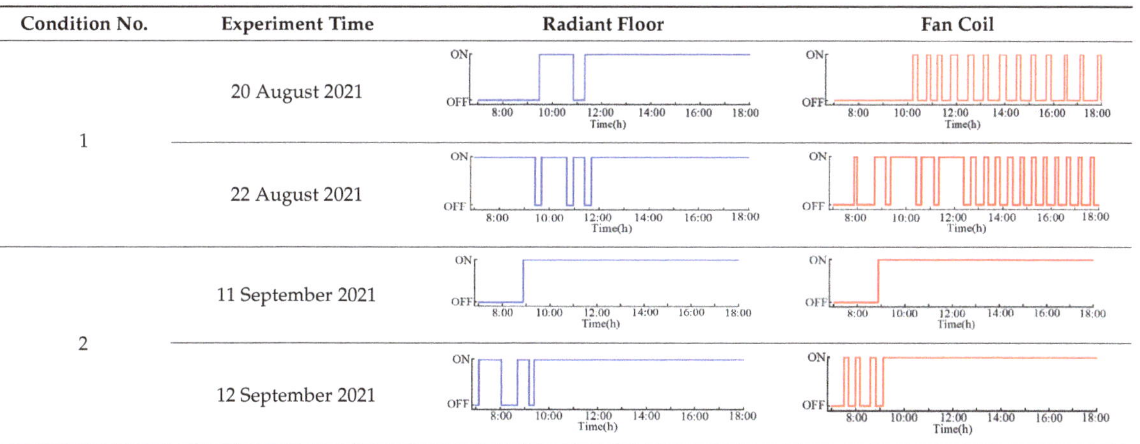

Condition No.	Experiment Time	Radiant Floor	Fan Coil
1	20 August 2021		
	22 August 2021		
2	11 September 2021		
	12 September 2021		

Table 6. *Cont.*

Condition No.	Experiment Time	Radiant Floor	Fan Coil
3	19 August 2021		
	26 August 2021		

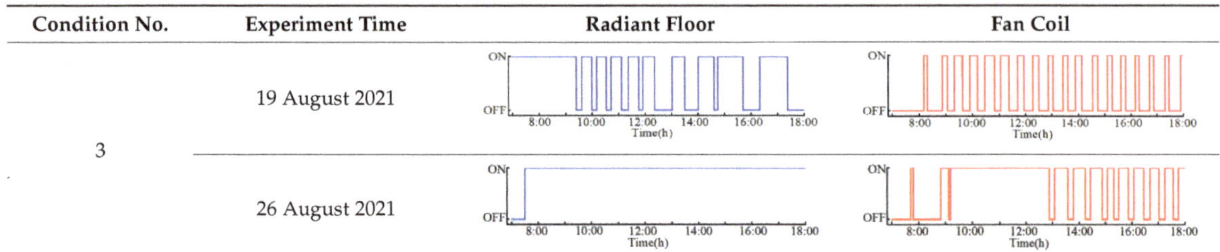

3.2. Evaluation Indices

3.2.1. Radiant Floor Surface Uniformity Temperature Coefficient

The average temperature of the surface of the radiant floor directly affects the heat exchange at the surface of radiant floor. This study used the S value to express the cooling capacity of the radiant floor. The closer the average temperature of the radiant floor surface to the lowest temperature of the radiant floor surface, the closer S is to 1. The better the S, the greater the cooling capacity of the radiant floor surface supply. The S calculation method is shown in Equation (3).

$$S = \frac{T_{s,max} - T_{s,min}}{T_g - T_h} \tag{3}$$

where $T_{s,max}$ is the maximum radiant floor surface temperature, °C; $T_{s,min}$ is the minimum radiant floor surface temperature, °C; T_g is the radiant floor water supply temperature, °C; and T_h is the radiant floor return water temperature, °C.

3.2.2. Time Constant

The time constant (τ) is used as a time scale to measure the time required from the start of the combined cooling system for the radiant floor to reach a relatively stable state. τ_{63} is defined as the time required for the temperature to change from the initial value T_s to T_e (approximately 63.2% of the total desired change). τ_{63} reflects the rapidity of the radiant floor surface temperature change. τ_{95} is defined as the time required for the radiant floor surface temperature to reach approximately 95% of the desired change. τ_{95} is a measure of the time required for the surface temperature of the radiant floor to reach a relatively stable state from the initial state. According to the physical meaning of τ, the τ calculation method is shown in Equations (4) and (5).

$$\tau_{63} = T_{im} - (T_{im} - T_{rs}) \times 63.2\% \tag{4}$$

$$\tau_{95} = T_{im} - (T_{im} - T_{rs}) \times 95.0\% \tag{5}$$

where τ_{63} is the time required for the radiant floor surface temperature to reach 63.2% of the total change, h; τ_{95} is the time required for the radiant floor surface temperature to reach approximately 95% of the total desired change, h; T_{im} is the initial radiant floor surface temperature, °C; and T_{rs} is the desired radiant floor surface temperature, °C.

3.2.3. Room Temperature Fluctuation

The sample standard deviation (S) was introduced in this experiment to evaluate the stability of the indoor thermal environment and characterize the degree of dispersion of the indoor temperature. Larger sample standard deviations indicate a greater degree of fluctuation in indoor temperature during the test period, which would indicate poor stability in the indoor thermal environment. The *s* calculation method is shown in Equation (6).

$$S = \sqrt{\frac{\sum_{i=1}^{11}\left(T_i - T_{age}\right)^2}{n}} \tag{6}$$

where S is the standard deviation of indoor temperature; T_i is the indoor temperature at the ith hour after the start of the combined system, °C; and T_{age} is the average indoor temperature during the test period, °C.

3.2.4. Thermal Comfort

The international standard ISO7730 uses the predicted mean vote (PMV) and predicted percentage dissatisfaction (PPD) to describe and assess the comfort of an indoor thermal environment. PMV represents the sensations experience by the majority of people in the same environment and can be used to evaluate the comfort or discomfort of a thermal environment. However, as a mean value, the PMV metric ignores individual variability, which means that it does not represent the thermal sensations of all people within a thermal environment. Therefore, researchers have proposed the PPD index which uses probability analysis to indicate the percentage of the population dissatisfied with a thermal environment. The recommended PMV value range in the international standard ISO7730 is -0.5 to $+0.5$, and the recommended value of PPD is $\leq 10\%$. The PMV calculation method is shown in Equation (7) and the PMV quantitative index is shown in Table 7.

$$\begin{aligned}
\text{PMV} = & \left(0.303e^{-0.036M} + 0.0275\right) \times \{M - W - 3.05[5.733 - 0.007(M - W) - P_a] - 0.42(M - W - 58.2) \\
& -0.0173M(5.867 - P_a) - 0.0014M(34 - t_a) - 3.96 \times 10^{-8}f_{cl}\left[(t_{cl} + 273)^4 - \left(t_{age} + 273\right)^4\right] - f_{cl}h_c(t_{cl} - t_a)\}
\end{aligned} \tag{7}$$

where M is the human energy metabolic rate, W/m^2; W is the mechanical work done by the human body, W/m^2; P_a is the partial pressure of water vapor around the human body, kPa; f_{cl} is the garment area coefficient; t_{cl} is the temperature of the outer surface of the garment, °C; t_{age} is the average radiation temperature, °C; h_r is the convective heat transfer coefficient, W/(m^2·K); and t_a is the air temperature around the human body, °C.

Table 7. Predicted mean vote index.

Thermal Sensation	Hot	Warm	Slightly Warm	Neutral	Slightly Cool	Cool	Cold
PMV value	+3	+2	+1	0	−1	−2	−3

3.2.5. System Cooling Capacity

In principle, when a radiant floor is used for summer cooling, it needs to be combined with an air supply system. However, most existing studies on the cooling capacity of RFC systems were based on experimental benches without air supply systems, which does not consider the influence of different air supply forms on the cooling capacity of a radiant floor, and also ignores the influence of air supply on the convective heat exchange of the radiant floor. The introduction of the air supply system not only affects the convective heat exchange between the surface of the radiant floor and the room air, but also changes the thermal environment condition of the room, which directly influences the RFC capacity.

The cooling capacity and operation status of the combined cooling system was evaluated by calculating the heat load carried at different air conditioning terminals. The operation of the combined cooling system at each stage of the day was analyzed based on the heat load capacities of different system components. The cooling capacity of the RFCAFC system can be calculated according to Equations (8) and (9).

$$Q_R = c_p m\left(T_h - T_g\right) \tag{8}$$

$$Q_F = G_n(h_2 - h_1) \tag{9}$$

where Q_R is the RFC capacity, W; Q_F is the fan coil cooling capacity, W; $c_p$is the specific heat at constant pressure, J/(kg·°C); m is the mass flow rate per unit area of circulating duct, kg/s; G_n is the fan coil air supply volume, kg/s; h_2 is the return air outlet enthalpy, kJ/(kg·K); and h_1 is the enthalpy at the supply air outlet, kJ/(kg·K).

3.2.6. System Operation Status and Energy Consumption

The EER_h, a measure of cooling capacity per unit power, was used to evaluate the energy efficiency of the combined cooling system during operation. The greater the EER_h, the better the energy conversion efficiency of the RFCAFC system. The EER_h of the combined system is calculated according to Equation (10).

$$EER_h = \frac{Q}{3.6 \times 10^6 E_h} \tag{10}$$

where EER_h is the hourly cooling performance coefficient; Q is the cooling capacity, W; and E_h is the system energy consumption, kWh.

4. Results

Experimental data under similar outdoor meteorological conditions were grouped to assess the different operating strategies for the combined radiant floor and fan coil cooling system in high humidity environments. The variations in indoor temperature (T_{in}), radiant floor surface temperature (T_f), and PMV-PPD values over time differed among different operation strategies and different outdoor meteorological conditions. The operational status and energy consumption of the combined cooling system under different operation strategies were analyzed using the EER_h and power consumption of the system. The analysis verified that the RFCAFC system was effective and energy efficient for different outdoor meteorological conditions in a high humidity environment. These results provide a strong theoretical basis for the promotion and application of the combined RFCAFC system in high humidity environments in Jinan, China.

4.1. Indoor Temperature and Humidity

Figures 8 and 9 show the variation of indoor humidity (φ), outdoor temperature (T_O), indoor temperature (T_{in}), and dew point temperature (T_d) over time using different operating strategies. It can be seen from Figure 8 that φ varied greatly during the test period. The φ tended to decrease significantly 0–3 h after turning on the combined cooling system, but after that the φ gradually reached a relatively stable state. This was because higher outdoor humidity resulted in higher indoor humidity as well, which affected the operation of the system. In general, the combined cooling system had a significant dehumidification effect in the high humidity indoor environment.

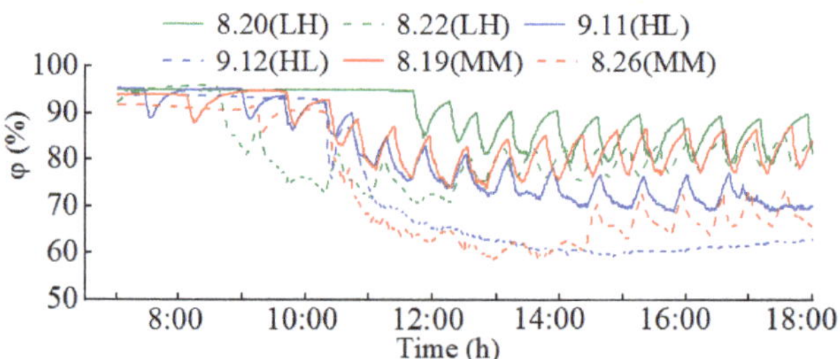

Figure 8. φ during the experimental period.

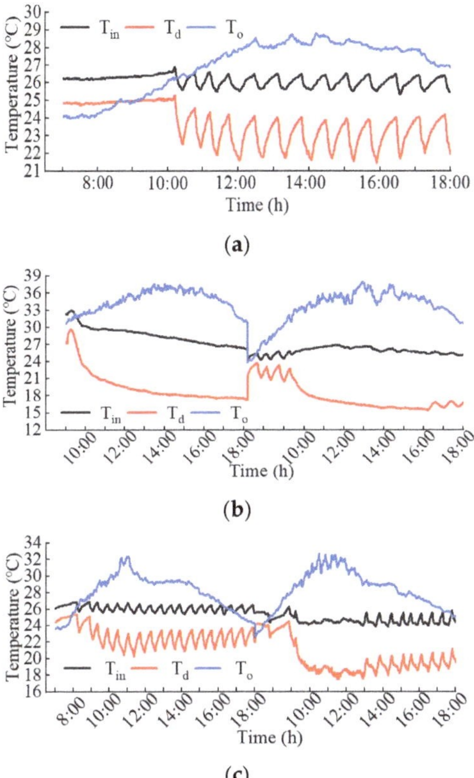

Figure 9. Temperature field showing (**a**) T_{in} change on the LH, (**b**) T_{in} change on the HL, and (**c**) T_{in} change on the MM.

As can be seen in Figure 9, the outdoor temperature fluctuated between 23 °C and 40 °C during the test period in Jinan, China. The average outdoor temperatures were 27.4 °C, 33.9 °C, and 28.7 °C during operating hours (8:00–18:00). The room temperature fluctuated between 25.5 °C and 27.5 °C during operation after the combined system reached a relative steady state. The average indoor temperatures were 26.1 °C, 26.6 °C, and 25.4 °C, respectively, during working hours. These values met the indoor temperature requirements for occupants. The standard deviations of T_r were low, i.e., 0.45 °C, 1.04 °C, and 0.58 °C, and the maximum standard deviation of T_{in} was only 1.04 °C. These values indicated that T_{in} was relatively stable and the operation strategy was effective in regulating T_{in} under different outdoor meteorological conditions.

Figure 9 shows that T_O was significantly lower than the indoor temperature during the test period 7:00–9:00. Therefore, the room temperature appeared to be reduced by natural cooling during that time period. By comparing the operating strategies for different outdoor meteorological conditions, it was found that the indoor temperature (T_{in}) under the operation strategy for HL showed a decreasing trend during the test period, while the T_{in} under the LH and MM operating strategies exhibited oscillating changes during the test period. The reason for this phenomenon was that there was a larger heat load entering the room through the envelope structure with continuously high outdoor temperatures. At the same time, the combined cooling system operated intermittently and the low outdoor temperature environment at night further improved the stability of the indoor thermal environment by reducing the internal surface temperature of the building envelope and offsetting part of the heat load accumulated by the building during the day. Therefore, to further improve the stability of the indoor thermal environment, intermittent operation of the combined cooling system at night can be considered. From the perspective of energy saving, natural cooling can effectively be used from 7:00 to 9:00.

4.2. Vertical Air Temperature Difference

Figure 10 shows the T_{in} change curves at different heights from the ground over time. It can be seen that the indoor vertical temperature fluctuated between 24.5 °C and 33.1 °C and at 1100 mm above the ground the temperature fluctuated between 23.4 °C and 26.3 °C under different operating strategies. Under the same operational strategy, the trends of T_{in} change were similar and relatively uniform among different heights. Under the LH operating strategy, the maximum temperature difference between different locations in the room was 0.3 °C at the same height, while under the LH and MM operating strategies the maximum temperature differences between locations in the room at the same height were 0.7 °C and 0.5 °C, respectively. The figure also shows that the room was colder at the lower points and hotter at higher points, and the temperature gradient was slightly more pronounced above 1.5 m. Under the LH, HL, and MM operating strategies, the maximum temperature differences at different heights in the room were about 1.7 °C, 1.6 °C, and 1.8 °C, respectively.

According to the ASHRAE552017 standard, the maximum temperature gradient should not exceed 3 °C in an area with major human activities. Therefore, the operating strategies used under all different outdoor meteorological conditions met the indoor temperature gradient requirements. Due to the influence of solar radiation, the indoor local maximum temperature occurred at the height of 1.1–1.5 m from 13:00 to 15:00, which could cause thermal discomfort at specific locations. Therefore, shading measures should be considered in the building construction process to reduce thermal discomfort caused by the local high temperature phenomenon.

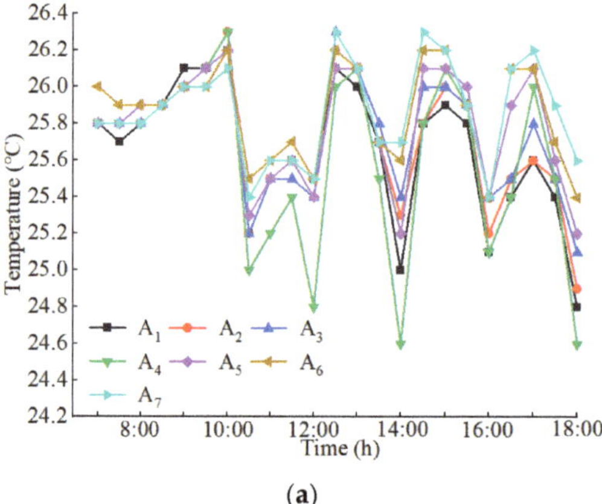

(**a**)

Figure 10. *Cont.*

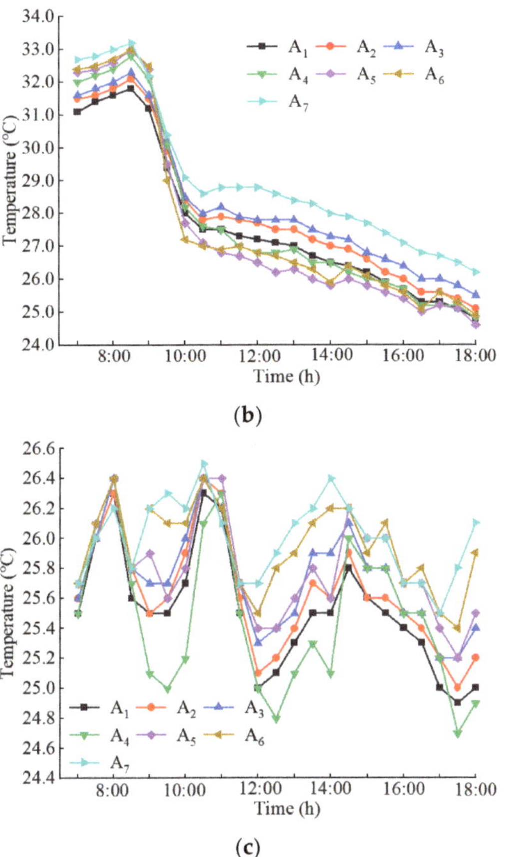

(b)

(c)

Figure 10. Vertical temperature differences under the (**a**) LH, (**b**) HL, and (**c**) MM operating strategies.

4.3. Radiant Floor Surface Temperature

The changes in radiation floor surface temperature under different operating strategies are shown in Figure 11. The black curve indicates the change trend of S and the red line indicates the fluctuation range of S. Under the HL operating strategy, S remained basically stable at about 1.0, which was a desirable level. Under the LH and MM operating strategies, the S was low, mostly within the 0.7–1.0 range, which can be considered the normal range. However, under the MM operating strategy, the s appeared to fluctuate more. So, under the MM operating strategy, the radiation floor had worse stability, but still remained within the normal range. This may have been because the indoor humidity was too high under this environmental condition. From the following Figure 11, it can be found that S was more stable when the radiant floor operated continuously than when it operated intermittently. Therefore, when the combined cooling system is artificially regulated, the radiant floor should be set to a constant open state to reduce fluctuation and improve comfort.

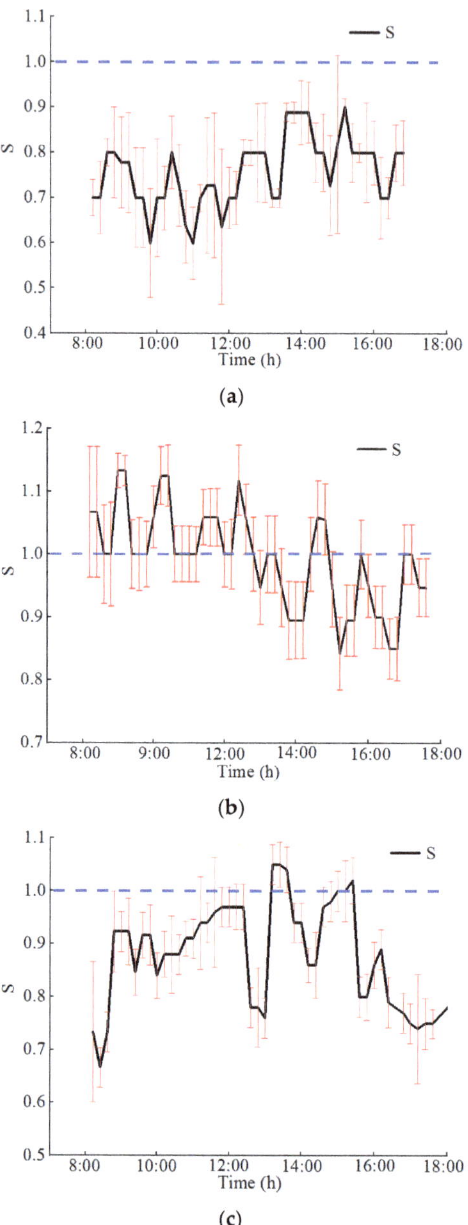

Figure 11. *S* range for the (**a**) LH, (**b**) HL, and (**c**) MM operating strategies.

4.4. Radiant Floor Time Constant

Figure 12 shows the radiant floor surface temperature change over time for different operating strategies for different outdoor meteorological conditions. It can be seen that the τ_{63} of different operating strategies ranged between 4.4 and 4.7 h.

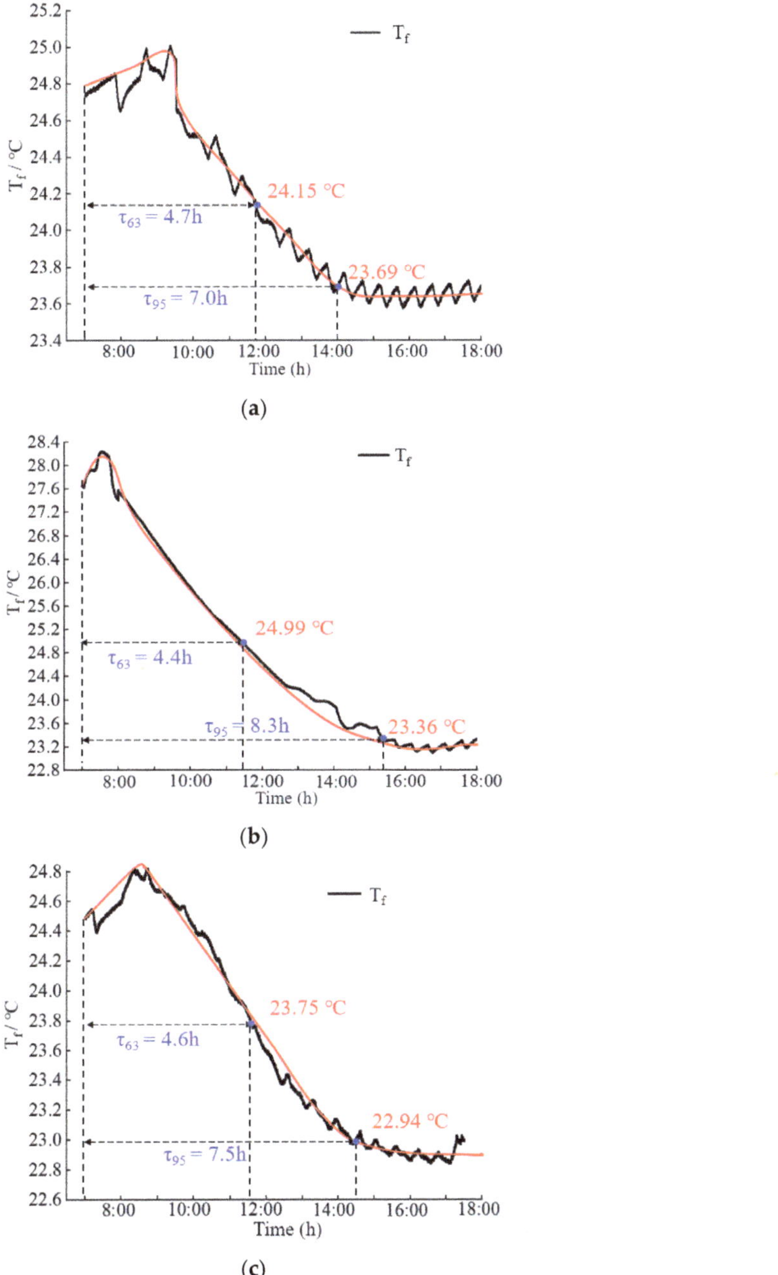

Figure 12. Temperature change of the radiant floor surface, T_f change during the (**a**) LH, (**b**) HL, and (**c**) MM operating strategies.

The differences in τ_{63} between operating strategies were small, and the final floor surface temperatures all stabilized at about 23 °C. This indicated that all the different RFCAFC operating strategies for different outdoor meteorological conditions were effective in the high humidity environment. In particular, when high outdoor temperatures led to high initial indoor temperatures, the τ_{63} value was only 4.4 h, which corresponds to a radiant floor temperature drop rate of 0.73 °C/h. The LH and MM operating strategies corresponded to radiant floor surface temperature drop rates of 0.19 °C/h and 0.21 °C/h. This may have been due to the high outdoor humidity levels, which required that the fan coil system be in continuous operation at all times and had a large effect on the drop in floor surface temperature. The τ_{95} values among operating strategies had large differences. Comparisons showed that the higher the outdoor humidity, the greater the τ_{95} value of the radiant floor, or the longer it takes the radiant floor surface to stabilize. Therefore, in cases of high outdoor humidity, we can consider activating the radiant floor system in advance to ensure the system reaches relative stability before being occupied.

4.5. Thermal Comfort

Figure 13 plots the PMV values over time for the different operation strategies. It can be seen that when the combined cooling system used the LH operating strategy, the PMV value fluctuated from −0.03 to 0.49, with the average value of 0.24. The PMV value fluctuated between −0.82 and 0.51 under the HL operating strategy, with an average of −0.08. The PMV value under the MM operating strategy fluctuated from −0.17 to 0.54, with an average of 0.14. The values of PMV under the LH and HL operating strategies were in accordance with ISO7730. However, the PMV was slightly higher than the recommended value during certain periods while using the MM operation strategy, but the comfort level was still acceptable. From the comfort point of view, the combined cooling system was able to effectively manage indoor thermal comfort.

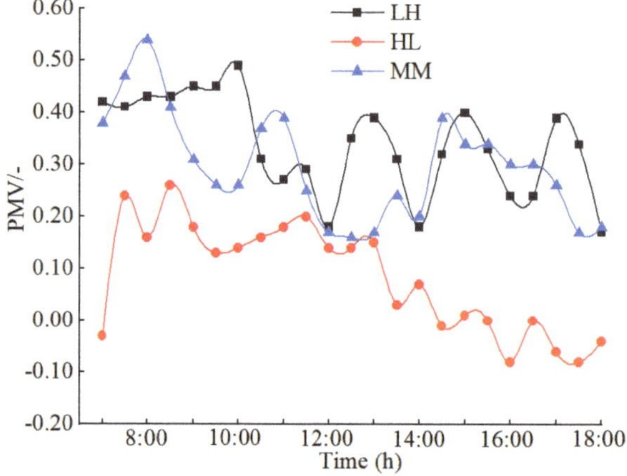

Figure 13. Thermal comfort under different operating strategies.

Figure 14 shows that indoor thermal comfort was better under different meteorological conditions. Therefore, in order to optimize indoor thermal comfort and energy saving, some interior design strategies can be considered. Due to the higher amount of solar radiation reaching the room through the south window, the combined cooling system was unable to completely eliminate the indoor heat load, which finally led to a certain reduction in indoor comfort from 13:00 to 15:00. Therefore, to improve thermal comfort, it is necessary to adjust the operation strategy of the combined cooling system in real time according to the changes in outdoor meteorological conditions.

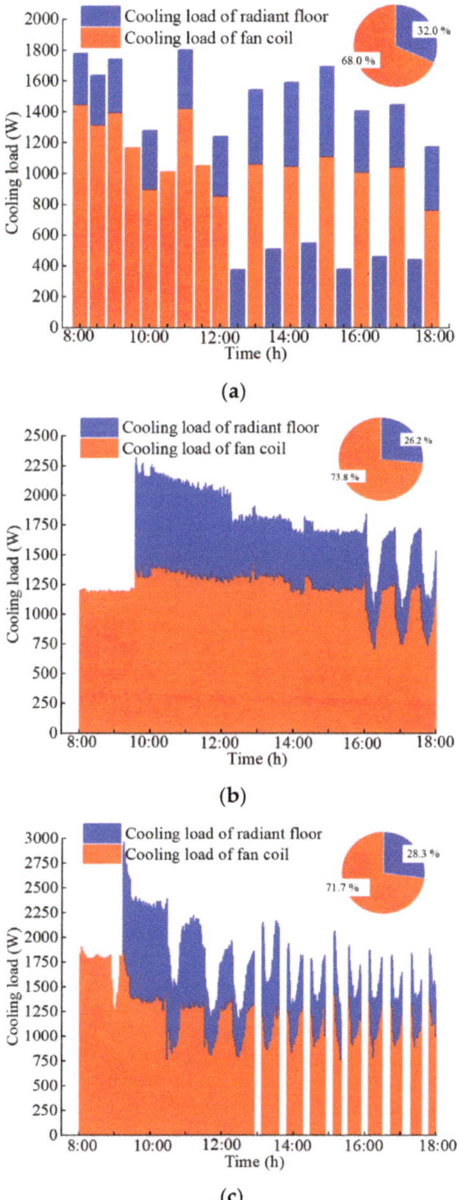

Figure 14. Cooling loads taken on by radiant floor and fan coil under the (**a**) LH, (**b**) HL, and (**c**) MM operating strategies.

4.6. System Cooling Capacity

Figure 14 shows the change in cooling capacity of the combined cooling system over time under different outdoor meteorological conditions. According to the figure, in the initially high humidity environment, the fan coil accounted for about 68.0% of the total cooling load under the LH operating strategy. Under the HL and MM operation strategies the fan coil made up about 73.8% and 71.7% of the total cooling capacity, respectively. In short, the fan coil of the combined system contributed more to the total cooling capacity than the radiant floor.

Comparing the proportion of fan coil cooling capacity to the total cooling capacity, we see that only during the LH operating strategy were fan coil system and floor radiation system turned on simultaneously to dehumidify and cool due to the high outdoor humidity. Under the HL and MM operating strategy, the fan coil system was turned on about 1 h before the radiant floor to compensate for the slow start and long thermal response time of the radiant floor system as well as to the eliminate indoor heat load. The cooling load taken on by the fan coil system was much higher than that of the radiant floor system because the outdoor temperature was higher under the HL and MM operating conditions.

In order to avoid thermal discomfort in the room, the fan coil was necessary for cooling when the outdoor temperature was high. Therefore, when the indoor and outdoor temperatures are high, in order to reduce indoor temperature and the risk of thermal discomfort, the fan coil system should be turned on in advance.

4.7. System Operation and Energy Consumption

Figure 15 shows the EER_h and system energy consumption over time for the combined cooling system the operation strategies for different outdoor meteorological conditions. It can be seen that when the combined cooling system was under the LH operating strategy, the EER_h fluctuated between 1.04 and 2.41, with an average of 1.98. During the HL and MM operating strategies, the EER_h of the combined cooling system fluctuated between 1.69–2.96 and 1.73–2.41, with averages of 2.52 and 2.14, respectively. The EER_h values of the combined system were slightly higher under HL outdoor meteorological conditions. This was because while using the HL operation strategy, the air source heat pump was operating at a high load when the room temperature reached a relatively stable state. In the MM operating strategy, the air source heat pump was maintained at a low load operation state. The combined cooling system consumed 13.43 kWh and 17.18 kWh when utilizing the LH and HL operating strategies, respectively, under the three different outdoor meteorological conditions. During the MM operating strategy, the daily energy consumption of the combined system was 11.36 kWh.

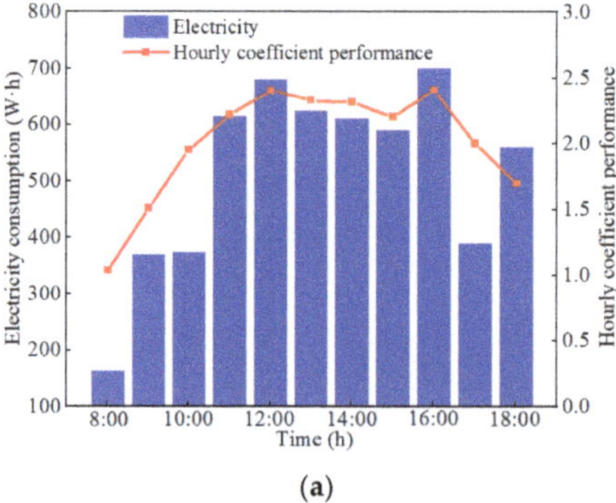

(a)

Figure 15. *Cont.*

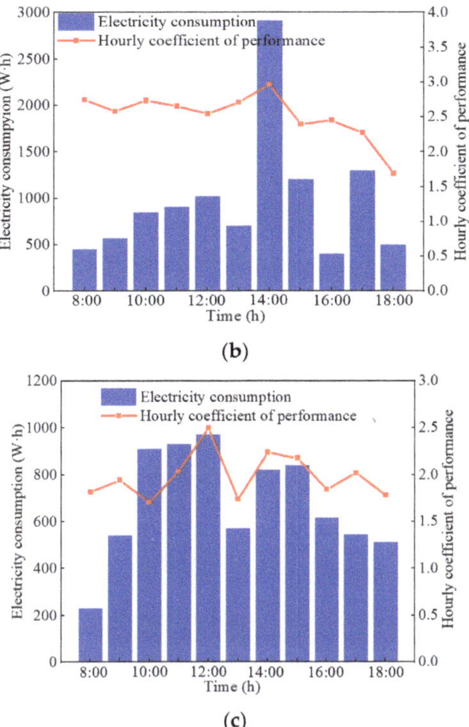

Figure 15. Energy consumption under the (**a**) LH, (**b**) HL, and (**c**) MM operating strategies.

Calculations showed that the energy consumption of the LH operating strategy was about 21.8% lower than that of the HL operating strategy. The energy consumption of the MM operating strategy was about 15.4% lower than that of the LH operating strategy. The energy consumption of the HL was much larger than those of the other two operating strategies. This showed that energy consumption was mainly influenced by the outdoor temperature. Thus, the higher the outdoor temperature, the more energy is consumed by the combined system. Furthermore, the EER_h was smaller during the initial operating period each day compared to other time periods. This was because the air source heat pump was operated at low load during the initial time period each day. In addition, the EER_h tended to increase as the cooling load taken by the radiant floor system increased. Therefore, from the perspective of reducing energy consumption and achieving efficient operation of the system, the radiant floor system should take on as much cooling capacity as possible and be continuously operated, while the fan coil can be operated intermittently.

5. Discussion

This experiment is one of the few that integrate an RFCAFC system into a standalone outdoor building in China. Previous experimental studies on the RFCAFC system have been based in office or residential buildings wherein not all walls are exterior walls. This experiment assessed the automatic operation of the RFCAFC system based on outdoor meteorological conditions, which freed the system from defects due to manual control present in previous studies. The subsequent results have created a strong theoretical basis for the promotion and application of combined cooling systems in the construction of intelligent buildings. They also provide new ideas for the operation of RFCAFC systems in actual building projects. The effectiveness of the operation strategies for different outdoor meteorological conditions can be used to guide future system operations and help to improve the reliability of the combined cooling system, especially in spaces that are frequently characterized by high humidity.

However, there were five limitations in this study that need to be further explored. First, this experimental study used only three different outdoor meteorological conditions (HL, LH, and MM) and did not consider others, such as high temperature and high humidity/low temperature and low humidity. Therefore, the operation strategies that were effective here may be different from those required for a combined cooling system in an actual project. In-depth studies will need to be conducted to consider more types of outdoor meteorological conditions in the future.

Second, the effect of indoor moisture dissipation on the combined cooling system was not considered in the study. Moreover, the study only assessed control strategies and did not study the control principles and the implementation process of the controls, which introduces certain limitations. A more in-depth study is needed to ensure the combined cooling system is widely applicable to different outdoor meteorological conditions.

Third, the experimental test was only conducted for six days, which is relatively short considering the length of the cooling season. The outdoor meteorological conditions were limited and cannot be considered representative of all outdoor meteorological conditions in Jinan, China during the cooling season. Therefore, when promoting the application of the combined cooling system, the influence of different outdoor meteorological conditions on the combined cooling system should be fully considered.

Fourth, this experiment was conducted in Jinan, China, which is in a hot summer and cold winter climate zone. Therefore, appropriate validation studies should be carried out before promoting the application of this combined cooling system in other climate zones to ensure that the RFCAFC system is the most effective and energy-saving option. The next steps will gradually address each of the above deficiencies to provide a stronger theoretical basis for the application of this type of the RFCAFC system in high humidity environments.

Fifth, this study will need to be augmented by developing numerical models and conducting simulation tests for validation. These methods will allow us to obtain results from different, more general case studies and under a wider variety of conditions.

6. Conclusions

In this study, a RFCAFC system for a single building was established and tested. The effects of different operation strategies on indoor temperature, vertical temperature difference, thermal comfort, cooling capacity, EER_h, and system energy consumption under different outdoor meteorological conditions were analyzed. The conclusions of the study are as follows.

(1) The level of radiation floor surface uniformity coefficient was low, mostly within 0.7–1.0. The range of τ_{63} of different operation strategies for different outdoor meteorological conditions was 4.4–4.7 h. The differences among operation strategies were small. The final floor surface temperature was stable at about 23 °C. In order for the radiant floor to reach its relatively stable state earlier, the radiant floor system could be turned on in advance.

(2) For all different operating strategies, the indoor thermal comfort was good and the temperature distribution was uniform in the indoor high humidity environment. Therefore, interior design approaches can be considered to further reduce the temperature and ensure indoor thermal comfort.

(3) The fan coil of the combined system took up a large proportion of the cooling load, accounting for about 68.0–73.8% of the total cooling capacity. The fan coil system plays a significant role in compensating for the long thermal response time of the radiant floor system.

(4) The energy consumption during the HL operating conditions was much larger than those of the other two outdoor meteorological conditions. The energy consumed by the LH operating strategy was about 21.8% lower than that of the HL operation strategy. The energy consumption of the MM operating strategy was about 15.4% lower than that of the LH operation strategy. From the energy saving point of view, use of the radiant floor

system for cooling should be prioritized and it should be operated continuously, while the fan coil can be operated intermittently.

Author Contributions: Conceptualization, X.Z. (Xuwei Zhu) and J.L.; methodology, J.L.; formal analysis, X.Z. (Xuwei Zhu), X.W. and J.L.; investigation, X.Z. (Xuwei Zhu), J.L., Y.D. and X.Z. (Xiangyuan Zhu); writing—original draft preparation, X.Z. (Xuwei Zhu); writing—review and editing, J.L., X.Z. (Xiangyuan Zhu), J.M., X.W. and X.Z. (Xuwei Zhu); supervision, J.L.; project administration, J.L.; funding acquisition, J.L. and J.M. All authors have read and agreed to the published version of the manuscript.

Funding: This work was funded by Natural Science Foundation of Shandong Province (ZR2021ME199) and Key Research and Development Project in Shandong Province (2018GSF121003).

Institutional Review Board Statement: Not applicable.

Informed Consent Statement: Not applicable.

Data Availability Statement: Not applicable.

Acknowledgments: This work acknowledges the support of the Plan of Introduction and Cultivation for Young Innovative Talents in Colleges and Universities of Shandong Province.

Conflicts of Interest: The authors declare no conflict of interest.

Nomenclature

c_p	specific heat of air, J/(kg·°C)
h	enthalpy, kJ/kg
T	Temperature, °C
Q	cooling capacity, W
S	radiant floor surface temperature uniformity coefficient
M	human energy metabolic rate, W/m^2
W'	mechanical work done by the human body, W/m^2
P_a	partial pressure of water vapor around the human body, kPa
m	mass flow rate per unit area of circulating duct, kg/s

Abbreviation	
RFCAFC	combined radiant floor and fan coil cooling
RFC	radiant floor cooling
LH	low temperature and high humidity
HL	high temperature and low humidity
MM	medium temperature and medium humidity
PMV	predicted mean vote
PPD	predicted percentage dissatisfaction
PVC	polyvinyl chloride
S	sample standard deviation

Greek letters	
τ	time constant, h
φ	relative humidity, %

Subscripts	
s,max	radiant floor surface maximum temperature, °C
s,min	radiant floor surface minimum temperature, °C
g	radiant floor water supply temperature, °C
h	radiant floor water return temperature, °C
s	initial temperature, °C
e	temperature at reaching 63% of total temperature change, °C
63	the time required for the radiant floor surface temperature change of the radiant floor to reach a total change of approximately 63.2%, h

95	the time required for the radiant floor surface temperature change of the radiant floor to reach a total change of approximately 95%, h
im	the initial moment radiant floor surface temperature, °C
rs	radiant floor surface temperature to reach a relatively stable state, °C
i	temperature after the i hour after the start of the combined system, °C
age	average indoor temperature during the test period, °C
f_{cl}	garment area coefficient
t_{cl}	temperature of the outer surface of the garment, °C
age	average radiant temperature, °C
r	convective heat transfer coefficient, $W/(m^2 \cdot K)$
a	air temperature around the human body, °C
R	RFC capacity, W
F	fan coil cooling capacity, W
n	fan coil air supply volume, kg/s
2	return air outlet enthalpy, $kJ/(kg \cdot K)$
1	enthalpy at the supply air outlet, $kJ/(kg \cdot K)$
E_h	system energy consumption, kWh
EER_h	hourly cooling performance coefficient
O	outdoor temperature, °C
in	indoor temperature, °C
d	dew point temperature, °C

References

1. Harish, V.S.K.V.; Kumar, A. A review on modeling and simulation of building energy systems. *Renew. Sust. Energy Rev.* **2016**, *56*, 1272–1292. [CrossRef]
2. Yang, L.; Yan, H.; Lam, J.C. Thermal comfort and building energy consumption implications—A review. *Appl. Energy* **2014**, *115*, 164–173. [CrossRef]
3. Hu, Y.; Xia, X.; Wang, J. Research on operation strategy of radiant cooling system based on intermittent operation characteristics. *J. Build. Eng.* **2022**, *45*, 103483. [CrossRef]
4. Qi, R.; Lu, L.; Yang, H. Investigation on air-conditioning load profile and energy consumption of desiccant cooling system for commercial buildings in Hong Kong. *Energy Build.* **2012**, *49*, 509–518. [CrossRef]
5. Chua, K.J.; Chou, S.K.; Yang, W.M.; Yan, J. Achieving better energy-efficient air conditioning—A review of technologies and strategies. *Appl. Energy* **2013**, *104*, 87–104. [CrossRef]
6. Rupp, R.F.; Vásquez, N.G.; Lamberts, R. A review of human thermal comfort in the built environment. *Energy Build.* **2015**, *105*, 178–205. [CrossRef]
7. Song, D.; Kim, T.; Song, S.; Hwang, S.; Leigh, S.-B. Performance evaluation of a radiant floor cooling system integrated with dehumidified ventilation. *Appl. Therm. Eng.* **2008**, *28*, 1299–1311. [CrossRef]
8. Feustel, H.E.; Stetiu, C. Hydronic radiant cooling—Preliminary assessment. *Energy Build.* **1995**, *22*, 193–205. [CrossRef]
9. Zhang, L.Z. Energy performance of independent air dehumidification systems with energy recovery measures. *Energy* **2006**, *31*, 1228–1242. [CrossRef]
10. Kim, M.K.; Liu, J.; Cao, S.-J. Energy analysis of a hybrid radiant cooling system under hot and humid climates: A case study at Shanghai in China. *Build. Environ.* **2018**, *137*, 208–214. [CrossRef]
11. Liu, J.; Li, Z.; Kim, M.K.; Zhu, S.; Zhang, L.; Srebric, J. A comparison of the thermal comfort performances of a radiation floor cooling system when combined with a range of ventilation systems. *Indoor Built Environ.* **2020**, *29*, 527–542. [CrossRef]
12. Liu, J.; Zhu, S.; Kim, M.K.; Srebricv, J. A review of CFD analysis methods for personalized ventilation (PV) in indoor built environments. *Sustainability* **2019**, *11*, 4166. [CrossRef]
13. Liu, J.; Kim, M.K.; Srebric, J. Numerical analysis of cooling potential and indoor thermal comfort with a novel hybrid radiant cooling system in hot and humid climates. *Indoor Built Environ.* **2022**, *31*, 929–943. [CrossRef]
14. Liu, J.; Ren, J.; Zhang, L.; Xie, X.; Kim, M.M.; Zhang, L. Optimization of Control Strategies for the Radiant Floor Cooling System Combined with Displacement Ventilation: A Case study of an Office Building in Jinan, China. *Int. J. Archit. Eng. Technol.* **2019**, *6*, 33–48. [CrossRef]
15. Ren, J.; Liu, J.; Zhou, S.; Kim, M.K.; Song, S. Experimental study on control strategies of radiant floor cooling system with direct-ground cooling source and displacement ventilation system: A case study in an office building. *Energy* **2022**, *239*, 122410. [CrossRef]
16. Li, N.; Chen, Q. Study on dynamic thermal performance and optimization of hybrid systems with capillary mat cooling and displacement ventilation. *Int. J. Refrig.* **2020**, *110*, 196–207. [CrossRef]
17. Hu, R.; Niu, J.L. A review of the application of radiant cooling & heating systems in Mainland China. *Energy Build.* **2012**, *52*, 11–19. [CrossRef]

18. Sourbron, M.; De Herdt, R.; Van Reet, T.; Van Passel, W.; Baelmans, M.; Helsen, L. Efficiently produced heat and cold is squandered by inappropriate control strategies: A case study. *Energy Build.* **2009**, *41*, 1091–1098. [CrossRef]

19. Romaní, J.; de Gracia, A.; Cabeza, L.F. Simulation and control of thermally activated building systems (TABS). *Energy Build.* **2016**, *127*, 22–42. [CrossRef]

20. Li, Y.; Zhuang, Z.; Zhu, Q.; Song, J.; An, H. Research on control methods of roof radiant cooling system. *Procedia Eng.* **2017**, *205*, 2149–2155. [CrossRef]

21. Schmelas, M.; Feldmann, T.; Bollin, E. Savings through the use of adaptive predictive control of thermo-active building systems (TABS): A case study. *Appl. Energy* **2017**, *199*, 294–309. [CrossRef]

22. Jia, H.; Pang, X.; Haves, P. Experimentally-determined characteristics of radiant systems for office buildings. *Appl. Energy* **2018**, *221*, 41–54. [CrossRef]

23. Bourdakis, E.; Kazanci, O.B.; Olesen, B.W. Load Calculations of Radiant Cooling Systems for Sizing the Plant. *Energy Procedia* **2015**, *78*, 2639–2644. [CrossRef]

24. Mikeska, T.; Svendsen, S. Dynamic behavior of radiant cooling system based on capillary tubes in walls made of high performance concrete. *Energy Build.* **2015**, *108*, 92–100. [CrossRef]

25. Mikeska, T.; Fan, J.; Svendsen, S. Full scale measurements and CFD investigations of a wall radiant cooling system integrated in thin concrete walls. *Energy Build.* **2017**, *139*, 242–253. [CrossRef]

26. Liu, J.; Zhu, X.; Kim, M.K.; Cui, P.; Kosonen, R. A Transient Two-dimensional CFD Evaluation of Indoor Thermal Comfort with an Intermittently-operated Radiant Floor Heating System in an Office Building. *Int. J. Archit. Eng. Technol.* **2020**, *7*, 62–87. [CrossRef]

27. Zarrella, A.; De Carli, M.; Peretti, C. Radiant floor cooling coupled with dehumidification systems in residential buildings: A simulation-based analysis. *Energy Convers. Manag.* **2014**, *85*, 254–263. [CrossRef]

28. Seo, J.-M.; Song, D.; Lee, K.H. Possibility of coupling outdoor air cooling and radiant floor cooling under hot and humid climate conditions. *Energy Build.* **2014**, *81*, 219–226. [CrossRef]

29. Zhao, K.; Liu, X.-H.; Jiang, Y. Application of radiant floor cooling in a large open space building with high-intensity solar radiation. *Energy Build.* **2013**, *66*, 246–257. [CrossRef]

30. Jin, W.; Jia, L.; Wang, Q.; Yu, Z. Study on Condensation Features of Radiant Cooling Ceiling. *Procedia Eng.* **2015**, *121*, 1682–1688. [CrossRef]

31. Kim, M.K.; Leibundgut, H. A case study on feasible performance of a system combining an airbox convector with a radiant panel for tropical climates. *Build. Environ.* **2014**, *82*, 687–692. [CrossRef]

32. Schmelas, M.; Feldmann, T.; Wellnitz, P.; Bollin, E. Adaptive predictive control of thermo-active building systems (TABS) based on a multiple regression algorithm: First practical test. *Energy Build.* **2016**, *129*, 367–377. [CrossRef]

33. Srivastava, P.; Khan, Y.; Bhandari, M.; Mathur, J.; Pratap, R. Calibrated simulation analysis for integration of evaporative cooling and radiant cooling system for different Indian climatic zones. *J. Build. Eng.* **2018**, *19*, 561–572. [CrossRef]

34. Feng, J.; Bauman, F.; Schiavon, S. Experimental comparison of zone cooling load between radiant and air systems. *Energy Build.* **2014**, *84*, 152–159. [CrossRef]

35. Fernández Hernández, F.; Cejudo López, J.M.; Fernández Gutiérrez, A.; Domínguez Muñoz, F. A new terminal unit combining a radiant floor with an underfloor air system: Experimentation and numerical model. *Energy Build.* **2016**, *133*, 70–78. [CrossRef]

36. Gu, X.; Cheng, M.; Zhang, X.; Qi, Z.; Liu, J.; Li, Z. Performance analysis of a hybrid non-centralized radiant floor cooling system in hot and humid regions. *Case Stud. Therm. Eng.* **2021**, *28*, 101645. [CrossRef]

37. Joe, J.; Karava, P. A model predictive control strategy to optimize the performance of radiant floor heating and cooling systems in office buildings. *Appl. Energy* **2019**, *245*, 65–77. [CrossRef]

38. Zhang, D.; Cai, N.; Cui, X.; Xia, X.; Shi, J.; Huang, X. Experimental investigation on model predictive control of radiant floor cooling combined with underfloor ventilation system. *Energy* **2019**, *176*, 23–33. [CrossRef]

39. Liu, D.; Zhou, H.; Hu, A.; Zhang, Q.; Liu, N.; Wen, J. Study on the intermittent operation mode characteristic of a convection-radiation combined cooling system in office buildings. *Energy Build.* **2022**, *255*, 111669. [CrossRef]

40. Atienza Márquez, A.; Cejudo López, J.M.; Fernández Hernández, F.; Domínguez Muñoz, F.; Carrillo Andrés, A. A comparison of heating terminal units: Fan-coil versus radiant floor, and the combination of both. *Energy Build.* **2017**, *138*, 621–629. [CrossRef]

41. Khan, Y.; Khare, V.R.; Mathur, J.; Bhandari, M. Performance evaluation of radiant cooling system integrated with air system under different operational strategies. *Energy Build.* **2015**, *97*, 118–128. [CrossRef]

42. Lin, B.; Wang, Z.; Sun, H.; Zhu, Y.; Ouyang, Q. Evaluation and comparison of thermal comfort of convective and radiant heating terminals in office buildings. *Build. Environ.* **2016**, *106*, 91–102. [CrossRef]

43. Sun, H.; Yang, Z.; Lin, B.; Shi, W.; Zhu, Y.; Zhao, H. Comparison of thermal comfort between convective heating and radiant heating terminals in a winter thermal environment: A field and experimental study. *Energy Build.* **2020**, *224*, 110239. [CrossRef]

44. Cen, C.; Jia, Y.; Liu, K.; Geng, R. Experimental comparison of thermal comfort during cooling with a fan coil system and radiant floor system at varying space heights. *Build. Environ.* **2018**, *141*, 71–79. [CrossRef]

45. Dong, J.; Zhang, L.; Deng, S.; Yang, B.; Huang, S. An experimental study on a novel radiant-convective heating system based on air source heat pump. *Energy Build.* **2018**, *158*, 812–821. [CrossRef]

46. Tian, Z.; Love, J.A. A field study of occupant thermal comfort and thermal environments with radiant slab cooling. *Build. Environ.* **2008**, *43*, 1658–1670. [CrossRef]

47. Liu, J.; Dalgo, A.D.; Zhu, S.; Zhang, L.; Srebric, J. Performance analysis of a ductless personalized ventilation combined with radiant floor cooling system and displacement ventilation. *Build. Simul.* **2019**, *12*, 905–919. [CrossRef]
48. ISO. *Guide to the Expression of Uncertainty in Measurements*; IO Standardization: Geneva, Switzerland, 1995.
49. Mathioulakis, E.; Voropoulos, K.; Belessiotis, V. Assessment of uncertainty in solar collector modeling and testing. *Sol. Energy* **1999**, *66*, 337–347. [CrossRef]

 *buildings*

Article

Prototyping a Lighting Control System Using LabVIEW with Real-Time High Dynamic Range Images (HDRis) as the Luminance Sensor

Aris Budhiyanto * and Yun-Shang Chiou

Department of Architecture, National Taiwan University of Science and Technology, Taipei 106, Taiwan; ychiou@mail.ntust.edu.tw
* Correspondence: aris.budhiyanto@gmail.com; Tel.: +866-986341427

Abstract: Lighting control systems (LCSs) play important roles in maintaining visual comfort and energy savings in buildings. This paper presents a prototype LCS using LabVIEW with real-time high dynamic range images and a digital multiplex controller to brighten lamps sequentially to provide visual comfort. The prototype is applied to a scaled classroom model with three schemes involving different activities and needs: writing and reading, requiring a uniform luminance of approximately 100 cd/m^2, teaching using a whiteboard, requiring an illuminance of approximately 120 cd/m^2 for the whiteboard and 60 cd/m^2 for the desks, and drawing and art activities focused on the center of the room, requiring an illuminance of approximately 100 cd/m^2 for the center area and 50 cd/m^2 for the background area. For each scheme, two conditions are presented: one in which the room is treated as a closed room without windows, and the one in which the room has a large window on one wall that enables daylight to penetrate the room. The prototype works well with both schemes and provides different combinations of lamp brightness levels, starting from 10% to 60%, based on the activities and required luminance, and can save around 73–82% of electricity. The presence of daylight does not always result in more energy savings, as the brightness contrast for visual comfort needs to be considered.

Keywords: lighting design; lighting control system; visual comfort; HDRi luminance analysis; LabVIEW; energy saving

Citation: Budhiyanto, A.; Chiou, Y.-S. Prototyping a Lighting Control System Using LabVIEW with Real-Time High Dynamic Range Images (HDRis) as the Luminance Sensor. *Buildings* **2022**, *12*, 650. https://doi.org/10.3390/buildings12050650

Academic Editor: Alessandro Cannavale

Received: 25 April 2022
Accepted: 9 May 2022
Published: 13 May 2022

Publisher's Note: MDPI stays neutral with regard to jurisdictional claims in published maps and institutional affiliations.

1. Introduction

Lighting is an essential factor for indoor environmental quality since it is related to occupants' visual comfort. In terms of building energy consumption, approximately 30% is dedicated to lighting. Therefore, it is important to find a way to reduce energy consumption without sacrificing occupants' visual comfort [1]. To achieve that purpose, lighting control systems (LCSs) are implemented in buildings. Implementing an LCS not only provides visual comfort for the occupants, but also reduces energy consumption by 60% [2].

The LCS receives information from the sensors, which is then processed by control rules and algorithms to determine the actions of the lighting fixtures. Based on the sensors, there are three lighting control schemes commonly used in buildings: occupancy-based sensors, daylight-linked systems, and scheduling schemes. Occupancy-based sensors depend on movement detection using passive infrared (PIR) sensors, ultrasonic sensors, or radio frequency identification (RFID), while daylight availability (illuminance) measured by a photosensor is used in the daylight-linked system, and a fixed schedule is used to control the lights in the scheduling system [3]. LCSs may use several types of sensors to achieve optimal results. An intelligent LCS research has developed a combination of lux sensors to monitor light intensity and motion sensors to sense human presence connected with LabVIEW (Laboratory Virtual Instrument Engineering Workbench) programming to control light intensity [4]. Based on the actions of the lights, there are two light-control methods:

switching and dimming methods. In the switching method, lights can be controlled to switch on or off, while in dimming, the system reduces or increases the light level and dims the lights.

There have been many studies on LCSs. Table 1 presents selected related research and techniques.

Table 1. Selected research and their techniques of LCSs.

Ref.	Objectives	Detection/Measurement		Method
		Tools/ Sensor	**Unit**	
[5]	Energy saving by applying daylight harvesting systems and lighting control	Daylight sensor	Illuminance (lux)	Simulation using DYSIM software based on the close loop control algorithm
[6]	Minimizing energy consumption	Occupancy sensor, ambient lux sensor, and scheduling	Illuminance (lux)	ZigBee protocol and illumination control rules
[7]	Providing visual comfort and energy saving for different uses and room layouts	Daylight sensor and user control	Illuminance (lux)	Home automation system
[8]	Adjusting lighting brightness, CCT, and illuminance distribution to meet various needs	Occupancy sensor, ambient lux sensor, scheduling, and user control	Illuminance (lux) and correlated color temperature (K)	IoT (Internet of Things) gateways and cloud platform
[9]	Providing illumination levels based on the users' needs and routines	PIR sensor, lux sensor, scheduling	Illuminance (lux)	Simulation using Dialux Evo software based on Fuzzy Logic and Artificial Neural Network
[10]	Proposing a control system for shop window lighting	Digital camera to produce HDRi	Luminance (cd/m^2)	DMX controller based on rules and algorithm
[11]	Developing a lighting system integrating users in the control loop	lux sensor, user survey	Illuminance (lux) and users' response	Q-learning algorithm
[12]	Comparing independent and integrated control strategies based on energy saving and lighting performance	Occupancy sensor, ambient lux sensor, and scheduling	Illuminance (lux)	Co-simulation platform consisting of BCVTB, EnergyPlus, and MATLAB

Illuminance sensors are common sensors used in LCSs [5–9,11,12], and working plane illuminance is mostly used to indicate lighting quantity because it is easily measured. However, as placing the illuminance sensor on the working plane is not practical, most illuminance sensors are mounted on walls or ceilings instead of the working plane, which can affect their performance [3]. In addition to illuminance, luminance also needs to be taken into consideration for providing visual comfort [13], as it is related to the light received by the human eye and human visual perception of brightness [14]. Luminance distribution can be measured by a spot luminance meter, but it is not effective to measure individual spots continuously. Current technology uses digital cameras to measure the luminance of a scene, called the high dynamic range (HDR), which captures luminance ranges within a scene and calculates the luminance based on pixel values [10,14]. The digital camera offers advantages over the light sensor in that the digital camera can be mounted on walls or ceilings without affecting the quality and performance of the HDRi. The method of using a digital camera as a luminance meter using high-dynamic-range images (HDRis) has been used for shop window LCSs to maintain a constant contrast between the shop window interior and its surroundings. The control system varies the brightness level of the lamps based on the level of exterior light to maintain this contrast [10].

Common LCSs are programmed to brighten or dim lamps simultaneously based on illuminance or occupancy detection [9,11,12]. However, this method mostly provides a uniform brightness, and for some activities that require brightness contrast, brightening or dimming lamps simultaneously is not able to achieve the desired brightness contrast, resulting in visual discomfort. In comparison, the occupancy detection sensor usually responds by turning on the lamps in a certain area where the occupants are detected and keeping the lamps in other areas turned off. This will lead to substantial brightness contrast in the room and cause visual discomfort [15,16]. In addition, the presence of daylight in the room should be considered in LCSs. Several LCSs are integrated with a blind/shade control system to block daylight. To fill this research gap, this study proposes an LCS based on HDRi for view sensing that uses dimmable LED lamps with various brightness levels as a solution. The lamps can be controlled to brighten in sequence in response to brightness contrast or daylight, so their brightness levels are varied.

The objective of this study is to develop a prototype of an LCS consisting of several dimmable LED lamps and a 360° internet protocol (IP) camera that serves as a viewing sensor for maintaining the visual comfort of the task area for various activities by brightening the lamps in sequence based on the activities and lighting needs. The 360° IP camera is used as a luminance sensor to produce HDRis and measure luminance. The LCS is developed based on HDRis for a room used for three activities, each requiring a different luminance. Different from the research conducted by [10], which focused only on the given contrast values of indoor and exterior environments without considering the various activities occurring there, this LCS is designed to provide visual comfort by maintaining the indoor luminance level at a given value and saving energy for both activities and needs.

Instead of using common programming languages, such as C++, Python, or MATLAB, which require advanced programming skills [5–12], the LabVIEW environment is used since it employs G programming, which can analyze several layers of data and execute according to the rules of data flow instead of a more traditional procedural approach. In addition, it enables parallel processing and performing multiple tasks at once, different from traditional and sequential languages such as C and C++ [17]. LabVIEW's biggest advantage is the rapid and simple construction of the graphical user interface (GUI), which makes it easier to create the algorithm and does not require advanced programming skills to operate [18]. Based on HDRi luminance values, LabVIEW sets the brightness levels of the LED lamp and communicates with a digital multiplex (DMX) controller to control and brighten the LED lamps. Each lamp is brightened in sequence to respond to the indoor luminance levels.

2. Materials and Methods

2.1. Model Prototype

The prototype of lighting control is composed of:

- 9 dimmable LED lamps with a dimmable DMX driver.
- 2 LED Fresnel lamps.
- An Arduino Uno as the DMX controller.
- A 360° IP camera.
- A luminance meter.
- A laptop installed with LabVIEW software.

The prototype is assumed to be a classroom environment that is used for three different types of activities: the first includes writing and reading (self-studying room), the second is teaching using a whiteboard (conventional classroom), and the last one is drawing and art activities focused on the center of the room (art activity classroom).

The prototype is set on a 1.5×1.5 m^2 measurement area divided into 25 squares (X_1,Y_1 to X_5,Y_5) as the working plane, and a piece of paper is placed in each square. Nine LED lamps arranged in three rows (A, B, C) and three columns (1, 2, 3) are hung 70 cm above the floor, so the proportion of room depth:height is approximately 2.15:1. Compared to a real classroom, the prototype is scaled down by four times. The LED lamps are covered with papers to reduce the brightness to levels in line with the scale of the prototype. To ensure the brightness scales as intended, an LED lamp is adjusted to various brightness levels, and an illuminance meter is used to measure the illuminance of the working plane in the prototype and in the real classroom for comparison. The illuminance measurements of the working plane in both the prototype and the real classroom are approximately 160 lux,

260 lux, and 500 lux as the lamp is brightened to 30%, 50%, and 100%, respectively. On one side of the wall, three papers are placed at a height of 45 cm (two-thirds of room height) as the whiteboard (X_2,Y_6 to X_4,Y_6). A 360° IP camera and a luminance meter are installed at the center of the experimental area, and two LED Fresnel lamps are placed outside of the measurement area, but their light is directed toward the measurement area to provide additional light to be assumed as daylight when the measurement scheme with daylight is presented (Figure 1). The IP camera captures images of the scene, and an algorithm is used to create an HDRi. The luminance value of each piece of paper placed inside the 28 squares is obtained from the HDRi and calibrated using the luminance meter. Based on these luminance values, LabVIEW is programmed to determine the brightness level of each LED lamp, and an Arduino Uno, used as a DMX controller, controls the brightness levels of the lamps [19,20] (Figure 2).

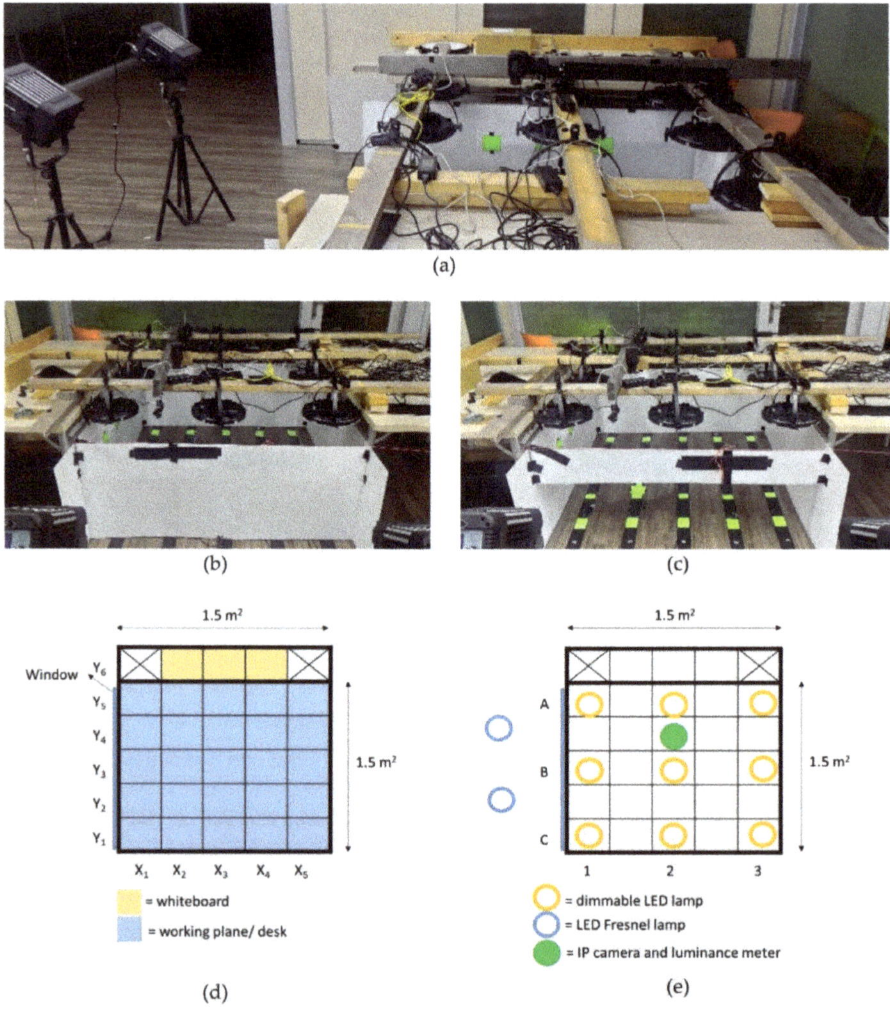

Figure 1. (**a**) The prototype is set on a 1.5×1.5 m² area; (**b**) The model for the closed room without daylight scheme; (**c**) The model for the room with a window on one side and daylight is presented; (**d**) The layout of the prototype; (**e**) The arrangement of lamps.

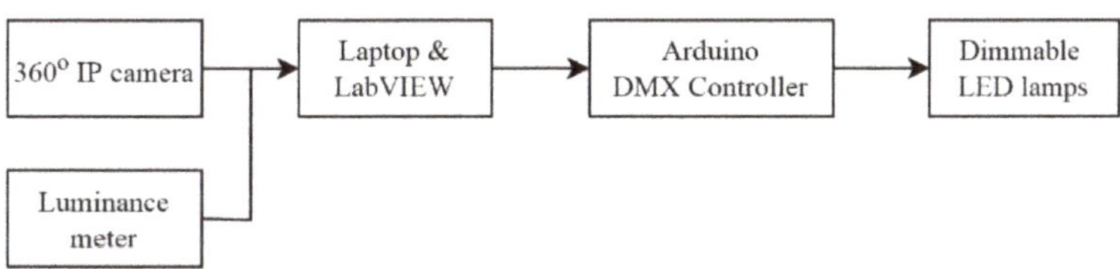

Figure 2. Control system framework.

2.2. HDRi Processing

The HDR technique has been widely accepted and applied for various purposes, such as digital photography, image editing, virtual reality, digital cinema and video, rendering, lighting simulation, and remote sensing [21]. In this study, an IP camera is used to measure the luminance of a scene using the HDR photography technique. The camera acquires images at multiple exposures to capture a wide luminance variation within a scene within one minute (Figure 3). The images are merged to create an HDRi based on the Debevec algorithm [10,22]. The luminance value is calculated based on the Radiance-based program using the following equation:

$$L = (0.265 \times R + 0.670 \times G + 0.065 \times B) \tag{1}$$

where L is the luminance value of the pixel (cd/m^2); R, G, and B are the spectrally weighted radiance values of the pixel (W/m^2 sr), and 0.265, 0.670, and 0.065 are calculated from CIE chromaticity used by Radiance [23,24]. A luminance meter is used to calibrate the 360° IP camera luminance value to obtain the correct luminance results [25].

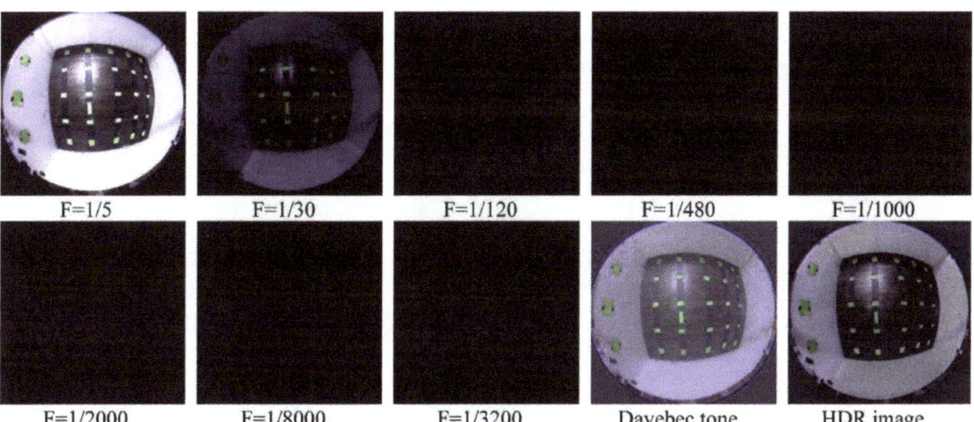

Figure 3. The images with different exposure and HDRi produced by 360° IP camera.

2.3. Calibration and Verification

2.3.1. Vignetting Effect and Correction of HDRi Luminance Results

The HDRi luminance measurement needs to be corrected because the light attenuation at the periphery of the lens results in luminance values for points away from the center of an HDRi image being systematically less than the actual luminance values in the scene. This is referred to as the vignetting effect, which is a nonlinear radial effect along the image radius of the lens and is often approximated by a polynomial function. To determine the vignetting effect, an HDRi of a uniform surface is captured, and the derived luminance values are compared to multiple spot measurements taken across the same surface [26,27]. Based on the vignetting effect, the vignetting correction factor with $R^2 = 0.98$ for each individual pixel is defined (Figure 4).

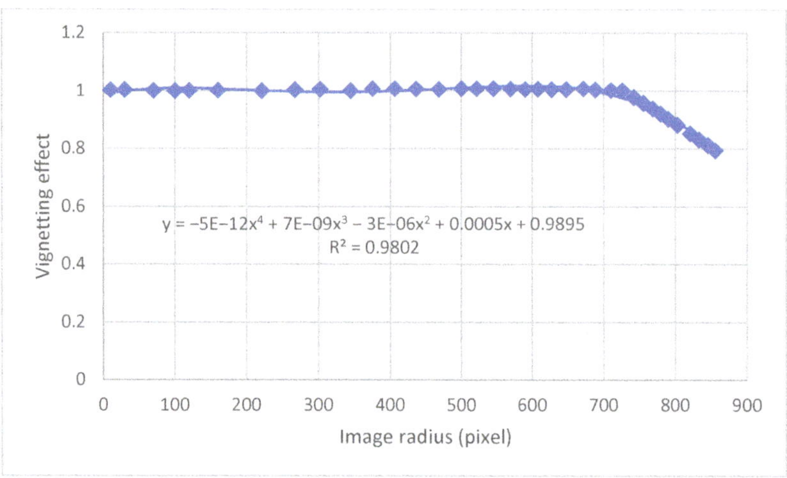

Figure 4. Measured luminance values and the vignetting function derived to fit the measured values.

To verify the HDRi luminance value results, multiple spot measurements using a luminance meter are acquired at various LED lamp brightness levels for comparison. Figure 5 shows the luminance values obtained from the HDRi and measurement. The lamp brightness levels for HDRi 1 and measurement 1, HDRi 2 and measurement 2, HDRi 3 and measurement 3, HDRi 4 and measurement 4, and HDRi 5 and measurement 5 are 10%, 20%, 40%, 60%, and 80%, respectively. The average difference between the HDRi result and the measurement is within the range of 5–15%, which is deemed acceptable for practical measurements [26].

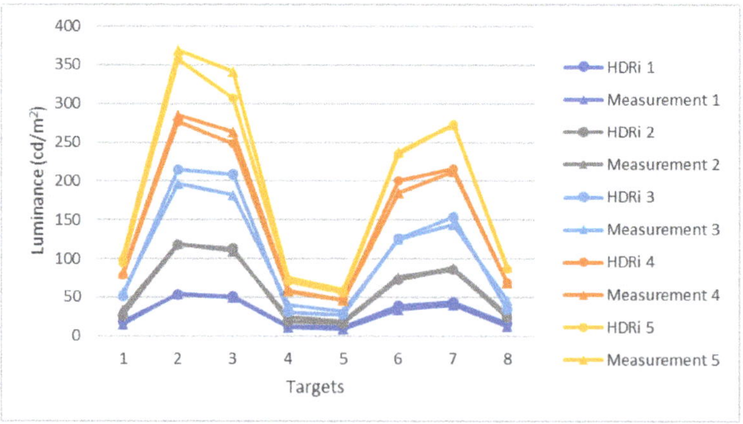

Figure 5. The luminance values obtained from the HDRis and measurements.

2.3.2. Luminance Requirement for Each Activity

Three activities are presented in this study: the first is writing and reading, which requires an illuminance level of approximately 500 lux for the entire room; the second is teaching using a whiteboard, which requires an illuminance level of approximately 600 lux on the whiteboard and approximately 300 lux on the desk area, with a brightness contrast level between the whiteboard and the desk area of approximately 1.5, and the last is drawing and art activities that focus on the center of the room, which require an illuminance level of approximately 500 lux on the center of the room and approximately 250 lux on the background area, with a contrast level between the center and the background area of approximately 1.5 [28]. Since the HDRi produces luminance, the illuminance must be transformed to luminance. To obtain the required luminance value for each activity, measurements using an illuminance meter, luminance meter, and the sample paper used in the prototype and measurement area were conducted in different spaces of the classroom according to the activities. Table 2 shows the luminance values required for each activity.

Table 2. The luminance values required for each activity.

Activity	Measurement Area	Illuminance (Lux)	Luminance (cd/m^2)
Reading and writing	Working plan/desks	500	100
Teaching	Whiteboard	600	120
	Working plan/desks	300	60
Drawing and art activity	Center area	500	100
	Background area	250	50

2.3.3. Brightness Level and Energy Consumption of Dimmable LED Lamp

Since nine dimmable LED lamps are used in this prototype, the brightness level and energy consumption of each LED lamp needs to be calibrated and verified so that all the lamps have the same performance. During that process, each lamp is turned on with various brightness levels, from 10% to 100%, and an energy meter is used to measure the energy consumption of the lamp at various brightness levels. Table 3 shows the energy consumption of an LED lamp in various brightness levels. The difference between the measured energy consumption and the product information is around 0.33–11.67%.

Table 3. The energy consumption of an LED lamp in various brightness levels.

Brightness Level (%)	Measured Energy Consumption (W)	Product Information Output Power (W)	Difference (%)
10	6.7	6	11.67
20	12.3	12	2.5
30	18.6	18	3.33
40	23.9	24	0.42
50	29.9	30	0.33
60	36.7	36	1.94
70	41.9	42	0.24
80	48.4	48	0.83
90	56	54	3.7
100	62	60	3.33

2.4. Measurement Schemes

Based on the three different activities mentioned in Section 2.1, three measurement schemes with different LCSs are presented. Each scheme is applied for two conditions: the first is the room assumed as a closed room without windows, and the second is the room assumed to have a large window. The LED Fresnel lamps placed outside the measurement area are turned on, and it is assumed that daylight penetrates the room. The control system algorithm is built in the LabVIEW environment. LabVIEW is a visual programming language developed by National Instruments and used for measuring, monitoring, controlling, and recording operating conditions [18,29–31]. LabVIEW supports thousands of hardware devices, including Arduino, and has been implemented in smart building control to control lighting electricity, shading devices, and home safety [32–34]. In LabVIEW, the brightness levels of each lamp are set by responding to the luminance values.

2.4.1. Measurement Scheme 1

In Measurement Scheme 1, the room functions as a self-studying room for writing and reading activities, which requires a uniform luminance value of approximately 100 cd/m^2 for the entire room. The whiteboard is not considered in this scheme. In Measurement Scheme 1A, the room is assumed to be a closed room without windows, and in Measurement Scheme 1B, the room is assumed to have a large window on one side such that daylight penetrates through the window. In Measurement Scheme 1B, the LED Fresnel lamps outside the measurement area are dimmed to 50% so that the luminance values in the measurement area do not exceed the requirement and shading devices do not need to be used. Figure 6 shows the LCS for Measurement Scheme 1. First, the room is divided into two areas, and the average luminance values of each area are compared to detect whether daylight is present. If the average luminance values of one area are much higher than those of the other, the lamps in the other area are brightened by 10% in sequence until the average luminance values of both areas are close to equal. Then, if the luminance values of the entire room are less than 100 cd/m^2, the lamps in columns 1 and 3 are simultaneously brightened by 10%, followed by the lamps in column 2 (L_1, L_3–L_2). This order is used instead of the order from column 1 to column 3 (L_1–L_2–L_3) because the latter order leads to all the lamps brightening to the same brightness level to meet the required luminance values, which has an effect no different from brightening all the lamps simultaneously (Appendix A). In addition, when the lamps placed on the rear sides brighten, they also illuminate the area in the center of the room. The measurement is repeated every three minutes per iteration because over one minute, the IP camera creates a series of images, and the lamp brightness levels should not change; otherwise, the created HDRis will not be accurate. Therefore, at least one additional minute after a change in lamp brightness levels is needed to create the next series of images.

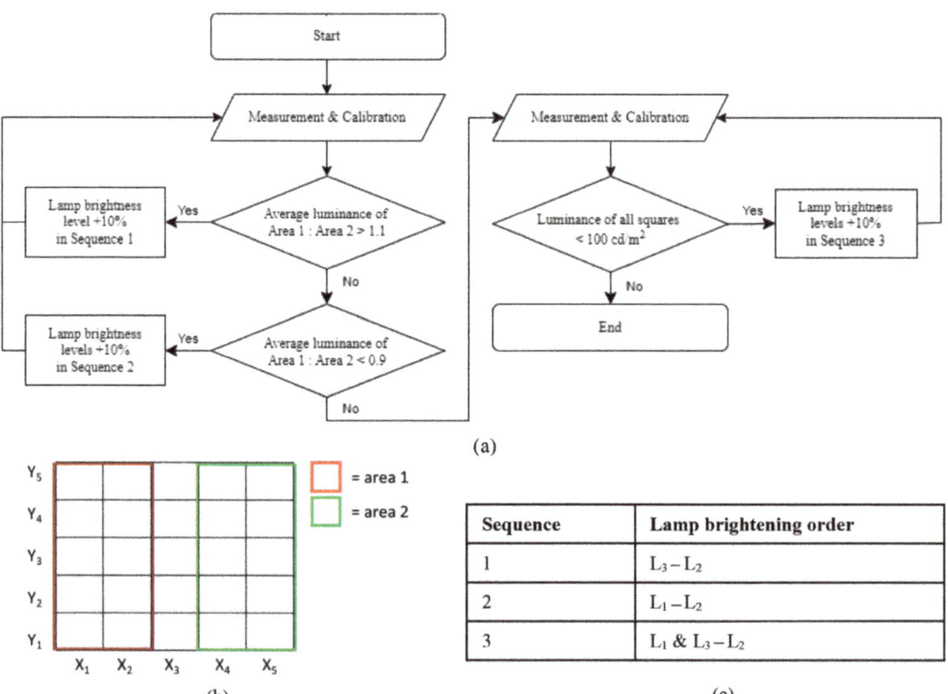

Sequence	Lamp brightening order
1	$L_3 - L_2$
2	$L_1 - L_2$
3	L_1 & $L_3 - L_2$

(c)

Figure 6. (**a**) The LCS flow diagram for scheme 1; (**b**) The layout of the room; (**c**) The sequence of the lamps brightened.

2.4.2. Measurement Scheme 2

Measurement Scheme 2 considers the room to function as a conventional classroom for teaching using a whiteboard, which requires a luminance value of approximately 120 cd/m^2 on the whiteboard (area A) and 60 cd/m^2 on the working plane/desks (area B); the brightness contrast between the whiteboard and the working plane should not be less than 1.5. This contrast value means that the average luminance values of whiteboards should be 50% higher than the average luminance value of the working plane/desks. Regarding the measurement accuracy of 5–15%, this value is more than three times higher than the error; thus, it can guarantee that the result of this measurement is correct. In Measurement Scheme 2A, the room is assumed to be a closed room without windows, and in Measurement Scheme 2B, the room is assumed to have a large window on one side so that daylight penetrates through the window. The LCS for Measurement Scheme 2 is presented in Figure 7. The first step of the LCS is similar to the LCS for Measurement Scheme 1, which is detecting whether there is daylight and making the luminance almost equal for both sides. If the luminance values of area A are less than 120 cd/m^2, the brightness levels of the lamps in row A (L_A) will increase by 10%. After the luminance values of area A meet the requirement, the luminance values of area B are examined. If the luminance values are less than 60 cd/m^2, the lamps in row C are brightened by 10%, followed by the lamps in row B (L_C–L_B). The L_C lamps are brightened first because the areas of rows Y_3 to Y_5 are already bright, resulting in the L_A brightness meeting the requirement for area A. The last step is calculating the contrast between area A and area B' (area B without row Y_5). If the ratio of the average luminance values of area A to those of area B' is less than 1.5, the L_A is brightened by 10%. To calculate the contrast, row Y_5 is not considered because in the classroom, the area in front of the whiteboard is not used as a working plane, and the desks are placed at a certain distance from the whiteboard.

2.4.3. Measurement Scheme 3

The art activity classroom is presented in Measurement Scheme 3. In this scheme, the drawing and art activities are focused on the center of the room (area A), which requires a luminance value of approximately 100 cd/m^2; the required luminance value on the background area (area B) is approximately 50 cd/m^2; the brightness contrast between the center of the room, where the art pieces are placed (area A'/X3,Y3), and area B should not be less than 1.5. The whiteboard is not considered in this scheme. A closed room without windows is used in Measurement Scheme 3A, and a room with a large window on one side and daylight penetrating through the window is used in Measurement Scheme 3B. Figure 8 shows the LCS for Measurement Scheme 3. Similar to Measurement Schemes 1 and 2, the first step of the LCS is similar to the LCS for Measurement Scheme 1, the presence of daylight is detected, and the luminance value is made almost equal for both sides. Then, if the luminance values of area A are less than 100 cd/m^2, the lamps are brightened by 10%, with the lamp in the center of the room (L2B) brightened first, followed by the four lamps L2C, L2A, L1B, and L3B simultaneously in the next iteration. Once the luminance values of area A meet the requirement, the luminance values of area B are examined. If the luminance values are less than 50 cd/m^2, the remaining lamps L_{1A}, L_{1C}, L_{3A}, and L_{3C} are brightened by 10% simultaneously. After all the required luminance values are met, the contrast between area A' and area B is calculated. If the ratio of the average luminance values of area A' to those of area B is less than 1.5, the lamp L_{2B} is brightened by 10%.

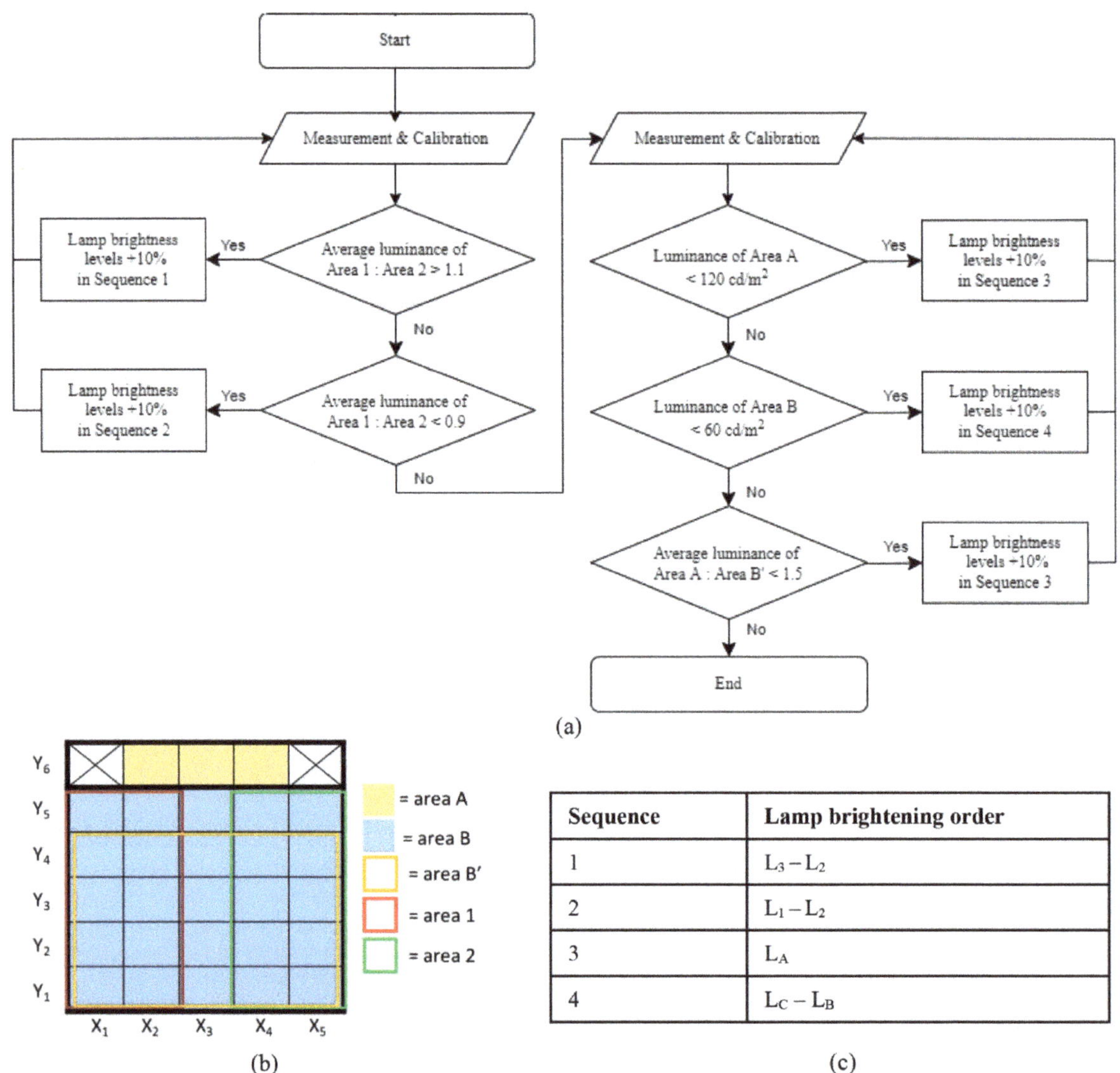

Figure 7. (**a**) The LCS flow diagram for scheme 2; (**b**) The layout of the room; (**c**) The sequence of the lamps brightened.

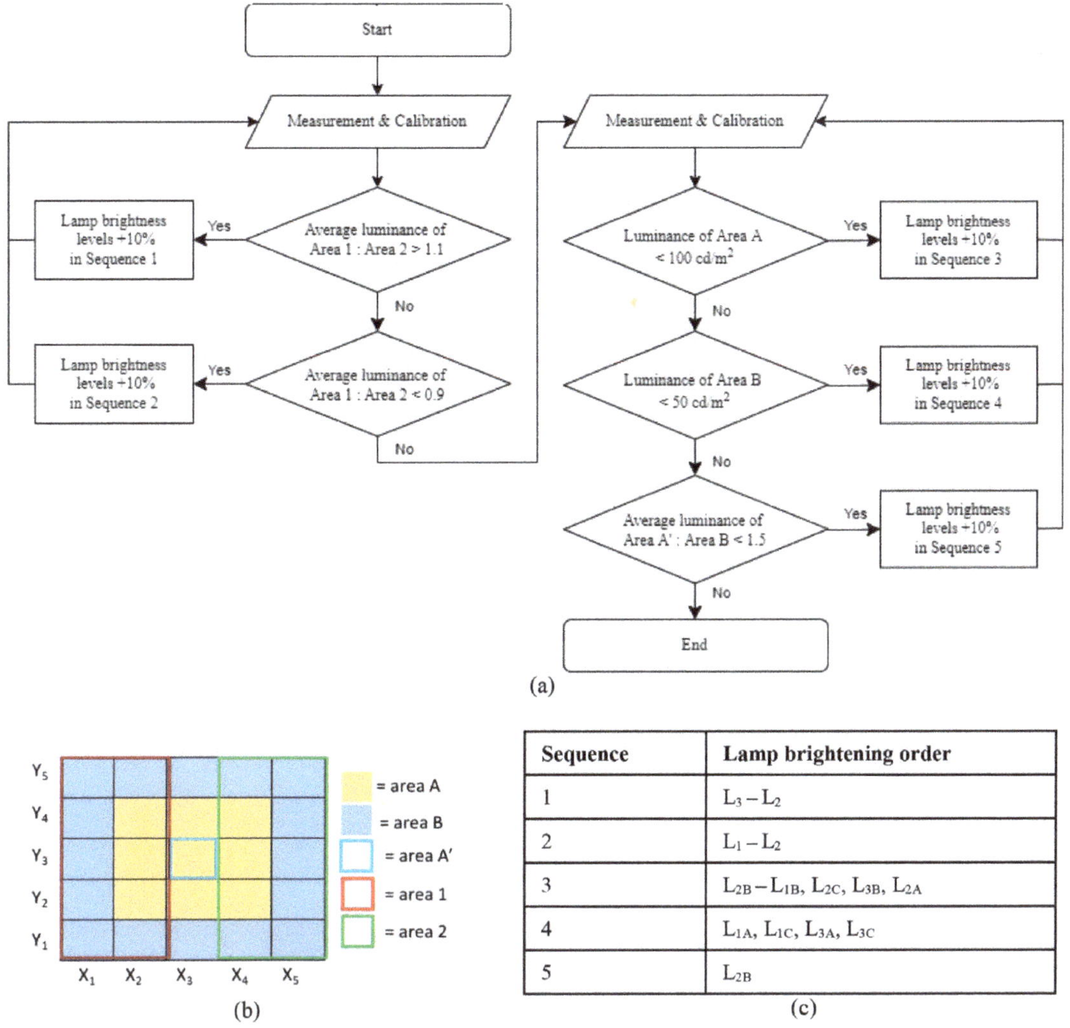

Sequence	Lamp brightening order
1	$L_3 - L_2$
2	$L_1 - L_2$
3	$L_{2B} - L_{1B}, L_{2C}, L_{3B}, L_{2A}$
4	$L_{1A}, L_{1C}, L_{3A}, L_{3C}$
5	L_{2B}

(c)

Figure 8. (**a**) The LCS flow diagram for scheme 3; (**b**) The layout of the room; (**c**) The sequence of the lamps brightened.

3. Results and Discussion

3.1. Luminance Values and Lamp Brightness Levels

In this section, the luminance values of the squares and the brightness levels of the LED lamps are presented and analyzed according to the results of the measurements.

3.1.1. Measurement Scheme 1A

In Measurement Scheme 1A, the room is assumed to be a closed self-studying room without windows. The initial condition shows that there is no difference between the luminance values of area 1 and area 2; thus, the lamps brightened by 10% in sequence 3. This sequence brightens the lamps in columns 1 and 3 (L_1 and L_3) simultaneously, followed by the lamps in column 2 (L_2) during the ensuing iterations. The required luminance values of all squares greater than 100 cd/m² are met after five iterations, with lamps in L_1 and L_3 at 30% brightness and L_2 at 20% brightness (Figure 9). After iteration 5, the brightness levels of the lamps remain steady and no longer change. This result shows that it is not necessary to brighten all lamps to the same level to provide uniform luminance values: some lamps can be brightened at a lower level depending on their positions. In this case, the lamps in L_2, in the middle column of the room, need to be brightened to only 20% because the lamps in L_1 and L_3 provide enough additional brightness to the middle area (columns X_2 to X_4).

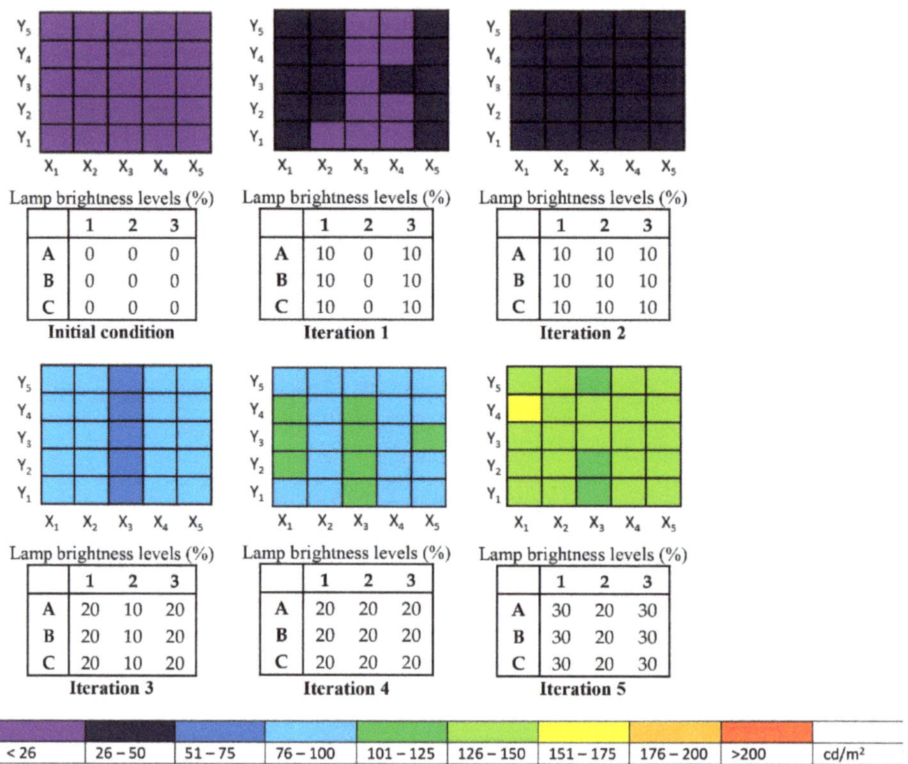

Figure 9. The luminance values and lamp brightness levels for Measurement Scheme 1A.

3.1.2. Measurement Scheme 1B

The room in Measurement Scheme 1B is assumed to be a self-studying room with a large window on one side of the wall and daylight penetrating the room. The initial conditions show that the luminance values of column X1 are higher than those of the other areas. This condition makes the luminance values of area 1 higher than those of area 2 and triggers the lamps to brighten by 10% in sequence 2: the lamps in column 3 (L3) brighten first, followed by the lamps in column 2 (L2), until the average luminance values of both areas are almost the same. To balance the luminance values of areas 1 and 2, the lamps in L_3 adjust to 10% brightness (iteration 1). After the luminance value of the measurement area is balanced, the lamps brighten in sequence 3 to satisfy the required luminance values. In this case, after five iterations, lamps in L_1 and L_2 brighten to 20%, and those in L_3 brighten to 30% to provide luminance values of more than 100 cd/m^2. Figure 10 shows the luminance values and brightness levels of the lamps at each iteration. Compared to those in Measurement Scheme 1A, the L1 lamps in Measurement Scheme 1B brighten to a level 10% lower since they are positioned near the window such that the luminance values of column X1 are higher than those of other areas in the initial condition. However, since the brightness levels of the L1 lamps are lower in Measurement Scheme 1B than in Measurement Scheme 1A, most of the luminance values of X1 to X4 are lower in Measurement Scheme 1B than in Measurement Scheme 1A, but they remain higher than the required luminance because of the daylight (the light from the LED Fresnel lamps).

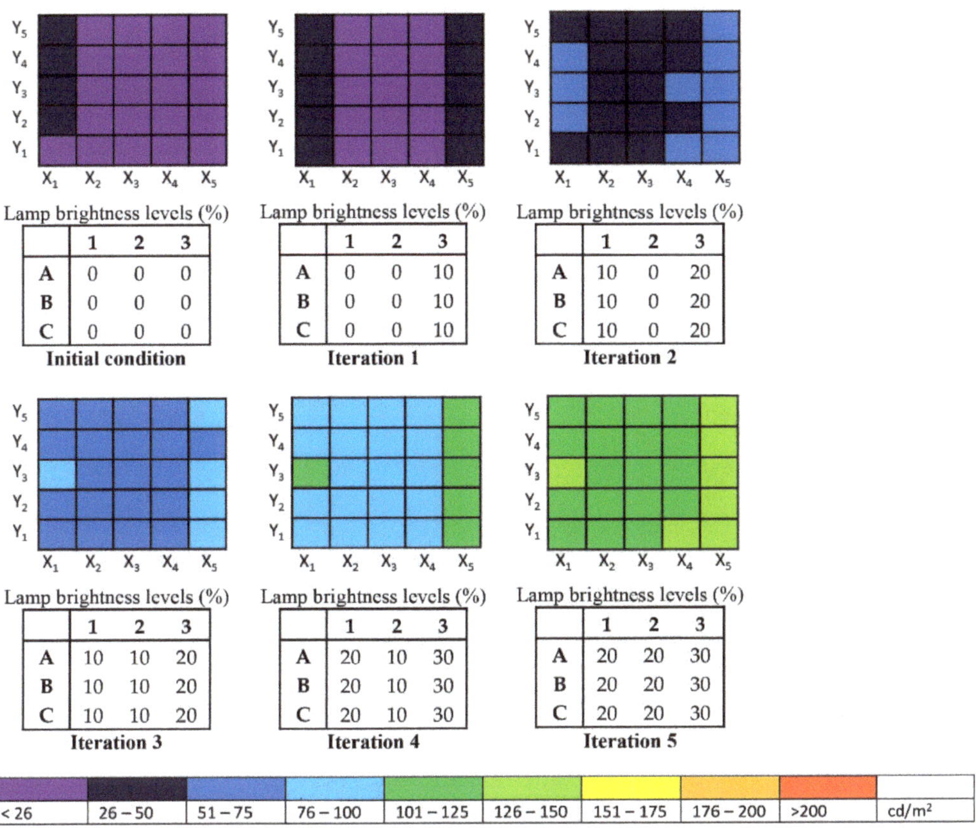

Figure 10. The luminance values and lamp brightness levels for Measurement Scheme 1B.

3.1.3. Measurement Scheme 2A

Measurement Scheme 2A considers that the room has functioned as a conventional classroom without windows, with the luminance required for area A being approximately 120 cd/m² and for area B being approximately 60 cd/m², with a brightness contrast of approximately 1.5 or more. As the initial condition shows that the luminance values of all areas are the same, the lamps brighten in sequence 3, and the lamps in row A (L_A) brighten by 10% until the luminance values of area A are 120 cd/m² or more, shown in iterations 1 to 5. That condition is achieved at iteration 5, with the brightness level of the lamps in L_A being 50%. After that condition is met, the other lamps brighten in sequence 4, which starts with the lamps in row C (L_C), followed by the lamps in row B (L_B). The required luminance levels and the brightness contrast between areas A and B' are met after eight iterations, when the lamps in row A, row B, and row C brighten at 50%, 10%, and 20%, respectively, and the brightness contrast between areas A and B' is approximately 1.7. The lamps in L_A are illuminated to a much greater extent than other lamps since they are placed near the whiteboard and need to provide luminance for the whiteboard, which is higher than the working plane/desks. This condition also provides higher luminance in the front area of the classroom (row Y_5), the area where the teacher stands (Figure 11).

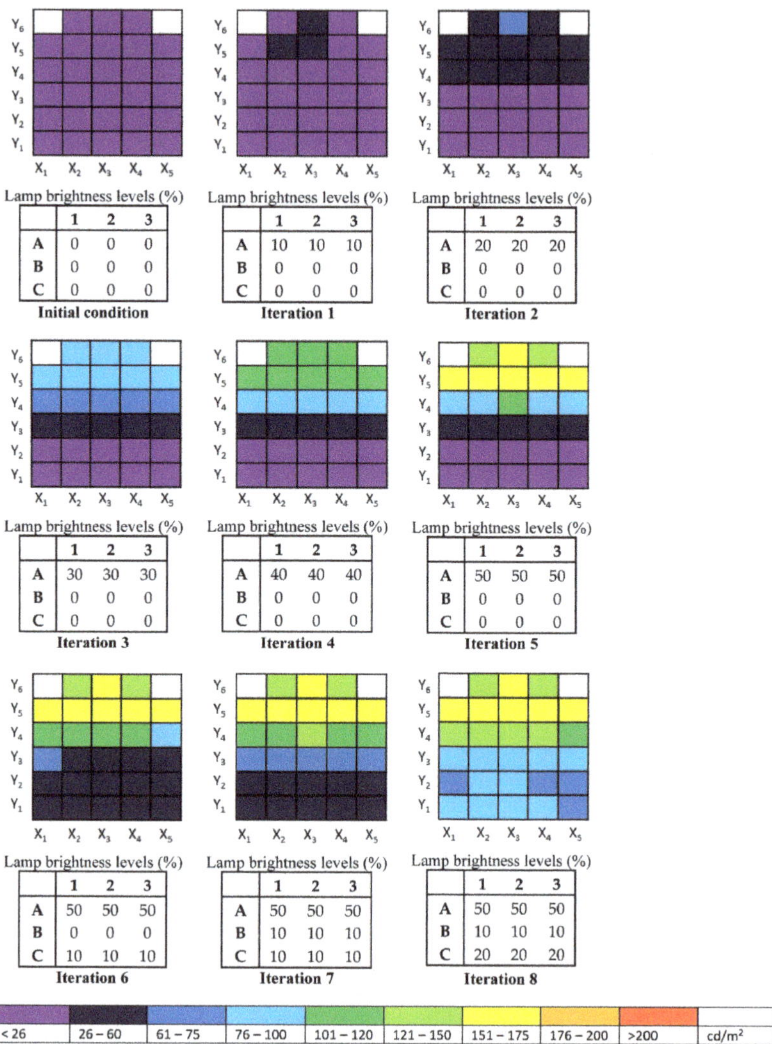

Figure 11. The luminance values and lamp brightness levels for Measurement Scheme 2A.

3.1.4. Measurement Scheme 2B

In Measurement Scheme 2B, the room is assumed to be a conventional classroom with a large window on one side of the wall, where daylight penetrates the room. Since the luminance values of column X1 are higher than those of other areas in the initial condition, the lamps brighten in sequence 2 at iteration 1. At that iteration, the lamps in L3 brighten to 10%, resulting in the luminance values of area 1 and area 2 being almost the same. This condition leads the lamps to brighten in sequence 3, in which the lamps in LA brighten by 10% until the luminance values of area A are approximately 120 cd/m^2 or more; this condition is met at iteration 6, when the lamps L1A and L2A brighten to 50% and the lamp L3A brightens to 60%. Lamp L3A brightens 10% higher than the other lamps in the same row because it already brightened 10% in iteration 1 to balance the luminance values of areas 1 and 2. After the luminance values of area A met the requirement, the lamps in L$_B$ and L$_C$ brightened in sequence 4. To meet the required luminance values and achieve the contrast between areas A and B', a lamp brightens at 60%, two lamps brighten at 50%, two lamps brighten at 20%, and the other lamps brighten at 10% after eight iterations (Figure 12). The brightness contrast between areas A and B' is approximately 1.61. Compared to the same lamps in Measurement Scheme 2A, lamps L3A and L3B brighten 10% higher, and L1C and L2C brighten 10% lower. This brightness level difference and the light from the LED Fresnel lamps are assumed to be due to daylight, resulting in the difference in luminance values of the same squares between Measurement Scheme 2A and Measurement Scheme 2B. The squares in row Y_5 mostly have higher luminance values in Measurement Scheme 2B than in Measurement Scheme 2A, while the squares in row Y4 mostly have higher luminance values in scheme 2B than in scheme 2A. Some squares in rows Y1 and Y2 have luminance values of approximately 51–76 cd/m^2, and some have luminance values of approximately 76–100 cd/m^2 in Measurement Scheme 2A. However, in Measurement Scheme 2B, all squares in the same rows have luminance values of approximately 51–76 cd/m^2.

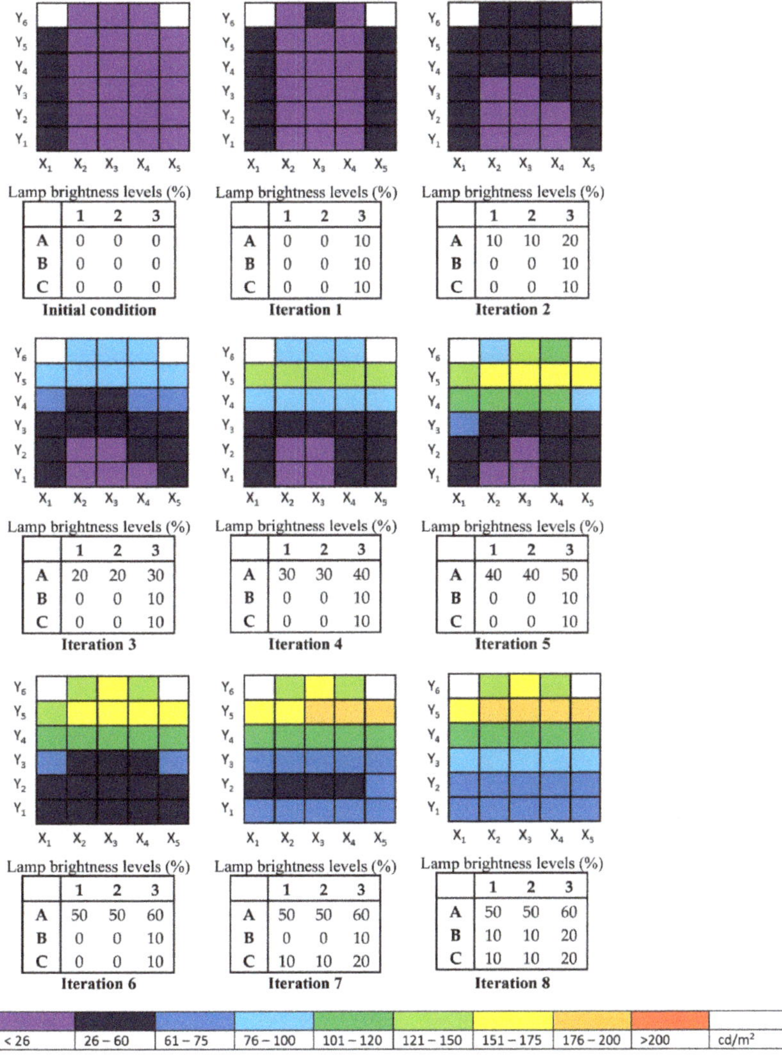

Figure 12. The luminance values and lamp brightness levels for Measurement Scheme 2B.

3.1.5. Measurement Scheme 3A

Measurement Scheme 3A presents the room as an art activity classroom without windows. The luminance required for area A and area B is approximately 100 cd/m^2 and 50 cd/m^2, respectively, and the brightness contrast should not be less than 1.5. As the initial condition shows that the luminance values of areas 1 and 2 are almost the same, the lamps brighten in sequence 3: L_{2B} in the center of the room, followed by lamps L_{1B}, L_{2C}, L_{3B}, and L_{2A} in the next iteration (by 10%) until the luminance values of area A are 100 cd/m^2 or more. In this scheme, the required luminance values are achieved after seven iterations, when lamp L_{2B}, in the center of the room, brightens to 40% and the four lamps L_{1B}, L_{2C}, L_{3B}, and L_{2A} brighten to 30%. At that iteration, the luminance values of area B are more than 50 cd/m^2, which is more than the required luminance, and the contrast between areas A' and B is approximately 1.6. Since the luminance values and the brightness contrast requirements are met, the lamps in sequence 4 do not need to brighten, and the rest of the lamps remain turned off. The process of increasing the brightness levels of the lamps and the luminance values at each iteration are shown in Figure 13. The luminance values and lamp brightness levels indicate that, for the room with activities focused on the center, the rear lamps do not need to be turned on because that area needs to remain darker than the center area of the room to create the desired brightness contrast.

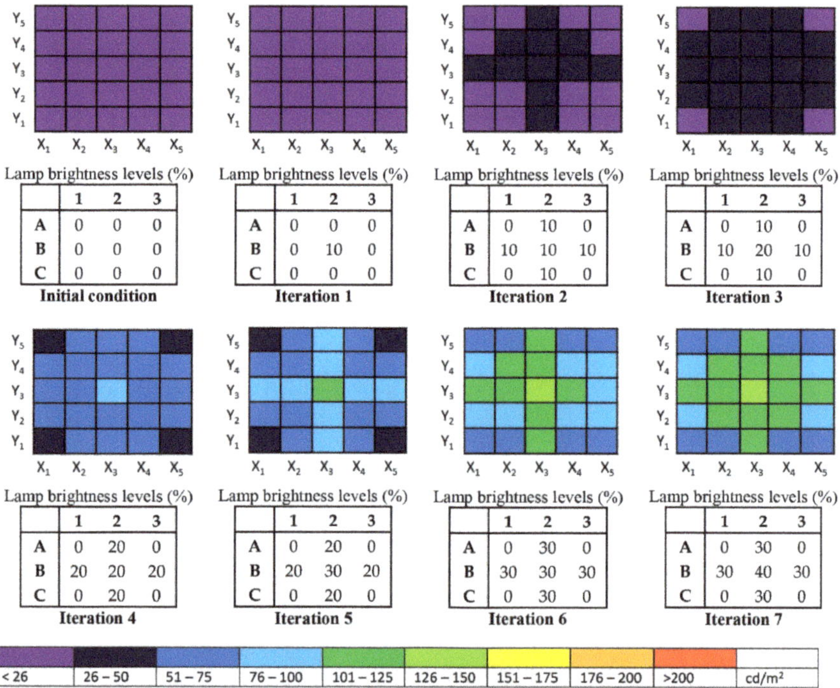

Figure 13. The luminance values and lamp brightness levels for Measurement Scheme 3A.

3.1.6. Measurement Scheme 3B

Measurement Scheme 3B considers the room as an art activity classroom with a large window on one side of the wall and daylight penetrating the room. As the initial condition shows that the average luminance values of area 1 are higher than those of area 2, the lamps brighten in sequence 2 to make the average luminance values of both areas almost the same, which happens at iteration 1, where the lamps in L3 brighten by 10%. After the average luminance values of areas 1 and 2 are almost the same, the lamps brighten in sequence 3 to ensure that the luminance values of area A are approximately 100 cd/m^2 or higher. In this measurement scheme, the required luminance values are achieved at iteration 7, in which lamp L3B is brightened to 40%, the four lamps L1B, L2B, L2A, and L2C are brightened to 30%, the two lamps L3A and L3C are brightened to 10%, and the two other lamps are off. However, the brightness contrast between areas A′ and B is only 1.4, which is still lower than the requirement. Therefore, at iteration 8, the brightness level of lamp L2B increases to 40%, leading two lamps to brighten to 40%, three lamps to brighten to 30%, and two lamps to brighten to 20%. At this iteration, the brightness contrast between areas A′ and B is approximately 1.6; thus, all requirements are met. This condition shows that in the presence of daylight, even the luminance values meet the requirement, but the lamp brightness levels still increase because the brightness contrast needs to be maintained at a given value. In this scheme, lamps L3A and L3B brighten 10% more than the same lamps in Measurement Scheme 3A because of the presence of daylight. Figure 14 shows the process of brightening the lamps and the luminance values at each iteration.

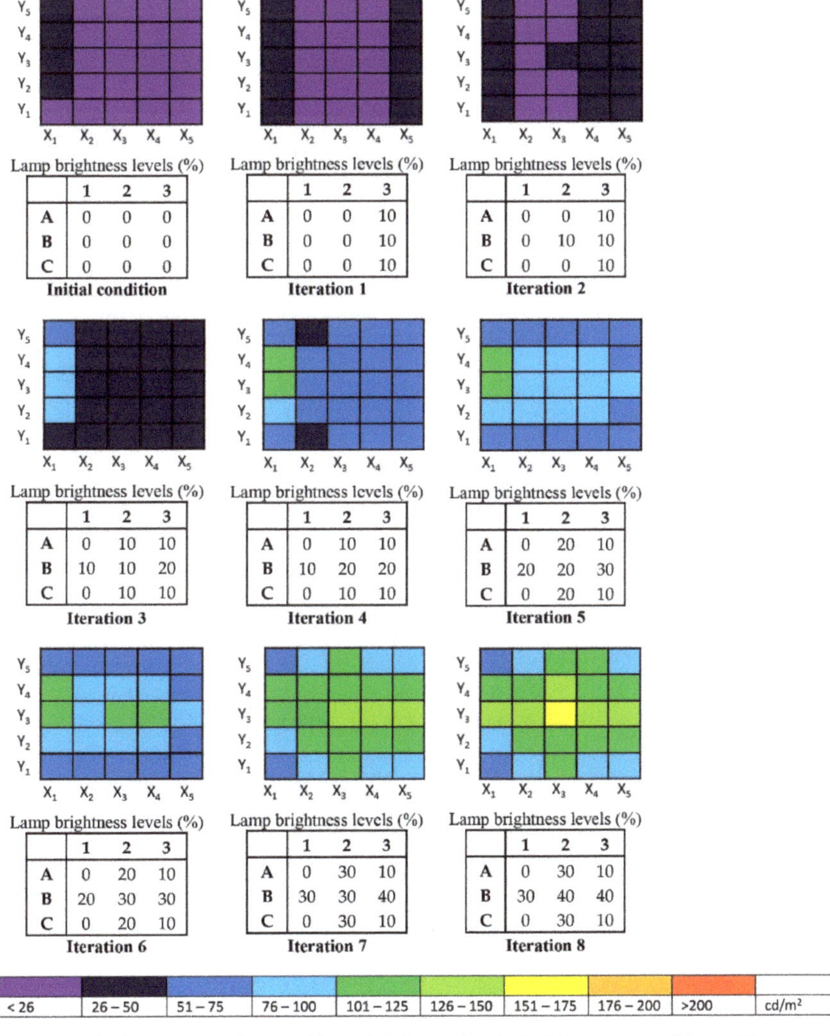

Figure 14. The luminance values and lamp brightness levels for Measurement Scheme 3B.

Table 4 summarizes the brightness levels of the lamps in each measurement scheme. The LCS resulted in different combinations of lamp brightness levels for different activities. In Measurement Schemes 1A and 1B, the presence of daylight resulted in a lower brightness level in Measurement Scheme 1B; in Measurement Schemes 2A and 2B, the daylight resulted in the brightness levels of some lamps being lower and those of other lamps being higher than the corresponding levels in the room without daylight, and in Measurement Schemes 3A and 3B, the presence of daylight increased rather than decreased the brightness levels of some lamps. These results show that daylight presence does not always have a great reducing impact on lamp brightness levels. Despite the presence of daylight in Measurement Scheme 3B, the brightness levels of some lamps are higher in Measurement Scheme 3B than in Measurement Scheme 3A because the activity is focused on the center of the room and the brightness contrast between the center of the room and the background area needs to be maintained at the given value.

Table 4. The brightness levels of each lamp in each measurement scheme.

Measurement Scheme	Number of Lamps Brightened at						
	60%	50%	40%	30%	20%	10%	off
1A	-	-	-	6	3	-	-
1B	-	-	-	3	6	-	-
2A	-	3	-	-	3	3	-
2B	1	2	-	-	2	4	-
3A	-	-	1	4	-	-	4
3B	-	-	2	3	-	2	2

The prototype successfully demonstrates the utilization of an IP cam as a luminance meter for simultaneously measuring several spots in a whole room and reveals that an IP cam can replace several traditional spot lighting sensors. In all measurement schemes, the HDRis produced by the IP cam prove successful in measuring the luminance values under various settings and functions of the room without changing the sensor or requiring additional sensors. In summary, the HDRi method supports IP cam utilization as a set of "field of view" sensors for a flexible space.

3.2. Energy Consumption

In this section, the energy consumption of the lamps in each scheme, wherein the equilibrium condition is met and the lamp brightness levels no longer change, is monitored by using an energy meter, and compared to the baseline condition of nine lamps, at 100% brightness. The energy consumption of the baseline, with nine LED lamps at 100% brightness, is approximately 558 W. Table 5 summarizes the energy consumption of each measurement scheme and the energy savings compared to the baseline. For Measurement Scheme 1, Measurement Scheme 1A consumes 14.58% more energy than Measurement Scheme 1B; however, for Measurement Scheme 2, Measurement Scheme 2A and Measurement Scheme 2B consume almost the same amount of energy, and for Measurement Scheme 3, Measurement Scheme 3B consumes 19% more energy than Measurement Scheme 3A. Compared to the measurement schemes without daylight, Measurement Schemes 2 and 3 with daylight do not save energy. This finding indicates that daylight does not always result in more energy savings because, for certain activities, the lamps still need to be turned on to maintain the brightness contrast and provide visual comfort. The applied LCS can save approximately 73.39%, 76.77%, 73.71%, 73.49%, 82.38%, and 79.03% of the energy consumed under baseline for Measurement Schemes 1A, 1B, 2A, 2B, 3A, and 3B, respectively, since most of the lamps are only brightened to between 10% and 60%.

Table 5. The comparison of the energy consumption.

Measurement Scheme	Energy Consumption (W)	Energy Saving (%)
1A	148.5	73.39
1B	129.6	76.77
2A	146.7	73.71
2B	147.9	73.49
3A	98.3	82.38
3B	117	79.03

4. Conclusions

This study presents an LCS prototype based on HDRi using a LabVIEW platform that consists of a 360° IP camera, a DMX controller, and several LED lamps to provide visual comfort and energy savings. The brightness levels of the LED lamps are changed in sequence to meet the luminance value requirement. The prototype is applied to the room assumed as a classroom with three different measurement schemes: the first is a self-studying room that requires a uniform luminance value of approximately 100 cd/m^2, the second is a conventional classroom with a whiteboard that requires luminance values of approximately 120 cd/m^2 for the whiteboard and 60 cd/m^2 for the working plane/desks, with a brightness contrast between the whiteboard and the working plane/desks of approximately 1.5, and the latter is an art activity classroom that requires different luminance values for the center area and background area, which are 100 cd/m^2 and 50 cd/m^2, respectively, with a brightness contrast between the center area and the background area of approximately 1.5. The findings of this study are summarized as follows:

1. The LCS based on HDRi that brightens the lamps in sequence can be a solution to be applied in a room with various functions and activities, and in the presence of daylight to provide visual comfort. This prototype proves that the LCS works well for a room that requires a uniform luminance value, and for a room that requires different luminance values and brightness contrast, with or without daylight.

2. The results of this study indicate that lamp brightness levels vary depending on the activities that occur in the room. The lamp brightness levels of the different schemes are as follows:

    a. Measurement Scheme 1A: 6 lamps at 30% brightness and 3 lamps at 20% brightness
    b. Measurement Scheme 1B: 3 lamps at 30% brightness and 6 lamps at 30% brightness
    c. Measurement Scheme 2A: 3 lamps at 50% brightness, 3 lamps at 20% brightness, and 3 lamps at 10% brightness
    d. Measurement Scheme 2B: 1 lamp at 60% brightness, 2 lamps at 50% brightness, 2 lamps at 20% brightness, and 4 lamps at 10% brightness
    e. Measurement Scheme 3A: 1 lamp at 40% brightness, 4 lamps at 30% brightness, and 4 lamps turned off
    f. Measurement Scheme 3B: 2 lamps at 40% brightness, 3 lamps at 30% brightness, 2 lamps at 10% brightness, and 2 lamps turned off

3. Compared to the electricity consumption of the baseline, with nine LED lamps brightened at 100%, which is approximately 558 W, the LCS prototype presented can achieve energy savings for scheme 1A, scheme 1B, scheme 2A, scheme 2B, scheme 3A, and scheme 3B of approximately 73.39%, 76.77%, 73.71%, 74.49%, 82.38%, and 79.03%, respectively.

4. The presence of daylight does not always result in more energy savings since the brightness contrast should be considered to achieve visual comfort.

5. This study demonstrates the advantages in supporting the flexible use of space of "field of view" sensors over traditional spot lighting sensors.

Buildings **2022**, 12, 650

Author Contributions: Conceptualization, A.B.; methodology, Y.-S.C.; software, A.B.; validation, A.B.; formal analysis, A.B. and Y.-S.C.; investigation, A.B.; resources, Y.-S.C.; data curation, A.B. and Y.-S.C.; writing—original draft preparation, A.B.; writing—review and editing, A.B.; visualization, A.B.; supervision, Y.-S.C.; project administration, A.B. and Y.-S.C.; funding acquisition, Y.-S.C. All authors have read and agreed to the published version of the manuscript.

Funding: This research was funded by the Taiwan Ministry of Science (Project No.: MOST 109-2221-E-011-033-) and by the Taiwan Ministry of Education.

Acknowledgments: The authors would like to thank Ben-Cheng Guo for assisting the HDRi processing using a 360° IP camera, Ronaldo Gabe Manik for assisting Arduino Uno utilization as DMX controller, and the ISAAC (Integrating Simulation and Analysis for Architectural Complexity) Lab at the National Taiwan University of Science and Technology (NTUST) of Taiwan (R.O.C.). for supporting this research.

Conflicts of Interest: The authors declare no conflict of interest.

Appendix A

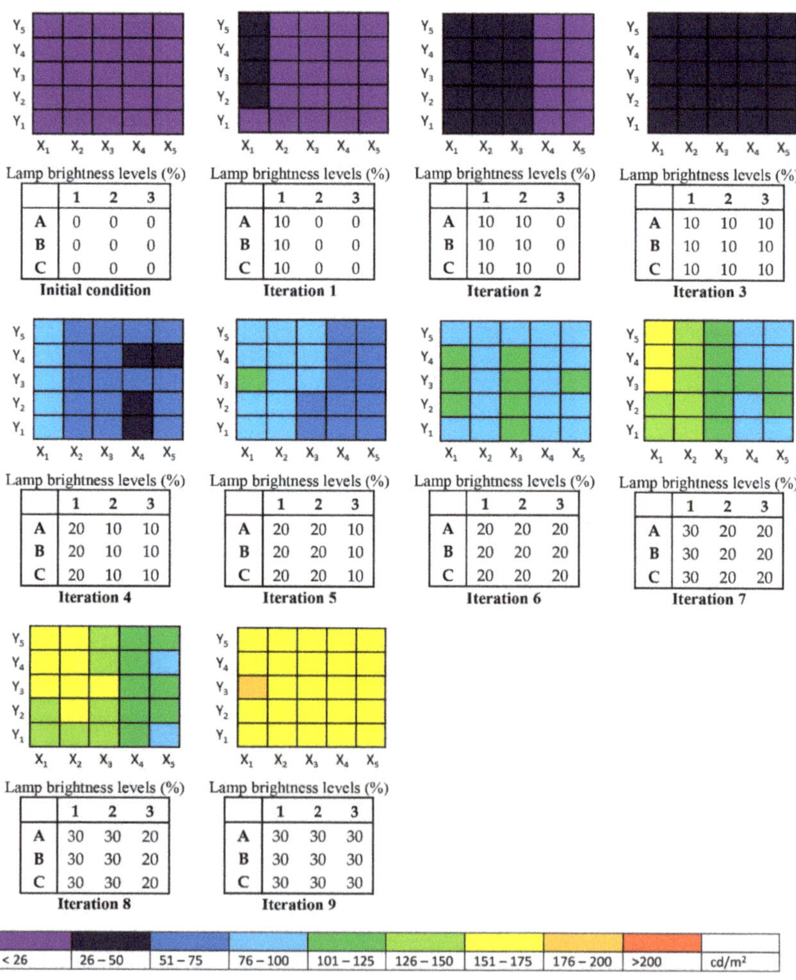

Figure A1. The luminance values and lamp brightness levels of Measurement Scheme 1A in each iteration. The lamps brighten in the sequence of L_1, L_2, and L_3.

References

1. Kruisselbrink, T.; Dangol, R.; van Loenen, E. A comparative study between two algorithms for luminance-based lighting control. *Energy Build.* **2020**, *228*, 110429. [CrossRef]
2. Kumar, A.; Kar, P.; Warrier, R.; Kajale, A.; Panda, S. Implementation of Smart LED Lighting and Efficient Data Management System for Buildings. *Energy Procedia* **2017**, *143*, 173–178. [CrossRef]
3. Haq, M.A.U.; Hassan, M.Y.; Abdullah, H.; Rahman, H.A.; Abdullah, P.; Hussin, F.; Said, D.M.; Haq, M.A.U.; Hassan, M.Y.; Abdullah, H.; et al. A review on lighting control technologies in commercial buildings, their performance and affecting factors. *Renew. Sustain. Energy Rev.* **2014**, *33*, 268–279. [CrossRef]
4. Mohamed, S.; Minhat, M.; Kasim, M.; Adam, M.; Sulaiman, M.; Rizman, Z. An Intelligent Lighting Control System (ILCS) using LabVIEW. *J. Fundam. Appl. Sci.* **2018**, *9*, 602. [CrossRef]
5. Doulos, L.T.; Kontadakis, A.; Madias, E.N.; Sinou, M.; Tsangrassoulis, A. Minimizing energy consumption for artificial lighting in a typical classroom of a Hellenic public school aiming for near Zero Energy Building using LED DC luminaires and daylight harvesting systems. *Energy Build.* **2019**, *194*, 201–217. [CrossRef]
6. Cheng, Y.; Fang, C.; Yuan, J.; Zhu, L. Design and Application of a Smart Lighting System Based on Distributed Wireless Sensor Networks. *Appl. Sci.* **2020**, *10*, 8545. [CrossRef]
7. Frascarolo, M.; Martorelli, S.; Vitale, V. An innovative lighting system for residential application that optimizes visual comfort and conserves energy for different user needs. *Energy Build.* **2014**, *83*, 217–224. [CrossRef]
8. Sun, B.; Zhang, Q.; Cao, S. Development and Implementation of a Self-Optimizable Smart Lighting System Based on Learning Context in Classroom. *Int. J. Environ. Res. Public Health* **2020**, *17*, 1217. [CrossRef]
9. Karyono, K.; Abdullah, B.; Cotgrave, A.; Bras, A. A Novel Adaptive Lighting System Which Considers Behavioral Adaptation Aspects for Visually Impaired People. *Buildings* **2020**, *10*, 168. [CrossRef]
10. Carrillo, C.; Diaz-Dorado, E.; Cidrás, J.; Bouza-Pregal, A.; Falcón, P.; Fernández, A.; Álvarez-Sánchez, A. Lighting control system based on digital camera for energy saving in shop windows. *Energy Build.* **2013**, *59*, 143–151. [CrossRef]
11. Cheng, Z.; Zhao, Q.; Wang, F.; Jiang, Y.; Xia, L.; Ding, J. Satisfaction based Q-learning for integrated lighting and blind control. *Energy Build.* **2016**, *127*, 43–55. [CrossRef]
12. Shen, E.; Hu, J.; Patel, M. Energy and visual comfort analysis of lighting and daylight control strategies. *Build. Environ.* **2014**, *78*, 155–170. [CrossRef]
13. Chiou, Y.-S.; Saputro, S.; Sari, D.P. Visual Comfort in Modern University Classrooms. *Sustainability* **2020**, *12*, 3930. [CrossRef]
14. Suk, J.Y. Luminance and vertical eye illuminance thresholds for occupants' visual comfort in daylit office environments. *Build. Environ.* **2018**, *148*, 107–115. [CrossRef]
15. Chraibi, S.; Creemers, P.; Rosenkötter, C.; Van Loenen, E.; Aries, M.; Rosemann, A. Dimming strategies for open office lighting: User experience and acceptance. *Light. Res. Technol.* **2018**, *51*, 513–529. [CrossRef]
16. Yu, T.-H.; Kwon, S.-Y.; Im, K.-M.; Lim, J.-H. An RTP-based dimming control system for visual comfort enhancement and energy optimization. *Energy Build.* **2017**, *144*, 433–444. [CrossRef]
17. Benefits of Programming Graphically in NI LabVIEW Table of Contents A Brief History of the Pursuit of Higher-Level Pro-Gramming. Available online: http://www.technologyreview.com/sites/default/files/legacy/benefits_of_programming_graphically_with_lv.pdf (accessed on 10 January 2022).
18. Tasner, T.; Lovrec, D.; Tasner, F.; Edler, J. Comparison of LabVIEW and MATLAB for scientific research. *Ann. Fac. Eng. Hunedoara* **2012**, *3*, 389–394.
19. Bogdan, M. Light Intensity Control Using Arduino Uno and Lab VIEW. In Proceedings of the 13th International Conference on Virtual Learning (ICVL), Alba Iulia, Romania, 26–28 October 2018; pp. 306–310.
20. Schwartz, M.; Manickum, O. *Programming Arduino with LabVIEW*.; Packt Publishing: Birmingham, UK, 2015.
21. Reinhard, E.; Wolfgang, H.; Debevec, P.; Pattanaik, S.; Ward, G.; Myszkowski, K. *High Dynamic Range Imaging: Acquisition, Display and Image-Based Lighting*, 2nd ed.; Morgan Kaufmann: Burlington, MA, USA, 2010.
22. Debevec, P.E.; Malik, J. Recovering high dynamic range radiance maps from photographs. In Proceedings of the SIGGRAPH '97: Proceedings of the 24th Annual Conference on Computer Graphics and Interactive Techniques, Los Angeles, CA, USA, 3–8 August 1997; pp. 369–378. [CrossRef]
23. Radiance: A Simulation Tool for Daylighting Systems. Available online: http://radsite.lbl.gov/radiance/refer/rc97tut.pdf (accessed on 10 January 2022).
24. Pierson, C.; Cauwerts, C.; Bodart, M.; Wienold, J. Tutorial: Luminance Maps for Daylighting Studies from High Dynamic Range Photography. *LEUKOS* **2020**, *17*, 140–169. [CrossRef]
25. Inanici, M. Evaluation of high dynamic range photography as a luminance data acquisition system. *Light. Res. Technol.* **2006**, *38*, 123–134. [CrossRef]
26. Kruisselbrink, T.; Aries, M.; Rosemann, A. A Practical Device for Measuring the Luminance Distribution. *Int. J. Sustain. Light.* **2017**, *19*, 75–90. [CrossRef]
27. Safranek, S.; Davis, R.G. Sources of Error in HDRI for Luminance Measurement: A Review of the Literature. *LEUKOS-J. Illum. Eng. Soc. N. Am.* **2020**, *17*, 187–208. [CrossRef]
28. *BS EN 12464-1:2021*; BSI Standards Publication Light and Lighting—Lighting of Work Places Part 1: Indoor Work Places. BSI Standards Limited 2021: London, UK, 2021.

29. Chinomi, N.; Leelajindakrairerk, M.; Boontaklang, S.; Chompoo-Inwai, C. Design and Implementation of a smart monitoring system of a modern renewable energy micro-grid system using a low-cost data acquisition system and LabVIEWTM program. *J. Int. Counc. Electr. Eng.* **2017**, *7*, 142–152. [CrossRef]
30. Essick, J. *Hands-On Introduction to LabVIEW for Scientists and Engineers*; Oxford University Press: New York, NY, USA, 2009.
31. Hamed, B. Design & Implementation of Smart House Control Using LabVIEW. *Int. J. Soft Comput. Eng.* **2012**, *1*, 98–106.
32. Angalaeswari, S.; Deepa, T.; Subbulekshmi, D.; Krithiga, S.; Reddy, M.N.; Siddartha, K.; Chaitanya, C. Smart House Control using LabVIEW. *J. Phys. Conf. Ser.* **2021**, *1716*, 012004. [CrossRef]
33. Taleb, M.; Mannsour, N. A self-controlled energy efficient office lighting system. *J. Assoc. Arab Univ. Basic Appl. Sci.* **2012**, *11*, 9–15. [CrossRef]
34. Chiou, Y.; Lin, Y. A Portable Testbed for Integrative Daylighting Design. In Proceedings of the PLEA 2016 Proceedings, Cities, Buildings People: Toward Regenerative Environments, Los Angeles, CA, USA, 11–13 July 2016.

 *buildings*

Article

Thermal Environment and Thermal Comfort in University Classrooms during the Heating Season

Jiuhong Zhang [1,*], Peiyue Li [1] and Mingxiao Ma [2]

[1] Jangho Architecture College, Northeastern University, Shenyang 110819, China; 2101400@stu.neu.edu.cn

[2] School of Automation Science and Electrical Engineering, Beihang University, Beijing 100191, China; mamingxiao@buaa.edu.cn

* Correspondence: zhangjiuhong@mail.neu.edu.cn

Abstract: In recent years, there has been increasing concern about the effects of indoor thermal environments on human physical and mental health. This paper aimed to study the current status of the thermal environment and thermal comfort in the classrooms of Northeastern University during the heating season. The indoor thermal environment was analyzed with the use of field measurements, a subjective questionnaire, regression statistics, and the entropy weight method. The results show that personnel population density is an important factor affecting the temperature and relative humidity variations in classrooms. The results also show that the temperature and relative humidity in a lecture state are respectively 4.2 °C and 11.4% higher than those in an idle state. In addition, in university classrooms in Shenyang, the actual thermal neutral temperature is 2.5 °C lower than the predicted value of the Predicted Mean Vote. It was found that increasing indoor relative humidity can effectively improve the overall thermal comfort of subjects. Furthermore, the temperature preference of women was higher than that of men. Therefore, when setting the initial heating temperature, the personnel population density and sufficient indoor relative humidity have been identified as the key factors for improving the thermal environment of the classroom.

Keywords: thermal environment; thermal comfort; cold regions; university classroom; heating season

Citation: Zhang, J.; Li, P.; Ma, M. Thermal Environment and Thermal Comfort in University Classrooms during the Heating Season. *Buildings* **2022**, *12*, 912. https://doi.org/10.3390/buildings12070912

Academic Editors: Yue Wu, Zheming Liu and Zhe Kong

Received: 10 May 2022
Accepted: 27 June 2022
Published: 28 June 2022

Publisher's Note: MDPI stays neutral with regard to jurisdictional claims in published maps and institutional affiliations.

1. Introduction

1.1. Overview

As an important part of university space, the classroom is the main place for students' daily study, and its indoor thermal environment directly affects students' comfort, physical and mental health, and learning efficiency.

The traditional way of creating a comfortable environment is by controlling the indoor environmental parameters within a specified range according to the relevant standards. However, the people's comforts are so different that these parameters cannot effectively meet individual states, feelings, and preferences. Additionally, individual comfort claims are not conveyed and feedback is not given during the creation of the thermal environment [1]. Therefore, it is necessary to propose reasonable and effective solutions in a targeted manner [2]. The present study provides a reasonable basis for the further improvement of indoor thermal comfort by summarizing and analyzing the results concerning thermal satisfaction through a subjective survey of student subjects.

1.2. Literature Review

The Predicted Mean Vote (PMV) metric proposed by Fanger has been commonly used in the field of thermal comfort to predict human thermal sensation in a steady-state environment [3]. However, deviations between the PMV and the Actual Mean Vote (AMV) of thermal sensation have been found in actual extensive field survey results [4]. To address these deviations, researchers have further revised and developed thermal adaptation models. For example, the ASHRAE Standard 55-2017 [5] contains the thermal adaptation model established by de Dear [6]. In 2012, China released the Standard GB/T 50785-2012 [7], which introduced a thermal adaptation model for naturally ventilated buildings based on Chinese field survey data. Compared with the steady-state model, the adaptive thermal comfort model can more accurately reflect the thermal sensation and thermal comfort of the human body in an actual building environment. Furthermore, the adaptive thermal comfort model can improve the internal thermal environment of buildings in a targeted way in order to reduce building energy consumption.

To date, the thermal comfort environment of college classrooms in different regions has been extensively studied [8–10]. Note that most of these studies have focused on summer or hot-summer and cold-winter regions, while the thermal comfort in college classrooms during the heating season in severe cold regions has been less studied [11–17]. Jung et al. [18] showed that students prefer a slightly cool indoor environment. Cao et al. [19] showed that in the heating season, the PMV of students in the classroom is lower than the actual thermal feeling. However, it should be noted that there are limited studies on the thermal comfort environment of classrooms as compared with other environments [10]. Kuru et al. [4] found that the thermal comfort of the indoor environment has a great impact on the health and well-being of users. Cognati et al. found that during the heating season in Italy, students prefer a warm environment in university classrooms [20]. Some scholars have also studied the influencing factor of gender. They found that females prefer to feel warmer and accept higher temperatures than males [2]. Wang [21] found that since females wear heavier clothes indoor than males, the neutral temperature of females remains higher than that of males. The results of some studies are listed in Table 1 [14,22–28]. Note that China has a vast territory, and there are great differences in the average temperature of the coldest month in winter within the same climate area. According to the thermal zoning in China, Shenyang belongs to severe cold zone C. Moreover, the standard temperature set for heating in Shenyang is different. Therefore, the present paper studies the indoor thermal environment of Shenyang during the heating season. To date, there is still a lack of indoor thermal environment standards for educational buildings with high indoor occupant density, which makes it difficult to meet the thermal environment requirements of the classroom by only relying on the existing norms. Hence, the objective of the present study is to find out how to improve the thermal comfort in the classroom so as to make students have a more comfortable learning environment.

Table 1. Research on human thermal comfort and adaptation in different regions.

Survey Time	Researcher	Survey Location	Architectural Environment	Neutral Temperature/°C
1984	de Dear [22]	Australia	Office buildings	24.2
2003–2005	Goto [23]	Japan	Office buildings	26.0
2010–2011	ZJ Wang [14]	China	Educational buildings	Spring: 21.7 Winter: 22.6
2011	Kim [24]	South Korea	Office buildings	23.5
2016	Ning [25]	China	Educational buildings	19.7–23.2
2018	Fang [26]	China	Educational buildings	24
2019	Liu [27]	China	Educational buildings	20.6
2020	Carolina [28]	Brazil	Educational buildings	23–24

2. Methods

2.1. Location and Climate

Shenyang is located in the northeast of China, which has a large temperature difference throughout the year and where winter is cold and dry. According to the Chinese standard GB 50178-1993 [29], Shenyang is a typical city in a severe cold climate zone. Figure 1 shows the temperature map of Shenyang in winter. It is also worth mentioning that the heating time is from November of the current year to March of the following year.

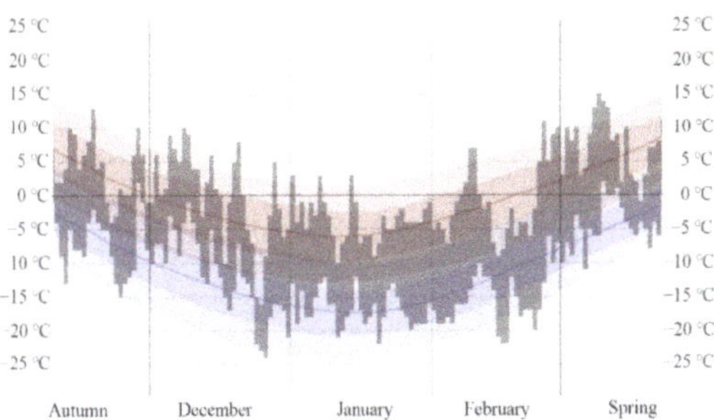

Figure 1. Winter temperature map of Shenyang for 2021 (data source: https://zh.weatherspark.com/ accessed on 1 February 2022).

The present study was conducted in the middle of the heating season, in November and December 2021, when the indoor thermal environment parameters were more stable and less affected by the outdoor environment. Note that teaching activities were from 8:30–20:30. Thus, the test period selected was in the middle, from 14:00 to 15:00, because the morning teaching activities could make the thermal environment in the classroom reach a stable state. Therefore, the afternoon measurement was more representative. The selected building was built in 2014. Additionally, the selected classroom was facing north in order to avoid the effects of direct sunlight. The classroom had an octagonal shape and 150 seats, with a floor height of 3.8 m and an area of 128 m^2. As shown in Figure 2, the classroom was equipped with heating radiators but without ventilation systems. Table 2 summarizes the thermal properties of the building envelopes. The thermal properties of the building envelopes met the requirements of the Design Standard for Energy Efficiency in public buildings [30].

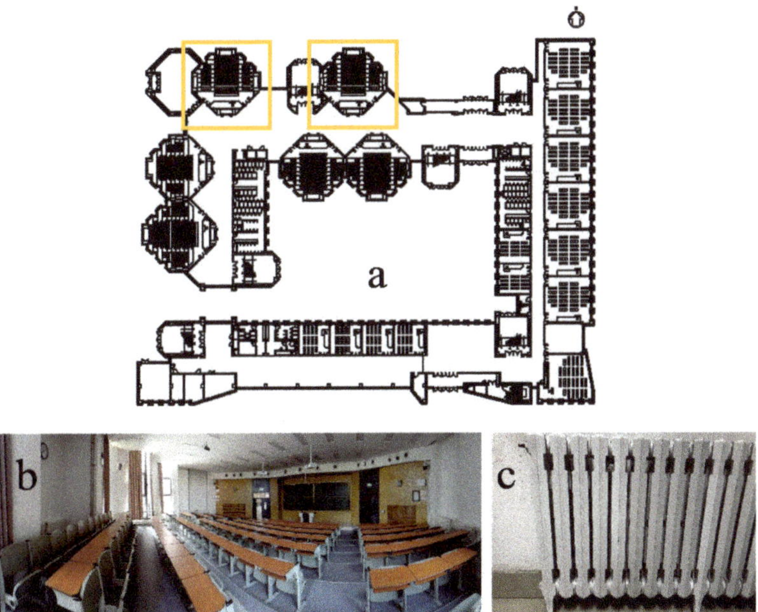

Figure 2. The studied classroom. (**a**) Floor plan of the location of the studied classroom, (**b**) interior view of the classroom, (**c**) heating radiator.

Table 2. Thermal properties of building envelopes.

Envelope	Material	Heat Transfer Coefficient (W/m^2·K)
Wall	Figure 3	0.29
Window	Hollow glass (air thickness 12 mm)	2.41

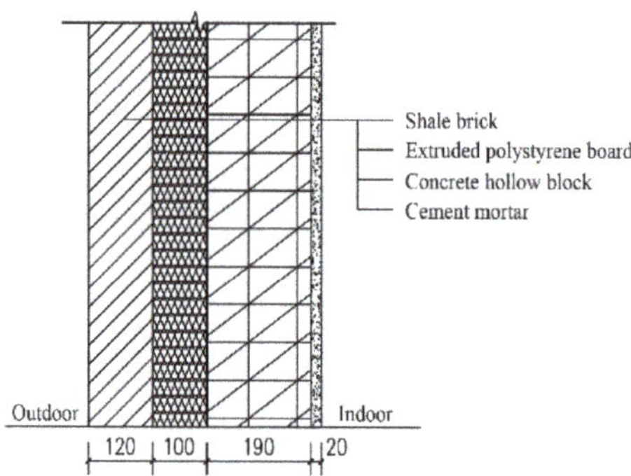

Figure 3. Wall structure.

2.2. Method

It should be noted that in the present study, both the objective evaluation and the subjective evaluation of the environment as well as the adaptability of people to their surroundings are reflected.

2.2.1. Testing Methods of Environmental Parameters

The testing parameters of this study included indoor air temperature, relative humidity, air velocity, and black-bulb temperature. The field test process was carried out using time, area, and situation to obtain reliable and relevant data.

Note that the selected classroom was analyzed in November and December 2021, each with a different number of users but with the same indoor heating temperature and measurement conditions that take into account both idle and full conditions. According to the standard "Ergonomics of Thermal Environment Physical Measurements" (GB/T 40233-2021) [31], the indoor test instruments were placed in three different areas in the front, middle, and back. In addition, four measurement points were arranged horizontally in each area. The final results were averaged and located, as shown Figure 4.

According to the Indoor Air Quality Standard (GB/T 18883-2002) [32], the height of the sampling point should be 1.1 m, which, in principle, is the same as the height of the human respiratory belt. The indoor thermal environment parameters were measured at the same time as the completion of the subjective questionnaires by the subjects in order to ensure that the measured indoor environmental parameters truly reflected the thermal environment conditions of the area at the time. The scheduled lecture time was 60 min with a 10 min break. Additionally, the environmental parameters were recorded by the investigators at 20 min intervals during the lectures. The 20 min interval was selected because the pre-experiment showed that the environmental parameters change significantly within 20 min during the classroom monitoring. Furthermore, the moment after the end of class was taken as an important time node. Thus, a data recording point at 14:50 was added. It is also worth mentioning that the test apparatus in the classroom was set 15 min before the class to ensure the stability of the measured data during the class period. Figure 5 shows pictures of the monitoring equipment used for measuring the indoor environmental parameters. In addition, Table 3 summarizes the specifications of the sensor probes used in this study.

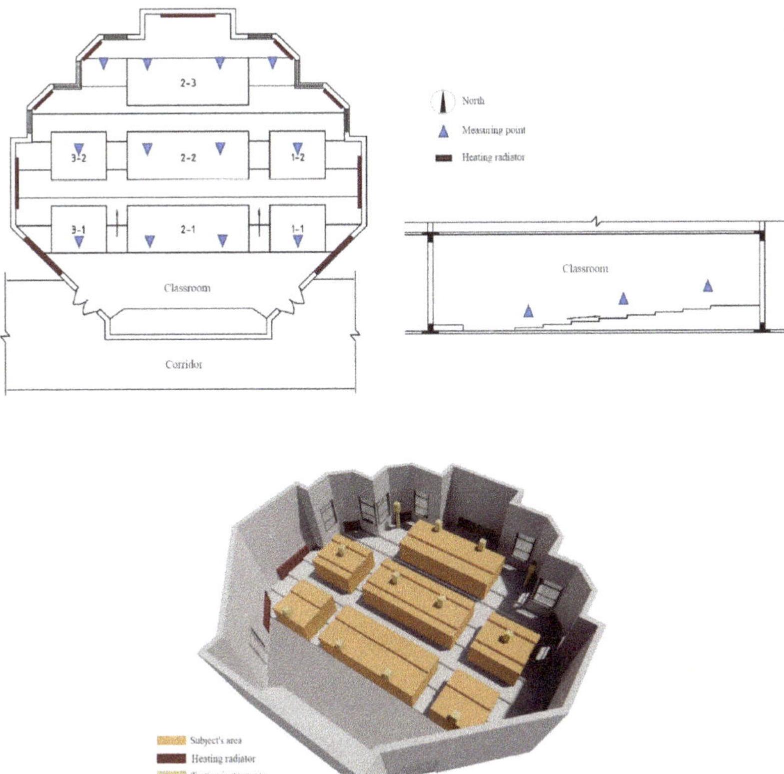

Figure 4. Location plan of the measuring points.

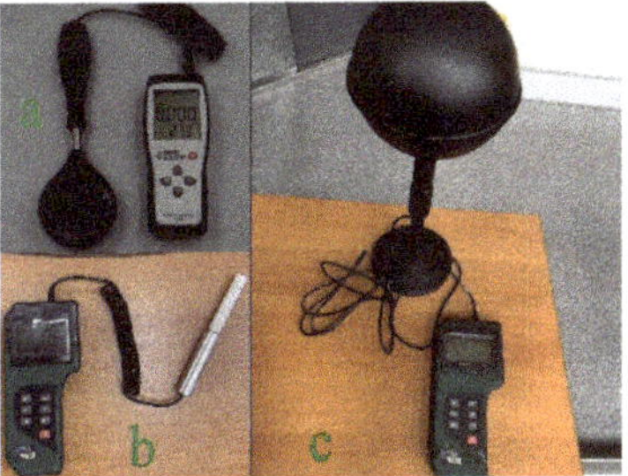

Figure 5. Test instruments: (**a**) Hot-wire anemometer, (**b**) temperature and humidity meter, (**c**) black globe thermometer.

Table 3. Specifications of sensors for measuring indoor climatic parameters.

Parameter	Sensor	Measurement Range	Accuracy
Air temperature	Temperature and	$-20{\sim}85\ °C$	$\pm0.3\ °C$
Relative humidity	humidity meter	$0{\sim}100\%$	$\pm1.5\%$
Global temperature	Black globe thermometer	$5{\sim}120\ °C$	$\pm0.5\ °C$
Air velocity	Hot-wire anemometer	$0.3{\sim}30\ m/s$	$\pm3\%$

2.2.2. Subjective Questionnaire

An electronic questionnaire was used in the field, and the subjects were 22–24-year-old students in good physical condition, who had fully adapted to the climate of Shenyang and had been informed of the survey in advance. In order to avoid the influence of the outdoor environment on the subjects' thermal sensation, the questionnaires were administered and filled in only after the subjects had been indoors for more than 20 min. The ASHRAE scale [33] was used in the subjective questionnaire, allowing the students to vote on thermal sensation, thermal comfort, temperature, and relative humidity expectations. Table 4 shows the voting scale for the subjective questionnaire, which can assess students' preferences. Table 4 also shows the basic information about the subjects such as gender, age, and clothing. A total of 135 questionnaires were distributed, and 133 questionnaires were completed and collected, thus facilitating a subjective evaluation of the thermal and humid environment in the classroom based on the content of the questionnaire.

Table 4. Scales used to measure subjective response to environmental variables.

Scale	Thermal Sensation Vote (TSV)	Thermal Comfort Vote (TCV)	Temperature/Relative Humidity Expectation	Wind Speed Perception
3	hot	-		-
2	warm	-		-
1	slightly warm	-	reduce	slightly perception
0	neutral	comfortable	unchanged	no perception
−1	slightly cool	slightly uncomfortable	rise	-
−2	cool	uncomfortable	-	
−3	cold	very uncomfortable	-	
−4	-	extremely uncomfortable	-	

2.2.3. Entropy Weighting Method

The thermal environment in the classroom is evaluated from two perspectives: in-door temperature and relative humidity using the "entropy weighting method". The steps for calculating the weights by the "entropy weighting method" are described below.

First, the temperature and relative humidity are standardized as follows:

$$y_{ij} = \frac{x_{ij} - min(x_{ij})}{max(x_{ij}) - min(x_{ij})}, 0 \leq y_{ij} \leq 1, j = 1,2 \tag{1}$$

The standardized matrix of the score is

$$Y = \{y_{ij}\}_{m \times n} \tag{2}$$

Then, the information entropy (H_j) of the two indexes is calculated as follows:

$$H_j = -\frac{1}{\ln(n)} \sum_{j=1}^{n} P_{ij} \ln(P_{ij}) \tag{3}$$

where P_{ij} is the proportion of each item in the total, i is the index, and n is the number of records.

The relationship between information utility value (E_j) and information entropy value (H_j) is expressed as

$$E_j = 1 - H_j \tag{4}$$

Note that the information utility value is related to the weight of the two indicators as follows:

$$Wj = \frac{Ej}{\sum_{j=1}^{n} Ej} \tag{5}$$

where W_1 and W_2 are the temperature and relative humidity, respectively.

3. Results

First, as mentioned before, Shenyang is located in cold region C. Additionally, the selected project was an indoor site and the actual measurement time was from 1 November 2021, the heating start time, to 19 December 2021. After 20 December, the school was closed for examination week, and in January 2022, the school was closed for the winter vacation. Thus, the site environment at the selected time interval represented the heating season for the college classrooms. Second, a comparison of the outdoor temperatures taken during the period of 1 November 2019–19 December 2021 shows that the average outdoor temperature at 14:00–15:00 during this interval was 4.7 °C, and the average outdoor relative humidity was 65.2%, as shown in Figures 6 and 7.

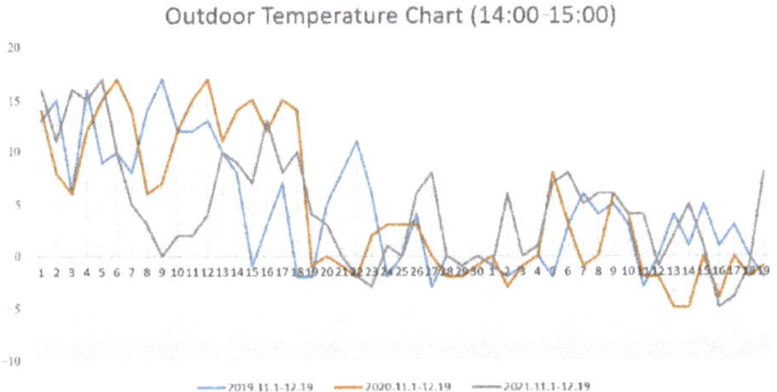

Figure 6. Outdoor temperature change from 14:00 to 15:00 on 1 November 2019–19 December 2021.

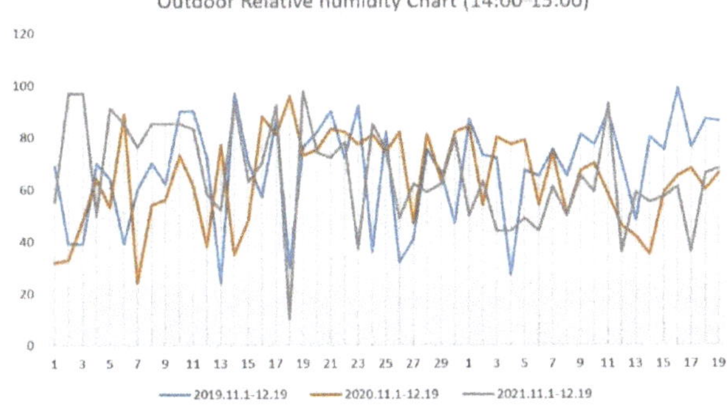

Figure 7. Outdoor relative humidity variation from 14:00 to 15:00 on 1 November 2019–19 December 2021.

Furthermore, some researchers have also used a three-day time volume for refinement analysis [34,35]. Thus, 7–9 December 2021 was selected for the specific analysis of the indoor environmental parameters because the outdoor temperature and relative humidity for these three days were closer to the average outdoor temperature and relative humidity in previous years and were more stable, with a standard deviation of 0.5 °C and 4.5%, as shown in Figures 8 and 9 and the box is the value of 7–9.

Finally, during the three-day measurement period, the indoor environment of the empty field classroom did not change much, and the trend was relatively consistent. The standard deviation was 1.15 °C for indoor temperature and 0.4% for relative humidity, as shown in Figures 10 and 11.

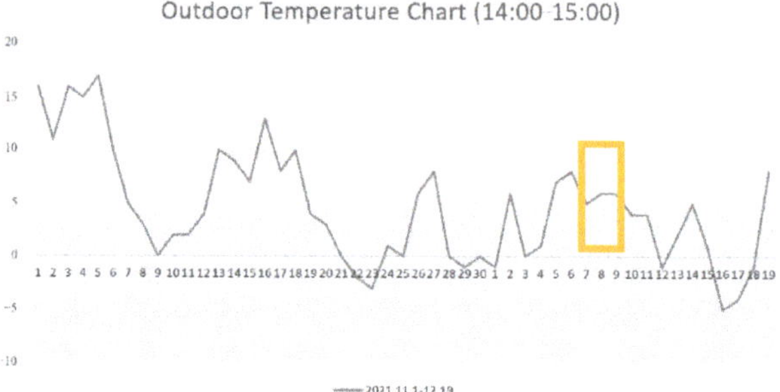

Figure 8. Outdoor temperature change from 14:00 to 15:00 on 1 November–19 December 2021.

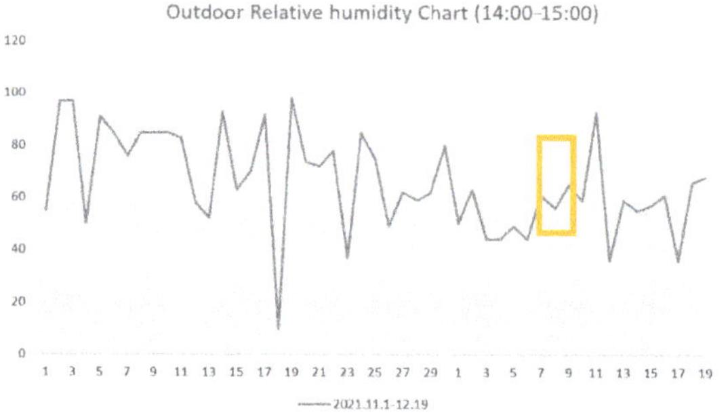

Figure 9. Outdoor relative humidity variation from 14:00 to 15:00 on 1 November–19 December 2021.

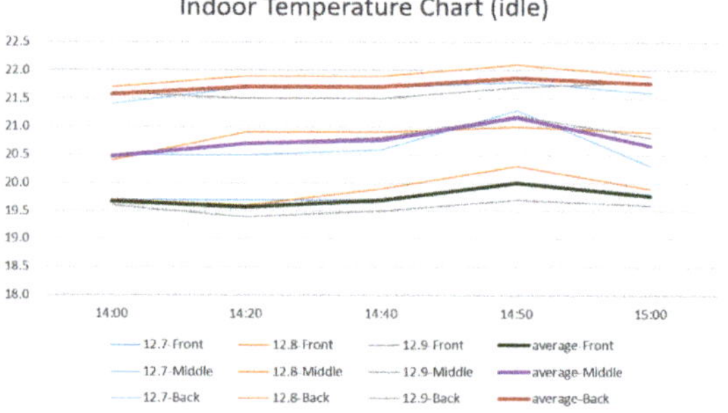

Figure 10. Temperature variation in the classroom (idle).

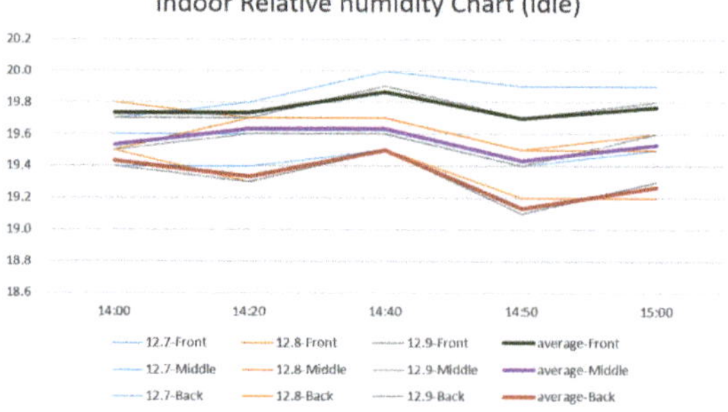

Figure 11. Relative humidity variation in the classroom (idle).

Note that the data were averaged and analyzed. The indoor environmental parameters in the full-field condition changed in a similar trend, with a standard deviation of 1 °C for indoor temperature and 2.7% for relative humidity, as shown in Figures 12 and 13. Thus, the selected day was the closest to the average, and the distributed questionnaire was selected for a detailed partitioning study (Table 5).

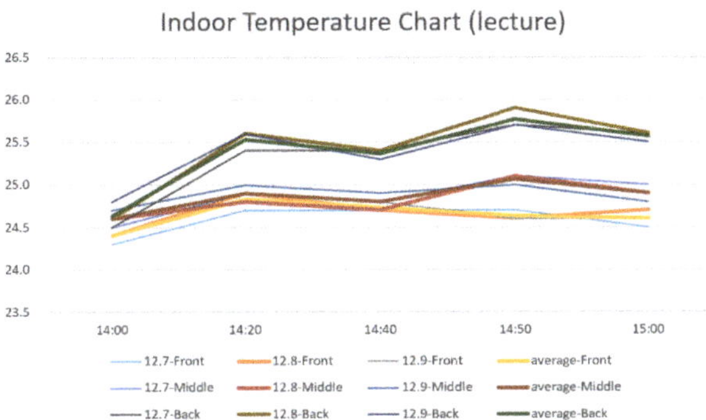

Figure 12. Temperature variation in the classroom (lecture state).

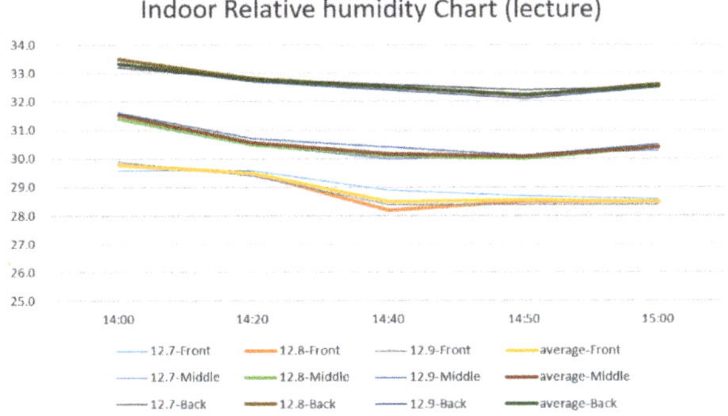

Figure 13. Relative humidity variation in the classroom (lecture state).

Table 5. Summary of the indoor physical environment.

Number	Time	$T_a/°C$	$T_g/°C$	RH/%	$V/m·s^{-1}$
1 (idle)	7 December 2021, 14:00–15:00	20.8	20.2	19.6	0.02
2 (idle)	8 December 2021, 14:00–15:00	20.5	20.1	19.7	0.03
3 (idle)	9 December 2021, 14:00–15:00	20.8	20.7	19.3	0.01
4 (lecture)	7 December 2021, 14:00–15:00	24.9	25.6	30.9	0.02
5 (lecture)	8 December 2021, 14:00–15:00	23.6	26.2	30.4	0.02
6 (lecture)	9 December 2021, 14:00–15:00	24.3	25.7	30.6	0.02

Table 6 shows the different environmental parameters in the classroom under idle and lecture conditions, according to the requirements of the international standard ISO 7726-1998 "Ergonomics of the Thermal Environment-Instruments for Measuring Physical Quantities" [36]. Note that for those subjects engaging in near-sedentary conditions (metabolic rate between 1.0 and 1.3 MET), the operating temperature (T_{op}) can be calculated by using an approximation of the average indoor air velocity and the average radiation temperature. The questionnaire statistics show that the average thermal resistance of students' indoor clothing in winter is $0.24/m^2{\cdot}K{\cdot}W^{-1}$.

Table 6. Indoor environment parameters.

Parameter	$t_{a(i)}/°C$	$RH_{(i)}/\%$	$T_{g(i)}/°C$	$V_{(i)}/m{\cdot}s^{-1}$	$T_{op(i)}/°C$	$R_{cl}/m^2{\cdot}K{\cdot}W^{-1}$
max(idle)	21.9	19.9	24.6	0.03	23.3	-
min(idle)	19.6	19.1	20.7	0.01	20.1	-
average(idle)	20.7	19.5	22.7	0.02	21.7	-
standard deviation(idle)	1.15	0.4	1.95	0.00	1.55	-
max(lecture)	25.9	33.5	29.5	0.02	27.7	0.32
min(lecture)	23.9	28.2	23.1	0.02	23.5	0.16
average(lecture)	24.9	30.8	26.3	0.02	25.6	0.24
standard deviation(lecture)	1	2.7	3.2	0.00	2.1	0.08

3.1. Analysis of Measurement Indicators (Heating Season and Idle State)

The temperature and relative humidity in the classroom were measured in the idle state. At this stage, the classroom is not used as a place for lecturing but only for students' independent study. Thus, the occupancy rate was below 5%, and the doors and windows were closed.

The temperature in the front, middle, and back of the classroom during the idle state remained relatively stable (Figure 14), with an average temperature of 19.8, 20.8, and 21.7 °C, respectively. Thus, all three areas meet the Chinese standard (GB50736-2012) [37], which stipulated the standard of 18–24 °C. Note that the back area is higher than the other two areas due to its relative position, and that the air with higher temperature is less dense and moves to the upper part of the space more easily. Thus, the temperature in the back is higher than in the other two areas.

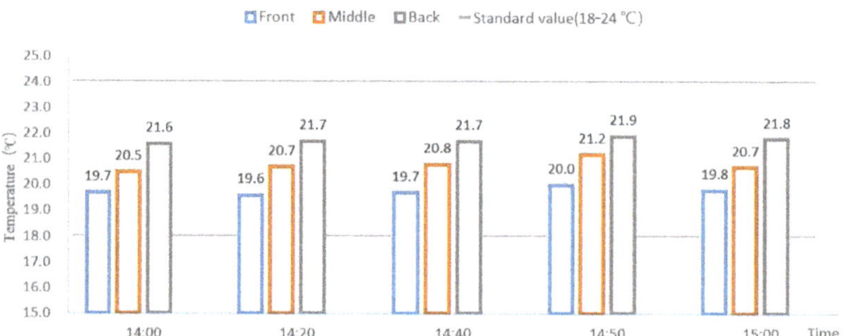

Figure 14. Temperature change in the idle state.

Figure 15 shows that the average relative humidity in the three areas within 60 min was 19.8, 19.5, and 19.3%, respectively. As shown in the figure, the back area has the lowest value because the relative humidity is the ratio of the current humidity to the saturation humidity. Thus, when the temperature rises, the ability of the air to carry moisture increases, and the saturation humidity rises accordingly; meanwhile, the relative humidity falls. Furthermore, during the test period, the difference between the average relative humidity of the three areas is small, not more than 1%, and none of them meet the Chinese standard (GB/T 18883-2002) [32]. In addition, the control range of indoor phase humidity during the winter heating period is 30–60%.

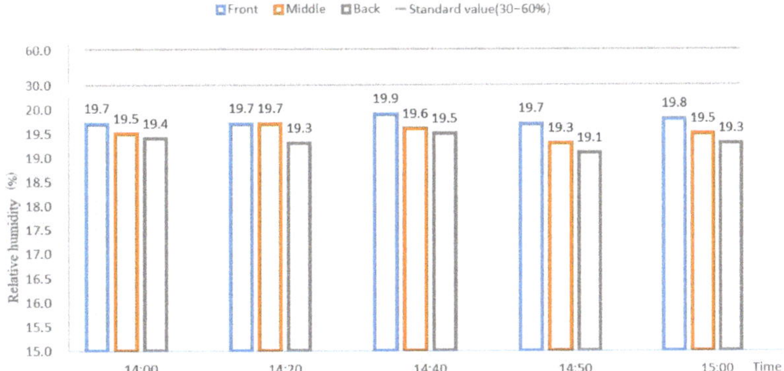

Figure 15. Relative humidity change in the idle state.

3.2. Analysis of Measurement Indicators (Heating Season and Lecturing)

During the lectures, the classroom occupancy rate was over 90%, the volume per person was close to 3.5 m^3/person, the doors and windows were closed, and the classroom door was only opened for 10 min during the class period.

Figure 16 shows that the overall temperature in the classroom during the lectures gradually increases over time. Only a small drop occurred between classes, with the largest drop in the front part. The influencing factors are the break time between classes, the opening of the classroom door, and cooling due to the small-scale ventilation caused by people entering and leaving the classroom. The average temperatures in the classrooms were 24.7 °C (front), 24.8 °C (middle), and 25.4 °C (back) during lectures. The temperature in the back of the classroom was the highest because the test site was a step classroom, which allowed for the heat dissipation of the human body to cause hot air to form and flow up. Moreover, the temperature in the upper space of the classroom was slightly higher than that in the lower space, while the back area had the highest relative position. Therefore, the temperature in the back space was relatively high. Note that all three areas exceeded the indoor temperature control range of 18–24 °C during the winter heating period, as stipulated in the Chinese standard (GB50736-2012) [37]. This shows that the heat released from the crowd in the classroom during the class period is not easily dissipated because of the large number of people. In addition, the design value of the heating space temperature in winter does not consider the influence of the heat generated by the activities of a large number of people during operation. Consequently, the indoor temperature exceeded the code value. Note that people engaging in mental work generally prefer a slightly cooler environment [38]. Thus, reducing the heat in the classroom during lessons can effectively improve human comfort and create a more conducive learning environment. Furthermore, this results in energy savings.

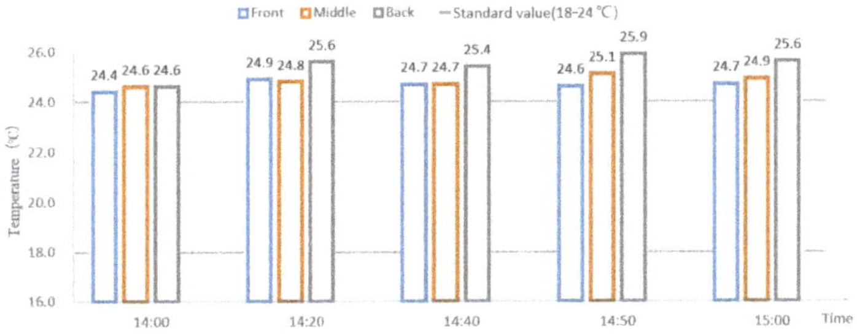

Figure 16. Temperature change during lectures.

Figure 17 shows that the overall relative humidity in the classroom increased significantly by 11.4% as compared to the idle state. This indicates that the dispersed humidity caused by a large number of people can lead to a more significant increase in the relative humidity of the room. However, the overall relative humidity tended to decrease over time in the classroom, indicating that the increase in temperature increases the rate of saturated water vapor much faster than the increase in absolute humidity, which in turn contributes to the decrease in relative humidity. The average relative humidity for the front area of the classroom was 28.9%, which is below the control range of 30–60% for indoor phase humidity during the heating period in winter, as stipulated in the Indoor Air Quality Standard (GB/T 18883-2002) [32]. The average relative humidity of the middle and back areas were 30.5 and 32.7%, respectively. Note that the water vapor is in the upper part of the space due to its lighter density as compared to the air density. This causes the relative humidity to be the largest in the back area, where the class seating position is relatively high. Although the temperature in the back area of the class is also the highest under this condition, the effect of temperature on the overall regional relative humidity is much less than the effect of water vapor.

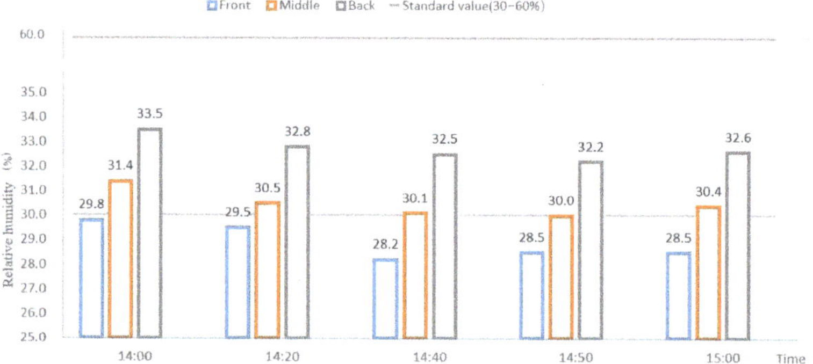

Figure 17. Relative humidity change during lectures.

The relative humidity in each zone of the classroom basically meets the standard requirements during lectures, but the values are low. Thus, self-regulating, i.e., humidifying the room can be a solution based on the subjective feelings of the students.

Figure 18 shows that all zones within the classroom during lectures do not simultaneously meet the two thermal environment parameters set by the specification. More specifically, nine points meet the standard for relative humidity but exceed the specification for air temperature. This implies that a relative reduction of heat within the classroom can effectively improve the thermal environment, meet the requirements of the standard, and result in energy savings and emission reduction.

3.3. Summary

In the horizontal direction of the field measurements, the temperature difference of each point was less than 0.1 °C, and the relative humidity difference was less than 0.2%. Additionally, there was no obvious difference, mainly because the test site was completed in 2015, the main body of the building and the insulation properties of the doors and windows were good, and the seating positions were far from the classroom boundary. Thus, the cold radiation generated by the doors and windows had less impact on the subjects. Based on this situation, the on-site measurements and analysis mainly used the different height areas in the front, middle, and back as the dependent variables.

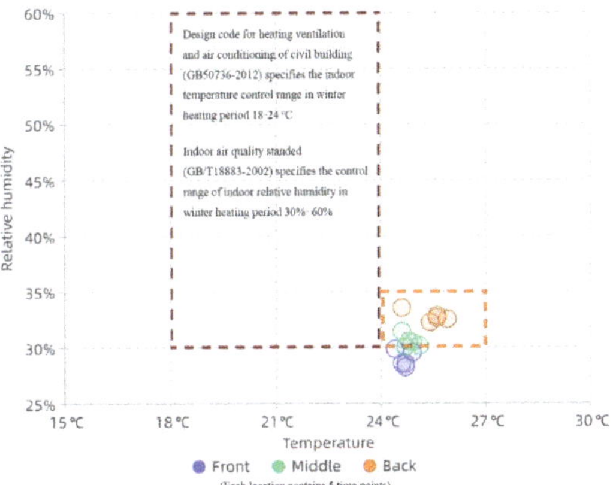

Figure 18. Thermal environment parameters of two kinds of heat exposure during lectures.

The temperature and relative humidity distributions during the idle state remained relatively constant and even, with only minor fluctuations, possibly related to the infiltration of cold air from the doors and windows and the low-temperature radiation from the three side windows. The temperature had a tendency to rise in the classroom and gradually reached a peak, indicating that the population density of people is an effective factor in the room's temperature. Likewise, individual heat dissipation, when reaching a certain value, affects the temperature change in the whole space. In terms of relative humidity, the comparison between the idle and lecture states shows that the factor more significantly affecting the relative humidity in the room is the body's own dissipation of moisture. At the end of the class, the temperature tended to drop, and the relative humidity slowly rebounded. This implies that opening the doors does not cause a significant convection of hot and cold air, but it has a significant effect on the reduction of excessive indoor temperatures and the increase in the amount of fresh air in the room to a certain extent.

4. Discussion

Human thermal sensation is influenced by indoor air temperature, average radiant temperature, air flow rate, air humidity, thermal resistance of clothing, and the different metabolic rates of the subjects. During the test period, the classroom doors and windows were tightly closed, and the threshold of the human perception of wind speed was 0.20 m/s (GB/T 28591-2012) [39]. The measured wind speed ($\leq$0.05 m/s) was much smaller than the threshold of human perception of wind speed (Table 5) and also comply with national standards [37]. Thus, the effect of indoor wind speed on the evaluation of the subject's thermal sensation could be ignored. In addition, the average radiation temperature and the thermal resistance of clothing were basically at constant values in this test. Therefore, this analysis focused on indoor air temperature and indoor relative humidity. A subjective survey was conducted in the classroom on the classroom's quality of temperature and humidity. Note that the sitting rate in the classroom was over 90%, and the per capita volume was nearly 3.5 m^3 per person. The results show that 85% of the classrooms had a sitting rate of over 50% during the class, which means that in most classroom situations, the classroom was not empty. Thus, the impact of the people population density in the classroom on the environment and the impact of indoor density on the environment are representative.

4.1. TSV and TCV

Figure 19 shows that the TCV value corresponds to each TSV voting value. The frequency of voting values between warm and hot was 49.6%, with nearly half of the subjects considering the temperature inside the classroom to be high during the test period. As the test was in the middle of the heating period, the subjects had gradually adapted to the colder climate outside and showed some rejection of the higher heating temperature inside the room under standard heating conditions. At the same time, the subjects felt most comfortable when the thermal sensory poll was thermoneutral. The thermal discomfort of the subjects corresponding to the cold and warm ends of the TSV also reached the peak, indicating that there is a very high correlation between thermal sensation and thermal comfort.

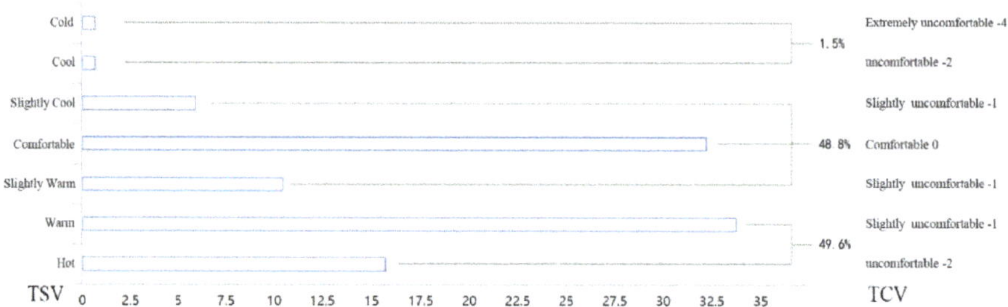

Figure 19. Relationship between TSV and TCV during lectures.

4.2. Comparison of Mean Thermal Sensation Vote (MTSV) and PMV

Some scholars [14] have proposed a linear regression equation for the average indoor thermal sensory vote in winter in severe cold regions versus air temperature; this equation can be expressed as follows:

$$V_{MTSV} = 0.2402t_a - 5.431 \quad R^2 = 0.86, \tag{6}$$

where V_{MTSV} is the average thermal sensation, and t_a (°C) is the room temperature.

The significance levels of regression of Equation (6) were calculated as 0.005 by the F-test. This value implies that the linear regression equation is appropriate.

Note that this regression equation does not take into account the effect of human heat dissipation on the room temperature when the people population density is high. Accordingly, this equation is further modified in the present paper.

It should be noted that the following relationships were obtained by linear regression using the Bin method (temperature frequency method) for a regression analysis of the indoor air temperature. Note that the indoor temperature, relative humidity, and personnel density did not change much during the three days of 7–9 December, which include the same day that was chosen for the subjective questionnaire, i.e., 7 December. In addition, these parameters only include the data obtained from the 1 h field test measurements. The reason for choosing only this time period is to enable an accurate correspondence between the subjective and objective investigations.

The MTSV of the subjects was calculated as follows:

$$V'_{MTSV} = 0.3418t_a - 7.3123 \quad R^2 = 0.85, \tag{7}$$

where V'_{MTSV} is the average thermal sensation, and t_a (°C) is the room temperature.

The significance levels of regression of Equation (7) were calculated as 0.008 by the F-test. This value implies that the linear regression equation is appropriate.

If V_{MTSV} and V'_{MTSV} are zero, the thermal neutral temperature is calculated as 22.6 °C with Equation (6) and 21.4 °C with Equation (7). It can be seen that when the people population density is high, the thermal neutral temperature is relatively low. Note that when the initial operating heating temperature is set for large classrooms, further consideration should be given to the impact of personnel heat dissipation on the overall thermal environment of the space, in addition to meeting code requirements.

According to ISO standard 7726 [38], in most practical cases where the relative velocity is small (<0.2 m/s) or where the difference between mean radiant and air temperature is small (<4 °C), the operative temperature can be calculated with sufficient approximation as the mean value of air and the mean radiant temperature. Furthermore, the corresponding predicted thermal sensory vote (PMV) values can be calculated by putting the corresponding parameters into the Chinese standard (GB/T 50785-2012) [7]

Using the temperature frequency method, regression curves were obtained between the PMV values and room temperature as follows:

$$PMV = 0.3008t_a - 7.2038 \quad R^2 = 0.99, \tag{8}$$

where PMV is the predicted thermal sensory vote value, and t_a (°C) is the room temperature.

The significance levels of regression of Equation (8) were calculated as 0.015 by the F-test. This value implies that the linear regression equation is appropriate.

Although the trends of the linear regression curves for the V'_{MTSV} model and the PMV model were generally consistent (Figure 20), the slope of the V'_{MTSV} curve was larger than the PMV curve, indicating that the actual thermal sensation of the subjects was hotter than the predicted thermal sensation for the same room temperature. This implies that 49.6% of the subjects perceived the classroom to be hotter despite the thermally neutral PMV value at this operating condition.

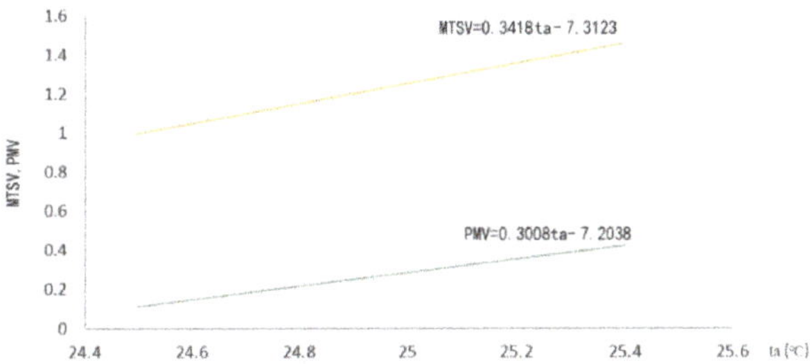

Figure 20. MTSV and PMV fitting curves.

The predicted thermoneutral temperature in Equation (8) was 23.9 °C, which is 2.5 °C higher than the thermoneutral temperature in Equation (7). This implies that the subjects had adapted to the colder outdoor climate at this stage and showed some rejection of the higher indoor heating temperature. It has been suggested in Ref. [40] that prolonged high temperatures can weaken students' ability to adapt to cold climates, and that the use of thermally neutral temperatures to set indoor heating temperatures can result in significant energy savings and improved thermal comfort.

4.3. Study of Coupled Temperature and Humidity Evaluation Models

The entropy weighting method mentioned in Section 2.2.3 is used to normalize the temperature and relative humidity to obtain the standard matrix, which can be expressed as follows:

$$\begin{bmatrix} 0 & 0 \\ 0.14 & 0.37 \\ 1 & 1 \end{bmatrix}. \tag{9}$$

According to Equation (3), the information entropy of temperature and relative humidity is $H_j = \{0.32, 0.51\}$.

According to Equations (4) and (5), the weight values of temperature and relative humidity are 0.58 and 0.42, respectively.

The predicted thermally neutral temperature of 23.9 °C can be calculated with Equation (8). Additionally, the corresponding relative humidity is 28% when PMV is zero. Note that according to the weighting value, the final evaluation score is 25.6 when PMV is zero, while according to the temperature and relative humidity of the front, middle, and rear areas, the evaluation scores are 26.6, 27.2, and 28.5, respectively. This implies that the thermal environment in the front area is the most suitable.

Figure 21 shows that the subjects in the front area had the highest thermal comfort, which is consistent with the final evaluation results of the entropy weighting method. This implies that the entropy weighting method can be used in the subjective analysis of thermal satisfaction.

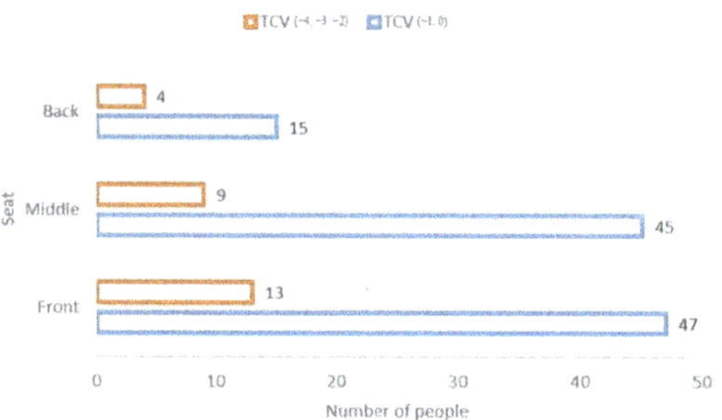

Figure 21. Relationship between TCV and seat area.

4.4. Temperature Expectation, Relative Humidity Expectation, and Thermal Comfort

There are six main factors that affect human thermal comfort: human energy metabolic rate, clothing thermal resistance, air temperature, air humidity, airflow velocity, and average radiation temperature. The human energy metabolic rate, clothing thermal resistance, and average radiation temperature varied slightly in this experiment. Thus, our analysis focused on the TCV voting values (from -4 to -2). The results of a precise statistics calculation is plotted in Table 7; it can be seen that 50% of the subjects felt that the room temperature could be lowered and that the relative humidity could be raised in subsequent improvement measures. However, higher relative humidity is not more beneficial, and some studies have shown [41] that relative humidity should not exceed 60% in order to avoid thermal discomfort, which is also in accordance with the relevant national upper limit [32]. Nevertheless, three of the subjects also chose not to change both environmental parameters, which may have been due to psychological reasons as teaching in the indoor environment is not autonomously controllable. Thus, the subjects might have had different degrees of additional thermal sensations, which would have affected the overall thermal comfort, although it is not related to the objective environmental parameters. It is also worth mentioning that some scholars have proved this point [42].

Table 7. Relationship between TCV, temperature, and relative humidity expectation.

	Extremely Uncomfortable (7 People)	Very Uncomfortable (4 People)	UNCOMFORTABLE (15 People)
T(h), RH(un)			1
T(h), RH(h)	1		1
T(h), RH(l)	1		2
T(l), RH(un)			1
T(l), RH(h)	4	3	6
T(l), RH(l)	1		1
T(un), RH(un)			3
T(un), RH(h)		1	
T(un), RH(l)			
total	7	4	15

T-temperature, RH-relative humidity, (h)-higher, (un)-unchanged

4.5. Gender and Thermal Comfort

Both temperature and relative humidity are influencing factors of thermal comfort. However, the environmental factor that had a greater impact on thermal comfort was mainly air temperature due to the relatively low sensitivity of humans to humidity, which fluctuates within a range of ±15% [43]. Furthermore, the relative humidity in the classroom during this test was generally low and relatively stable.

Information about the students' sex is summarized in Table 8. As can be seen in Figure 22, in terms of gender influence, 81.8% of females and 76.5% of males voted for thermal comfort concentrated between −1 and zero. The average temperature in the classroom in the previous field test was 24.9 °C, indicating that at room temperatures above the prescribed value of 24 °C [39], thermal comfort values were higher for females than for males in the same thermal environmental conditions. It should be noted that some studies on the same severe cold climate zone used a linear relationship between AMV (actual mean vote) and room temperature for different gender groups [44]. These studies calculated the thermal neutral temperature of male and female subjects as 21.6 and 22 °C, respectively. Through the field measurements, the present study also proves that women prefer a warmer environment.

Table 8. Summary of the students' sex in the field survey.

Subjects		Number	%
Total		133	
Sex	male	34	25.6%
	female	99	74.4%

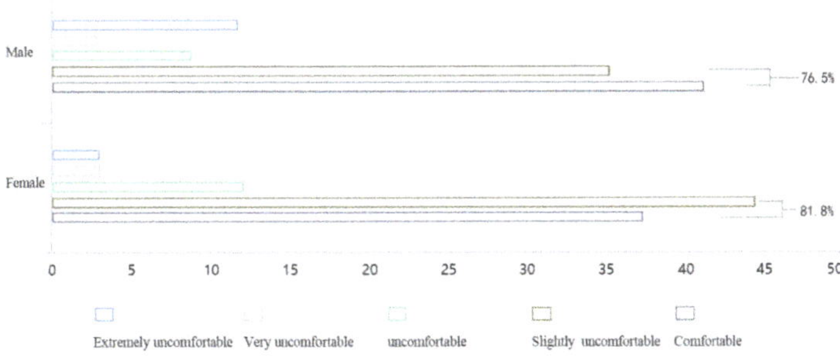

Figure 22. Relationship between gender and thermal comfort.

4.6. Summary

Note that the majority of subjects in this test considered the room temperature to be high, and the thermally neutral temperature was calculated as relatively low. However, when the people population density was high, the majority of subjects were satisfied with the current state of the air temperature in the classroom, indicating a certain level of thermal tolerance. When the two types of expectations were counted, more subjects wanted the temperature to remain the same and the relative humidity to increase, indicating a stronger desire to improve the dry environment in the classroom. In terms of gender, the thermal comfort values for females were relatively higher than those for males in classroom conditions, indicating women's preference for a warmer indoor environment. Note that the entropy weighting method can also be used as a supplementary verification tool in the subjective analysis of the thermal environment.

4.7. Limitations

First, the test time set in the experiment was in the afternoon. Thus, the present study did not cover indoor thermal environment conditions in the morning and evening. Second, the present study focused on a certain type of young people and did not cover the whole age group. In the future, it would be possible to further study the personnel density thresholds corresponding to the appropriate temperature and relative humidity in the classroom. Additionally, the test site chosen was north-facing in order to avoid the effects of sunlight exposure, which is relatively simple. The air quality in the classroom, which would have affected the overall comfort level in the classroom, was also not studied. Finally, in the context of the 2019 coronavirus pandemic, the impact of wearing masks on thermal comfort and the indoor environment should be considered. However, there was no epidemic in the Shenyang campus during the test time of this project, and the students were not wearing masks. Therefore, further research can be carried out to address these limitations.

5. Conclusions

This study aimed at analyzing the characteristics of the indoor thermal environment and personnel activities in university classrooms in Shenyang. Through objective and subjective analyses, combined with the auxiliary analysis of relevant linear equations and the entropy weight method, the following conclusions can be drawn:

1. Through actual measurement and analyses under two scenarios, it was found that the size of the personnel density during lectures is an important factor affecting the change of indoor temperature and relative humidity. Furthermore, the individual's own heat dissipation and moisture dissipation have some influence on the overall thermal environment.
2. PMV overestimates the thermal neutral temperature of students, while the thermal neutral temperature of the subjects was calculated as relatively low. Therefore, when the density of the indoor personnel population is large, we can consider reducing the amount of indoor heating appropriately to enhance thermal comfort and save energy at the same time.
3. Women are more satisfied in a relatively warm environment, with women voting 5.3% higher than men with regard to thermal comfort.
4. The entropy method can be introduced into the subjective analysis of the thermal environment.
5. Thermal comfort can be improved by improving indoor relative humidity. Opening doors can slow down the thermal discomfort, and it does not cause a significant convection phenomenon with hot and cold air. Clothing storage areas can be set up in classrooms to allow students to add or remove clothing and to provide storage.

Author Contributions: Conceptualization, J.Z., P.L. and M.M.; methodology, J.Z., P.L. and M.M.; software, P.L. and M.M.; validation, P.L. and M.M.; formal analysis, J.Z., P.L. and M.M.; investigation, J.Z. and P.L.; resources, M.M.; data curation, J.Z., P.L. and M.M.; writing—original draft preparation, P.L.; writing—review and editing, J.Z. and P.L.; visualization, P.L. and M.M.; supervision, J.Z., P.L. and M.M. All authors have read and agreed to the published version of the manuscript.

Funding: This research received no external funding.

Institutional Review Board Statement: The study was conducted in accordance with the Declaration of Helsinki, and approved by the Ethics Committee of School of Architecture, Northeastern University.

Informed Consent Statement: Informed consent was obtained from all subjects involved in the study.

Data Availability Statement: Data are not publicly available due to restrictions regarding the privacy of the participants.

Acknowledgments: We are grateful to Northeastern University for providing the equipment used in this study.

Conflicts of Interest: The authors declare no conflict of interest.

References

1. Cao, B.; Zhu, Y.; Hou, Y.; Wu, Y.; Li, S.; Liu, S. Ergonomics in built environments: Prospects of human thermal comfort research. *Chin. Sci. Bull.* **2022**, *67*, 1757–1770. [CrossRef]

2. Wang, Z.; de Dear, R.; Luo, M.; Lin, B.; He, Y.; Ghahramani, A.; Zhu, Y. Individual difference in thermal comfort: A literature review. *Build. Environ.* **2018**, *138*, 181–193. [CrossRef]

3. Fanger, P.O. *Thermal Comfort-Analysis and Application in Environment Engineering*; Danish Technology Press: Copenhagen, Denmark, 1970.

4. Kuru, M.; Calis, G. Understanding the Relationship between Indoor Environmental Parameters and Thermal Sensation of users Via Statistical Analysis. *Procedia Eng.* **2017**, *196*, 808–815. [CrossRef]

5. *ASHRAE Standard 55-2017*; Thermal Environmental Conditions for Human Occupancy. ASHRAE: Atlanta, GA, USA, 2017.

6. De Dear, R.; Brager, G.S. Developing an adaptive model of thermal comfort and preference. *ASHRAE Trans.* **1998**, *1*, 145–167.

7. *GB/T 50785-2012*; Evaluation Standard for Indoor Thermal and Humid Environment in Civil Buildings. China Mohan Huapai Art Seminar: Beijing, China, 2012.

8. De Dear, R.; Kim, J.; Candido, C.; Deuble, M. Adaptive thermal comfort in Australian school classrooms. *Build. Res. Inform.* **2015**, *43*, 383–398. [CrossRef]

9. Zomorodian, Z.S.; Tahsildoost, M.; Hafezi, M. Thermal comfort in educational buildings: A review article. *Renew. Sustain. Energy Rev.* **2016**, *59*, 895–906. [CrossRef]

10. Singh, M.K.; Ooka, R.; Rijal, H.B.; Kumar, S.; Kumar, A.; Mahapatra, S. Progress in thermal comfort studies in classrooms over last 50 years and way forward. *Energy Build.* **2019**, *188–189*, 149–174. [CrossRef]

11. Haverinen-Shaughnessy, U.; Moschandreas, D.J.; Shaughnessy, R.J. Association between substandard classroom ventilation rates and students' academic achievement. *Indoor Air* **2011**, *21*, 121–131. [CrossRef]

12. Lee, S.C.; Chang, M. Indoor air quality investigations at five classrooms. *Indoor Air* **1996**, *9*, 134–138. [CrossRef]

13. Wang, Z.; Li, A.; Ren, J.; He, Y. Thermal adaptation and thermal environment in university classrooms and offices in Harbin. *Energy Build.* **2014**, *77*, 192–196. [CrossRef]

14. Wang, Z.; Zhang, X.; Ning, H.; Ji, Y. Human thermal comfort and thermal adaptability in Harbin. *J. Harbin Inst. Technol.* **2012**, *8*, 48–52.

15. Sui, X.; Liu, Q.; Li, M.; Liu, J. Analysis on indoor air quality of university classrooms in Xi'an city during heating season. *Energy Conserv.* **2019**, *38*, 153–158. [CrossRef]

16. Jiang, J.; Li, W.; Di, Y.H.; Qin, S.L.; Yun, X.X. Evaluation of indoor air quality of teaching buildings in colleges and universities in cold regions. *J. Xi'an Polytech. Univ.* **2021**, *35*, 1–8. [CrossRef]

17. Yan, X.; Lei, Y.; Jing, S.; Yin, H. Research of winter indoor thermal environment and thermal comfort of classrooms in colleges and universities in cold regions. *J. Huaqiao Univ.* **2022**, *2*, 198–205.

18. Jung, G.J.; Song, S.K.; Ahn, Y.C.; Oh, G.S.; Bin Im, Y. Experimental research on thermal comfort in the university classroom of regular semesters in Korea. *J. Mech. Sci. Technol.* **2011**, *25*, 503–512. [CrossRef]

19. Cao, B.; Zhu, Y.; Ouyang, Q.; Zhou, X.; Huang, L. Field study of human thermal comfort and thermal adaptability during the summer and winter in Beijing. *Energy Build.* **2011**, *43*, 1051–1056. [CrossRef]

20. Corgnati, S.P.; Filippi, M.; Viazzo, S. Perception of the thermal environment in high school and university classrooms: Subjective preferences and thermal comfort. *Build. Environ.* **2007**, *42*, 951–959. [CrossRef]

21. Wang, Z. A field study of the thermal comfort in residential buildings in Harbin. *Build. Environ.* **2006**, *41*, 1034–1039. [CrossRef]

22. de Dear, R.J.; Auliciems, A. Validation of the Predicted Mean Vote model of thermal comfort in six Australian field studies. *ASHRAE Trans.* **1985**, *91*, 452–468.

23. Goto, T.; Mitamura, T.; Yoshino, H.; Tamura, A.; Inomata, E. Long-term field survey on thermal adaptation in office buildings in Japan. *Build. Environ.* **2007**, *42*, 3944–3954. [CrossRef]

24. Kim, J.T.; Lim, J.H.; Cho, S.H.; Yun, G.Y. Development of the adaptive PMV model for improving prediction performances. *Energy Build.* **2015**, *98*, 100–105. [CrossRef]

25. Ning, H.; Wang, Z.; Zhang, X.; Ji, Y. Adaptive thermal comfort in university dormitories in the severe cold area of China. *Energy Build.* **2016**, *99*, 161–169. [CrossRef]

26. Fang, Z.; Zhang, S.; Cheng, Y.; Fong, A.M.; Oladokun, M.O.; Lin, Z.; Wu, H. Field study on adaptive thermal comfort in typical air conditioned classrooms. *Energy Build.* **2018**, *133*, 73–82. [CrossRef]

27. Liu, J.; Yang, X.; Jiang, Q.; Qiu, J.; Liu, Y. Occupants' thermal comfort and perceived air quality in natural ventilated classrooms during cold days. *Build. Environ.* **2019**, *158*, 73–82. [CrossRef]

28. Buonocore, C.; De Vecchi, R.; Scalco, V.; Lamberts, R. Thermal preference and comfort assessment in air-conditioned and naturally-ventilated university classroom under hot and humid conditions in Brazil. *Energy Build.* **2020**, *211*, 109783. [CrossRef]

29. *GB50178-93*; Building Climate Zoning Standards. MOC: Beijing, China, 1993.

30. *GB50189-2015*; Design Standard for Energy Efficiency of Public Buildings. MOC: Beijing, China, 2015.

31. *GB/T40233-2021*; Instruments for the Measurement of Physical Quantities in the Ergonomics of Thermal Environments. State Administration of Market Supervision: Beijing, China, 2021.

32. *GB/T18883-2002*; Indoor Air Quality Standard. State Administration of Market Supervision: Beijing, China, 2002.

33. Berglund, L.G. Thermal acceptability. *ASHRAE Trans.* **1979**, *85*, 825–834.

34. Jing, S.; Lei, Y.; Wang, H.; Song, C.; Yan, X. Thermal comfort and energy-saving potential in university classrooms during the heating season. *Energy Build.* **2019**, *202*, 109390. [CrossRef]

35. Guoqiang, Z. *Study on the Thermal Environment of University Classroom in Severe Cold Region: Taking Harbin Institute of Technology as an Example*; Harbin Institute of Technology: Harbin, China, 2020.
36. *ISO 7726-1998*; Ergonomics of the Thermal Environment-Instruments for Measuring Physical Quantities. International Organization for Standardization: Geneva, Switzerland, 1998.
37. *GB50736-2012*; Design Code for Heating Ventilation and Air Conditioning of Civil Buildings. China Mohan Huapai Art Seminar: Beijing, China, 2012.
38. Kosonen, R.; Tan, F. Assessment of productivity loss in air-conditioned buildings using PMV index. *Energy Build.* **2004**, *36*, 987–993. [CrossRef]
39. *GB/T 28591-2012*; Wind Scale. State Administration of Market Supervision: Beijing, China, 2012.
40. Wang, Z.; Ning, H.; Zhang, X.; Ji, Y. Human thermal adaptation based on university students in China's severe cold area. *Sci. Technol. Built Environ.* **2017**, *23*, 413–420. [CrossRef]
41. Miao, P. Humidity effect on comfort. *Contam. Control. Air-Cond. Technol.* **2003**, *4*, 13–16.
42. Yu, J.; Ouyang, Q.; Zhu, Y.; Shen, H.; Cao, G.; Cui, W. A comparison of the thermal adaptability of people accustomed to air-conditioned environments and naturally ventilated environments. *Indoor Air* **2012**, *22*, 110–118. [CrossRef] [PubMed]
43. Ji, W.; Cao, B.; Zhu, Y. Indoor thermal and humidity environment and thermal adaptation in different heating periods of North China. *Heat. Vent. Air Cond.* **2019**, *49*, 103–107.
44. Haoran, N. Research on Human Thermal Comfort and Thermal Adaptation in Heating Building Environments in Severe Cold Area. Ph.D. Thesis, Harbin Institute of Technology, Harbin, China, 2017.

Article

A VR Experimental Study on the Influence of Chinese Hotel Interior Color Design on Customers' Emotional Experience

Jian Xu [1,2], Muchun Li [1,*], Kaizhong Cao [3], Fangqi Zhou [1], Boyi Lv [1], Ziqi Lu [4], Zihan Cui [5] and Kailiang Zhang [6]

[1] Department of Tourism Management, South China University of Technology, Guangzhou 510006, China; jianxu@scut.edu.cn (J.X.); 202020151408@mail.scut.edu.cn (F.Z.); tdbylyu@mail.scut.edu.cn (B.L.)
[2] State Key Laboratory of Subtropical Building Science, Guangzhou 510006, China
[3] The Art Department, Communication University of China, Beijing 100024, China; ckz.ckz@cuc.edu.cn
[4] Media Art Department, Communication University of China, Beijing 100024, China; cuc_lu@cuc.edu.cn
[5] Drama-Light and Shadow Space Art, Communication University of China, Beijing 100024, China; guoyueqian@cuc.edu.cn
[6] Landscape Architecture, Communication University of China, Beijing 100024, China; zhangsuzhe@cuc.edu.cn
* Correspondence: limch@scut.edu.cn; Tel.: +86-139-2882-0357

Citation: Xu, J.; Li, M.; Cao, K.; Zhou, F.; Lv, B.; Lu, Z.; Cui, Z.; Zhang, K. A VR Experimental Study on the Influence of Chinese Hotel Interior Color Design on Customers' Emotional Experience. *Buildings* **2022**, *12*, 984. https://doi.org/10.3390/buildings12070984

Academic Editor: Ricardo M. S. F. Almeida

Received: 1 June 2022
Accepted: 4 July 2022
Published: 11 July 2022

Publisher's Note: MDPI stays neutral with regard to jurisdictional claims in published maps and institutional affiliations.

Abstract: As an important part of a hotel's internal environment, color design affects not only customers' hotel stay experiences, but also their check-in experiences. However, how hotel guests' emotional experiences are affected by interior color design is understudied in China. Drawing on the theory of color psychology, we designed a Virtual Reality (VR) experiment and a questionnaire to explore how hotel guests' emotional experience can be influenced by the color scheme of hotel interior color design. The results show that hotel rooms decorated in yellow have a pleasurable effect, those decorated in gray a calming effect, and those decorated in blue a relatively neutral effect. Young participants have more negative emotional responses to rooms decorated in dark yellow. The emotional impact of both gray and yellow with higher grayscale values shifts from positive to negative with the improvement of customers' educational background. Low grayscale color schemes are preferred over high grayscale ones, and indoor environments with synergistic colors are preferred over contrasting colors. It is also found that male subjects tend to have more positive emotional reactions to all color schemes than females. For most subjects, age and education have no effect on their emotional reactions to different color schemes. These findings have important implications for hotel interior environment color design.

Keywords: color psychology; hotel indoor environment; emotion; subjective evaluation; VR experiment

1. Introduction

Color plays an important role in influencing human perception. About 83% of human perceptual information comes from vision, and visual information is dominated by color [1]. As a sensory attribute, the impact of color on the service industry has been of interest to some scholars, and some studies have been carried out on customers' perceptions of the wall color of barber shops [2], clothing stores [3], restaurants [4], etc. Studies of this type have examined customers' visit intentions or attitudes by changing the color of a space [5] and have traced the changes in their psychological and physical perception of the space [6]. Hotel rooms are the spaces where customers spend most of their stay time [7]. Therefore, compared with other parts of a hotel, the decoration and design of rooms have a more substantial impact on customers' emotional experiences. As color is one of the dominant visual elements in hotel internal environments [8], many scholars have conducted in-depth research on it. Color has been proven to significantly affect customers' aesthetic perceptions, emotional experiences, behavioral intentions, and other subjective reactions [9,10]. Although subjective evaluation methods have been used in most existing studies, the contents evaluated in them vary from one to another. For example, though

Selin [8] and Lee [11] both adopted subjective evaluation methods to investigate hotel guests' emotional experiences of interior color design, the former focused on the subjects' states of pleasure in blue, yellow, and gray rooms of urban hotels, whereas the latter concerned itself with the influence of cold color-themed rooms, especially those dominated by green, on customers' emotional well-being. Virtual digital image experiments have been designed in some recent studies, which have found that the influence of warm color schemes on arousal, stimulation, and excitement is stronger than that of cold color and colorless schemes, and cold color schemes can produce a better ambience of spaciousness and tranquility than warm color and colorless schemes [12]. Notably, previous research has mainly dealt with the emotional effect of the singular cold/warm dimension of colors. Hence, further research is needed to fully reveal the impact of different color systems and different color schemes of different grayscale values in the hotel indoor environment on customers' emotional experiences, and whether the synergistic and contrasting colors of such items as bed flags and pillows in the room will cause changes in customers' emotional responses.

Color psychology is the study of human psychological activities in relation to colors. Drawing on traditional psychological theories, it scientifically studies the relationship between color, humans, and the environment. Different colors convey different visual information to the brain, where different associations are formed, and different psychological and emotional reactions are generated, including the perception of temperature and distance as well as the feeling of excitement, depression, irritability, and stability [13]. Research on the relationship between color and psychological function can be traced back to W. Goethe, a German poet and erudite who intuitively speculated on the relationship between color perception and emotional experience and divided colors into positive and negative categories. Positive colors, namely, yellow, red yellow, and yellow red, are considered to be stimulants of positive emotions, such as liveliness, ambition, and warmth, and blue, red blue, and blue red are regarded as the ideal colors that trigger negative emotions, such as anxiety and cold [14]. Psychological research has proved that different colors have different meanings [15] and can affect human perceptions [16]. For example, red represents excitement and passion [17], blue is related to openness and peace [18], yellow means happiness, while green generally symbolizes safety and environmental awareness [10], and so forth. Existing research on color psychology in China mainly focuses on packaging design [19], interior decoration design [20], architectural design [21], etc. As is widely accepted by color psychologists, the meaning of colors is culturally bounded [22], and due to the influence of traditional Chinese culture, high purity red and yellow are favored by most people in China. However, with the influx of foreign cultures, the color preferences of the younger generation are more diverse, and the color choice of the living environment is related to the decoration style, the location of the city, and the purpose of living. According to the authoritative data analysis of "Nippon CCM Color Management System", red and yellow are not among the 13 Chinese favorite colors released in 2022, which indicates that the color issue in hotel interior design is in need of further investigation. While the hotel indoor environment is under-researched in the realm of color psychology, the rise of boutique hotels in recent years, especially those with more diverse and bold colors, has further pushed the boundaries of the research field.

Previous studies have used field experiments or combined virtual digital images with questionnaires to investigate the impact of different colors on subjects' emotional responses. Field experiments can obtain objective and practical information and eliminate the deviation of subjective estimation. Nonetheless, experimental studies are time-consuming, costly, and highly demanding of the experimental conditions. By contrast, subjective evaluation of virtual digital images based on 2D screens is more cost-effective. Nevertheless, such studies may well be problematic in that the subjects cannot be completely immersed and are susceptible to external interference. In a word, current research on the emotional experience of hotel indoor environment is either confronted with the challenge of variable control and color classification, or it is defective due to the inadequate immersion of the hotel indoor environment and the diversity of color collocation. Therefore, this study was intended not only to focus on the impact of gray, blue, and yellow hotel indoor environments on customers' emotional experiences, but also to take a further step to explore the emotional effect of gray, blue, and yellow rooms of different grayscale values. Despite the demanding requirements of the colors of hotel indoor environments, the virtual environment makes it possible to combine the control of laboratory measures with the fidelity of daily experience [23] and hence ensures the reliability of subjective evaluations. Virtual reality (VR) can evaluate cognition, emotion, and behavior in an ecologically effective environment, and individuals can experience it in an immersive VR environment [24]. Moreover, the experimental cost of VR technology is relatively low, and the experimental process is relatively simple and convenient. In this study, VR modeling was used to build virtual hotel rooms, and experiments were carried out through VR technology to overcome the challenges of demanding requirements for hotel color and difficult control of the external environment. When the experiment was over, the subjects were immediately invited to fill out a questionnaire to report their authentic feelings in a direct and timely manner. The questionnaire was designed in line with the most updated scale of existing studies and has high reliability and validity.

To sum up, this study first collected the main color systems of contemporary hotels through Ctrip and summarized the three main color systems of hotel indoor environments, i.e., yellow, blue, and gray, in reference to the color wheel. Drawing on color psychology theories, it was intended to provide a new perspective for the influence mechanism of color on customers' emotional experience of hotel indoor environments by addressing the following four specific questions:

Q1: Whether and how hotel rooms decorated in gray, blue and yellow have a significantly different emotional impact on customers' hotel stay experience.
Q2: Whether and how changing the grayscale value of gray, blue, and yellow leads to a change in customers' emotional experience.
Q3: Whether synergistic colors or contrasting colors have a significantly different emotional impact on customers' emotional experiences, and what color preferences customers have.
Q4: Are the impacts indicated in Questions 1–3 associated with demographic characteristics ?

In accordance with the principle of positivist philosophy, this study applied VR technology and conducted a questionnaire to ensure the authenticity and objectivity of the subjects' reported emotional responses. From the perspective of color psychology, it is a useful attempt to explore the impact of different colors in the hotel indoor environment on customers' emotional experience. It can not only complement with existing research on potential factors affecting the mood of hotel guests, but also provide new insights into the interior color design of hotels.

2. Methodology

2.1. Research Framework

This study was comprised of a VR experiment and a questionnaire. When the experiment on each participant was completed, the participant was invited to fill out the questionnaire, which was intended to collect their emotional perception of the hotel internal environment decorated in different colors. Subsequently, the relationship between different colors and respondents' mood, age, education background, and gender were explored through repeated measures analysis of variance, one-way analysis of variance, and an independent samples t-test. Finally, the interior color design of hotels, particularly the choice of specific color systems and the combination of different colors, was discussed, and implications were provided for the design of hotel interior environments. The research framework of this paper is demonstrated in Figure 1.

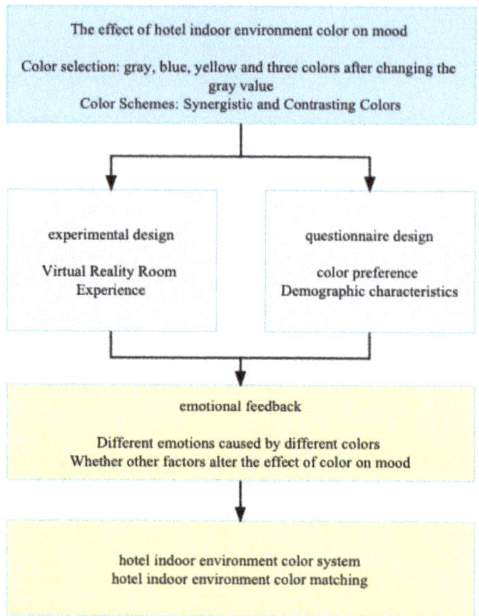

Figure 1. Experimental simulation diagram and their relevant parameters.

2.1.1. Experimental Design

(1) Experimental instrument

VR technology has proven to have an incomparable edge over two-dimensional planes in subjective evaluation. It can overcome the limitation of time and space; integrate immersion, interactivity, and imagination; and bring immersive life experience to users through virtual environment maps, videos, audios, etc., so that users can immerse themselves in it and have a sense of intuition substitution or intuition enhancement [25–28]. In addition, the virtual environment fully reflects the complexity of the real world, and the value obtained by virtual space is almost the same as the value obtained from the real world [4]. Stamp's meta-analysis gives full support to the use of virtual space in studying the effectiveness of responses to architectural features [29].

The VR device used in this study was Pico/Bird Look, which has the merits of simple setup steps, better clarity, and more stable connection. The specific parameters of the device are displayed in Table 1.

Table 1. Equipment parameters.

Brand Name	Pico/Bird Look
Model	Pico Neo3 Xiaomang joint edition
Type	Pico Neo3 128G Xiaomang Pioneer Edition
Production enterprise	Goertek Inc.
Intelligent type	Other intelligence
Time to market	1 February 2022
Product details and display	

(2) VR scene building

In order to measure the emotional effect of color schemes on hotel guests' stay experiences, this study showed the participants a series of digital images of virtual rooms that differed only in the color of indoor walls, curtains, pillows, and bed-end cloth. According to the auditory scale in environmental psychology, the convenient distance of conversation is less than 3 m, and the most effective distance of hearing is less than 6 m [30]. As the standard rooms of express hotels are usually occupied by two people, the length is 6 m. As for the width, according to the human body scale in ergonomics, the best width of express hotels is 3.6 m [30]. Therefore, the virtual room designed in this study was a 21.6 m^2 large bedroom, including a 1.8 m × 2.2 m bed, two bedside cabinets of 0.42 m × 0.55 m × 0.5 m each, a 2.4 m × 0.6 m × 2.1 m wardrobe, a 0.4 m × 1.6 m × 0.54 m TV table, and a 1.05 m × 0.4 m × 0.79 m desk. In this study, Sketch-up was used to model, and Enscape was used to obtain a surreal attempt. As regards the color scheme of the virtual space, the three color systems of yellow, blue, and gray were adopted to compare indoor items with the five color schemes of yellow, green, blue, green, purple, and gray. In light of previous research [8,31] and the color wheel, we chose these five color schemes. The reason is that on the color wheel, the radial areas of green and yellow form a rectangle, and purple is opposite to yellow, which makes them complementary. In other words, the five color schemes can encompass all the color systems contained in the color wheel. Twelve 15 W white LED downlights, a 10.2 m white 3000 K LED strip, and two 8 W white LED bedside chandeliers were used for the lighting of the virtual space. In order to improve the reliability of the experiment, the furniture and decoration of each experimental VR room remained unchanged, and only two colors were contrasted. For the first contrast, the lightness and saturation of yellow, blue, green, and purple were fixed, and only the hue of the four colors was changed. For the second contrast, the hue and saturation of yellow, blue, and gray were fixed, and the grayscale value of each color was reduced by 40% (as shown in Figure 2a–c).

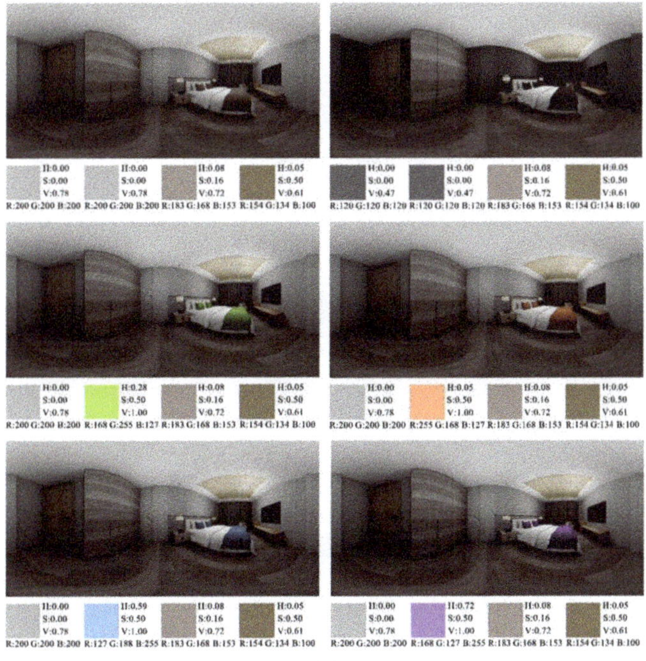

(a)

Figure 2. *Cont.*

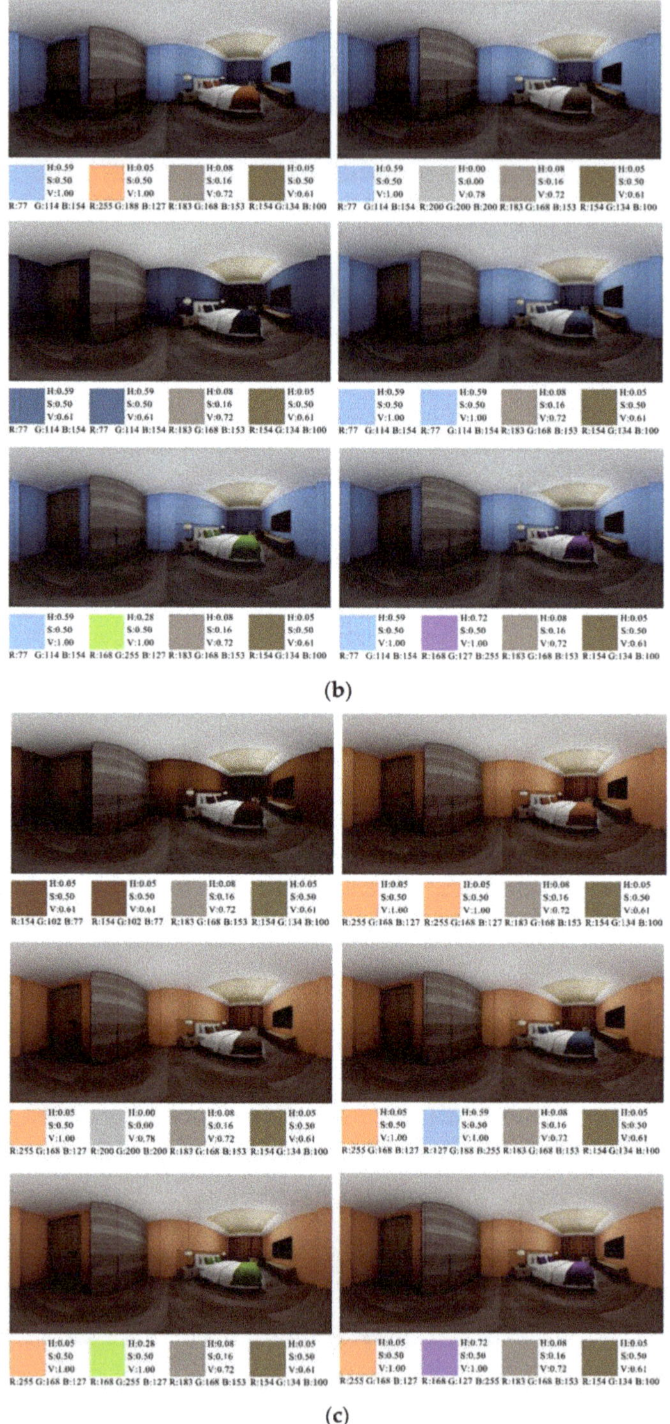

(b)

(c)

Figure 2. *Cont.*

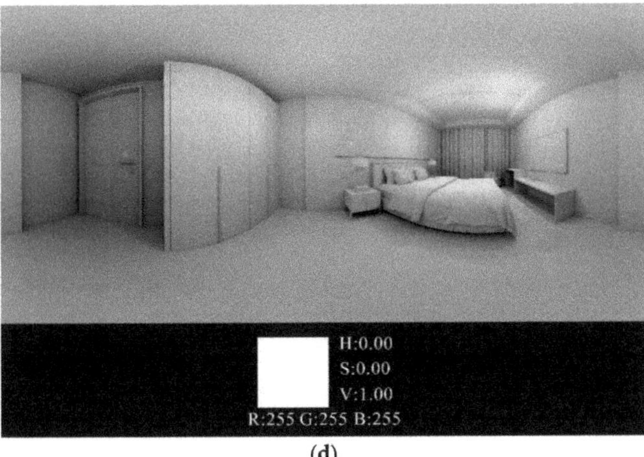

(d)

Figure 2. Experimental simulations and relevant parameters: (**a**) gray schemes and relevant parameters; (**b**) blue schemes and relevant parameters; (**c**) yellow schemes and relevant parameters; (**d**) colorless scheme and relevant parameters.

In addition to the six virtual experimental settings mentioned above, a colorless (pure white) auxiliary setting was set up separately (Figure 2d). The major consideration to do so was that the subjects needed to observe different color combinations within a certain period of time, which might cause visual fatigue, which would in turn lead to inconsistency in their feedback. The color-free auxiliary setting was designed to reduce this type of error. Before the formal experiment, a 3–5 min transition period was arranged to allow the subjects to wear VR glasses, enter the colorless auxiliary setting for observation, and to become familiar with the color as well as the spatial environment of the virtual room. Furthermore, the subjects were required to wear noise-reducing headphones that played light music so as to alleviate experiment-induced anxiety and tension. Throughout the experiment, the subjects always wore noise-reducing headphones to ensure that there was no external auditory interference.

This experiment was carried out from late February to March 2022 in Jian Xu's laboratory of South China University of Technology, Guangzhou, Guangdong, China. February to March is the time for teachers and students to return to school, and they have a low level of work-related pressure and are in a relaxed state of mind. The experimental time was intentionally set between two o'clock and five o'clock in the afternoon, for the temperature at this time is also suitable. Teachers' laboratories are bright, quiet, and dull in color, which can minimize external interference.

2.1.2. Questionnaire Design

The questionnaire used in this experiment consisted of three parts and contained 61 questions in total. The first part of the questionnaire was designed as a color blindness and color weakness test with a view to screening out the subjects insensitive to color. For that purpose, three test charts were provided, and the subject was required to identify the numbers or images in the charts. If the subject failed to identify the symbols in all the three charts, which meant that he/she could not accurately distinguish the colors, the VR experiment would not be conducted on the subject. Only the qualified subjects would proceed to the second part, that is, to choose a favorite color system from the color wheel. The third part was designed to obtain the subjects' subjective feelings while observing the virtual rooms. The subjects would enter different virtual rooms that were randomly ordered. When observing each virtual room, subjects were requested to answer questions concerning the emotional experience of brightness satisfaction and color combination, etc. In light of existing research, a seven-level semantic gradient scale was adopted, and the subjects were asked to report their emotional responses to a given color scheme simply by marking the semantic level. The specific measurement items of the scale were optimized in accordance with existing research [12] and are listed as follows: happy/unhappy, warm/cold, vivid/lackluster, stimulating/non-stimulating, calm/restless, peaceful/unpeaceful. Questions concerning demographic information, including gender, age, education, occupation, income, etc., were also placed in the third part in order to ensure that the subjects' willingness to participate in the experiment would not be affected due to privacy concerns.

Prior to the formal recruitment of experimental subjects, a host of male and female testers with different educational backgrounds aged between 18 and 52 were invited to evaluate the comprehensibility of the questionnaire items, and all the testers were able to correctly understand and answer the questions. Excluding individual factors (such as reaction to and acceptance of the VR device) and equipment debugging, the total amount of time each subject spent wearing VR equipment was within 5 min [31], the duration of the experiment was reasonable enough, and no sign of fatigue or impatience was detected in each subject. During the experiment, the researchers asked the questions orally, the subjects answered each of them orally, and the researchers recorded the answers.

2.2. Participants

From January to February 2022, 112 participants, who were born and brought up in different regions of China but all living in Guangzhou, were recruited to participate in the experiment through online and offline promotion. The researchers promised to the recruited participants that they would be presented a carefully prepared prize if they could take active participation in the experiment.

A color blindness and weakness test was then administered on the recruited participants. Those who were color blind or unable to accurately distinguish colors were identified, and the data of the subsequent experiment on them were to be excluded. In addition, those who had received color education or engaged in professional color research were also screened out. In the end, a total of 104 valid experimental samples was obtained in this study, and the effective response rate reached 92.86%. As displayed in Figure 3, the gender ratio was roughly balanced, with females slightly outnumbering males; the age distribution was dominated by the group aged 24 and below; and most of the participants were mainly from young and middle-aged groups. As for education level, the majority was bachelors, junior college students, and students of master programs or above, and the proportion of bachelor/junior college students/graduates was the highest. With respect to occupation, the proportion of students was the highest, followed by enterprise employees and freelancers. With regard to average monthly income, all income segments were covered, but because the subjects were mainly students, the average monthly income was dominantly 3000 yuan or less.

2.3. Data Analysis

SPSS 26 was applied to analyze the statistical data in this study. First, the reliability and validity of the dependent variables were examined, the existence of a relationship between the dependent variables was evaluated by correlation tests, and then the average and standard deviation of the data were calculated. The repeated measures ANOVA was used to test the effect of the interior color design on the participants' emotional experience. One-way analysis of variance was used to test the effect of age and education. An independent-samples T test was used to test whether there was a significant difference between different age groups' responses. The data obtained were presented in graphical or tabular form.

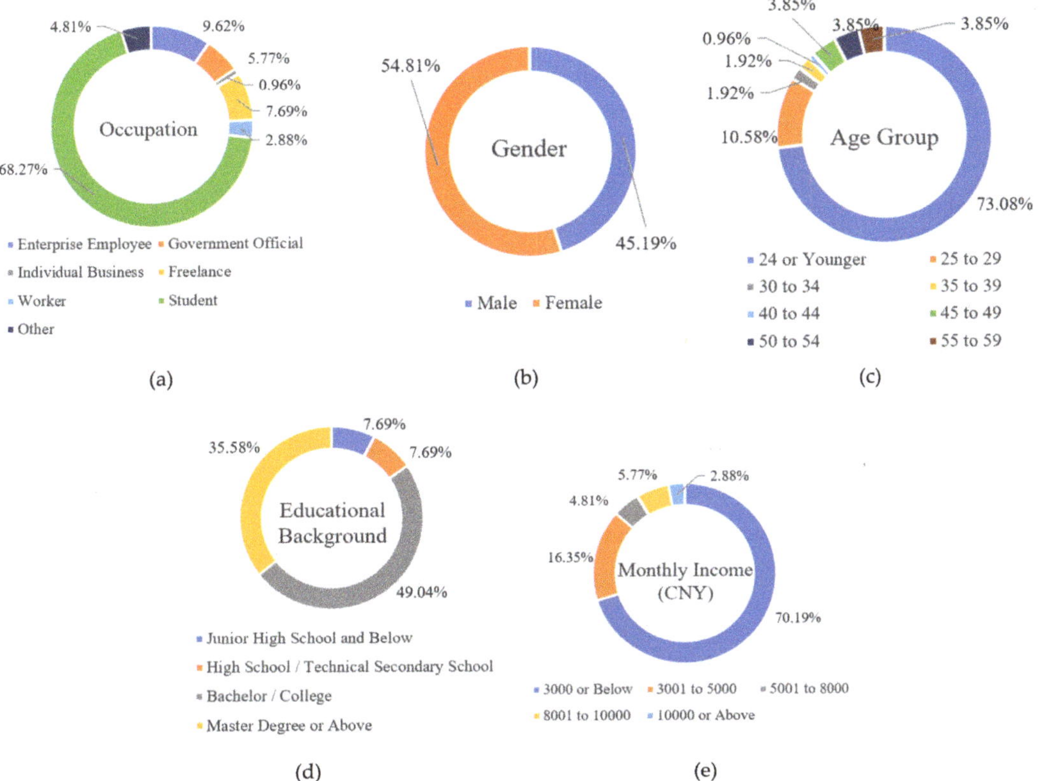

Figure 3. Demographic characteristics of the sample: (**a**) occupation; (**b**) gender; (**c**) age group; (**d**) educational background; (**e**) monthly income.

2.4. Reliability and Validity Test

Cronbach's consistency coefficient α value in the alpha analysis model was used to analyze the reliability of the scale. As can be seen from Table 2, the α coefficient of the scale was 0.956, much higher than the standard value of 0.7, indicating that the scale had high internal consistency; the results of the KMO and Bartlett tests, as displayed in Table 3, showed that the significance was less than 0.05, and the KMO statistic was 0.839, higher than 0.7, indicating that the scale had construct validity [32,33]. Initially, 112 participants were recruited, and when the invalid experimental results were excluded, 104 valid experimental samples were finally obtained, which met the basic requirements of the current research.

Table 2. Reliability analysis.

Project	Index
Cronbach's alpha	0.956
Number of items	45

Table 3. KMO and Bartlett's Test.

KMO (Kaiser–Meyer–Olkin)		0.839
Bartlett test of sphericity	Approx. Chi-Square	4737.332
	Degrees of freedom	990
	Significance	0.000

3. Results

3.1. Color Preference Statistics

The results of the first part of the questionnaire revealed that the participants had the highest preference for yellow (47.11%), followed by blue (28.84%) and gray (7.69%). Of the five color systems, the aforementioned three were favored by more than 80% (83.64%) of all participants (Figure 4). It was also found that synergistic and gray colors were preferred over the others, which indicates that in interior decoration, synergistic and gray color schemes are more pleasurable and welcoming to individuals.

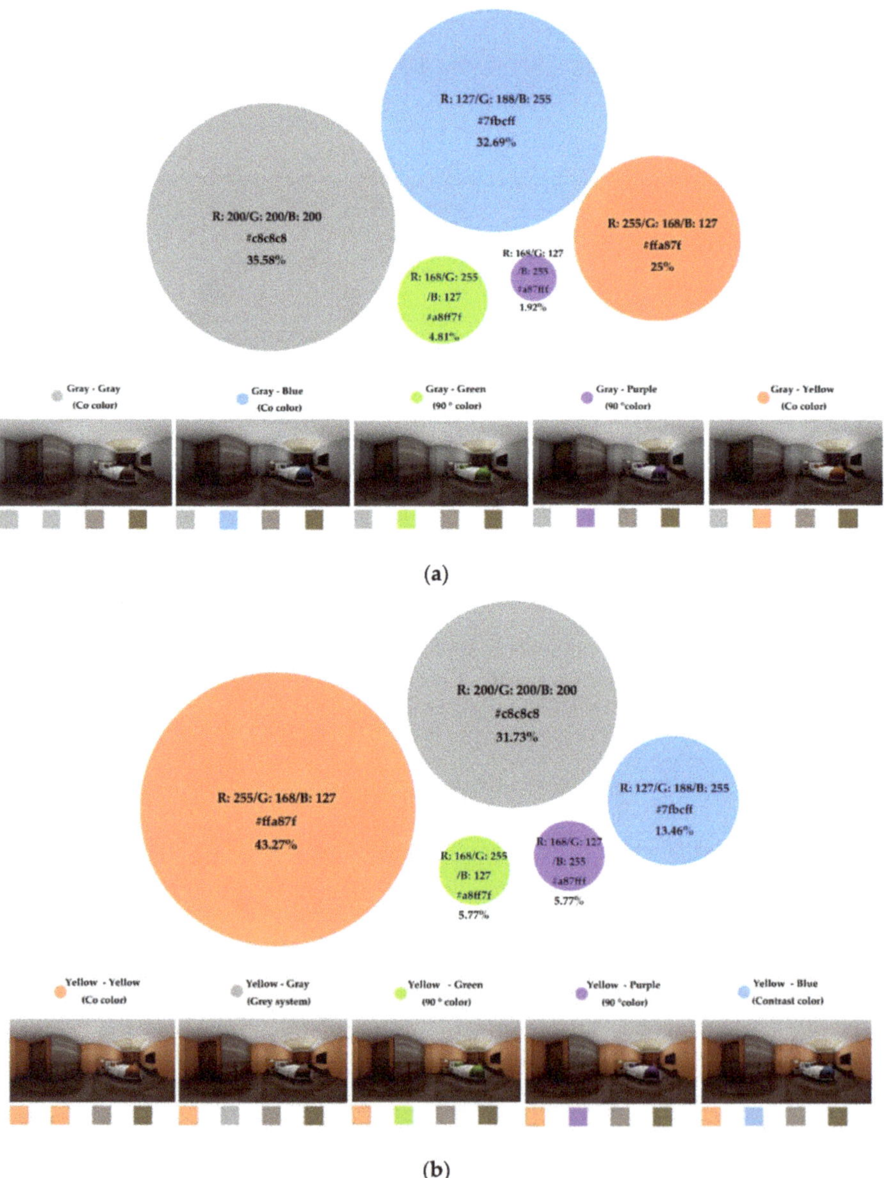

Figure 4. *Cont.*

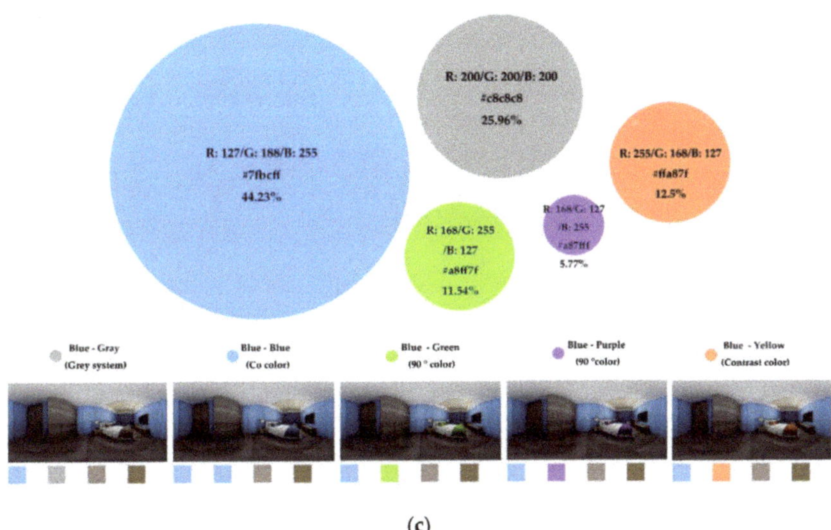

Figure 4. Preference statistics of color combination in hotel internal environment: (**a**) color preference statistics of the gray room; (**b**) color preference statistics of the yellow room; (**c**) color preference statistics of the blue room.

3.2. Effect of Color on Emotional Experience

The results of the six experimental settings were analyzed to reveal the statistical relationship between different color schemes and the participants' emotional responses. The means and standard deviations of the participants' ratings of their emotional responses to the six virtual rooms are shown in Table 4. Differences between participants' emotional responses to the color schemes of the virtual room are shown in Figure 5. The following findings could be drawn from the statistical analysis:

(1) Among the three color systems of gray, yellow, and blue, yellow could evoke more positive individual emotions than gray and blue, which is supportive of Jonauskaite et al.'s conclusion [34] that yellow is more related to positive emotions such as happiness.

(2) Hotel rooms decorated in gray and blue were more effective in bringing customers a sense of calmness, which is also consistent with previous research, indicating that cool colors can give people a more calm and quiet mood [34]. For example, Ayash et al.'s [16] experimental study on the effect of color on students' mood, heart rate, and performance in the learning environment suggests that blue makes people feel more relaxed and calm.

(3) As the grayscale values of the three color systems were increased, the participants' emotional responses become less strong, and yet the positive impact of the yellow room remained higher than that of the gray room and the blue room. In addition to the fact that the yellow system chosen in the experiment was a warm color, and warm colors can bring more positive emotions, there was another reason that the dark yellow hotel internal environment was still brighter than the dark gray one and the dark blue one. Bright colors are more likely to generate positive emotions, which is in accord with Ulusoy et al.'s findings [35]. In other words, lighter and less saturated colors evoke more positive meanings than darker and more saturated colors [36,37].

Table 4. Means and standard deviations of participants' ratings of emotional experience.

Dependent Variable	Color											
	Gray		Gray (Grayscale Increase 40%)		Blue		Blue (Grayscale Increase 40%)		Yellow		Yellow (Grayscale Increase 40%)	
	M	SD	M	SD	M	SD	M	SD	M	SD	M	SD
Happy/Unhappy	4.587	1.251	3.567	1.413	4.933	1.100	3.606	1.210	5.298	0.934	3.856	1.347
Warm/Cold	4.375	1.323	3.490	1.246	4.279	1.218	3.337	1.235	5.558	0.943	4.135	1.322
Vivid/Lackluster	3.808	1.337	3.317	1.248	4.712	1.155	3.452	1.299	5.404	1.010	3.933	1.279
Stimulating/Non-Stimulating	3.856	1.280	3.423	1.236	4.423	1.163	3.471	1.262	5.106	1.114	3.865	1.285
Calm/Restless	4.817	1.305	3.817	1.406	4.635	1.183	3.702	1.253	4.404	1.162	3.885	1.324
Peaceful/Unpeaceful	4.856	1.280	3.817	1.328	4.673	1.226	3.673	1.310	4.817	1.189	3.933	1.279

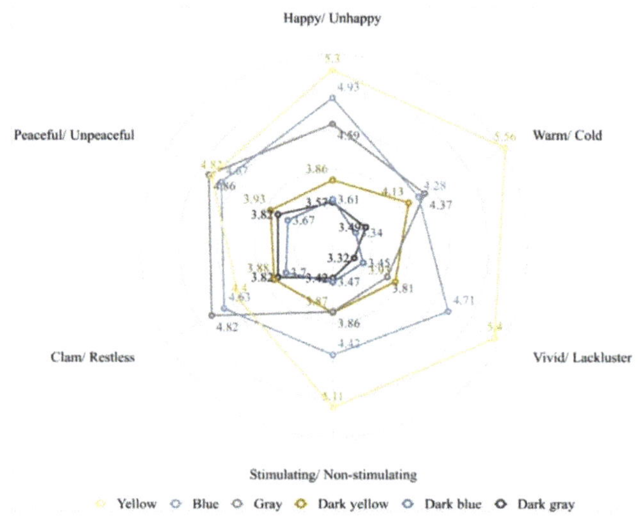

Figure 5. Influence of different color schemes in hotel internal environment on mood.

As the *p* value of the Mauchly's test of sphericity was less than 0.05 (Table 5), the null hypothesis of sphericity was rejected. Combined with the results obtained by the multivariate test ($p = 0.000$) (Table 6), it was found that there was a statistically significant difference ($p < 0.001$) between the emotional effects of different colors.

Table 5. Mauchly's test of sphericity [a].

Measure:	Emotion						
Within Subjects Effect	Mauchly's W	Approximate Chi-Square	Df.	Sig.	Greenhouse–Geisser	Epsilon [b] Huynh–Feldt	Lower–bound
Color	0.446	81.596	14	0	0.753	0.785	0.2

Test the null hypothesis that the error covariance matrix of the dependent variable after the orthogonal transformation is proportional to the identity matrix. [a] Design: Intercept: Main body design: color; [b] The degrees of freedom that can be used to adjust the mean significance test. Modified: tests are displayed in the Tests for Within-Subject Effects table.

Table 6. Multivariate test [a].

Effect		Value	F	Hypothesis df	Error df	Sig.
Color	Pillai's trace	0.681	42.314 [b]	5.000	99.000	0.000
	Wilks' lambda	0.319	42.314 [b]	5.000	99.000	0.000
	Hotelling's trace	2.137	42.314 [b]	5.000	99.000	0.000
	Roy's largest root	2.137	42.314 [b]	5.000	99.000	0.000

[a] Design: Intercept: Main body design: color; [b] Precise Statistics.

As shown in Figure 6, it is apparent that different color schemes had different effects on emotional responses. Among the color systems of gray (color 1), blue (color 3), and yellow (color 5), gray had the least impact on emotions, followed by blue, and yellow was most likely to cause emotional fluctuations. Among the adjusted schemes of dark gray (color 2), dark blue (color 4), and dark yellow (color 6), the effect of dark gray and dark blue on human emotions was not significantly different; compared with the other two schemes, dark yellow was still more likely to cause emotional fluctuations. It can also be found that changing the grayscale value of each color system could change their emotional influence: the average rating of the gray system was reduced from 4.383 to 3.572, the blue system from 4.609 to 3.540, and the yellow system from 5.098 to 3.933, indicating that changing the grayscale value could effectively stabilize the emotion. The comparison between the different color systems showed that changing the grayscale value of the gray system resulted in the least emotional fluctuation; the marginal average rating of the blue system, with the grayscale changed, was closer to the median of 3.540, indicating that the blue system with a lowered grayscale value was more stabilizing; the effect of the yellow system with a lowered grayscale on emotion was accordingly weakened, but it was still statistically significant. The following three aspects were further elaborated in accordance with the results of the questionnaire content and findings of existing research [8,38,39].

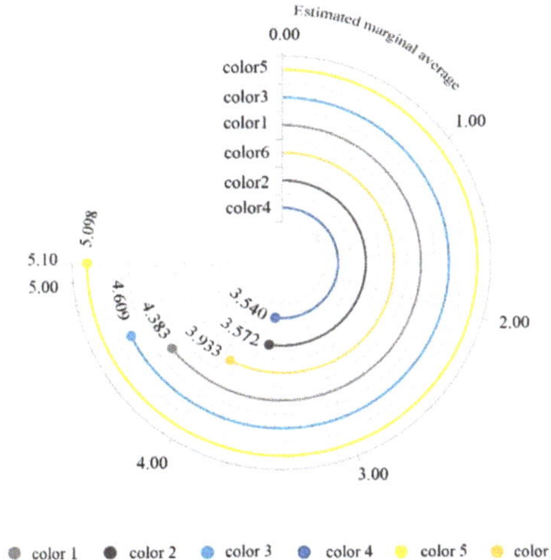

Figure 6. Estimated marginal average of emotional responses.

3.2.1. Pleasing

Yellow received the highest rating in the happy/unhappy item with a mean of 5.298, and no obvious difference was shown between the ratings of blue and gray. The average ratings of the three colors with adjusted grayscale values were all within the range of 3.5–4, and there was no significant difference between them. It was found that yellow had a role to play in generating pleasurable emotions, and colors with lower grayscale values could also make people enter into a pleasant state of mind. In the warm/cold item, the yellow system had a very significant warming effect, with a mean of 5.557. Even the warming effect of the yellow system with a lower grayscale value was not substantially different from that of the blue and gray systems. However, the blue and gray systems with lower grayscale values had a negative effect in this respect, and the mean was lower than the median of 3.5.

3.2.2. Evoking

In the vivid/lackluster item, the yellow system still received the highest average rating, followed by the blue system—there was a moderate difference between the two—and the rest of the colors had a weak wake-up effect. The average ratings of dark blue and dark gray were both lower than the median of 3.5, indicating that they were counterproductive to emotional evocation. It was the same with the results of the stimulating/non-stimulating item. These results demonstrated that yellow and blue systems had arousing and stimulating effects on emotions, while other colors were rather weak or even negative in this respect.

3.2.3. Calming

When it comes to the calm/restless item, the gray, blue, and yellow systems were slightly different from each other, with gray receiving the highest average rating, followed by blue. This shows that color systems with lower grayscale values had a moderately calming effect on people's emotions, especially the gray one. The emotional influence of the three color systems after changing the grayscale values was relatively weak, and there was no obvious difference between them. In the peaceful/unpeaceful item, the ratings of the gray and yellow systems were close to the average, followed by blue, indicating that gray and blue were more placating than yellow. With the grayscale values changed, the ratings of the three color systems were similar to the results of the previous item, and their emotional influence was relatively weak.

In a word, the three colors with lower grayscale values had a higher influence on emotions than the three colors with higher grayscale values.

3.3. Impact of Demographic Variables on the Results

3.3.1. The Effect of Gender on the Results

From the results of the six sets of experiments, it can be seen that there were differences between the responses of male and female participants to the six virtual hotel rooms respectively decorated in the color schemes of gray, blue, yellow, dark gray, dark blue, and dark yellow (Figures 7 and 8). The results of "one-way ANOVA" analysis (Table 7) showed that there was no gender difference in the participants' responses to the gray, blue, and yellow systems, with the mere exception of the calm/restless rating of the yellow system, but the means of the male participants' ratings of all items were higher than those of the female participants, which means that males were more positive towards each of the color schemes than females. When it came to the hotel interior color design, the yellow system, whether its grayscale value was changed or not, could bring greater emotional pleasure to both male and female subjects than the other color schemes. In terms of the strength of emotional impact, the three basic color systems followed the order of: yellow > blue > gray.

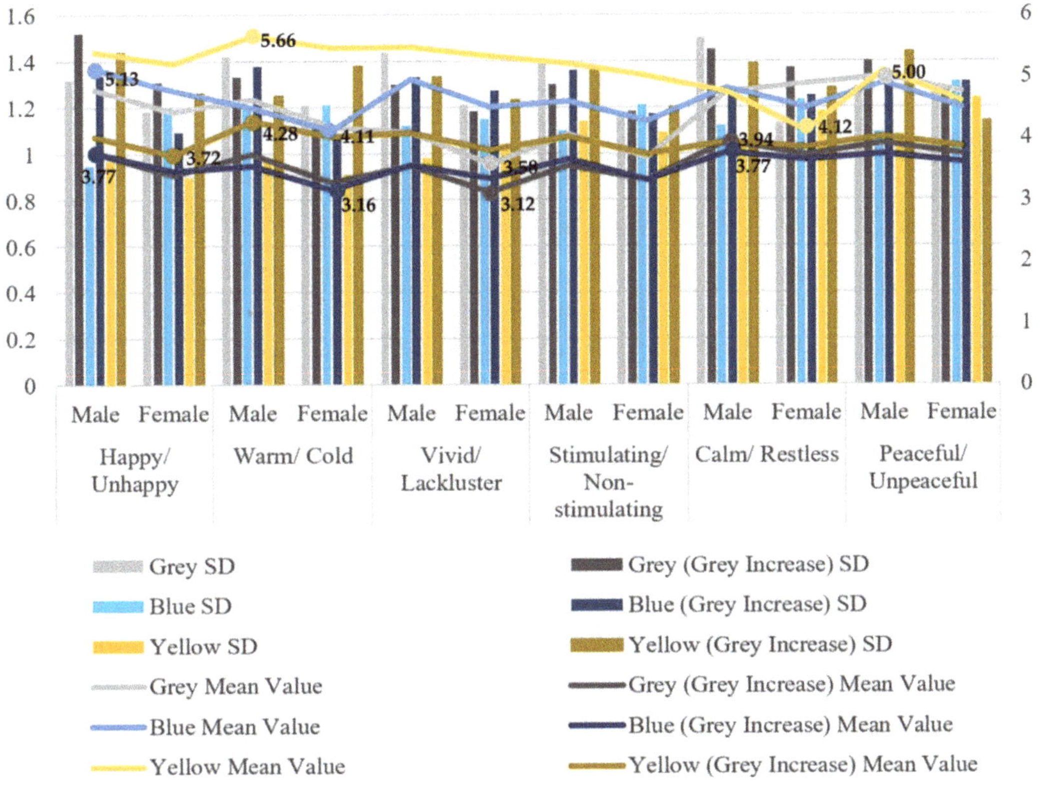

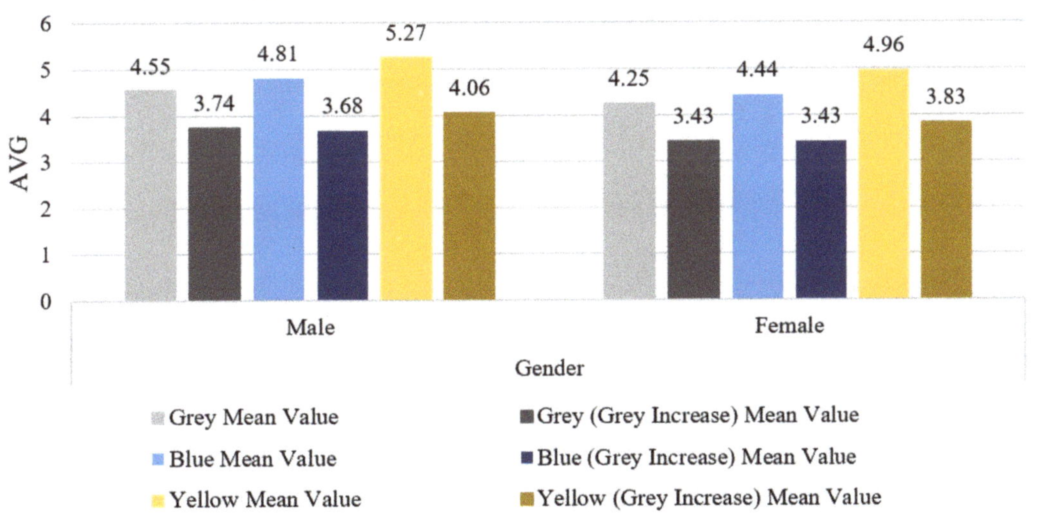

Figure 7. The influence of gender on color perception.

Figure 8. Gender differences in emotional perception of color.

Table 7. Perception of different gender groups of the color of hotel indoor environment.

Happy/Unhappy			Warm/Cold			Vivid/Lackluster			Stimulating/Non-Stimulating			Calm/Restless			Peaceful/Unpeaceful				
BG	WG	Sum	BG	WG	Sum	BG	WG	Sum	BG	WG	Sum	BG	WG	Sum	BG	WG	Sum		
3.454	157.767	161.221	5.023	175.352	180.375	6.600	177.554	184.154	3.712	165.124	168.837	0.452	175.077	175.529	1.784	167.053	168.837	SS	
3.454	1.547		5.023	1.719		6.600	1.741		3.712	1.619		0.452	1.716		1.784	1.638	103	MS	Gray Virtual Room
2.233			2.922			3.791			2.293			0.264			1.089			F	
0.138			0.090			0.054			0.133			0.609			0.299			Sig.	
3.384	202.145	205.529	5.545	154.445	159.990	4.771	155.757	160.529	1.452	155.933	157.385	1.212	202.317	203.529	0.817	180.712	181.53	SS	
3.384	1.982		5.545	1.514		4.771	1.527		1.452	1.529		1.212	1.984		0.817	1.772		MS	Gray Virtual Room
1.708			3.662			3.125			0.950			0.611			0.461			F	(Grayscale Increase)
0.194			0.058			0.080			0.332			0.436			0.499			Sig.	
3.260	121.269	124.529	3.800	149.113	152.913	5.186	132.161	137.346	2.557	136.828	139.385	2.593	141.522	144.115	4.171	150.714	154.885	SS	
3.260	1.189		3.800	1.462		5.186	1.296		2.557	1.341		2.593	1.387		4.171	1.478		MS	Blue Virtual Room
2.742			2.600			4.002			1.906			1.869			2.823			F	
0.101			0.110			0.048			0.170			0.175			0.096			Sig.	
2.200	148.636	150.837	4.025	153.196	157.221	1.288	172.472	173.760	3.044	160.869	163.913	0.352	161.408	161.760	0.440	176.445	176.885	SS	
2.200	1.457		4.025	1.502		1.288	1.691		3.044	1.577		0.352	1.582		0.440	1.730		MS	Blue Virtual Room
1.510			2.680			0.762			1.930			0.222			0.254			F	(Grayscale Increase)
0.222			0.105			0.385			0.168			0.638			0.615			Sig.	
0.967	88.793	89.760	0.890	90.764	91.654	0.627	104.411	105.038	0.982	126.855	127.837	9.962	129.077	139.038	6.150	139.379	145.529	SS	
0.967	0.871		0.890	0.890		0.627	1.024		0.982	1.244		9.962	1.265		6.150	1.366		MS	Yellow Virtual
1.111			1.000			0.613			0.789			7.872			4.501			F	Room
0.294			0.320			0.436			0.376			0.006			0.036			Sig.	
2.349	184.487	186.837	1.729	178.387	180.115	1.992	166.537	168.529	2.084	168.031	170.115	0.228	180.387	180.615	0.673	167.856	168.529	SS	
2.349	1.809		1.729	1.749		1.992	1.633		2.084	1.647		0.228	1.769		0.673	1.646		MS	Yellow Virtual
1.299			0.988			1.220			1.265			0.129			0.409			F	Room
0.257			0.322			0.272			0.263			0.720			0.524			Sig.	(Grayscale Increase)
1	102	103	1	102	103	1	102	103	1	102	103	1	102	103	1	102	103	df	All

3.3.2. The Effect of Education on the Results

The results of "one-way ANOVA" analysis (Table 8) suggested that there was no significant difference between the emotional responses of participants from different educational backgrounds to the color systems of gray, blue, and yellow ($p > 0.05$), with the only exception of the stimulating/non-stimulating rating of the gray system, which indicated that education level was not a factor that had an effect on people's emotional responses to different color schemes. Although the stimulating/non-stimulating rating of the gray system was significantly higher than that of the others ($p = 0.03 < 0.05$), it can be drawn from multiple sets of data that this set of data was not representative, and its significance belonged to a small probability event. Thus, it can be concluded that the color effect on emotional experience of hotel internal environment was generally independent of educational background when the grayscale values were not changed. Nonetheless, when the grayscale values of the above three color systems were adjusted, the change of the participants' emotional responses was dependent on their educational backgrounds ($p < 0.05$), with the only exception being the dark blue color scheme, which was also an event of small probability, indicating that education had a significant effect on people's emotional perception of the grayscale of different colors.

3.3.3. The Effect of Age on the Results

As shown in Table 9, there was no significant difference between the emotional responses to the gray system of participants of all age groups ($p > 0.05$), which means that age had no significant effect on the emotional perception of the gray system. No significant difference was shown in their emotional responses to the dark gray color scheme, except for the vivid/lackluster and stimulating/non-stimulating ratings. As verified by multiple sets of data, the p values showed that the ratings of these two items were also small probability events, indicating that age had no significant effect on the emotional perception of dark gray. No statistically significant difference was shown between different age groups' emotional responses to blue and dark blue ($p > 0.05$ in), indicating that age had no significant effect on the emotional perception of these two color schemes. Notably, this was not the case with yellow and dark yellow. While the p values of different age groups' emotional responses to yellow were all above 0.05, indicating that age had no significant effect on the emotional perception of this color system, the overall p value of their responses to dark yellow was below 0.05, indicating that age had a significant impact on the emotional perception of the color with an adjusted grayscale value. To sum up, it can be generally argued that the emotional effect of different color schemes is not affected by age, with the exception of dark yellow.

Table 8. Perception of different educational groups of the color of hotel internal environment.

	Happy/Unhappy			Warm/Cold			Vivid/Lackluster			Stimulating/Non-Stimulating			Calm/Restless			Peaceful/Unpeaceful				
	BG	WG	Sum	BG	WG	Sum	BG	WG	Sum	BG	WG	Sum	BG	WG	Sum	BG	WG	Sum		
	6.02	155.2	161.22	3.12	177.25	180.38	5.73	178.42	184.15	9.55	159.29	168.84	10.92	164.61	175.53	2.23	166.6	168.84	SS	
	0.86	1.62		0.45	1.85		0.82	1.86		1.36	1.66		1.56	1.71		0.32	1.74		MS	Gray Virtual Room
	0.53			0.24			0.44			0.82			0.91			0.18			F	
	0.81			0.97			0.87			0.57			0.5			0.99			Sig.	
	23.45	182.08	205.53	17.29	142.7	159.99	24.62	135.91	160.53	24.6	132.79	157.38	12.5	191.03	203.53	20.19	161.33	181.53	SS	
	3.35	1.9		2.47	1.49		3.52	1.42		3.51	1.38		1.79	1.99		2.88	1.68		MS	Gray Virtual Room
	1.77			1.66			2.48			2.54			0.9			1.72			F	(Grayscale Increase)
	0.1			0.13			0.02			0.02			0.51			0.11			Sig.	
	10.64	113.89	124.53	17.35	135.57	152.91	4.92	132.43	137.35	6.72	132.66	139.38	8.04	136.08	144.12	7.37	147.51	154.88	SS	
	1.52	1.19		2.48	1.41		0.7	1.38		0.96	1.38		1.15	1.42		1.05	1.54		MS	Blue Virtual Room
	1.28			1.75			0.51			0.69			0.81			0.69			F	
	0.27			0.11			0.83			0.68			0.58			0.68			Sig.	
	9.46	141.37	150.84	10.34	146.88	157.22	11.15	162.61	173.76	13.15	150.76	163.91	13.02	148.74	161.76	16.85	160.03	176.88	SS	
	1.35	1.47		1.48	1.53		1.59	1.69		1.88	1.57		1.86	1.55		2.41	1.67		MS	Blue Virtual Room
	0.92			0.97			0.94			1.2			1.2			1.44			F	(Grayscale Increase)
	0.5			0.46			0.48			0.31			0.31			0.2			Sig.	
	4.24	85.52	89.76	4.43	87.23	91.65	4.14	100.9	105.04	10.87	116.97	127.84	11.95	127.08	139.04	13.78	131.75	145.53	SS	
	0.61	0.89		0.63	0.91		0.59	1.05		1.55	1.22		1.71	1.32		1.97	1.37		MS	Yellow Virtual
	0.68			0.7			0.56			1.27			1.29			1.43			F	Room
	0.69			0.68			0.78			0.27			0.26			0.2			Sig.	
	34.24	152.59	186.84	30.01	150.11	180.12	20.38	148.15	168.53	24.93	145.19	170.12	25.85	154.76	180.62	19.45	149.08	168.53	SS	
	4.89	1.59		4.29	1.56		2.91	1.54		3.56	1.51		3.69	1.61		2.78	1.55		MS	Yellow Virtual Room
	3.08			2.74			1.89			2.35			2.29			1.79			F	(Grayscale Increase)
	0.01			0.01			0.08			0.03			0.03			0.1			Sig.	
	7	96	103	7	96	103	7	96	103	7	96	103	7	96	103	7	96	103	df	All

Table 9. Perception of different age groups of the color of hotel internal environment.

| Happy/Unhappy | | | Warm/Cold | | | Vivid/Lackluster | | | Stimulating/Non-Stimulating | | | Calm/Restless | | | Peaceful/Unpeaceful | | | | |
BG	WG	Sum	BG	WG	Sum	BG	WG	Sum	BG	WG	Sum	BG	WG	Sum	BG	WG	Sum		
6.94	154.29	161.22	6.05	174.33	180.38	11.54	172.61	184.15	14.15	154.68	168.84	7.07	168.46	175.53	5.61	163.22	168.84	SS	
2.31	1.54		2.02	1.74		3.85	1.73		4.72	1.55		2.36	1.68		1.87	1.63		MS	Gray Virtual Room
1.5			1.16			2.23			3.05			1.4			1.15			F	
0.22			0.33			0.09			0.03			0.25			0.33			Sig.	
37.63	167.89	205.53	26.9	133.09	159.99	32.87	127.66	160.53	20.56	136.83	157.38	19.32	184.21	203.53	16.44	165.09	181.53	SS	
12.54	1.68		8.97	1.33		10.96	1.28		6.85	1.37		6.44	1.84		5.48	1.65		MS	Gray Virtual Room
7.47			6.74			8.58			5.01			3.5			3.32			F	(Grayscale Increase)
0			0			0			0			0.02			0.02			Sig.	
4.05	120.48	124.53	1.59	151.32	152.91	2.78	134.56	137.35	2.11	137.27	139.38	5.56	138.55	144.12	0.37	154.52	154.88	SS	
1.35	1.2		0.53	1.51		0.93	1.35		0.7	1.37		1.85	1.39		0.12	1.55		MS	Blue Virtual Room
1.12			0.35			0.69			0.51			1.34			0.08			F	
0.34			0.79			0.56			0.67			0.27			0.97			Sig.	
13.94	136.89	150.84	8.73	148.49	157.22	5.95	167.81	173.76	7.6	156.31	163.91	10.61	151.15	161.76	9.23	167.65	176.88	SS	
4.65	1.37		2.91	1.48		1.98	1.68		2.53	1.56		3.54	1.51		3.08	1.68		MS	Blue Virtual Room
3.39			1.96			1.18			1.62			2.34			1.84			F	(Grayscale Increase)
0.02			0.12			0.32			0.19			0.08			0.15			Sig.	
0.43	89.33	89.76	0.95	90.7	91.65	0.56	104.48	105.04	2.33	125.51	127.84	10.97	128.07	139.04	6.01	139.52	145.53	SS	
0.14	0.89		0.32	0.91		0.19	1.04		0.78	1.26		3.66	1.28		2	1.4		MS	Yellow Virtual
0.16			0.35			0.18			0.62			2.86			1.44			F	Room
0.92			0.79			0.91			0.6			0.04			0.24			Sig.	
39.06	147.78	186.84	21.05	159.07	180.12	24.95	143.57	168.53	25.87	144.24	170.12	26.04	154.57	180.62	15.93	152.6	168.53	SS	
13.02	1.48		7.02	1.59		8.32	1.44		8.62	1.44		8.68	1.55		5.31	1.53		MS	Yellow Virtual
8.81			4.41			5.79			5.98			5.62			3.48			F	Room
0			0.01			0			0			0			0.02			Sig.	(Grayscale Increase)
3	100	103	3	100	103	3	100	103	3	100	103	3	100	103	3	100	103	df	All

4. Discussion

4.1. Theoretical Contributions

This study is both linked to and different from existing research, such as the studies by Bilal et al. [8] and Kim et al. [40]. In Bilal et al.'s study, the color of the bedding was changed, while the color of the walls was fixed, and it was concluded that the subjects preferred yellow and blue to gray. In Kim et al.'s study, only the color of the walls in the room was changed, and the color of the bedding remained constant, and it was found that the subjects preferred brighter colors. This study, however, is a progression of these studies by changing not only the color of the walls in the room, but also the color of the bedding.

Color psychologists believe that yellow can activate thinking but will cause emotional instability; and that blue can eliminate tension, reduce the pulse rate and calm the mood. People have less emotional fluctuations in a non-colored environment, and it is difficult-to-produce pleasant feelings through color design. The conclusions of this study are supportive of the basic theory of color psychology and are in agreement with the findings of Yildirim et al. [12]. Other studies also suggest that blue is the most popular color [41,42], and Marco et al.'s research even shows that yellow is one of the most unpopular colors for college students [43].

Labkin, a famous color scientist from the former Soviet Union, found that the colors in the middle of the spectrum, that is, light yellow, light green, light blue, white, etc., cause the least fatigue to the eyes [44]. From the perspective of color psychology, bright colors will bring people a positive and energetic feeling, while dark colors will bring people a sense of heaviness and depression. In short, this study confirms that bright light colors are more popular, which is consistent with existing research [43,45].

The combination of synergistic colors will increase the sense of light and shade and regularity and give people a sense of the beauty of order. The emotional impact of such color schemes is milder than the combination of contrasting colors, and it is the most conservative and safe method to use synergistic colors in interior color design [46]. Yildirim et al. believe that changing the spatial characteristics (such as furniture and decoration) of an indoor environment where the color remains the same will also lead to the change of emotional responses [12]. In comparison, this study explores people's preferences of different schemes of color combination in an indoor environment where the spatial characteristics remain unchanged and is an endeavor to expand the scope of research in this direction.

4.2. Innovations of Research Design

Although the emotional influence of color shades on hotel guests has been dealt with in some previous studies [43], this study has pushed the boundary of this line of research by introducing the variable of grayscale. In most existing studies, only gender has been utilized to distinguish groups of respondents. This study, however, differs from previous studies by introducing the previously ignored variables of age and educational background and offering new insights into the potential influence of age and education on the emotional experiences of different color schemes. In addition, this study added color blindness and weakness tests to the experimental design and hence avoided errors caused by color-blind subjects. In order to make customers have a good accommodation experience, it is common practice that a stable, gentle, and homelike environment be created to soothe their nerves and guarantee a good night's sleep. In light of this practice, the indoor environment of the virtual hotel room built in the current study is decorated with more high-end items so as to simulate an emotional experience that is better than feeling at home. Overall, this study created a fine-grained virtual hotel internal environment and has convincingly validated numerous basic theories of color psychology.

Despite the dominant effect of the main interior color, this study innovatively presented to the participants virtual hotel rooms decorated with a range of color combination schemes, including the combination of synergistic colors, contrasting colors, and 90° colors, and found that the combination of synergistic colors was preferable over that of all others, which gives further evidence to clarify the emotional effect of different color schemes in an indoor environment where the dominant color remains unchanged.

4.3. Limitations and Future Directions

Considering that prolonged experimental processes and questionnaires may cause respondents to lose patience, only three color systems were selected in the current study to minimize the duration of the experiment and the questionnaire and to reduce fatigue-induced errors. It is advisable for future research to take more colors into account through innovative research design. Marco et al. [43] stretched the timespan of their study and recorded respondents' overall perceptions of interior color designs longitudinally over a year. Such a design may be recommendable for practitioners in the field of long-term rentals (such as timeshares), but it should be noted that respondents may develop attachments to the environment in which they have lived for a year [47,48] and hence give biased responses.

This study was based on a largely homogeneous cultural background, and most of the subjects were young and middle-aged. The researchers believed that although a multicultural background might help yield results with higher generalizability, there would be an overwhelming number of experimental variables, and consequently the researchers would be deterred from progressing the research agenda. However, we suggest that participants from multicultural backgrounds be recruited in future research so as to identify the similarities and differences across different cultures in the emotional experience of hotel internal environment colors.

To better control the experimental variables and ensure the accuracy of the results, this study fixed the lighting conditions and did not involve the discussion of the emotional effect of lighting. In future research, lighting can be added as a variable. Jingyi et al. examined the effect of different lights and concluded that limited light has a negative emotional effect on elderly subjects [49]. Natural light, or artificial light that is close to natural light, can be pleasant, reduce fatigue, and improve sleep [50,51]. Therefore, different lighting environments may also have an emotional impact. Future research can also take materials and texture into consideration, because the color of materials is a key factor in interior design or architectural design. Kim et al. mentioned that different indoor building materials have different degrees of reflection, which can affect people's emotional experience of the color scheme in an indoor environment [52].

This study merely focused on the subjective evaluations of the participants, so the data collected comprised the participants' emotional responses to different color schemes. Yilin et al. used eye-tracking glasses to record the subjects' eye movements as affected by the lighting environment and simultaneously applied E-Prime 3.0 tools to obtain subjects' psychological reactions [53]. We suggest that behavioral data and psychological data be collected to triangulate participants' subjective evaluations in future research.

5. Conclusions and Suggestions

5.1. Conclusions

This study set out to investigate how hotel rooms decorated with different color schemes affect customers' emotional experiences through experiments and questionnaires, and the answers to the four research questions can be summarized as follows:

1. Without changing the grayscale value, the three color systems of yellow, blue, and gray all have impacts on customers' emotional responses, and males give more positive emotional responses than females, which is consistent with the results of existing studies [25,26]. Yellow rooms evoke more positive emotions than blue and gray rooms. The yellow system has obvious advantages in creating pleasure and arousing senses and produces the greatest emotional fluctuations; the gray system has a prominently calming effect and causes less emotional fluctuations, and the blue system is relatively neutral. Warm colors can bring people a higher sense of pleasure and arousal, while

cool colors make people calmer and more peaceful [45], which is in agreement with the findings of existing research [8].

2. Light colors cause more pronounced mood changes in customers, as opposed to dark colors; room colors that are less gray are more preferable. When the grayscale value is increased (i.e., the color becomes darker), all three color systems are not only less likely to bring positive emotions to customers but also more likely to weaken arousal and to stimulate "cold" and "uneasy" emotions, etc.

3. Compared with hotels rooms decorated with contrasting colors, customers prefer hotel rooms that are decorated with synergistic colors.

4. In general, there is no significant difference in the color preferences of different age groups, but in the case of the yellow-colored room, if the grayscale value is increased, different age groups may have different emotional responses. Respondents under the age of 49 had slightly more negative emotional responses to the high grayscale yellow rooms, and there was no obvious emotional bias among them. However, the group aged 35 to 39 might be an exception, as they obviously gave more positive emotional responses. Respondents aged 50 to 59 also had significantly more positive emotional reactions. Due to the limited number of samples, this study did not collect sufficient data concerning customers' emotional responses to high grayscale yellow rooms, but it could be tentatively concluded that there could be a positive correlation between age and positive emotional experience as the participants under the age of 24 were not satisfied with the high grayscale yellow indoor environment.

Educational background does not have an obvious impact on the emotional experience of blue rooms, but for gray and yellow rooms with higher grayscale values, the emotional responses of different educational groups are different. This study found that the participants' emotional responses to these two high grayscale color systems gradually shifted from positive to negative with the improvement of educational background. Considering that the majority of the subjects were college students and under the age of 24, it would be questionable to claim a definite connection between age and educational background, but those dissatisfied with the high grayscale yellow color were all undergraduates/college graduates or above and under the age of 24.

5.2. Suggestions

Taken together, the findings of this study have a number of implications.

1. Light colors with lower grayscale values are preferable over other colors in hotel interior color design.

2. Color psychologists believe that the combination of more than three main colors in a room will stimulate optic nerves and easily cause excitement and insomnia. Therefore, it is suggested that synergistic colors be of primary consideration in hotel interior color design.

Specifically, different types of hotels need to apply different color schemes. Hotel image is an important factor affecting customer expectations [54]. Yaoqi et al. introduced hotel image perception as an intermediary variable between decoration style and customer purchase intention [55] and found that the hotel image will affect customers' evaluations and expectations of room decoration. In that light, the following recommendations are put forward:

1. Hotels with social attributes, such as social hotels, youth hostels, home-stays, and inns should choose yellow as a pleasant color.

2. For budget hotels that focus on sleep, it is suggested to use blue as the primary color for hotel indoor environments.

3. High-end luxury hotels and designed hotels can choose gray as the main color for hotel indoor environments to create an ambiance of elegance, tranquility, and profundity for customers.

Author Contributions: Supervision and revision, J.X., M.L. and K.C.; investigation, F.Z., B.L. and Z.L.; resources, F.Z. and B.L.; draft, F.Z., B.L., Z.L., Z.C. and K.Z.; proof-reading and editing, F.Z., B.L., Z.L. and K.Z.; data analysis, F.Z.; formal analysis, Z.L. and Z.C.; methodology, Z.L.; visualization, Z.L. and Z.C.; modeling, K.Z. All authors have read and agreed to the published version of the manuscript.

Funding: This study was part of the project called "The study on rural tourism revitalization planning and sustainable living coordination mechanism" funded by the International Cooperation Open Project of State Key Laboratory of Subtropical Building Science, South China University of Technology, the funding number is 2019ZA02. "Image construction and branding output of rural landscape in Beijing–Tianjin–Hebei region: based on spatial communication perspective", the funding number is CUC220B024. And "Guangzhou philosophy and social science planning project: Research on Guangzhou Higher Education Reform Promoting the coordinated development of Guangdong Hong Kong Macao Greater Bay Area", the funding number is 2022GZGJ131.

Data Availability Statement: Data are available on request from the authors.

Acknowledgments: We would especially like to thank Jian Xu and Muchun Li, for their assistance in proofreading the manuscript, and the International Cooperation Open Project of State Key Laboratory of Subtropical Building Science, South China University of Technology. Furthermore, we are grateful to Guangwei Chen for his contribution to Python programming.

Conflicts of Interest: The authors declare no conflict of interest.

References

1. Jianming, S. Color theory, design, professional and experimental psychology. *Zhuangshi* **2020**, *4*, 21–26. (In Chinese)
2. Yildirim, K.; Capanoglu, A.; Cagatay, K.; Hidayetoglu, M.L. Effect of wall colour on the perception of hairdressing salons. *Int. Colour Assoc.* **2012**, *7*, 51–63.
3. Cho, J.Y.; Lee, E.-J. Impact of Interior Colors in Retail Store Atmosphere on Consumers' Perceived Store Luxury, Emotions, and Preference. *Cloth. Text. Res. J.* **2017**, *35*, 33–48. [CrossRef]
4. Wardono, P.; Hibino, H.; Koyama, S. Effects of interior colors, lighting and decors on perceived sociability, emotion and behavior related to social dining. *Procedia Soc. Behav. Sci.* **2012**, *38*, 362–372. [CrossRef]
5. Tantanatewin, W.; Inkarojrit, V. The influence of emotional response to interior color on restaurant entry decision. *Int. J. Hospit. Manag.* **2018**, *69*, 124–131. [CrossRef]
6. Uluçay, N.Ö. An interior design exhibition: An assessment of color scheme preferences and the emotional states of students. *Color Res. Appl.* **2019**, *44*, 132–138. [CrossRef]
7. Jones, P.; Lockwood, A. *The Management of Hotel Operations*; Cassell: London, UK, 1989.
8. Bilal, S.Y.; Aslanolu, R.; Olguntürk, N. Colour, emotion, and behavioral intentions in city hotel guestrooms. *Color Res. Appl.* **2021**, *47*, 771–782. [CrossRef]
9. Alfakhri, D.; Harness, D.; Nicholson, J.; Harness, T. The role of aesthetics and design in hotelscape: A phenomenological investigation of cosmopolitan consumers. *J. Bus. Res.* **2018**, *85*, 523–531. [CrossRef]
10. Labrecque, L.I.; Patrick, V.M.; Milne, G.R. The Marketers' Prismatic Palette: A Review of Color Research and Future Directions. *Psychol. Mark.* **2013**, *30*, 187–202. [CrossRef]
11. Lee, A.H.; Guillet, B.D.; Law, R. Tourists' emotional wellness and hotel indoor environment colour. *Curr. Issues Tour.* **2018**, *21*, 856–862. [CrossRef]
12. Yildirim, K.; Hidayetoglu, M.L.; Capanoglu, A. Effects of Interior Colors on Mood and Preference: Comparisons of Two Living Rooms. *Percept. Mot. Skills* **2011**, *112*, 509–524. [CrossRef] [PubMed]
13. Shishan, H.; Xu, A.; Aihua, Z. Research progress of color psychology in adjuvant treatment of diseases. *Chin. Gen. Pract. Nurs.* **2020**, *18*, 288–290. (In Chinese)
14. Goethe, W. *Theory of Colors*; Frank Cass: London, UK, 1810.
15. Cyr, D.; Head, M.; Larios, H. Colour appeal in website design within and across cultures: A multi-method evaluation. *Int. J. Hum. Comput. Stud.* **2010**, *68*, 1–21. [CrossRef]
16. AL-Ayash, A.; Kane, R.T.; Smith, D.; Green-Armytage, P. The influence of color on student emotion, heart rate, and performance in learning environments. *Color Res. Appl.* **2016**, *41*, 196–205. [CrossRef]
17. Labrecque, L.I.; Milne, G.R. Exciting red and competent blue: The importance of color in marketing. *J. Acad. Mark. Sci.* **2012**, *40*, 711–727. [CrossRef]
18. Hsieh, Y.C.; Chiu, H.C.; Tang, Y.C.; Lee, M. Do colors change realities in online shopping? *J. Interact. Mark.* **2018**, *41*, 14–27. [CrossRef]
19. Jian, Z. Packaging design of edible fungus products based on color psychology. *Edible Fungi China* **2020**, *39*, 129–131. (In Chinese)
20. Yin, F.; Wei, W. Research on the application of color psychology in interior design of tea house. *Tea Fujian* **2017**, *39*, 51. (In Chinese)
21. Yiwei, C.; Ye, H.; Lei, L. Research on color optimization of Lingnan campus architecture based on color psychology. *Design* **2021**, *34*, 142–145. (In Chinese)

22. Xuan, C. Application of color psychology in interior design. *China National Exhib.* **2020**, *18*, 192–193. (In Chinese)
23. Parsons, T.D.; Courtney, C.G.; Arizmendi, B.; Dawson, M. Virtual Reality Stroop Task for Neurocognitive Assessment. *Stud Health. Technol. Inform.* **2011**, *163*, 433–439. [PubMed]
24. Valmaggia, L.R.; Latif, L.; Kempton, M.J.; Rus-Calafell, M. Virtual reality in the psychological treatment for mental health problems: An systematic review of recent evidence. *Psychiatry Res.* **2016**, *236*, 189–195. [CrossRef] [PubMed]
25. Qing, Y.; Shuhua, Z. A Study on the Development of Virtual Reality Technology in China: Review and Prospect. *Sci. Manage. Res.* **2020**, *38*, 20–26. (In Chinese)
26. Qin, L.; Lili, Q.; Tianyu, Y.; Yewei, C. A Review of Virtual Tourism Research: Bibliometrics and Content Analysis Based on Scopus Database. *Tourism Sci.* **2022**, *36*, 16–35. (In Chinese)
27. Sherman, W.R.; Craig, A.B. *Understanding Virtual Reality: Interface, Application, and Design*; Morgan Kauffman: San Francisco, CA, USA, 2018; pp. 441–442.
28. Wang, Y.; Yu, Q.; Fesenmaier, D.R. Defining the virtual tourist community: Implications for tourism marketing. *Tour. Manag.* **2002**, *23*, 407–417. [CrossRef]
29. Stamps, A.E. Use of static and dynamic media to simulate environments: A meta-analysis. *Percept. Mot. Skills.* **2010**, *111*, 355–362. [CrossRef]
30. Yudi, Z. Study on Convenience of Room Space Design in Express Hotel—Take Standard Room of Beijing Chain Hotel as an Example. Master's Thesis, North China University of Science and Technology, Beijing, China, 2020. (In Chinese).
31. Leiqing, X.; Ruoxi, M.; Shuqing, H.; Zheng, C. Healing Oriented Street Design: Experimental Explorations via Virtual Reality. *Urban. Plan. Int.* **2019**, *34*, 38–45. (In Chinese)
32. Gieyong, A.; Pearce, S. A beginner's guide to factor analysis: Focusing on exploratory factor analysis. *Tutor. Quant. Methods Psychol.* **2013**, *9*, 79–94.
33. Williams, B.; Onsman, A.; Brown, T. Exploratory factor analysis: A five-step guide for novices. *Australas. J. Paramed.* **2010**, *19*, 42–50. [CrossRef]
34. Jonauskaite, D.; Althaus, B.; Dael, N.; Dan-Glauser, E.; Mohr, C. What color do you feel? Color choices are driven by mood. *Color Res. Appl.* **2019**, *44*, 272–284. [CrossRef]
35. Ulusoy, B.; Olguntürk, N.; Aslanoğlu, R. Pairing colours in residential architecture for different interior types. *Color Res. Appl.* **2021**, *46*, 1079–1090. [CrossRef]
36. Kaya, N.; Crosby, M. Color associations with different building types: An experimental study on American college students. *Color Res. Appl.* **2006**, *31*, 67–71. [CrossRef]
37. Ulusoy, B.; Olguntürk, N.; Aslanoğlu, R. Colour semantics in residential interior architecture on different interior types. *Color Res. Appl.* **2020**, *45*, 941–952. [CrossRef]
38. Valdez, P.; Mehrabian, A. Effects of colour on emotions. *J. Exp. Psychol.* **1994**, *123*, 394–409. [CrossRef]
39. Mehrabian, A.; Russell, J.A. *An Approach to Environmental Psychology*; MIT Press: Cambridge, UK, 1974.
40. Kim, D.; Hyun, H.; Park, J. The effect of interior color on customers' aesthetic perception, emotion, and behavior in the luxury service. *J. Retail. Consum. Serv.* **2020**, *57*, 102252. [CrossRef]
41. Singh, S. Impact of color on marketing. *Manag. Decis.* **2006**, *44*, 783–789. [CrossRef]
42. Amsteus, M.; Al-Shaaban, S.; Wallin, E.; Sjöqvist, S. Colors in marketing: A study of color associations and context (in)dependence. *Int. J. Bus. Soc. Sci.* **2015**, *6*, 32–45.
43. Costa, M.; Frumento, S.; Nese, M.; Predieri, I. Interior Color and Psychological Functioning in a University Residence Hall. *Front. Psychol.* **2018**, *9*, 1580. [CrossRef]
44. Peng, D.; Yushu, C. Analysis on Color Matching Methods of Panel Furniture. *Furnit. Inter. Design* **2017**, *2*, 26–27. (In Chinese)
45. Qingqing, Z.; Lina, Z. Color matching of interior design and its influence on people's psychology. *Art Panorama* **2018**, *9*, 96–97. (In Chinese)
46. Fangmin, L.; Chenyang, Z.; Haiyong, L. Application of physiological and psychological effects of color in interior design. *Dazhongwenyi* **2018**, *8*, 128–129. (In Chinese)
47. Tognoli, J. Leaving home: Homesickness, place attachment, and transition among residential college students. *J. Coll. Stud. Psychother.* **2003**, *18*, 35–48. [CrossRef]
48. Rioux, L.; Scrima, F.; Werner, C.M. Space appropriation and place attachment: University students create places. *J. Environ. Psychol.* **2017**, *50*, 60–68. [CrossRef]
49. Jingyi, M.; Shanshan, Z.; Wu, Y. The Influence of Physical Environmental Factors on Older Adults in Residential Care Facilities in Northeast China. *Herd-Health Env. Res. Des. J.* **2022**, *15*, 131–149. [CrossRef]
50. Cajochen, C.; Freyburger, M.; Basishvili, T.; Garbazza, C.; Rudzik, F.; Renz, C.; Kobayashi, K.; Shirakawa, Y.; Stefani, O.; Weibel, J. Effect of daylight LED on visual comfort, melatonin, mood, waking performance and sleep. *Lighting Res. Technol.* **2019**, *51*, 1044–1062. [CrossRef]
51. Agnes, D.; Acacia, A.; Jean, F.; Philippe, G.; Anne, K.; Odile, B. Bright light affects alertness and performance rhythms during a 24-h constant routine. *Physiol. Behav.* **1993**, *53*, 929–936.
52. Kim, I.T.; Choi, A.S.; Sung, M.K. Development of a Colour Quality Assessment Tool for indoor luminous environments affecting the circadian rhythm of occupants. *Build. Environ.* **2017**, *126*, 252–265. [CrossRef]

53. Yilin, L.; Shanshan, Z.; Yue, W.; Da, Y. Studies on visual health features of luminous environment in college classrooms. *Build. Environ.* **2021**, *205*, 108184.
54. Flavian, C.; Guinaliu, M.; Torres, E. The influence of corporate image on consumer trust: A comparative analysis in traditional versus internet banking. *Internet Res.* **2005**, *15*, 447–470. [CrossRef]
55. Yaoqi, L.; Hui, F.; Songshan, S.H. Does conspicuous decoration style influence customer's intention to purchase? The moderating effect of CSR practices. *Int. J. Hosp. Manag.* **2015**, *51*, 19–29.

Article

Study on a New Type of Ventilation System for Rural Houses in Winter in the Severe Cold Regions of China

Baogang Zhang [1],*, Xianglu Cai [1] and Ming Liu [2]

[1]   Faculty of Infrastructure Engineering, Dalian University of Technology, Dalian 116081, China; caixianglu@mail.dlut.edu.cn

[2]   School of Architecture and Fine Art, Dalian University of Technology, Dalian 116081, China; liuming@dlut.edu.cn

*   Correspondence: zhangbaogang@dlut.edu.cn or zhangbaogangtj@163.com

Abstract: The weather in the high latitudes of China is cold in winter. The pollution caused by the burning of biomass fuels used in rural individual heating is a great threat to human health. This study finds that the amounts of CO_2, CO, $PM_{2.5}$, and PM_{10} in the bedroom exceed the standard and the temperature does not meet the standard based on indoor air measurements in rural residential buildings in Liaoning Province in winter. In this study, a mechanical ventilation method which uses flue gas to preheat fresh air is proposed, for the purpose of simultaneously improving the indoor air quality and the thermal environment of rural houses. The flue gas–fresh air heat exchange (FGFAHE) experiment shows that biomass combustion flue gas can increase the outdoor air temperature by 23.7 °C on average. The ventilation experiment shows that the method of mechanical ventilation combined with external window penetration can increase the dilution rate of indoor CO by more than 1 times. The simulation results of six different ventilation schemes show that the ventilation mode of the diagonal opposite side upper air supply and lower exhaust air (DOUSLE) has the best effect on indoor CO prevention and control, and the mode of median air supply is the most beneficial to the indoor thermal environment.

Keywords: rural houses; winter; ventilation; waste heat recover; air quality

Citation: Zhang, B.; Cai, X.; Liu, M. Study on a New Type of Ventilation System for Rural Houses in Winter in the Severe Cold Regions of China. *Buildings* **2022**, *12*, 1010. https://doi.org/10.3390/buildings12071010

Academic Editors: Yue Wu, Zheming Liu and Zhe Kong

Received: 19 May 2022
Accepted: 11 July 2022
Published: 14 July 2022

Publisher's Note: MDPI stays neutral with regard to jurisdictional claims in published maps and institutional affiliations.

1. Introduction

In the rural areas of China's severe cold regions, it is difficult to apply central heating due to factors such as population density and living habits [1]. Traditional individual heating is still the main means of heating houses. Inexpensive biomass fuels such as straw and firewood have been widely used until now [2]. According to the statistics, the total amount of bioenergy produced in rural areas of China is about 6 million tons per year, of which about 60% is used for heating and cooking [3]. The heating and cooking activities of individual farmers often produce a large quantity of pollutants, which seriously pollute the indoor air and pose a great threat to human health [4]. Studies have shown that long-term exposure to the environment containing biomass combustion fumes can cause coughing, asthma and other respiratory diseases [5], and it is especially harmful to pregnant women and children [6]. There are also studies showing that $PM_{2.5}$ produced by biomass burning in farmhouses can increase blood pressure in residents [7]. According to the statistics, in Europe, PM pollution shortens human life expectancy by 8.6 months [8]. Therefore, it is necessary for residents living in rural houses to adopt appropriate ventilation strategies to improve indoor air quality.

In recent years, the problem of indoor air quality in rural residential buildings has attracted great attention. Relevant scholars have carried out a lot of research. Wang et al. [9] conducted a measurement on the indoor air quality in the heating season of farmhouses around Harbin City, Heilongjiang Province, China, and found that the air pollutants over the standard level included $PM_{2.5}$, PM_{10}, CO_2, CO, SO_2 and NO_x, etc. Additionally, there is a linear correlation between the mass concentrations of different pollutants, as observed through statistical analysis. There are many factors affecting the indoor air quality of farmhouses. Zhang et al. [10] found that the indoor pollution level of rural houses is related to the living habits of people, indoor temperature, relative humidity, biomass fuel combustion and ventilation methods, etc. Gu et al. [11] found that $PM_{2.5}$ directly generated by burning straw and wood accounted for 52% and 37% of the air in the room, and the indirectly generated $PM_{2.5}$ accounted for 7% and 15%. The indoor air pollutants in farmhouses also show different concentrations in different time and spaces. Yang [12] found that the indoor temperature, humidity and the concentration of air pollutants in rural residential buildings demonstrate similar regular changes every day during measurement. Liu [13] found that the concentration of indoor air pollutants in rural residential buildings is significantly different in winter and summer, and also in rooms with different functions. The above studies show that the indoor air pollution in rural residential buildings is characterized by a wide variety of air pollutants and is influenced by complex factors.

Ventilation is the most economical and practical way of removing indoor air pollutants [14]. A report on the indoor air quality of 85 buildings in the United States, Canada and Europe showed that improper ventilation is the biggest problem causing indoor air pollution [15]. A low ventilation rate can also cause a buildup in indoor moisture and provide growth conditions for mold and other microorganisms [16]. According to different air flow dynamics, ventilation can be divided into natural ventilation and mechanical ventilation [17]. Natural ventilation driven by thermal pressure and wind pressure [18] is the most important ventilation method in Chinese buildings at this stage. However, in the winter in severe cold regions, the temperature difference between indoor and outdoor environments can reach more than 30 °C, and improper ventilation can cause cold air intrusion, causing discomfort. Therefore, conventional natural ventilation methods are not suitable for farmhouses in cold regions. Passive ventilation based on solar energy and building waste heat, which can simultaneously increase ventilation rate and supply air temperature, has received extensive attention in recent years. A number of achievements has been made, including roof-mounted solar chimneys, wall-mounted solar chimneys [19], combination system of trombe walls, geothermal air pipes and a solar chimney [20], and a combination system of rotating heat recovery devices and rooftop wind catchers [21]. However, passive ventilation is often unstable due to uncontrollable solar radiation and outdoor air movement [22]. Several studies have shown that natural ventilation is not effective in removing pollutants from indoor air [23], and improper natural ventilation can even worsen air quality [24].

Compared with passive ventilation, mechanical ventilation can control the volume of the wind, and the starting and stopping of the ventilation, so as to regulate the microenvironment such as indoor airflow and air pressure [25]. Studies have shown that mechanical ventilation is better than natural ventilation at controlling pollutants such as particulate matter and fungi [26]. However, conventional mechanical ventilation can only control ventilation volume, and cannot change the air supply load. Indoor circulating air with supplementary air together can reduce the heat load of ventilation to a certain extent, but if the proportion of supplementary air is too low, the air quality of the supply air cannot be guaranteed [27]. Jinkyu et al. [28] found that when the volume ratio of supplementary air to recirculation air is adjusted to 1.2, the energy consumption can be reduced by about 22.1%, improving the indoor air quality at the same time. Underground duct ventilation uses the temperature of the soil to preheat the outdoor cold air, which has been widely used in Europe [29]. Zhao et al. [30] studied the application of a horizontal multi-row parallel underground pipeline ventilation system, and found that the system can preheat the outdoor air from −8~−10 °C to 0~15 °C. Wang et al. [31] conducted a survey of several houses in Sweden that had been reformed with combined ventilation heat recovery and low-temperature heating systems, and found that the comprehensive energy consumption and primary energy consumption were reduced by 55% and 22%, respectively, and the indoor air quality dissatisfaction rate dropped from 16.2% to 14.5%. Nonetheless, the above methods can improve the supply air temperature to some extent, but the preheating effect is not obvious.

The average annual heating fuel consumption in rural areas in China is as high as 600 million tons, of which 60% is used for heating kang or stove-kang [32]. However, the heating efficiency of the kang system is low, the heat gain rate of the kang body is often lower than 50%, and the temperature at the exit of the chimney is as high as 343–363 K [33], which is a serious waste of energy. Some scholars have launched research to recover this part of the waste heat. Yu et al. [34] found that the wall can effectively absorb the waste heat of the flue gas and increase the wall temperature by conducting an experimental study on the chimney combination wall. Jin et al. [35] conducted research on a combined heating system of hanging kang and soil heating. The results show that the efficiency of this system is 21.26% higher than that of traditional floor-standing kang. Zhao [36] proposed a tunnel heating system with flue gas waste heat, and it was found that this system can save 9.17% of the heating energy consumption. The above studies all use the building envelope construction to absorb the waste heat of the combustion flue gas of farmhouses. Up to now, few people have carried out research on the heat exchange between flue gas and fresh air.

This study intends to explore the appropriate ventilation strategy according to the law of change of the indoor air pollutant concentration in the farmhouse using the waste heat of the flue gas emitted by the farmhouse stove and kang system (SKS) to preheat the outdoor fresh air and reduce the heat load at the same time, and. This scheme can simultaneously improve the indoor air quality and indoor thermal environment of rural houses in severe cold areas in China, and can recycle the waste heat resource, which has far-reaching significance for improving the living comfort and health level of rural residents and achieving sustainable development. In this study, we conduct a field measurement of indoor air in rural houses in severe cold areas in China, and summarizes the concentration changes; we also conduct a flue gas–fresh air heat exchange (FGFAHE) experiment to explore the preheating effect of smoke exhaust from the SKS of farmhouses on outdoor fresh air and the dilution effect of mechanical ventilation on indoor air pollutants in farmhouses. This study also adopts the simulation method to explore the best air supply and exhaust outlet layout scheme suitable for farmhouses in severe cold areas.

2. Indoor Air Measurement of Rural Houses in Western Liaoning, China

2.1. Materials and Methods

Reasonable ventilation should achieve maximum energy saving on the basis of effectively improving indoor air quality. For rural houses in severe cold areas, the concentration state of indoor air pollutants can be divided into different levels, and the purpose of precise air supply can be achieved by adjusting the ventilation volume according to the concentration level of pollutants. Therefore, it is necessary to grasp the concentration state and the changing regularities of indoor air pollutants in farmhouses. In this study, CO_2, CO, PM_{10}, $PM_{2.5}$, formaldehyde, TVOC, temperature and relative humidity are measured for five consecutive days in the bedroom of three rural houses in Chaoyang City, Liaoning Province, China. The test instruments and the introductions are shown in Table 1.

Table 1. Introduction of test instrument.

Equipment	Model	Metrics	Range	Accuracy
Environmental monitor	Environmental monitor EVM serious	CO_2	0~5000 ppm	±100 ppm
		CO	0~1000 ppm	±5%
		Temperature	−20~60 °C	±1.1 °C
		Relative humidity	0~100%	±5%
Particulate Monitor	LIGHTHOUSE3016	$PM_{2.5}$ PM_{10}	0~4,000,000/ft^3	±5%
Formaldehyde Monitor	AKBT-PID-CH2O	Formaldehyde	0~100 ppm	±0.01 ppm
TVOC Monitor	AKBT-PID-TVOC	TVOC	0~100 ppm	±0.01 ppm

The test time was from January 27 2021 to February 10 2021. The test method refers to "Indoor Air Quality Standards (GBT18883-2002)".

2.2. Results and Discussion

The test results are shown in Figure 1. The white area in the figure represents the personnel activity period (PAP, 7:00–22:00), the purple area represents the nighttime sleep period (NSP, 22:00–7:00), and the red horizontal dotted line represents the safety limit in relevant standards.

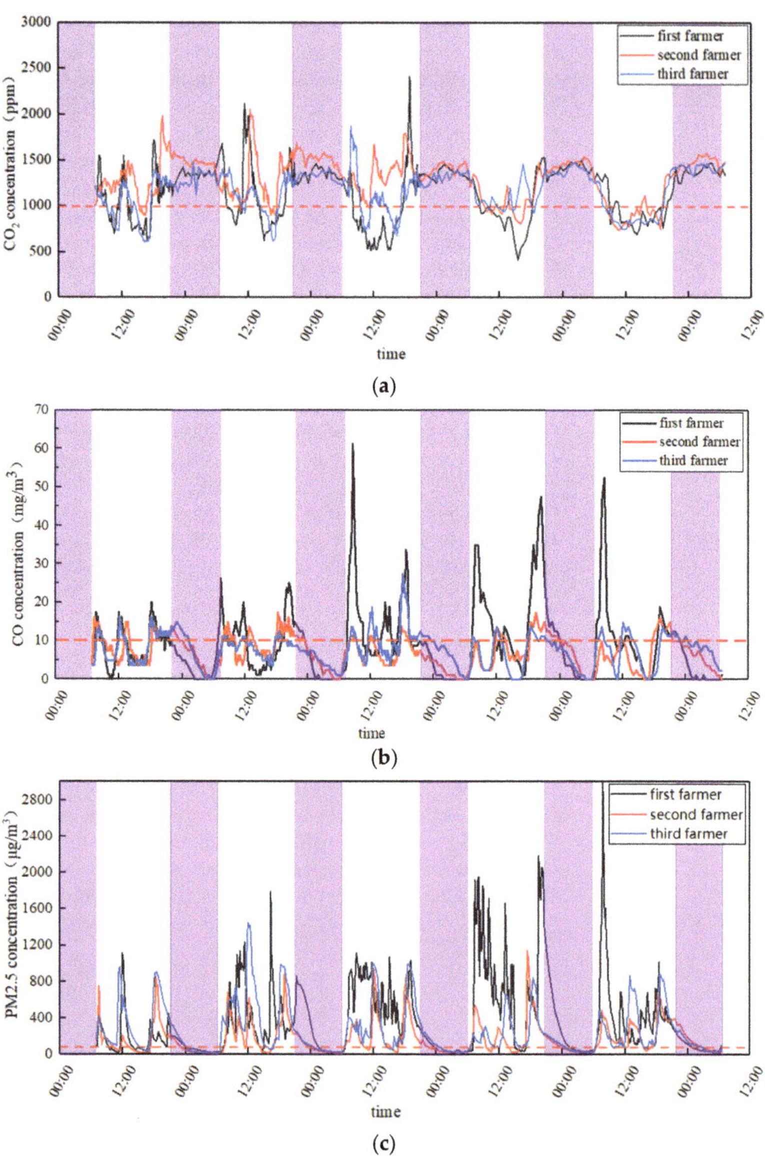

(a)

(b)

(c)

Figure 1. *Cont.*

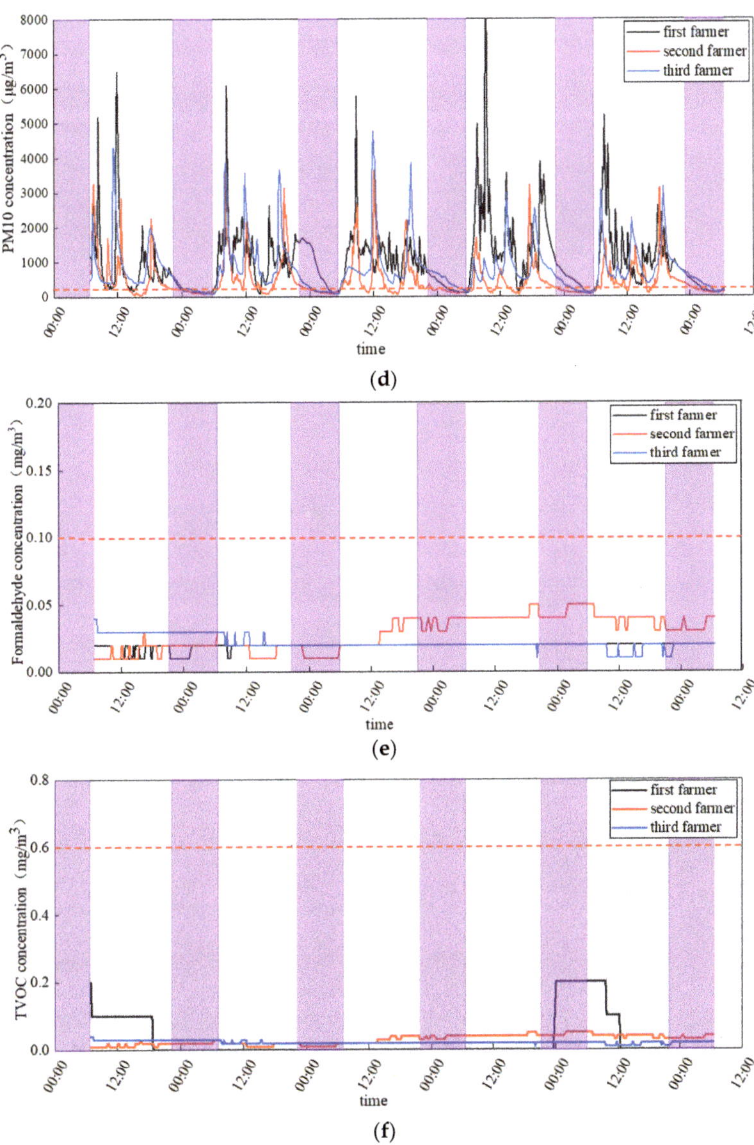

Figure 1. *Cont.*

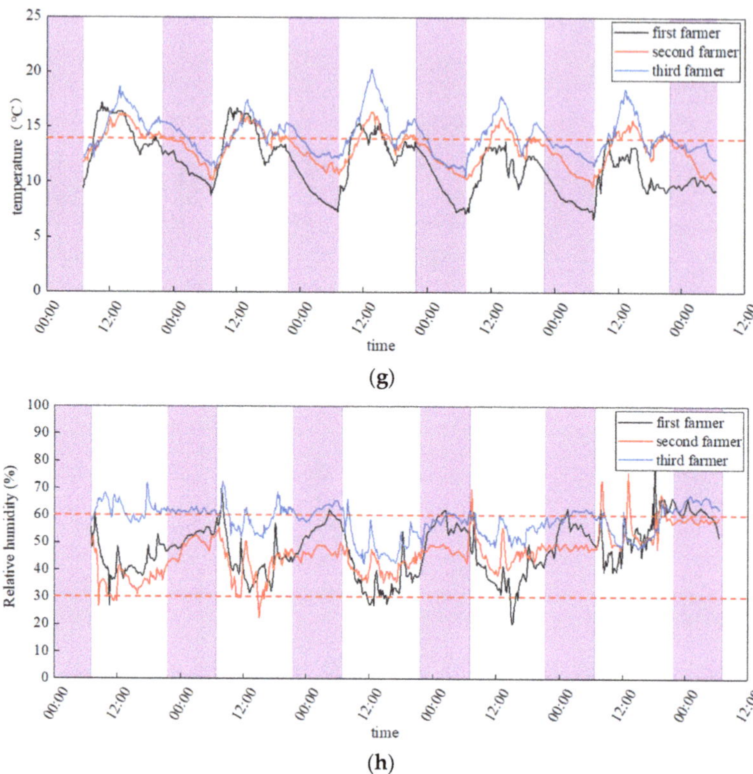

Figure 1. Field test of air quality in bedrooms of three farmhouses for 5 consecutive days: (**a**) CO_2, (**b**) CO, (**c**) $PM_{2.5}$, (**d**) PM_{10}, (**e**) Formaldehyde, (**f**) TVOC, (**g**) temperature, (**h**) relative humidity.

According to the above test, it can be found that during the test period, the relative humidity of the three farmhouse bedrooms is within the comfortable range most of the day, and it slightly deviates from the comfort zone for a short period of time. The concentrations of formaldehyde and TVOC are far below the standard values throughout the day. The concentration of CO_2 exceeds the standard value most of the time throughout the day, and it continues to be in a high concentration state during NSP, generally between 1000 ppm and 2000 ppm. The concentrations of CO, $PM_{2.5}$, and PM_{10} exceed the standard to varying degrees during PAP, and they gradually decrease during NSP and are within the standard value most of the time. The indoor temperature shows a trend of rising first and then falling during PAP, and gradually decreases during NSP. The average temperature throughout the day is lower than the winter calculation temperature of farmhouses stipulated in the "Design Standards for Energy Efficiency of Rural Residential Buildings (GBT 50824-2013)".

Studies have shown that a high CO_2 concentration can reduce the cognitive performance and attention of building occupants [37]. The studies show that 10,000 ppm exposure for 30 min can cause respiratory toxicity in an adult [38]. An indoor CO_2 concentration in the range of 1000–2000 ppm generally causes drowsiness but no short-term health effects [12]. Therefore, the impact of CO_2 on people during bedtime may not be considered. The main factor that affects residents' thermal comfort during NSP is the small-scale microenvironment around kang, and so the impact of comprehensive indoor air temperature at night may not be focused on. The critical period is PAP, and the key indicators are CO_2, CO, $PM_{2.5}$, PM_{10} and temperature.

Observing the above curves, it can be found that the changes in CO_2, CO, $PM_{2.5}$ and PM_{10} concentrations in the bedrooms of the three farmhouses are disorderly, which is caused by the irregular activities of indoor residents. However, in general, there is a common trend, in that there are some peaks exceeding the standard value during the daily cooking period. In addition, many local residents also report that the bedroom often produces a choking feeling of smoke during these periods.

It can be found that there are two main reasons for the severe indoor air pollution in rural houses observed through the on-site investigation. The first reason is the unreasonable layout of the rooms. The kitchen is the most polluted place in the farmhouses. However, most of the rural houses in this area have adjacent bedrooms and kitchens, which are connected by bedroom doors that are open all day. The door curtain made from cotton cloth on the bedroom doors cannot effectively prevent the spread of kitchen pollution into the room. The second reason is the poor smoke resistance of the building. The kang body is mainly composed of clay, which is different from the surrounding wall material and can easily cause improper connection, resulting in smoke leakage. The extremely high concentration of pollutants in the flue gas is bound to have an adverse effect on the indoor air. Cooking is the most important period of rural kitchen activities. The main reason for the serious pollution during the three meal periods is the large quantity of pollutants directly or indirectly produced by cooking activities that are released into the room.

According to statistics, the average CO_2 over-standard rates in the bedrooms of the three farmhouses are 45.7%, 84.9%, and 68.6%, respectively, and the average excess multiples are 1.01, 1.26, and 1.17. The average CO over-standard rates are 54.8%, 46.1%, and 48.1%, respectively. The average $PM_{2.5}$ over-standard rates are 85.8%, 77.4% and 82.6%, and the average exceeding multiples are 5.5, 2.89 and 4.2, respectively. The average PM_{10} over-standard rates are 95.1%, 78.0% and 86.4%, and the average exceeding multiples were 7.5, 4.2 and 5.3, respectively. In view of the poor air quality in the bedrooms of farmhouses, it is necessary to take reasonable ventilation measures. The cooking periods should be taken as the key ventilation periods.

3. Experiment of the New Ventilation System

3.1. Materials and Methods

Observing Figure 1g, it can be seen that the temperature level of the bedrooms of the three farmhouses is poor. In severe cold areas where the outdoor temperature is low in winter, introducing outdoor fresh air directly into the room would further damage the indoor thermal environment. This study seeks to use the flue gas emitted by the SKS in the farmhouses to preheat the outdoor air. The design of the flue gas–fresh air heat exchange (FGFAHE) experiment bench is shown in Figure 2. It is not advisable to set up exhaust vents in the experiment room because the room is surrounded by walls on three sides, and one side is adjacent to the kitchen. In this study, infiltration exhaust is selected. The schematic graph for the pipeline paths of the fresh air system is shown in Figure 3.

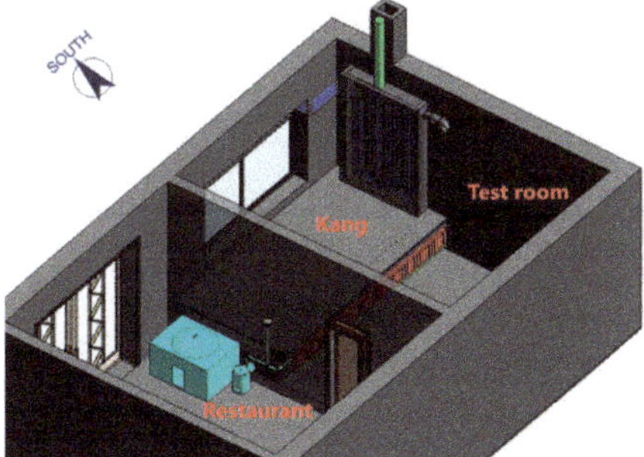

Figure 2. Design of FGFAHE experiment bench.

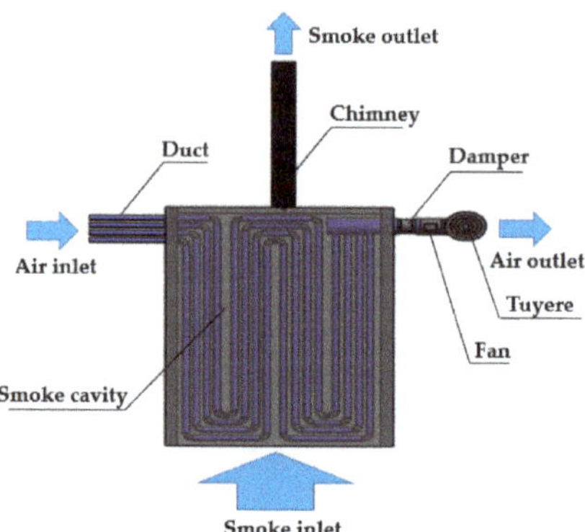

Figure 3. Schematic graph for the pipeline paths of the fresh air system.

The feasibility of the design scheme was demonstrated before the experiment bench was built. Considering the price, melting point, thermal conductivity and other aspects comprehensively, we chose a 304 stainless steel telescopic tube as the heat exchange air tube. Considering the size of the smoke cavity, the diameter of the air duct was preset to 50 mm. The size of the smoke cavity was 1.2 m × 1.5 m × 0.3 m, and the air ducts were arranged in parallel in a serpentine shape in the smoke cavity.

Fan selection is a very critical step. This study sought to check the resistance of the test system with a ventilation frequency of 3 h^{-1}. The size of the experiment room was 4.5 m × 3.5 m × 3 m. The calculation result of the ventilation volume in this state was about 141.3 m^3/h, and the air velocity was about 5 m/s. This was the state with the largest system resistance. The relevant local resistance coefficients refer to the second edition of the "Practical Heating and Air Conditioning Design Handbook" [39]. The resistance check results are shown in Table 2 below. It can be found from the table that the maximum resistance of the system was about 274.6 Pa. Taking the wind pressure correction coefficient as 1.2, the total resistance of the system was obtained as 329.72 Pa. A HF150 variable frequency duct fan with a rated power of 57 w was selected. Under the rated power, it could provide a static pressure value of 330 pa and a ventilation volume of 650 m^3/h, ensuring the normal operation of the system.

The outer wall of the smoke chamber of the experiment bench was made of red clay brick, and the connection between the air duct and the wall of the smoke chamber is smoothed with local high-viscosity laterite mud to avoid the infiltration of smoke. In order to avoid the influence of cold air entry caused by the poor sealing of the air duct and the outer window, this position was sealed with foam material, as shown in Figure 4.

(a)

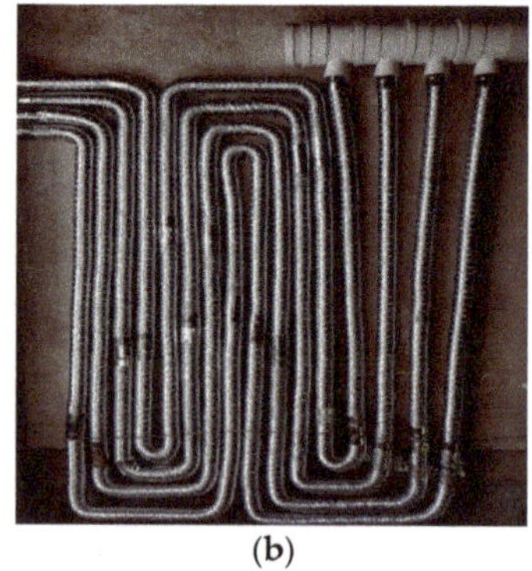

(b)

Figure 4. FGFAHE experiment bench: (**a**) experiment bench's appearance, (**b**) duct arrangement.

Table 2. Hydraulic calculation of fresh air system.

Pipe Number	Air Flow Q_v/(m³/h)	Length L/m	Width a/mm	Height b/mm	Actual Cross-Sectional Area S/m²	Tube Wind Speed V_s/(m/s)	Dynamic Pressure P_d/Pa	Coefficient of Locall Resistance $\sum\xi$	Local Resistance ΔP_j/Pa	Specific Frictional Resistance R_m/(Pa/m)	Frictional Drag $R_m L$/Pa	Section Resistance $(R_m L + \Delta P_j)$/Pa	Total System Resistance ΔP/Pa	Allow for the Wind p_f/Pa
Main pipe	141.3	0.6	100	100	0.00785	5.00	15.00	1.77	26.55	1.20	0.72	27.27		
First branch	141.3	0.25	100	100	0.00785	5.00	15.00	0.20	3.00	1.20	0.30	3.30		
	35.33	5.53	50	50	0.0019625	5.00	15.00	1.40	21.00	6.00	33.15	54.15		
Second branch	105.98	0.08	100	100	0.00785	3.75	8.44	1.52	12.83	3.20	0.26	13.08	274.76	329.72
	35.33	5.36	50	50	0.0019625	5.00	15.00	1.40	21.00	6.00	32.16	53.16		
Third branch	70.65	0.08	100	100	0.00785	2.50	3.75	5.40	20.25	1.30	0.10	20.35		
	35.33	5.17	50	50	0.0019625	5.00	15.00	1.40	21.00	6.00	31.02	52.02		
Fourth branch	35.33	0.08	100	100	0.00785	1.25	0.94	0.20	0.19	1.00	0.08	0.27		
	35.33	5.03	50	50	0.0019625	5.00	15.00	1.40	21.00	6.00	30.16	51.16		

This study carried out a three-day experiment from 15 to 17 February 2021. The weather was sunny during the three days. The experiment time was in the morning, and the outdoor average air temperatures on the three days were: 15 February: $-17\,°C$, 16 February: $-14\,°C$, 17 February: $-10\,°C$.

The content of this experiment can be divided into two parts. The first part mainly discussed the heat exchange effect between flue gas and fresh air. The experiment was divided into two working conditions: the time of burning and one hour after the burning stopped. The flue gas temperature was measured with the TM902c thermometer. The average temperature of the flue gas at the entrance of the kang was $150\,°C$ in the first working condition and $50\,°C$ in the second working condition. During the experiment, the air velocity was controlled by adjusting the fan gear position and the air valve, and the change in the ventilation volume was reflected by the change in the air velocity. The TES1341 hot-wire anemometer was used to measure the temperature of fresh air outlet.

The second part of the experiment mainly discussed the dilution effect of fresh air on indoor CO. In this part, the stove was first fired to create a CO polluted environment in the room, and the environmental monitor (an EVM series environmental detector) was turned on to test the CO concentration. Before starting the experiment, the door was closed tightly and the fan was turned on for ventilation. Exhaust could be eliminated by the penetration of external windows. The air velocity was controlled to be 5 m/s, 4 m/s and 3 m/s, respectively (the corresponding fresh air volumes were $141\ m^3/h$, $113\ m^3/h$ and $85\ m^3/h$, respectively) in 3 groups of experiments.

3.2. Results and Discussion

As shown in Figure 5, the results show that the flue gas can preheat the outdoor air from $-17\,°C\sim-10\,°C$ to $6\,°C\sim30\,°C$ during the time of burning. When the experiment ventilation volume reached the maximum ($141\ m^3/h$), the system raises the outdoor air temperature by $7.2\,°C$ on average. When the experiment ventilation rate was the minimum ($29\ m^3/h$), the system could raise the outdoor air temperature by an average of $23.7\,°C$. In the working period of one hour after the burning stopped, the residual heat in the smoke chamber preheated the outdoor cold air to $-6\,°C\sim10\,°C$.

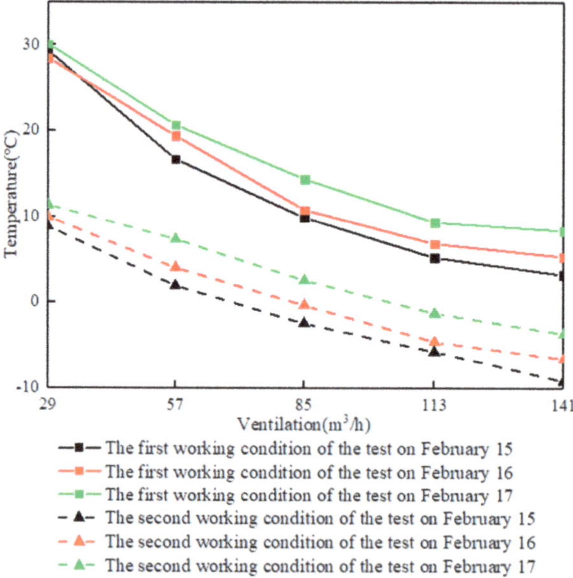

Figure 5. Fresh air preheating experiment results.

The change in indoor CO concentration of the three groups without any ventilation measures was selected as a comparison, and the curve of CO concentration with time was drawn, as shown in Figure 6. The experiment results show that the combination of mechanical ventilation and external window penetration could effectively dilute indoor air pollutants, and the dilution rate was more than double that of natural penetration. When the ventilation rate was 85~141 m^3/h, it only took about 17–30 min to reduce the CO concentration of 20~25 mg/m^3 to below the standard value (10 mg/m^3).

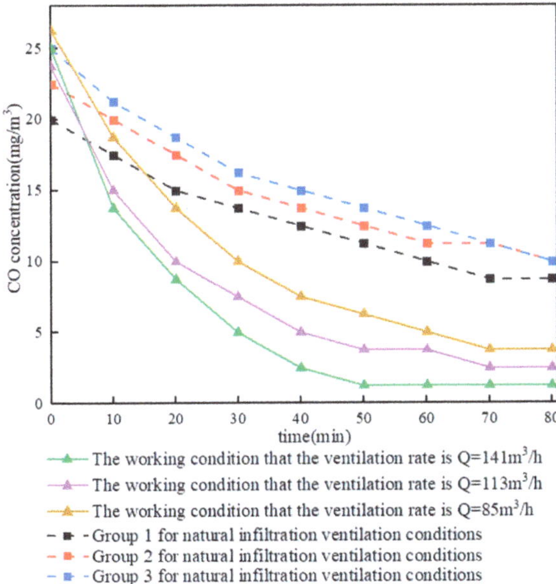

Figure 6. Fresh air dilution experiment results.

For mechanical ventilation, reasonable airflow organization design is the key to achieving efficient ventilation. The above experiments only explore the ventilation effect of one scheme (combination of mechanical ventilation and exterior window penetration). According to the characteristics of the experiment room and the bench, it is not appropriate to adjust the position of the air supply port or add additional exhaust ports to conduct more experiments. Therefore, it is necessary to adopt the method of simulation to discuss more schemes.

4. Simulation of the New Ventilation System

4.1. Methods

In this study, ANSYS FLUENT software was selected for the simulation of pollutant dispersion and flow field analysis. A CFD model was established first, and it was necessary to verify the accuracy of the model with the above experiments. The exhaust method of the above ventilation experiment was the penetration of external window, and so it was necessary to clarify the size of the penetration gap. In this study, CO was used as the tracer gas. The volume of the gap was determined by the CO natural penetration experiment, the procedure of which is as follows.

First, we opened the door to connect the experiment room with the kitchen, and started a fire on the stove to create a high-concentration CO environment in the room. Then, we turned on the environmental detector to test the CO concentration and air temperature in the experiment room. We turned off the flame completely and closed the stove door when the CO concentration in the experiment room reached 25 ppm. We prepare the experiment when the CO concentration of the kitchen and the experiment room were similar for the purpose of eliminating the influence of mutual penetration on the experiment. We recorded the concentration of CO at the time when the experiment started. After about 2–3 h, the experiment ended when the CO in the room was sufficiently diluted. We carried out three experiments and derived the results. Then, we plotted the change curve of CO concentration with time, as shown in Figure 6 for natural infiltration ventilation conditions.

The time taken for the CO concentration to decrease from 20 mg/m^3 (16 ppm) to 10 μg/m^3 (8 ppm) in the three experiments was 61 min, 66 min and 71 min, respectively. The average dilution time of 66 min was taken to calculate the permeation ventilation. The size of the kang was 3.5 m × 2.0 m × 0.7 m, and the effective space size of the experiment room was 42.35 m^3. According to the above conditions, the natural penetration ventilation volume was obtained as Q = 38.5 m^3/h.

The main driving force of infiltration ventilation is the thermal pressure formed by the temperature difference between indoor and outdoor. The formula for the calculation of hot pressure [40] is as follows:

$$\Delta P_t = (\rho_w - \rho_n)gh \tag{1}$$

where ΔP_t is the level of the hot pressure, Pa; ρ_w is the outdoor air density at the experiment time, kg/m^3; ρ_n is the indoor air density at the experiment time, kg/m^3; and h is the net height difference between the supply and exhaust air, m.

The calculation of natural ventilation dynamic pressure is shown as follows [40].

$$\Delta P_v = \xi \frac{v^2}{2} \rho \tag{2}$$

where ΔP_v is the pressure difference between the two sides of the window hole, Pa; v is the flow velocity of the air flowing through the window hole, m/s; ρ is the density of the air passing through the window hole, kg/m^3; and ξ is the local resistance coefficient of the window hole.

The size of the outer window of the experiment room is 1.5 m × 1.5 m. Taking the half plane of the window height as the neutral plane, then the average height of air supply and exhaust is 0.75 m. In the three natural infiltration experiments, the average outdoor temperature was −12 °C and the average indoor temperature was 10 °C. Air density refers to Appendix 5 in the textbook "Heat Transfer" published by China Higher Education Press. The above data are substituted into (1) to obtain the hot pressure size as ΔP_t = 0.57 Pa. The thermal pressure is converted into dynamic pressure of natural infiltration, that is, $\Delta P_t = \Delta P_v$. The local resistance coefficient of the window hole refers to the vertical axis plate window in the textbook "Ventilation Engineering" published by China Machinery Industry Press. The above data are substituted into (2) to obtain the infiltration velocity as v = 0.6 m/s. Then, the gap area of the outer window can be obtained as 0.018 m^2. The total length of the slit (openable part) is 6.6 m through measurement, so the average width of the outer window slit is 5.39 mm.

The natural infiltration process is slow. Due to the lack of room insulation measures, the indoor temperature will gradually drop during the experiment. Therefore, the result obtained by the above method is greater than the actual value. Then, a certain degree of correction should be carried out. Taking the convenience of calculation into consideration, the gap width of this model is set to 5 mm after correction. The average width of sliding window gaps in China are about 2–4 mm, which is reasonable for rural buildings with poor construction technology.

The experiment platform is geometrically modeled by ICEM CFD software, as shown in Figure 7. The following assumptions are made to facilitate problem analysis. First, air is an incompressible fluid, and the viscous dissipation in the flow process should be ignored; second, the physical parameters of air do not change with temperature, and the density satisfies the Boussinesq assumption; third, the temperature of the envelope structure is uniform and stable; fourth, there are no other sources of CO pollution and infiltration in the room; fifth, the infiltration of the room only occurs in the external windows. In the experiment room, the air flows under the disturbance of mechanical supply air, which is turbulent flow under forced convection. The control equations established in this study are as follows:

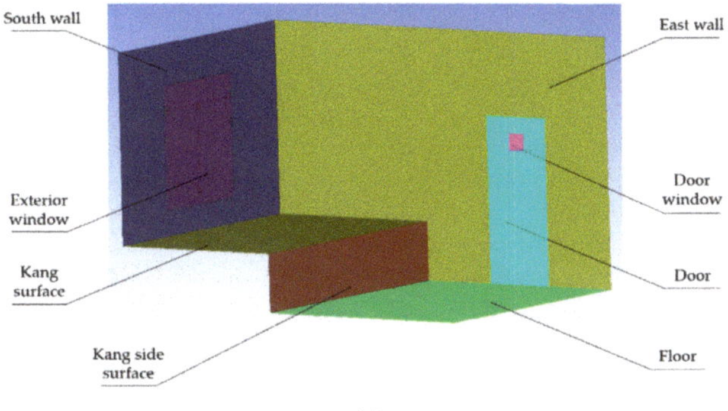

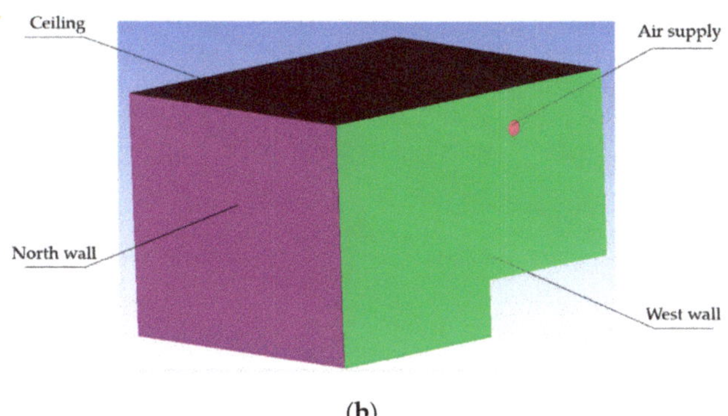

Figure 7. The geometric model of the experiment room, (**a**) southeast view, (**b**) Northwest view.

Continuity Equation

$$\frac{\partial u}{\partial x} + \frac{\partial v}{\partial y} + \frac{\partial w}{\partial z} = 0 \tag{3}$$

Momentum conservation equation

$$\rho u \frac{\partial u}{\partial x} + \rho v \frac{\partial u}{\partial y} + \rho w \frac{\partial u}{\partial z} = \mu\left(\frac{\partial^2 u}{\partial x^2} + \frac{\partial^2 u}{\partial y^2} + \frac{\partial^2 u}{\partial z^2}\right) - \frac{\partial p}{\partial x} \tag{4}$$

$$\rho u \frac{\partial v}{\partial x} + \rho v \frac{\partial v}{\partial y} + \rho w \frac{\partial v}{\partial z} = \mu\left(\frac{\partial^2 u}{\partial x^2} + \frac{\partial^2 u}{\partial y^2} + \frac{\partial^2 u}{\partial z^2}\right) - \frac{\partial p}{\partial y} - \rho g \tag{5}$$

$$\rho u \frac{\partial w}{\partial x} + \rho v \frac{\partial w}{\partial y} + \rho w \frac{\partial w}{\partial z} = \mu\left(\frac{\partial^2 u}{\partial x^2} + \frac{\partial^2 u}{\partial y^2} + \frac{\partial^2 u}{\partial z^2}\right) - \frac{\partial p}{\partial z} \tag{6}$$

Energy conservation equation

$$\rho c_p\left(u \frac{\partial T}{\partial x} + v \frac{\partial T}{\partial y} + w \frac{\partial T}{\partial z}\right) = \lambda\left(\frac{\partial^2 T}{\partial x^2} + \frac{\partial^2 T}{\partial y^2} + \frac{\partial^2 T}{\partial z^2}\right) + S_H \tag{7}$$

Species mass-conservation equation

$$\frac{\partial(\rho c_s)}{\partial t} + \frac{\partial(c_s u)}{\partial x} + \frac{\partial(c_s v)}{\partial y} + \frac{\partial(c_s w)}{\partial z} = \frac{\partial}{\partial x}\left(D_S \frac{\partial(c_s)}{\partial x}\right) + \frac{\partial}{\partial y}\left(D_S \frac{\partial(c_s)}{\partial y}\right) + \frac{\partial}{\partial z}\left(D_S \frac{\partial(c_s)}{\partial z}\right) \tag{8}$$

In the formula, u, v, and w represent the velocity of air in the X, Y, and Z directions, respectively, m/s; ρ is the density of the air, kg/m^3; p is the pressure of the air, Pa; μ is the dynamic viscosity, Pa·s; cp is the specific heat capacity at constant pressure, J/(kg·K); T is the temperature of the gas, °C; λ is the thermal conductivity of the gas, W/(m·K); S_H is the radiation heat transfer term; c_s is the volume concentration of the component; and D_S is the diffusion coefficient of the component.

When choosing a turbulence model, it is necessary to consider its calculation accuracy, calculation time and the computing power of the CPU. In this study, three turbulence models, namely the standard k-ε model, the RNG k-ε model and the realizable k-ε model, are used to numerically simulate the ventilation process. The calculation results show that the convergence rate of the standard k-ε model is slow, and the error is large when calculating the non-uniform turbulence problem. The residual curve of the RNG k-ε model oscillates seriously. Using the realizable k-ε model can improve the convergence speed of the calculation and reduce the error. Therefore, the realizable k-ε model is selected for simulation calculation.

The boundary condition of the air inlet adopts the velocity inlet, and the inlet air velocity and temperature refer to the above experiment conditions, as shown in Table 3 below. The boundary condition of the air outlet adopts the pressure outlet. The boundary condition of the wall adopts the first type. The wall temperature is tested by the TM902C thermometer, and the average value of the experiment results is taken as shown in Table 4. The measured results of initial CO concentration and indoor air temperature are shown in Table 5. The SIMPLE algorithm is used to solve the problem, the non-steady time step is 0.1, and the maximum time step is 10.

Table 3. Inlet boundary conditions for ventilation dilution simulations.

Test Conditions	Velocity (m/s)	Temperature (°C)
Q = 141 m^3/h	5	8.40
Q = 113 m^3/h	4	9.38
Q = 85 m^3/h	3	14.44

Table 4. Wall temperature boundary condition for ventilation dilution simulation.

Project	Exterior Wall	Exterior Window	Floor	Roof	Kang Surface	Kang Side
Temperature (°C)	8.1	1.8	4.5	7.2	42.6	28.7

Table 5. Initial conditions for ventilation dilution simulations.

Test Conditions	CO Concentration (mg/m^3)	Temperature (°C)
Q = 141 m^3/h	25	10.6
Q = 113 m^3/h	23.75	10.9
Q = 85 m^3/h	26.25	11.2

The geometric model is shown in Figure 7. The type of grid is a structured grid. Five sets of structured grid schemes are established to verify the grid independence, and the grid numbers are 43,624, 61,470, 156,408, 428,400 and 1,051,612, respectively. The five schemes are simulated under the test conditions of Q = 141 m^3/h. The inlet velocity and temperature are shown in Table 3, the temperature of each wall of the experiment room is shown in Table 4, and the initial CO concentration and initial temperature of the test room are shown in Table 5. The curve of indoor CO concentration with time is drawn as shown in Figure 8. It can be seen from the figure that there is a small difference between the simulation results when the grid number is 156,408 and 428,400, the maximum error is 0.4 mg/m^3, and the average error is only 0.3 mg/m^3. Therefore, when the number of grids is greater than

156,408, the accuracy of numerical calculation is no longer significantly improved, and 156,408 grids can be used for calculation. The result of grid division is shown in Figure 9.

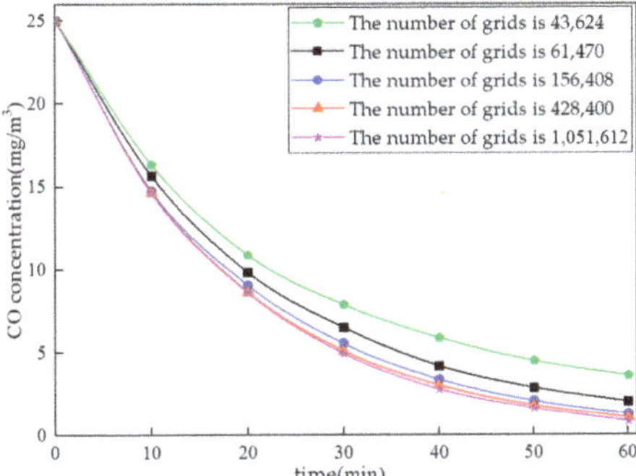

Figure 8. Grid independence verification.

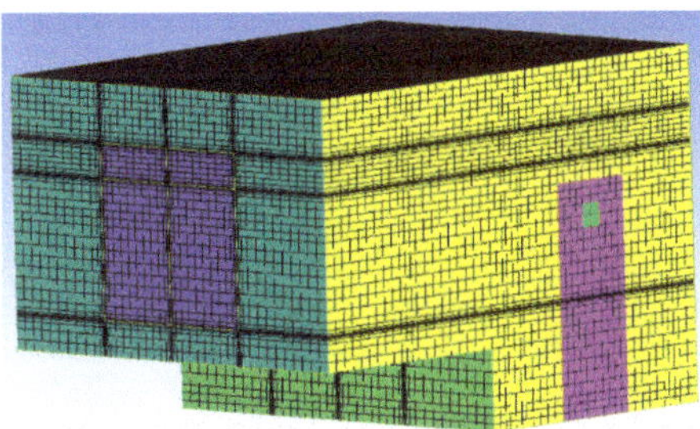

Figure 9. Structural meshing of room models.

The reliability of the model is verified by experiments with ventilation volume of $Q = 141$ m^3/h, $Q = 113$ m^3/h and $Q = 85$ m^3/h, respectively. The boundary conditions and initial conditions of each test are shown in Tables 3–5. The variation curve of the average CO concentration over time at a height of 1.2 m in the room is as shown in Figure 10. It can be seen from the figure that the errors between the simulation result and the experiment are small, of which the maximum error is 0.98 mg/m^3, and the average error is 0.53 mg/m^3. Therefore, it can be considered that the CFD model established in this study can represent the experimental results.

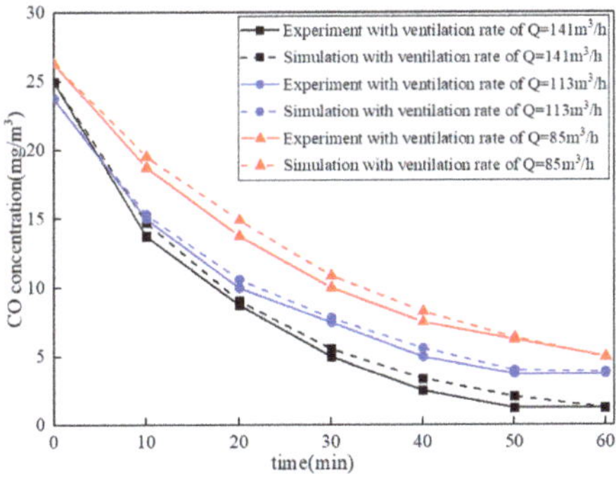

Figure 10. Model reliability verification.

The above model is now used to further explore the ventilation of different airflow organizations. According to the characteristics of rural houses and the relevant recommendations of "Design Code for Heating Ventilation and Air Condition (GB50736-2012)", this study proposes six layout schemes of air supply and exhaust vents as follows: median opposite side upper air supply and lower exhaust air (MOUSLE), median same side upper air supply and lower exhaust air (MSUSLE), median upper air supply and exterior window penetration (MUSEWP), diagonal opposite side upper air supply and lower exhaust air (DOUSLE), diagonal same side upper air supply and lower exhaust air (DSUSLE), and diagonal upper air supply and exterior window penetration (DUSEWP). As shown in Figure 11. The simulation results of the six ventilation schemes are used as examples for discussion and analysis.

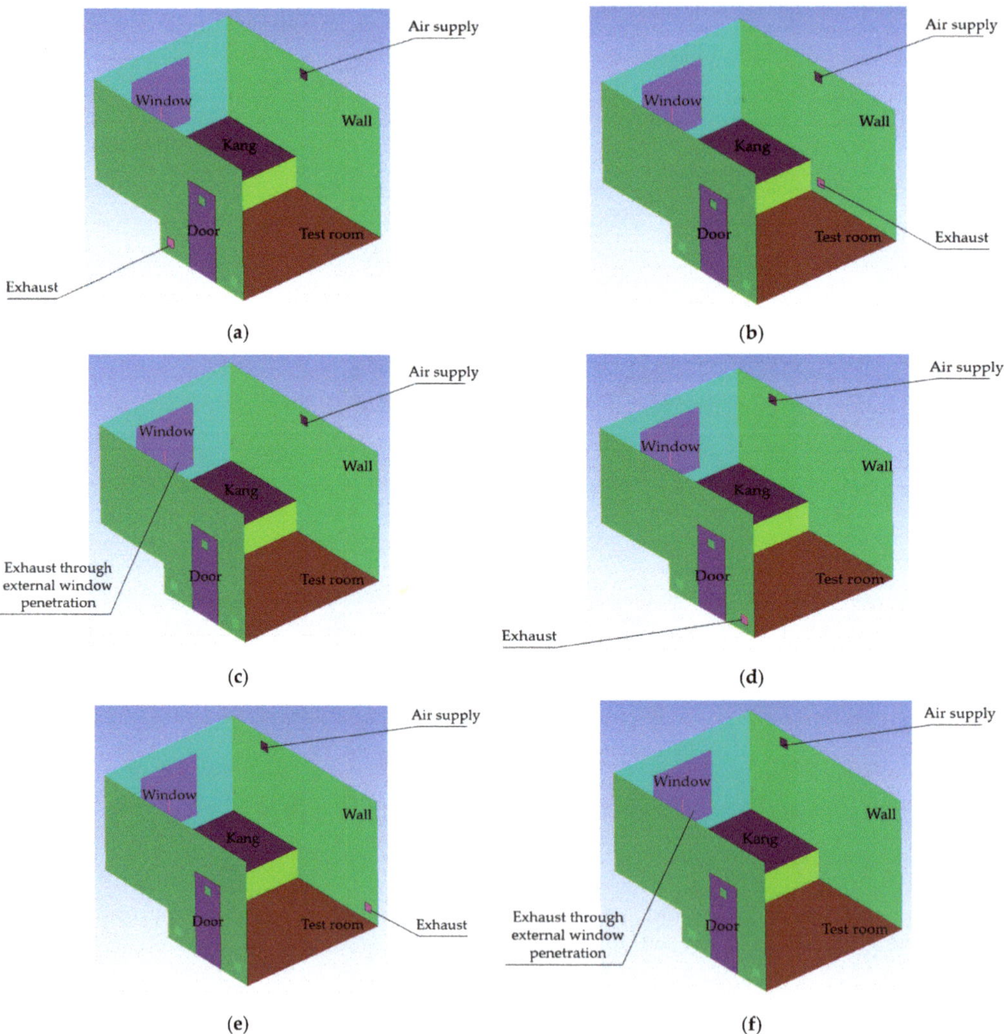

Figure 11. Six layout schemes of air supply and exhaust vents: (**a**) MOUSLE, (**b**) MSUSLE, (**c**) MUSEWP, (**d**) DOUSLE, (**e**) DSUSLE, (**f**) DUSEWP.

4.2. Results and Discussion

The initial concentration of CO is set to 25 mg/m^3, and the data of the Q = 141 m^3/h working condition of the above experiments are used as the boundary conditions to simulate the above six air supply and exhaust vents layout schemes. The calculation of average plane CO concentration at a height of 1.2 m in the room in the simulation results, and the CO concentration variation curves with time are as shown in Figure 12.

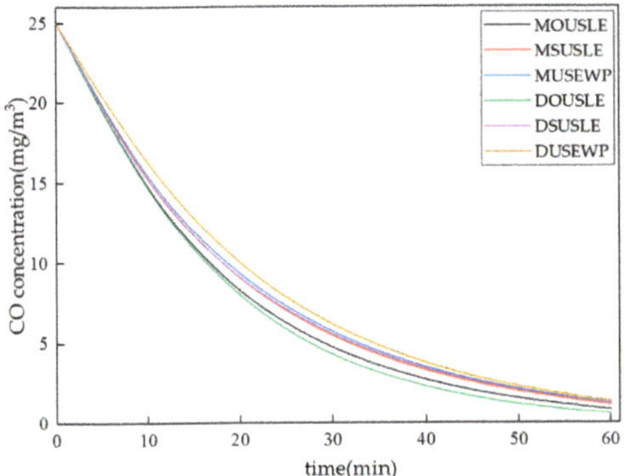

Figure 12. The curve of CO concentration changing with time in the room at a height of 1.2 m.

From the curve, we can see that all of the six schemes have a good dilution effect on indoor CO. The CO concentration in the room can be reduced from 25 mg/m^3 to below the standard value (10 mg/m^3) within 20 min. The dilution rate is the fastest when the ventilation scheme is adopted with the DOUSLE, and the dilution time is 16.6 min. The dilution rate is the slowest when the ventilation is adopted with the DUSEWP, and the dilution time is 20.0 min.

Rural residents spend most of their time on the kang in the bedroom, so the environmental conditions around the kang should be focused. The CO concentration distribution contour is generated by intercepting the plane at a height of 1.2 m, as shown in Figure 13. From the contour, it can be found that when the ventilation mode of DOUSLE is adopted, the CO concentration level in the area where the kang is located is relatively low, and the area with a high CO concentration is mainly on the floor side. When adopting the mode of DUSEWP for ventilation, the CO concentration in the area where the kang is located is relatively high, which may be due to the fact that the air supply port is too close to the external window.

It can be seen from the FGFAHE experiment that with the increasing of ventilation volume, the supply air temperature gradually decreases. During the time of burning, when the fresh air volume is 29 m^3/h, the fresh air outlet temperature can reach can reach 3.24~8.4 °C. According to the previous measurement, the average indoor temperature of rural house during PAP is 13.2 °C. Therefore, if the FGFAHE system is used for ventilation, when the supply air volume is small, the supply air temperature can be higher than indoor air temperature, which is beneficial to the thermal environment of the bedroom; when the air supply volume is large, the air supply temperature can be lower than the indoor air temperature, which will destroy the thermal environment of the bedroom. In the following section, the indoor thermal environment is discussed by means of simulation in two cases of high ventilation (Q = 141 m^3/h) and low ventilation (Q = 29 m^3/h).

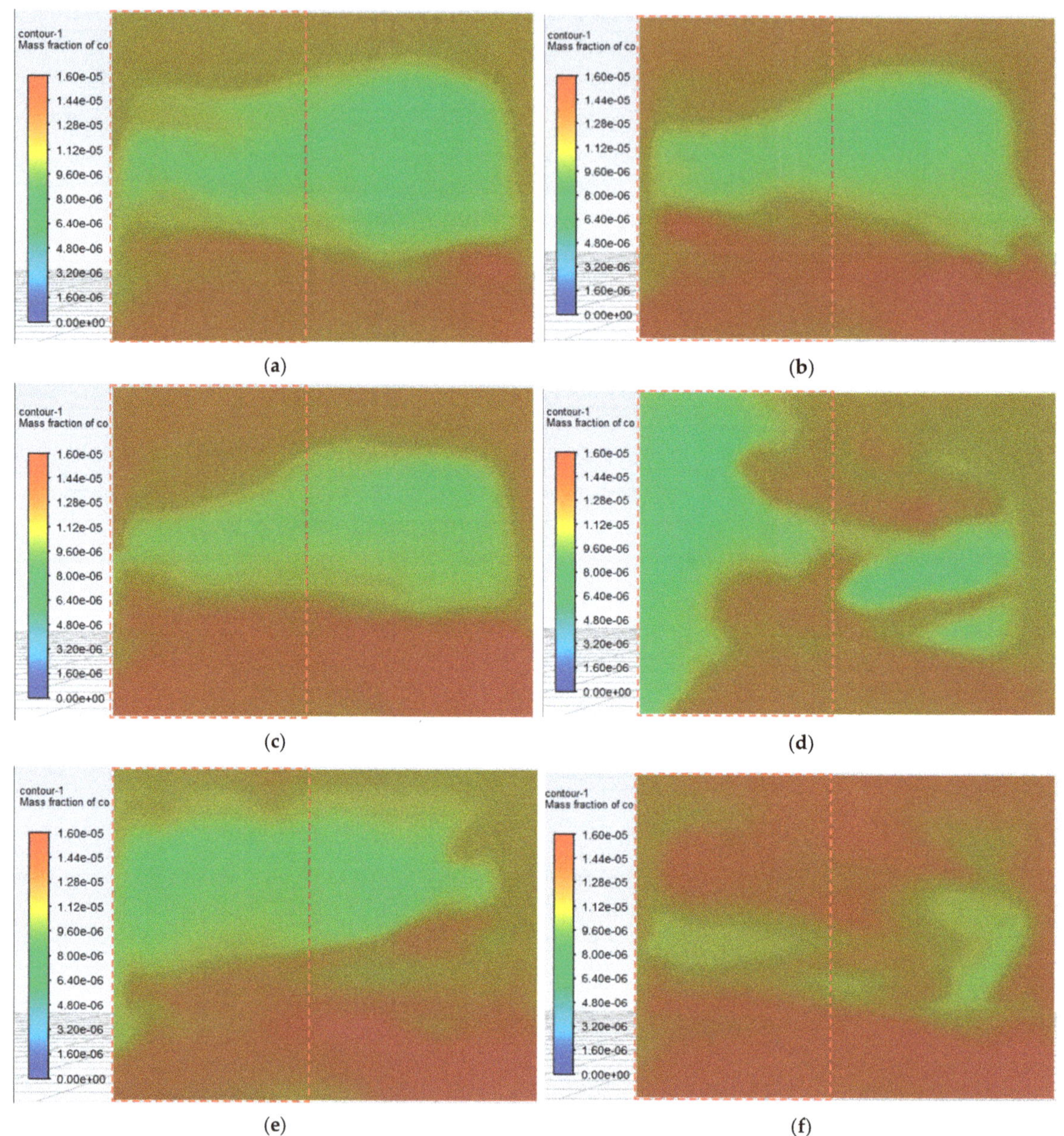

Figure 13. Contours of indoor 1.2 m high plane CO concentration distribution under six schemes: (**a**) MOUSLE, (**b**) MSUSLE, (**c**) MUSEWP, (**d**) DOUSLE, (**e**) DSUSLE, (**f**) DUSEWP.

When the air supply volume is Q = 141 m³/h, the supply air temperature is set as 8.4 °C according to the FGFAHE experimental results. The ventilation simulation of the room is carried out in the above six ventilation schemes. The temperature distribution contours are generated by intercepting the plane at the height of 1.2 m in the room, and the area where the kang is located is marked with a red frame, as shown in Figure 14.

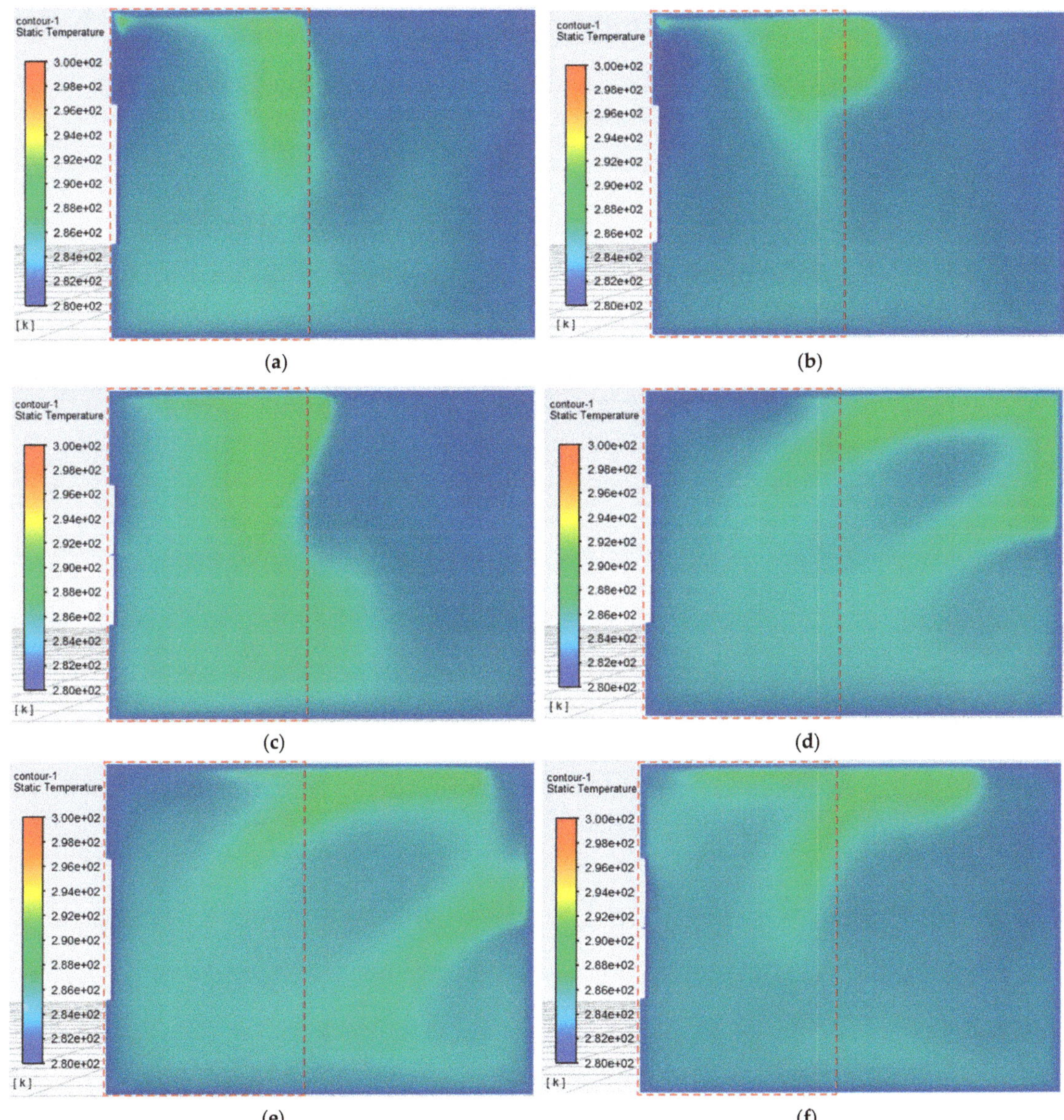

Figure 14. Contours of indoor 1.2 m high plane temperature distribution under six schemes: (**a**) MOUSLE, (**b**) MSUSLE, (**c**) MUSEWP, (**d**) DOUSLE, (**e**) DSUSLE, (**f**) DUSEWP.

It can be found through calculation that the indoor air temperature is reduced by an average of 1.3 °C when the above six ventilation schemes are used for ventilation. It can be seen from Figure 14 that when the three ventilation methods of median air supply are used, the areas with higher temperature are mainly distributed on the kang side. Calculating the air temperature of the area where the kang is located and the area where the floor is located, a table is drawn, as shown in Table 6. It can be found that the temperature of the area where the kang is located is 0.63 °C higher when the three median air supply schemes are used than when the three diagonal air supply schemes are used. That is to say, the use of these three ventilation methods of median air supply can reduce the impact of insufficient preheated fresh air on the occupants' comfort.

Table 6. The average temperature of the air on the kang side and the floor side when the ventilation volume is Q = 141 m^3/h.

	MOUSLE	MSUSLE	MUSEWP	DOUSLE	DSUSLE	DUSEWP
Kang side	285.65 K	285.95 K	285.40 K	284.92 K	285.06 K	285.14 K
Floor side	283.94 K	283.79 K	284.69 K	285.00 K	285.27 K	284.88 K

The air flow rate should not be too high in the areas where people need to stay indoors for a long time, otherwise it will cause discomfort to the people and even affect the daily activities of residents. It is required in the "Design Code for Heating Ventilation and Air Condition of civil buildings (GB50736-2012)" that the average air velocity of indoor activity areas of the buildings with hot air heating in winter should not be greater than 0.3 m/s, and the average indoor air velocity of air conditioning in indoor long-term stay areas in winter should not be greater than 0.2 m/s.

A plane with a height of 1.2 m to generate a contour of air velocity distribution is shown in Figure 15. The speed range in the contour is 0–1.3 m/s. It can be seen from the figure that under the six ventilation modes, most of the area where the kang is located has an air velocity exceeding 0.3 m/s, and the air velocity near the south side wall of the bedroom reaches nearly 1.3 m/s. The air velocity of the area is calculated for where the kang is located and where the floor is located, and a table is drawn, as shown in Table 7. It can be found that the average air velocity in the area where the kang is located reaches more than 0.35 m/s. Therefore, when high air volume is required for ventilation, it is recommended to turn off or switch the ventilation mode as soon as possible after diluting the indoor air pollutants.

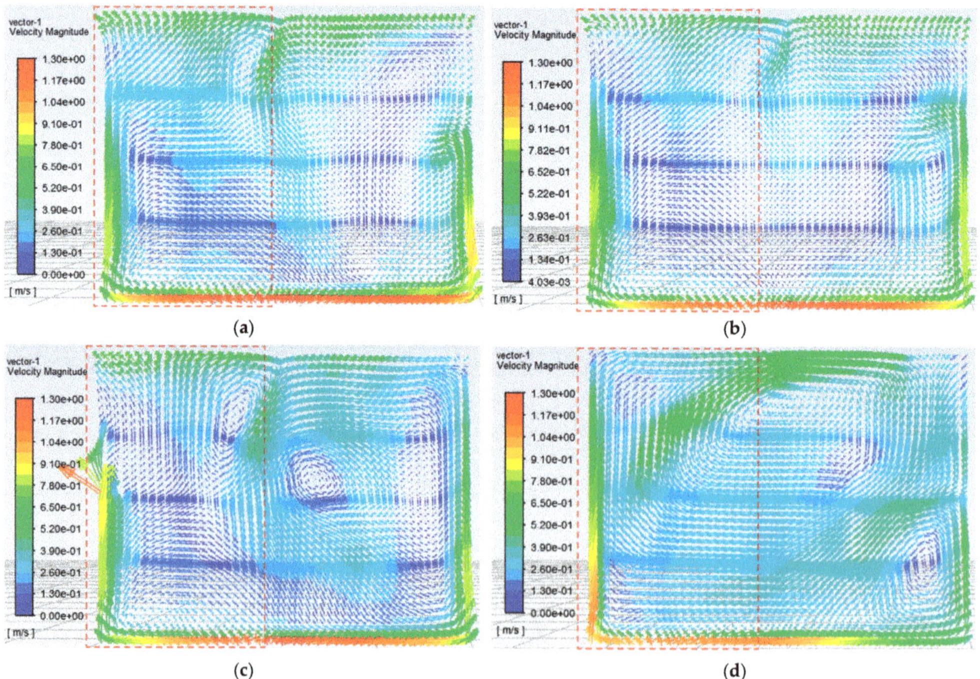

(a)

(b)

(c)

(d)

Figure 15. *Cont.*

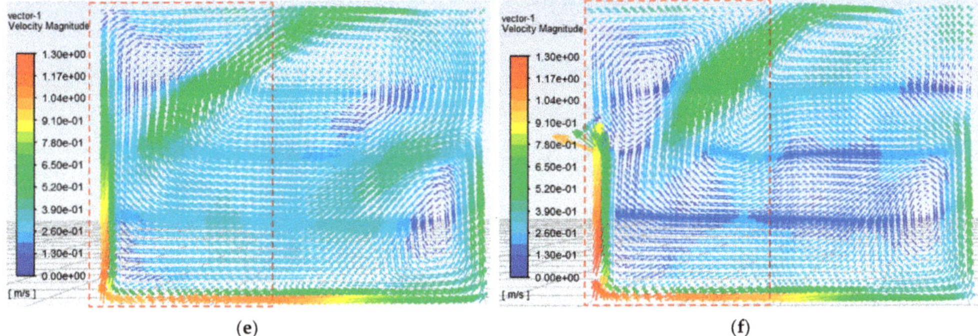

Figure 15. Contours of indoor 1.2 m high plane air velocity distribution under six schemes: (**a**) MOUSLE, (**b**) MSUSLE, (**c**) MUSEWP, (**d**) DOUSLE, (**e**) DSUSLE, (**f**) DUSEWP.

Table 7. The average air velocity on the kang side and the floor side when the ventilation volume is $Q = 141$ m^3/h.

	MOUSLE	MSUSLE	MUSEWP	DOUSLE	DSUSLE	DUSEWP
Kang side	0.35 m/s	0.35 m/s	0.37 m/s	0.35 m/s	0.35 m/s	0.31 m/s
Floor side	0.38 m/s	0.37 m/s	0.38 m/s	0.33 m/s	0.31 m/s	0.26 m/s

When the air supply volume is $Q = 29$ m^3/h, referring to the FGFAHE experimental results, the air supply temperature is set to 30.12 °C. The ventilation simulation of the room is carried out as in the above six ventilation schemes. The temperature distribution contours are generated by intercepting the plane at a height of 1.2 m in the room, and the area where the kang is located is marked with a red frame, as shown in Figure 16.

The calculation shows that the indoor temperature will increase by 1.0 °C on average when the above six ventilation schemes are used for ventilation. It can be seen from Figure 16 that when the three ventilation methods of median air supply are adopted, the areas with higher temperature are mainly distributed on the kang side. The air temperature of the area is calculated where the kang is located and where the floor is located, and a table is drawn, as shown in Table 8. It can be found that the temperature of the area where the kang is located is 2.07 °C higher when the three median air supply schemes are used than when the three diagonal air supply schemes are used. That is, the use of these three ventilation methods can improve the thermal comfort of the occupants to a greater extent. A plane with a height of 1.2 m is taken to generate contours of air velocity distribution, as shown in Figure 17.

Table 8. The average temperature of the air on the kang side and the floor side when the ventilation volume is $Q = 29$ m^3/h.

	MOUSLE	MSUSLE	MUSEWP	DOUSLE	DSUSLE	DUSEWP
Kang side	291.12 K	290.75 K	291.67 K	288.97 K	288.89 K	289.45 K
Floor side	286.88 K	286.59 K	286.04 K	287.00 K	286.99 K	286.44 K

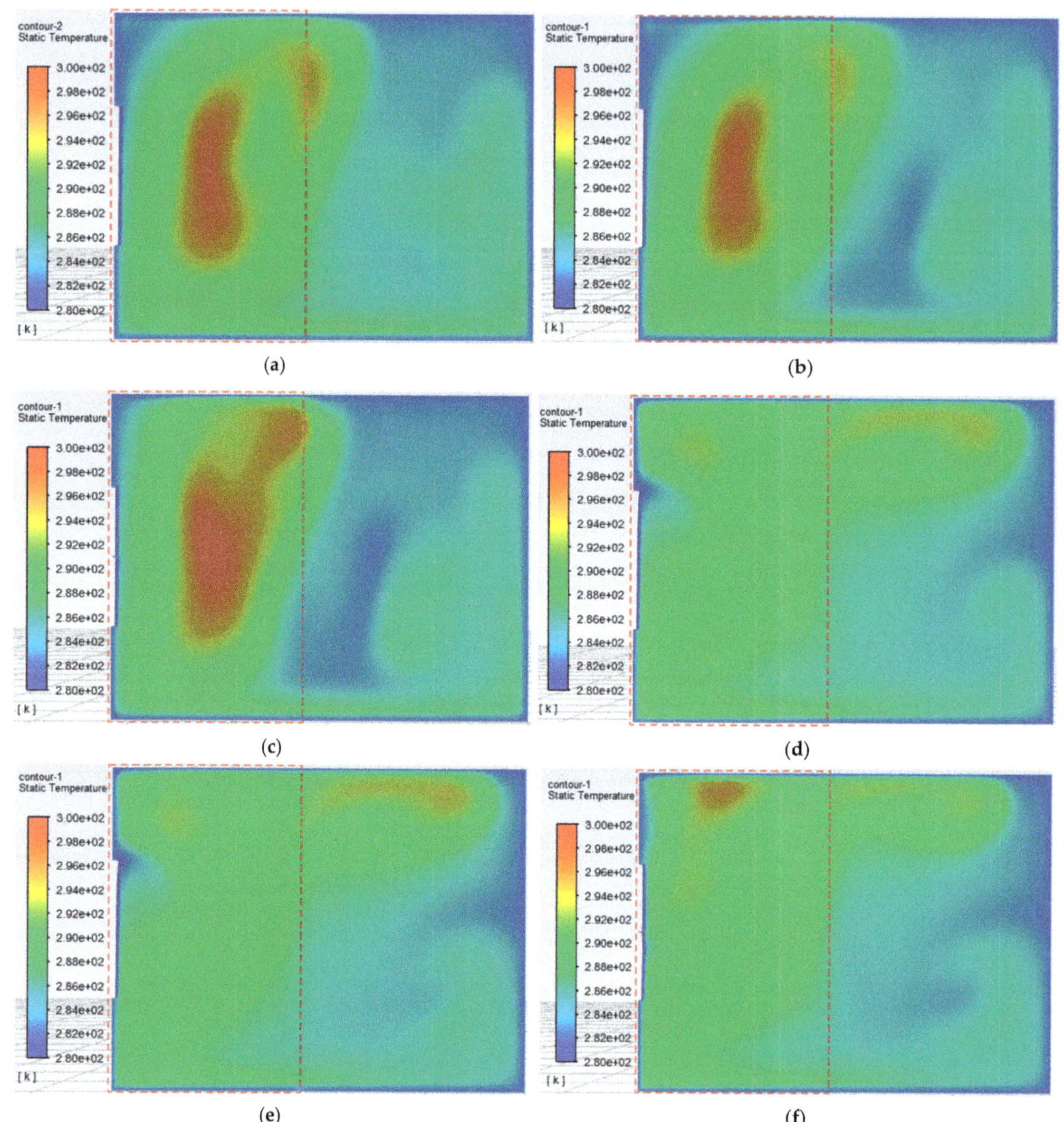

Figure 16. Contours of indoor 1.2m high plane temperature distribution under six schemes: (**a**) MOUSLE, (**b**) MSUSLE, (**c**) MUSEWP, (**d**) DOUSLE, (**e**) DSUSLE, (**f**) DUSEWP.

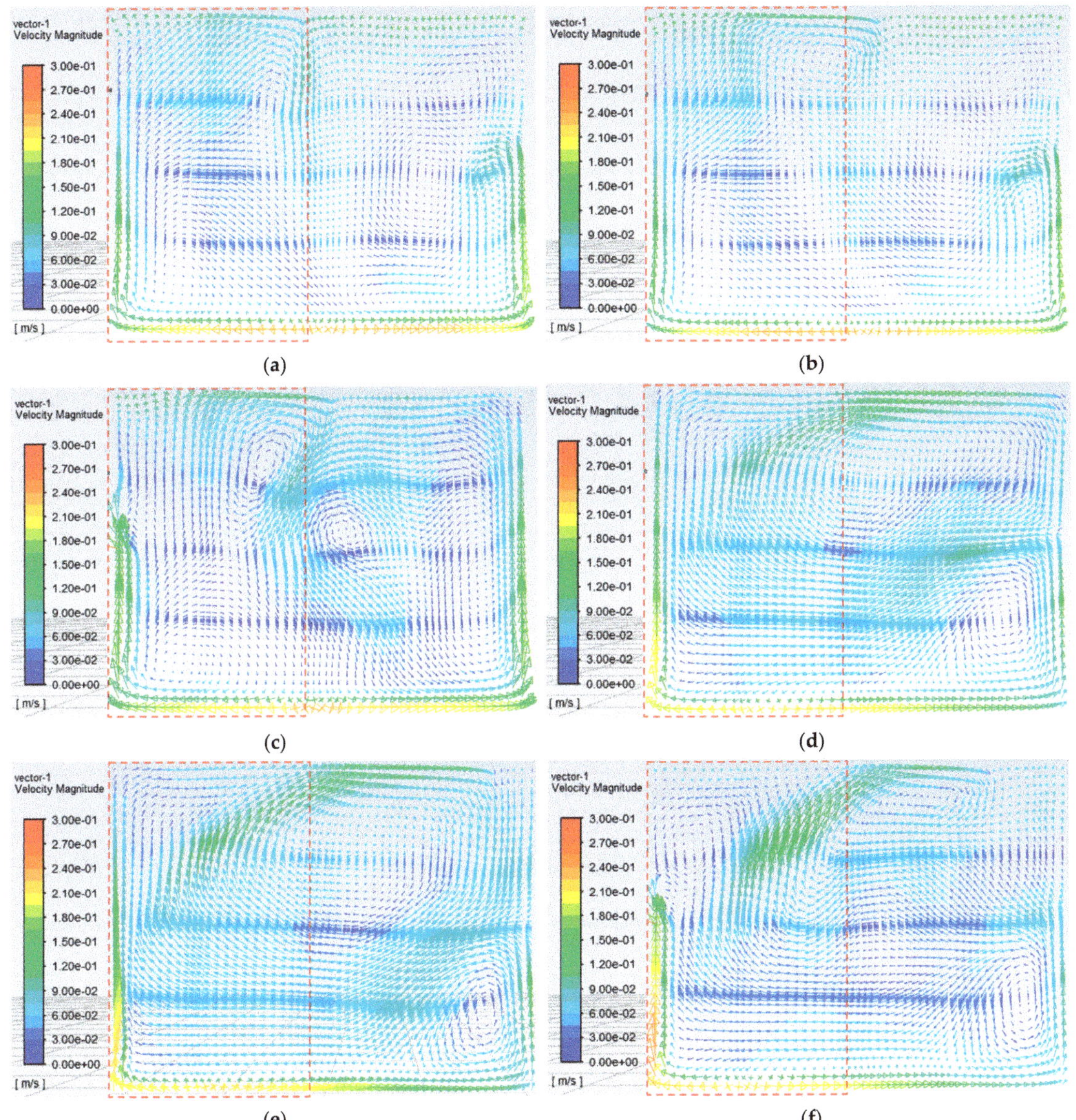

Figure 17. Contours of indoor 1.2 m high plane air velocity distribution under six schemes: (**a**) MOUSLE, (**b**) MSUSLE, (**c**) MUSEWP, (**d**) DOUSLE, (**e**) DSUSLE, (**f**) DUSEWP.

The velocity range in the contours is 0–0.3 m/s. It can be seen from Figure 17 that under the six ventilation modes, there are some areas where the air velocity is higher than 0.2 m/s, but lower than 0.3 m/s, mainly distributed near the inner wall on the south side of the bedroom. Most of the air velocity in the area where the kang is located is lower than 0.2 m/s. Such air velocity would not cause discomfort to indoor occupants. The air velocity of the area where the kang is located and the area where the floor is located is calculated, and a table is drawn, as shown in Table 9. It can be found that the average air velocity in the area where the kang is located reaches about 0.06 m/s. It can be ventilated for a long time in this state.

Table 9. The average air velocity on the kang side and the floor side when the ventilation volume is $Q = 29$ m^3/h.

	MOUSLE	MSUSLE	MUSEWP	DOUSLE	DSUSLE	DUSEWP
Kang side	0.05 m/s	0.04 m/s	0.03 m/s	0.08 m/s	0.06 m/s	0.09 m/s
Floor side	0.03 m/s	0.02 m/s	0.01 m/s	0.08 m/s	0.1 m/s	0.08 m/s

5. Conclusions

In this study, indoor CO_2, CO, PM_{10}, $PM_{2.5}$, formaldehyde, TVOC, temperature and relative humidity are measured in three typical houses in a rural area of Chaoyang City, Liaoning Province, China. A FGFAHE experiment bench is built and ventilation experiments are carried out. Six ventilation schemes, namely MOUSLE, MSUSLE, MUSEWP, DOUSLE, DSUSLE, DUSEWP, are proposed, and their effects on indoor CO dilution and indoor thermal environment in farmhouses are explored by means of simulation. The conclusions of the study are as follows.

(1) The air pollutants such as CO_2, CO, $PM_{2.5}$ and PM_{10} in the bedrooms of the farmhouses seriously exceed the standard. Between 7:00 and 22:00, the concentrations of CO_2 and CO are in a state of exceeding or close to the standard most of the time. The average excess multiples of $PM_{2.5}$ and PM_{10} are 4.2 and 5.7. Between 22:00 and 7:00, the concentration of CO_2 continued to exceed the standard, and the concentrations of CO, $PM_{2.5}$ and PM_{10} are within the standard value for most of the time. Formaldehyde and TVOC levels do not exceed the standard throughout the day, and their average concentrations are far lower than the standard value. Relative humidity is within the standard range for most of the day. The bedroom temperature shows a trend of rising first and then falling between 7:00 and 22:00, and the average temperature is lower than the design temperature of farmhouses.

(2) The biomass combustion flue gas can preheat the outdoor air from $-17\,°C\!\sim\!-10\,°C$ to $6\,°C\!\sim\!30\,°C$ during the time of burning. The mechanical air supply and external window infiltration can effectively dilute the indoor CO of farmhouses, and the dilution rate is more than double that of natural infiltration. When the ventilation volume is 29 m^3/h$\sim$141 m^3/h, it only takes about 17–30 min to reduce the CO concentration of 20 mg/m$^3\!\sim\!25$ mg/m^3 to below the standard value (10 mg/m^3).

(3) All six of the ventilation schemes have a good dilution effect on indoor CO. When the ventilation is carried out with DOUSLE, the indoor CO dilution is the fastest and the CO concentration in the area near the kang is the lowest. When the bedroom is ventilated with three central air supply schemes, the thermal environment in the area where the kang is located is the best. Ventilation with a high ventilation volume will cause the indoor air temperature to drop and the flow velocity to increase, causing discomfort to indoor occupants. When the ventilation is performed with low air volume, the supply air temperature is higher than the indoor air temperature and has little influence on the indoor velocity field, which can improve indoor thermal environment.

Author Contributions: Conceptualization, B.Z. and X.C.; methodology, X.C.; formal analysis, M.L.; investigation, X.C.; writing—original draft preparation, X.C.; writing—review and editing, B.Z.; supervision, M.L. All authors have read and agreed to the published version of the manuscript.

Funding: This research received no external funding.

Institutional Review Board Statement: Not applicable.

Informed Consent Statement: Not applicable.

Data Availability Statement: Not applicable.

Conflicts of Interest: The authors declare no conflict of interest.

Nomenclature

SKS	Stove and Kang System
TVOC	Total Volatile Organic Compounds
PAP	Personnel Activity Period (7:00–22:00)
NSP	Nighttime Sleep Period (22:00–7:00)
FGFAHE	Flue Gas-Fresh Air Heat Exchange
MOUSLE	Median Opposite Side Upper Air Supply and Lower Exhaust Air
MSUSLE	Median Same Side Upper Air Supply and Lower Exhaust Air
MUSEWP	Median Upper Air Supply and Exterior Window Penetration
DOUSLE	Diagonal Opposite Side Upper Air Supply and Lower Exhaust Air
DSUSLE	Diagonal Same Side Upper Air Supply and Lower Exhaust Air
DUSEWP	Diagonal Upper Air Supply and Exterior Window Penetration

References

1. Zhang, Y.; Zhang, Y.; Wang, Z.C. Analysis of indoor air pollution in rural residences in northern China. *Chem. Adhes.* **2018**, *40*, 461–463.
2. Ji, W.H.; Luo, Q.; Zhang, J.L.; Wang, H.H.; Du, T.; Heiselberg, P.K. Investigation on thermal performance of the wall-mounted attached ventilation for night cooling under hot summer conditions. *Build. Environ.* **2018**, *146*, 268–279. [CrossRef]
3. Gao, X.X.; Liu, J.P.; Hu, R.R.; Akashi, Y.; Sumiyoshi, D. A simplified model for dynamic analysis of the indoor thermal environment of rooms with a Chinese kang. *Build. Environ.* **2017**, *111*, 265–278. [CrossRef]
4. Huo, H.M.; Xu, W.; Li, A.G. Comparison and analysis of envelope structure retrofitting schemes for rural residential building in Northern China based on entropy method. *Build. Sci.* **2019**, *35*, 57–64.
5. Guo, Y.M.; Zeng, H.M.; Zheng, R.S.; Li, S.S.; Barnett, A.G.; Zhang, S.W.; Zou, X.N.; Huxley, R.; Chen, W.Q.; Williams, G. The association between lung cancer incidence and ambient air pollution in China: A spatiotemporal analysis. *Environ. Res.* **2016**, *144*, 60–65. [CrossRef]
6. Kilabuko, J.H.; Matsuki, H.; Nakai, S. Air quality and acute respiratory illness in biomass fuel using homes in Bagamoyo, Tanzania. *Int. J. Environ. Res. Public Health* **2007**, *4*, 39–44. [CrossRef] [PubMed]
7. Chen, Y.C.; Fei, J.; Sun, Z.; Shen, G.F.; Du, W.; Yang, L.Y.; Wu, R.X.; Chen, A.; Zhao, M.R. Household air pollution from cooking and heating and its impacts on blood pressure in residents living in rural cave dwellings in Loess Plateau of China. *Environ. Sci. Pollut. Res.* **2020**, *27*, 36677–36687. [CrossRef]
8. Xue, Q.W.; Wang, Z.J.; Liu, J.; Dong, J.K. Indoor $PM_{2.5}$ concentrations during winter in a severe cold region of China: A comparison of passive and conventional residential buildings. *Build. Environ.* **2020**, *180*, 106857. [CrossRef]
9. Wang, Z.J.; Xie, D.D.; Tang, R. Indoor air pollutions and their correlation at rural houses in severe cold region in winter. *J. Harbin Inst. Technol.* **2014**, *46*, 60–64.
10. Zhang, X.Y.; Chen, B. Analysis on the $PM_{2.5}$ Concentration During Heating and Cooking in Cold Rural Areas of Northern China. In Proceedings of the 4th International Conference on Sustainable Energy and Environmental Engineering, Shenzhen, China, 20–21 December 2013; Atlantis Press: Shenzhen, China; pp. 1083–1086.
11. Gu, Q.P.; Gao, X.; Chen, Y.; Yu, Q.; Zhang, Y.; Chen, L.M. The Mass Concentration Characters of Indoor $PM_{2.5}$ in Rural Areas in Jiangsu Province. *J. Fudan Univ.* **2009**, *48*, 593–597.
12. Yang, Z. Reserach on the Improvement of Indoor Air Quality and Strategies of Ventilation for Rural Houses in Cold Region. Master's Thesis, Harbin Institute of Technology, Harbin, China, 2009.
13. Liu, C.L. Research on Indoor Air Quality at Rural Residential House in Northeast. Master's Thesis, Harbin Engineering University, Harbin, China, 2007.
14. Yang, X.B.; Jin, X.Q.; Du, Z.M.; Fan, B.; Chai, X.F. Evaluation of four control strategies for building VAV air-conditioning systems. *Energy Build.* **2011**, *43*, 414–422. [CrossRef]
15. Collett, C.W.; Ross, J.A.; Sterling, E.M. Quality assurance strategies for investigating IAQ problems. *ASHRAE J. Am. Soc. Heat. Refrig. Air Cond. Eng.* **1994**, *36*, 42.
16. Hagerhed-Engman, L.; Sigsgaard, T.; Samuelson, L.; Sundell, J.; Janson, S.; Bornehag, C.G. Low home ventilation rate in combination with moldy odor from the building structure increase the risk for allergic symptoms in children. *Indoor Air* **2009**, *19*, 184–192. [CrossRef]
17. Wang, H.Q. *Ventilation Engineering*, 1st ed.; China Machine Press: Beijing, China, 2007; p. 17.
18. Zhang, K.; Ren, L.; Wang, N.X. Analysis on nature ventilation design of residential buildings in northern towns. *J. Shenyang Jianzhu Univ. (Soc. Sci.)* **2011**, *13*, 274–277.
19. Zhai, X.Q.; Song, Z.P.; Wang, R.Z. A review for the applications of solar chimneys in buildings. *Renew. Sustain. Energy Rev.* **2011**, *15*, 3757–3767. [CrossRef]

20. Elghamry, R.; Hassan, H. Experimental investigation of building heating and ventilation by using Trombe wall coupled with renewable energy system under semi-arid climate conditions. *Sol. Energy* **2020**, *201*, 63–74. [CrossRef]
21. Calautit, J.K.; O'Connor, D.; Tien, P.W.; Wei, S.Y.; Pantua, C.A.J.; Ben, H. Development of a natural ventilation windcatcher with passive heat recovery wheel for mild-cold climates: CFD and experimental analysis. *Renew. Energy* **2020**, *160*, 465–482. [CrossRef]
22. Ye, W.; Zhang, X.; Gao, J.; Gao, G.Y.; Zhou, X.; Su, X. Indoor air pollutants, ventilation rate determinants and potential control strategies in Chinese dwellings: A literature review. *Sci. Total Environ.* **2017**, *586*, 696–729. [CrossRef]
23. Mijakowski, M.; Sowa, J. An attempt to improve indoor environment by installing humidity sensitive air inlets in a naturally ventilated kindergarten building. *Build. Environ.* **2017**, *111*, 180–191. [CrossRef]
24. Leung, D.Y.C. Outdoor-indoor air pollution in urban environment: Challenges and opportunity. *Front. Environ. Sci.* **2015**, *2*, 69. [CrossRef]
25. Zhao, X.Y. Research on the Natural Ventilation Design of the Residence in Nangjing. Master's Thesis, Southeast University, Nanjing, China, 2005.
26. Niculita-Hirzel, H.; Yang, S.; Jörin, C.H.; Perret, V.; Licina, D.; Pernot, J.G. Fungal Contaminants in Energy Efficient Dwellings: Impact of Ventilation Type and Level of Urbanization. *Int. J. Environ. Res. Public Health* **2020**, *17*, 4936. [CrossRef] [PubMed]
27. Elkilani, A.; Bouhamra, W. Estimation of optimum requirements for indoor air quality and energy consumption in some residences in Kuwait. *Environ. Int.* **2001**, *27*, 443–447. [CrossRef]
28. Joo, J.; Zheng, Q.; Lee, G.; Kim, J.T.; Kim, S. Optimum energy use to satisfy indoor air quality needs. *Energy Build.* **2012**, *46*, 62–67. [CrossRef]
29. Shen, H.T. Research on Suitable Ventilation Mode in Rural Housing of Northern Region in Heating Season. Master's Thesis, Harbin Institute of Technology, Harbin, China, 2014.
30. Zhao, J.X.; An, B.Y.; Ma, Y.F.; Liu, X.Y.; Liu, S.N.; Liu, L. Research on the design of energy-saving ventilation system for farm buildings. *Sci. Technol. Innov.* **2020**, *5*, 90–91.
31. Wang, Q.; Ploskic, A.; Song, X.Q.; Holmberg, S. Ventilation heat recovery jointed low-temperature heating in retrofitting-An investigation of energy conservation, environmental impacts and indoor air quality in Swedish multifamily houses. *Energy Build.* **2016**, *121*, 250–264. [CrossRef]
32. Chen, C.Y.; Yao, C.S.; Li, M. Analysis of rural residential energy consumption and its carbon emission in China, 2001–2010. *Renew. Energy Resour.* **2012**, *30*, 121–127.
33. Jin, X. Flow and Heat Transfer Analysis of Flue Gas in Flue Composite Wall in Northern Rural Areas. Master Thesis, Harbin Institute of Technology, Harbin, China, 2018.
34. Yu, K.C.; Tan, Y.F.; Zhang, T.T.; Jin, X.; Zhang, J.D.; Wang, X.M. Experimental and simulation study on the thermal performance of a novel flue composite wall. *Build. Environ.* **2019**, *151*, 126–139. [CrossRef]
35. Jin, X.; Tan, Y.F.; Yu, K.C. Test and analysis of the thermal performance of combined elevated kang and radiator heating system in northern rural areas. *J. Harbin Inst. Technol.* **2019**, *51*, 179–186.
36. Zhao, Y.B. Research on Tunnel-Type Heating System Using Fuel Gas Heat in Northern Rural House. Master's Dissertation, Harbin Institute of Technology, Harbin, China, 2014.
37. Kim, J.; Hong, T.; Lee, M.; Jeong, K. Analyzing the real-time indoor environmental quality factors considering the influence of the building occupants' behaviors and the ventilation. *Build. Environ.* **2019**, *156*, 99–109. [CrossRef]
38. Guyot, G.; Sherman, M.H.; Walker, I.S. Smart ventilation energy and indoor air quality performance in residential buildings: A review. *Energy Build.* **2018**, *165*, 416–430. [CrossRef]
39. Lu, Y.Q. *Practical Heating and Air Conditioning Design Handbook*, 2nd ed.; China Architecture & Building Press: Beijing, China, 2007; pp. 1099–1138.
40. Wang, H.Q. *Ventilation Engineering*; Machinery Industry Press: Beijing, China, 2007; pp. 41–42.

 *buildings*

Article

Assessment of Indoor Climate for Infants in Nursery School Classrooms in Mild Climatic Areas in Japan

Kahori Genjo

Graduate School of Engineering, Nagasaki University, Nagasaki 852-8521, Japan; genjo@nagasaki-u.ac.jp; Tel.: +81-95-819-2598

Abstract: In Japan, the standard of indoor climate in nursery school classrooms has not been established, and the control and maintenance of indoor climate in the classrooms are entrusted to individual childminders. Therefore, indoor climate in nursery school classrooms was measured to prepare fundamental information for proper environmental design and environmental control, considering infants' comfort and health. The climate of 0-year-old and 1-year-old children's rooms in 15 nursery schools located in mild climatic areas in Japan were measured in the summer and winter over four years. Consequently, a lower average temperature was found during winter at lower heights at which infants spend time and indoor air quality was found to be poor in both summer and winter due to a lower ventilation rate in some classrooms with a smaller area per infant compared to the minimum standards for child welfare institutions. One classroom with an average CO_2 concentration of over 1500 ppm was found in both summer and winter due to less ventilation. Illumination less than 300 lx in one-third of the studied classrooms and high equivalent noise level in most classrooms were measured. The need for indoor environmental standards was indicated in terms of infants' comfort and health.

Keywords: nursery; classrooms; vertical temperature difference; humidity; CO_2; air stuffiness index; particle matter; illumination; noise level

Citation: Genjo, K. Assessment of Indoor Climate for Infants in Nursery School Classrooms in Mild Climatic Areas in Japan. *Buildings* **2022**, *12*, 1054. https://doi.org/10.3390/buildings12071054

Academic Editors: Yue Wu, Zheming Liu and Zhe Kong

Received: 24 June 2022
Accepted: 18 July 2022
Published: 20 July 2022

Publisher's Note: MDPI stays neutral with regard to jurisdictional claims in published maps and institutional affiliations.

1. Introduction

In Japan, with the advancement of women in society, the number of children who use nursery schools is increasing, and the proportion of young children is increasing [1]. Infants exhibit less thermoregulatory function than adults and are more affected by the surrounding thermal environment. Childcare time increases up to 11 h, and children who go to the nursery spend most time in nursery schools' indoor environments. Therefore, the environmental improvement of the nursery school classrooms is becoming increasingly important in maintaining children's mental and physical health. Guidelines for the indoor environment have been defined by the school environmental health standard for kindergartens, which is under the jurisdiction of the Ministry of Education, Culture, Sports, Science and Technology [2]. However, no specific guidelines have been developed for the classrooms in nursery schools, which fall under the jurisdiction of the Ministry of Health, Labor and Welfare. Consequently, control and maintenance of the indoor environment in nursery schools are entrusted to individual childminders. The indoor environments in nursery schools are controlled based on the sense of comfort for adults, and the comfort and health of infants who spend time at lower heights tend to be overlooked. If an indoor environment standard for the nursery school classrooms is prepared, childminders could adjust indoor environment appropriately to make the indoor climate in the classrooms more conducive to comfort and health. Infants are particularly susceptible to infectious diseases such as influenza; hence, controlling indoor temperature and humidity in nursery classrooms is necessary for infants' wellbeing. Additionally, ventilation is important to prevent the spread of coronavirus disease 2019 (COVID-19)—not only in office and school buildings but also in nursery schools. The infectious disease control guidelines for nursery schools were established in 2018, which was before the COVID-19 pandemic; thus, the standards for thermal environment were demonstrated specifically for the first time [3]. According to the guidelines, the temperature should be set at 26–28 °C in the summer and at 20–23 °C in the winter, and relative humidity should be set to 60%. Although the basics of air infection control clearly include the isolation of the affected person and management of room ventilation, the guideline for indoor air environment is not clearly indicated. The aforementioned school environmental health standard provides the following guidelines: carbon dioxide concentration should be set to less than 1500 ppm as an index of ventilation; temperature should be set to more than 18 °C and less than 28 °C (the lower limit of the temperature standard was 17 °C until the end of March 2022) [2]; relative humidity should be set to more than 30% and less than 80%, illumination should be set to more than 300 lx; noise level should be set to less than 50 dBA when the window is closed and 55 dBA when the window is open. The humidity ratio (HR) is not specified in Japan; the current ASHRAE Standard 55-2017 [4] only specifies the upper limit of HR as 0.012 kg/kg (DA) and does not specify the lower limit. By contrast, recently, noise has been acknowledged as a social problem in nursery schools in Japan. Kawai et al. [5] investigated the conditions of the acoustic environment of six nursery schools of the Kanto area that were opened during 2005–2012 and conducted a review of overseas standards and guidelines. They revealed the problems of acoustic environment specific to nursery schools, such as the possibility of health effects by indoor high sound pressure level, securing quietness during nap time, and the importance of speech intelligibility for children in the developmental stage. On the contrary, in the United Kingdom, Building Bulletin 93 [6], published by the Institute of Acoustics and the Association of Noise Consultants in 2015, provided design criteria for acoustics and lighting design and guidance on fulfilling these criteria. The guidelines on ventilation, thermal comfort, and indoor air quality in schools have been updated in Building Bulletin 101 (BB101) in 2018 [7], wherein thermal comfort criteria are classified into four categories, and a higher level of expectation for thermal comfort may be needed for very young pupils. For example, the normal maintained operative temperature during the heating season is set at 25 °C, and the strictest draught criteria to provide thermal comfort are applied to the space utilized by young children less than 5 years of age. The control of the ventilation rate in nursery school classrooms is as important as that of the indoor thermal environment. In the case of natural and hybrid ventilation systems, the control set points on carbon dioxide concentration for the ventilation system in teaching spaces, which include opening windows, should be set to achieve less than 1000 ppm whenever possible; however, although carbon dioxide concentration in school classrooms should be kept below 1500 ppm according to the school environmental hygiene standards in Ministry of Education, Culture, Sports, Science and Technology, Japan (as mentioned earlier), no standards for carbon dioxide concentration have been established in Japanese nursery classrooms. The recommended illumination in nursery schools is 300 lx in BB101 [7], but no standards for illumination have been established in Japanese nursery classrooms.

Thus far, compared to the numerous studies on the indoor environment in school classrooms [8–21], few studies have been conducted on the indoor environment in nursery schools [22–26]. Particularly in Japan, research focusing on the environmental humidity in the winter has been conducted for kindergartens from the perspective of influenza prevention by Aoki et al. [27]; however, recently, Taneichi et al. [28] investigated the indoor environment of nursery schools. To date, however, few survey cases on indoor environment in nursery schools have been conducted. Therefore, for the present study, we have conducted questionnaire and measurement surveys focusing on the indoor environment of nursery schools since before the onset of the COVID-19 pandemic [29,30].

As mentioned earlier, except for the infectious disease control guidelines, there have not been any indoor environmental standards—particularly indoor air environmental standards—in nursery schools in Japan, until now [3]. Therefore, indoor climate in nursery school classrooms was measured to prepare fundamental information for proper environmental design and environmental control considering infants' comfort and health. Indoor environments were measured and questionnaire surveys were administered in 15 different nursery schools in mild climatic areas in Japan during the summer and winter from 2016 to 2019. This paper summarizes the results of the measurements thus far and, further, proposes the indoor environmental standards that should be established for nursery school classrooms.

2. Materials and Methods

The survey was administered in 15 different nursery schools located in Nagasaki City, which has a population of around 400,000 inhabitants in 406 km^2 and is located in western Kyushu Island, with mild climate in Japan (32°45′ N, 129°53′ E). Tables 1 and 2 present the description of the investigated nursery school buildings. The investigated nursery schools included 14 authorized nursery schools—comprising four public schools (Nursery A, B, C, and D), 10 private nursery schools, and one unlicensed in-house nursery school (Nursery O). Thirteen nursery schools, excluding Nursery K and Nursery L, correspond to Certified Childcare Centers, which fall under the jurisdiction of the Cabinet Office and are supervised by the Ministry of Health, Labor and Welfare together with the Ministry of Education, Culture, Sports, Science and Technology. Nurseries K and L fall under the jurisdiction of the Ministry of Health, Labor and Welfare. The newest nursery is Nursery O, which was constructed in 2017, while Nursery A is an extremely old building, with a completion date in 1949. Three public nursery schools, excluding Nursery A, were completed around 1970. The airtightness values of observed buildings were unknown because the measurement of airtightness itself is not obligatory in Japan, although the value influences draught, CO_2 concentrations, and noise level. Air conditioners were installed in all the nurseries for space cooling and space heating. Along with air conditioners, floor heating was installed in Nursery I, J, and N. Ventilation systems were installed in 11 nursery schools, excluding three public schools and one private school. Humidifiers were used in 13 nursery schools, excluding Nurseries L and O, and air cleaners were used in 13 nursery schools, excluding Nurseries B and E. In the two nursery schools, Nurseries J and M, the floor area did not meet the standard of 1.65 m^2 per child, which was defined by the standards on facilities and operation of child welfare institutions [31], because, due to a shortage of nursery schools in Nagasaki City at the time of this research, it was supposed to accept children up to 120% of its capacity.

Thirteen nursery schools excluding A and F were evaluated in both summer and winter. Nursery A was evaluated during winter and nursery F was evaluated during summer. Each school was studied during one full school week—from Monday to Friday.

Table 3 shows the description of measurement items and measuring equipment used. We administered a survey on indoor thermal environment, indoor air environment, illumination environment, and acoustic environment for 0-year-old and 1-year-old children's rooms used by infants, and the outdoors. The vertical temperature was measured at the following four points: 0.1 m from the floor, which is the height at which infants crawl; 0.3 m from the floor surface, which is the height at which they sit on the floor; 0.6 m, which is the height of their head when they stand; and 1.1 m, which is the height of a head when adults sit in a chair. In Nurseries H and L, no partition exists between the 0-year-old and the 1-year-old children's rooms; hence, measurements were conducted at one point, considering the two spaces as one room. The measurement survey was administered for about two weeks in Nurseries A, B, C, E, F, and J and for about one week in the other nine nursery schools. The sensors continuously collected data while they were installed in the classroom. Additionally, measurements on the acoustic environment were conducted since the winter of 2016. $PM_{2.5}$ concentration and equivalent noise level were evaluated once by a 10 min measurement per classroom. Outdoor air temperature and relative humidity were obtained from the nearest meteorological station [32]. In addition to the measurement survey, a questionnaire survey on activities in the nursery classroom was also administered for each nursery classroom. In the questionnaire survey, investigated items included the following: the number of infants and nursery teachers in each nursery room, usage of cooling and heating equipment, usage of ventilation equipment, window opening, and the weekday and Saturday timeline.

Table 1. Description of the investigated nursery school buildings.

Name	Nursery A	Nursery B	Nursery C	Nursery D	Nursery E	Nursery F	Nursery G	Nursery H
Classification	public	public	public	public	private	private	private	private
Building form	single	composite	composite	composite	single	single	single	single
Completion	1949	1969	1970	1972	1972	1988	1999	2012
Structure [1,2]	RC	S	RC	RC	RC	RC	S	RC
Floor number	2 floors	2 floors	3 floors	1 floor	3 floors	2 floors	3 floors	2 floors
Gross floor area (m^2)	665.28	483.29	667.56	827.52	662.56	1014.18	408.67	763.68
Space cooling system [3]	AC	AC	AC	AC	AC	AC	AC	AC
Space heating system [3,4,5]	AC, KH	AC, EC	AC	AC, EC	AC	AC	AC	AC
Ventilation system	natural	natural	natural	mechanical	natural	mechanical	mechanical (not used)	mechanical
Other equipment	air cleaner, humidifier	humidifier	air cleaner, humidifier	air cleaner, humidifier	humidifier	air cleaner, humidifier	air cleaner, humidifier	air cleaner, humidifier
Lighting equipment [6,7]	FL	FL, IL	LED, FL	FL	LED, FL	LED	FL	FL
Opening time	7:30–18:00	7:30–18:00	7:30–18:00	7:00–18:00	7:00–19:00	7:00–18:00	7:00–18:00	7:00–18:00
Capacity	120	90	120	140	130	90	135	80
Area of infants' room (m^2/head) summer/winter [8]	0-year-old room:-/5.78 1-year-old room: -/3.20	0-year-old room:6.4/4.8 1-year-old room: 4.0/4.0	0-year-old room:5.3/4.0 1-year-old room:3.1/2.9	0-year-old room:4.5/4.5 1-year-old room:5.8/5.3	0-year-old room:5.5/3.6 1-year-old room:2.7/2.8	0-year-old room:6.8/- 1-year-old room:1.9/-	0-year-old room:4.3/4.1 1-year-old room:1.7/1.7	0- and 1-year-old room: 5.1/4.6
Survey year	2016	2017	2017	2019	2017	2017	2018	2018

[1] RC, reinforced concrete. [2] S, steel frame. [3] AC, air conditioner. [4] KH, kerosene fan heater. [5] EC, electric heating carpet. [6] FL, fluorescent light. [7] IL, incandescent lamp. [8] '-', 'not measured'.

Table 2. Description of the investigated nursery schools.

Name	Nursery I	Nursery J	Nursery K	Nursery L	Nursery M	Nursery N	Nursery O
Classification	private	private	private	private	private	private	unlicensed in-house
Building form	single	composite	single	single	single	composite	single
Completion	2013	2013	2013	2014	2015	2016	2017
Structure [1,2,3]	RC	RC	W	S	RC	RC	S
Floor number	2 floors	3 floors	2 floors	2 floors	2 floors	2 floors 1 basement	1 floor
Gross floor area	968.24	1310	unknown	1186.93	963.17	1233.41	252
Space cooling system [4]	AC	AC	AC	AC	AC	AC	AC
Space heating system [4,5,6,7]	AC, FH	AC, FH	AC, KH	AC	AC	AC, FH	AC
Ventilation system	mechanical	mechanical	mechanical	mechanical	mechanical	mechanical	mechanical
Other equipment	air cleaner, humidifier	humidifier	air cleaner, humidifier	air cleaner	air cleaner, humidifier	air cleaner, humidifier	air cleaner
Lighting equipment [8,9]	LED	LED	FL	LED, FL	LED	LED	LED
Opening time	7:00–18:00	7:00–18:00	7:00–19:00	7:00–20:00	7:00–19:00	7:00–18:00	7:00–18:00
Capacity	100	119	60	140	140	190	30
Area of infants' room (m^2/head) summer/winter	0-year-old room: 2.1/2.3 * 1-year-old room: 3.4/2.3 *	0-year-old room: 2.6/1.6 1-year-old room: 1.5/1.5	0-year-old room: 3.7/3.7 1-year-old room: 2.7/2.7	0- and 1-year-old room: 4.5/3.2	0-year-old room: 1.6/1.5 1-year-old room: 1.3/1.3	0-year-old room: 3.2/3.2 1-year-old room: 2.8/2.8	0-year-old room: 5.5/3.9 1-year-old room: 4.6/4.6
Survey year	2018	2016	2019	2019	2019	2018	2019

[1] RC, reinforced concrete. [2] W, wooden. [3] S, steel frame. [4] AC, air conditioner. [5] KH, kerosene fan heater. [6] EC, electric heating carpet. [7] FH, floor heating. [8] FL, fluorescent light. [9] IL, incandescent lamp. * 0- and 1-year-old room.

Table 3. Description of measurement items and measurement equipment used.

Measurement Items	Measurement Height (Number of Sensors)	Measurement Interval	Measurement Equipment	Range and Accuracy
Temperature	0.1, 0.3, and 0.6 m above the floor (1 for each height)		small data logger for temperature and humidity measurement (RSW-21S)	range: 0 to 55 °C, 10 to 95%RH
Humidity				accuracy: ±0.5 °C, ±5%RH
Carbon dioxide concentration, temperature, humidity	1.1 m above the floor (1)	10 or 5 min (as for CO_2 measured in 2016 and 2017)	data logger for carbon dioxide concentration, temperature, and humidity measurement (TR-76Ui)	range: 0 to 9999 ppm accuracy: ±50 ppm of range ± 5% of reading range: 0 to 55 °C, 10 to 95%RH accuracy: ±0.5 °C, ±5%RH
Globe temperature	1.1 m above the floor (1)		small data logger for temperature measurement (RTW-31S), globe thermometer with a diameter of 75 mm	range: −20 to 80 °C accuracy: ±0.3 °C
$PM_{2.5}$ concentration (particle number)	1.1 m above the floor (1)	1 min for 10 min measurement	air quality monitor (laser light scattering particle counter, DC170, sampling flow rate: 1.7 L/min)	particle size sensitivity: 0.5, 2.5 μm
Illumination	0.1–1.1 m above the floor (1)	10 min	data logger for illumination measurement (TR-74Ui)	range: 0 to 130 klx accuracy: ±5%
Equivalent noise level	1.1 m above the floor (1)	10 min measurement	normal noise level meter (NL-27)	range: 30 to 137 dB resolution: 0.1 dB (class 2)

An index called ICONE [22] was calculated from the measurement results of CO_2 concentration and used to evaluate indoor air quality (IAQ). The ICONE is an air stuffiness index used in the air quality evaluation of the classroom environment in France. Occupancy periods of less than 5 h are discarded. CO_2 values are classified according to their levels: n_0—values < 1000 ppm, n_1—values between 1000 and 1700 ppm, and n_2—values > 1700 ppm. The ICONE air stuffiness index is then calculated by applying Equation (1), where f_1 is the proportion of CO_2 values between 1000 and 1700 ppm ($f_1 = n_1/(n_0 + n_1 + n_2)$ and f_2 is the proportion of CO_2 values above 1700 ppm ($f_2 = n_2/(n_0 + n_1 + n_2)$).

$$\text{ICONE} = \left(\frac{2.5}{\log_{10}(2)} \right) \log_{10}(1 + f_1 + 3f_2), \tag{1}$$

The final results are rounded to the nearest integer. The air stuffiness level of the room is then expressed by a score ranging from 0–5. ICONE scores from 0–5 correspond to an air stuffiness gradient: 0, none; 1, low; 2, average; 3, high; 4, very high; and 5, extreme stuffiness [22].

3. Results and Discussion

The median outdoor temperatures and relative humidity (RH) during the measuring period for the summer and winter were 27.8 °C and 80%, and 8.5 °C and 69%, respectively. The results are evaluated with reference to the infectious disease control guidelines for nursery schools established in 2018 [3]. Further, the school environmental health standard [2] and ASHRAE Standard 55-2017 [4] are referenced as needed.

3.1. Temperature and Humidity

Results of temperature at 1.1 m above the floor during opening hours in the summer and winter are presented in Figure 1. Figures 2 and 3 present a part of the vertical temperature distributions during the summer and winter opening hours. The temperature difference between 1.1 m above the floor and 0.1 m above the floor is termed the vertical temperature difference, and whether the temperature difference is above or below the comfortable range of 3 °C is evaluated [4]. Results of RH, HR, and globe temperature during opening hours in the summer and winter are presented in Figures 4–6, respectively. The mean values of temperature, globe temperature, RH, and HR during the measuring period were 27.4 °C, 26.5 °C, 62%, and 0.014 kg/kg (DA) in the summer; and 19.2 °C, 18.6 °C, 46%, and 0.006 kg/kg (DA) in the winter, respectively. Compared to the infectious disease control guideline mentioned above [3], the summer mean temperature was between 26 and 28 °C, but the winter mean temperature was lower than the lower limit of 20–23 °C. Mean RH was close to 60% of the standard in the summer, but lower than the standard in the winter.

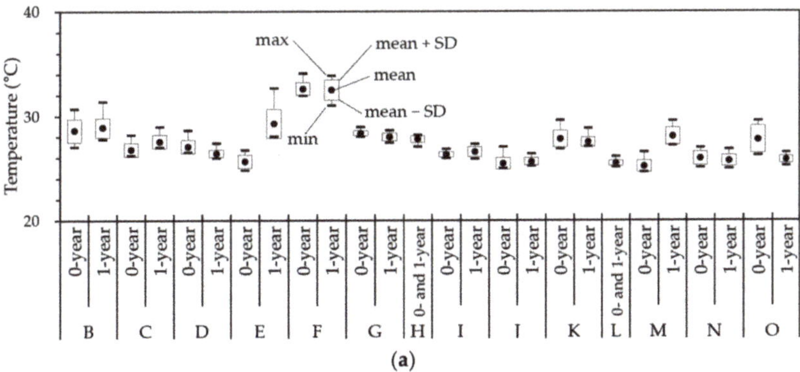

(a)

Figure 1. *Cont.*

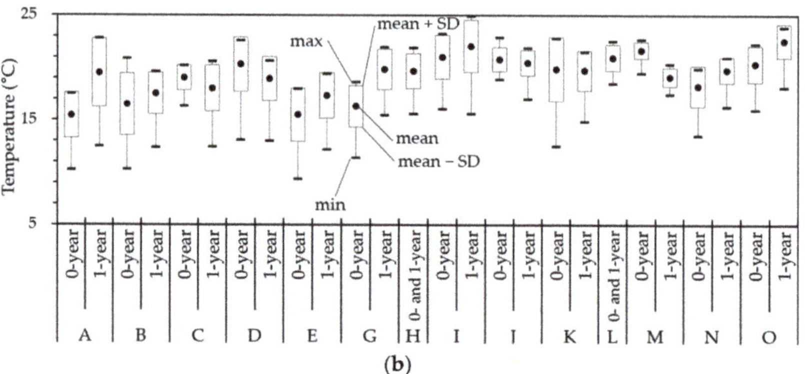

(b)

Figure 1. Temperature at 1.1 m from floor surface during opening hours in (**a**) summer and (**b**) winter.

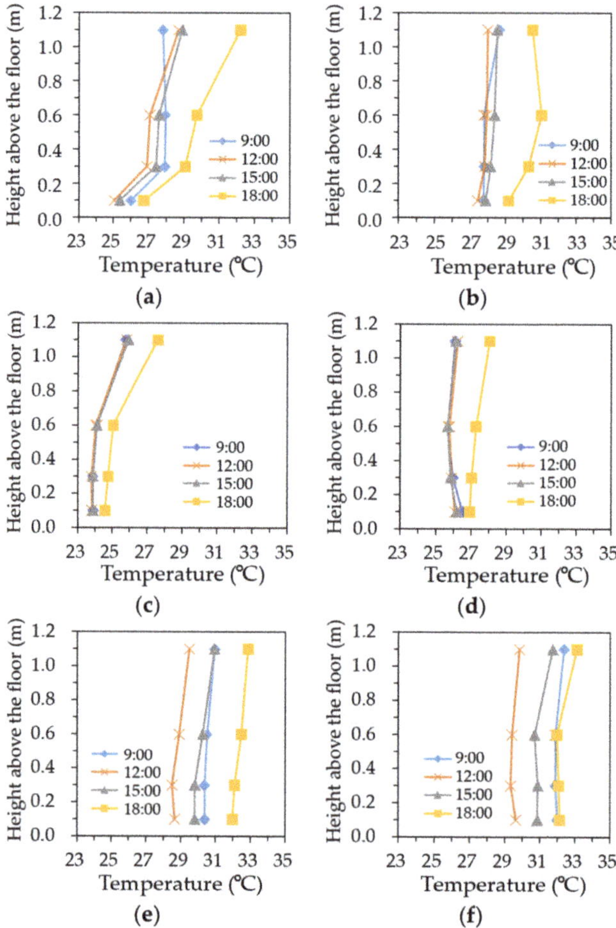

(a) **(b)** **(c)** **(d)** **(e)** **(f)**

Figure 2. *Cont.*

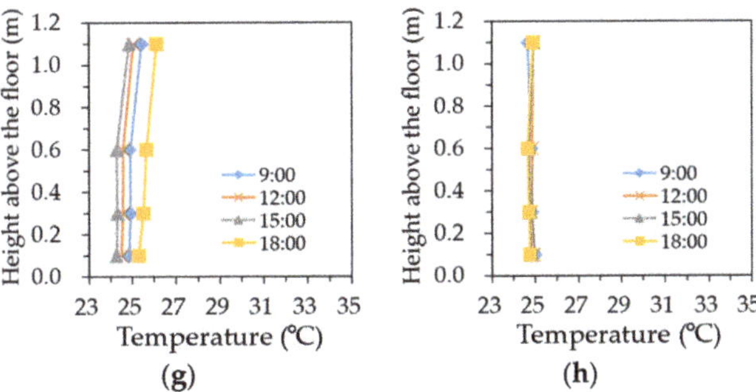

Figure 2. Vertical temperature distributions during opening hours in summer: (**a**) 0-year-old room in Nursery B; (**b**) 1-year-old room in Nursery B; (**c**) 0-year-old room in Nursery C; (**d**) 1-year-old room in Nursery C; (**e**) 0-year-old room in Nursery F; (**f**) 1-year-old room in Nursery F; (**g**) 0-year-old room in Nursery N; (**h**) 1-year-old room in Nursery N.

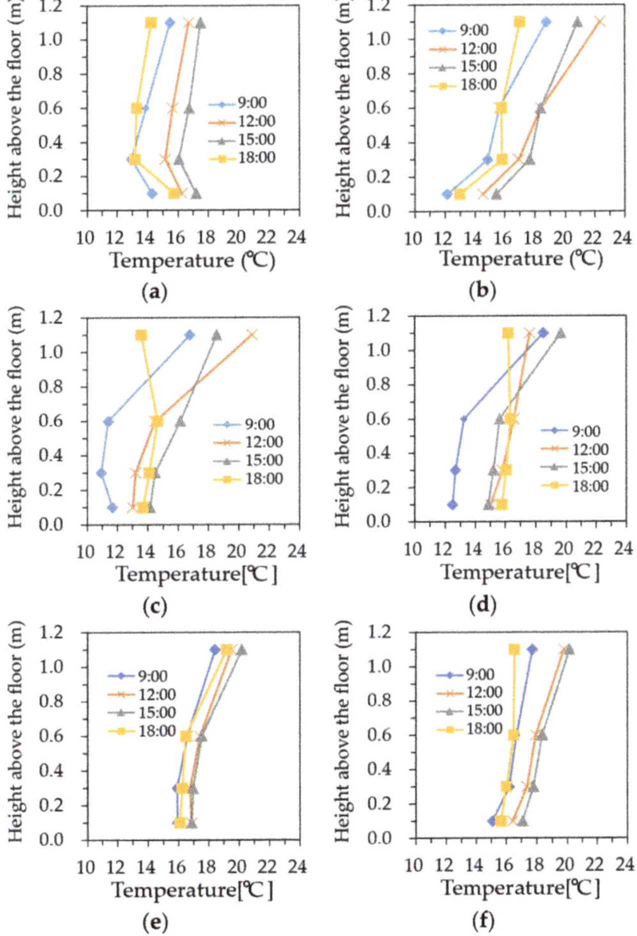

Figure 3. *Cont.*

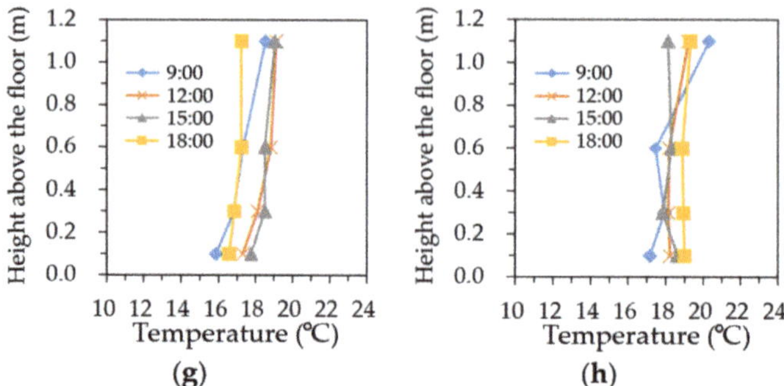

Figure 3. Vertical temperature distributions during opening hours in winter: (**a**) 0-year-old room in Nursery A; (**b**) 1-year-old room in Nursery A; (**c**) 0-year-old room in Nursery B; (**d**) 1-year-old room in Nursery B; (**e**) 0-year-old room in Nursery C; (**f**) 1-year-old room in Nursery C; (**g**) 0-year-old room in Nursery N; (**h**) 1-year-old room in Nursery N.

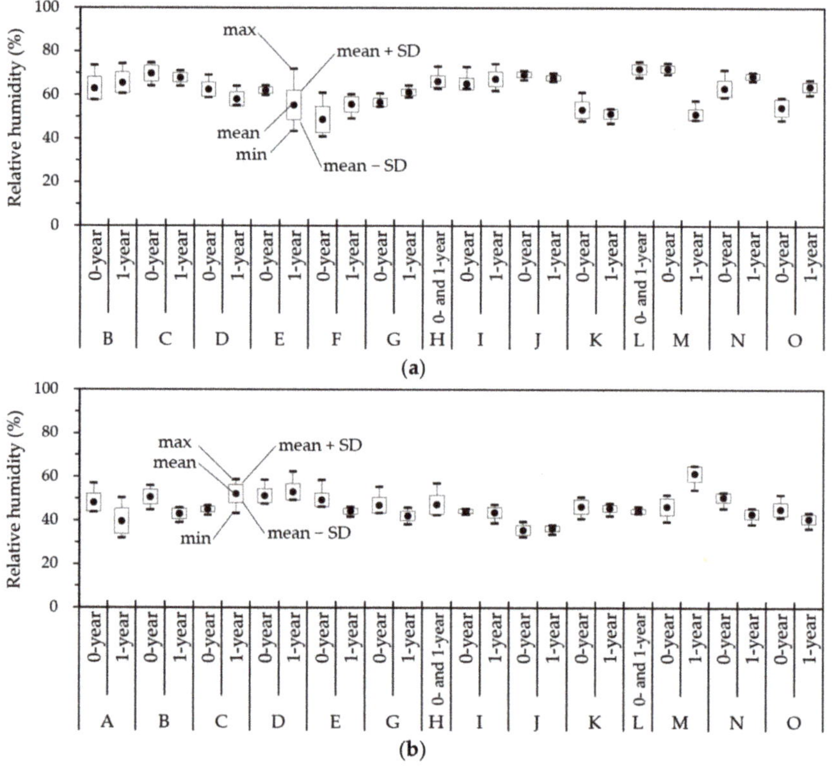

Figure 4. Relative humidity at 1.1 m from floor surface during opening hours in (**a**) summer and (**b**) winter.

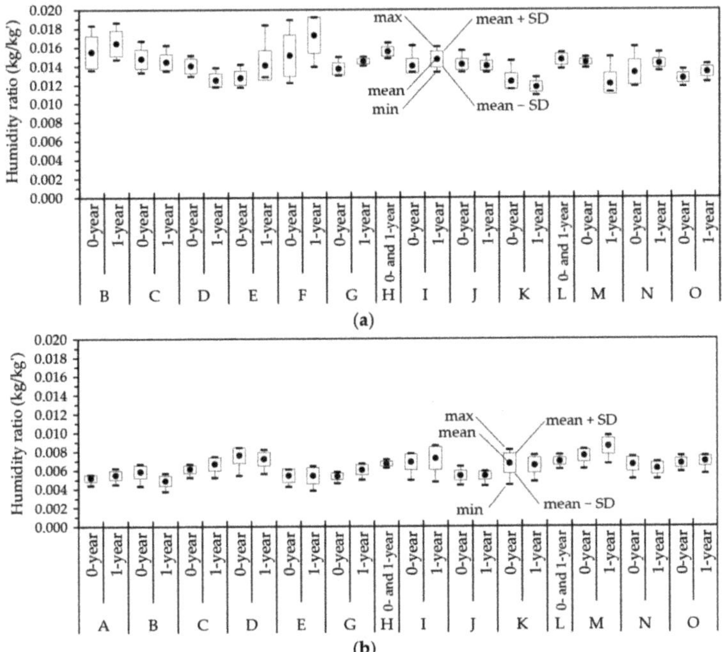

Figure 5. Humidity ratio at 1.1 m from floor surface during opening hours in (**a**) summer and (**b**) winter.

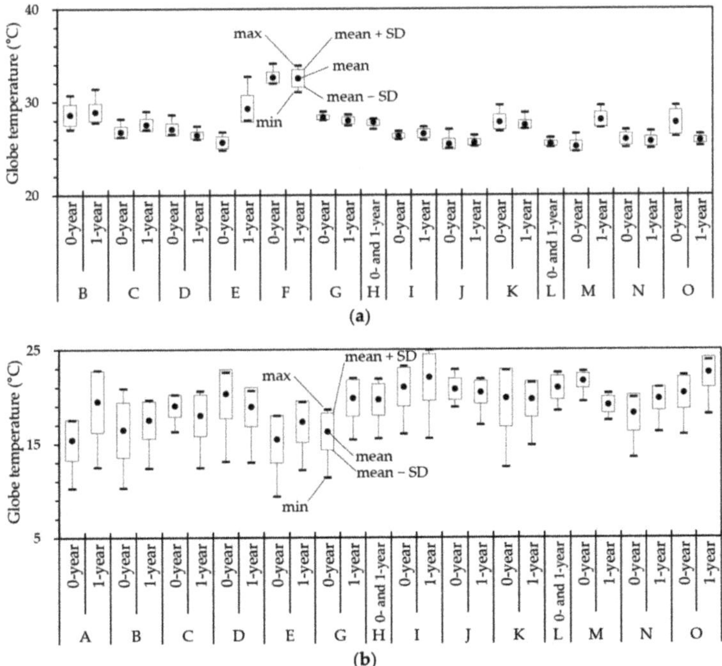

Figure 6. Globe temperature at 1.1 m from floor surface during opening hours in (**a**) summer and (**b**) winter.

In the summer, the temperatures of the following nursery classrooms nearly satisfied the above guideline [3]: both classrooms in nursery B, 0-year-old room in nursery E, both rooms in nursery G, 0-year-old room in nursery K, 1-year-old room in nursery M, and 0-year-old room in nursery O. In most other nursery rooms, the temperature was below the lower limit of the guideline and met the requirement except in Nursery F, where the temperature was too high and failed to fulfill the guidelines. Nursery F refrained from using an air conditioner in accordance with the policy of childcare regarding sweating, and thus, the temperature in the 1-year-old room at any height except at 12:00 was higher than 30 °C, as illustrated in Figure 2f [30]. From the perspective of preventing heat stroke, using air conditioning is necessary in this nursery school. Overall, the temperature was lower at a height of 0.1 m above the floor than that at 1.1 m above the floor in most nursery rooms, for example, in the 0- and 1-year-old room in nursery H, where cold air was circulated by a ceiling fan.

As shown in Table A2, in the winter, the temperature at 1.1 m above the floor generally meets the aforementioned standard [3] of 20–23 °C, except for some nursery schools, such as those that are older; however, the lower the height above the floor, the lower the temperature tended to be (Figure 3b–d) [29,30]. Therefore, the temperature at 0.1 m above the floor did not meet the standards in any nursery room. This may be due to poor insulation performance or poor air circulation, even in recently constructed nursery schools. In nursery schools I, J, and N, where floor heating was installed, the temperature at 0.1 m above the floor was kept relatively high; hence, the vertical temperature difference was within 3 °C, as is shown in Figure 3g,h. Since infants spend their time at a lower height than adults, raising the temperature at 0.1, 0.3, and 0.6 m above the floor in the winter is necessary to ensure infants' comfort and health.

Figures 2 and 3 show that the vertical temperature difference was over 3 °C in two (8% of the total) out of 26 rooms in the summer and 11 (42% of the total) out of 26 rooms in the winter. Reportedly, the vertical temperature difference was larger in the winter than in the summer. Except for the 1-year-old room in nursery E and 0-year-old room in nursery F in the summer, temperature was evaluated to be comfortable in terms of the vertical temperature difference because it was less than 1.0 °C. On the contrary, most nursery rooms studied in the winter were evaluated to be less comfortable. Figure 3b,c show that the largest and second largest vertical temperature differences were found to be 6.8 °C in the 0-year-old room in nursery B and 5.9 °C in the 1-year-old room in nursery A, respectively. As illustrated in Figure 3, compared to the vertical temperature distribution at 9:00, 12:00, 15:00, and 18:00 in each studied room, the largest vertical temperature difference was found at 9:00 in the morning in most nursery rooms. In the winter, the room temperature was under 17 °C in addition to the large vertical temperature difference; hence, numerous nursery rooms were evaluated to be less comfortable for both infants and childminders. The reason for this is the low insulation performance of the older buildings; thus, the temperature at 0.1 m above the floor did not rise easily even if the temperature at 1.1 m above the floor rose when heated. Most nursery schools heat the rooms only with air conditioners, but when nursery rooms are heated only with air conditioners, especially in old buildings, the temperature near the floor does not increase enough; hence, devising ways to ensure comfort is necessary, for example, through the use of auxiliary heating. Additionally, since the insulation performance in mild climatic areas in Japan is lower than that of Europe or the United States, efforts such as raising the insulation performance of buildings and repairing the insulation of existing buildings can also be considered necessary.

Figure 4 demonstrates that according to the infectious disease control guideline [3], mean RH did not satisfy the value of 60% in any nursery classrooms through the summer and winter measurements. However, according to the school environmental health standard [2], mean RH satisfied the value of 30–80% in all studied nursery classrooms. The mean RH ranged 49–72% in the summer and 36–62% in the winter, respectively. According to Tables A3 and A4, the 95th percentile RH exceeded the upper limit of 80% in three classrooms in the summer, and the 5th percentile RH was below the lower limit of 30% in six rooms in the winter. Figure 5 shows that the mean HR in most studied nursery classrooms except one classroom (1-year-old room in nursery K) ranged from 0.012 to 0.017 kg/kg (DA) and did not satisfy the upper limit of 0.012 kg/kg (DA) specified by ASHRAE Standard 55-2017 [4] in the summer, though the mean HR in all studied nursery classrooms, which ranged from 0.005 to 0.009 kg/kg (DA), satisfied the standard value in the winter. Since it is hot and humid during the summer in Japan, humidity control from the viewpoint of microbial contamination is vital. On the contrary, the mean HR across eight classrooms (31% of the total) of 26 nursery classrooms in the winter was less than 0.006 kg/kg (DA), which was proposed by Shoji [33] as the index of influenza epidemic warning level despite using humidifiers. HR measured in the winter in this study was higher than that measured in the winter in Aoki et al.'s survey [27] for kindergartens in Gifu prefecture in the Tokai area, which is located in central Japan, with the range 0.004–0.005 kg/kg (DA). Indoor HR values studied in this survey in the winter were similar to the abovementioned survey [27].

From Figures 1 and 6, globe temperature difference obtained by subtracting the temperature at 1.1 m above the floor from the globe temperature of the same height in each nursery room ranged from −2.6 to 0.0 °C in the summer and −5.1 to 0.4 °C in the winter. Although the globe temperature differences were negative in all nursery classrooms except one classroom (0-year-old room in nursery B) in the summer, it was negative in the studied nursery classrooms in the winter except for some nursery classrooms (0-year-old room in nursery A, 1-year-old room in nursery D, 0-year-old-room in nursery K, and 0-year-old room in nursery M). The globe temperature differences were negative in most classrooms both in the summer and winter, and nursery classrooms were affected by cold radiation.

3.2. CO_2 Concentration

The results of the measurements of CO_2 concentration during opening hours in the summer and winter are presented in Figure 7. Tables A5 and A6 show that the median CO_2 concentrations calculated for opening hours on weekdays were 922 ppm in the summer and 990 ppm in the winter. According to the results of a survey of nursery schools in Denmark [23], the median CO_2 concentration during opening hours was 579 ppm, which was lower than that of the nursery classrooms in this study. On the other hand, according to the results of a survey of nursery schools in France [24], the mean CO_2 concentration during opening hours was 1200 ppm, which was higher than that of the nursery classroom in this study. In both summer and winter, the mean CO_2 concentration in one classroom was found to be over 1500 ppm, which was specified by the school environmental health standard [2]. The mean CO_2 concentration during opening hours in the summer exceeded 1500 ppm in the 0-year-old room in nursery K, probably because of a lower ventilation rate, despite meeting the area standard per infant, as shown in Table 2. The mean CO_2 concentrations during opening hours in the winter were over 1500 ppm in the 1-year-old room in nursery M, where the ventilation rate was probably low because it did not meet the area standard per infant as mentioned earlier.

Figure 7a shows that during opening hours in the summer, of the 26 measured classrooms, one classroom had a mean value of more than 1500 ppm, and the mean values exceeded 1000 ppm in six classrooms, which is 23% of the measured classrooms. Figure 7b shows that during opening hours in the winter, of the 26 measured classrooms, one classroom had a mean value of more than 1500 ppm, and the mean values exceeded 1000 ppm in eight classrooms, which is 31% of the measured classrooms. In this way, the CO_2 concentration in the nursery classrooms tended to be higher in the winter than in the summer. Additionally, except for some nursery schools (nursery I in summer and nursery D, I, and K in winter), there was a significant difference in CO_2 concentration between the 0-year-old room and 1-year-old room on weekdays in both the summer and winter. In the summer and winter (nurseries B and K), in the summer (nurseries E, F, and N), and in the winter (nursery J), the CO_2 concentration in the 0-year-old room was significantly higher than that in the 1-year-old room. In other nursery schools, the 1-year-old room was found to exhibit a significantly higher CO_2 concentration than the 0-year-old room (Tables A5 and A6). The high concentrations of CO_2 occurred when numerous people gathered in one classroom at the time of pick-up or during activities according to its fluctuations (data not shown). However, there was no relationship between CO_2 concentration and the areas per infant for both classrooms (Tables 1 and 2).

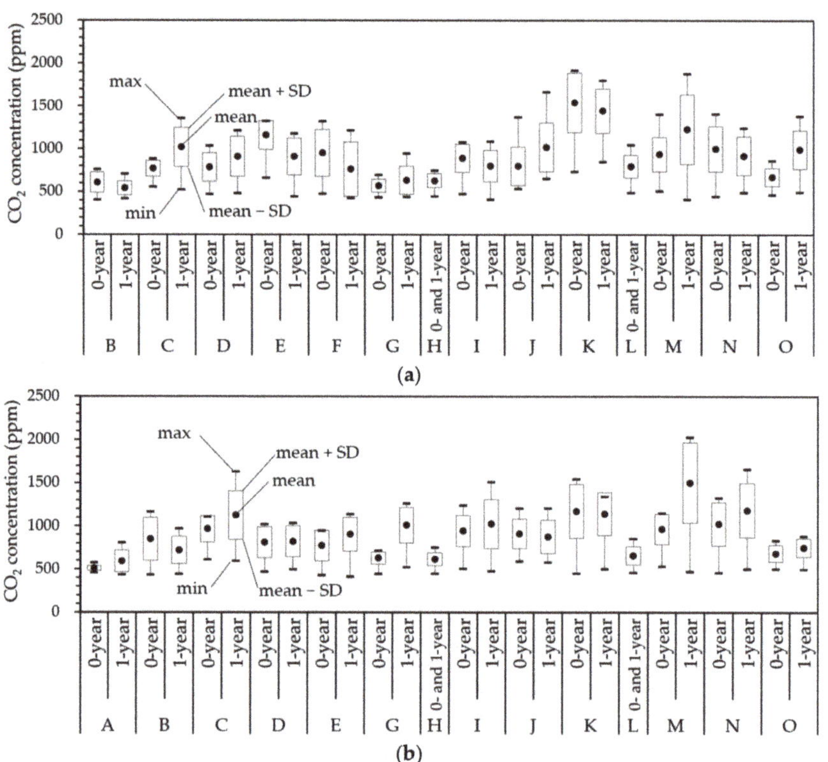

Figure 7. CO_2 concentrations in the nursery classrooms during opening hours in (**a**) summer and (**b**) winter.

Figure 8 shows the evaluation results of the IAQ during the opening hours for each nursery classroom in the summer and winter using ICONE [22]. No extreme air stuffiness (ICONE score = 5) was found in any of the nursery classrooms. Of 52 studied classrooms in the summer and winter, only one classroom (2%) had very high air stuffiness (ICONE score = 4), three (6%) had high air stuffiness (ICONE score = 3), 16 (31%) had average air stuffiness (ICONE score = 2), 20 (39%) had low air stuffiness (ICONE score = 1), and 12 (23%) had no air stuffiness (ICONE score = 0). With the exception of some nursery schools, the 0-year-old rooms and 1-year-old rooms had lower percentages (8%) of high air stuffiness (ICONE score $\geq$ 3) compared with that of nursery schools in France (38% [24]). Both 0-year-old and 1-year-old rooms in nursery K exhibited high or very high air stuffiness scores in both rooms in the summer, and the 1-year-old room in nursery M exhibited high scores in both summer and winter. Table 2 showed that in the 1-year-old room of nursery M, the area per infant did not meet the minimum standard, and the ventilation rate was insufficient due to the high density of the classroom. In nursery K, the small room volume contributed to the insufficient ventilation rate, even though the area per infant met the minimum standard.

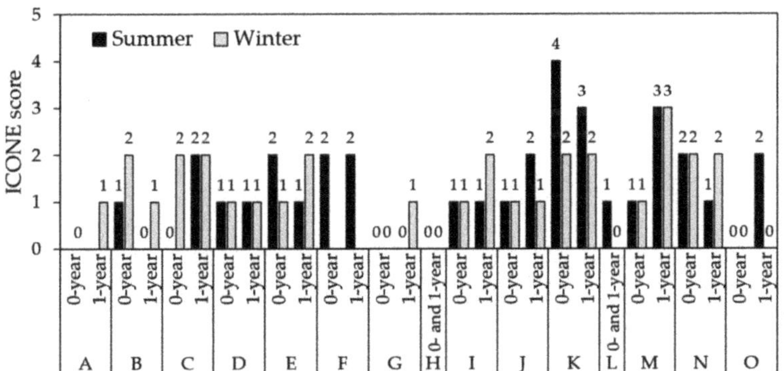

Figure 8. ICONE air stuffiness index scores in the studied nursery rooms during opening hours in summer and winter.

Figure 9 shows the relationship between the infants' room area and weekly median CO_2 concentration in the summer and winter. In the winter, there is a moderate correlation between the area per infant and the weekly median value of CO_2 concentration. It can be seen that in the winter, the larger the area per infant, the lower the weekly median CO_2 concentration, although there is no similar correlation in the summer.

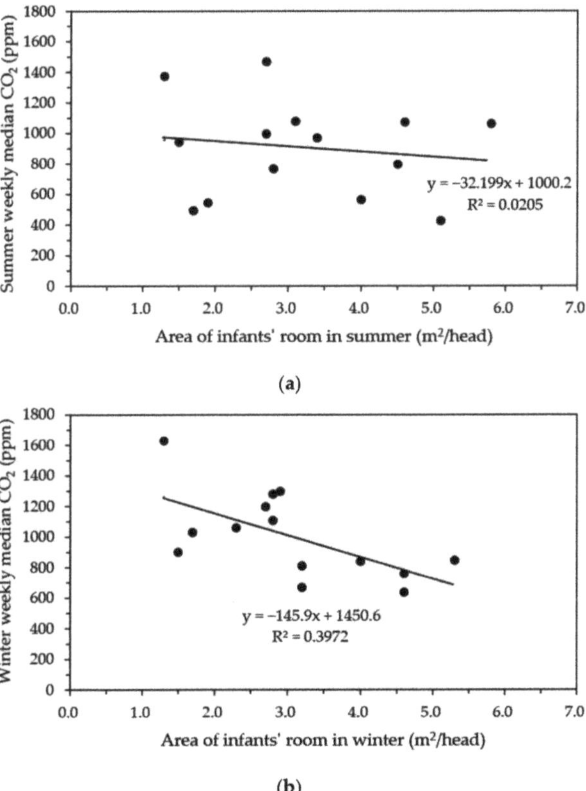

(**a**)

(**b**)

Figure 9. Relationship between infants' room area and weekly median CO_2 concentration in (**a**) summer and (**b**) winter.

3.3. Particulate Matter

The results of $PM_{2.5}$ measurements during the 10 min spot measuring period in nursery classrooms and outside are presented in Table 4, which shows the number of small particles with a diameter of less than 0.5 μm, indoor-to-outdoor ratios (I/O ratios) that represent the ratios of the number of small particles between the indoor air and outdoor air, and IAQ determined based on the number of the small particles during measuring period in the summer and winter. I/O ratios can determine the impact of indoor/outdoor sources on indoor environments. The small particles were detected in all of the studied classrooms. IAQ in each nursery classroom was rated to be fair, poor, or very poor in most nursery schools, according to an IAQ judgment criteria based on the number of small particles. The air quality criteria were evaluated as follows on a five-point scale according to the number of small particles: 0–75, excellent; 75–150, very good; 150–300, good; 300–1050, fair; 1050–3000, poor; and >3000: very poor. Of the 24 classrooms measured in both summer and winter, 12 had a lower IAQ in the winter than in the summer, five had a lower IAQ in the summer than in the winter, and there was no change in IAQ between summer and winter in the other seven classrooms. The time that the windows are open tends to be shorter in the winter than in the summer, and the lack of ventilation affects the low IAQ result based on the number of small particles. I/O ratios ranged from 0.11 to 4.47 in 0-year-old rooms, and ranged from 0.10 to 2.01 in the 1-year old rooms, throughout the year. Compared by season, the I/O ratios ranged from 0.48 to 4.47 in 0-year-old rooms and ranged from 0.39 to 2.01 in 1-year old rooms in the summer, though the I/O ratios ranged from 0.11 to 1.89 in 0-year-old rooms and from 0.10 to 1.21 in 1-year-old rooms in the winter. In 1-year-old rooms, the I/O ratio was significantly higher in the summer than in the winter ($p < 0.05$), but there was no significant seasonal difference in the 0-year-old rooms. From the viewpoint of I/O ratio, the number of the small particles in the classroom was larger than outside in seven out of 26 classrooms (one-third of the measured rooms) measured in the summer and four out of 26 classrooms (one-sixth of the measured rooms) measured in the winter. In these 11 classrooms in both summer and winter, there is a source of small particles in the classroom. It is believed that small particles detected indoors were influenced by activities such as playing with sand in the garden. Children might come and go more frequently when playing outside in the playground in the summer than in the winter; hence, it may be easier to bring dust from the outside air into the classroom. In a Portuguese preschool environment, values of I/O ratios showed that a considerable part of small particle matter, including $PM_{2.5}$, originated indoors and carcinogenic risks due to exposure indoors were 10 and 4 times higher than for exposure outdoors, respectively, for younger (3 years) and older (4–5 years) pupils [25]. This was due to the fact that younger children tended to stay indoors for longer periods of time [25]. Therefore, focusing on the health effects of small particles in the classroom is necessary.

Table 4. Small particles (count/ft^3), I/O ratios, and IAQ during the summer and winter measurement periods.

Nursery, Room	0-Year-Old Room	1-Year-Old Room	Outdoor
A, in the summer	Not measured	Not measured	Not measured
A, in the winter	5107, 0.90, very poor	4580, 0.81, very poor	5673, very poor
B, in the summer	1302, 0.48, poor	1373, 0.51, poor	2685, poor
B, in the winter	1391, 0.31, poor	1529, 0.34, poor	4447, very poor
C, in the summer	6596, 1.37, very poor	6051, 1.25, very poor	4823, very poor
C, in the winter	2541, 1.40, poor	1299, 0.71, poor	1820, poor
D, in the summer	2562, 1.97, poor	2610, 2.01, poor	1301, poor
D, in the winter	1802, 0.25, poor	4700, 0.64, very poor	7340, very poor
E, in the summer	1005, 0.75, fair	872, 0.65, fair	1348, poor
E, in the winter	1005, 0.25, fair	1351, 0.34, poor	3992, very poor

Table 4. *Cont.*

Nursery, Room	0-Year-Old Room	1-Year-Old Room	Outdoor
F, in the summer	2813, 0.85, poor	4857, 1.47, very poor	3295, very poor
F, in the winter	Not measured	Not measured	Not measured
G, in the summer	1219, 1.04, poor	1275, 1.08, poor	1176, poor
G, in the winter	1347, 0.87, poor	1345, 0.87, poor	1543, poor
H, in the summer	2502, 4.47, poor		560, fair
H, in the winter	7515, 0.94, poor		7998, very poor
I, in the summer	776, 0.70, fair	944, 0.85, fair	1108, poor
I, in the winter	4907, 0.66, very poor	5778, 0.78, very poor	7450, very poor
J, in the summer	2863, N.A., poor	1645, N.A., poor	Not measured
J, in the winter	6805, 1.89, very poor	4375, 1.21, very poor	3610, very poor
K, in the summer	545, 0.84, fair	514, 0.80, fair	645, fair
K, in the winter	3448, 0.66, very poor	1647, 0.32, poor	5195, very poor
L, in the summer	5031, 0.79, very poor		6355, very poor
L, in the winter	1208, 1.07, poor		1133, poor
M, in the summer	797, 0.70, fair	445, 0.39, fair	1146, poor
M, in the winter	1292, 0.11, poor	1166, 0.10, poor	12,239, very poor
N, in the summer	820, 0.73, fair	1069, 0.95, poor	1121, poor
N, in the winter	2893, 0.54, poor	4070, 0.76, very poor	5375, very poor
O, in the summer	1469, N.A., poor	1374, N.A., poor	Not measured
O, in the winter	699, 0.48, fair	611, 0.42, fair	1444, poor

3.4. Illumination

Figure 10 shows the mean values of illumination in the studied nursery classrooms during opening hours in the summer and winter, including 95% confidence intervals. It shows a fluctuation from the lighting being temporarily turned off at any time from 12:00 to 14:00 due to the children napping in the nursery classroom, and the illumination is reduced to the minimum (results not shown). Notably, the measurement location may be partially affected by solar radiation due to the convenience of measurement installation. The numbers of the nursery classrooms where the average illumination during opening hours was less than the standard value of 300 lx established by the school environmental health standard [2] were eight (31% of the total) out of the 26 studied classrooms in the summer, and nine out of 26 (35% of the total) in the winter. There was no significant difference in the average illumination between summer and winter. The average illumination in the two classrooms was lower than 100 lx. In particular, the average illumination of the 1-year-old room in nursery F in the summer was under 77 lx, which was significantly lower because the lights were turned off to save energy for space cooling. Infants do not study as do school students, but in consideration for the needs of infant visual development, they do need to maintain minimal illumination for reading picture books and other indoor activities in the classrooms.

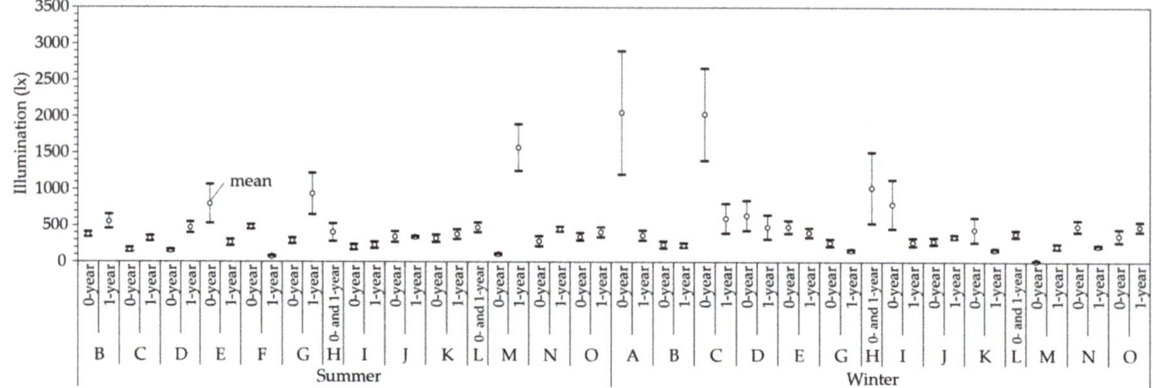

Figure 10. Mean values of illumination in the studied nursery rooms during opening hours in summer and winter, including 95% confidence intervals.

3.5. Noise Level

In this study, for nursery schools that could not be measured during naps, the noise levels during various activities were measured and treated as reference values. Results provide evidence showing how noisy it is; noise in the vicinity of nursery schools in Japan is considered to be a social problem. Strictly speaking, the noise level during nap time is not equal to the background noise level, but perhaps close to background noise and therefore useful. The equivalent noise level during various activities ranged from 50.4 to 70.6 dBA with an average of 60.1 dBA in 0-year-old rooms, and ranged from 48.9 to 84.3 dBA with an average of 64.1 dBA in 1-year-old rooms (results not shown). The equivalent noise level during the nap time ranged from 50.4 to 59.3 dBA with an average of 54.4 dBA in 0-year-old rooms and ranged from 49.3 to 58.5 dBA with an average of 53.1 dBA in 1-year-old rooms. According to the guidelines of WHO, the background equivalent noise level in classrooms of preschools should not exceed 35 dB Laeq during class [34]. Compared to this, the noise level in Japanese nursery classrooms is considerably higher even during naps. Considering the development of hearing of infants, establishing appropriate noise standards for nursery school classrooms is necessary.

3.6. Examination of Indoor Environmental Standards

Indoor environmental standards for nursery school classrooms where children are under 2 years old and who cannot easily wear a mask, cannot receive a COVID-19 vaccination, and spend a long time, will become increasingly necessary in Japan in the future. From the results of this study, first, concerning the thermal environment, the temperature should be 26–28 °C in the summer and 20–23 °C in the winter, as given in the infectious disease control guidelines for nursery schools [3]. Considering the time that children spend at lower heights close to the floor, this study's findings recommend assessing whether the temperatures at the heights of 0.1, 0.3, and 0.6 m above the floor meet the standard value. If the nursery school buildings are old, their insulation is also important.

Second, regarding relative humidity, the infectious disease control guideline only stipulates a relative humidity of 60% [3], but it is believed that the 40–70% standard with a range applied to office buildings under the Act on Maintenance of Sanitation in Buildings [35] should be applied to nursery school classrooms.

Third, regarding the indoor air environment, it became clear that the CO_2 concentrations in the nursery classrooms are high during activities or the waiting time for pick-up; hence, it is proposed that the daily average concentration of CO_2 should be less than 1000 ppm and the maximum concentration should not exceed 1500 ppm, with reference to BB101 established in the UK [7], though the infectious disease control guidelines do not set ventilation standards [3]. Additionally, adhering to the standard for the area of the nursery classroom per infant, which is specified in the minimum standard for child-welfare-facility minimum criteria [28], will lead to satisfying the required ventilation rate per child, and provide a well-ventilated indoor air environment. As for $PM_{2.5}$ concentration, as mentioned earlier, the spot measurement was only for 10 min; hence, a concrete standard could not be proposed in this study.

Fourth, regarding the illumination environment, a minimum illumination of 300 lx, which is specified in BB101 [7], is required for infants' visual development. The nursery school classrooms are used by people of all ages, including infants' grandparents who pick-up the children instead of working parents. Older adults need twice as much illumination as adults [36]; hence, they need more than 300 lx from the viewpoint of universal design.

Fifth, regarding the acoustic environment, the indoor background noise level in the nursery school classrooms should ideally be restricted to 35 dBA, as determined by WHO. According to the results of inspections of nursery facilities in Munich, Germany, where regulations of acoustic design for daycare facilities exist, all the classrooms in the inspected facilities were equipped with sound absorptive material for sound insulation and the reduction in reverberation [37]. By contrast, in Japan, without any standards or regulations, meeting these strict requirements is difficult for nursery schools. Therefore, firstly, as stipulated in the school environmental health standard [2], fulfilling the minimum indoor background noise level of 50 dBA when windows are closed is necessary; however, establishing the indoor equivalent noise level slightly lower to consider the development of hearing in infants is also vital.

4. Conclusions

The measurements of indoor environments and questionnaire surveys in 15 different nursery schools in mild climatic areas in Japan were conducted in the summer and winter of 2016–2019. The results of the measurements were summarized and the indoor environmental standards that should be set in the nursery school classrooms were considered.

The study revealed that summer mean temperature was between 26 and 28 °C, but the winter mean temperature was lower than the lower limit of 20–23 °C. Mean RH was close to 60% of the standard in the summer, but lower than the standard in the winter, compared to the infectious disease control guideline. In addition, in the summer, there was one nursery school where the temperature in the 1-year-old room at any height was higher than 30 °C at almost time during the day in the summer, which was too high and did not meet the guidelines, although the temperature was lower compared to the lower limit of the guideline in most of the other nursery classrooms. It is necessary to use air conditioning in this nursery school to prevent heat stroke. In order to avoid an extremely hot indoor environment in the summer, the necessity of indoor environmental standards in the nursery classroom was shown. In the winter, it was found that the temperature at 1.1 m above the floor generally met the infectious disease control guidelines for nursery schools (20–23 °C), except for some nursery schools such as those in older buildings, in contrast, the temperature at 0.1 m above the floor in all of the nursery classrooms did not meet the standards (20–23 °C). This may be due to poor insulation performance or poor air circulation, even in nursery schools that were built more recently. Since infants spend their time at lower heights than adults, it is necessary to raise the temperatures at 0.1, 0.3, and 0.6 m from floor surface in the winter to ensure the comfort and health of infants. It can be said that efforts such as improving the insulation performance of buildings and repairing the insulation of existing buildings are also necessary to solve the problems of lower temperature at the lower height from the floor surface in the nursery classrooms during winter.

The mean RH satisfied the value of 30–80% specified by the school environmental health standard in all studied nursery classrooms both in the summer and winter. The mean HR in most of the studied nursery classrooms did not satisfy the upper limit of 0.012 kg/kg (DA) in the summer, specified by ASHRAE Standard 55-2017, although the mean HR in all studied nursery classrooms satisfied the requirement in the winter. Since it is hot and humid during the summer in Japan, focusing on humidity control in the summer is important from the viewpoint of microbial contamination.

Nursery school classrooms were affected by cold radiation because the globe temperature differences were negative in most classrooms both in the summer and winter.

The mean CO_2 concentrations during opening hours exceeded 1500 ppm, specified by the school environmental health standard, in one classroom both in the summer and winter. The reasons for this were probably because of the low ventilation rate despite meeting the area standard per infant or insufficient ventilation rate because of not meeting the area standard per infant. The evaluation results of the IAQ during the opening hours of each nursery room in the summer and winter using ICONE showed that the 0-year-old and 1-year-old rooms, except some nursery schools, exhibit lower percentages (8%) of high air stuffiness (ICONE score $\geq$ 3).

Small particles with a diameter of less than 0.5 μm were detected in all studied classrooms. IAQ in each nursery classroom was rated to be fair, poor, or very poor in most nurseries, according to an IAQ judgment criteria based on the number of small particles. Half of the nursery school classrooms exhibited lower IAQ in the winter than the summer. The opening time of the windows is probably shorter in the winter than in the summer, and the lack of ventilation affects the low IAQ in the winter. As a result of I/O ratios, the number of small particles in the classrooms was larger than outside in one-third of the studied classrooms in the summer and one-sixth of the studied classrooms in the winter; therefore, sources of small particles may be present indoors. It is believed that small particles detected indoors were influenced by activities such as playing with sand in the garden.

The mean illumination during opening hours in one-third of the studied nursery classrooms was less than the standard value of 300 lx established by the school environmental health standard—both in the summer and winter. Considering the needs of infant visual development, minimal illumination for indoor activities for infants should be maintained.

The equivalent noise level by applying the A-weighting filter in studied nursery classrooms was considerably high even during nap times. Therefore, considering the development of infant hearing, the need for an acoustic standard for nursery school classrooms was shown.

To summarize, some nursery school classrooms were found to have large vertical temperature differences, while other classrooms had poor IAQ because they held more children than their capacities allowed. Compliance with the specified area per child would lead to the maintenance of a desirable IAQ. In this study, the need for indoor environmental standards in nursery schools was strongly indicated in terms of infants' comfort and health. It is necessary to continue to evaluate the actual conditions of indoor environment in nursery school classrooms and establish suitable indoor environment standards as soon as possible. Future indoor environmental standards for nursery classrooms—where children under 2 years old spend considerable time, who cannot wear a mask easily, and cannot receive COVID-19 vaccinations—will become increasingly necessary in Japan.

5. Limitations to the Study and Future Research

This study had the following limitations:

(1) The installation location of the measurement equipment tends to be restricted, especially in 1-year-old rooms, to ensure that children do not touch the measurement equipment.
(2) It is necessary to consider the metabolic rate of the children to set the temperature standard.
(3) In this study, particle matter and equivalent noise level were evaluated by the results of a spot measurement by sampling for only 10 min for each visit. Therefore, evaluating particle matter and equivalent noise level by lengthening the sampling time will be necessary in the future. Further, quantitatively evaluating $PM_{2.5}$ concentration and assessing its health risk to infants and childminders will be crucial.

(4) In this study, indoor air environment was evaluated by CO_2 concentration and particulate matter, but investigating the ventilation rate for each nursery classroom will be vital in the future.

Funding: This research was partially funded by LIXIL JS Foundation in 2016, grant number 16-13.

Informed Consent Statement: Informed consent was obtained from all subjects involved in the study.

Acknowledgments: I deeply appreciate the assistance of all the people involved in the nursery schools who cooperated in the survey. I am extremely grateful to all the students at Nagasaki University who supported me throughout the preparation and administration of the survey.

Conflicts of Interest: The author declares no conflict of interest.

Appendix A

Table A1. Vertical temperature during opening hours in summer.

Nursery, Room	Temperature at 0.1 m from Floor Surface (°C) Median (5th, 95th Percentile)	Temperature at 0.3 m from Floor Surface (°C) Median (5th, 95th Percentile)	Temperature at 0.6 m from Floor Surface (°C) Median (5th, 95th Percentile)	Temperature at 1.1 m from Floor Surface (°C) Median (5th, 95th Percentile)
A, 0-year-old room A, 1-year-old room	Not measured	Not measured	Not measured	Not measured
B, 0-year-old room B, 1-year-old room	27.4 (25.1, 29.2) 27.7 (25.1, 29.7)	27.5 (24.8, 29.3) 27.8 (24.8, 30.4)	27.7 (24.6, 29.6) 27.9 (24.8, 30.9)	28.1 (25.0, 31.7) 27.9 (24.6, 31.1)
C, 0-year-old room C, 1-year-old room	25.6 (25.0, 26.7) 25.8 (24.9, 27.3)	25.5 (24.8, 26.7) 25.3 (24.0, 27.4)	25.8 (25.1, 27.1) 25.2 (23.7, 26.8)	26.4 (25.8, 28.0) 27.3 (26.3, 28.6)
D, 0-year-old room D, 1-year-old room	25.7 (23.8, 27.0) 25.9 (23.0, 26.6)	25.5 (23.8, 26.8) 25.8 (23.3, 26.7)	25.5 (24.0, 27.1) 25.8 (23.5, 26.8)	27.0 (24.6, 28.7) 26.3 (24.9, 27.9)
E, 0-year-old room E, 1-year-old room	26.8 (25.8, 28.8) 24.9 (23.8, 26.3)	26.5 (25.6, 28.8) 24.7 (23.5, 26.1)	26.4 (25.4, 28.9) 24.7 (23.6, 26.2)	25.6 (24.1, 27.1) 28.5 (27.0, 32.9)
F, 0-year-old room F, 1-year-old room	28.6 (27.1, 32.8) 31.2 (28.6, 32.9)	28.4 (26.9, 33.0) 31.1 (28.2, 32.9)	28.7 (27.0, 33.4) 31.0 (28.3, 32.8)	31.7 (29.6, 34.3) 31.9 (29.2, 34.8)
G, 0-year-old room G, 1-year-old room	27.8 (26.4, 29.7) 26.4 (24.1, 28.7)	27.7 (26.4, 29.7) 26.5 (24.2, 29.0)	27.6 (26.3, 29.6) 26.6 (24.6, 29.2)	28.2 (27.0, 30.0) 28.0 (26.1, 30.1)
H, 0- and 1-year-old room	26.6 (26.2, 27.4)	26.5 (26.0, 27.4)	26.6 (26.1, 27.6)	27.8 (26.9, 29.5)
I, 0-year-old room I, 1-year-old room	24.8 (24.4, 25.9) 25.1 (24.6, 26.1)	24.8 (24.4, 25.9) 25.2 (24.6, 26.3)	24.8 (24.4, 25.9) 25.2 (24.5, 26.5)	24.8 (24.3, 25.8) 25.2 (24.6, 26.8)
J, 0-year-old room J, 1-year-old room	25.0 (23.9, 26.7) 25.2 (24.0, 26.6)	24.8 (23.6, 26.3) 24.9 (23.6, 26.3)	24.7 (23.4, 26.4) 25.1 (23.8, 26.6)	25.2 (24.0, 27.5) 25.5 (24.2, 27.1)
K, 0-year-old room K, 1-year-old room	26.5 (25.6, 28.8) 25.3 (24.8, 26.3)	26.7 (25.8, 29.1) 25.3 (24.7, 26.4)	27.1 (25.9, 29.3) 25.5 (25.0, 26.7)	27.4 (26.1, 30.1) 27.3 (26.5, 28.4)
L, 0- and 1-year-old room	24.2 (23.2, 25.1)	24.4 (23.5, 25.3)	24.4 (23.5, 25.3)	25.5 (24.5, 26.3)
M, 0-year-old room M, 1-year-old room	24.6 (23.9, 26.1) 25.9 (25.1, 27.3)	24.7 (23.8, 26.2) 26.1 (25.2, 27.8)	24.7 (23.8, 27.5) 26.2 (25.3, 28.1)	24.9 (24.2, 26.6) 27.8 (26.7, 30.1)
N, 0-year-old room N, 1-year-old room	24.8 (23.9, 25.9) 24.8 (24.4, 25.9)	24.8 (23.9, 26.1) 24.7 (24.2, 25.9)	24.8 (23.9, 26.1) 24.7 (24.2, 25.9)	25.7 (24.7, 27.1) 25.3 (24.5, 28.0)
O, 0-year-old room O, 1-year-old room	27.0 (26.0, 29.5) 24.5 (23.7, 25.4)	25.9 (24.8, 29.4) 24.4 (23.6, 25.4)	26.4 (25.2, 29.7) 24.5 (23.7, 25.7)	26.9 (24.8, 31.5) 25.8 (24.6, 27.3)

Table A2. Vertical temperature during opening hours in winter.

Nursery, Room	Temperature at 0.1 m from Floor Surface (°C) Median (5th, 95th Percentile)	Temperature at 0.3 m from Floor Surface (°C) Median (5th, 95th Percentile)	Temperature at 0.6 m from Floor Surface (°C) Median (5th, 95th Percentile)	Temperature at 1.1 m from Floor Surface (°C) Median (5th, 95th Percentile)
A, 0-year-old room	16.0 (11.6, 18.4)	14.8 (8.4, 18.2)	15.3 (8.5, 19.1)	16.7 (9.1, 20.5)
A, 1-year-old room	14.1 (8.2, 17.4)	16.6 (11.5, 19.0)	17.6 (11.5, 20.1)	20.0 (12.3, 25.8)
B, 0-year-old room	13.7 (9.7, 16.0)	14.7 (9.9, 16.9)	16.8 (10.5, 18.7)	20.5 (12.0, 22.6)
B, 1-year-old room	15.7 (11.0, 17.3)	16.4 (11.0, 18.4)	17.1 (11.6, 19.5)	18.7 (14.0, 21.2)
C, 0-year-old room	17.1 (14.9, 18.3)	17.2 (14.8, 18.2)	17.9 (15.0, 19.5)	20.0 (16.6, 21.1)
C, 1-year-old room	17.5 (12.4, 20.0)	18.6 (13.1, 21.0)	19.2 (12.9, 22.6)	18.8 (14.5, 24.1)
D, 0-year-old room	18.6 (11.4, 21.2)	19.3 (12.3, 21.6)	20.0 (13.2, 22.5)	21.6 (15.3, 24.3)
D, 1-year-old room	16.8 (10.8, 19.9)	18.8 (13.6, 21.2)	19.6 (14.0, 21.9)	19.8 (13.5, 22.1)
E, 0-year-old room	11.8 (7.6, 15.2)	13.3 (8.6, 17.0)	16.0 (9.7, 19.1)	11.8 (10.9, 19.2)
E, 1-year-old room	14.6 (10.8, 18.0)	15.2 (11.2, 18.8)	16.1 (12.0, 19.7)	13.5 (12.9, 21.7)
F, 0-year-old room F, 1-year-old room	Not measured	Not measured	Not measured	Not measured
G, 0-year-old room	12.2 (10.8, 14.2)	13.4 (10.8, 16.4)	14.7 (10.9, 18.9)	16.5 (10.8, 20.2)
G, 1-year-old room	16.4 (12.4, 18.2)	17.1 (12.6, 19.4)	17.8 (12.6, 20.5)	20.5 (14.9, 22.7)
H, 0- and 1-year-old room	16.7 (14.5, 18.4)	16.8 (14.3, 18.9)	17.3 (14.4, 19.8)	19.7 (15.9, 24.2)
I, 0-year-old room	Not measured	Not measured	Not measured	21.6 (15.9, 23.9)
I, 1-year-old room	19.9 (14.2, 22.8)	20.8 (14.3, 23.6)	21.1 (14.2, 24.0)	22.7 (15.7, 25.5)
J, 0-year-old room	17.4 (15.1, 18.9)	17.8 (15.5, 20.2)	18.5 (15.9, 21.6)	20.4 (17.6, 24.0)
J, 1-year-old room	18.5 (15.4, 19.6)	19.1 (15.5, 20.9)	19.8 (15.8, 22.4)	20.7 (17.0, 24.2)
K, 0-year-old room	17.5 (10.5, 20.5)	18.3 (10.8, 21.5)	19.3 (11.5, 22.2)	20.8 (12.7, 23.7)
K, 1-year-old room	17.5 (12.9, 19.8)	18.1 (13.2, 20.1)	18.7 (13.9, 20.5)	20.3 (15.4, 22.0)
L, 0- and 1-year-old room	18.4 (15.2, 20.2)	19.3 (16.7, 20.9)	20.3 (17.6, 21.7)	20.8 (18.2, 23.1)
M, 0-year-old room	19.0 (16.7, 19.6)	20.2 (17.5, 22.0)	21.9 (18.7, 24.2)	22.2 (19.2, 24.0)
M, 1-year-old room	17.9 (15.3, 19.2)	18.0 (15.4, 19.3)	18.3 (15.9, 19.7)	19.2 (17.1, 21.5)
N, 0-year-old room	16.9 (13.2, 18.7)	17.6 (13.2, 19.5)	17.9 (13.1, 19.9)	19.4 (13.6, 21.4)
N, 1-year-old room	17.8 (14.8, 19.8)	17.9 (14.7, 20.1)	17.9 (14.3, 19.9)	20.1 (15.9, 21.8)
O, 0-year-old room	16.1 (11.9, 18.3)	17.1 (11.8, 19.3)	18.7 (12.3, 20.7)	20.9 (15.2, 23.3)
O, 1-year-old room	18.7 (12.3, 20.7)	20.4 (15.7, 21.9)	20.9 (15.8, 22.4)	23.1 (18.2, 24.3)

Table A3. Relative humidity, humidity ratio, and globe temperature during opening hours in summer.

Nursery, Room	Relative Humidity at 1.1 m from Floor Surface (%) Median (5th, 95th Percentile)	Humidity Ratio at 1.1 m from Floor Surface (kg/kg (DA)) Median (5th, 95th Percentile)	Globe Temperature at 1.1 m from Floor Surface (°C) Median (5th, 95th Percentile)
A, 0-year-old room A, 1-year-old room	Not measured	Not measured	Not measured
B, 0-year-old room	60 (50, 82)	0.015 (0.011, 0.020)	28.1 (25.0, 31.7)
B, 1-year-old room	66 (52, 82)	0.015 (0.012, 0.020)	27.9 (24.6, 31.1)
C, 0-year-old room	66 (56, 77)	0.014 (0.012, 0.017)	25.4 (24.6, 27.4)
C, 1-year-old room	59 (49, 69)	0.013 (0.011, 0.016)	25.6 (24.3, 27.4)
D, 0-year-old room	63 (47, 74)	0.014 (0.011, 0.017)	25.6 (24.1, 27.2)
D, 1-year-old room	59 (41, 70)	0.013 (0.009, 0.015)	26.3 (24.1, 28.1)

Table A3. *Cont.*

Nursery, Room	Relative Humidity at 1.1 m from Floor Surface (%) Median (5th, 95th Percentile)	Humidity Ratio at 1.1 m from Floor Surface (kg/kg (DA)) Median (5th, 95th Percentile)	Globe Temperature at 1.1 m from Floor Surface (°C) Median (5th, 95th Percentile)
E, 0-year-old room	61 (55, 72)	0.013 (0.011, 0.016)	25.1 (23.6, 26.7)
E, 1-year-old room	56 (40, 72)	0.013 (0.011, 0.018)	26.4 (25.4, 29.2)
F, 0-year-old room	46 (34, 62)	0.013 (0.010, 0.019)	29.5 (27.7, 33.6)
F, 1-year-old room	55 (47, 69)	0.017 (0.012, 0.021)	31.6 (28.5, 34.1)
G, 0-year-old room	56 (44, 66)	0.013 (0.010, 0.017)	27.5 (26.4, 29.3)
G, 1-year-old room	62 (50, 70)	0.015 (0.012, 0.018)	26.3 (24.3, 28.9)
H, 0- and 1-year-old room	65 (52, 76)	0.015 (0.013, 0.018)	26.8 (26.2, 28.1)
I, 0-year-old room	65 (57, 74)	0.013 (0.011, 0.015)	24.8 (24.3, 25.8)
I, 1-year-old room	67 (57, 78)	0.014 (0.012, 0.016)	25.2 (24.6, 26.8)
J, 0-year-old room	69 (63, 76)	0.014 (0.012, 0.016)	25.1 (24.1, 27.2)
J, 1-year-old room	68 (62, 75)	0.014 (0.012, 0.016)	25.4 (24.2, 26.9)
K, 0-year-old room	52 (41, 65)	0.012 (0.011, 0.016)	26.7 (25.6, 29.1)
K, 1-year-old room	52 (44, 60)	0.012 (0.010, 0.014)	26.1 (24.0, 29.1)
L, 0- and 1-year-old room	73 (64, 79)	0.015 (0.013, 0.017)	24.4 (23.5, 25.3)
M, 0-year-old room	73 (65, 77)	0.014 (0.013, 0.015)	24.8 (24.1, 26.3)
M, 1-year-old room	51 (43, 63)	0.012 (0.010, 0.016)	27.5 (26.0, 31.1)
N, 0-year-old room	62 (55, 73)	0.013 (0.011, 0.016)	25.3 (24.4, 26.7)
N, 1-year-old room	69 (54, 82)	0.014 (0.011, 0.017)	24.4 (23.7, 27.3)
O, 0-year-old room	55 (42, 63)	0.012 (0.011, 0.016)	26.1 (23.9, 31.0)
O, 1-year-old room	64 (57, 70)	0.013 (0.012, 0.015)	25.2 (23.8, 26.5)

Table A4. Relative humidity, humidity ratio, and globe temperature during opening hours in winter.

Nursery, Room	Relative Humidity at 1.1 m from Floor Surface (%) Median (5th, 95th Percentile)	Humidity Ratio at 1.1 m from Floor Surface (kg/kg (DA)) Median (5th, 95th Percentile)	Globe Temperature at 1.1 m from Floor Surface (°C) Median (5th, 95th Percentile)
A, 0-year-old room	48 (35, 58)	0.006 (0.003, 0.007)	16.6 (9.2, 20.5)
A, 1-year-old room	39 (25, 52)	0.005 (0.004, 0.007)	20.0 (12.3, 26.0)
B, 0-year-old room	48 (32, 56)	0.007 (0.004, 0.008)	20.5 (12.0, 22.6)
B, 1-year-old room	39 (32, 47)	0.005 (0.003, 0.007)	18.7 (14.0, 21.2)
C, 0-year-old room	44 (39, 49)	0.006 (0.005, 0.007)	19.4 (15.9, 21.1)
C, 1-year-old room	52 (40, 60)	0.007 (0.005, 0.008)	20.1 (14.6, 25.0)
D, 0-year-old room	49 (42, 66)	0.008 (0.006, 0.011)	20.5 (14.7, 23.1)
D, 1-year-old room	50 (40, 70)	0.007 (0.005, 0.011)	20.0 (15.3, 22.5)
E, 0-year-old room	59 (51, 67)	0.005 (0.004, 0.008)	18.5 (11.4, 20.7)
E, 1-year-old room	48 (38, 61)	0.005 (0.004, 0.007)	18.1 (13.9, 21.5)
F, 0-year-old room F, 1-year-old room	Not measured	Not measured	Not measured
G, 0-year-old room	46 (33, 64)	0.005 (0.004, 0.007)	16.5 (10.8, 20.2)
G, 1-year-old room	43 (32, 52)	0.006 (0.004, 0.008)	19.6 (14.1, 21.7)
H, 0- and 1-year-old room	46 (29, 66)	0.006 (0.004, 0.009)	18.2 (15.2, 22.2)
I, 0-year-old room	45 (35, 53)	0.007 (0.005, 0.009)	21.4 (14.6, 25.2)
I, 1-year-old room	44 (35, 52)	0.008 (0.005, 0.009)	22.0 (15.6, 24.8)

Table A4. *Cont.*

Nursery, Room	Relative Humidity at 1.1 m from Floor Surface (%) Median (5th, 95th Percentile)	Humidity Ratio at 1.1 m from Floor Surface (kg/kg (DA)) Median (5th, 95th Percentile)	Globe Temperature at 1.1 m from Floor Surface (°C) Median (5th, 95th Percentile)
J, 0-year-old room	36 (25, 45)	0.006 (0.004, 0.007)	19.8 (17.1, 23.7)
J, 1-year-old room	36 (27, 44)	0.005 (0.004, 0.007)	20.3 (16.3, 23.8)
K, 0-year-old room	47 (37, 53)	0.007 (0.005, 0.008)	20.1 (11.8, 31.9)
K, 1-year-old room	46 (36, 53)	0.007 (0.005, 0.008)	19.3 (14.5, 21.6)
L, 0- and 1-year-old room	41 (26, 64)	0.006 (0.004, 0.010)	21.0 (18.1, 22.4)
M, 0-year-old room	46 (29, 75)	0.007 (0.005, 0.011)	22.3 (19.0, 24.5)
M, 1-year-old room	63 (43, 79)	0.009 (0.006, 0.011)	19.1 (16.7, 21.3)
N, 0-year-old room	51 (41, 60)	0.007 (0.005, 0.009)	18.7 (13.4, 20.5)
N, 1-year-old room	44 (33, 51)	0.006 (0.004, 0.008)	19.4 (14.8, 21.4)
O, 0-year-old room	44 (37, 55)	0.007 (0.005, 0.009)	20.6 (15.3, 23.1)
O, 1-year-old room	41 (33, 51)	0.007 (0.005, 0.009)	22.5 (17.9, 23.6)

Table A5. CO_2 concentrations (ppm) in nursery classrooms during opening hours in summer.

Nursery, Room	Weekday Median (5th, 95th Percentile)	Saturday Median (5th, 95th Percentile)	Holiday Median (5th, 95th Percentile)
A, 0-year-old room A, 1-year-old room	Not measured	Not measured	Not measured
B, 0-year-old room	692 (394, 1424)	432 (378, 630)	388 (375, 411)
B, 1-year-old room	564 (417, 1126)	414 (402, 518)	407 (382, 437)
C, 0-year-old room	801 (515, 1000)	758 (493, 891)	503 (476, 554)
C, 1-year-old room	1078 (525, 1523)	993 (471, 1451)	486 (445, 556)
D, 0-year-old room	827 (464, 1210)	532 (480, 807)	527 (521, 532)
D, 1-year-old room	1061 (457, 1359)	536 (485, 808)	463 (451, 476)
E, 0-year-old room	1268 (669, 1758)	1080 (622, 1322)	429 (416, 467)
E, 1-year-old room	996 (441, 1518)	497 (424, 1108)	440 (417, 465)
F, 0-year-old room	1157 (426, 1815)	654 (411, 1894)	421 (404, 431)
F, 1-year-old room	545 (406, 1803)	455 (404, 1657)	411 (396, 428)
G, 0-year-old room	551 (419, 1163)	444 (421, 512)	410 (398, 414)
G, 1-year-old room	494 (407, 1208)	479 (434, 687)	430 (419, 432)
H, 0-year-old room	657 (444, 829)	457 (429, 850)	419 (409, 433)
H, 1-year-old room	426 (401, 448)	430 (408, 940)	406 (403, 419)
I, 0-year-old room	967 (454, 1268)	860 (463, 1097)	554 (533, 606)
I, 1-year-old room	969 (464, 1332)	876 (438, 1101)	473 (450, 495)
J, 0-year-old room	668 (408, 1282)	866 (494, 1490)	427 (404, 443)
J, 1-year-old room	944 (393, 1875)	913 (435, 1752)	406 (401, 422)
K, 0-year-old room	1708 (652, 2172)	1298 (800, 1791)	574 (521, 651)
K, 1-year-old room	1468 (843, 2003)	1293 (866, 1754)	585 (532, 669)
L, 0- and 1-year-old room	795 (469, 1121)	662 (505, 902)	412 (401, 422)
M, 0-year-old room	985 (495, 1306)	708 (534, 878)	398 (393, 407)
M, 1-year-old room	1373 (410, 2224)	843 (410, 1170)	370 (366, 382)
N, 0-year-old room	922 (439, 1790)	813 (698, 904)	425 (419, 456)
N, 1-year-old room	766 (455, 1689)	812 (615, 1015)	446 (441, 477)
O, 0-year-old room	675 (458, 1014)	492 (444, 569)	442 (438, 451)
O, 1-year-old room	1072 (490, 1836)	634 (495, 739)	430 (423, 448)

Table A6. CO_2 concentrations (ppm) in nursery classrooms during opening hours in winter.

Nursery, Room	Weekday Median (5th, 95th Percentile)	Saturday Median (5th, 95th Percentile)	Holiday Median (5th, 95th Percentile)
A, 0-year-old room	598 (429, 876)	457 (399, 691)	471 (458, 531)
A, 1-year-old room	810 (442, 1151)	687 (442, 1192)	421 (406, 458)
B, 0-year-old room	1144 (451, 1746)	N.A.	390 (381, 433)
B, 1-year-old room	839 (467, 1421)	N.A.	419 (404, 450)
C, 0-year-old room	1084 (585, 1303)	N.A.	543 (461, 672)
C, 1-year-old room	1296 (549, 1773)	N.A.	532 (440, 644)
D, 0-year-old room	880 (468, 1261)	592 (447, 788)	428 (422, 448)
D, 1-year-old room	845 (496, 1290)	677 (486, 920)	461 (457, 486)
E, 0-year-old room	950 (439, 1397)	733 (446, 890)	423 (411, 446)
E, 1-year-old room	1108 (660, 1436)	942 (516, 1222)	411 (406, 426)
F, 0-year-old room F, 1-year-old room	Not measured	Not measured	Not measured
G, 0-year-old room	665 (435, 879)	474 (419, 630)	418 (404, 428)
G, 1-year-old room	1029 (449, 1482)	1121 (470, 2388)	444 (429, 458)
H, 0- and 1-year-old room	635 (429, 850)	478 (395, 593)	416 (402, 427)
I, 0-year-old room	1033 (510, 1386)	814 (495, 946)	474 (459, 499)
I, 1-year-old room	1059 (455, 1585)	963 (466, 1298)	456 (448, 461)
J, 0-year-old room	943 (503, 1403)	719 (482, 1012)	404 (399, 418)
J, 1-year-old room	900 (539, 1394)	785 (482, 1116)	414 (403, 435)
K, 0-year-old room	1236 (452, 1885)	941 (440, 1304)	412 (408, 450)
K, 1-year-old room	1196 (509, 1926)	877 (472, 1213)	413 (408, 457)
L, 0- and 1-year-old room	667 (453, 964)	595 (432, 664)	380 (378, 390)
M, 0-year-old room	1042 (547, 1386)	716 (507, 1055)	405 (401, 429)
M, 1-year-old room	1629 (473, 2670)	1169 (478, 1817)	400 (393, 418)
N, 0-year-old room	1126 (449, 1685)	543 (431, 660)	415 (406, 438)
N, 1-year-old room	1279 (441, 1992)	1076 (420, 1468)	416 (410, 435)
O, 0-year-old room	692 (487, 1074)	529 (430, 566)	408 (404, 415)
O, 1-year-old room	757 (469, 1188)	693 (459, 830)	417 (415, 424)

References

1. Ministry of Health, Labour and Welfare, Japan. Summary of Related Situations such as Nursery School in 1 April 2021. Available online: https://www.mhlw.go.jp/stf/newpage_20600.html (accessed on 25 October 2021). (In Japanese)
2. Ministry of Education, Culture, Sports, Science and Technology, Japan. The School Environmental Health Standard. 2009. Available online: https://www.mext.go.jp/a_menu/kenko/hoken/1353625.htm (accessed on 25 October 2021). (In Japanese)
3. The Infectious Disease Control Guidelines for Nursery Schools (2018 Revised Edition). Available online: https://www.mhlw.go.jp/file/06-Seisakujouhou-11900000-Koyoukintoujidoukateikyoku/0000201596.pdf (accessed on 25 October 2021). (In Japanese)
4. *ANSI/ASHRAE Standard 55-2017*; Thermal Environmental Conditions for Human Occupancy. American Society of Heating, Refrigerating and Air-Conditioning Engineers, Inc.: Atlanta, GA, USA, 2017.
5. Kawai, K.; Ueno, K. Overview of present situation in Japan and international standards and guidelines of acoustic environment of nursery facilities: Toward the expansion of the AIJES for sound environment in school buildings. In Proceedings of the Annual Meeting of AIJ, Kobe, Japan, 14 September 2014; Volume D-1, pp. 301–304. (In Japanese)
6. *Acoustic Design of Schools: Performance Standards*; Building Bulletin 93; Education & Skills Funding Agency: Coventry, UK, 2015. Available online: https://www.gov.uk/government/publications/bb93-acoustic-design-of-schools-performance-standards (accessed on 21 October 2021).
7. *Guidelines on Ventilation, Thermal Comfort and Indoor Air Quality in Schools Version 1 in 2018*; Building Bulletin 101; Education & Skills Funding Agency: Coventry, UK, 2018. Available online: https://www.gov.uk/government/publications/building-bulletin-101-ventilation-for-school-buildings (accessed on 25 October 2021).
8. Wargocki, P.; Wyon, D.P. Providing better thermal and air quality conditions in school classrooms would be cost-effective. *Build. Environ.* **2013**, *59*, 581–589. [CrossRef]

9. Almeida, R.M.S.F.; de Freitas, V.P. Indoor environmental quality of classrooms in Southern European climate. *Energy Build.* **2014**, *81*, 127–140. [CrossRef]

10. Hou, Y.; Liu, J.; Li, J. Investigation of indoor air quality in primary school classrooms. *Procedia Eng.* **2015**, *121*, 830–837. [CrossRef]

11. Stabile, L.; Dell'Isola, M.; Frattolillo, A.; Massimo, A. Effect of natural ventilation and manual airing on indoor air quality in naturally ventilated Italian classrooms. *Build. Environ.* **2016**, *98*, 180–189. [CrossRef]

12. Deng, S.; Lau, J. Seasonal variations of indoor air quality and thermal conditions and their correlations in 220 classrooms in the Midwestern United States. *Build. Environ.* **2019**, *157*, 79–88. [CrossRef]

13. Korsavi, S.S.; Montazami, A.; Mumovic, D. Indoor air quality (IAQ) in naturally-ventilated primary schools in the UK: Occupant-related factors. *Build. Environ.* **2020**, *180*, 106992. [CrossRef]

14. Kurabuchi, T.; Iino, Y.; Kawase, T. Air quality, thermal environment and thermal environmental evaluations in elementary school classrooms equipped with cooling systems in mild climate areas. *J. Environ. Eng. AIJ* **2009**, *74*, 893–899. (In Japanese) [CrossRef]

15. Iwashita, G.; Yoshino, H. The effect of natural ventilation on the indoor air quality in classroom of the elementary school without heating equipments during winter season. In Proceedings of the 2nd PALENC Conference and 28th AIVC Conference on Building Low Energy Cooling and Advanced Ventilation Technologies in the 21st Century, Crete, Greece, 27–29 September 2007; pp. 446–449.

16. Iwashita, G. Indoor air quality and thermal environment in classrooms of an elementary school during summer season before/after installing air conditioner. *J. Environ. Eng. AIJ* **2009**, *74*, 877–882. (In Japanese) [CrossRef]

17. Sudo, M.; Ito, K.; Sasaki, H.; Iwashita, G.; Ueno, K.; Hiwatashi, K.; Nakae, T.; Goto, T. The effect of academic performance on classroom environment for junior high school students. *J. Environ. Eng. AIJ* **2011**, *76*, 201–209. (In Japanese) [CrossRef]

18. Yanai, Y.; Ikaga, T.; Kawakubo, S. A field survey of classroom environmental quality in terms of students' physical conditions and concentration. *J. Environ. Eng. AIJ* **2012**, *77*, 533–539. (In Japanese) [CrossRef]

19. Iwashita, G. Air environmental quality during winter period in classrooms of elementary/secondary schools in Tokyo metropolitan district X based on the environmental hygienic audit in schools, Study on air environmental quality in schools using administrative data as an evidence (Part 1). *J. Environ. Eng. AIJ* **2014**, *79*, 865–870. (In Japanese)

20. Iwashita, G. Time series analysis on the incidence of influenza in elementary schools during winter season using temperature and humidity. *J. Environ. Eng. AIJ* **2017**, *82*, 257–264. (In Japanese) [CrossRef]

21. Iwashita, G. Study on evaluation method for indoor air environment in classrooms with open-space using index of air stuffiness "ICONE". *AIJ J. Technol. Des.* **2020**, *26*, 185–190. (In Japanese) [CrossRef]

22. Ramalho, O.; Mandin, C.; Ribéron, J.; Wyart, G. Air stuffiness and air exchange rate in French schools and daycare centres. *Int. J. Vent.* **2013**, *12*, 175–180. [CrossRef]

23. Kolarik, B.; Andersen, Z.J.; Ibfelt, T.; Engelund, E.H.; Møller, E.; Bräuner, E.V. Ventilation in day care centers and sick leave among nursery children. *Indoor Air* **2016**, *26*, 157–167. [CrossRef]

24. Canha, N.; Mandin, C.; Ramalho, O.; Wyart, G.; Ribéron, J.; Dassonville, C.; Hänninen, O.; Almeida, S.M.; Derbez, M. Assessment of ventilation and indoor air pollutants in nursery and elementary schools in France. *Indoor Air* **2016**, *26*, 350–365. [CrossRef]

25. Oliveira, M.; Slezakova, K.; Deleru-Matos, C.; Pereira, M.C.; Morais, S. Assessment of air quality in preschool environments (3–5 years old children) with emphasis on elemental composition of PM10 and PM2.5. *Environ. Pollut.* **2016**, *214*, 430–439. [CrossRef]

26. Kawai, K. Current situation of acoustic environment in childcare facilities in Japan: A comprehensive survey in Kumamoto City region. *J. Phys. Conf. Ser.* **2018**, *1075*, 012004. [CrossRef]

27. Aoki, T.; Mizutani, A. Field survey on the actual conditions and improvement of indoor temperature and humidity in kindergartens in Gifu area during winter. *J. Environ. Eng. AIJ* **2017**, *82*, 821–830. (In Japanese) [CrossRef]

28. Taneichi, S.; Aoki, T.; Sudo, M.; Mizutani, A. Effects of influenza prevention on indoor temperature and humidity environment in children's facilities during winter (part 1): Examination using questionnaire and field survey in the southern Tohoku region. *J. Environ. Eng. AIJ* **2021**, *86*, 59–67. (In Japanese) [CrossRef]

29. Genjo, K. Indoor climate for infants in nursery school classrooms. In Proceedings of the Annual Meeting of SHASE, Kochi, Japan, 14 September 2017; Volume 6, pp. 121–124. (In Japanese)

30. Genjo, K. Indoor climate in nursery school in mild climatic area. In Proceedings of the 42nd Symposium on Human-Environment System, Osaka, Japan, 9 December 2018; pp. 187–190. (In Japanese)

31. Ministry of Health, Labour and Welfare. *Standards on Facilities and Operation of Child Welfare Institution*; Ministry of Health, Labour and Welfare: Tokyo, Japan, 2011. (In Japanese)

32. Japan Meteorological Agency. Available online: https://www.data.jma.go.jp/obd/stats/etrn/ (accessed on 25 April 2022). (In Japanese)

33. Shoji, M. Influenza Epidemic Forecast in Japan. *Glob. Environ. Res.* **2003**, *8*, 165–174. (In Japanese)

34. Berglund, B.; Lindvall, T.; Schwela, H.D. *Guidelines for Community Noise WHO*; WHO: Geneva, Switzerland, 1999; Available online: https://apps.who.int/iris/handle/10665/66217?show=full (accessed on 3 June 2022).

35. Ministry of Health, Labour and Welfare. *Act on Maintenance of Sanitation in Buildings*; Ministry of Health, Labour and Welfare: Tokyo, Japan, 1970.

36. *JIS Z 9110: 1979*; Recommended Levels of Illumination. Japanese Industrial Standard: Tokyo, Japan, 1979.
37. Kawai, K.; Sato, M.; Noguchi, S.; Funaba, H. A survey on the current situation of acoustic design in child daycare facilities in Munich city region. *AIJ J. Technol. Des.* **2018**, *24*, 1083–1086. (In Japanese) [CrossRef]

Article

"In-Between Area" Design Method: An Optimization Design Method for Indoor Public Spaces for Elderly Facilities Evaluated by STAI, HRV and EEG

Haining Wang [1], Keming Hou [2], Zhe Kong [1,*], Xi Guan [2], Songtao Hu [2], Mingli Lu [2], Xun Piao [2] and Yuchong Qian [1]

1   School of Architecture, Southeast University, Nanjing 210096, China
2   School of Environmental and Municipal Engineering, College of Architecture and Urban Planning, Qingdao University of Technology, Qingdao 266033, China
*   Correspondence: kongzhe@seu.edu.cn

Abstract: The indoor public spaces of most elderly facilities in China have a monotonous space form, which, thus, causes low comprehensive performance and is less likely to satisfy participants' various requirements. This study proposes an optimization design method of "In-Between Area" for a space form operation to improve the performance of indoor public spaces. First, two models were established: Model A to reflect current indoor public spaces and Model B to represent the indoor public spaces designed by using the "In-Between Area" method. Second, a walk-through video was created from each model, with a duration of 196 s. Subjective assessment (STAI) data and objective physiological data (HRV and EEG), were collected from 40 participants while they were watching walk-through videos. The comparison analysis showed statistically significant differences between Model A and Model B. The results of STAI, HRV and EEG proved that the "In-Between Area" method, as an optimization design method, created a more pleasant and comfortable environment for the elderly and improved the overall efficiency of the indoor space.

Keywords: elderly facility; in-between area; EEG; HRV; STAI; architectural scheme design

Citation: Wang, H.; Hou, K.; Kong, Z.; Guan, X.; Hu, S.; Lu, M.; Piao, X.; Qian, Y. "In-Between Area" Design Method: An Optimization Design Method for Indoor Public Spaces for Elderly Facilities Evaluated by STAI, HRV and EEG. *Buildings* **2022**, *12*, 1274. https://doi.org/10.3390/buildings12081274

Academic Editor: Morten Gjerde

Received: 16 July 2022
Accepted: 15 August 2022
Published: 19 August 2022

Publisher's Note: MDPI stays neutral with regard to jurisdictional claims in published maps and institutional affiliations.

1. Introduction

People's daily life is inseparable from the built environment [1] and the quality of the built environment profoundly affects people's living experience [2,3]. Space is one of the most critical elements in the built environment and also the only carrier of other built environment elements [4]. Therefore, space plays a special and critical role in the built environment. As an important topic of space design, the discussion of form has always been carried out, because it not only stipulates the basic physical form of the space [5], but also decisively affects the performance of all aspects of the space.

At present, the indoor public activity space design of most elderly facilities in China has significant disadvantages and cannot fully meet the needs of various participants [6]. The participants of the elderly facilities were mainly composed of the elderly and their children who come to visit, as well as various staff. Among them, the elderly need to be particularly considered in designing the interior space, because they spend most of their time in interior space [7], conducting their social, leisure and sports activities in the indoor public spaces of these elderly facilities [8]. Therefore, good indoor public space design is critical to their positive experience, such as comfort, health and happiness [9], and the success of elderly facilities design [10]. However, the indoor public spaces of most elderly facilities in China are short of transitional connection with other spaces due to the lack of effective design-method guidance. Hence, a single, rigid, oppressive and dull spatial form is generated, which results in overall performance deterioration, as shown in Figure 1.

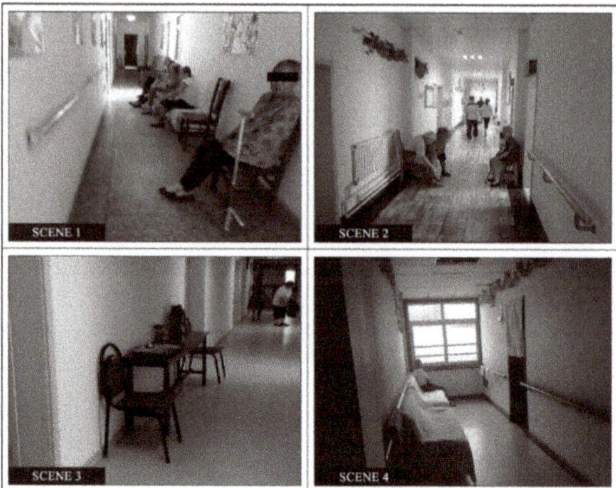

Figure 1. Status of indoor public activity space in elderly facilities.

Some effective design methods were proposed by enriching indoor public spaces, including promoting the functional carrying capacity and visual pleasure. Hertzberger believed that the semi-public and semi-private space environments were more interesting and more likely to stimulate the occurrence of behavioral events [11]. Newman mentioned that it was vital to create a buffer area between residential and urban streets, which could improve the comfort and vitality of urban public spaces [12]. Zhou proposed to set up a certain number of open spaces along the corridors of elderly facilities, which was conducive to improving the willingness of the elderly to communicate [13]. These researchers advocated the idea that "gray" spaces with ambiguous attributes need to be established in designing space form. Inspired by these works, this study proposes an "In-Between Area" design method to improve the current low performance of indoor public spaces in most Chinese elderly facilities.

At the heart of the "In-Between Area" design approach is the idea to enhance the user's sense of control over the space and increase their participation. The boundary of the corridor space is richly processed and some leisure and entertainment areas of different sizes are set up, so that the elderly walking around them can feel the interesting and rhythmic spatial changes. These areas provide more choices for the elderly, thereby enhancing the possibility and flexibility of corridor spaces for interaction and increasing the user participation. What deserves special attention is the entry space, which can be flexibly arranged according to the user's own preferences and mood and strengthen the user's sense of control over the space. At the same time, this area is good in terms of sight and interaction and the most relaxing and casual social environment.

Due to the fact that the method is experience based, its effectiveness was tested using subjective and objective evaluation methods for space performance. A large number of studies used subjective voting methods to evaluate the comprehensive performance of various living spaces [14,15]. However, studies solely based on subjective methods are limited by the ambiguity of the methods, for they lack a complete, systematic and reliable evaluation standard [16]. To overcome the limitation imposed by subjective methods, some other studies employed objective methods. Studies have shown that emotional changes are usually produced under the stimulation of the external environment, accompanied by physiological and psychological reactions [17,18]. In terms of heart rate variability (HRV), the variation in heartbeats is suppressed when subjects are anxious and depressed and returns to normal in a relaxed state [19,20]. As an important test method, Electroencephalography (EEG) can reflect the cognitive state of subjects comprehensively [21,22]. Maryam et al. used EEG to study the influence of interior morphology on emotions [23]. Parastou et al. used

EEG to observe the emotional changes in people facing different building facades [24]. Li et al. explored the correlation between subjective questionnaires and EEG and established a method to use EEG to characterize people's feedback on building space [25,26]. Using physiological indicators can well measure the physiological changes in users in different spatial environments. At the same time, the scientificity of the subjective architectural design can be verified using physiological indicators, which are more objective and authentic and can better clear up the ambiguity caused by subjective evaluation. Subjective evaluation facilitates qualitative analysis, while objective evaluation can achieve quantitative analysis of emotions. Therefore, the combination of subjective and objective evaluation can be used to better evaluate the architectural space design scheme.

Aiming at the low performance and its causes of indoor public space in most Chinese elderly facilities, this study proposed an optimal design method, "In-Between Area", in terms of spatial form operation. The subjective and objective evaluation method was used to verify the effectiveness of the method, as well as to try and achieve two research goals and value significance. One is to propose a design method that can effectively improve the performance of space and provide assistance for improving the indoor public space of Chinese elderly facilities. The other is to provide a set of feasible evaluation methods for spatial performance under the influence of space form differences. We also aimed to explore the scientific evaluation theory of subjective design in order to promote the development of the research field of built environment evaluation.

2. Materials and Methods

2.1. Laboratory Environment

The experiment was carried out in an artificial climate chamber (5.10 m × 3.50 m × 2.80 m), as shown in Figure 2, at Qingdao University of Technology. The chamber was equipped with a constant temperature and humidity air conditioning system.

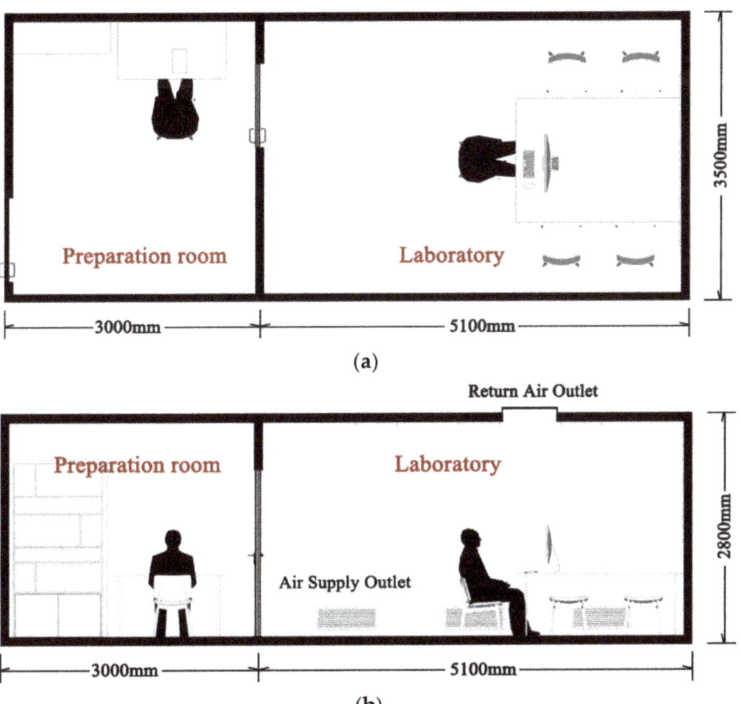

Figure 2. (**a**) Layout of artificial climate chamber; (**b**) section of artificial climate chamber.

Previous studies have shown that environmental factors, such as indoor light environment [27,28], temperature [29], air quality and ventilation rate [30], are important, because they affect the environmental quality of building spaces. The air temperature, relative humidity, horizontal illuminance on a work plane, color temperature and sound pressure levels were controlled in order to reduce the influence of physical environment (Table 1).

Table 1. Physical environment parameter settings.

Parameter	Range
Air temperature/°C	24 ± 0.5
Relative Humidity/%	60 ± 5
Wind Speed/m·s^{-1}	<0.2
Illuminance/lux	500 ± 50
Color Temperature/K	4000 ± 100
Sound pressure level/dB	<40

2.2. Participants

G*power (Version 3.1, Düsseldorf, Germany) [31] was used to calculate the sample size required for the study, the effect size was set to 0.5 and α was set to 0.05. The calculation results showed that to achieve a statistical power of 0.8, the study required a total of 34 subjects. Different age groups have different needs for living space [32], so we recruited the participants for this experiment that were young, middle aged and elderly. Forty participants were recruited from volunteers: the elderly, their families and staff of pension institutions. The participants included 22 men and 18 women (21 young people, 9 middle aged and 10 elderly). Due to the greater impact of COVID-19 on the elderly, it is more difficult to organize the elderly to participate in the experiment and, thus, the numbers of participants in the three groups were adjusted. All participants were informed of the aim of the study, experiment procedure and confidentiality issues before the experiment began. The participants were all physically and mentally healthy. Smoking and alcohol consumption were prohibited for 48 h before the experiment. Table 2 presents the demographic information of 40 participants.

Table 2. Participants' demographic information.

	Number (Male/Female)	Maximum Age	Minimum Age	Average Age	Standard Deviation
Young people	21 (15/6)	26	20	22.51	1.64
Middle age	9 (7/2)	37	30	33.84	2.61
Elderly	10 (5/5)	80	59	66.52	5.63
Total	40 (27/13)	-	-	36.06	18.45

2.3. Experimental Design

Three-dimensional software makes it possible to do near-reality modeling and visualize changing and complex environments [33]. In this study, Google SketchUp (Version 2020, Trimble, Sunnyvale, CA, USA) [34] was used to set up the experimental scene and Enscape (Version 3.1, Enscape GmbH, Karlsruhe, Germany) [35] was used to render the video as the emotion-evoking environment.

First, a survey was conducted in elderly facilities located in the urban area of Qingdao. There was a total of 265 elderly facilities registered in Qingdao in 2022, among which 141 facilities were in the urban area. The survey was carried out on the spatial environment and spatial form of these 141 elderly facilities. To a large extent, the data collected about these urban elderly facilities of Qingdao can presumably represent the situation of elderly facilities in the city. The survey results show that "Inner corridor" and "Stub-end type" are the main modes of corridor space in these facilities. In addition, typical spatial patterns, floor plans, colors, interior materials and appearance of facilities are also summarized.

Based on this prior investigation, Model A was designed. The time of video A was set to 196 s, considering the duration of the experiment and subjective fatigue [26]. Since the normal walking speed of adults is around 1.1 m/s, the corridor length of Model A was determined as 112 m. The corridor of Model A was designed as two long and straight parallel lines. There was no buffer area between the elderly living room and the corridor, and the activity spaces for dining or games were closed independent rooms along the corridor. Figure 3a shows the layout and nine interior perspectives for Model A.

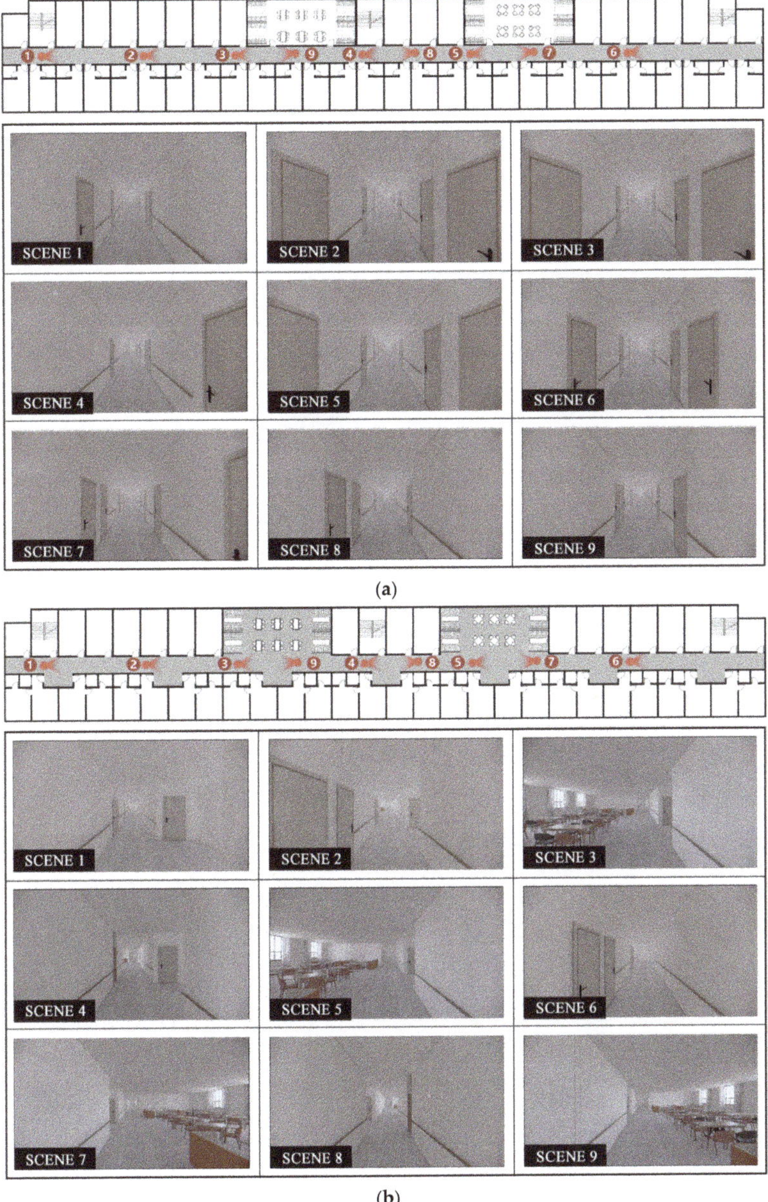

(a)

(b)

Figure 3. *Cont.*

Figure 3. (**a**) Model A of a corridor without centralized activity spaces; (**b**) Model B of a corridor with centralized activity space; (**c**) axonometric drawing comparisons between Models A and B.

Compared to Model A, Model B was optimized in the following way: (1) reducing bedroom spaces and expanding the corridor space to create common spaces in the corridor and (2) eliminating the walls of the dining rooms and game rooms to create common spaces in the corridor. The increase in the accessibility of space leads to a higher level of living for all [36]. Apart from the two points, Model B and Model A were identical. Figure 3b shows the layout and nine interior perspectives for Model B. Video A and video B entailed a round trip of a walking experience in the two models.

2.4. Measures

2.4.1. Survey Questionnaire

Many systematic and in-depth studies are carried out on the issue that the space environment of elderly facilities affects the physical and mental health of the elderly. The results of the study show that lack of personal privacy in indoor public spaces makes the elderly feel that they are losing control over the personal field, which may lead to negative emotions of anxiety and stress [37]. The State–Trait Anxiety Inventory (STAI), developed by Spielberger et al., is a commonly used measure of trait and state anxiety and can provide a more reliable psychological state test for clinical practice [38,39]. Therefore, this study selected the State–Trait Anxiety Inventory (STAI) to explore the effects of different spatial forms on the emotions of the elderly.

The scale is self-rating and consists of 40 descriptive items divided into two subscales. Among them, items 1–20 are the State Anxiety Inventory (S-AI) to measure short-term unpleasant emotional experiences such as tension, fear and worry. Items 21–40 are the Trait Anxiety Inventory (T-AI) to describe underlying and long-term anxiety tendencies. For more details, please refer to Appendix A.

The scales vary from 1 to 4 and the subjects were required to choose the most appropriate grade according to their feelings. The scores of S-AI and T-AI were obtained by adding up all the relevant questions, respectively. Table 3 presents the STAI scores.

Table 3. Scores for STAI.

Question Number	Not at All or Almost Never	Somewhat or Sometimes	Moderately So or Often	Very Much So or Almost Always
3, 4, 6, 7, 9, 12, 13, 14, 17, 18, 22, 25, 28, 29, 31, 32, 35, 37, 38, 40	1	2	3	4
1, 2, 5, 8, 10, 11, 15, 16, 19, 20, 21, 23, 24, 26, 27, 30, 33, 34, 36, 39	4	3	2	1

2.4.2. Heart Rate Variability

Heart rate variability (HRV) refers to the variation in heartbeat cycle differences and is an important indicator to reflect the activity of the autonomic nervous system and the balance of sympathetic and parasympathetic nerves [40]. HRV is more suitable for characterizing the health, resilience and stress resistance of a human body in a resting state [41]. Generally speaking, a low HRV value indicates that a human being is under stress from exercise, psychological events or other stressors, while a high HRV value indicates that a human being has a strong ability to tolerate stress or a human being is in the process of stress recovery. In a resting state, a high HRV value is generally favorable, compared to a low HRV value. The main analysis methods of HRV include time-domain analysis and frequency domain analysis [42]. Time-domain analysis is more intuitive, while frequency-domain analysis is more comprehensive. In terms of analysis time, there are two main types of HRV analysis in clinical medicine: short-range analysis (5 min) and long-range analysis (24 h). The short-range analysis method has a short recording time and the experimental conditions are easy to control. The commonly used parameters and characteristics are shown in Table 4. It should be noted that HRV is referred to as SDNN in this paper, unless otherwise stated.

Table 4. Common parameters and characteristics of HRV.

Parameter	Meaning	Domain	Characteristic
SDNN	Standard deviation of beat interval	Time Domain	The lower the value, the more active the sympathetic nerves, manifested as fatigue and tension. On the contrary, the more active the parasympathetic nerves, manifested as relaxation.
RMSSD	RMS of consecutive beat intervals	Time Domain	Decreases when fatigued and increases when recovering from fatigue.
LF/HF	The ratio of low frequency to high frequency	Frequency domain	Increases when fatigued and decreases when recovering from fatigue.

Polar® H10 chest strap heart rate sensor (Polar, Kempele, Finland) was used to monitor HRV and Polar® Unite (Polar, Kempele, Finland) for recording and storing data [43,44] (Figure 4). Before the experiment, the Polar® H10 and Polar® Unite were worn on the chest and left wrist separately with the sensors fitting snugly on the skin. The raw data stream was preprocessed with ARTiiFACT software (T. Kaufmann, Würzburg, Germany) [45,46] and the parameters, such as LF, HF, LF/HF, SDNN and RMSSN, were calculated using Kubios-HRV-Standard (Kubios Oy, Kuopio, Finland) [47,48].

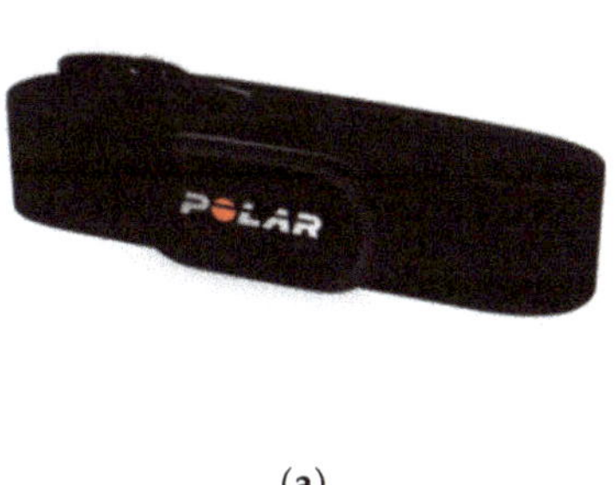

(a)	(b)

Figure 4. (a) Polar H10; (b) Polar Unite.

2.4.3. EEG Signals

Cognitive electrophysiology studies cognitive functions at the neuron population level, including perception, memory, emotion, language and behavior. EEG is an important part of cognitive electrophysiology which is a method of collecting and recording the activity in the cerebral cortex using sophisticated instruments [49,50]. The collected data are an overall reflection of nerve cells on the surface of the cerebral cortex or scalp.

Many neurological studies have shown that EEG rhythms are closely related to the working state of the brain [51]. In the process of excitation conduction, the "passive diffusion" and "active entry" of Na^+ and K^+ cause EEG to produce oscillations and the level of oscillation frequency is related to the state of the human body. Generally, EEG signals can be divided into 5 bands [52]. For adults, a range of 0.5–4 Hz is denoted as the Delta band, which is mainly related to deep sleep. A range of 4–8 Hz is denoted as Theta band, which is considered the transition between drowsiness and consciousness. A range of 8–13 Hz is denoted as the Alpha band, which is prominent in relaxed awareness but attenuates or disappears with concentration or attention. A range of 13–30 Hz is denoted as the Beta band, which is related to active thinking, attention and solving specific problems. A range of 30–50 Hz is denoted as the Gamma band, whose amplitude is often smaller than those of the other bands.

Higher temporal resolution and lower spatial resolution are two significant features in EEG [53]; therefore, the following considerations were taken in this study. First, the high temporal resolution allows us to monitor millisecond-level signals, which also increases the amount of data. Considering that the subjects were active for a long time, the excessive temporal resolution may cause data redundancy. Second, the low spatial resolution means that it is difficult to pinpoint exactly where the signal source is generated. Usually, the data collected by an electrode from the EEG is the sum of the voltages in the vicinity of the point and this experiment did not involve traceability.

Based on the above two considerations, Emotiv Epoc (Emotiv Systems, San Francisco, CA, USA) [54] was selected and the sampling rate was set to 256 Hz. The device had a total of 16 metal electrodes, corresponding to 16 standard points in the 10–20 system developed by the International Society of Electroencephalography. Among them, P3 and P4 are located at the mastoid as reference electrodes. Figure 5a is a physical view of the device and Figure 5b shows the relative positions of the 16 electrodes on the scalp.

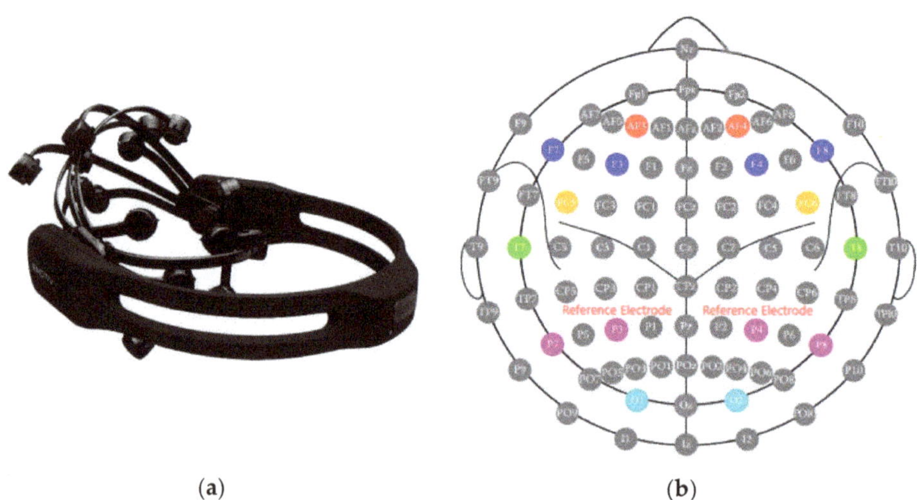

(a) (b)

Figure 5. (a) Physical map of Emotiv Epoc; (b) channel location.

Before the test, the experimenter made sure that the equipment was fully charged and the sponge pad was immersed in physiological saline. The fully wetted pad acted as a physical bridge between the scalp and the metal electrodes to enhance the conductivity. Subjects were required to wash and blow their hair dry to avoid the effects of oil and skin keratin on conductivity. When the equipment was worn, the measuring points fully touched the scalp. The subject was required to sit still on the seat to minimize unnecessary activities. The test could be started only after the impedance was stable and met the requirements.

MATLAB® 2019b (MathWorks, Natick, MA, USA) and the EEGLAB plugin [54] were used to process the EEG data. Figure 6 shows the basic processing process. First, apparent spikes or oscillations outside the normal range of EEG signals were manually rejected, which were due to abnormal point contact or slight movement of the subject. Second, the AC working frequency caused the EEG spectrum sag at 50 Hz. For this reason, the 1–48 Hz signal was retained by bandpass filtering. Third, the EEG signals contained noises such as eye movement artifacts, ECG artifacts, EMG artifacts and skin artifacts. To identify and remove them, the independent component analysis function was used in the EEGLAB plugin. Finally, the time domain signal was converted to the frequency domain by Fast Fourier Transform (FFT), which well separated the Delta, Theta, Alpha and Beta bands. The total power of a band was the sum of the squares in the frequency-domain sequences of the band, as shown in Formula (1).

$$P_{total} = \sum_{n=i}^{j} \left| X_n^2 \right| \tag{1}$$

where P_{total} is the total power of EEG, i is the lower limit of frequency, j is the upper limit of frequency and X_n is frequency domain sequence.

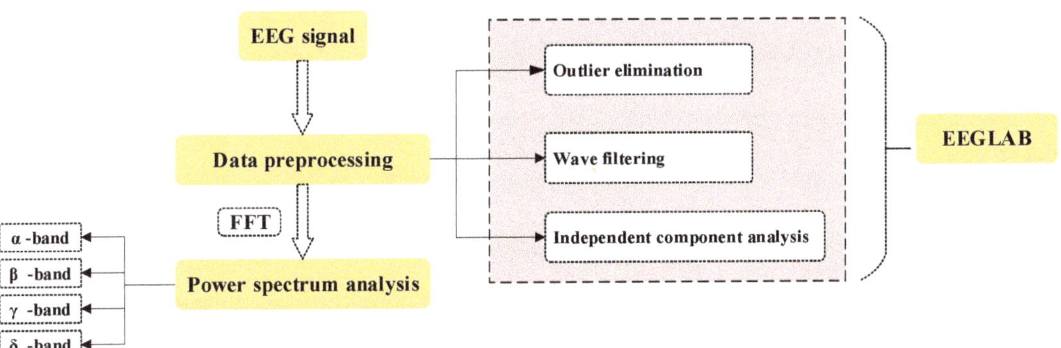

Figure 6. EEG data processing.

Based on Formula (1), the total EEG power of 14 channels for each subject was obtained. Due to the large magnitude of the value, Formula (2) was used for logarithmic processing.

$$P_{total}' = log_{10}(P_{total}) \tag{2}$$

2.5. Experiment Procedure

Before the experiment, the basic situation was introduced to the participants in the preparation room. Then, the participants took the STAI test and wore the equipment with the assistance of staff. After the experiment began, the participants had 10 min to rest and adapt to the physical environment. They were required to successively take Model A test and STAI test, each for about 20 min. Model B test was the same as the Model A. The experiment totally took about 40 min. The experimental procedure is shown in Figure 7.

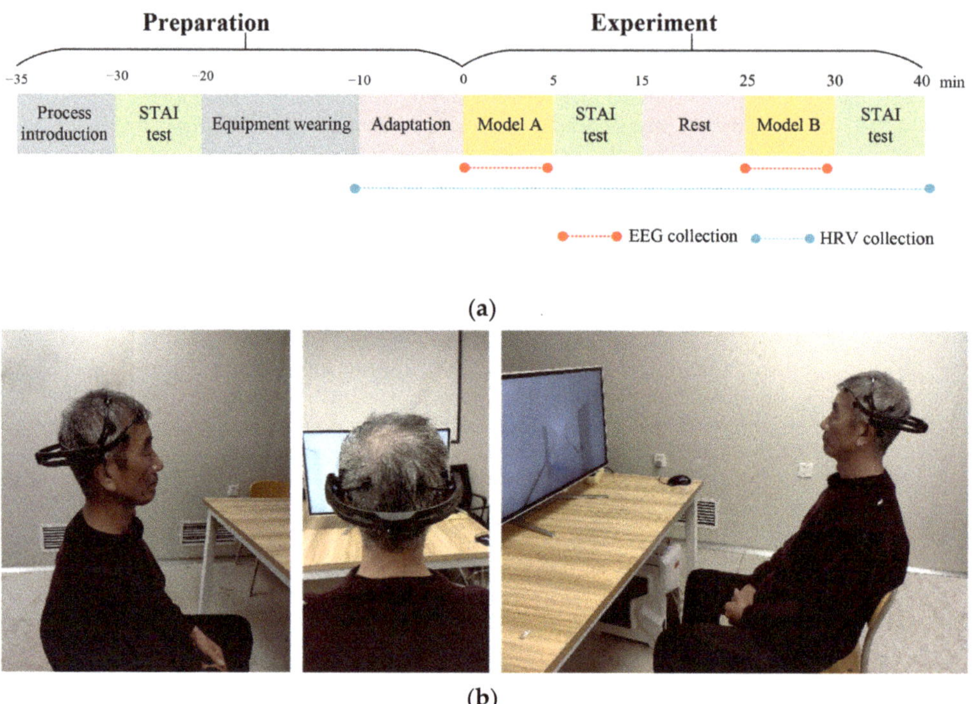

Figure 7. (**a**) Experimental procedure; (**b**) experimental photos.

3. Results

3.1. Subjective Evaluation

Each subject conducted two experiments, Model A and Model B. According to the scoring rules in Table 3, the STAI scores of 40 subjects in the preparation stage (pre-experiment) and the experimental stage (post-experiment) were calculated. Further, the post-experiment value minus the pre-experiment value is the outcome variable (Δx_1) for the paired samples *t*-test. A Kolmogorov–Smirnov test was performed using SPSS and the results showed that the data conformed to a normal distribution. The exploratory data analysis showed a group of outliers, which, thus, was eliminated, resulting in 39 groups of valid data. The results of the paired samples *t*-test show that there is a significant difference in Δx_1 between Model A and Model B, as shown in Table 5.

Table 5. STAI paired sample *t*-test results.

	T	*p*
S-AI	2.270	0.029 *
T-AI	3.530	0.001 **

Note: * means $p < 0.05$. ** means $p < 0.01$.

The comparison of S-AI and T-AI is shown in Figure 8a,b. The Δx_1 of Model A is positive, indicating that Model A with a single spatial form is likely to cause psychological pressure to participants. Because the activity space in Model A is almost a closed room, lacking the connection and interaction with other functional areas, which is less likely to stimulate the participant's communication behavior, the subjects tended to be more nervous and anxious in this kind of environment.

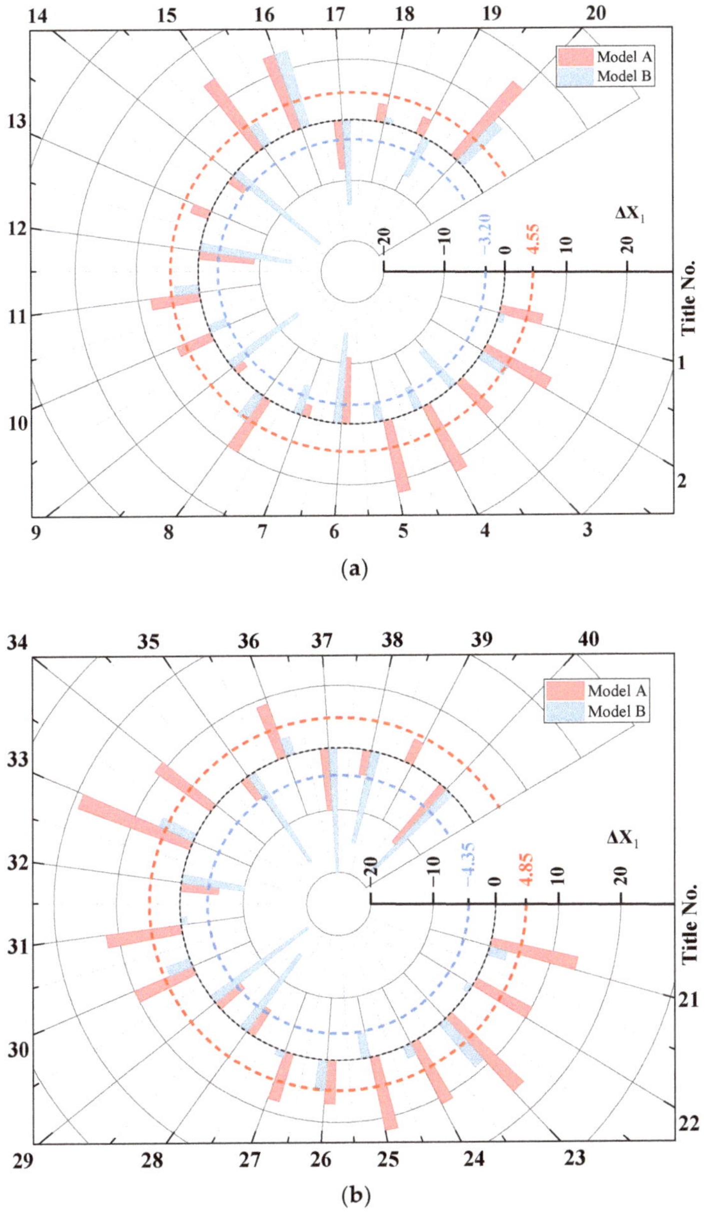

Figure 8. (**a**) Comparison of S-AI between Model A and Model B; (**b**) comparison of T-AI between Model A and Model B.

On the contrary, the Δx_1 of Model B is negative. It shows that the "In-Between Area" design relieved the visual fatigue, enriched the space form, improved the overall efficiency of the indoor public activity space and used the space in a more abundant and diverse way; the optimized design provided a more pleasant and comfortable environment for the elderly at the psychological level, which would be beneficial to physical and mental development and disease prevention in the elderly.

3.2. Heart Rate Variability

The post-experiment value minus the pre-experiment value is the outcome variable (Δx_2) and a paired samples t-test was performed on the heart rate, LF, HF, LF/HF, SDNN and RMSSD of Model A and Model B. The normal distribution test generated 39 groups of valid data and the results are shown in Table 6. The results are not statistically different in heart rate and RMSSD, but are in LF, LF/HF and SDNN.

Table 6. STAI paired samples t-test results.

	Heart Rate	**HRV**				
		LF	**HF**	**LF/HF**	**SDNN**	**RMSSD**
p	0.643	0.046 *	0.077	0.002 **	0.011 *	0.086

Note: * means $p < 0.05$. ** means $p < 0.01$.

The SDNN difference (Δx_2) between Model A and Model B is shown in Figure 9. It is not difficult to find that the SDNN mean value of Model A is lower than that of pre-experiment, while the SDNN mean value of Model B is slightly higher than that of pre-experiment. On the one hand, SDNN is an important indicator to measure the activity in the autonomic nervous system. Decreased SDNN is found in patients with anxiety, depression, insomnia, etc. On the other hand, SDNN reflects the ability to withstand stress and an increase in SDNN value indicates an increase in the complexity of HRV and an increase in the body's ability to adapt to environmental changes. This result indicates that the abundant space form in Model B creates a transitional space with dual properties: (1) the participant's sense of security is enhanced by strengthening control over the territory and (2) the penetration and interaction of the space improves the user's participation, which, in turn, enhances the participant's ability to adapt to the environment.

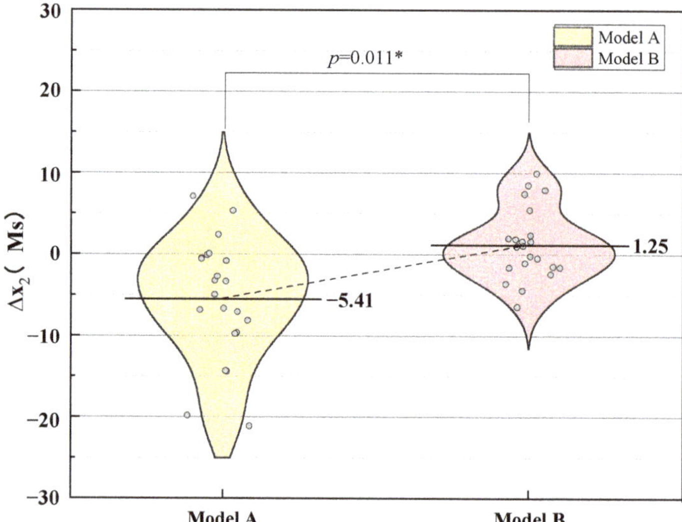

Figure 9. Comparison of Δx_2 between Model A and Model B.

3.3. EEG Signal

Emotiv Epoc provides 14 channels, corresponding to 14 points on the cerebral cortex distributed in different functional areas of the left and right brain. Power spectral density (PSD) is a measure of EEG energy per unit frequency and the PSD of a subject with typical characteristics is shown in Figure 10. The energy of the frontal lobe is higher than that of other regions and the high-frequency components are more distributed in the right brain, which is in line with Sperry's theory [55]. Further, maximum values of 2.0 Hz and 6.0 Hz in Model A are higher than in Model B, indicating that the energy of EEG in Model A is higher.

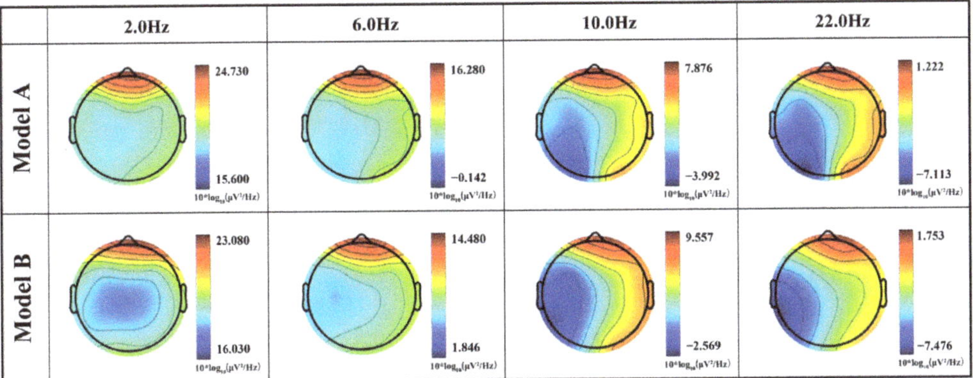

Figure 10. PSD of a subject between Model A and Model B.

Considering the functional division of the left and right brain is not precise enough while the mutual influence between functional areas is ubiquitous, the paired samples *t*-test was used to find the eigenvalues that can significantly distinguish the difference between Model A and Model B under 14 channels and four bands. The data generated by the Kolmogorov-Smirnov test conformed to a normal distribution. The exploratory data analysis showed three groups of outliers, which, thus, were eliminated, resulting in 37 groups of valid data. The results are shown in Table 7.

Table 7. Paired samples *t*-test results for EEG logarithmic power.

Channel	*p*-Value				
	Delta	Theta	Alpha	Beta	Total
AF3	0.069	0.064	0.308	0.379	0.004 **
AF4	0.393	0.371	0.962	0.704	0.030 *
F3	0.223	0.485	0.928	0.972	0.480
F4	0.120	0.105	0.349	0.111	0.081
F7	0.687	0.991	0.925	0.307	0.721
F8	0.391	0.674	0.623	0.443	0.629
FC5	0.243	0.375	0.529	0.735	0.323
FC6	0.956	0.474	0.979	0.099	0.505
T7	0.184	0.205	0.276	0.225	0.213
T8	0.482	0.132	0.457	0.032 *	0.141
P7	0.415	0.622	0.524	0.622	0.279
P8	0.676	0.340	0.891	0.091	0.392
O1	0.434	0.891	0.111	0.447	0.736
O2	0.466	0.548	0.573	0.731	0.683
Total	0.152	0.029 *	0.307	0.110	0.108

Note: * means $p < 0.05$. ** means $p < 0.01$.

Among them, AF3 and AF4 are located in the frontal lobe, which are the closest to the eyes among the 16 electrodes in Emotiv Epoc. Since the frontal lobe is mainly responsible for learning, language, decision-making, abstract thinking, emotions and other functions, we analyzed the logarithmic average power of AF3 and AF4, as shown in Figure 11.

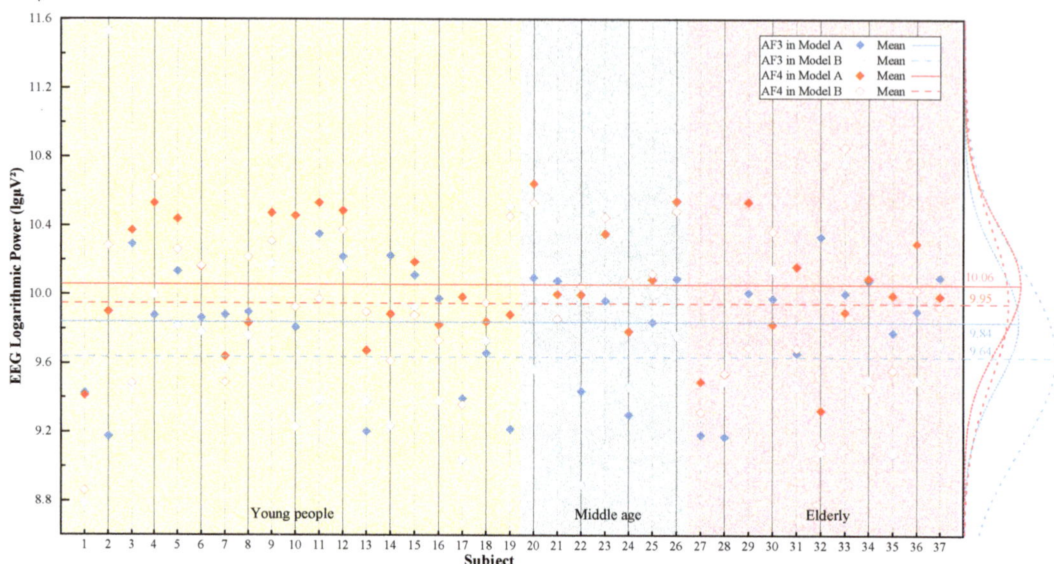

Figure 11. Comparison of AF4 logarithmic average power between Model A and Model B.

The logarithmic average power of Model B is lower than that of Model A: 28.8% in AF4 and 58.5% in AF3. Among the 37 subjects, 23 had this pattern, accounting for 62.2% (young people account for 63.2%, middle aged account for 42.9% and elderly account for 80%).

Low power means low neuronal activity, that is, weak intensity of EEG signal oscillations. Our findings showed that the total EEG power of subjects in an uncomfortable environment is higher than that in a comfortable environment, which is consistent with Guan et al. [56]. Compared to Model A, subjects were more relaxed and comfortable in Model B.

4. Discussion and Conclusions

This study proposes a design method for improving indoor public spaces in elderly facilities. Aiming at the disadvantages of corridor space in most elderly facilities in China, this study clarified the core reasons for the problem first and then proposed a design strategy. The psychological and physiological data of the participants were collected to obtain a more comprehensive and accurate evaluation. Based on the data analysis, the following conclusions were drawn:

(1) The STAI results show that the anxiety and psychological stress of the participants in Model A were higher than those of the pre-experiment, while those in Model B were lower. That indicates that Model B is better than Model A in terms of the psychological dimension.

(2) The HRV results show that the SDNN was statistically different in the two models. Autonomic nervous system was more active in Model B, which means that Model B improved the participants' ability to withstand the stress while adapting to the environment.

(3) The EEG results show that the EEG power of Model B is significantly lower than that of Model A. From a neurophysiological perspective, Model B consumes less energy and is more comfortable for participants.

In conclusion, the "In-Between Area" design method could enrich the spatial form and relieve the visual fatigue caused by separated monotonous spatial forms by enhancing the connection between activity spaces and other functional areas, as well as im-proving the overall efficiency of interior spaces. Moreover, this design method also strengthens user sense of control over the space and enhances user participation.

Since the impact of COVID-19 has dramatically increased the difficulty of organizing more participants for each age group, future research will increase the number of subjects to achieve comparative analysis between different populations. Compared with the actually built scene or the virtual reality (VR) scene, the video-based scene used in the current study lacked the immersion experience, which may cause insufficiency in the emotional arousal of the subjects. Therefore, future research may consider implementing VR to increase immersion.

Author Contributions: Conceptualization, K.H.; methodology, M.L.; software, X.G.; validation, S.H.; formal analysis, H.W.; investigation, Z.K.; resources, H.W.; data curation, X.G.; writing—original draft preparation, X.G.; writing—review and editing, K.H., X.G., M.L., H.W., Y.Q., Z.K. and X.P.; visualization, X.G.; supervision, K.H.; project administration, K.H.; funding acquisition, K.H. All authors have read and agreed to the published version of the manuscript.

Funding: This research was funded by National Natural Science Foundation of China Youth Science Foundation, grant number 51908301 and 51908111, General project of Shandong social science planning and research project, grant number 21CSHJ06, Shandong Natural Science Foundation Training Fund, grant number ZR2019PEE034 and Qingdao social science planning and research project, grant number QDSKL2101178.

Institutional Review Board Statement: The study was conducted in accordance with the Declaration of Helsinki, and approved by the *Research Ethics Committee of Qingdao Innovation Center for CIM + Urban Regeneration* (protocol code 2022-08-10-01 and 10 August 2022).

Informed Consent Statement: Informed consent was obtained from all subjects involved in the study.

Data Availability Statement: The data presented in this study are available from the corresponding author upon reasonable request.

Conflicts of Interest: The authors declare no conflict of interest.

Nomenclature

HRV	Heart rate variability	LF	Low frequency
EEG	Electroencephalography	HF	High frequency
STAI	State–Trait anxiety inventory	LF/HF	The ratio of low frequency to high frequency
S-AI	State Anxiety Inventory	SDNN	Standard deviation of beat interval
T-AI	Trait Anxiety Inventory	RMSSD	RMS of consecutive beat intervals
EMG	Electromyography	VR	Virtual reality
FFT	Fast Fourier Transform	Δx_1	The post-experiment value minus the pre-experiment value of STAI
PSD	Power spectral density	Δx_2	The post-experiment value minus the pre-experiment value of SDNN

Appendix A

Table A1. (**a**) Self-evaluation questionnaire: STAI Form Y-1; (**b**) self-evaluation questionnaire: STAI Form Y-2.

(**a**)

Self-Evaluation Questionnaire—Stai Form Y-1

Please provide the following information:

Name:	Date:	Age:	Gender (*Circle*):M F

Directions:

A number of statements that people have used to describe themselves are given below. Read each statement and then circle the appropriate number to the right of the statement to indicate how you feel right now, that is, at this moment. There are no right or wrong answers. Do not spend too much time on any one statement but give the answer that seems to describe your present feelings best.

Meaning of options:

1: Not At All; 2: Sometimes; 3: Moderately So; 4: Very Much So

1. I feel calm	1	2	3	4
2. I feel secure	1	2	3	4
3. I am tense	1	2	3	4
4. I feel strained	1	2	3	4
5. I feel at ease	1	2	3	4
6. I feel upset	1	2	3	4
7. I am presently worrying over possible misfortunes	1	2	3	4
8. I feel satisfied	1	2	3	4
9. I feel frightened	1	2	3	4
10. I feel comfortable	1	2	3	4
11. I feel self-confident	1	2	3	4
12. I feel nervous	1	2	3	4
13. I am jittery	1	2	3	4
14. I feel indecisive	1	2	3	4
15. I am relaxed	1	2	3	4
16. I feel content	1	2	3	4
17. I am worried	1	2	3	4
18. I feel confused	1	2	3	4
19. I feel steady	1	2	3	4
20. I feel pleasant	1	2	3	4

Table A1. *Cont.*

(b)

Self-Evaluation Questionnaire—Stai Form Y-2

Please provide the following information:

Name: Date: Age: Gender (*Circle*):M F

Directions:

A number of statements that people have used to describe themselves are given below. Read each statement and then circle the appropriate number to the right of the statement to indicate how you *generally* feel.

Meaning of options:

1: Almost Never; 2: Sometimes; 3: Often; 4: Almost Always

21. I feel pleasant	1	2	3	4
22. I feel nervous and restless	1	2	3	4
23. I feel satisfied with myself	1	2	3	4
24. I wish I could be as happy as others seem to be	1	2	3	4
25. I feel like a failure	1	2	3	4
26. I feel rested	1	2	3	4
27. I am "calm, cool, and collected"	1	2	3	4
28. I feel that difficulties are piling up so that I cannot overcome them	1	2	3	4
29. I worry too much over something that really doesn't matter	1	2	3	4
30. I am happy	1	2	3	4
31. I have disturbing thoughts	1	2	3	4
32. I lack self-confidence	1	2	3	4
33. I feel secure	1	2	3	4
34. I make decisions easily	1	2	3	4
35. I feel inadequate	1	2	3	4
36. I am content	1	2	3	4
37. Some unimportant thought runs through my mind and bothers me	1	2	3	4
38. I take disappointments so keenly that I can't put them out of my mind	1	2	3	4
39. I am a steady person	1	2	3	4
40. I get in a state of tension or turmoil as I think over my recent concerns and interests	1	2	3	4

References

1. Shan, X.; Yang, E.H.; Zhou, J.; Chang, V.W.C. Neural-signal electroencephalogram (EEG) methods to improve human-building interaction under different indoor air quality. *Energy Build.* **2019**, *197*, 188–195. [CrossRef]
2. Miao, Y.; Ding, Y. Indoor environmental quality in existing public buildings in China: Measurement results and retrofitting priorities. *Build. Environ.* **2020**, *185*, 107216. [CrossRef]
3. Alkabashi, B.H.A.; Dokmeci Yorukoglu, P.N. Evaluating Indoor Environmental Quality of a Wellness Center Through Objective, Subjective and Architectural Criteria. *Megaron* **2019**, *14*, 483–494. [CrossRef]
4. Lee, K. The Interior Experience of Architecture: An Emotional Connection between Space and the Body. *Buildings* **2022**, *12*, 326. [CrossRef]
5. Han, Y.J.; Kotnik, T. A Tomographic computation of Spatial Dynamics. In *Ecaade 2020: Anthropologic—Architecture and Fabrication in the Cognitive Age*; Werner, L.C., Koering, D., Eds.; Ecaade-Education & Research Computer Aided Architectural Design Europe; eCAADe: Berlin, Germany, 2020; Volume 2, pp. 89–94. Available online: https://www.webofscience.com/wos/alldb/full-record/WOS:000651200700009 (accessed on 5 August 2022).
6. Su, X.; Wang, S. Research on functional space of community-based medical-nursing combined facilities for the aged in small cities. *J. Asian Archit. Build. Eng.* **2022**, *21*, 1399–1410. [CrossRef]
7. Zhan, H.; Yu, J.; Yu, R. Assessment of older adults' acceptance of IEQ in nursing homes using both subjective and objective methods. *Build. Environ.* **2021**, *203*, 108063. [CrossRef]
8. Li, M.; Liu, Q.; Li, X. Research on the Influence of Environmental Art Design Based on Environmental Science on the Indoor Environment of Old-age Buildings. *IOP Conf. Ser. Earth Environ. Sci.* **2020**, *617*, 012053. [CrossRef]
9. Leung, M.Y.; Wang, C.; Famakin, I.O. Integrated model for indoor built environment and cognitive functional ability of older residents with dementia in care and attention homes. *Build. Environ.* **2021**, *195*, 107734. [CrossRef]
10. Wang, C.; Lu, W.; Ohno, R.; Gu, Z. Effect of Wall Texture on Perceptual Spaciousness of Indoor Space. *Int. J. Environ. Res. Public Health* **2020**, *17*, 4177. [CrossRef]
11. Herman, H. *Lessons for Students in Architecture*; NAi010 Publishers: Rotterdam, The Netherlands, 2017.
12. Newman. *Defensible Space: Crime Prevention through Urban Design*; Macmillan: New York, NY, USA, 1972.
13. Zhou, Y.; Cheng, X. *Housing for the Elderly*; China Architecture & Building Press: Beijing, China, 2011.
14. Xu, H.; He, P.; Hsu, W.L.; Lee, Y.C.; Ho, C.Y. Establishment of Assessment Indexes for Intelligent Elderly Residential Space by Analytic Hierarchy Process. In Proceedings of the 2019 IEEE Eurasia Conference on Biomedical Engineering, Healthcare and Sustainability (IEEE Ecbios 2019), Okinawa, Japan, 31 May–3 June 2019; pp. 69–72. Available online: https://www.webofscience.com/wos/alldb/full-record/WOS:000539483300019 (accessed on 5 August 2022).
15. Song, Y.; Li, J.; Wang, J.; Hao, S.; Zhu, N.; Lin, Z. Multi-criteria approach to passive space design in buildings: Impact of courtyard spaces on public buildings in cold climates. *Build. Environ.* **2015**, *89*, 295–307. [CrossRef]
16. Li, J.; Lu, S.; Wang, Q. Graphical visualisation assist analysis of indoor environmental performance: Impact of atrium spaces on public buildings in cold climates. *Indoor Built Environ.* **2016**, *27*, 331–347. [CrossRef]
17. Ojha, V.K.; Griego, D.; Kuliga, S.; Bielik, M.; Buš, P.; Schaeben, C.; Treyer, L.; Standfest, M.; Schneider, S.; König, R.; et al. Machine learning approaches to understand the influence of urban environments on human's physiological response. *Inf. Sci.* **2019**, *474*, 154–169. [CrossRef]
18. Kong, Z.; Zhang, R.; Ni, J.; Ning, P.; Kong, X.; Wang, J. Towards an integration of visual comfort and lighting impression: A field study within higher educational buildings. *Build. Environ.* **2022**, *216*, 108989. [CrossRef]
19. Tarvainen, M.P.; Cornforth, D.J.; Jelinek, H.F. Principal Component Analysis of Heart Rate Variability Data in assessing Cardiac Autonomic Neuropathy. In Proceedings of the 2014 36th Annual International Conference of the IEEE Engineering in Medicine and Biology Society (Embc), Chicago, IL, USA, 26–30 August 2014; pp. 6667–6670. Available online: https://www.webofscience.com/wos/alldb/full-record/WOS:000350044706160 (accessed on 5 August 2022).
20. Alvarez-Gonzalez, M.; Vila, X.A.; Lado, M.J.; Mendez, A.J.; Rodriguez-Linares, L. Web Site on Heart Rate Variability: HRV-Site. In *Computing in Cardiology 2010*; Murray, A., Ed.; IEEE: New York, NY, USA, 2010; Volume 37, pp. 609–612. Available online: https://www.webofscience.com/wos/alldb/full-record/WOS:000328956000153 (accessed on 5 August 2022).
21. Shemesh, A.; Leisman, G.; Bar, M.; Grobman, Y.J. A neurocognitive study of the emotional impact of geometrical criteria of architectural space. *Archit. Sci. Rev.* **2021**, *64*, 394–407. [CrossRef]
22. Hu, L.; Shepley, M.M. Design Meets Neuroscience: A Preliminary Review of Design Research Using Neuroscience Tools. *J. Inter. Des.* **2022**, *47*, 31–50. [CrossRef]
23. Banaei, M.; Hatami, J.; Yazdanfar, A.; Gramann, K. Walking through architectural spaces: The impact of interior forms on human brain dynamics. *Front. Hum. Neurosci.* **2017**, *11*, 477. [CrossRef]
24. Rad, P.N.; Shahroudi, A.A.; Shabani, H.; Ajami, S.; Lashgari, R. Encoding pleasant and unpleasant expression of the architectural window shapes: An ERP study. *Front. Behav. Neurosci.* **2019**, *13*, 186. [CrossRef]
25. Li, J. Building environment information and human perceptual feedbackcollected through a combined virtual reality (VR) andelectroencephalogram (EEG) method. *Energy Build.* **2020**, *224*, 110259. [CrossRef]
26. Li, J.; Wu, W.; Jin, Y.; Zhao, R.; Bian, W. Research on environmental comfort and cognitive performance based on EEG plus VR plus LEC evaluation method in underground space. *Build. Environ.* **2021**, *198*, 107886. [CrossRef]

27. Hattar, S.; Liao, H.W.; Takao, M.; Berson, D.M.; Yau, K.W. Melanopsin-Containing Retinal Ganglion Cells: Architecture, Projections, and Intrinsic Photosensitivity. *Science* **2002**, *295*, 1065–1070. [CrossRef]

28. Kong, Z.; Jakubiec, J.A. Evaluations of long-term lighting qualities for computer labs in Singapore. *Build. Environ.* **2021**, *194*, 107689. [CrossRef]

29. Tao, Y.; Gou, Z.; Yu, Z.; Fu, J.; Chen, X. The challenge of creating age-friendly indoor environments in a high-density city: Case study of Hong Kong's care and attention homes. *J. Build. Eng.* **2020**, *30*, 101280. [CrossRef]

30. Hormigos-Jimenez, S.; Padilla-Marcos, M.Á.; Meiss, A.; Gonzalez-Lezcano, R.A.; Feijó-Muñoz, J. Assessment of the ventilation efficiency in the breathing zone during sleep through computational fluid dynamics techniques. *J. Build. Phys.* **2019**, *42*, 458–483. [CrossRef]

31. Faul, F.; Erdfelder, E.; Lang, A.G.; Buchner, A. G*Power 3: A flexible statistical power analysis program for the social, behavioral, and biomedical sciences. *Behav. Res. Methods* **2007**, *39*, 175–191. [CrossRef]

32. Katunsky, D.; Brausch, C.; Purcz, P.; Katunska, J.; Bullova, I. Requirements and opinions of three groups of people (aged under 35, between 35 and 50, and over 50 years) to create a living space suitable for different life situations. *Environ. Impact Assess. Rev.* **2020**, *83*, 106385. [CrossRef]

33. Barriuso, A.; De la Prieta, F.; Villarrubia González, G.; De La Iglesia, D.; Lozano, Á. MOVICLOUD: Agent-Based 3D Platform for the Labor Integration of Disabled People. *Appl. Sci.* **2018**, *8*, 337. [CrossRef]

34. Macris, V.; Weytjens, L.; Geyskens, K.; Knapen, M.; Verbeeck, G. Design Guidance for Low-Energy Dwellings in Early Design Phases Development of a simple design support tool in SketchUp. In *Ecaade 2012, Digital Physicality*; Ecaade-Education & Research Computer Aided Architectural Design Europe; Achten, H., Pavlicek, J., Hulin, J., Matejovska, D., Eds.; eCAADe: Brussels, Belgium, 2012; Volume 1, pp. 691–699. Available online: https://www.webofscience.com/wos/alldb/full-record/WOS:000330322400073 (accessed on 25 June 2022).

35. Dhanda, A.; Fai, S.; Graham, K.; Walczak, G. Leveraging Existing Heritage Documentation for Animations: Senate Virtual Tour. In *Icomos/Isprs International Scientific Committee on Heritage Documentation (Cipa) 26th International Cipa Symposium—Digital Workflows for Heritage Conservation*; Hayes, J., Ouimet, C., Quintero, M.S., Fai, S., Smith, L., Eds.; Copernicus Gesellschaft Mbh; Copernicus GmbH: Allegheny County, PA, USA, 2017; Volume XLII-2-W5, pp. 171–175. [CrossRef]

36. Alonso, F. The benefits of building barrier-free: A contingent valuation of accessibility as an attribute of housing. *Eur. J. Hous. Policy* **2002**, *2*, 25–44. [CrossRef]

37. Hou, K. *Space Evaluation and Optimal Design of Pension Facilities*; Southeast University Press: Nanjing, China, 2020.

38. Rossi, V.; Pourtois, G. Transient state-dependent fluctuations in anxiety measured using STAI, POMS, PANAS or VAS: A comparative review. *Anxiety Stress Coping* **2012**, *25*, 603–645. [CrossRef]

39. Knowles, K.A.; Olatunji, B.O. Specificity of trait anxiety in anxiety and depression: Meta-analysis of the State-Trait Anxiety Inventory. *Clin. Psychol. Rev.* **2020**, *82*, 101928. [CrossRef]

40. Zhu, J.; Ji, L.; Liu, C. Heart rate variability monitoring for emotion and disorders of emotion. *Physiol. Meas.* **2019**, *40*, 064004. [CrossRef]

41. Yang, Y.; Hu, L.; Zhang, R.; Zhu, X.; Wang, M. Investigation of students? short-term memory performance and thermal sensation with heart rate variability under different environments in summer. *Build. Environ.* **2021**, *195*, 107765. [CrossRef]

42. Salo, M.A.; Huikuri, H.V.; Seppanen, T. Ectopic Beats in Heart Rate Variability Analysis: Effects of Editing on Time and Frequency Domain Measures. *Ann. Noninvasive Electrocardiol.* **2001**, *6*, 5–17. [CrossRef] [PubMed]

43. Hinde, K.; White, G.; Armstrong, N. Wearable Devices Suitable for Monitoring Twenty Four Hour Heart Rate Variability in Military Populations. *Sensors* **2021**, *21*, 1061. [CrossRef]

44. Müller, A.M.; Wang, N.X.; Yao, J.; Tan, C.S.; Low, I.C.C.; Lim, N.; Tan, J.; Tan, A.; Müller-Riemenschneider, F. Heart Rate Measures from Wrist-Worn Activity Trackers in a Laboratory and Free-Living Setting: Validation Study. *JMIR mHealth uHealth* **2019**, *7*, e14120. [CrossRef]

45. Kaufmann, T.; Suetterlin, S.; Schulz, S.M.; Voegele, C. ARTiiFACT: A tool for heart rate artifact processing and heart rate variability analysis. *Behav. Res. Methods* **2011**, *43*, 1161–1170. [CrossRef]

46. Martino, P.F.; Miller, D.P.; Miller, J.R.; Allen, M.T.; Cook-Snyder, D.R.; Handy, J.D.; Servatius, R.J. Cardiorespiratory Response to Moderate Hypercapnia in Female College Students Expressing Behaviorally Inhibited Temperament. *Front. Neurosci.* **2020**, *14*, 588813. [CrossRef]

47. Lundell, R.V.; Tuominen, L.; Ojanen, T.; Parkkola, K.; Raisanen-Sokolowski, A. Diving Responses in Experienced Rebreather Divers: Short-Term Heart Rate Variability in Cold Water Diving. *Front. Physiol.* **2021**, *12*, 649319. [CrossRef] [PubMed]

48. Pino-Ortega, J.; Rico-Gonzalez, M.; Gantois, P.; Nakamura, F.Y. Level of agreement between sPRO and Kubios software in the analysis of R-R intervals obtained by a chest strap. *Proc. Inst. Mech. Eng. Part P J. Sports Eng. Technol.* **2021**, 17543371211031144. [CrossRef]

49. Azzazy, S.; Ghaffarianhoseini, A.; Ghaffarianhoseini, A.; Naismith, N.; Doborjeh, Z. A critical review on the impact of built environment on users' measured brain activity. *Arch. Sci. Rev.* **2021**, *64*, 319–335. [CrossRef]

50. Park, H.-R.; Lim, D.-Y.; Choi, E.-H.; Lee, K.-S. Analysis on the Domestic and International Illumination Research Trends in 2012 and Further Study Suggestion. *J. Korean Inst. Illum. Electr. Install. Eng.* **2013**, *27*, 28–37. [CrossRef]

51. Verma, G.K.; Tiwary, U.S. Multimodal fusion framework: A multiresolution approach for emotion classification and recognition from physiological signals. *Neuroimage* **2014**, *102*, 162–172. [CrossRef] [PubMed]

52. Stach, T.; Browarska, N.; Kawala-Janik, A. Initial Study on Using Emotiv EPOC+ Neuroheadset as a Control Device for Picture Script-Based Communicators. *IFAC-PapersOnLine* **2018**, *51*, 180–184. [CrossRef]
53. Ahn, M.; Ahn, S.; Hong, J.H.; Cho, H.; Kim, K.; Kim, B.S.; Chang, J.W.; Jun, S.C. Gamma band activity associated with BCI performance: Simultaneous MEG/EEG study. *Front. Hum. Neurosci.* **2013**, *7*, 848. [CrossRef]
54. Martínez-Cancino, R.; Delorme, A.; Truong, D.; Artoni, F.; Kreutz-Delgado, K.; Sivagnanam, S.; Yoshimoto, K.; Majumdar, A.; Makeig, S. The open EEGLAB portal Interface: High-Performance computing with EEGLAB. *Neuroimage* **2021**, *224*, 116778. [CrossRef]
55. Pearce, J.M.S. The "split brain" and Roger Wolcott Sperry (1913–1994). *Rev. Neurol.* **2019**, *175*, 217–220. [CrossRef] [PubMed]
56. Guan, H.; Hu, S.; Lu, M.; He, M.; Zhang, X.; Liu, G. Analysis of human electroencephalogram features in different indoor environments. *Build. Environ.* **2020**, *186*, 107328. [CrossRef]

Article

Is the Shortest Path Always the Best? Analysis of General Demands of Indoor Navigation System for Shopping Malls

Hui Deng [1], Yiwen Xu [1] and Yichuan Deng [2,*]

1   School of Civil Engineering and Transportation, South China University of Technology, Guangzhou 510335, China
2   Pazhou Lab, Guangzhou 510335, China
*   Correspondence: ctycdeng@scut.edu.cn

Abstract: Indoor navigation systems are basic services for shopping malls. However, the design and implementation of such systems are seldom studied, with most current indoor navigation systems showing the static route for the shortest distance, which causes confusion or even danger for users. Therefore, this paper analyzes the general demand for indoor navigation systems for shopping malls based on 498 questionnaires and the Kano model. The results of the study unveil three important functions, as outlined by "Congestion/emergency section avoidance", "Vertical elevator first", and "Passing by a particular type of store". The relationship between users' characteristics and shopping behavior is also discovered. Comparing this with the existing literature in terms of user demands research for indoor navigation, the general demand analysis method based on the Kano model of this paper is able to reveal the user accreditation degree towards different functions of indoor navigation systems in shopping malls. The findings of this paper provide insight into users' behaviors and preferences, which will benefit further studies on indoor navigation systems for shopping malls.

Keywords: general demand analysis; Kano model; indoor navigation; system design

Citation: Deng, H.; Xu, Y.; Deng, Y. Is the Shortest Path Always the Best? Analysis of General Demands of Indoor Navigation System for Shopping Malls. *Buildings* **2022**, *12*, 1574. https://doi.org/10.3390/buildings12101574

Academic Editor: Ricardo M. S. F. Almeida

Received: 6 September 2022
Accepted: 26 September 2022
Published: 30 September 2022

Publisher's Note: MDPI stays neutral with regard to jurisdictional claims in published maps and institutional affiliations.

1. Introduction

Indoor navigation systems realize the output of navigation paths in buildings through road network construction [1], indoor positioning [2], and path planning [3]. It is necessary and important to develop an indoor navigation system that reflects the general demands of users in shopping malls. Recently, many scholars have carried out independent research on the three important content factors of an indoor navigation system. In road network construction, IFC is a commonly used road network extraction method, combined with the BIM model [4,5]. In indoor positioning, Liu et al. [6] fused magnetic and visual sensors to study indoor localization without infrastructure, while Farahsari et al. [7] studied Internet of Things (IoT)-based indoor localization. In path planning, Deng et al. [8] studied path planning under fire evacuation scenarios, while Lee and Medioni [9] used an improved D* algorithm to plan paths. However, at present, there is no mature indoor navigation system [10–12] that combines the three independent parts, and in particular, the existing indoor navigation systems do not really understand the users' demand, and there is no indoor navigation system that combines the users' general demand and dynamic environmental changes. At present, the visual navigation service only uses the fixed point where the LED display screen is located as the navigation starting point, and the user cannot navigate in real time on their mobile phone. The navigation service only provides the resulting output of the shortest path, and only a few shopping malls provide a personalized choice of priority vertical elevators or priority horizontal escalators, without considering environmental changes (such as congestion and emergencies) and their impact on user navigation experience and satisfaction. This leads to user dissatisfaction with indoor navigation deficiency. Users often find that the positioning of the navigation is inaccurate, the planned shortest route is congested, the traffic time is wasted, and navigation in the vertical direction is not clear. The reason for these existing affairs and problems is that the indoor navigation system has not undergone an adequate demand functional analysis. As a functional service system for space management and operation and maintenance management, the indoor navigation system is used by "people" with independent will. Therefore, it is not enough for designers to only satisfy the planning of the shortest path between the start point and the end point. Only by conducting sufficient user research and demand analysis can the indoor navigation system achieve the greatest practical role and meet the individual needs of users.

In general, existing indoor navigation systems of shopping malls have problems such as ignoring the general demand of users and ignoring the dynamic changes of the scene. The lack of a complete survey for demand analysis of indoor navigation in shopping malls is, thus, the key issue to be resolved. General demand analysis has been proven to be the key starting point of system design, with successful examples such as industrial energy demand models [13], cooling demand models to design cooling systems for large office buildings [14], and hotel room demand analysis, which changed the direction of the hospitality industry [15]. However, there is a gap in the effective analysis of the general demand for indoor navigation systems.

There are several mainstream models for demand analysis, including Maslow's hierarchy of needs [16], SWOT analysis [17], the Boston matrix [18], and PEST analysis [19]. Among them, Maslow's Hierarchy of Needs method divides people's needs into five levels, physiology, safety, emotional belonging, respect, and self-realization, which is suitable for determining the macro function of products, such as designing internet products with the function of making friends out of emotional belonging. SWOT analysis divides events into four dimensions, strengths, weaknesses, opportunities, and threats, and constructs a matrix to facilitate the design of product positioning and competitive strategies. The Boston Matrix divides products into four categories, star category, thin dog category, problem category, and golden bull category, which is helpful for sales strategy adjustment. The PEST analysis method obtains the macro-environmental analysis of the product through the analysis of the political environment, economic environment, social environment, and technical environment. Compared with these methods, the Kano model is another important model that better suits the purpose of the study. The Kano model is a theory invented and proposed by Professor N. Kano in 1984. With its essence of reflecting the nonlinear relationship between satisfaction and performance [20], the model can classify and rank the demand of users. The Kano model helps to increase the value of a system or product, and will focus on the service design, development and verification phases, and functional design by real customers in the development design phase [21]. Integrating the Kano model into existing design methods can improve users' satisfaction with product design [22]. At present, the Kano model has been fully applied and studied in various aspects such as product development and the healthcare industry. Asian et al. [20] used this model to study the effective variables of third-party logistics providers in the automobile manufacturing industry. Hashim and Dawal [23] improved ergonomic design with the help of the Kano model. Li et al. [3] studied the user needs of an eco-city based on the Kano model. Materla et al. [24] summarized the application of the Kano model in the healthcare industry. However, no report on the application of the Kano model in the building operation and maintenance management stage has been found, especially in the design of indoor navigation products in shopping malls.

In order to fill the gap in which there is a lack of effective general demand analysis for indoor navigation, which causes deviation between the functions of the indoor navigation system and the user's general demand, this paper uses the Kano model with 498 questionnaires to determine the priority of different general demands in the functional design for indoor navigation in shopping malls. The design of a shopping mall indoor navigation system based on users' general demand and dynamic environmental changes is also proposed to inspire future designers for related products. The main contribution of this paper is to apply the Kano model, for the first time, to determine the general demand and functions of indoor navigation in shopping malls. Existing research (Table 1) on user demand analysis of indoor navigation systems has mostly focused on the navigation needs of special populations [25,26], and mobility needs in the navigation process [27,28]. Compared to the existing literature in terms of user demand research for indoor navigation, the general demand analysis method based on the Kano model in this paper is able to reveal the user accreditation degree of the different functions of indoor navigation systems in shopping malls and meet the general demand of most people. Furthermore, the findings of this paper provide insight into users' behaviors and preferences from questionnaire research, which will benefit further studies on indoor navigation systems for shopping malls.

Table 1. Comparison between the method in this paper and the existing research.

Research Content	Advantages	Limitations	Author	References
An indoor navigation system for the blind and visually impaired (PERCEPT)	Focused on the subjective needs of special populations	The study only focused on the blind population	Ganz et al.	[25]
A speech-based, infrastructure-free indoor navigation system (MagNav)		The study only focused on validating the importance of voice navigation	Giudice et al.	[26]
Navigation aid based on mobile projector	Demonstrates the advantages of mobile screen navigation	Developed only for navigation pages/apps	Li et al.	[27]
Using smartphones and WiFi-based indoor navigation system (SWiN)	Consider the mobility and dynamic properties of the navigation system	Only focused on the use process	Zhang et al.	[28]

The following is a summary of the framework of this paper. Section 2 describes the Kano model and the related evaluation indicators. Section 3 presents the results of the questionnaire survey, including the analysis of the basic information of the questionnaire, the correlation analysis, and the related indicators of the Kano model. Section 4 is a practical implementation of the general demand for an indoor navigation system in shopping malls, which is based on the results of Section 3.

2. Materials and Methods

The method used in this paper is shown in Figure 1, which mainly includes literature research, offline interviews, and questionnaire research. Through literature research and offline interviews, several major functions of general demand can be initially considered in the indoor navigation of shopping malls, as well as related qualitative indicators. Except for a few customers who are particularly familiar with shopping malls, most customers have high indoor navigation demands. Zhou et al. [29] considered path complexity, congestion, and blocking events when planning indoor paths. Basu et al. [30] believe that the Pedestrian Route Choice (PRC) needs to consider the relationship between perceptual factors and objective factors. The qualitative indicators mentioned here are mainly crowd-density indicators and traffic-speed indicators. Based on the above discussion, this paper identifies five general demands (2.1 Identify 5 general demands), namely "Avoid crowded/emergency roads", "Passing by specific types of shops", "Bypass specific types of shops", "Vertical elevator first", and "Escalator first". Next, we outline the questionnaire design based on these five general demands. The questionnaire is divided into three parts, including the basic personal information of users, objective data of users related to the indoor navigation of shopping malls, and subjective data of users. Due to the COVID-19 pandemic, this questionnaire was collected online. After obtaining the results of the questionnaire research, the analysis work was carried out. The reliability, validity, and correlation were tested, and the Kano model and related evaluation indicators were introduced. The types of general demands and the priority of consideration were determined through Kano model classification, mixed class, and coefficient analysis. A design of the indoor navigation system of a shopping mall incorporating general demands identified from the results is presented to show how the conclusions drawn by the Kano model can be applied to the system design.

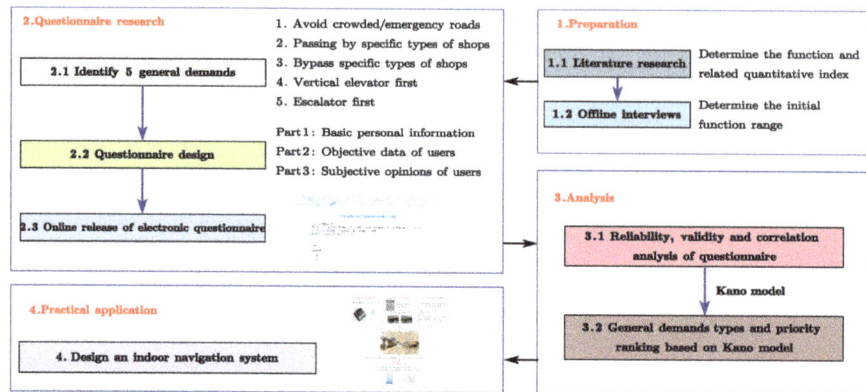

Figure 1. Method of this paper.

2.1. The Kano Model

The Kano model divides demand attributes into 5 types, as shown in Table 2 and Figure 2 [31,32]. The X-axis represents the level of quality performance (from insufficient to sufficient) and the Y-axis represents the level of user satisfaction (from dissatisfaction to satisfaction), which can be divided into five categories. The must-be quality (M) means that when this factor is applied or improved, the user's satisfaction with the product will not be improved. If this factor is not considered or is weakened, the user's satisfaction with the product will drop significantly. This factor must be considered in product design. The indifferent quality (I) means that regardless of whether this factor is applied or not, the user's satisfaction with the product does not fluctuate, and this factor does not need to be considered in product design. The one-dimensional quality (O) means that when the factor is applied or improved, the user's satisfaction with the product is greatly improved. If the factor is not considered or is weakened, the user's satisfaction with the product will decrease accordingly. This factor is a competitive attribute and is an important factor in product design. The considered part is different from other conventional products and reflects unique, special, and high-quality characteristics. The attractive quality (A) means that when the factor is applied or improved, the user's satisfaction with the product is greatly improved. If the factor is not considered or is weakened, the user's satisfaction with the product does not change, and the factor can be developed within the scope of the cost. The reversal quality (R) means that when the factor is applied or improved, the user's satisfaction with the product does not rise but falls, and the user has no demand for the factor, which should be eliminated in the design.

Table 2. Definitions of Kano types.

Kano Type	Definition
Must-be quality (M)	When this factor is applied or improved, the user's satisfaction of the product will not be improved.
Indifferent quality (I)	No matter whether this factor is applied or not, the user's satisfaction of the product does not fluctuate.
One-dimensional quality (O)	when the factor is applied or improved, the user's satisfaction of the product is greatly improved.

Table 2. *Cont.*

Kano Type	Definition
Attractive quality (A)	When the factor is applied or improved, the user's satisfaction of the product is greatly improved.
Reversal quality (R)	When the factor is applied or improved, the user's satisfaction of the product does not rise but falls.

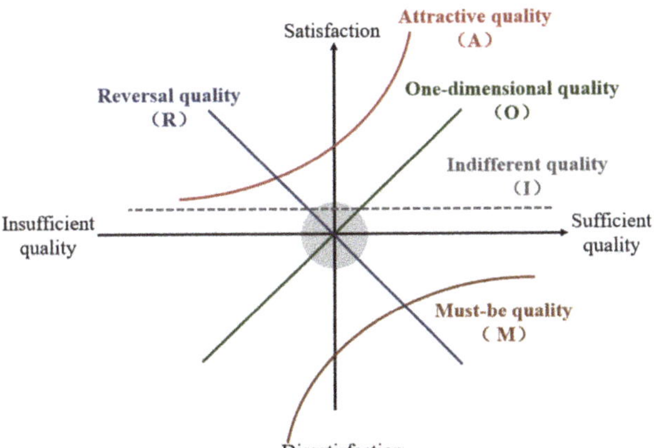

Figure 2. Relationship between quality performance and user satisfaction of Kano types [31,32].

2.2. Questionnaire Design Based on Kano Model

The premise of the Kano model is questionnaire research. Functional questions and dysfunctional questions were set in the questionnaire. Functional questions aim to ask whether the user is satisfied if this demand is considered in the product design. The purpose of the dysfunctional question is to ask if the user would be satisfied if the product design left this demand out. Therefore, in the questionnaire design, a maximum of five demand/function questions should be set in the same questionnaire, and the keywords of the functional question and dysfunctional question should be bolded in different colors to prevent the questionnaire results from being affected by unclear questions. In addition, multiple-choice questions were utilized when setting the question type, avoiding using an array of questions and preventing the respondents from answering questions in confusion due to the small degree of distinction.

Therefore, the questionnaire developed in this study adopted the form of "Single choice + Multiple choice", combined with the interviews of pedestrians in shopping malls, to obtain three elements, including user's basic information, objective data, and subjective data about users related to indoor navigation in shopping malls. The questionnaire included the following three main parts, and the specific item settings are shown in Figure 3.

Part1: User's personal basic information, including gender, age, and occupation.

Part2: Objective data of users related to the indoor navigation of shopping malls, including the frequency of visiting shopping malls, the purpose of visiting shopping malls, the types of shopping malls visited, whether users are familiar with shopping mall maps, the types of shops they often visit, whether they can accurately find the shortest path, etc.

Part3: Subjective data of users related to the indoor navigation of shopping malls, including opinions and demands on indoor navigation in shopping malls, and satisfaction with existing indoor navigation signs/systems. Among them, the functional and dysfunctional questions for indoor navigation in shopping malls were combined with 5 general demands, including "Avoid crowded/emergency roads", "Passing by specific types of shops", "Bypass specific types of shops", "Vertical elevator first", and "Escalator first".

The sources references of the 5 general demands are shown in Table 3, which prove that the investigated functions cover most shopping mall users. The setting of functional and dysfunctional questions is shown in Table 4. The questionnaire respondents are allowed to fill in personalized needs in addition to the 5 general demands. Among them, because "Avoid congestion/emergency roads" involves quantitative analysis, the choice of indicators for the congestion environment was added to the questionnaire. We adopted results from Zhou et al. [29], which divided the congestion index into the traffic density index and the traffic speed index. The traffic density index (person/m^2) was divided into four grades ([0, 0.75], (0.75, 2.00), (2.00, 3.50), and (3.5, +∞]). The traffic speed index (m/s) was divided into four levels ((1.40, +∞], (1.08, 1.40], (0.30, 1.08], and [0, 0.30]). Based on this, the questionnaire for this study was divided into multiple levels, and the traffic density index was divided into six grades (0.25 people/m^2, 0.5 people/m^2, 0.75 people/m^2, 1.25 people/m^2, 2 people/m^2, and ≥2 people/m^2). The traffic speed index was divided into four grades (1.5 m/s, 0.75 m/s, 0.3 m/s, and ≤ 0.3 m/s). In order to improve people's engagement with the questionnaire, pictures of the scene under different traffic densities were simulated (shown in Figure 4), which is analogous to the normal walking speed.

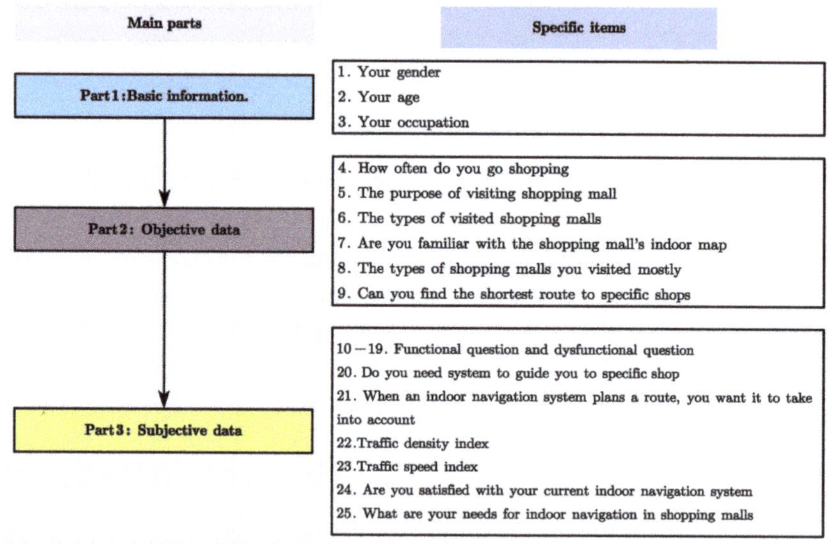

Figure 3. Main parts and specific items of questionnaire.

Table 3. Source of 5 general demands.

Items	Corresponding Source	References
Avoiding crowded/emergency roads	Literature research	Zhou et al. [29]
Passing by specific types of shops	Offline interviews	Randomly interviewed 5 male customers and 5 female customers, the age of the customers involved "old", "middle" and "youth" generations
Bypassing specific types of shops		
Vertical elevators first	Shopping mall field research	Hankyu Ningbo
Escalator first		

Table 4. Sample functional and dysfunctional questions of Kano questionnaire.

	Like	Must-Be	Neutral	Live with	Dislike
Do you accept avoiding crowded/emergency roads on the way to your destination (Functional question)	√				
Do you accept not avoiding crowded/emergency sections on the way to your destination (Dysfunctional question)					√

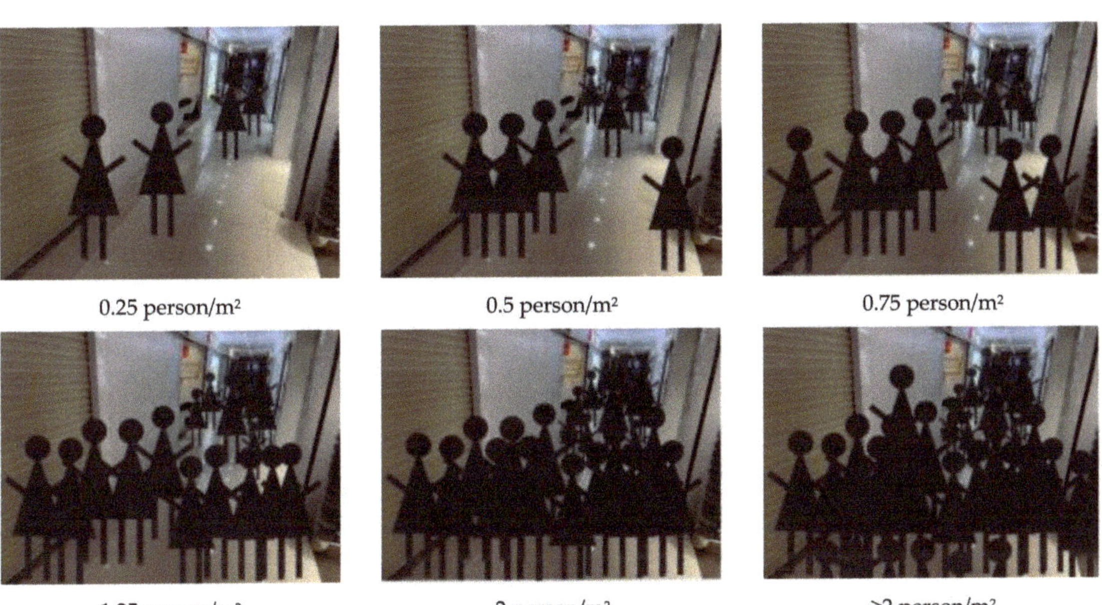

Figure 4. Pictures with the scene under different crowd densities.

2.3. Determination of Kano Model by Berger

There were five options for functional and dysfunctional questions, including "Like", "Must-be", "Neutral", "Live with", and "Dislike". According to the choices of the people investigated, the Kano type of this general demand was obtained, as shown in Table 5. Among them, the Kano types obtained according to the dysfunctional and functional questions correspond to the Kano type in Table 2. Then, the principle of relative majority was adopted to summarize all Kano types of each demand item and determine the final Kano type of each demand item. Berger [33] proposed an improved Kano category, which is defined as follows:

Table 5. Dysfunctional and functional questions [33,34].

Scale		Dysfunctional Question Evaluation				
		Like	Must-Be	Neutral	Live with	Dislike
	Like	Q [1]	A	A	A	O
	Must-be	R	I	I	I	M
Functional question evaluation	Neutral	R	I	I	I	M
	Live with	R	I	I	I	M
	Dislike	R	R	R	R	Q [1]

[1] Note: Q is a suspicious result due to incorrectly set question or misunderstanding of respondent.

If f (O+A+M) > f (I+R+Q), the Kano type is the highest frequency type among O, A, and M;

If f (O+A+M) < f (I+R+Q), the Kano type is the highest frequency type among I, R, and Q;

If f(O+A+M) = f (I+R+Q), the Kano type is the highest frequency type among O, A, M, and I;

where f (X) is the frequency of the demand type.

2.4. Determination of TS and CS of the Kano Model

The second part is mixed class analysis, proposed by Newcomb [35], which includes two indicators (as shown in Equations (1) and (2)), TS (Total Strength) and CS (Category Strength). If the TS value ≥ 0.6 and the CS value ≤ 0.06, it indicates that the demand belongs to the mixed class H, and the two mixed types, namely the two types with the highest frequency, should be explained.

$$TS = f(O + A + M)/f(O + A + M + I + R + Q) \tag{1}$$

$$CS = \frac{\max\{f(O),\ f(A),\ f(M),\ f(I),\ f(R),\ f(Q)\} - \text{second } \max\{f(O),\ f(A),\ f(M),\ f(I),\ f(R),\ f(Q)\}}{f(O + A + M + I + R + Q)} \tag{2}$$

2.5. Determination of Better-Worse Coefficient of Kano Model

The third part is the analysis of the Better–Worse coefficient, which is calculated by the percentage of each demand to the classification. The Better–Worse coefficient is used to indicate the influence degree of whether the demand is satisfied or not for the respondents. The better coefficient is an "increased satisfaction coefficient", indicating that when the demand/function is satisfied, the user's satisfaction will be improved. The closer the coefficient is to 1, the more obvious the improvement in satisfaction will be. The Worse coefficient is "dissatisfaction coefficient after elimination", indicating that when the demand/function is eliminated, the user's satisfaction will decrease. The closer the coefficient is to -1, the more obvious the decrease in satisfaction will be. The Better coefficient and Worse coefficient are calculated as follows (Equations (3) and (4)):

$$\text{Better coefficient} = \frac{f(O) + f(A)}{f(O + A + M + I)} \tag{3}$$

$$\text{Worse coefficient t} = -\frac{f(O) + f(M)}{f(O + A + M + I)} \tag{4}$$

The evaluation of the Average Satisfaction Coefficient (ASC) based on the Better–Worse coefficient was proposed by Park [36], which is strongly correlated with the Better and Worse coefficients and can reflect the priority degree in functional design. The higher the ASC value (as shown in Equation (5)), the higher priority the change demand/function should be.

$$ASC = \frac{|\text{Better}| + |\text{Worse}|}{2} \tag{5}$$

If we plot a four-quadrant graph based on the Better–Worse coefficients of the demand factors, the "Better" coefficient is along the X-axis and the absolute value of the "Worse" coefficient is along the Y-axis. The origin is the average value of the absolute value of the "Better" coefficient and the "Worse" coefficient at each point.

3. Findings from the General Demand Analysis

As shown in Figure 5, 78.51% of respondents indicated that they needed software/ systems to guide them to specific shops. As shown in Figure 6, 47.19% of respondents indicated that they were not satisfied with the current indoor navigation signage/system.

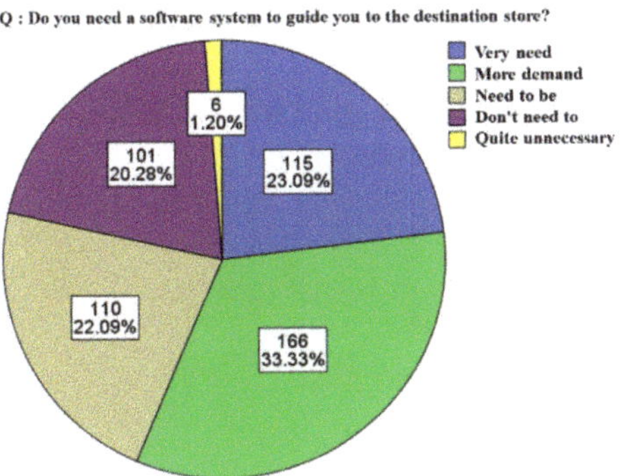

Figure 5. "Do you need indoor navigation?" fan chart.

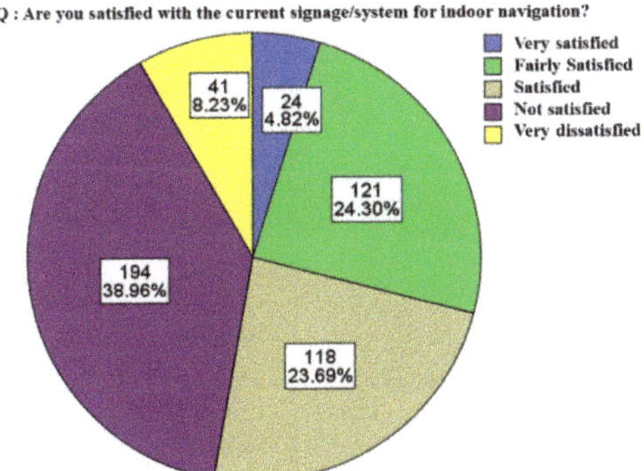

Figure 6. "Satisfaction with existing indoor navigation" fan chart.

3.1. Reliability and Validity Analysis

In this survey, 502 questionnaires were issued in total, of which 498 were effective, accounting for 99.20%. SPSS 22.0 software (IBM SPSS Statistics, Norman H. Nie & C. Hadlai (Tex) Hull & Dale H. Bent Chicago, IL, USA) was used for reliability and validity statistics. The gender, age, and occupation of the respondents are shown in Table 6 and Figures 7–9. The gender, age, and occupation of the respondents in the questionnaire conform to the actual access situation of the shopping mall, and also conform to the conclusion of Wu [37] on customers of different ages in user surveys of a shopping mall. Due to economic ability and desire, users are mainly concentrated between 18 and 49 years old. Professional office workers with stable economic incomes accounted for the largest proportion. Therefore, the questionnaire is in line with the actual situation and the general results of previous studies.

Table 6. Sex and age cross-comparison table.

Distribution of Options	Male		Female	
	Number	Proportion	Number	Proportion
<18	0	0.00%	5	2.01%
18~29	95	38.15%	87	34.94%
30~39	61	24.50%	70	28.11%
40~49	43	17.27%	50	20.08%
50~59	40	16.06%	31	12.45%
≥60	10	4.02%	6	2.41%
Total	249	100.00%	249	100.00%

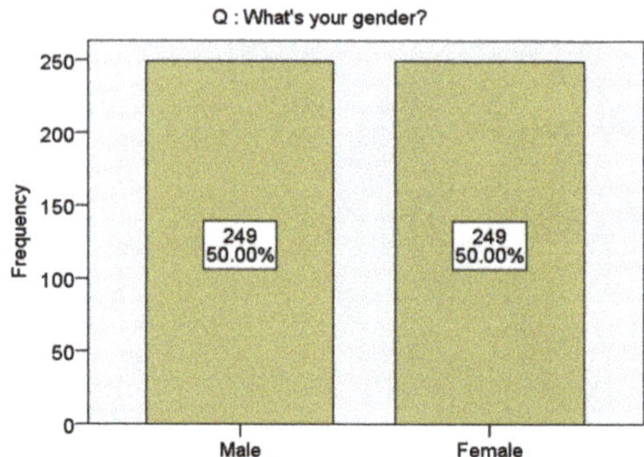

Figure 7. Gender frequency.

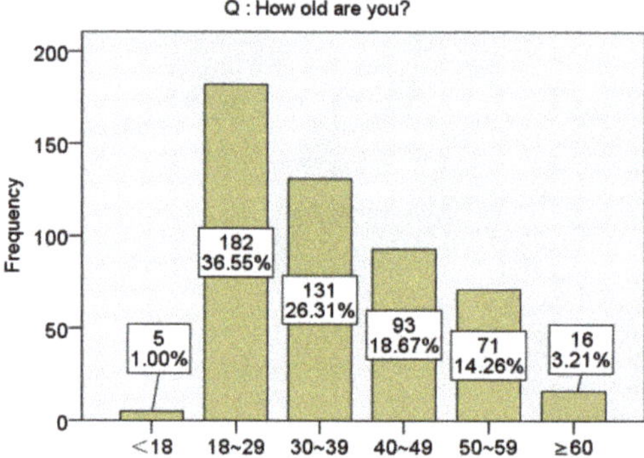

Figure 8. Age frequency.

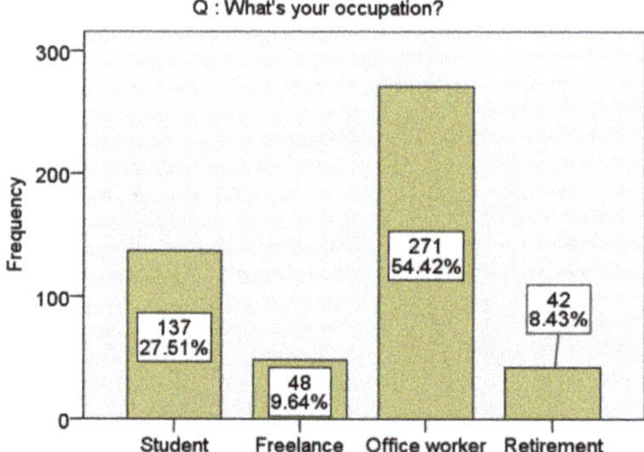

Figure 9. Occupation frequency.

Finding 1: *The user profile of the indoor navigation system of the shopping mall is 18–49-year-old non-retired users with a certain economic ability and frequent access to different shopping malls.*

As a basic index of a questionnaire survey, reliability and validity are divided into two levels. Reliability is used to test the internal stability and consistency of the evaluation results, while validity is used to judge whether the results obtained from the questionnaire can accurately represent the evaluated functional requirements. The reliability measurement indexes include Cronbach's Alpha value, the halved coefficient, retest reliability, etc. The validity measurement is factor analysis, and the KMO value and Bartlett sphericity test value are used in this paper. As shown in Table 7, Cronbach's Alpha of the functional questions is between 0.7 and 0.8 with good reliability, the KMO value is between 0.7 and 0.8, and Bartlett's sphericity test corresponds to a p value < 0.05, indicating that the results of functional questions in this questionnaire are suitable for factor analysis. Cronbach's Alpha of dysfunctional questions was between 0.7 and 0.8, with good reliability, the KMO value was between 0.7 and 0.8, and Bartlett's sphericity test corresponded to a p value < 0.05, indicating that the results of dysfunctional questions in this questionnaire were suitable for factor analysis.

Table 7. Reliability and validity analysis of forward and reverse problems of Kano model.

Type of Test	Cronbach's Alpha	KMO Value	Bartlett's Sphericity Test Sig Value
Functional question	0.710	0.767	0.000
Dysfunctional question	0.707	0.739	0.000

3.2. Correlation Analysis

Correlation analysis can determine the degree of correlation between variables and judge the degree of correlation between basic personal information and objective data and subjective data in the questionnaire, using the Pearson correlation coefficient and significance SIG value. The correlation analysis framework of this questionnaire is shown in Figure 10.

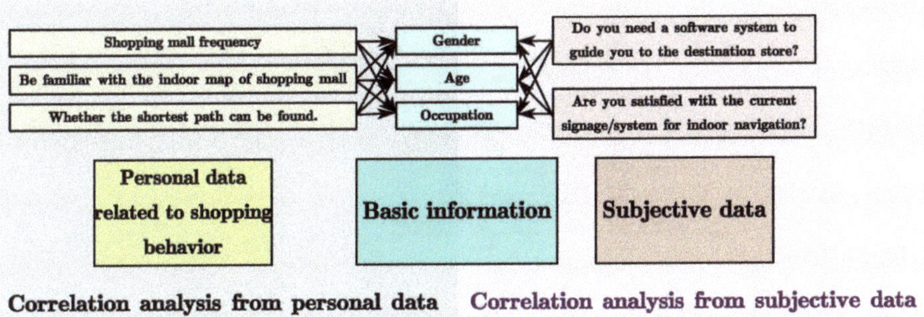

Figure 10. Framework for correlation analysis.

3.2.1. Correlation Analysis from Objective Data

As shown in Table 8, there is a certain correlation between respondents' basic information and objective data. There is a significant difference between "shopping mall frequency" and "age" at 0.05 (double tail), and an extremely significant difference between "occupation" at 0.01 (double tail). The correlation between "familiar with indoor map of shopping mall", "age", and "occupation" is significantly different at 0.05 (double tail). The correlation between "whether can find the shortest path", "gender", and "age" is significant at 0.01 (double tail), and the correlation between "occupations" is significant at 0.05 (double tail). In addition to the correlation between characteristics (personal basic information) and state variables (objective data) in Table 7, there is also a correlation between status variables (objective data). The r value of "frequency of shopping mall" and "familiarity with indoor map of shopping mall" is 0.377, with a SIG value of 0.000, and the correlation is very significant at 0.01 (double tail). The r value and SIG value of "are you familiar with the indoor map of the shopping mall" and "can you find the shortest path to the destination" are 0.297 and 0.000, showing a very significant difference in the correlation at 0.01 (double tail).

Table 8. Correlation analysis between characteristics (personal basic information) and status variables (objective data) [1].

	Gender		Age		Occupation	
	r	p	r	p	r	p
Shopping mall frequency	−0.012	0.787	−0.095	0.034	−0.115	0.010
Be familiar with the indoor map of shopping mall	−0.075	0.093	−0.103	0.022	−0.099	0.028
Whether the shortest path can be found	0.189	0.000	−0.151	0.001	−0.109	0.015

[1] r is Pearson correlation coefficient. p is significance sig value. Blue indicates significant difference. Red indicates very significant difference.

Figure 11 shows the basic information of "gender" and the frequency of the personal data contrast, according to the Pearson correlation coefficient and the significant SIG value of "gender" and "if you can find the shortest path". Males' perceptions of finding the shortest path were evenly balanced, while more than half of females felt they were unsure whether they would find the shortest path.

Finding 2: *Males feel that they have a clearer grasp of pathfinding, while females are not sure.*

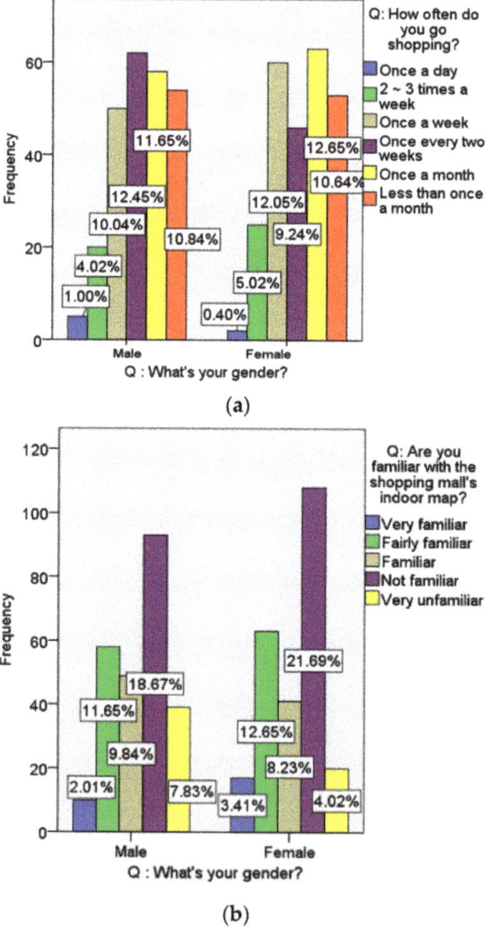

Figure 11. *Cont.*

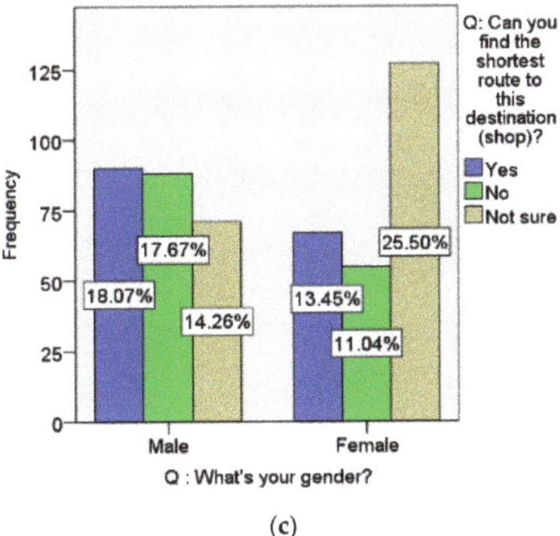

(c)

Figure 11. Frequency comparison between "Gender" and three objective data. (**a**) Shopping mall frequency; (**b**) Be familiar with the indoor map of shopping mall; (**c**) Whether the shortest path can be found.

Figure 12 shows the comparison between the frequency of "age" and personal data. According to the Pearson correlation coefficient and significance SIG value, it can be seen that "age" is correlated with personal data related to shopping behavior. Although the proportion of respondents aged 30–49 is 1.23 times higher than that of respondents aged 18–29, their frequency of going to the mall more than once a week is 1.97 times higher than that of respondents aged 18–29, indicating that respondents aged 30–49 visit shopping malls more frequently. The frequency of respondents aged 18 to 29 years old visiting shopping malls mainly ranges from once a week (8.83%) to once every two weeks (9.03%) to less than once a month (10.04%), which is related to different work and rest schedules for different age groups. Young users need to consider academic and economic pressures, while middle-aged users have more disposable time and consumption demands. Secondly, in terms of familiarity with the indoor map of the shopping mall, all age groups are concentrated in "not very familiar", and the proportion of "very familiar" with the indoor map is less than 2%. This again highlights the need for a properly designed indoor navigation system. Similar to "whether familiar with store indoor map", the 18-to-29-year-old group of respondents' answers to "if you can find the shortest path" focus on "no" (15.66%) and "uncertainty" (14.26%). Respondents in the 30–49 age group, however, concentrated on "yes" (15.26%) and "not sure" (19.48%), which were related to lifestyle fixation and familiarity with frequent mall visits. Respondents generally reflected that they could not find the shortest route in a new shopping mall. In addition, 55.18% of the respondents aged over 50 answered "yes", indicating that the elderly tend to go to a fixed shopping mall. They often go to the nearest shopping mall to buy daily necessities and are quite familiar with the route.

Finding 3: *Users of different age groups go to shopping malls with different frequencies, which is closely related to the lifestyle, economic strength, and work schedule of different age groups.*

Finding 4: *Users that frequently go to particular malls think they are familiar with the indoor map of the shopping mall, but still need a navigation system when they go to a new mall.*

Finding 5: *Users over the age of 50 tend to go to particular shopping malls due to habits and physical limitations, etc., and they do not use new navigation systems. Therefore, they are not the target users of indoor navigation systems in shopping malls.*

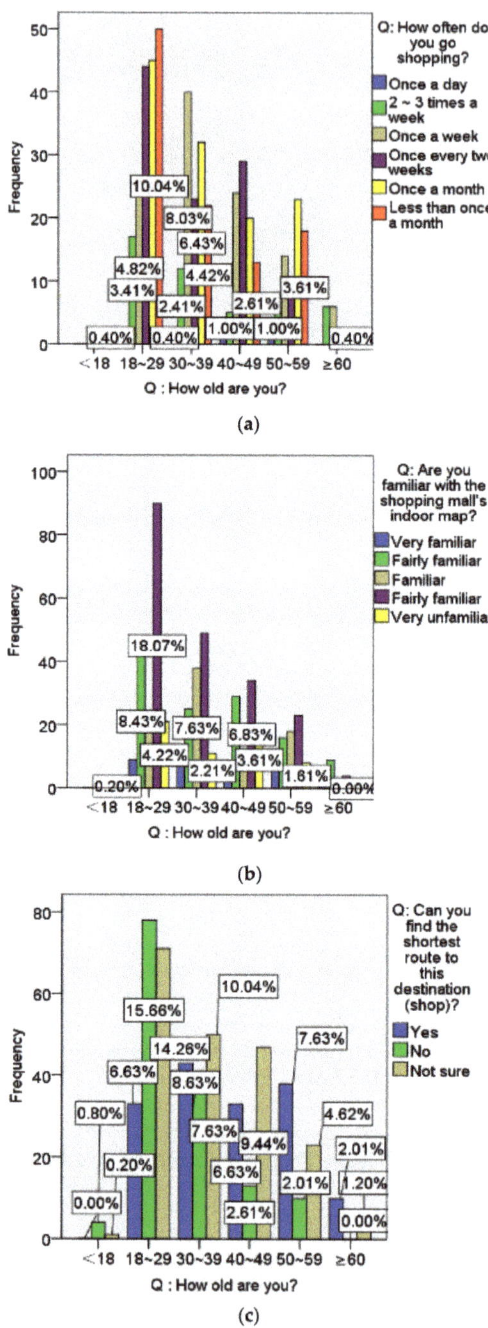

(a)

(b)

(c)

Figure 12. Frequency comparison between "Age" and personal data related to shopping behavior. (**a**) Shopping mall frequency; (**b**) Be familiar with the indoor map of shopping mall; (**c**) Whether the shortest path can be found.

Figure 13 shows the frequency comparison between "occupation" and personal data related to shopping behavior. According to the Pearson correlation coefficient and significance SIG value, it can be seen that "occupation" is correlated with personal data related to shopping behavior. The frequency of students visiting shopping malls is concentrated at least once every two weeks, which is similar to the related results of "age". The frequency of visiting shopping malls of the self-employed and retired respondents was relatively balanced. Office workers accounted for the largest proportion of respondents, with a frequency of mainly once a week. Secondly, in terms of whether they are familiar with the indoor map of shopping malls, students and office workers answered "not very familiar"; in contrast, 26% of office workers chose "relatively familiar", which also matched the middle-aged survey respondents' "regular life" and "regular shopping malls" mentioned above. In the part concerning "whether the shortest path can be found", 64.23% of the students believe that they cannot find it or are not sure, while 75.28% of the office workers focus on "yes" and "not sure", which also matches the previous conclusion.

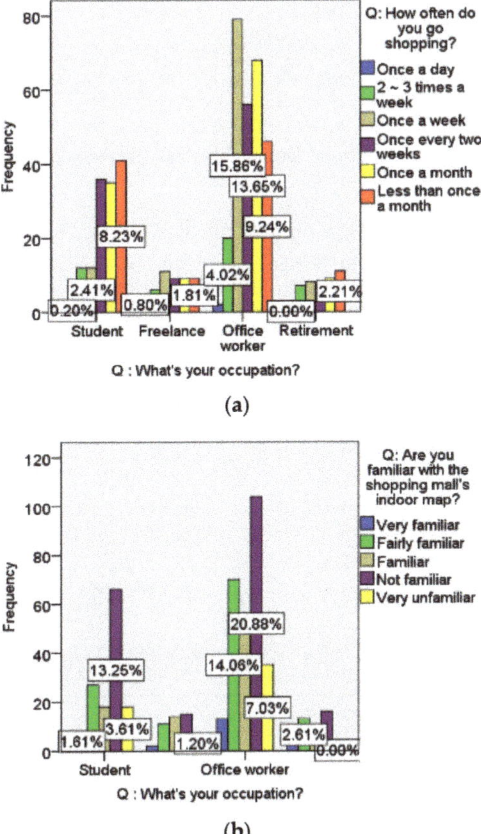

(a)

(b)

Figure 13. *Cont.*

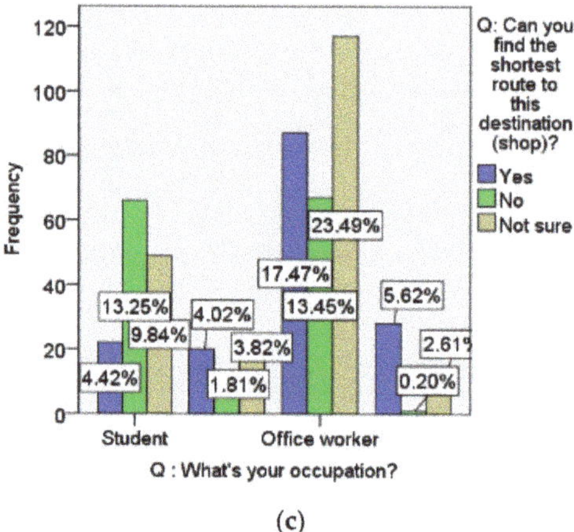

(**c**)

Figure 13. Frequency comparison between "Occupation" and personal data related to shopping behavior. (**a**) Shopping mall frequency; (**b**) Be familiar with the indoor map of shopping mall; (**c**) Whether the shortest path can be found.

Finding 6: *Similar results of "occupation" and the personal data related to shopping behavior can be corroborated with "age".*

3.2.2. Correlation Analysis from Subjective Data

As shown in Table 9, there is a certain correlation between people's basic information and data related to shopping behavior. The correlation between "need navigation software/system" and "occupation" is significant at 0.05 (two tails). In addition to the correlation between characteristics (personal basic information) and demand variables (subjective data) in Table 8, there is also a correlation between demand variables (subjective data). The r value of "whether you need navigation software/system" and "Yes or no you are satisfied with the current navigation system" is −0.156, the SIG value is 0.000, and the correlation is significant at 0.01 (double tail).

Table 9. Correlation analysis between characteristics (personal basic information) and demand variables (subjective data) [1].

	Gender		Age		Occupation	
	r	p	r	p	r	p
Do you need a software system to guide you to the destination store?	0.028	0.538	0.057	0.205	0.105	0.019
Are you satisfied with the current signage/system for indoor navigation?	0.010	0.832	−0.075	0.097	0.039	0.383

[1] r is Pearson correlation coefficient. p is significance sig value. Blue is significant difference.

Figure 14 shows the frequency comparison of "gender", a person's basic information, and the personal data related to shopping behavior, while Figure 15 shows the frequency comparison of "age", a person's basic information, and personal data related to shopping behavior. According to the Pearson correlation coefficient and significance sig value, it can be seen that "gender" and "age" are not correlated with them.

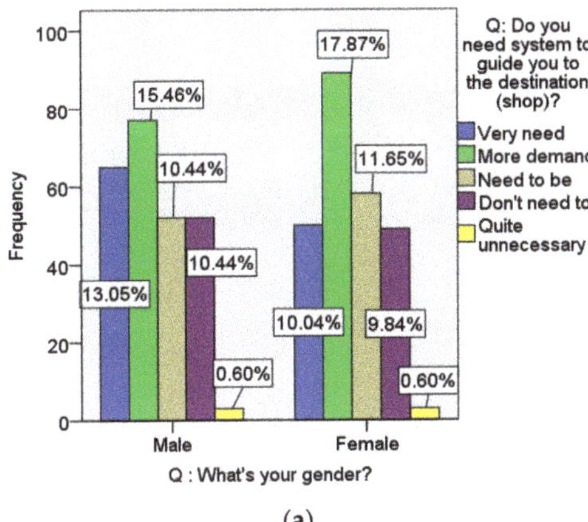

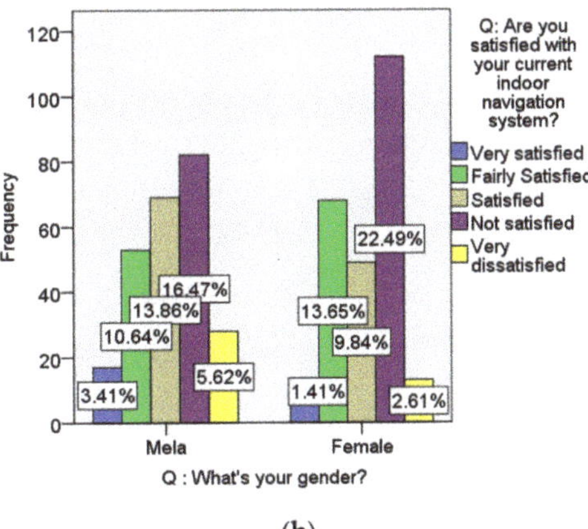

(a)   (b)

Figure 14. Frequency comparison between "Gender" and personal data related to shopping behavior. (**a**) Do you need a software system to guide you to the destination store? (**b**) Are you satisfied with the current signage/system for indoor navigation?

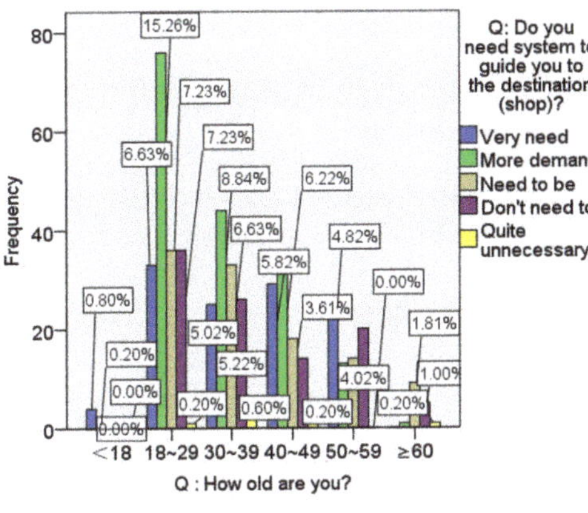

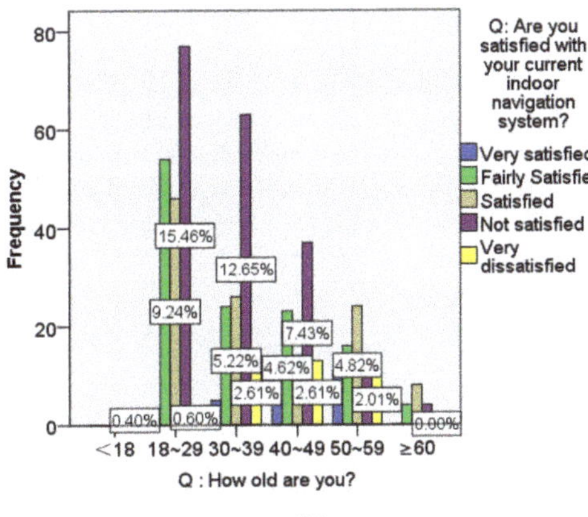

(a)   (b)

Figure 15. Frequency comparison between "Age" and personal data related to shopping behavior. (**a**) Do you need a software system to guide you to the des-tination store; (**b**) Are you satisfied with the current signage/system for indoor navigation.

Finding 7: *"Gender" and "Age" are not correlated with personal data related to shopping behavior.*

Figure 16 shows the basic information of "professional" respondents and the frequency of personal data related to shopping behavior contrast, according to the Pearson correlation coefficient and the significant sig value of "professional" and "whether need navigation software/system". Except for the retired group, the respondents of other occupational types are not satisfied with the current indoor navigation signage or system, especially students (10.84%) and office workers (22.69%).

Finding 8: *Satisfaction with indoor navigation systems is low among students and office workers, while retirement groups do not care. This also verifies that the retired group is not the target user of the indoor navigation system.*

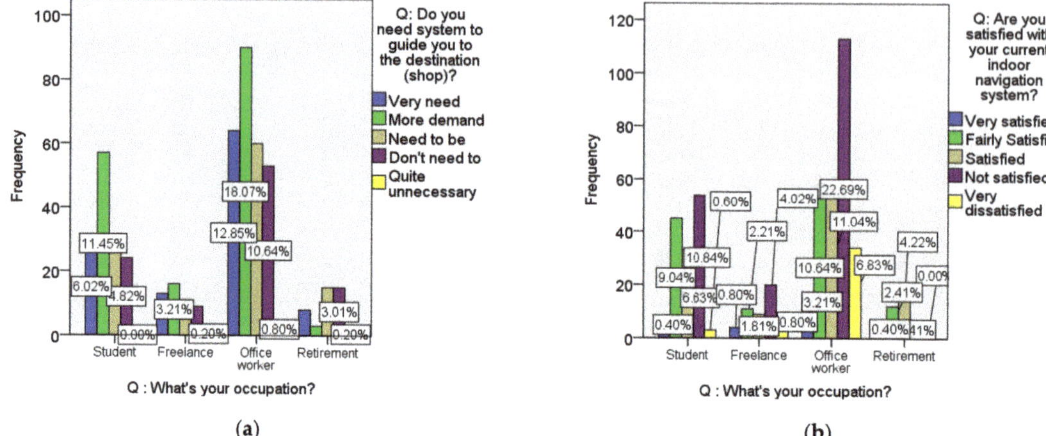

(a)   (b)

Figure 16. Frequency comparison of "Occupation" and personal data related to shopping behavior. (**a**) Do you need a software system to guide you to the destination store? (**b**) Are you satisfied with the current signage/system for indoor navigation?

Students and office workers are selected as the main users. Young people, limited by their academic and economic abilities, tend to choose uncertain shopping malls, and they are more willing to try different shopping malls, having a greater demand for indoor navigation. Middle-aged people have a more fixed life trajectory and are very familiar with the indoor map of a specific shopping mall, but they also have a greater demand for indoor navigation when going to a shopping mall they have never been to.

Finding 9: *There is a certain correlation between the characteristics of users and their behavior in shopping malls.*

3.3. Kano Model Classification

As shown in Table 10, the Kano type of the five demands is consistent with the modified Kano types of Berger.

Table 10. Frequency statistics and classification of indoor navigation demand factors in shopping malls.

Demand Item	Demand Type Frequency f(X)						Kano Category	Berger Improved the Kano Category
	O	A	M	I	R	Q		
Avoid crowded/emergency roads	226	58	27	151	18	18	O	O
Passing by specific types of shops	40	263	3	142	23	27	A	A
Bypass specific types of shops	11	111	3	306	31	36	I	I
Vertical elevator first	50	260	8	148	10	22	A	A
Escalator first	48	89	14	311	11	25	I	I

3.4. Mixed Classification

As shown in Table 11, combined with the Kano model classification and mixed classification, the five demands of this questionnaire are all non-mixed types, which are consistent with the Kano model classification. "Vertical elevator first" belongs to the Attractive quality (A), while "Escalator first" belongs to the Indifferent quality (I). Both vertical elevators and horizontal escalators are important means of vertical transportation in shopping malls. Users prefer to take the vertical elevator because they think the vertical elevator is faster and more convenient while consuming less energy. Some users stated that they are afraid of the height of the horizontal escalator.

Table 11. Mixed analysis of indoor navigation demand factors in shopping malls.

Demand Item	TS Value	CS Value	Mixed Type
Avoid crowded/emergency roads	0.624498	0.150602	O
Passing by specific types of shops	0.614458	0.242972	A
Bypass specific types of shops	0.251004	0.391566	I
Vertical elevator first	0.638554	0.224900	A
Escalator first	0.303213	0.445783	I

Finding 10: *Although vertical elevators are more difficult to find than horizontal escalators, users prefer to use vertical elevators, which may be related to the user's belief that the vertical elevator is faster.*

"Passing by a particular type of store" belongs to the Attractive quality (A), while "Bypass specific types of shops" belongs to the Indifferent quality (I). This shows that users do not accept unnecessary detours, but only detours due to congestion.

Finding 11: *The users prefer that the navigation system passes the desired space when reaching a certain destination, but users who are not interested will tend to avoid certain spaces.*

3.5. Better–Worse Coefficient Analysis

Table 12 shows the Better–Worse coefficient and ASC values of indoor navigation demand factors in shopping malls. According to AEC values, the priority of the five demands can be obtained: "Avoid crowded/emergency sections" > "Vertical elevator first" > "passing by specific types of shops" > "Escalator first" > "Bypass specific types of shops".

Table 12. Analysis of Better–Worse coefficient and ASC value of indoor navigation demand factors in shopping malls.

Demand Item	Better Coefficient	Worse Coefficient	ASC Value	Importance Ranking
Avoid crowded/emergency roads	0.614719	−0.547619	0.581169	1
Passing by specific types of shops	0.676339	−0.095982	0.386161	3
Bypass specific types of shops	0.283063	−0.032483	0.157773	5
Vertical elevator first	0.665236	−0.124464	0.394850	2
Escalator first	0.296537	−0.134199	0.215368	4

The four-quadrant diagram is drawn according to the Better–Worse coefficient of the demand factors for indoor navigation in shopping malls, as shown in Figure 17. "Avoid crowded/emergency roads" is located in the first quadrant, which is an expected requirement and should be given first priority in functional design. "Passing by specific types of shops" and "Vertical elevator first" are in the second quadrant, which should also be given priority. "Bypass specific types of shops" and "Escalator first" are located in the third quadrant, which are undifferentiated requirements and can be ignored in functional design.

In terms of functional design, it can be concluded that users prefer vertical elevators for vertical transportation. In addition, it is generally recognized as necessary to avoid congested road sections or emergencies. In terms of going to a certain destination and passing the rest of the space, the users tend to think that it is necessary to pass a certain type of store that they want to pass through, but it is not necessary to make a detour to avoid passing through a certain type of store.

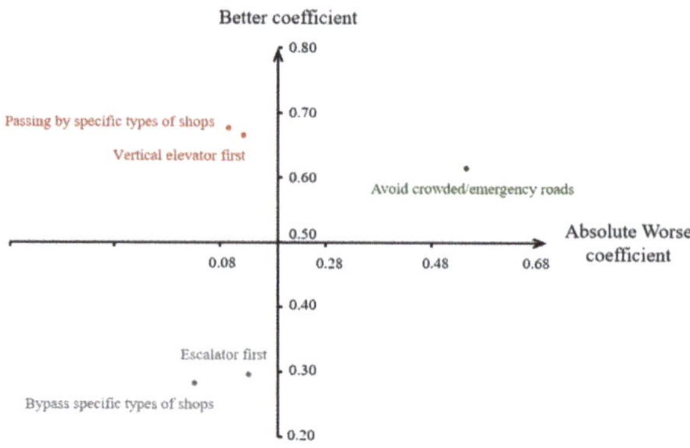

Figure 17. Better–Worse coefficient quadrant diagram.

Finding 12: *The three indoor navigation system functions identified in this paper are "Avoid crowded/emergency roads", "Vertical elevator first", and "Passing by a particular type of store".*

4. Practical Implementations of Findings from the Survey

Based on literature research and the questionnaire analysis of the Kano model, it can be concluded that "Avoid crowded/emergency roads" is an important demand, while "Vertical elevator first" and "Passing by specific types of shops" are expected general demands when designing the indoor navigation system of shopping malls. "Bypass specific types of shops" and "Escalator first" are undifferentiated requirements and may not be considered. Table 13 shows answers to "When planning the path of indoor navigation system, you want it to consider: _______" in the questionnaire. The survey results are similar to those of the Kano model. "Avoiding crowded/emergency sections", "Vertical elevator first", and "Passing by specific types of shops" account for the largest proportion, while "Bypassing specific types of shops" and "Escalator first" account for less than 40%.

Table 13. Percentage of indoor navigation considerations (multiple choice) in the questionnaire.

Demand Item	Number	Percentage	Percentage of the Observed Value
Avoid crowded/emergency roads	274	24.3%	55.0%
Passing by specific types of shops	338	29.9%	67.9%
Bypass specific types of shops	114	10.1%	22.9%
Vertical elevator first	231	20.5%	46.4%
Escalator first	159	14.1%	31.9%
Else	13	1.2%	2.6%

In addition to the analysis of the five general demands, the questionnaire also set the following question: "What are your requirements for indoor navigation in shopping malls/what do you think the most convenient and effective indoor navigation system should have: _______". A total of 141 valid answers were collected. The users' expectation of the system is being clear and easy to use, having detailed precision, and real-time accuracy. In terms of specific needs, some users mentioned the demand to identify shopping mall activities, indicate the location of toilets, and provide a variety of options and non-graphic designs by floor.

This paper determines demand functions of three levels, as shown in Table 14, including the function level, use level, and effect level. The function level is to realize three exciting needs and expectation needs and can choose whether to implement specific functions through a personalized interface. The interface should be simple and clear, with navigation made easy enough even for people with a weak sense of direction. At the effect level, new demands are put forward for the specific environment of shopping malls. For example, the path congestion index needs to change in real time, and even predict future congestion based on past long-term data. It is necessary to solve the three-dimensional navigation results of the shopping mall from the two aspects of positioning and navigation, rather than a two-dimensional plane. More consideration should be given to special spaces such as toilets, vertical elevators, and horizontal escalators. Multiple paths can be provided if the user desires.

Table 14. Classification and design of shopping mall interior navigation requirements.

Level	Specific Requirements
Function	Avoid crowded/emergency roads Vertical elevator first Passing by specific types of shops
Use	Convenient, portable use Simple and clear
Effect	Updates environment data in real time and prompts for specific activities Positioning and navigation are three-dimensional and accurate Multiple paths are available Special calibration for toilet, vertical elevator, horizontal escalator and other shopping mall special space

Based on the determined system functions, the shopping mall indoor navigation system as shown in Figure 18 was designed. The dotted box in Figure 16 is the preparatory work for the indoor navigation system of the shopping mall. First, based on the BIM model, the 3D road network is obtained through IFC and stored in the form of a matrix. Meanwhile, a database of indoor images of shopping malls is established, and 200 photos were selected for each of the four different types of scenes of elevators, atriums, gates, and ordinary indoor space. Subsequently, computer vision was used to determine whether a location in the mall is a congested route. Then, the multi-source heterogeneous information such as the congestion situation, shops on the intended route, three-dimensional coordinates, and vertical elevators are integrated into the shortest path routing algorithm. The parameters are continuously adjusted to obtain the most suitable route evaluation algorithm. Among them, the login page of the indoor navigation system of the shopping mall is shown in Figure 19, which provides the basic information of the user. In addition, the functional page design of the shopping mall indoor navigation system is shown in Figure 20. The user scans the QR code of the nearest store through the mobile phone, and their three-dimensional coordinates can be obtained in the background. On the mobile phone terminal, the user enters the destination, the shops they intend to pass through, whether to avoid crowded road sections, and whether vertical elevators are preferred, and then the optimal path calculated by the path planning algorithm can be obtained. The mobile terminal of the mobile phone presents the visual effect of the path, and displays the important store nodes of the path to improve readability and understandability.

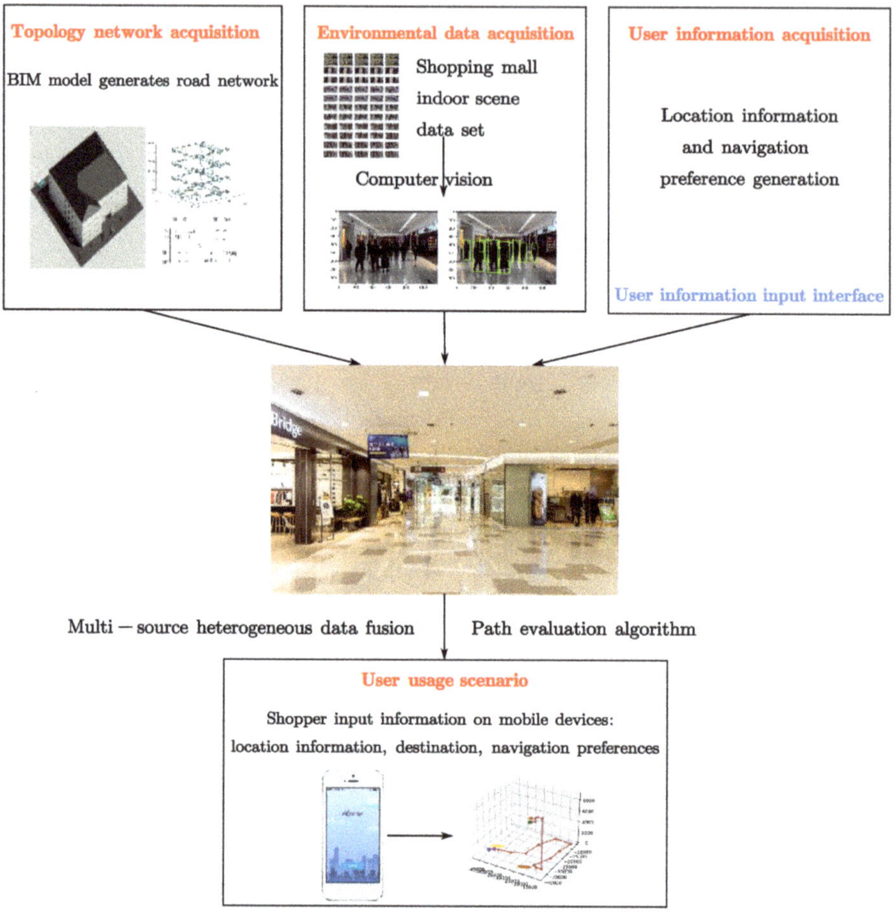

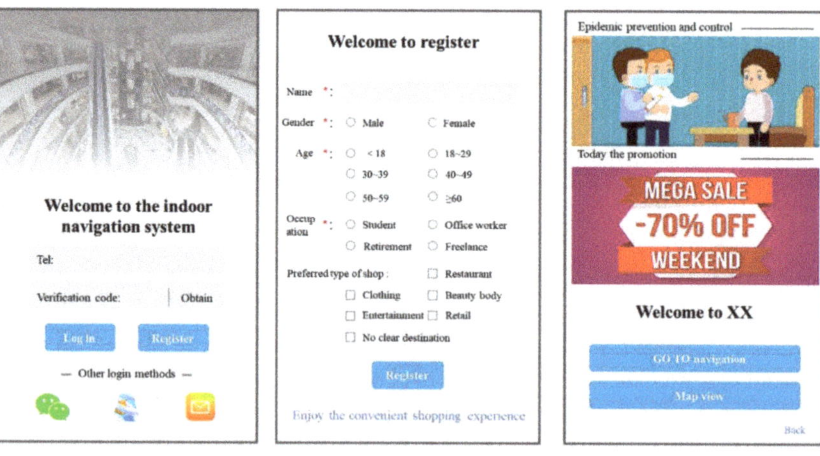

Figure 18. Shopping mall indoor navigation system.

Figure 19. Shopping mall indoor navigation system login page.

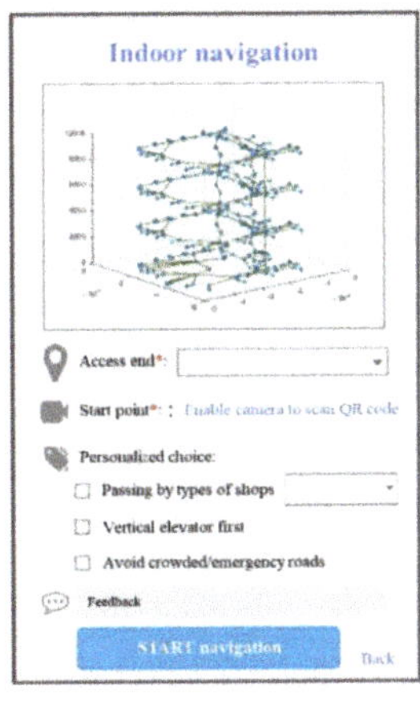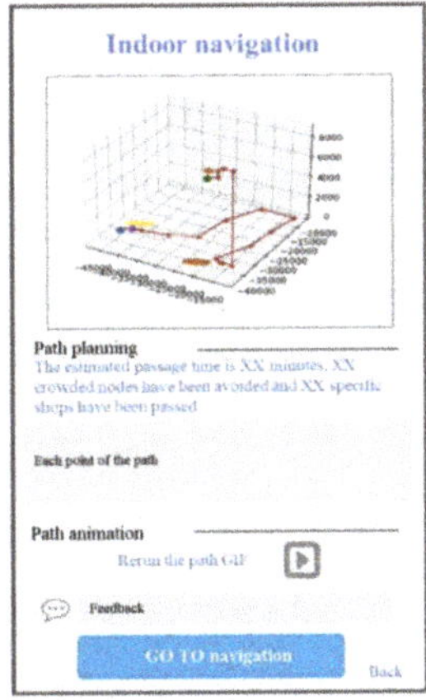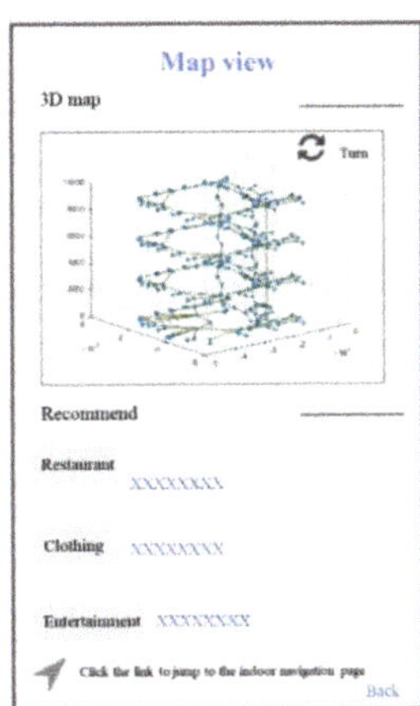

Figure 20. Shopping mall indoor navigation system function page.

5. Conclusions

This paper provides a detailed analysis of users' general demand for indoor navigation systems for shopping malls. Three important functions that need to be considered in the design of shopping mall indoor navigation system are obtained. According to the questionnaire, this paper also drew some interesting conclusions, which will benefit further studies on indoor navigation systems for shopping malls: (1) Both vertical elevators and horizontal escalators are important means of vertical transportation in shopping malls. Although vertical elevators are more difficult to find than horizontal escalators, users prefer to use vertical elevators, which may be related to users' beliefs that the vertical elevator is faster. (2) Users prefer that the navigation system can indicate the passage to the desired space when reaching a certain destination, such as buying milk tea and other beverages along the way when going to the cinema, but users who are not as interested will avoid certain spaces. (3) This paper finds the correlation between "gender", "age", "occupation", and users' behaviors in shopping malls. For example, due to the singularity of moving lines, retired groups do not care about the indoor navigation signs of shopping malls. (4) Retirees over the age of 50 are not the target users of indoor navigation systems in shopping malls.

There are still some deficiencies in this paper. The functions formulated in the questionnaire survey are still insufficient, and the multi-source heterogeneous data fusion and application in the design of shopping mall indoor navigation systems are not fully explored. In future work, more in-depth research will be carried out on these two aspects.

Author Contributions: Y.X.: Methodology (lead), formal analysis, investigation, writing—original draft. H.D.: Resources, validation, writing—review and editing, funding acquisition. Y.D.: Conceptualization (lead), methodology, writing—review and editing, funding acquisition. All authors have read and agreed to the published version of the manuscript.

Funding: This research was funded by Guangdong Science Foundation (Grant No. 2022A1515010174), support from the State Key Lab of Subtropical Building Science, South China University of Technology (No. 2022ZB19), and the support from the Guangzhou Science and Technology Program (No. 202201010338).

Data Availability Statement: Not applicable.

Conflicts of Interest: The authors declare no conflict of interest.

References

1. Geoinformation; Reports from International Institute of Information Technology Describe Recent Advances in Geoinformation. Politics & Government Week 2020. Available online: https://geoinformation.com/ (accessed on 11 August 2022).
2. Luo, H.; Hu, X.; Zou, Y.; Jing, X.; Song, C.; Ni, Q. Research on a reference signal optimisation algorithm for indoor Bluetooth positioning. *Appl. Math. Nonlinear Sci.* **2021**, *6*, 525–534. [CrossRef]
3. Li, J.; Wang, Q.; Xuan, Y.; Zhou, H. User demands analysis of Eco-city based on the Kano model—An application to China case study. *PLoS ONE* **2021**, *16*, e0248187. [CrossRef] [PubMed]
4. Mirahadi, F.; McCabe, B.; Shahi, A. IFC-centric performance-based evaluation of building evacuations using fire dynamics simulation and agent-based modeling. *Autom. Constr.* **2019**, *101*, 1–16. [CrossRef]
5. Liu, L.; Li, B.; Zlatanova, S.; van Oosterom, P. Indoor navigation supported by the Industry Foundation Classes (IFC): A survey. *Autom. Constr.* **2021**, *121*, 103436. [CrossRef]
6. Liu, Z.; Zhang, L.; Liu, Q.; Yin, Y.; Cheng, L.; Zimmermann, R. Fusion of Magnetic and Visual Sensors for Indoor Localization: Infrastructure-Free and More Effective. *IEEE Trans. Multimed.* **2016**, *19*, 874–888. [CrossRef]
7. Farahsari, P.S.; Farahzadi, A.; Rezazadeh, J.; Bagheri, A. A Survey on Indoor Positioning Systems for IoT-Based Applications. *IEEE Internet Things J.* **2022**, *9*, 7680–7699. [CrossRef]
8. Deng, H.; Ou, Z.; Zhang, G.; Deng, Y.; Tian, M. BIM and Computer Vision-Based Framework for Fire Emergency Evacuation Considering Local Safety Performance. *Sensors* **2021**, *21*, 3851. [CrossRef] [PubMed]
9. Lee, Y.H.; Medioni, G. RGB-D camera based wearable navigation system for the visually impaired. *Comput. Vis. Image Underst.* **2016**, *149*, 3–20. [CrossRef]
10. Schuldt, C.; Shoushtari, H.; Hellweg, N.; Sternberg, H. L5IN: Overview of an Indoor Navigation Pilot Project. *Remote Sens.* **2021**, *13*, 624. [CrossRef]
11. Yao, D.; Park, D.-W.; An, S.-O.; Kim, S.K. Development of Augmented Reality Indoor Navigation System based on Enhanced A* Algorithm. *KSII Trans. Internet Inf. Syst.* **2019**, *13*, 4606–4623. [CrossRef]
12. Bian, W.; Guo, Y.; Qiu, Q. Research on Personalized Indoor Routing Algorithm. In Proceedings of the 2014 13th International Symposium on Distributed Computing and Applications to Business, Engineering and Science, Xi'an, China, 24–27 November 2014; pp. 275–277. [CrossRef]
13. Maaouane, M.; Zouggar, S.; Krajačić, G.; Zahboune, H. Modelling industry energy demand using multiple linear regression analysis based on consumed quantity of goods. *Energy* **2021**, *225*, 120270. [CrossRef]
14. Zeferina, V.; Wood, F.R.; Edwards, R.; Tian, W. Sensitivity analysis of cooling demand applied to a large office building. *Energy Build.* **2021**, *235*, 110703. [CrossRef]
15. Ozdemir, O.; Kizildag, M.; Dogru, T.; Madanoglu, M. Measuring the effect of infectious disease-induced uncertainty on hotel room demand: A longitudinal analysis of U.S. hotel industry. *Int. J. Hosp. Manag.* **2022**, *103*, 103189. [CrossRef]
16. Hopper, E. Maslow's Hierarchy of Needs Explained. *ThoughtCo ThoughtCo* 2020, 24. Available online: https://www.thoughtco.com/maslows-hierarchy-of-needs-4582571 (accessed on 11 August 2022).
17. Leigh, D. SWOT Analysis. In *Handbook of Improving Performance in the Workplace: Volumes 1–3*; Wiley Online Library: Hoboken, NJ, USA, 2009; pp. 115–140. [CrossRef]
18. Tao, Z.-Q.; Shi, A.-M. Application of Boston matrix combined with SWOT analysis on operational development and evaluations of hospital development. *Eur. Rev. Med. Pharmacol. Sci.* **2016**, *20*, 2131–2139. [PubMed]
19. Sammut-Bonnici, T.; Galea, D. PEST Analysis. 2014. Available online: https://www.um.edu.mt/library/oar/bitstream/123456789/21816/1/sammut-bonnici%20pest.pdf (accessed on 11 August 2022).
20. Asian, S.; Pool, J.K.; Nazarpour, A.; Tabaeeian, R.A. On the importance of service performance and customer satisfaction in third-party logistics selection: An application of Kano model. *Benchmarking Int. J.* **2019**, *26*, 1550–1564. [CrossRef]
21. Yao, M.-L.; Chuang, M.-C.; Hsu, C.-C. The Kano model analysis of features for mobile security applications. *Comput. Secur.* **2018**, *78*, 336–346. [CrossRef]
22. Chen, C.-C.; Chuang, M.-C. Integrating the Kano model into a robust design approach to enhance customer satisfaction with product design. *Int. J. Prod. Econ.* **2008**, *114*, 667–681. [CrossRef]
23. Hashim, A.; Dawal, S.Z. Kano Model and QFD integration approach for Ergonomic Design Improvement. *Procedia-Soc. Behav. Sci.* **2012**, *57*, 22–32. [CrossRef]
24. Materla, T.; Cudney, E.A.; Antony, J. The application of Kano model in the healthcare industry: A systematic literature review. *Total Qual. Manag. Bus. Excel.* **2019**, *30*, 660–681. [CrossRef]
25. Ganz, A.; Schafer, J.; Gandhi, S.; Puleo, E.; Wilson, C.; Robertson, M. PERCEPT Indoor Navigation System for the Blind and Visually Impaired: Architecture and Experimentation. *Int. J. Telemed. Appl.* **2012**, *2012*, 894869. [CrossRef] [PubMed]
26. Giudice, N.A.; Whalen, W.E.; Riehle, T.H.; Anderson, S.M.; Doore, S.A. Evaluation of an accessible, real-time, and infrastructure-free indoor navigation system by users who are blind in the mall of america. *J. Vis. Impair. Blind.* **2019**, *113*, 140–155. [CrossRef]
27. Li, M.; Arning, K.; Sack, O.; Park, J.; Kim, M.-H.; Ziefle, M.; Kobbelt, L. Evaluation of a Mobile Projector-Based Indoor Navigation Interface. *Interact. Comput.* **2013**, *26*, 595–613. [CrossRef]
28. Zhang, Z.; He, S.; Shu, Y.; Shi, Z. A Self-Evolving WiFi-based Indoor Navigation System Using Smartphones. *IEEE Trans. Mob. Comput.* **2019**, *19*, 1760–1774. [CrossRef]

29. Zhou, Y.; Chen, H.; Zhang, Y.; Huang, Y.; Zhang, P.; Yang, W. Multi-objective indoor path planning method with dynamic environment awareness. *J. Southwest Jiaotong Univ.* **2019**, *54*, 611–618. Available online: https://en.cnki.com.cn/Article_en/CJFDTotal-XNJT201903022.htm (accessed on 11 August 2022).
30. Basu, N.; Haque, M.; King, M.; Kamruzzaman; Oviedo-Trespalacios, O. A systematic review of the factors associated with pedestrian route choice. *Transp. Rev.* **2021**, *42*, 672–694. [CrossRef]
31. Noriaki, K.; Seraku, N.; Takahashi, F.; Tsuji, S. Attractive Qaulity and Must-Be Quality. *J. Jpn. Soc. Qual. Control* **1984**, *14*, 39–48. [CrossRef]
32. Ivarsson, L.; Nilsson, A.; Rimfält, T. *Customer Satisfaction—An Investigation of Trivselhus*; LAP LAMBERT Academic Publishing: Chisinau, Moldova, 2010.
33. Berger, C.; Blauth, R.; Boger, D. Kano's Methods for Understanding Customer-Defined Quality 1993. Available online: https://www.semanticscholar.org/paper/KANO (accessed on 11 August 2022).
34. Wang, T.; Ji, P. Understanding customer needs through quantitative analysis of Kano's model. *Int. J. Qual. Reliab. Manag.* **2010**, *27*, 173–184. [CrossRef]
35. Lee, M.C.; Newcomb, J.F. Applying the Kano methodology to meet customer requirements: NASA's microgravity science program. *Qual. Manag. J.* **1997**, *4*, 95–106. [CrossRef]
36. Jang, H.Y.; Song, H.G.; Park, Y.T. Determining the importance values of quality attributes using ASC. *J. Korean Soc. Qual. Manag.* **2012**, *40*, 589–598. [CrossRef]
37. Wu, C.X. Experience-Based Customer Requirement Study in Emporium. 2008. Available online: https://www.dissertationtopic.net/doc/1398157 (accessed on 11 August 2022).

Article

A Comparative Study on the Influence of Different Decoration Styles on Subjective Evaluation of Hotel Indoor Environment

Jian Xu [1,2,3], Muchun Li [1,3,4,*], Dandan Huang [1], Yuxin Wei [1] and Sijia Zhong [1]

[1] Department of Tourism Management, South China University of Technology, Guangzhou 510006, China
[2] State Key Laboratory of Subtropical Building Science, Guangzhou 510006, China
[3] Key Laboratory of Digital Village and Sustainable Development of Culture and Tourism, Guangzhou 510006, China
[4] Guangdong Tourism Strategy and Policy Research Center, Guangzhou 510006, China
* Correspondence: limch@scut.edu.cn; Tel.: +86-181-4288-6885

Abstract: The purpose of this study is to explore the occupants' subjective evaluation of the indoor environmental quality (IEQ) of hotels with the same physical environment and different decoration styles, and to reveal the influence of different decoration styles on the subjective evaluation of the indoor environmental quality. The study found a hotel with three mainstream styles of modern simple style, British pastoral style, and modern Japanese style, and adopted standard rooms with the same area, pattern, lighting, orientation, and decoration cost. The only variable controlled was the decoration style, and the subjective feelings of customers on the physical environment were investigated. Based on the literature and 604 online comments, the researchers designed a questionnaire and collected 710 effective questionnaires for empirical analysis. The analysis results of KH coder and SPSS software (Chicago, IL, USA) show that ① the light environment in the indoor environment (including indoor natural lighting, lighting and other influencing factors) and non-light visual factors (including indoor color matching, plant layout, closeness to nature, decoration texture, space materials, decoration atmosphere and other factors) has the greatest impact on the subjective evaluation of decoration style, especially on the subjective evaluation of modern simple indoor environment. ② Light environment, air quality and non-light visual factors play a key role in the subjective evaluation of the indoor environment of the British pastoral-style hotels. ③ The light environment, thermal environment and non-light visual factors are the most sensitive to the subjective evaluation of the indoor environment of modern Japanese-style hotels. ④ Thermal environment, light environment, acoustic environment, air quality environment and non-light visual factors have the greatest impact on the subjective evaluation of the hotel indoor environment. Based on the findings, this study puts forward some suggestions to improve the interior environment of the hotel with different decoration styles to improve the quality and attractiveness of the hotel.

Keywords: hotel indoor environment; decoration style; text analysis; questionnaire survey; subjective evaluation; comparative study; China

Citation: Xu, J.; Li, M.; Huang, D.; Wei, Y.; Zhong, S. A Comparative Study on the Influence of Different Decoration Styles on Subjective Evaluation of Hotel Indoor Environment. *Buildings* **2022**, *12*, 1777. https://doi.org/10.3390/buildings12111777

Academic Editors: Yue Wu, Zheming Liu and Zhe Kong

Received: 14 August 2022
Accepted: 24 September 2022
Published: 24 October 2022

Publisher's Note: MDPI stays neutral with regard to jurisdictional claims in published maps and institutional affiliations.

1. Introduction

With the rapid development of China's tourism industry, the number of hotels has also increased rapidly. From 2016 to 2019, the number of hotel rooms in China increased to 17.62 million, with an annual growth rate of 7.71% [1]. Subsequently, user experience has become the focus of competition among hotels. With the increasing demand of consumers for hotels, hotels with different decoration styles are widely popular to meet the aesthetic needs of consumers. Customers' perception of the overall value of the hotel mainly comes from the indoor environmental conditions of the hotel. The physical environment is a key function of consumer psychology, which links consumers with tangible products and services [2]. The physical environment has a positive impact on the mood, satisfaction, and behavioral intentions of residents [3]. With the increasing demand for hotel characteristics in the tourism market, hotel rooms with various decoration styles came into being. Different decoration styles mean different decoration materials, furniture layouts, and so on. Therefore, we get inspiration on whether different decoration styles affect customers' subjective evaluation of the hotel. This research question can provide valuable suggestions for hotel interior environment design, which is beneficial to the development of China's tourism industry. The National Institute for Occupational Safety and Health of the United States has defined indoor environmental quality (IEQ), including the comprehensive psychological and physiological effects of five aspects of acoustic, thermal, light environment, air quality, and non-light visual factors on the residents [4]. Studies have shown that the values hotel customers care about most are seven dimensions, including thermal environment comfort, visual environment comfort, and indoor air quality [5,6]. Existing studies have preliminarily explored the impact of hotel aesthetic design on the subjective feelings of hotel consumers [7,8], but few scholars have paid attention to the impact of hotel decoration style on the subjective feelings of consumers. Therefore, the purpose of this study is to explore the residents' subjective evaluation of hotel rooms with the same physical environment but different decoration styles and to reveal the impact of different decoration styles on the subjective evaluation of various indoor environments (Table 1).

In terms of theoretical research on hotel indoor design, hotel decoration style, i.e., "atmosphere", is considered to be extremely important by hotel operators, hotel designers, and architects. Similarly, it also has a crucial impact on customer satisfaction [9]. Many studies have explored the research field of indoor landscape, including the hotel indoor environment transformation through different decoration materials, decoration environment, and plant configuration [10–12]. With the refined development of the hotel industry and people's continuous pursuit of experience, some scholars have provided ideas on the space layout, furniture and materials, theme styles, and green environmental protection of the hotel interior design from the perspective of subjective evaluation [13–16]. In terms of the indoor environment, there are many professional studies concerning the general architecture of the physical space, but the hotel indoor decoration style rarely becomes the objective of study [17–19] (Table 1).

Table 1. Literature review framework.

Topics	Content	Authors and Published Year	Limitations
	"atmosphere" has a crucial impact on customer satisfaction.	Heide, M. et al. (2007) [9]	
Hotel indoor design	The hotel indoor environment transformation through different decoration materials, decoration environment and plant configuration.	Song, J.H. (2005) [10]; Chen, Y.X. (2015) [11]; Han, Y.Y. et al. (2013) [12]	There are many professional studies concerning the general architecture of the physical space, but the hotel indoor decoration style rarely becomes the objective of study.
	Providing ideas on the space layout, furniture and materials, theme styles and green environmental protection of the hotel interior design from the perspective of subjective evaluation.	Nanu, L. et al. (2020) [13]; Xu, Z.X. et al. (2020) [14]; Han, H. (2020) [15]; Kim, J.J. et al. (2022) [16]	

Table 1. *Cont.*

Topics	Content	Authors and Published Year	Limitations
Subjective evaluation of hotel rooms	Indoor air quality and thermal environment have the greatest impact on the online rating of hotels.	Shen, Z.F. et al. (2021) [20]	Single index of IEQ is concerned, but IEQ has rarely been paid attention to as a whole in the research of hotel evaluation systems.
	Location and accessibility, employee performance and room quality are the most important factors of customer satisfaction.	Xu, X. et al. (2016) [21]	
	The impact of online comments on customer satisfaction from the technical level of comments.	Zhao, Y.B. et al. (2019) [22]	
	Providing a new idea for the air quality management of the hotel.	Zanni, S. et al. (2021) [23]; Cociorva, S. et al. (2017) [24]	
	Measuring the acoustic environment of Okura Hotel in Tokyo and explaining the reasons for its reputation.	Nojima, R. et al. (2020) [25]	
	The thermal environment and air quality level of the hotel have an impact on the hotel room satisfaction.	Han, H. et al. (2019) [26]	
Hotel indoor environment evaluation system	A hotel lighting quality evaluation system and a lighting technology system for controlling lighting effects.	Zhang, Y.F. (2005) [27]	The indoor environment evaluation system is proposed, but the subjective evaluation of customers is not paid attention to.
	A TAIL rating scheme to obtain the overall quality rating of IEQ according to the quality of four indicators.	Wargocki, P. et al. (2021) [28]	
	Putting forward a "Feng Shui" evaluation system through experimental methods, and verifying that the evaluation system of "Feng Shui" in China is scientific.	Jin, Z. et al. (2021) [29]	
	The average contribution of thermal, acoustic, lighting environment and air quality parameters to the overall IEQ rating of buildings was 27%, 17%, 22%, and 34%, respectively.	Wei, W. et al. (2020) [30]	

In the aspect of subjective evaluation of hotel rooms, many studies explore the influencing factors of customers' satisfaction with hotel IEQ through methods of analysis of online word-of-mouth. Shen et al. evaluated the IEQ of the five major chain economy hotels in China through online comment text mining and found that for economy hotels, indoor air quality and thermal environment have the greatest impact on the online rating of hotels [20]. Xu et al. explored the factors of hotel customer satisfaction and dissatisfaction and their importance ranking and found that location and accessibility, employee performance and room quality are the most important factors of customer satisfaction [21]. Zhao et al. investigated the impact of online comments on customer satisfaction from the technical level of comments [22]. At the same time, some scholars have paid attention to the influence of various indicators of the hotel indoor environment on customer satisfaction or comfort degree. Zanni et al. tested the air quality of hotel rooms under the background of epidemic prevention and control, thus providing a new idea for the air quality management of the hotel [23]; Cociorva et al. monitored and regulated the indoor air quality of the hotel utilizing technical operation [24]; Nojima et al. measured the acoustic environment of Okura Hotel in Tokyo and explained the reasons for its reputation [25]; Han et al. concluded through experiments that the thermal environment and air quality level of the hotel have an impact on the hotel room satisfaction [26]. In the research of hotel indoor environment evaluation systems, many researchers have innovated and put forward new IEQ evaluation systems based on existing ones. Zhang proposed a hotel lighting quality evaluation system and a lighting technology system for controlling lighting effects through field research [27]. Wargocki et al. proposed a TAIL rating scheme to obtain the overall quality rating of IEQ according to the quality of four indicators [28]. Jin et al. put forward a "Feng Shui" evaluation system through experimental methods and verified that the evaluation system of "Feng Shui" in China is scientific [29]. Wei found that the average contribution of thermal, acoustic, lighting environment, and air quality parameters to the overall IEQ rating of buildings were 27%, 17%, 22%, and 34%, respectively, through the parameters of IEQ evaluation in the green building certification plan [30]. Although IEQ has been paid attention to as a whole in the research of hotel evaluation systems, its influence on the subjective evaluation of hotel consumers for decoration style still needs to be supplemented (Table 1).

According to the existing literature, IEQ mainly includes five factors: acoustic, light, thermal environment, air quality and non-light factors. This study explores the satisfaction of consumers with the above five factors through questionnaires. The acoustic environment comfort refers to the existence of a comfortable acoustic environment without any uncomfortable noise. The noise of the hotel is mainly affected by the noise of the external environment of the hotel, the noise generated by the occupants, and the noise generated by the operation of the hotel equipment. Therefore, the measurement of the acoustic environment mainly includes the road traffic noise, the noise of other occupants, and the noise generated by the operation of various equipment. The light environment comfort is related to illuminance uniformity, brightness distribution, and other factors. The natural lighting and lighting of the hotel mainly affect the light environment of the hotel. Therefore, the light environment mainly focuses on the natural lighting and lighting of the hotel [31]. The thermal environment comfort is expressed as "psychological conditions that are satisfied with the thermal environment", which is related to the physical factors in the natural ventilation and regulating environment. Therefore, the thermal environment is the measurement of the room temperature, indoor wind speed, humidity, and electrical heat dissipation of the hotel. The indoor air quality has a great impact on the health of the residents. It is directly related to the ventilation rate and the concentration of pollutants, and the supply of outdoor air can provide acceptable indoor air quality. The hotel occupants primarily judge the air quality of the hotel by the odor and ventilation of the hotel air. Therefore, the measurement of air quality mainly includes odor, air purification facilities, and indoor ventilation [32]. According to the analysis of the hotel review text, the non-light visual factors in this study refer to the visual factors that can affect the subjective evaluation of customers excluding the light environment in the hotel indoor environment, mainly including the internal color matching, plant layout, closeness to nature, decoration texture, space design, decoration atmosphere, and other factors.

Based on the above literature, it can be concluded that in the research on the hotel interior environment, many research objects focus on the hotel interior building materials, plant furnishings, furniture, etc. Additionally, some attach importance to "atmosphere", but few scholars quantify the difference in subjective feelings brought by the hotel decoration style. In terms of the research on the subjective evaluation of customers, scholars mainly use the network text as the research data source to explore the factors that affect hotel satisfaction and do not give a detailed description of the index impact of each dimension. Therefore, the research on the relationship between hotel interior decoration style and customers' subjective feelings needs to be supplemented and improved.

The innovations of this study are as follows:① although the literature has explored the influencing factors of customer satisfaction, decoration style is rarely paid attention to as an influencing factor. Therefore, the innovation of this study is to link the decoration style with the subjective evaluation of customers, collect the customer evaluation data using the questionnaire method, and explore the impact of decoration style on the subjective evaluation of customers. ② To date, many studies have focused on the indicators of the indoor environment, such as air quality and acoustic environment. However, these studies only focus on the individual indicators of the indoor environment. This study takes the five indicators of the indoor environment as a whole to explore the overall evaluation of the indoor environment by customers under different decoration styles. This research is very lucky to find a popular multi-style hotel in Guangzhou that has just been renovated, which fully meets our research needs to explore the impact of different decoration styles on the subjective feelings of occupants in the same physical environment. The research first obtains the online evaluation text of the hotel through the website and uses KH coder software to conduct word frequency analysis and semantic network analysis on the online evaluation text, puts forward hypotheses, and designs online questionnaires. A questionnaire survey on the subjective evaluation of 856 guests who have stayed in the hotel was conducted. SPSS quantitative analysis software was used for empirical analysis to reveal the impact and difference of different decoration styles on the subjective evaluation of the guests.

The problem definitions of this study are: firstly, whether the hotel indoor environment (including acoustic environment, thermal environment, light environment, air quality, and non-light visual factors) will have a positive impact on the subjective evaluation of customers; second, whether there are differences in customers' evaluation of IEQ factors under different decoration styles; thirdly, in the hotel environment with different decoration styles, what is the most influential factor in the subjective evaluation of customers.

Our research contributions have three points: ① exploring the relationship between different decoration styles and customers' subjective evaluation, which to a certain extent makes up for the research gap on the impact of decoration styles on customer satisfaction. ② Starting from the IEQ factors, we explore the subjective evaluation of customers on five factors, namely, acoustic environment, thermal environment, light environment, air quality, and non-light visual factors, under the environment of different decoration styles, and then explore whether there are differences in the subjective evaluation of customers under different decoration styles. ③ Through the questionnaire survey, the most important factors for customers in each decoration style of hotel rooms are obtained, which provides theoretical guidance for hotels to improve service quality and customer satisfaction.

2. Methodology

2.1. Method Design

To make the results of this study more objective and specific, the study used the online survey data to conduct text analysis to prepare for the questionnaire design. First, to make the final questionnaire more suitable for the hotel in Guangzhou, we adopted an online text analysis method. A questionnaire was designed in combination with a literature review. At the same time, combined with on-the-spot research, use the Questionnaire Star to conduct a questionnaire survey (the research framework of this paper is shown in Figure 1).

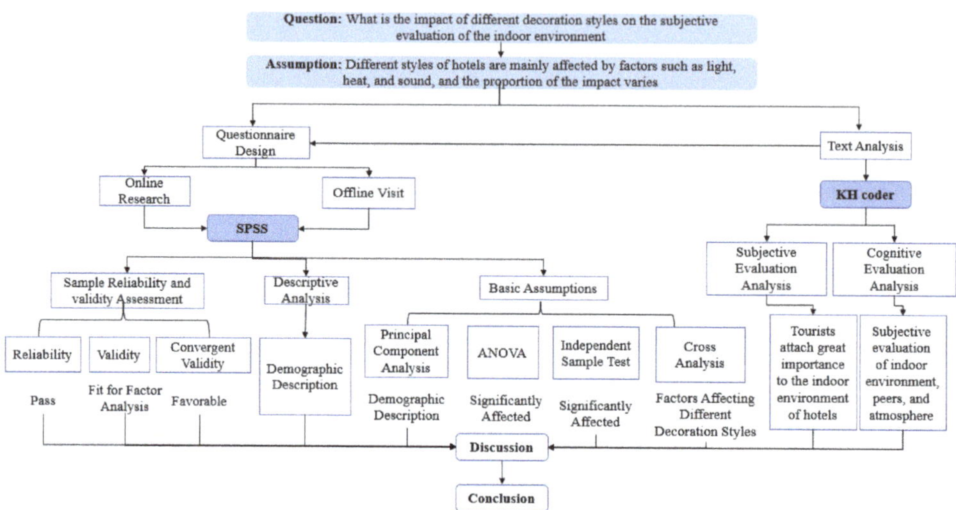

Figure 1. The technical route framework for research.

To highlight the influence of decoration style on guest room evaluation and reduce the influence of other physical factors on the subjective evaluation of the indoor environment of hotels with different decoration styles, the Shiyu Hotel in Guangzhou was selected as the main body of the investigation through network search and field inspection. Project case is located in the city center of China's first-tier cities. It has 278 rooms and possesses a superior geographical environment and a large passenger flow. The renovation time of the hotel is one year, and the complaint rate of management and equipment services is low (Figure 2). Hotel room decoration styles include three mainstream decoration styles: modern simple, British pastoral style, and modern Japanese style (Figure 2). Modern simple style is one of the simple styles. In addition, simple styles include Chinese simple style, Mediterranean simple style, American simple style, European simple style, and Japanese simple style. Pastoral styles include British pastoral style, American pastoral style, Chinese pastoral style, French pastoral style, South Asian pastoral style, and Korean pastoral style. We chose a British pastoral style to differentiate it from the sample's modern simple and modern Japanese styles. Japanese style is roughly divided into three types, simple Japanese style, modern Japanese style, and traditional Japanese style. At the same time, the three styles are relatively close. Then, define the hotel's three styles in terms of color, material, lighting, etc. (Table 2). Selecting standard rooms with the same area (21.24 m^3), layout, lighting, orientation, and decoration cost, the purpose is to control the only variable to conduct a subjective evaluation and comparison study for the decoration. At the same time, the outdoor sunny and cloudy weather is selected for indoor sampling, and the sampling time is from 12:00 noon to 4:00 p.m. When the curtains are drawn, the sampled temperature is basically 26.1–26.3 °C (average 26.2 °C), the noise level is 42–44 decibels (average 43 decibels), and the indoor wind speed is 2.5–2.7 m/s (average 2.6 m/s), the indoor light intensity is 70–75 xl (the light intensity of different decoration styles is different).

Ctrip, as the No. 1 hotel customer review website in China, is highly authoritative and representative. The online text research collected Chinese reviews of the hotel on Ctrip during the Spring Festival from 1 January to 1 March 2022. This research is based on network text materials, using the word frequency analysis and semantic network analysis functions of KH Coder, the analysis conclusions are used as the main basis for designing the questionnaire, and the SPSS results are used for auxiliary verification.

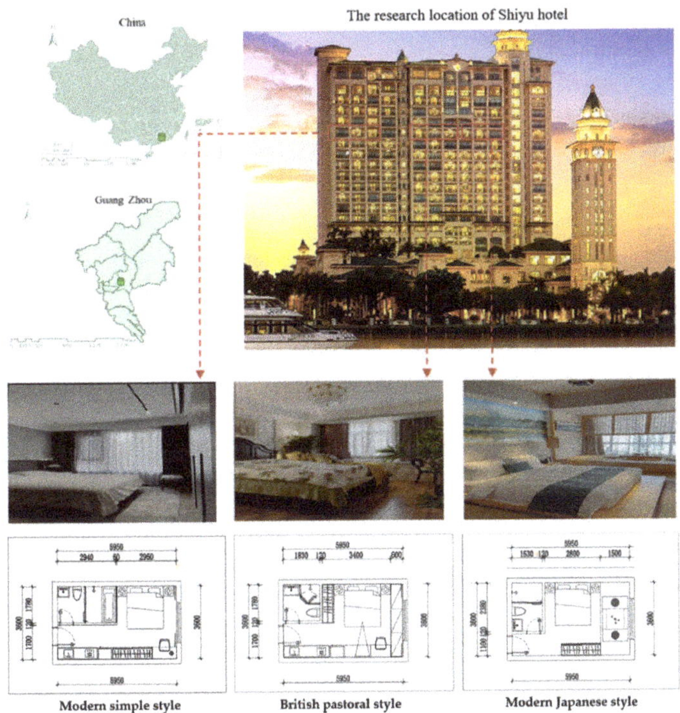

Figure 2. Location, room type map and plans of the case area. Source: bzdt.ch.mnr.gov.cn (accessed on 25 April 2022).

Table 2. Distinguish different hotel decoration styles.

Style	Features			The Case of This Study
	Color	**Material**	**Light**	
Modern simple style	Black, white, gray tones, strong color contrast. Space color matching usually pays attention to the overall unity of space color, people's psychological feelings and the needs of space composition, so as to create a harmonious and comfortable environment [33].	Emphasizing the texture of decorative materials, the choice of materials is also relatively wide.	The light and shadow part is dominated by natural light from glass windows, supplemented by a mix of artificial light and artificial natural light.	
British pastoral style	It pays great attention to the delicacy of the pattern and the softness of the color. Gives a fresh and comfortable feeling [34].	Rustic murals, luxurious chandeliers, lots of floral fabrics and more. Most of the furniture is handcrafted and the material is mainly solid wood.	The principle of asymmetry, the light is softer. Lamp lines are generally curved.	

Table 2. *Cont.*

Style	Features			The Case of This Study
	Color	**Material**	**Light**	
Modern Japanese style	The colors are simple, restrained and elegant [35].	Use more wood materials, such as natural wood, stone, bamboo, etc. Metal is rarely used.	The light is softer, and the interior space environment brought by it tends to be elegant and soft due to the blending of light and shadow, with a hazy beauty.	

2.2. Text Analysis

KH Coder is an unstructured text analysis and data mining software. The main functions of the software include word frequency, cluster analysis, etc. Restricted by statistical laws, a few reviews cannot objectively and truly reflect hotel customer satisfaction. Therefore, decoration styles below 200 are not involved in hotel customer satisfaction rankings or analysis. By eliminating invalid reviews and choosing the hotel's three decoration styles: modern simple, British pastoral, and modern Japanese, a total of 604 physiological and psychological review data were obtained. Through KH Coder, word frequency analysis and semantic network analysis are carried out on the obtained reviews, then the environmental factors that may affect the subjective evaluation of residents are extracted, and subjective evaluation analysis is carried out from the aspects of the overall evaluation and cognitive evaluation.

① Overall subjective evaluation analysis

KH Coder extracts the text and finds that the total distribution of all high-frequency words evaluated by customers is shown in Table 3. By setting stop words and eliminating invalid prepositions and pronouns, high-frequency words are mostly nouns and adjectives, and nouns are room, hotel, environment, clean, service, etc., adjectives mainly reflect the indoor environment and hotel services. In addition, the hotel appeared the most frequently, appearing 417 times in 604 comments. Regarding the decoration style of the hotel, there are adjectives Japanese (51), pastoral (33), idyllic (18), rustic (9), modern simple (36), and the noun style (105), which confirms the customer's attention and inclination to the decoration style.

Table 3. Total distribution table of all high-frequency words for customer reviews.

Class	Word	Frequency	Word	Frequency	Word	Frequency
	Clean	198	Bad	81	Strong	51
	Comfortable	141	Free	75	Hot	48
Adj	Front	123	Beautiful	69	Sound	48
	Convenient	87	Nice	63	Quiet	45
	Poor	87	Japanese	51	Suitable	45
	Hotel	417	Lot	96	Morning	63
	Service	267	Price	87	Attitude	60
Noun	Time	189	Door	84	Decoration	57
	Water	183	Child	81	Sea	57
	Desk	159	Facility	81	Bathroom	54

Table 3. *Cont.*

Class	Word	Frequency	Word	Frequency	Word	Frequency
	Night	138	Floor	81	Bit	54
	Environment	132	Pool	75	Lighting	54
	Style	105	Smell	75	Bed	51
Noun	Homestay	102	Location	72	Parking	51
	Air	99	People	72	House	48
	Day	99	Experience	66	Money	48
	Breakfast	96	Insulation	66	Friend	45
	Stay	147	Choose	69	Play	57
Verb	Recommend	78	Live	69	Check	54
	Feel	72	Eat	63	Change	48
	Sleep	72	Book	57	Shower	45

② Cognitive subjective evaluation analysis

Through the cluster analysis of the co-occurrence network of the high-frequency words, modular processing is selected, and a total of fourteen groups of relational networks are obtained (Figure 3). Comparing and analyzing the co-occurrence network graph and word frequency, it can be seen that the influencing factors of subjective evaluation are mainly manifested in the indoor environment of the destination, travel companions, and the environment and services provided by the hotel:

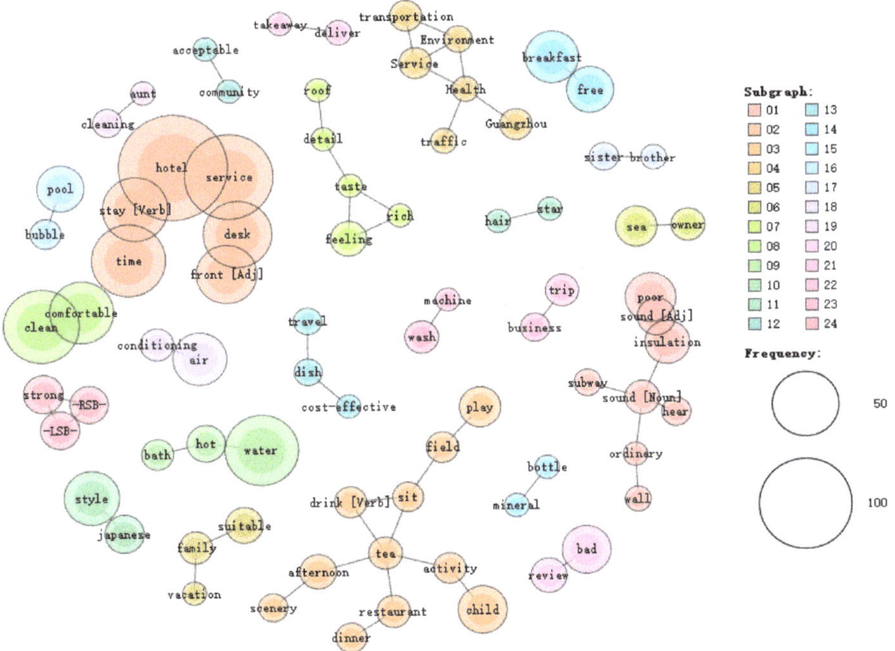

Figure 3. Co-existing semantic network distribution map.

The 24 groups of rings were sorted into three groups: First, cognitive evaluation of the indoor environment. With the hotel as the core, the words such as room and homestay are expanded as a circle. The above vocabulary is a group, including facilities, location, sound insulation, air conditioning, taste, comfort, etc., which reflects the influence of the indoor environment on subjective evaluation. Second, traveler's cognitive evaluation of fellow travelers. Most tourists travel together, and the related high-frequency words are children, family members, friends, etc. Third, cognitive evaluation of the hotel environment atmosphere. Most people think that the hotel's service is friendly, and the overall accommodation is comfortable. The most frequent texts include comfort, caring, completeness, thoughtfulness, and details. The above three points combined with the relevant literature provide the design direction for the questionnaire design.

2.3. Questionnaire Design

By combing the results of online texts and analyzing the evaluation dimensions of existing research [36], the questionnaire has 8 dimensions, 29 attributes, and 52 questions. The research mainly adopts Choi, K.'s indoor environmental quality (IEQ) evaluation system to establish the subjective evaluation dimension.

In addition to the hotel's factor variables required for this research, there are also differences between individual tourists. Appendix A summarizes the measurement properties of the relevant variables. In the questionnaire, the satisfaction selection was used to evaluate the indoor environmental quality, including the overall environment, heat, air quality, light, sound, and non-light visual factors. The overall customer satisfaction with the comprehensive evaluation of the hotel was also investigated. In each choice, a score scale or option is provided, through specific questions and their corresponding scale notes, the chosen option expresses the guest's subjective feelings.

The importance and satisfaction of subjective evaluations of different types of hotels were assessed using the Likert scale, ranging from extremely unimportant/unsatisfactory (1) to extremely important and extremely satisfactory (5). To ensure the validity and scientific nature of the questionnaire, this study set up a pre-investigation, randomly selecting 30 people among consumers and giving incentives in the form of gift certificates. Then, combined with expert consultation, the questions were supplemented with revisions and explanations to determine the questionnaire.

The questionnaire consists of 4 parts (as shown in Figure 4): (1) screening the respondents' questions, only the respondents who have stayed in the case hotel can continue the investigation, otherwise, they will be guided to an ending. (2) Personal information, such as the respondent's gender, age, occupation, and monthly income. (3) Other relevant information of the respondent, such as the hotel room type, decoration style, traveler, departure place, and reasons for choosing the hotel. (4) Respondents' subjective evaluation of hotel rooms, based on the results of IEQ theory and text analysis, and then the overall, and refinement of satisfaction assessments.

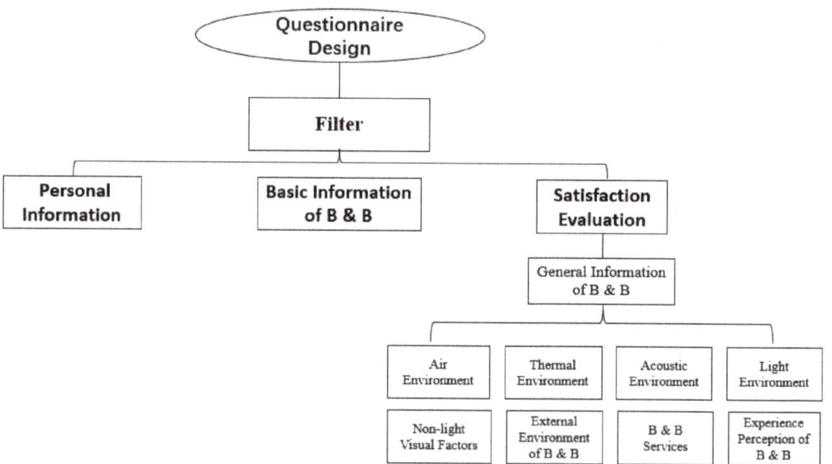

Figure 4. Questionnaire design framework.

To ensure the quality of the questionnaire and prevent invalid questionnaires, options unrelated to the research topic were set in the topic selection. This filters out respondents who did not answer the question. At the same time, the questionnaire is set to answer more than 3 min to be effective.

To avoid order effects [37], different sections of the questionnaire will be randomly adjusted in order.

2.4. Investigation Method

To avoid the limitations and differences between online research and offline research [38], the research adopts online random and offline fixed-point methods for sampling and obtains samples scattered in hotels with different decoration styles. The study selects the Canton Fair in April 2022 (China Import and Export Fair, held in Guangzhou in the spring and autumn every year, a comprehensive international trade fair with the longest history, the highest level, the largest scale, the largest number of buyers and the widest distribution in China) to investigate. The study selects the Canton Fair in April 2022 (China Import and Export Commodities Fair, held in Guangzhou every spring and autumn, as the comprehensive international trade fair with the longest history, highest level, largest scale, and most buyers) and the most widely distributed in China) to investigate. Conduct offline surveys during trade events (peak hotel occupancy). This can not only ensure the richness of customers, but also ensure the diversity of regions and ages of customer samples. Tourists who have stayed in a hotel at least once in the past 12 months and are currently living to constitute the main body of this sample and were selected and invited to participate in the survey. The online survey tool is Questionnaire Star (China's most professional online questionnaire, examination, evaluation, and voting platform, focusing on providing users with services such as online questionnaires, data research, and result analysis), and then the questionnaires are randomly distributed in the hotel private domain fan official account and We Chat (the most popular social software in China) group; offline research, after obtaining the assistance of hotel management personnel, will conduct rewarded interviews for guests who are checking out and who have already checked in. Finally, a total of 856 questionnaires were recovered, including 710 valid questionnaires (620 online and 90 offline), and the questionnaire response rate was 83%.

3. Result

3.1. Assumption Analysis

Based on IEQ theory, and also using KH Coder software to perform text analysis and network visualization analysis on network evaluation texts, the following hypothesis analysis model was proposed (Figure 5):

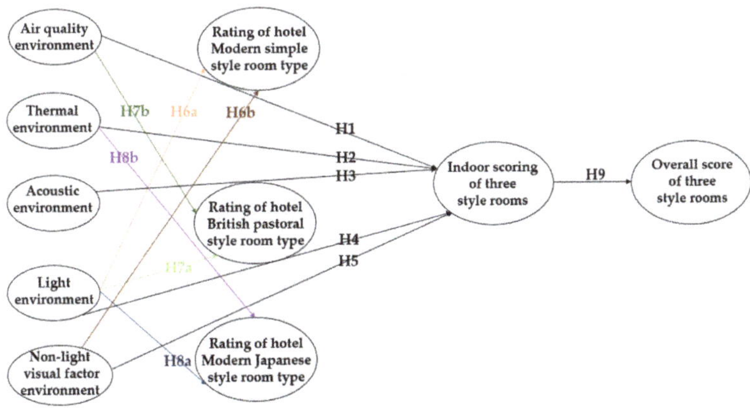

Figure 5. Analysis model.

① Analysis of influencing factors of hotel indoor environment

H1. *Hotel air quality environment will have a positive impact on customers' subjective evaluation of the hotel's indoor environment.*

H2. *Hotel thermal environment will have a positive impact on customers' subjective evaluation of the hotel's indoor environment.*

H3. *Hotel acoustic environment will have a positive impact on the subjective evaluation of the hotel's indoor environment by customers.*

H4. *Hotel light environment will have a positive impact on customers' subjective evaluation of the hotel's indoor environment.*

H5. *Hotel non-light visual factor environment will have a positive impact on customers' subjective evaluation of the hotel's indoor environment.*

② Hotel modern simple style room type and indoor environment influencing factors

Many scholars have found that modern simple style achieves harmony in visual effects with simple and smooth lines, black, white, gray, and log colors, and other non-light visual factors; in terms of light environment factors such as space lighting, it combines overall and local lighting to satisfy the owner. Many studies have confirmed that non-light visual factors and light environment have a greater impact on customers' subjective ratings of the hotel's modern simple style room type C [39,40]. Based on the above research results and questionnaire survey results, the following research hypotheses are proposed:

H6a. *The hotel's light environment will have a positive impact on customers' subjective evaluation of the hotel's modern simple style room type.*

H6b. *The non-light visual factors of the hotel will have a positive impact on customers' subjective evaluation of the hotel's modern simple style room type.*

③ Hotel British pastoral-style room type and indoor environment influencing factors

Many studies have confirmed the important influence of non-light visual factors on customers' subjective evaluation of British pastoral-style room type [41], but few studies have studied the effect of light factors and air quality on the creation of a quiet, natural, pleasant, and relaxed atmosphere in British pastoral-style room type [42]. Through the analysis of KH Coder text of online evaluation, this study finds that the two factors of light and air quality have a great influence on the hotel's British pastoral-style room type, and then put forward the following assumptions:

H7a. *The light environment of the hotel will have a positive impact on customers' subjective evaluation of the hotel's British pastoral-style room type.*

H7b. *The air quality environment of the hotel will have a positive impact on customers' subjective evaluation of the hotel's British pastoral-style room type.*

④ Hotel modern Japanese-style room type and indoor environmental influencing factors

Studies have shown that modern Japanese-style room type uses a variety of retro furniture such as tatami and fan-shaped windows as decorations to highlight elegance and simplicity and go deep into the realm of Zen [43,44]. Therefore, the adjustment of factors such as room temperature and wind speed included in the thermal environment may be of important significance and the Japanese-style house type pays special attention to the use of "light", most of them are mainly warm-toned lights. So, putting forward the following assumptions:

H8a. *The light environment of the hotel will have a positive impact on customers' subjective evaluation of the hotel's modern Japanese-style room type.*

H8b. *The thermal environment of the hotel will have a positive impact on customers' subjective evaluation of the hotel's modern Japanese-style room type.*

⑤ Hotel indoor environment and customer subjective evaluation

Based on the analysis of five influencing factors in IEQ theory, hypotheses are put forward:

H9. *The indoor environment of the hotel will have a positive impact on the overall customer subjective evaluation of the hotel.*

3.2. Data Analysis Method

The framework of this research is shown in Figure 6 below. First, the reliability, validity, and factor analysis of the questionnaire were carried out. After testing the feasibility of factor analysis of the questionnaire, a demographic descriptive analysis of customer reviews was carried out. Then, principal component analysis (PCA) was carried out to determine the influence of various factors of the hotel's indoor environment on the subjective evaluation of tourists [43,45]. The study further applied factor analysis to eliminate the redundancy of correlated variables to reduce the number of available composite data [46]. To reveal the significant influence, significant difference, and joint distribution of different decoration styles on the subjective evaluation of the hotel, the study used SPSS to carry out single factor ANOVA, independent sample *t*-test, and further analysis of the data using the custom table to obtain in-depth data analysis conclusions and verify the hypothesis.

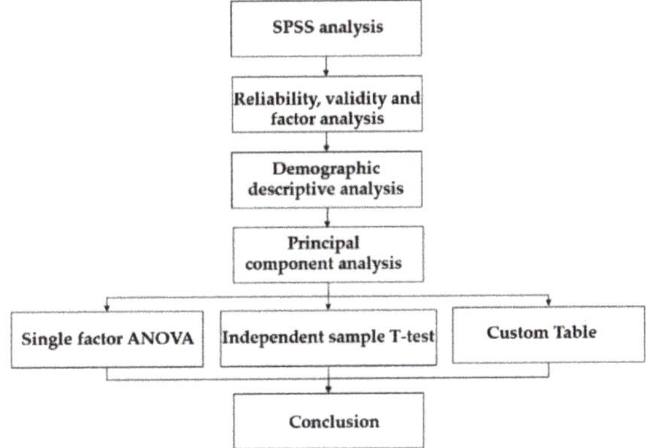

Figure 6. SPSS analysis frame diagram.

3.3. Evaluation of the Reliability and Validity of the Scale

(1) Reliability test

This study mainly used SPSS 25.0 to test the reliability of the scale. The larger the Cronbach's Alpha value, the greater the reliability of the scale. The coefficient should not be less than 0.6, and the questionnaire should be readjusted if it is below 0.6. The test results showed that the coefficient of the actual satisfaction scale was 0.962, and the coefficient of each sub-scale must be greater than 0.8, and most of them were greater than 0.9; then according to the criterion that the overall correlation coefficient (CITC) of the project proposed by the scholar Churchill should not be less than 0.5, the model was trusted. The results showed that the CITC values were all above 0.7, which meant that the scale of this research had a high degree of internal consistency and a small error value in the survey of hotel user groups, that is, the reliability of the questionnaire was high, and the reliability was strong.

(2) Validity test

The KMO test and Bartlett's test of sphericity were used to analyze whether factor analysis was possible. The results of the KMO test showed that the value of KMO was 0.958. At the same time, the results of the Bartlett sphericity test showed that the significant p value was 0.000 ***, infinitely close to 0, and the level was significant, rejecting the null hypothesis, indicating that this questionnaire had validity, and there was the correlation between variables, which was suitable for factor analysis. The factor analysis results showed that there was at least one value greater than 0.4 in the loading values of all the variables in the row, so the variables were reasonable and did not need to be deleted.

(3) Test the convergent validity of the model

Confirmatory factor analysis is performed on the convergent validity of the measurement model. Further evaluation of the structural model can only be performed if the fit of the measurement model meets acceptable standards. The test results showed that the standardized factor loadings of all dimensions were above 0.7 and were significant; the compositional reliability CR values were all higher than 0.8, and the average variance extraction (AVE value) was greater than 0.5. Therefore, the model had good convergent validity.

3.4. Demographic Description of the Sample

First, study participants were screened according to the inclusion criteria; second, questionnaires were collected through screening questions, 856 questionnaires were initially collected, and after excluding outliers, a total of 710 valid questionnaires were used for data analysis. Among them, the majority of respondents were women, accounting for 62%; the age group accounted for the largest proportion of the 18–29-year-old group (93%) because the way such as giving gifts was more attractive to young women; most respondents had a bachelor's degree (88%). For profession, middle school students (84%) accounted for the majority; 77% of all valid participants had a monthly income of less than 3000; this study was mainly based on college students. To be realistic, there are no barriers to the survey respondents, so there are some non-college student groups of respondents. The specific results can be seen in the following Figure 7:

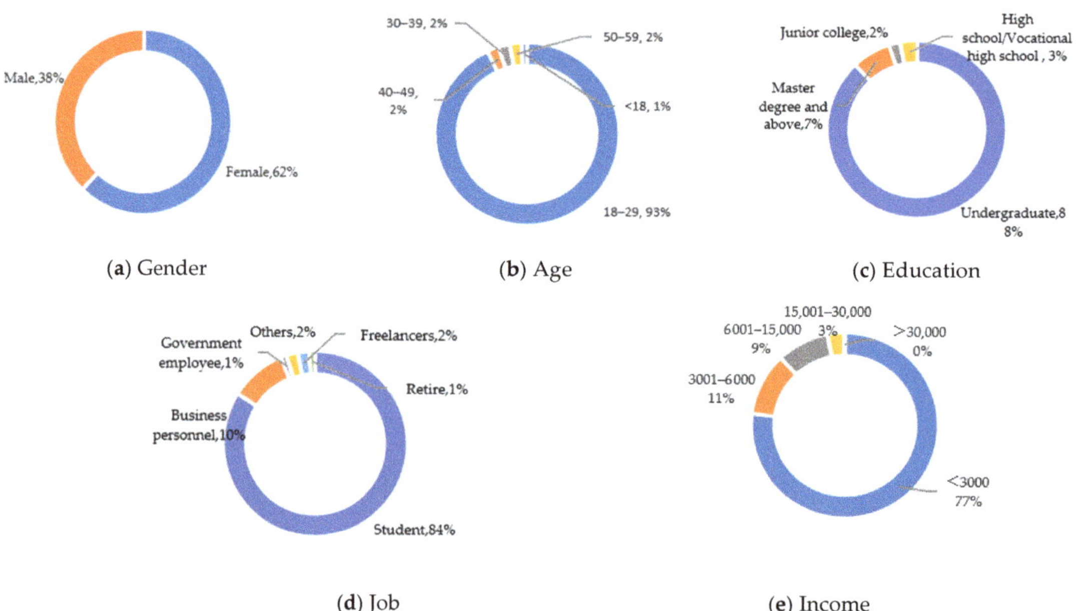

Figure 7. Demographic basic data indicator chart.

Additional relevant demographic information (Figure 8) was measured to further understand respondents. Respondents mostly chose the modern simple style (60.6%), followed by the British pastoral style (22.7%) and the modern Japanese style (16.7%); most respondents traveled with friends (40.8%); leisure travel was the reason why most respondents stayed in hotels (79.3%).

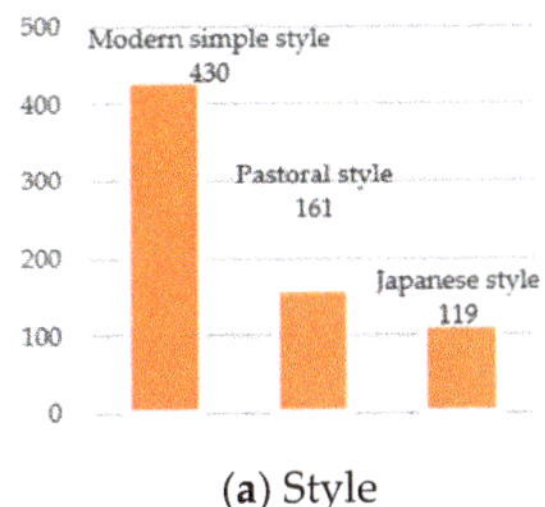

(**a**) Style

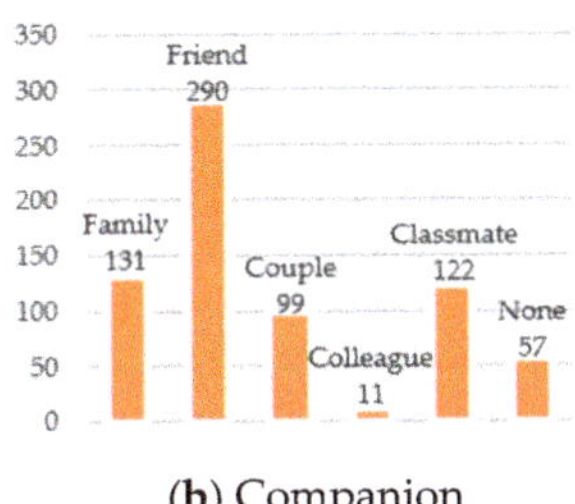

(**b**) Companion

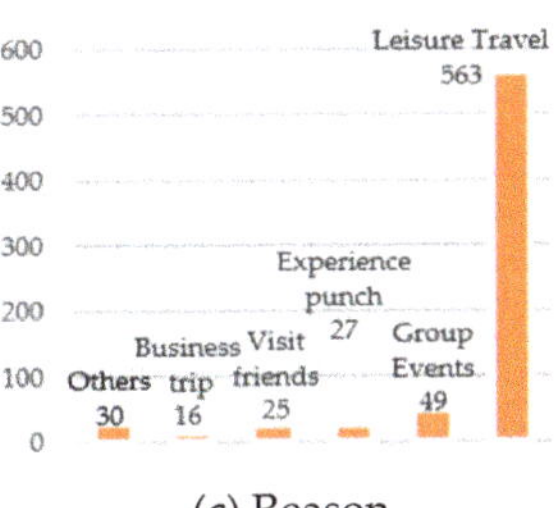

(**c**) Reason

Figure 8. Chart of other population data indicators.

3.5. Basic Assumptions

(1) Principal Component Analysis

Perform principal component analysis and select the maximum variance orthogonal method for rotation, as shown in Tables 4 and 5. For the five principal components extracted from Table 4, the cumulative variance contribution rate reached 73.796%, and the contribution rate of variable explanation was greater than 70%, indicating that the original data can be fully reflected; Table 5 revealed the indicators and proportions of the major components.

Table 4. Subjective evaluation and principal component analysis of hotel room types with different decoration styles.

Constituent	Initial Eigenvalue			Extract the Sum of Squares of the Load			Sum of Squares of Rotating Loads		
	Total	Percent Variance	Cumulative%	Total	Percent Variance	Cumulative%	Total	Percent Variance	Cumulative%
1	18.693	47.932	47.932	18.693	47.932	47.932	8.231	21.106	21.106
2	5.934	15.215	63.147	5.934	15.215	63.147	7.356	18.861	39.967
3	1.859	4.766	67.913	1.859	4.766	67.913	5.931	15.209	55.176
4	1.237	3.173	71.086	1.237	3.173	71.086	3.908	10.020	65.196
5	1.057	2.710	73.796	1.057	2.710	73.796	3.354	8.600	73.796

Table 5. Index and proportion of each component after rotation.

Index	Constituent 1	Index	Constituent 2	Index	Constituent 3	Index	Constituent 4	Index	Constituent 5
Botany arrangement	0.847	The general condition of the thermal environment	0.806	Hotel experience perception	0.892	Surrounding environment	0.518	Noise from other occupants	0.679
Closeness to nature	0.785	Room temperature	0.806	General information about the hotel	0.890	Traffic convenience	0.649	Noise in public areas	0.658
Decorative atmosphere	0.771	Heat dissipation of electrical appliances	0.799	Hotel indoor environment	0.875	Infrastructure	0.623	Overall acoustic environment	0.637
Decoration texture	0.769	Wind velocity	0.779	Overall air quality	0.864	Personal and property safety	0.578	Road traffic noise	0.556
Space design	0.723	Humidity	0.754	Air purification facilities	0.855	Sanitary condition	0.574	Operating noise of various equipment (air conditioner, fan)	0.553
Personalized service	0.718	Lighting	0.601	Odor condition	0.843	Cost performance of the hotel	0.563		

Table 5. *Cont.*

Index	Constituent 1	Index	Constituent 2	Index	Constituent 3	Index	Constituent 4	Index	Constituent 5
Color	0.699	General conditions of the light environment	0.571	Hotel service quality	0.829				
Catering	0.668	Ventilation	0.558	External environment of the hotel	0.824				
General situation of non-light visual factors	0.667	Operating noise of various equipment	0.513						
Experience special activities	0.632								
Surrounding environment	0.569								
Employee service level	0.513								

(2) Data analysis

This time, single factor analysis of variance, independent sample *t*-test, and custom table analysis were performed on the collected scores of three types of decoration styles: modern simple style, British pastoral style, and modern Japanese style, to study the following issues:

First: Whether the subjective evaluation of the hotel's different decoration styles is significantly affected by the factors set in this questionnaire.

Second: Based on the IEQ theory to explore the most important factors affecting the subjective evaluation of guests in hotel room types with different decoration styles.

Third: The factors that have a greater impact on the judgment factors of the guests' subjective evaluation of most hotel room types.

① Single Factor ANOVA

The single factor ANOVA study was used to determine the significant impact of different hotel decoration styles on the subjective evaluation of the guests. The results were shown in Table 6:

Table 6. Results of single factor ANOVA on the rating of hotel 3 decoration style by various influencing factors.

Type		ANOVA Score				
		Sum of Squares	Degree of Freedom	Mean Square	F	Conspicuousness
Modern simple Style	Intergroup	37.191	37	1.005	1.804	0.002
	In group	1460.557	2622	0.557		
	Total	1497.748	2659			
British pastoral style	Intergroup	39.449	37	1.066	1.663	0.008
	In group	1242.500	1938	0.641		
	Total	1281.949	1975			
Modern Japanese style	Intergroup	105.782	37	2.859	4.517	0.000
	In group	4473.572	7068	0.633		
	Total	4579.354	7105			

From the results shown in the above table, it can be seen that different factors have a significant impact on the scores of different styles of room types.

② Independent Sample *t*-test

The test was mainly carried out by taking the average of all relevant factors and constructing a new dimension (for example, using the average of the overall scores of the four factors that measure air quality to form a test variable and named it the air dimension); The test value was subjected to an independent sample test to verify whether the new five dimensions were significantly different from the test value.

Then, the average score of the overall situation of each room type of the hotel decoration style was used as the test value, the average score of the indoor environment of each room type of decoration style was used as the test variable, and an independent sample test was carried out to determine whether the hotel's indoor environment score was significantly different from the hotel's overall score.

The structural relationship between variables, T value, and hypothesis test results were shown in Table 7. All hypotheses had passed the *t*-test, and the path coefficients supported the hypothesis at the level of confidence $\alpha = 0.005$.

Table 7. Hypothesis test results.

Hypothesis	Relationship	T	Sig.	Conclusion
H1	Luminous environment→Hotel indoor environment	3.300	0.974	Support
H2	Acoustic environment→Hotel indoor environment	1.232	0.219	Support
H3	Thermal environment→Hotel indoor environment	−7.150	0.475	Support
H4	Air quality environment→Hotel indoor environment	7.286	0.100	Support
H5	Non-light visual factor environment→Hotel indoor environment	−3.133	0.202	Support
H6a	Luminous environment→Modern simple style	1.232	0.219	Support
H6b	Non-light visual factor environment→Modern simple style	−3.132	0.315	Support
H7a	Luminous environment→British pastoral style	−4.007	0.308	Support
H7b	Air quality environment→British pastoral style	6.614	0.308	Support
H8a	Luminous environment→Modern Japanese style	−4.063	0.389	Support
H8b	Thermal environment→Modern Japanese style	−6.216	0.202	Support
H9	Hotel indoor environment→Customers' subjective overall score	7.117	0.154	Support

It was further verified that the five dimensions of the IEQ theory had a significant impact on the subjective evaluation of customers.

③ Custom Table

In addition, the author used a custom table to further verify the factors that had a greater impact on the different decoration styles of the hotel. The analysis results were shown in Figure 9 below:

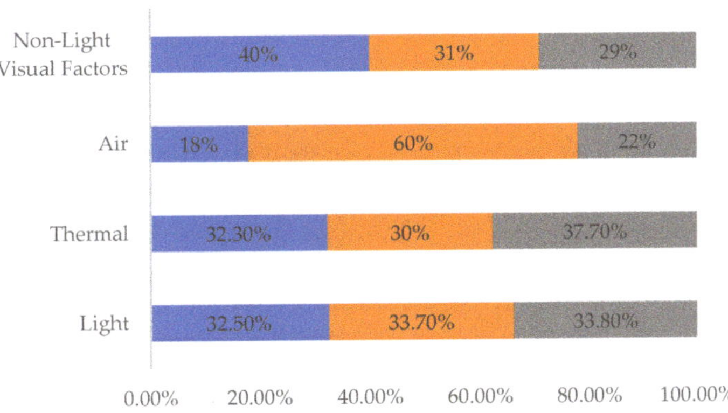

Figure 9. The influence of four factors on the subjective evaluation of three decoration styles.

The subjective scores of customers in modern simple style rooms were mainly affected by light environment and non-light visual factors, which were in line with the hypothesis. The factors of light environment and air quality played an important role in the subjective ratings of customers of British pastoral-style rooms, which were in line with the hypothesis;

the subjective scores of customers of British pastoral-style rooms were greatly affected by non-light visual factors, which was attributed to the British pastoral style, showing the characteristics of the pastoral atmosphere through decoration, and paid more attention to non-light visual factors to create a natural space that met customer expectations. For modern Japanese-style hotel room types, light and the thermal environment were the core influencing factors of customers' subjective ratings, which was in line with the hypothesis; the results showed that the subjective ratings of modern Japanese-style hotel room types were also greatly affected by non-light visual factors, mainly due to modern Japanese style was mainly created by non-light visual factors such as decoration atmosphere and home decoration such as tatami.

4. Discussion

4.1. Research Innovation

The innovation of this study is that starting from the hotel indoor environment, according to the IEQ definition of indoor environmental quality by the National Institute of Occupational Safety and Health, the hotel indoor environment is subdivided into acoustic, light, thermal, air quality, and non-light visual factors five dimensions to explore the impact of each dimension on the subjective evaluation of customers, innovating in subdivision angle. Furthermore, this study takes room types with different decoration styles of hotels as the research object and explores the similarities and differences in customers' subjective evaluations of hotels with different decoration styles. The study found that the influencing factors of the indoor environment of three different decoration styles were different. Existing studies have not incorporated the hotel's decoration style into the influencing factors of customers' subjective evaluation, which is another innovation of this study. The study also found that for the same hotel with different decoration styles, the factors that mainly affect the subjective evaluation of its customers are not the same. The study further found that the hotel's indoor environment plays a key role in influencing the subjective evaluation of customers. In particular, the five aspects of acoustic, light, thermal, air quality, and non-light visual factors have a great impact on the comprehensive psychological and physiological impact of the occupants, thus affecting the subjective evaluation of the hotel by customers.

4.2. Comparison of Similar Studies

At present, many studies on hotel customer satisfaction have found that hotel indoor environments can have an impact on customer satisfaction by mining online review texts [20,47,48].

Through this research, we deeply explored the influence of the hotel indoor environment on customers' subjective evaluation and found that the hotel indoor environment has a positive impact on customers' subjective evaluation. The test shows that all the hypotheses H1–H9 have passed the t-test, and the path coefficient is significant in the confidence level $\alpha = 0.001$ level, which further verifies the significant influence of the hotel indoor environment on customers' subjective evaluation, which is consistent with previous research conclusions.

This study uses the method of case study, selects a hotel with different decoration styles as the research object, and collects the subjective evaluation data of customers through questionnaires. The case study method can provide a real environment for us to collect relevant data. Based on the real environment, we can explore our research problems more deeply. However, the case study should also consider the particularity of the selected case hotel. Recently, some studies have selected specific hotels as research objects to explore issues related to customer satisfaction, such as selecting international hotels in Pyongyang to explore customer satisfaction under the condition of limited choice [49]. Case studies can make the problem more in-depth, but there are still limitations in case selection and conclusion promotion. In future research, multiple case hotels can be selected for the investigation to increase the universality of the research conclusions [26]. In addition, in terms of exploring customers' subjective evaluation of the hotel, many studies have adopted the method of online review analysis, which can make up for the lack of representativeness of case study data to a certain extent.

4.3. Significance

Modern people spend most of their time working and living indoors [21], and customers' perception of the overall value of the hotel is mostly derived from the status quo of the hotel's indoor environment. To improve the driving factors of customer satisfaction and thus improve the subjective evaluation of customers for the hotel, the major decoration styles of the hotel should ensure that the acoustic, light, thermal, air quality, and non-light visual factors of the indoor environment are higher than the hotel standards. For modern simple style rooms, more attention should be paid to their light and non-light visual factors. The British pastoral-style rooms should pay more attention to the influence of light environment, air quality, and non-light visual factors; the modern Japanese-style rooms should focus on creating their light, heat, and non-light visual environment. In addition, the individual needs of customers cannot be ignored, to provide customers with personalized services.

4.4. Limitations and Future Work

This study investigates the customer satisfaction evaluation of the indoor environment quality of Shiyu Hotel in Guangdong Province, adopting the method of text analysis combined with questionnaires. However, there are still shortcomings in this study that need to be improved.

The object of this study (Shi Yu Hotel) has seven major decoration styles, including modern simple, industrial, Japanese, and pastoral. However, to control the decoration style as the only variable, the hotel room type is finally reduced to three mainstream decoration styles: modern simple, British pastoral, and modern Japanese. The research results cannot fully represent various decoration styles of hotel room types.

Among the samples used for data analysis in this questionnaire survey, due to the different preferences of customers' hotel room styles, there is a large difference in the number of samples collected for the three hotel room styles in this study. This problem will also have a certain impact on the data analysis results to a certain extent.

This survey mainly focuses on the acoustic, light, thermal, air quality, and non-light visual factors stipulated by IEQ. It is a shortcoming of this study not to consider the subdivided influencing factors such as material type, texture, color, pattern, decorative element, and plant form in non-optical visual factors.

In future research, it is possible to expand the research on different hotels, room types with different decoration styles, and groups of different ages and occupations, so that the conclusion can more comprehensively determine the factors of different decoration styles and different hotels that affect the subjective evaluation of different guests and the importance of different factors affecting the subjective evaluation of guests in different hotel decoration styles. To establish the subjective evaluation system of the hotel indoor environment suitable for the hotel's multi-decoration style room type and expand the scope of application of the research conclusions, and can deeply study the impact of the subdivision factors of these five major influencing factors on the subjective evaluation of guests.

5. Conclusions and Suggestions

5.1. Conclusions

This research combines the text analysis method and questionnaire survey method, using KH Coder and SPSS to process and analyze the data. It also studies the influence of hotel indoor environment on customers' subjective evaluation, and draws the following conclusions:

(a) The light environment and non-light visual factors in the indoor environment of different decoration styles of the hotel are the key factors affecting the subjective evaluation of customers' indoor environment, and their impact is the largest and most obvious.

The light environment and non-light visual factors have a great influence on the subjective evaluation of customers in each decoration style room type, and the average satisfaction score is high, indicating that the respondents are relatively satisfied with the light conditions and non-light visual conditions of the Shiyu Hotel in Guangdong Province, which should continue.

(b) For the hotel's modern simple style room type's indoor environment, the light environment and non-light visual factors have a greater impact on the subjective evaluation of customers and are the most important.

From the hypothesis test t-values of H6a and H6b, 1.232 and -3.132, and the two-tailed probability p-values of the t-test are 0.219 and 0.315, respectively, which are greater than 0.05 of the significance level α. From the results of the hypothesis, it can be seen that the light environment and non-light visual factors are the main factors affecting the customer's satisfaction rating of the hotel's modern simple style room type. In addition, from the result data in Figure 9, it can be further verified that the light environment and non-light visual factors have an important influence on the customer satisfaction score of the hotel's modern simple style room type, and the non-light visual factors have an influence of up to 40%.

(c) Light environment, air quality, and non-light visual factors play a central role in the subjective evaluation of the indoor environment of the British pastoral style room type, and have a greater impact on it.

For the British pastoral style room type, it can be concluded from the test results in Table 7 that the assumptions H7a and H7b are established, that is, the light environment and air quality have a greater impact on customer satisfaction with the hotel British pastoral style room type. From Figure 9, visibly light environment and air quality have an important impact on customer satisfaction with the British pastoral style of the hotel, which is as high as 60% and 33.7%, respectively. At the same time, from Figure 9, it can be concluded that non-light visual factors have a great influence on the score of British pastoral-style room type, reaching 31%, which is consistent with the existing research results.

(d) The hotel's modern Japanese-style room type is significantly affected by the light environment, thermal environment, and non-light visual factors. This decoration style is also the most sensitive to light, heat, and non-light visual factors.

From the test result data in Table 7, it can be concluded that the H8a and H8b hypotheses are true, that is, the light environment and thermal environment are the most sensitive factors in the customer's satisfaction score for the hotel's modern Japanese-style room type, which are the most sensitive factors affecting the score. At the same time, from the result data in Figure 9, it can be further verified that the light environment and thermal environment have an important influence on the customer satisfaction score of the hotel's modern Japanese-style room type, and the thermal environment influence is as high as 37.7%. However, for modern Japanese-style hotels, there is no cliff-like distinction between the influences of various factors, but compared with other factors, the light and thermal environment have a greater impact on the subjective ratings of modern Japanese-style hotel customers to a certain extent.

(e) To sum up, the factors included in the five IEQ theories of indoor acoustic, light, thermal, air quality, and non-light visual factors have the greatest impact on customers' subjective evaluation of the hotel's indoor environment.

On the whole, the observed value of each F statistic in the results of single factor ANOVA is greater than 1, and the corresponding probability p-values is approximately 0. It is believed that the different levels of influencing factors have a significant impact on the score of each decoration style hotel. Additionally, the *p*-values of the results of hypothesis testing H1, H2, H3, H4, and H5 are 0.974, 0.219, 0.475, 0.100, and 0.202, respectively. The path coefficient supports the hypothesis at the level of confidence $\alpha = 0.005$; it is further demonstrated that the hotel studied in this paper is established. The five dimensions of the indoor environment are the main factors that affect the subjective evaluation of customers.

The research further shows that among the five factors that affect the subjective evaluation of customers, the importance of each factor is ranked as follows: light and non-light visual factors are the most important, followed by thermal environment, air quality, and finally the acoustic environment.

5.2. Suggestions

On this basis, we put forward the following suggestions to the hotel: each type of decoration style of the hotel should ensure that the acoustic, light, thermal, air quality, and non-light visual factors of the indoor environment are all higher than the hotel standard; for modern simple style rooms, more attention should be paid to the influence of light and non-light visual factors; the British pastoral-style rooms should pay more attention to the influence of light environment, air quality and non-light visual factors; the modern Japanese-style rooms should focus on creating its light, thermal and non-light visual environment. In addition, the individual needs of customers cannot be ignored, to provide customers with personalized services.

① Modern simple style hotels should strengthen lighting design and space design.

Light environment and non-light visual factors are the most important factors in the subjective evaluation of modern simple style hotels; for modern simple style hotels, high-quality lighting design can create a comfortable lighting environment and meet people's visual needs. In the process of lighting design, especially with the continuous development of LED technology, LED lighting products continue to emerge. The modern simple style hotels' lighting design can use more LED lighting products to improve the quality of the light environment and meet the health, energy saving, and cost reduction. In addition, modern simple style hotels can improve the quality of non-light visual elements of hotels by arranging vegetation that complements each other, focusing on decoration color matching, and matching furniture with high decorative texture.

② British pastoral-style hotels should strengthen the HVAC system.

The above conclusions show that the light environment, air quality, and non-light visual factors play a central role in the subjective evaluation of the British pastoral-style hotel, and have a greater impact; in addition to improving the lighting design space design, indoor air quality deserves attention, it is a very complex indicator, various time-dependent physical and chemical parameters are components of indoor air quality, which are in turn affected by outdoor conditions (climate), building conditions (materials, structure, and construction), the hotel's HVAC system (heating, ventilation, and air conditioning) systems), interior space arrangement (furniture, furniture, equipment), and the impact of occupant productivity patterns [50]. At this stage, many hotels have not done a good job of building ventilation to save costs. The indoor ventilation is completely carried out by air conditioners. The occupants often smell a musty smell when they enter the room, which will not only affect the occupancy experience of the occupants and the subjective evaluation of the straight line. It will also cause hidden dangers to the health of the guests. For the British pastoral style, it is very important to strengthen the HVAC system of the hotel: let the indoor air quality meet the standard, improve the occupant's check-in experience, strengthen the health protection, and finally maximize the subjective evaluation of the occupants

③ Modern Japanese-style hotels should improve heat and humidity treatment capabilities.

Through investigation, it is found that hotel guests are very sensitive to the thermal environment in modern Japanese-style hotels; therefore, for modern Japanese-style hotels: how to dehumidify and maintain the ideal thermal environment of the hotel is very important, such as ventilation and dehumidification, heating and dehumidification, Dew point dehumidification, rotary dehumidification, membrane dehumidification, solution dehumidification, and other common dehumidification methods.

Author Contributions: Supervision and revision, J.X., M.L. and D.H.; investigation, D.H., Y.W. and S.Z.; resources, Y.W.; draft, D.H., Y.W. and S.Z.; proofreading and editing, D.H., Y.W. and S.Z.; data analysis, D.H.; text analysis, Y.W.; methodology, Y.W.; visualization, D.H. and Y.W. All authors have read and agreed to the published version of the manuscript.

Funding: This study was part of the project called "The study on rural tourism revitalization planning and sustainable living coordination mechanism" funded by the International Cooperation Open Project of State Key Laboratory of Subtropical Building Science, South China University of Technology, the funding number is 2019ZA02; China youth fund of national natural science foundation projects "Research on visual landscape evaluation method of traditional villages in Guangdong based on deep learning", the funding number is 52208057 ; And "Guangzhou philosophy and social science planning project: Research on Guangzhou Higher Education Reform Promoting the coordinated development of Guangdong Hong Kong Macao Greater Bay Area", the funding number is 2022GZGJ131.

Acknowledgments: We would especially like to thank Jian Xu Muchun Li and Zhicai Wu, for their assistance in proofreading the manuscript, and the International Cooperation Open Project of State Key Laboratory of Subtropical Building Science, South China University of Technology.

Conflicts of Interest: The authors declare no conflict of interest.

Appendix A

Questionnaire on investigation on actual customer satisfaction of different decoration style hotel indoor environment. (This questionnaire is used to present the investigation process and facilitate readers' understanding).

Dear friends: this questionnaire is mainly to understand the actual satisfaction of different decoration style hotel indoor environment. Please recall your last experience of HOTEL and fill in the questionnaire. This questionnaire is filled in anonymously, and the questionnaire information is only for academic research. Thank you very much for your support!

Have you ever stayed in a HOTEL (single choice) *

☐ yes

☐ no (please skip to the end of the questionnaire and submit the answer sheet)

HOTEL city: (fill in the blank) *

- **1: Basic information of hotel**

1. Room type (single choice) *

☐ standard king room ☐ deluxe king room ☐ double bed room ☐ parent–child room ☐ deluxe suite ☐ complete building

2. Style (single choice) *

☐ modern simple style ☐ British pastoral style ☐ modern Japanese style

3. Price/day (single choice) *

☐ CNY 0–500 ☐ CNY 501–1000 ☐ CNY 1001–1500 ☐ CNY 1501–2000 ☐ CNY 2001–2500 ☐ more than CNY 2500

4. Companion (single choice) *

☐ none ☐ family ☐ lovers ☐ friends ☐ colleagues ☐ classmates ☐ others___________

5. Reason for check-in (single choice) *

☐ business trip ☐ leisure tourism ☐ visiting relatives and friends ☐ experience punch in ☐ league building activities ☐ others________________

6. Reason for selection (multiple choice) *

☐ convenient transportation ☐ unique theme ☐ beautiful appearance design ☐ high praise rate of the platform ☐ high-cost performance ☐ advanced facilities ☐ experience punch in ☐ recommended by friends ☐ others________________

- **2: Survey of customers' actual satisfaction with the elements of HOTEL how satisfied are you with the elements of B and B?**

Satisfaction score of overall situation of all dimensions of HOTEL (matrix scale) *

	Very dissatisfied	Dissatisfied	Neutral	Satisfied	Very satisfied
General condition of hotel	☐	☐	☐	☐	☐
Hotel indoor environment	☐	☐	☐	☐	☐
Hotel L external environment	☐	☐	☐	☐	☐
Hotel service quality	☐	☐	☐	☐	☐
Hotel experience perception	☐	☐	☐	☐	☐

Investigation on customers' actual satisfaction with various elements of HOTEL (matrix scale) *
How satisfied are you with the elements of B and B?

	Very dissatisfied	dissatisfied	neutral	satisfied	Very satisfied
Air environment					
Overall air quality	☐	☐	☐	☐	☐
Odor condition	☐	☐	☐	☐	☐
Air purification facilities	☐	☐	☐	☐	☐
Ventilation	☐	☐	☐	☐	☐
General condition of thermal environment	☐	☐	☐	☐	☐
Wind speed	☐	☐	☐	☐	☐
Humidity	☐	☐	☐	☐	☐
Heat dissipation of electrical appliances	☐	☐	☐	☐	☐
Room temperature	☐	☐	☐	☐	☐
Thermal environment					
Acoustic environment					
Overall acoustic environment	☐	☐	☐	☐	☐
Road traffic noise	☐	☐	☐	☐	☐
Noise from other occupants	☐	☐	☐	☐	☐
Noise in public areas	☐	☐	☐	☐	☐
Operating noise of various equipment (air conditioner, fan)	☐	☐	☐	☐	☐
Light environment					
General conditions of light environment	☐	☐	☐	☐	☐
Natural lighting	☐	☐	☐	☐	☐
Lighting	☐	☐	☐	☐	☐
Non-light visual factors					
General condition of non-light visual factors	☐	☐	☐	☐	☐
Color matching	☐	☐	☐	☐	☐
Plant arrangement	☐	☐	☐	☐	☐
Closeness to nature	☐	☐	☐	☐	☐

	Very dissatisfied	dissatisfied	neutral	satisfied	Very satisfied
Decoration texture (quality of decoration materials)	☐	☐	☐	☐	☐
space design	☐	☐	☐	☐	☐
Decorative atmosphere (local characteristics and "home" atmosphere)	☐	☐	☐	☐	☐
External environment of Hotel					
Surrounding environment	☐	☐	☐	☐	☐
Traffic convenience	☐	☐	☐	☐	☐
Hotel services					
Infrastructure	☐	☐	☐	☐	☐
Sanitary Conditions	☐	☐	☐	☐	☐
Food and Beverages	☐	☐	☐	☐	☐
Staff service level (service attitude, communication ability)	☐	☐	☐	☐	☐
Experience perception of hotel					
Personal and property safety	☐	☐	☐	☐	☐
Cost performance of hotel	☐	☐	☐	☐	☐
Experience special activities/customs	☐	☐	☐	☐	☐

- **3: Personal information**

1. Gender (single choice) *

☐ male ☐ female

2. Age (single choice) *

☐ under 18 ☐ 18–29 years old ☐ 30–39 years old ☐ 40–49 years old ☐ 50–59 years old ☐ 60 years and over

3. Educational background (single choice) *

☐ junior high school and below ☐ high school/vocational school ☐ junior college ☐ undergraduate ☐ master's degree or above

4. Occupation (single choice) *

☐ enterprise personnel ☐ public servants ☐ freelancer ☐ students ☐ retirement ☐ unemployed ☐ others________________

5. Monthly income (single choice) *

☐ less than CNY 3000 ☐ CNY 3001–6000 ☐ CNY 6001–15000 ☐ CNY 15001–30000 ☐ more than CNY 30000

Thank you for your support! I wish you good health and good luck in the year of the tiger!

References

1. Summary of China's Hotel Industry Development Report 2021. Available online: http://www.199it.com/archives/1235849.html (accessed on 13 August 2022).
2. Song, C.L.; Ali, F.; Cobanoglu, C.; Nanu, L.; Lee, S.H.J. The effect of biophilic design on customer's subjective well-being in the hotel lobbies. *J. Hosp. Tour. Manag.* **2022**, *52*, 264–274. [CrossRef]
3. Ali, F.; Amin, M. The influence of physical environment on emotions, customer satisfaction and behavioural intentions in Chinese resort hotel industry. *J. Glob. Bus. Adv.* **2014**, *7*, 249–266. [CrossRef]
4. Li, Q.D.; You, R.Y.; Chen, C.; Yang, X.D. A field investigation and comparative study of indoor environmental quality in heritage Chinese rural buildings with thick rammed earth wall. *Energy Build.* **2013**, *62*, 286–293. [CrossRef]
5. Yang, T.; Zhao, L.; Li, W.; Wu, J.; Zomaya, A.Y. Towards Healthy and Cost-Effective Indoor Environment Management in Smart Homes: A Deep Reinforcement Learning Approach. *Appl. Energy* **2021**, *300*, 117335. [CrossRef]
6. Zhao, Z.; Amasyali, K.; Chamoun, R.; El-Gohary, N. Occupants' Perceptions about Indoor Environment Comfort and Energy Related Values in Commercial and Residential Buildings. *Procedia Environ. Sci.* **2016**, *34*, 631–640. [CrossRef]

7. Baek, J.; Michael Ok, C. The Power of Design: How does Design Affect Consumers' Online Hotel Booking? *Int. J. Hosp. Manag.* **2017**, *65*, 1–10. [CrossRef]

8. Hou, L.; Pan, X. Aesthetics of hotel photos and its impact on consumer engagement: A computer vision approach. *Tour. Manag.* **2023**, *94*, 104653. [CrossRef]

9. Heide, M.; Laerdal, K.; Gronhaug, K. The Design and Management of Ambience—Implications for Hotel Architecture and Service. *Tour. Manag.* **2007**, *28*, 1315–1325. [CrossRef]

10. Song, J.H. Research on the Thematic Construction of Indoor Environment of Russian-Style Kasbah Hotel. Master's Thesis, Harbin Normal University, Harbin, China, 2005.

11. Chen, Y.X. Characteristics of Indoor Landscape Environment of Binhai Resort Hotel. Master's Thesis, Shanxi University, Taiyuan, China, 2015.

12. Han, Y.Y.; Meng, Y. Research on the Interior Design of the Atrium of the Chinese Hotel. *Sci. Educ. Lit.* **2013**, *10*, 148.

13. Nanu, L.; Ali, F.; Berezina, K.; Cobanoglu, C. The effect of hotel lobby design on booking intentions: An intergenerational examination. *Int. J. Hosp. Manag.* **2020**, *89*, 102530. [CrossRef]

14. Xu, Z.X.; Fang, H. Research on humanized design of indoor environment in hotel rooms. *Art Sci. Technol.* **2020**, *33*, 25–28.

15. Han, H.; Ariza-Montes, A.; Giorgi, G.; Lee, S. Utilizing Green Design as Workplace Innovation to Relieve Service Employee Stress in the Luxury Hotel Sector. *Int. J. Environ. Res. Public Health* **2020**, *17*, 4527. [CrossRef]

16. Kim, J.J.; Han, H. Redefining in-Room Amenities for Hotel Staycationers in the New Era of Tourism: A Deep Dive into Guest Well-Being and Intentions. *Int. J. Hosp. Manag.* **2022**, *102*, 103168. [CrossRef]

17. Wi, S.; Kang, Y.; Yang, S.; Kim, Y.U.; Kim, S. Hazard evaluation of indoor environment based on long-term pollutant emission characteristics of building insulation materials: An empirical study. *Environ. Pollut.* **2021**, *285*, 117223. [CrossRef]

18. Qaraman, A.F.A.; Elbayoumi, M.; Khadoura, K.J. Indoor environmental quality and sick building syndromes in naturally ventilated education laboratories: Case study from Israa University-Gaza. *Glob. Environ. Chang.* **2021**, *11*, 37–48.

19. Sastry, J.; Agawane, S.; Rajan, M.; Black, K.; Laumbach, R.; Ramagopal, M. The effect of the indoor environment on wheeze- and sleep-related symptoms in young Indian children. *Lung India* **2021**, *38*, 307–313.

20. Shen, Z.F.; Yang, X.R.; Liu, C.L.; Li, J.J. Assessment of Indoor Environmental Quality in Budget Hotels Using. *Sustainability* **2021**, *13*, 4490. [CrossRef]

21. Xu, X.; Li, Y.B. The Antecedents of Customer Satisfaction and Dissatisfaction toward Various Types of Hotels: A Text Mining Approach. *Int. J. Hosp. Manag.* **2016**, *55*, 57–69. [CrossRef]

22. Zhao, Y.B.; Xu, X.; Wang, M.S. Predicting Overall Customer Satisfaction: Big Data Evidence from Hotel Online Textual Reviews. *Int. J. Hosp. Manag.* **2019**, *76*, 111–121. [CrossRef]

23. Zanni, S.; Motta, G.; Mura, M.; Longo, M.; Caiulo, D. The Challenge of Indoor Air Quality Management: A Case Study in the Hospitality Industry at the Time of the Pandemic. *Atmosphere* **2021**, *12*, 880. [CrossRef]

24. Cociorva, S.; Iftene, A. Indoor Air Quality Evaluation in Intelligent Building. *Energy Procedia* **2017**, *112*, 261–268. [CrossRef]

25. Nojima, R.; Sugie, N.; Taguchi, A.; Kokubo, J. Understanding the Legendary Sound Environment in the Lobby of Hotel Okura Tokyo. *Appl. Sci.* **2020**, *10*, 4552. [CrossRef]

26. Han, H.; Moon, H.; Hyun, S.S. Indoor and Outdoor Physical Surroundings and Guests' Emotional Well-being. *Int. J. Contemp. Hosp. Manag.* **2019**, *31*, 2759–2775. [CrossRef]

27. Zhang, Y.F. Research on Environmental Quality Evaluation and Technical System of Hotel Lobby Lighting. Master's Thesis, Chongqing University, Chongqing, China, 2005.

28. Wargocki, P.; Wei, W.; Bendzalova, J.; Espigares-Correa, C.; Gerard, C.; Greslou, O.; Rivallain, M.; Sesana, M.M.; Olesen, B.W.; Zirngibl, J.; et al. TALT, a new scheme for rating indoor environmental quality in offices and hotels undergoing deep energy renovation. *Energy Build.* **2021**, *244*, 111029. [CrossRef]

29. Jin, Z.; Juan, Y.K. Is Fengshui a science or superstition? A New Approach Combining The Physiological and Psychological Measurement of Indoor Environments. *Build. Environ.* **2021**, *201*, 107992. [CrossRef]

30. Wei, W.; Wargocki, P.; Zirngibl, J.; Bendzalova, J.; Mandin, C. Review of Parameters Used to Assess the Quality of the Indoor Environment in Green Building Certification Schemes for Offices and Hotels. *Energy Build.* **2020**, *209*, 109683. [CrossRef]

31. Zuhaib, S.; Manton, R.; Griffin, C.; Hajdukiewicz, M.; Keane, M.M.; Goggins, J. An Indoor Environmental Quality (IEQ) assessment of a partially-retrofitted university building. *Build. Environ.* **2018**, *139*, 69–85. [CrossRef]

32. Xu, H.; Liu, Y.C.; Wu, H.J. Indoor environmental quality evaluation model of green office buildings. *Build. Sci.* **2021**, *37*, 149–155+179.

33. Su, Z.; Wang, D.Z.; Liu, Y.Z. Interpreting the embodiment of modern simplicity in interior design. *Hous. Ind.* **2016**, *9*, 59–61.

34. Hou, J.W. Discussion on pastoral decoration style based on low carbon concept. *Jiangxi Build. Mater.* **2015**, *2*, 24.

35. Tang, Y. Comparative Study on Minimalism in Modern Nordic and Japanese Interior Design. Master's Thesis, Hefei University of Technology, Hefei, China, 2016.

36. Choi, K.; Meng, B.; Lee, T.J. An investigation into the segmentation of Japanese traditional 'Ryokan' hotels using selection attributes. *J. Vacat. Mark.* **2018**, *24*, 324–339. [CrossRef]

37. Chan, J. Response-Order Effects in Likert-Type Scales. *Educ. Psychol. Meas.* **1991**, *51*, 531–540. [CrossRef]

38. Gosling, S.D.; Vazire, S.; Srivastava, S.; John, O.P. Should We Trust Web-Based Studies? A Comparative Analysis of Six Preconceptions about Internet Questionnaires. *Am. Psychol.* **2004**, *59*, 93–104. [CrossRef] [PubMed]

39. Jiao, X.L.; Wei, Y.Y. On The Application of Minimalism in Interior Design. *J. Wuzhou Univ.* **2017**, *27*, 30–34.
40. Liu, Y.F. Minimalist Design of Indoor light environment. *Build. Struct.* **2020**, *50*, 141–142.
41. Liu, L.W.; Wu, W.T. A Brief Discussion on the Embodiment of Pastoral Style in Contemporary Interior Design and Suggestions. *Pop. Lit. Art* **2019**, *24*, 87–88.
42. Liu, X.J. Pastoral Style in Contemporary Interior Design Reflects and Recommendations. *Guide Fam. Life* **2021**, *2*, 151–152.
43. Zhou, X. Discussion on Japanese Izakaya Interior Design. *Mod. Decor.* **2016**, *1*, 54.
44. Zhang, Y.F.; Ruan, Z.H. Japanese Style in Home Design—A Case Study of Chongqing Longhu District Type. *Art Ref.* **2019**, *32*, 265–266+269.
45. Sohrabi, B.; Vanani, I.R.; Qorbani, D.; Forte, P. An Integrative View of Knowledge Sharing Impact on E-Learning Quality: A Model for Higher Education Institutes. *Int. J. Enterp. Inf. Syst.* **2012**, *8*, 14–29. [CrossRef]
46. Lewis, E. How to Write and Publish A Scientific Paper. *Am. J. Nurs.* **1984**, *84*, 272–273.
47. Miao, Z.W.; Lu, Y.L. Research on Hotel Customer Satisfaction Evaluation System—Review Data of High-star Hotels in Hangzhou Based on tripadvisor.cn. *Tour. Res.* **2021**, *13*, 73–84.
48. Feng, X.B.; Huang, Z.Q.; Zheng, Q.W. Research on Hotel Service Quality Evaluation Based on Network Text Analysis: A Case Study of Crowne Plaza Chengdu Poly Park. *J. Guangxi Econ. Manag. Cadre Coll.* **2020**, *32*, 69–73+101.
49. Li, F.Y.; Ryan, C. Western guest experiences of a Pyongyang international hotel, North Korea: Satisfaction under conditions of constrained choice. *Tour. Manag.* **2020**, *76*, 103947. [CrossRef]
50. Al Horr, Y.; Arif, M.; Kaushik, A.; Mazroei, A.; Katafygiotou, M.; Elsarrag, E. Occupant productivity and office indoor environment quality: A review of the literature. *Build. Environ.* **2016**, *105*, 369–389. [CrossRef]

Article

Evaluation and Optimization of Daylighting in Heritage Buildings: A Case-Study at High Latitudes

Farimah Piraei [1], Barbara Matusiak [2] and Valerio R. M. Lo Verso [3,*]

[1] Department of Energy 'Galileo Ferraris', Politecnico di Torino, Corso Duca degli Abruzzi, 24, 10129 Turin, Italy
[2] Department of Architecture and Technology, Norwegian University of Science and Technology, Sentralbygg 1, 447, Gløshaugen, Alfred Getz vei 3, 7034 Trondheim, Norway
[3] Department of Energy 'Galileo Ferraris', TEBE Research Group, Politecnico di Torino, Corso Duca degli Abruzzi, 24, 10129 Turin, Italy
* Correspondence: valerio.loverso@polito.it; Tel.: +39-011-090-4508

Abstract: Transforming historical listed buildings into workplaces is a serious challenge, particularly for buildings with relatively small windows in the façades, which determine scarce daylighting indoors. This paper studied how daylighting can be significantly increased in a case-study historical building through rooflighting systems, as the façade cannot be modified. The case-study was a historic and iconic warehouse built-in 1681 in Trondheim, Norway. The optimized configuration was analyzed in terms of daylight amount and view analysis, according to EN 17037 and to LEED v4.1 protocol. A critical evaluation of the actual applicability of the optimized Scenario in the real building was carried out along with the constructors. A 3D model was built in Rhinoceros, and daylighting simulations of the base-case (the building in the existing configuration) and for 6 alternative Scenarios were run through Climate Studio. The following metrics were calculated: Daylight Factor (DF), Spatial Daylight Autonomy (sDA), Annual Sunlight Exposure (ASE), and views. An optimized configuration was eventually identified through the Galapagos component in Grasshopper, with an average DF value of 2.7% (against 0.9% in the base-case configuration), higher than the target DF_m of 2.4% for Norway), and a sDA value of 50.2% (14.2% in base-case configuration).

Keywords: daylight simulation; atrium; corelighting; heritage retrofit; historical building; indoor environmental quality; view analysis; Nordic latitude

---

Citation: Piraei, F.; Matusiak, B.; Lo Verso, V.R.M. Evaluation and Optimization of Daylighting in Heritage Buildings: A Case-Study at High Latitudes. *Buildings* **2022**, *12*, 2045. https://doi.org/10.3390/buildings12122045

Academic Editors: Yue Wu, Zheming Liu and Zhe Kong

Received: 14 October 2022
Accepted: 15 November 2022
Published: 22 November 2022

Publisher's Note: MDPI stays neutral with regard to jurisdictional claims in published maps and institutional affiliations.

1. Introduction

Historic, listed, or unlisted buildings account for 30% of the European building stock [1]. Heritage is regarded as a highly precious wealth, and its preservation for future generations is a big challenge. Neglecting the heritage buildings may expose the heritage to further deterioration. Therefore, adaptive reuse of heritage buildings is considered favorable to their preservation. However, adapting heritage buildings to new users with new occupancy conditions is a challenging task, concerned with implementing the new functions while keeping the historical originality of the building with only minimal alterations and in a sustainable way [2]. This implies a demand for retrofit solutions able to improve indoor environmental conditions while reducing energy use and preserving heritage significance. Improving the energy performance of the building envelope, such as roofs, walls, or windows, requires a more comprehensive analysis to balance out built heritage conservation, technical compatibility, health and comfort of occupants, and energy efficiency [1]. Daylighting particularly plays a crucial role in this process: with the goal of improving the positive experience for the occupants, it is crucial to provide internal spaces that promote daylight conditions without sacrificing the identity of the place. Moreover, a suitable amount of natural light is beneficial to the health and productivity of the occupants, as it regulates the circadian entrainment of individuals [3].

Heritage building reuse has been widely studied in previous research [4–8], which mainly dealt with green retrofitting and energy conservation whilst only partly including comfort-oriented strategies for the occupants [9,10]. The International Energy Agency IEA set up a dedicated project within the SHC Task 59, titled "Renovating Historic Buildings Towards Zero Energy", which was active over the quadrennium 2017–2021 [11]. It was mainly aimed at preserving the historic and aesthetic value of heritage buildings while increasing comfort, lowering energy bills, and minimizing the environmental impact [12].

Consequently, the energy issues have received the highest attention, but the visual comfort of the occupants must be guaranteed as well, to assure the successful reuse of heritage buildings. In a heritage building, evolving the original quality of daylighting through the extensive use of electric light can critically impact the visual character and the sense of the place [13,14]. Research in this regard strongly emphasizes the intersection of cultural heritage preservation and environmental engineering to change building usage and improve indoor daylighting. Thus, adaptive heritage reuse is considered beneficial for the well-being of the occupants [15,16]. Furthermore, a high level of daylight sufficiency leads to lower operating expenses and reduces the energy demand for electric lighting while increasing the indoor environmental quality for the occupants. Issues about daylighting should be addressed in the early design phase [17]. However, it should be stressed on the other hand that heritage-listed buildings present strong constraints that limit the opportunity to modify the façade layout, for instance by increasing the window area. In this regard, rooflighting is a promising strategy to admit daylight into a building by opening skylights on the roof [18,19]. As an alternative, the use of atria, attractive architectural elements that allow corelighting to be achieved, thus admitting daylighting into the core of the building, could be considered.

Over the last 40 years, atria have become a quite recurring architectural form [20,21]. They were used in a range of modern building types worldwide, to create the experience of openness and spaciousness inside buildings. An atrium is typically a large, multi-story, glass-roofed space that brings daylight into large buildings where sidelighting alone cannot penetrate enough [22]. Besides daylighting, atria can also provide other practical functions for a building, such as allowing natural ventilation to help maintain thermal comfort [23] or acting as a buffer space to reduce energy losses.

Nowadays, the usability of atria is usually linked to commercial and public buildings as they are commonly used as significant architectural features to emphasize main entrances, increase the usability of public circulation spaces, or highlight specific destinations within a structure [24]. In fact, many large-scale buildings are presently designed with atria.

Successful daylighting design of an atrium mainly depends on (i) the roofs fenestration system; (ii) the geometry of the atrium, typically expressed through the Aspect Ratio (height-to-section ratio); (iii) window-to-wall ratio WWR all along the vertical section of the atrium (typically higher at lower floors and smaller at top floors); and (iv) light reflectance of the atrium surfaces. The daylight availability can be evaluated in terms of daylight illuminance levels achieved in the atrium and, even more, in terms of daylight illuminance levels in spaces facing the atrium [21]. This in turn is a function of the daylight availability at the location [25].

Modern buildings use highly-glazed atriums more frequently because they offer pleasant architectural aesthetics and daylight [26]. In a work on atrium building design, Calcagni & Paroncini [27] found that increasing light reflectance R_v values of atrium opaque surfaces in atria with large daylight openings and high light transmittance T_v does not generate a substantial improvement in the daylight factor levels on the atrium ground floor. Differently, Matusiak et al. [21] examined different strategies for improving daylighting in buildings with atria. They found that with the variation of glazing reflectance across the atrium according to its height, a significant increase in daylight reaching the bottom of the atrium. Younis et al. [28] found that the light from the sky and the light reflected off the atrium walls and floor are the essential components for daylighting in rooms adjacent to an atrium well. In a similar study, Matusiak [29] found that the mean reflectance of the atrium surfaces had a significant impact on the distribution of daylight in the atrium: increasing atrium floor reflectance was a very efficient approach for improving daylighting in first-floor rooms, as well as increasing façade reflectance influences lower-floor daylighting very moderately but were quite successful for daylighting at the upper floors. As the view angle to the direct sky is lower at the lower levels of an atrium, reflected light becomes the most important component. Therefore, high reflectivity ratings on opaque surfaces are a crucial factor.

Most research has been focused on the performances of side windows [30–33], rather than on rooflighting systems. Low sun angles are a feature of high latitude regions, where electric lighting is essential for both visual and thermal comfort and daylighting needs to be optimized [34,35]. A reasonable illuminance level able to contribute to an impression of daylight presence in building cores and rooms without vertical windows was found to be equal to 50 lux [36]. Office buildings use about 40% of their energy for electric lighting, but this rate becomes even higher for other building types, such as commercial or industrial [34]. Therefore, providing daylight is vital and needs to be noticed, especially in high latitudes (greater than 55°). Daylighting is a particularly challenging topic to address in Nordic countries, due to its peculiar typical characteristics: (i) low solar elevation angles are prevailing during the year: up to 1/3 of the whole daytime during the year solar rays are nearly horizontal; and (ii) the frequency of sunny skies during the year is quite low, especially in winter. As an example, Figure 1 reports a map of the Scandinavian area, taken from Satel-Light, where the frequency of sunny skies is shown: it can be observed that the occurrence of sunny days is less than 25% on annual basis in Norway [37].

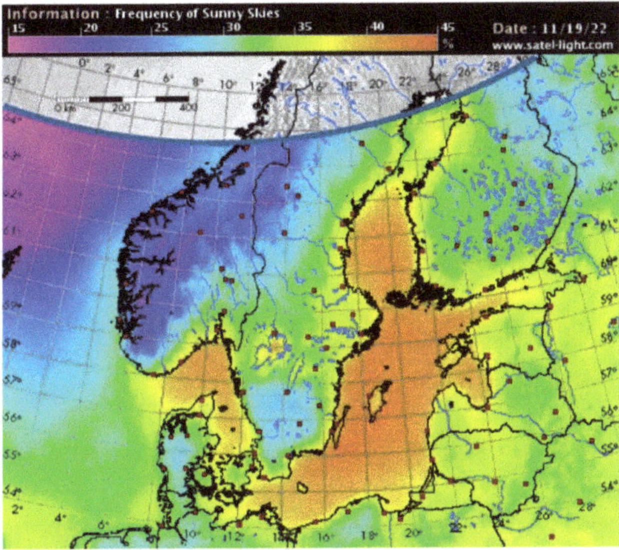

Figure 1. Frequency of sunny skies in Northern Europe (image taken from [37] with the permission of Barbara Matusiak).

In the retrofit process of existing buildings, additional elements are usually applied to windows to harvest, redirect, or even block the solar rays [38,39]. However, skylights make it possible to significantly increase daylighting without changing the visual character of the building since thy are hardly visible from the street level. The refurbishment of skylights in the NTNU university building (63° N) is an example showing how low sunlight can be both scattered and redirected down to the occupied room creating comfortable visual conditions with high daylight level during the year [37].

Within this context, this article focuses on how to improve daylighting in a heritage building through a specific case-study, which has poor daylighting conditions and presents strong constraints regarding the façades. Accordingly, the main research question was set as follows:

"How can it be possible to guarantee sufficient daylighting (according to international standards or protocols) in a heritage building located in a Nordic climate, where sidelighting cannot be incremented, so through corelighting?"

To address this question, the study analyzed an existing building, located in Trondheim, Norway. Different alternatives of skylights and atria were explored to improve daylighting inside the building, quantifying its distribution into the adjacent areas (which are all dark in the original layout of the building), so as to find how daylighting conditions can be optimized since the early stage of the restoration process of historical buildings.

2. Materials and Methods

2.1. Case-Study

This research studied one of the oldest and biggest wharves, namely Huitfeldtbrygga, in Trondheim, Norway, at a latitude of 63.4° N and longitude of 10.4° E. Figure 2 shows the solar paths of Trondheim: it can be observed that at the winter solstice there are 4 h only of sunlight (from 11 until 14), with a maximum sun elevation angle as low as 3.35°. In the period from the 21st of December to the beginning of February, the sun elevation angle never reaches 10°; this value occurs for the first time on the 3rd of February. After this day, the number of hours when the sun is over 10° increases rapidly, but 30° sun elevation angle cannot be observed before the 30th of March.

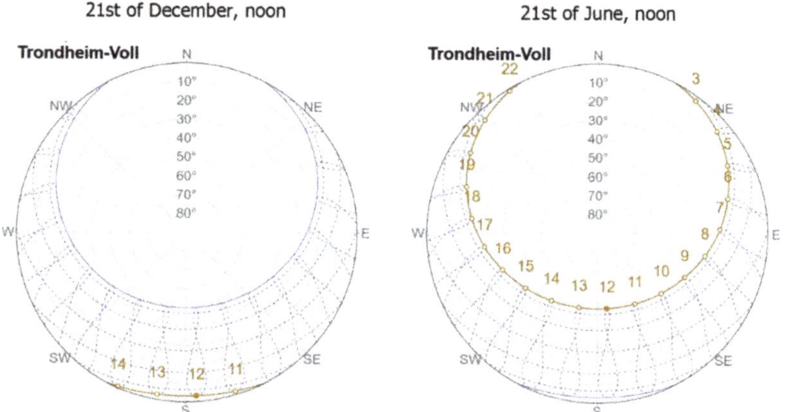

Figure 2. Solar paths for Trondheim.

In summer, the sunlight hours are 20 (from 3 until 22); the highest position of the sun is 50.1°, but this happens for only four days during the year, the 19th, 20th, 21st, and 22nd of June, and for a very short time: for instance, sun elevation angles higher than 50° lasts for 12 min only on 21st of June.

The Huitfeldtbrygga building strongly suffers from 40–50 years of lack of maintenance and use, and it is listed as a protection class A building, which indicates that it is regarded as one of Trondheim's most worthy of preservation. It is an outstanding building as it consists of three wharves built under the same roof, which also appears to be the most skewed one. The wharf has a total area of about 1900 m^2 and a very deep plan with small windows. Figure 3 shows some images of the case-study building.

Figure 3. Views of Huitfeldtbrygga in Trondheim, Norway (personal images by the Authors).

According to studies [40–42], a room with a depth of 4–6 m needs a window-to-wall ratio WWR of 60%. Therefore, WWR and WFR (window-to-floor ratio) play a crucial role in indoor space's daylight quality and quantity. In the building analyzed as case-study, the windows in the façades are relatively small, which results in low values of both WWR (18.4%) and WFR (2.4%).

The building consists of a basement, four stories, and an attic. The West wall is facing a street: they are separated by a sloping terrain, where trees were planted to transform it into a park. The building is currently undergoing a huge renovation, and the Authors of this paper support the optimization of daylighting inside the building, considering the constraint due to its architectural and historical character. The daylighting project was carried out through a continuous confrontation with the design team who realized the full renovation project of the building.

2.2. Research Framework

This study relied on empirical research and a quantitative strategy, which included modeling the building and investigating through simulations various alternative retrofitting Scenarios to optimize daylighting inside the building through core lighting (skylights and atria). The project was carried out through a sequence of steps, as shown in Figure 4.

The first step was the data collection and in-situ measurements. The plan and section drawings were provided by the architectural design team [43]. The architectural company working on the retrofit proposal for this building was interviewed, while luminance measures were taken inside the building to calculate the light reflectance of the materials (assumed as Lambertian), through the following equation:

$$\frac{R_{test}}{R_{ref}} = \frac{L_{test}}{L_{ref}} \tag{1}$$

In-situ measurement was taken to calculate the light reflectance (R_{test}) of the various surfaces inside the building. The luminance values of each surface (L_{test}) and of a reference grey card (L_{ref}) with known light reflectance (R_{ref}) were measured (Figure 5). The grey card was positioned in the same place and illuminated equally as the test surface. All measurements were done with using a luminance meter Konica Minolta LS-110.

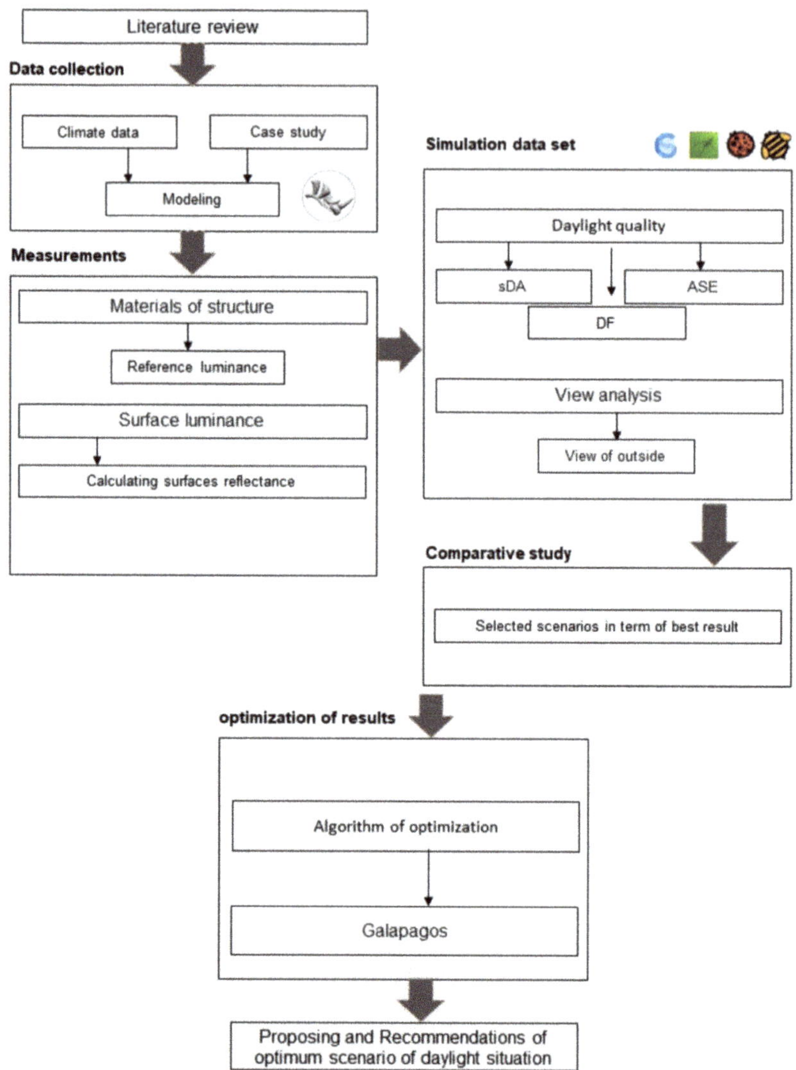

Figure 4. Conceptual framework of the study.

Figure 5. Field luminance measurement is used to calculate the light reflectance of real materials for simulations [44].

As the next step, the historical structure of the building was studied and a discussion was carried out with the design team on which acceptable Scenarios could be implemented into this heritage building: as a result, six alternatives were conceived, taking views, daylighting, and the solar exposure into consideration on an annual basis.

All alternatives were simulated using the Climate Studio (CS) simulation software package. This is a plug-in for the CAD modeler Rhinoceros and uses the validated Radiance algorithm [45] to calculate daylighting conditions in a space. Rhino was used to build the 3D model of both the existing building (base-case) and of the six Scenarios. Radiance-compatible materials were attributed through CS, by creating new materials with the light reflectances value that was measured in-situ for the various materials [46]. The EnergyPlus Weather File (.epw) of Trondheim was used as a source of meteorological data. The output of simulations was:

1. The daylight factor values in the building (at all floors), in accordance with what is required by the recent European standard EN 17037 [47] (mean and median DF); particularly, a mean Daylight Factor $DF_m \geq 2.4\%$ is required for the latitude and the climate of Norway.
2. The spatial daylight autonomy $sDA_{300,50\%}$ and the annual sunlight exposure $ASE_{1000,250}$; sDA quantifies the fraction of the regularly occupied area within a space for which the daylight autonomy (calculated for a threshold illuminance of 300 lx) exceeds a specified value (50%); ASE quantifies the fraction of the horizontal workplane that exceeds a specified direct sunlight illuminance level (1000 lx) for more than a specified number of hours per year (250 h) over a specified daily schedule with all operable shading devices retracted (excluding the sky) [48].
3. At all floors; the following targets were assumed, in accordance with the requirements set in the LEED protocol v4.1 [49]:

   - $sDA_{300,50\%} \geq 75\%$ for the regularly occupied floor area: 3 points are granted according to the EQ Credit: daylight
   - $sDA_{300,50\%} \geq 55\%$ for the regularly occupied floor area: 2 points are granted
   - $sDA_{300,50\%} \geq 40\%$ for the regularly occupied floor area: 1 point is granted
   - for any regularly occupied spaces with $ASE_{1000,250}$ greater than 10%, a strategy must be identified to address glare.

The software outcome from all Scenarios was then compared and the two most promising alternatives were identified based on DF_m, sDA, and ASE results. These two optimal Scenarios were further analyzed according to the requirements set by EN 17037 [47] (see Section 2.5).

After a comparison, the optimum Scenario was chosen using the Honeybee plugin tool [50] in Grasshopper [51]. Finally, some recommendations for future strategies were presented based on sDA and ASE values.

All metrics were calculated for a grid of sensors positioned 0.8 m above the floor finishing and with a spacing of 0.25 m.

2.3. Simulation Data Set

Annual simulations were run to analyze the daylighting conditions inside Huitfeldtbrygga, by calculating the following metrics: spatial daylight autonomy sDA, annual sunlight exposure ASE, daylight factor, and view out.

To understand the daylight behavior due to the changing geometry of atriums, all simulations were conducted in various Scenarios of atria space with considering the whole building of Huitfeldtbrygga. In order to do the simulation in ClimateStudio, laminated double-pane glazing was chosen for both the vertical windows in the façades and for the rooflighting systems (skylights and atrium). The window materials were selected based on conventional construction window material in Norway. Following up on the instructions received from the renovation design team, the light transmittance value T_v of a double-pane clear + clear glazing was used (clear Float glass 6mm, Krypton 13 mm, clear glazing 6mm), namely T_v of 0.70, with a corresponding U-value of 1.26 W/m²K. It is worth mentioning that laminated glazing was chosen for its protective function.

2.4. Design Configurations

The shape of the atrium in the building fits the design of the existing plan and, at the same time, aims to provide a visual connection between the floors internally through an atrium. Figure 6 shows the plan of the third floor: the various design strategies to reach the standard level of daylight were defined taken the interior layout of the building into account. There are two long interior walls (side-to-side) that divide the building into three separated parts. As a result, separate atriums were implemented on both sides of the building to improve daylighting in these areas; besides, the middle part of the building was not blocked by the floors except in the second floor which is consistently connected to both sides. The six Scenarios that were designed to improve daylighting in the building relied on using as much secure glazing as suitable for transferring daylight as possible. Scenarios are shown in Figure 7. For all configurations, the atrium consisted of laminated glazing panes (1 m × 1 m in size) mounted on a galvanized steel supporting frame structure that is 0.06 m × 0.06 m in section. Also, the upper glazing that closes the atrium was conceived as an openable surface to allow ventilation to get activated (stack-effect) to reduce solar and internal gains in summer, with positive effects on thermal comfort for the occupants and reduced energy demand for cooling. It is worth pointing out, though, that the thermal performance was not in the scope of this paper. The six Scenarios are highlighted in orange in Figure 6. The main atrium in the various configurations was placed on the North-facing sloped roof rather than in the South-facing one to avoid the penetration of direct solar rays in the spaces adjacent to the atrium itself, especially at the top floors.

The first Scenario considered one massive atrium with plan sizes of 3.34 m × 8.50 m in the middle of the building. Similarly, to Scenario 1, Scenarios 2 and 3 had central atriums in precisely the same position and sizes, but in Scenario 2 there were three smaller separated atria with plan sizes 1.84 m × 1.70 m. In Scenario 3, there were united side atriums with side plans of 1.84 m × 10.10 m. In Scenario 4 there was a square atrium in the middle of the building with plan sizes of 3.34 m × 3.34 m, while Scenarios 5 and 6 had the same atrium plan sizes for both lateral wharves as atria 2 and 3.

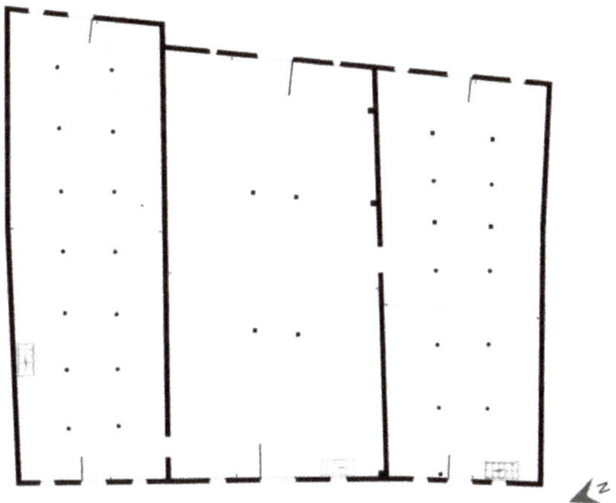

Figure 6. Plan of the 3rd floor of the case-study building, with the position of the structural elements.

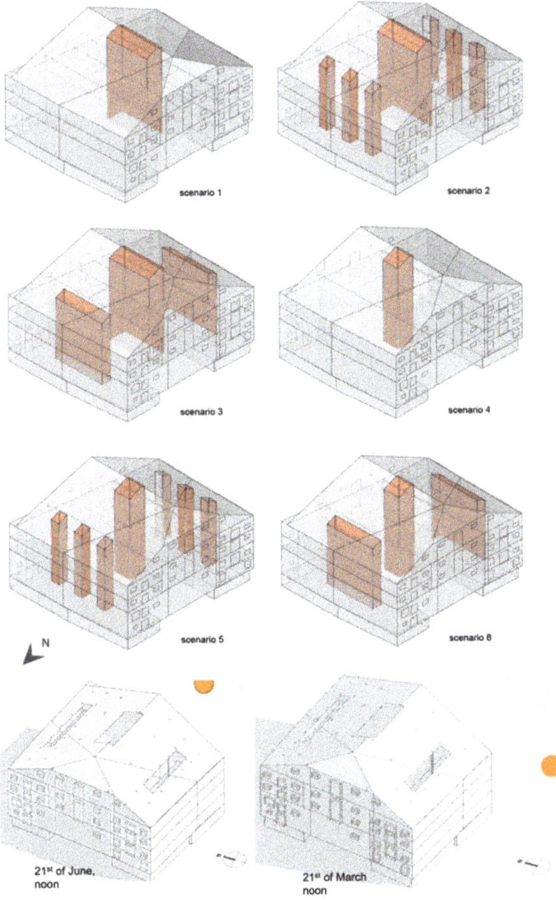

Figure 7. Schematic view of the six configurations that were defined for corelighting of the building through skylights and light atria. The position of the sun hitting the building is also shown.

Table 1 summarizes the following geometrical information: atrium volume, atrium roof area, percentage of the volume of each Scenario compared to the whole volume, and percentage of the roof surface of each Scenario compared to the total roof area. As shown in the Table, the atrium in Scenario 3 with 915.6 m^3 has the largest volume compared to the other Scenarios., whilst the atrium in Scenario 4, with a volume of 180.9 m^3, has the least space in the building. In between, Scenarios 1, 2, 5, and 6 had 462.1 m^3, 690.5 m^3, 408.9 m^3, and 634.4 m^3, respectively. It is worth mentioning that even if the spaces related to the atrium in each Scenario were quite large, each atrium geometry was considered functional spaces that could be used for different purposes, based on confrontation with the renovation design team.

Table 1. The detailed characteristics of considered atria area and volume in each Scenario.

Scenario	Atrium Volume (m^3)	Percentage of Atrium Volume Compared to the Building Volume (%)	Atrium Roof Area (m^2)	Percentage of Atrium Roof Area Compared to the Building Roof Area (%)
Scenario 1	462.1	4.9	31.1	4.4
Scenario 2	690.5	7.4	51.8	7.3
Scenario 3	915.6	9.8	72.6	10.3
Scenario 4	180.9	1.9	12.2	1.7
Scenario 5	408.9	4.4	32.9	4.7
Scenario 6	634.4	6.8	502	7.1

The total volume and total roof area of the case-study building (Huitfeldbrygga) were 9394.7 m^3 and 708.2 m^2, respectively.

2.5. View Analysis

The method evaluated the views for the occupants and determined eligibility for the EN 17037 European standard. View factors and distance to specific model layers or items of interest were also calculated for Scenarios with better daylighting (in terms of DF, sDA, and ASE), as well as for the optimized Scenario (see next section).

The results were provided at sensor places defined on an analysis grid distance of 1.20 m above the floor of each building level. The analysis grid size determined spacing 0.60 m to acquire precise findings. EN 17037 [47] covered four aspects of daylight in buildings, the second of which—View Out–was included in ClimateStudio view analysis workflow (as of Climate Studio v1.5).

Accordingly, in view result, CS shows the proportion of the building floor space that falls into each of four compliance categories: Failing, Minimum, Medium, and High. The compliance levels are determined by three assessments, which are performed for each point of view such as Horizontal Sight Angle, Outside View Distance, and the number of view levels (three, sky, ground, and landscape). A view position must observe:

1. at least the landscape layer, to achieve 'minimum' compliance
2. landscape layer as well as one additional layer (either ground or sky), to achieve 'medium' compliance
3. all three layers (landscape, ground, and sky), to achieve 'high' compliance
4. a point that does not benefit from any of the three view levels is labeled as 'failing' in the view assessment criterion.

2.6. Optimization

Grasshopper plug-in Ladybug and Honeybee were used to parametrize a Rhino model for daylighting simulation [52], and Galapagos was used to achieve the optimal solution [53,54]. Honeybee and Ladybug were validated and used in several studies [55–58] for daylight simulations. The simulation procedure started with creating the geometry and setting the parametric design variables. Throughout the parametric simulation process, each material of the real building was coupled to a specific Radiance materials component, by setting the material light transmittance T_V or reflectance R_V. After that, the building materials were linked to the daylighting simulation component, which takes the Trondheim's weather file, the position of daylighting sensors, and other simulation variables into consideration. Radiance creates a .rad file and simulates daylighting. Ladybug reads the daylight performance data and provides as output an annual lighting schedule, which is used by Honeybee to run an annual daylighting simulation and generate results in terms of daylighting metrics (DF, sDA, and ASE). The simulation results are then imported back into Grasshopper. Galapagos was utilized in the optimization process to determine the optimal atrium shape with the maximum sDA. To this end, different dimensions in the bottom and upper parts of each atrium were considered to test the variety of shapes in the optimization process. The boundary condition for maximum and minimum dimensions of atriums was also assigned to avoid unacceptable shapes. The assigned range of width was between 1.20 m and 3.60 m, and the length between 1.20 m and 15.80 m was adopted to achieve the optimum outcomes. These domains were set to control the optimal solutions according to the building floor area and the actual structure positions. The design variables were related to the Genetic input in Galapagos, and the sDA output was related to the Fitness input equal to 50. Each generation had a population of 100, with a population boost of twice the size of the first generation.

3. Results

Two key features distinguish the retrofit of a heritage building. The first component was related to the renovation process itself, as the original material asset should be preserved in the case of a historical building. The second issue was concerned with the historical feature buildings, which should be maintained in such a way that the authenticity of the asset was preserved with minimal modifications to the original construction.

This section reports all annual results from the Climate Studio and Grasshopper simulations. First, an inquiry was conducted to ensure that the final design tool with all iterations was given. Secondly, Scenario-based outcomes were evaluated using metrics such as mean and median DF, sDA, and ASE, highlighting the grid regions that were compliant with the LEED v4.1 protocol, option1, and the European standard EN17037:20218. Therefore, the general interaction with parameters and the effect of each variable input were explained using comparison and correlation studies. Finally, optimization was tailored to the parameters under consideration.

3.1. Analysis of DF, sDA, and ASE in the Base-Case

The simulation was initially carried out using two simulation tools, Climate Studio and Honeybee, to analyze DF, sDA, and ASE values within the Huitfildbrygga building. In order to improve daylight conditions in the existing building, it was necessary to quantify the daylight amount that was admitted in the base-case Scenario. Therefore, daylighting was evaluated at each floor by using mean and median DF, sDA, and ASE.

Figure 8 shows the simulation results for the base-case: it can be observed that daylighting levels in the building were quite limited at all floors, due to the reduced area of the vertical openings in the façades. This is testified by both DF (Figure 8a) and sDA (Figure 8b) values: the mean DF was 0.9%, while as for sDA, none of the floors in the base-case Scenario showed sDA value over 18%. The lowest daylighting level was observed on the ground floor, as expected: DF = 0.02% and sDA = 12.8%. The first floor and the second floor showed the same sDA results (17.1%), while the third floor showed the highest of sDA value (10.1%). On the other hand, the limited daylight penetration into the building yields ASE values that can be considered negligible (Figure 8c). The highest and lowest ASE values were observed on the lower floors (first and ground floors), with values of ASE = 5.1% and 3.1%, respectively.

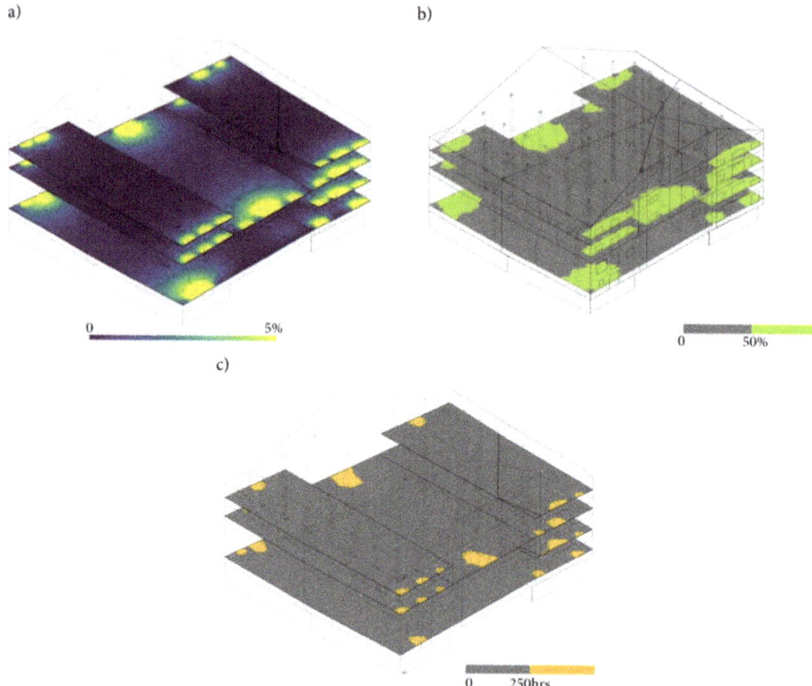

Figure 8. Simulation results in the base-case; (**a**) DF; (**b**) sDA$_{300,50\%}$; (**c**) ASE$_{1000,250}$.

The amount of sDA in winter was less than 10%. More specifically, sDA values were lower than 2% and close to zero in November, December, and January. On the other hand, sDA values exceeded 30% in May, June, and July. ASE values during a year were negligible in the base-case Scenario, especially in winter. The mean DF was quantified with a value of 0.9% and a median DF of 0.3%. As it typically happens in sidelighting configurations, point DF values decrease significantly as the distance from windows increases; however, they remain quite low also in the proximity of the façades, due to the reduced window area.

3.2. Comparison of DF, sDA and ASE in the Six Scenarios

Changing the atrium configuration affected how daylight penetrates deep down into the building. The simulation of the six Scenarios was intended to identify which Scenario resulted in the highest and most uniform daylight autonomy in the building spaces facing the atrium. Six Scenarios based on the number, shape, and size of atria were thus compared in the following sections.

Figure 9 illustrates and compares the DF, sDA, and ASE results that were obtained for each Scenario: the highest sDA values were observed for Scenario 3 (sDA = 28%), followed by Scenario 6 (sDA = 23%) and Scenario 2 (sDA = 21%). Scenario 4 offered a quite similar distribution of daylight autonomy to the base-case. Scenarios 1 and 4 were therefore the least favorable of the three shapes. The zones that received no light from the atrium were the largest for these atrium configurations, meaning that more electric lighting will be needed.

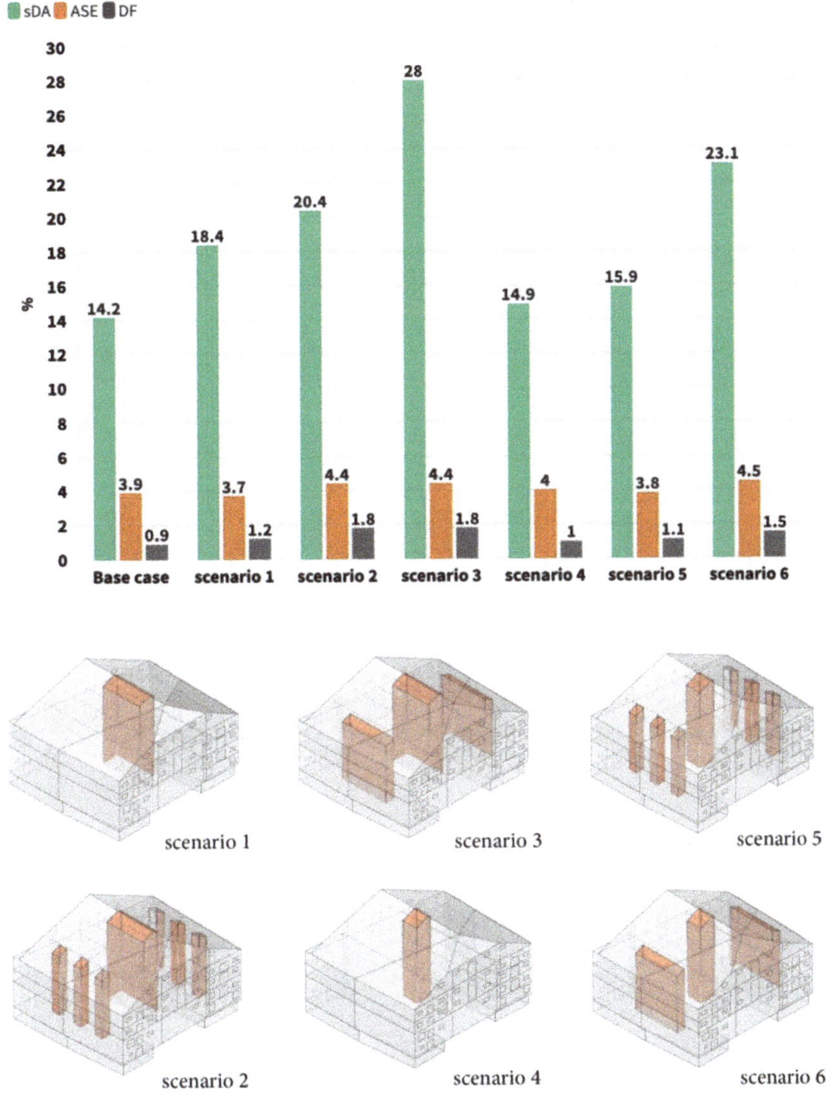

Figure 9. Results of $sDA_{300,50\%}$, $ASE_{1000,250}$, and DF in the six Scenarios.

As for the sDA metric, it was clear that the Scenarios with a united area of the atrium on the top and bottom had more advantages due to larger areas that face the external environment. It can also be noticed that Scenario 4 had the lowest sDA and DF values compared to the other Scenarios (absolute difference = 14.9% and 1% less, respectively). Scenarios 3 and 2 performed the best in achieving sDA with values of 28% and 20.4%, respectively. In contrast, DF in Scenarios 4 and 5 was 1% and 1.1%, and the sDA was 14.9% and 15.9%, respectively. These Scenarios yielded similar performances, thus being the worst-case Scenario among the six configurations, being closer to the base-case. In Scenarios 6 and 1, sDA was 23.1% and 18.4%, respectively. However, DF for Scenarios 6 and 1 was 1.5% and 1.2%. This means that these levels were not compliant with the requirement specified in EN 17037 and LEED v4.1.

Based on the comparative analysis of the six configurations, Scenarios 2 and 3 were thus identified as the best results, both in terms of sDA and of DF values, of daylight distribution inside the floor plan. Scenario 6 also showed high sDA, lower than Scenario 3 but even higher than Scenario 2. However, Scenario 6 was excluded due to a lower DF value and ASE value than Scenario 2. Note that Scenarios 2 and 3 showed the same DF value of 1.8%, which precedes the demand for daylight factor uniformity. Therefore, Scenarios 2 and 3 have been selected for more in-depth analysis.

3.3. Comparative Study of Selected Scenarios

According to the results, all the cases behaved similarly. Scenarios 2 and 3 demonstrate a great potential to become autonomous in terms of daylight, as their grids on the floor surface have more exposure to sunlight due to the atrium shape of the building.

3.4. Optimization of Scenario 3

Due to a large amount of data and time limitations, Scenario 3 was selected to investigate how the improvements affected the daylighting condition on each floor. Moreover, the amount of ASE was investigated to analyze glare amount.

The optimization process was carried out through the Galapagos component in Grasshopper [59]. Firstly, all building parameters, including walls, floors, windows, and atriums, were modeled in Grasshopper. Secondly, the materials related to each opaque and transparent surface were modeled. Finally, the algorithms of daylighting were designed in terms of optimization.

Galapagos was used to identify optimum fitness values and three parameters for each atrium through genetic algorithm optimization methods. The evolutionary algorithm was chosen since it was the only method to identify optimal solutions using Rhino and Grasshopper. Furthermore, this program was commonly used for architectural designs [60]. Galapagos was configured with three-parameter values and four variables to achieve circumstances similar to manual methods: Maximum Stagnant = 50; Population = 20; Maintain = 20%; Inbreeding = 50%. The Population value of 20 indicates that Honeybee simulates evolution 20 times for each generation, while the Maximum stagnant value of 50 computes up to 50 generations.

Because this study focuses on sDA as an objective for optimizing the atrium, most probably the highest area of the atrium on their boundaries could have opted. Therefore, two rectangles were considered to create an atrium for each atrium. The length and width of each rectangle are connected to the Galapagos component as variables. The highest value for the length and width of atriums based on their positions among columns was also considered to control the atrium size. It is worth mentioning that three steps were considered to decrease the number of simulations for the lengths and widths of each atrium as variables. As a result, the total number of simulations should be 1000; the skip and filtering method technique was designed to eliminate duplicates by 20% of the population that has remained stagnant. However, each atrium had 36 examples in this study, and the highest fitness value may be found within 30 generations.

Figure 10 shows the optimized shape of the atrium for Scenario 3. The shape of the optimized atriums had two rectangles consisting of bottom and top rectangles. After optimization, the amount of each variable was defined, and the final and optimized shape of the atria was achieved. The atrium sizes yielded the bottom rectangle bigger than the top one. The atrium sizes atrium from North-West (left to right in Figure 10) were: 1.8 m × 12.21 m in the top rectangle, 3 m × 15 m in the bottom rectangle, for the central atrium 10.30 m × 1.92 m in the top and 15.10 m × 1.92 m for bottom one, another side of the building the dimension was 3.60 m × 9.00 m in the top, 11.25 m × 3.60 m for the bottom. The optimized slope shape of the atria was defined to compensate for the decreased daylighting at the bottom of the atria. This atrium shape allows daylighting to be guided into the building, thus improving the daylighting across the building. This rule was the same for all three atrium shapes; however, the optimized sizes of each atrium were different.

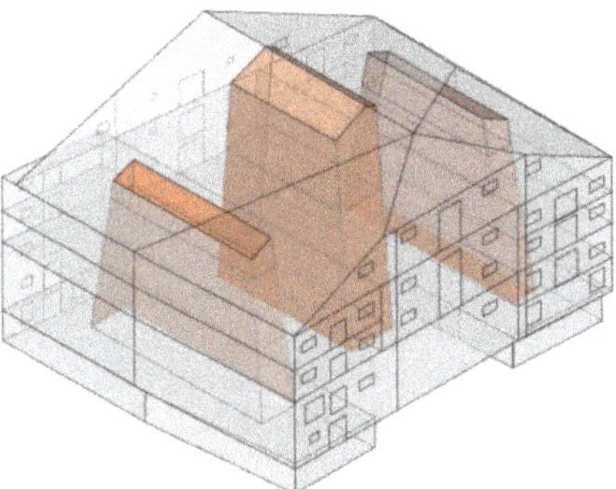

Figure 10. Configuration of atriums after optimization in Scenario 3.

In order to compare the daylight condition after the optimization, the optimized atrium was simulated, and the results of sDA, ASE, and DF were extracted.

Besides, the results of simulations of the optimized model, base-case Scenario, and Scenario 3 were compared in Figure 11, to better understand the effect of optimized atria on the building daylight condition.

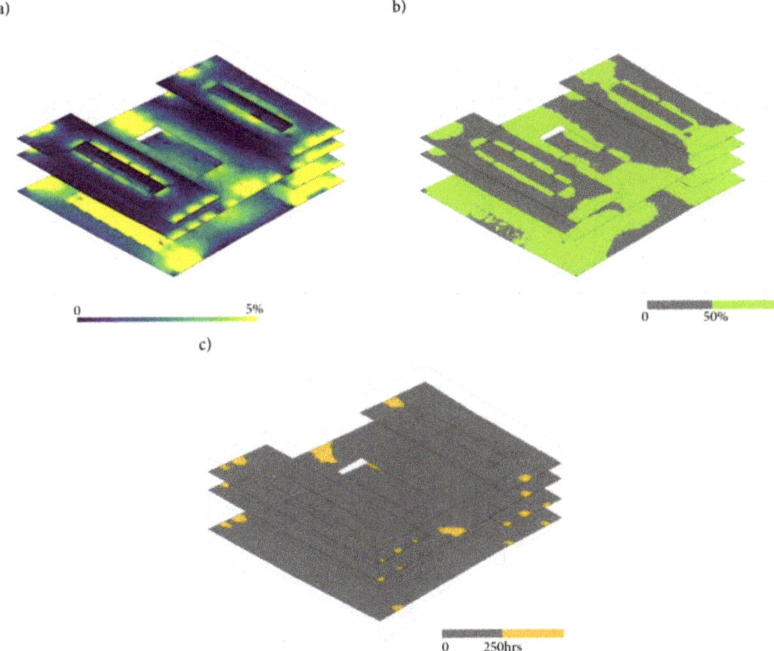

Figure 11. Simulation results in the optimized Scenario; (**a**) DF; (**b**) sDA$_{300,50\%}$; (**c**) ASE$_{1000,250}$.

The results shown in Figure 11 confirm that the value of sDA and ASE was 50.2% and 5.0%, respectively. The DF of optimized Scenario 3 was 2.7%, while a median daylight factor of 2.4% was obtained. Values of sDA and DF are higher on the ground floor than at the top floor. One reason for this is that the windows at the ground floor are much bigger than the equivalent ones at the top floor and the height of the ground floor is double as high. The second reason relates to the narrow shape of the atrium. High specularity of glass for light coming from slanted angles makes that the light from the sky is mainly reflected between glass with the vertical direction, which makes that the atrium glazing transports light from zenith effectively down to the bottom of the building. The light from zenith is a huge contributor at the location dominated by overcast sky condition with the yearly frequency of sunlight of about 20%. Additionally, the top floor was encompassed by the walls which blocked, and daylighting penetrations form the central part of building and the timber that was used for the top floor was much darker than the timber on the ground floor.

As indicated by the results, sDA in the base-case was 14.2%, increasing to 28% in Scenario 3 (Figure 12). After the optimization, the sDA exponentially increased and reached the value of 50.2%, with an increment of more than 32%. The ASE value in the base-case was 3.9%, while it slightly increased in Scenario 3, reaching the value of 4.4%. The ASE did not increase significantly even after optimization, reaching the value of 5%. According to the LEED protocol, the ASE should be lower than 10% to avoid glare occurrence based on the standard. After optimization, the low amount of ASE = 5% can guarantee a glare-free space within the interior space.

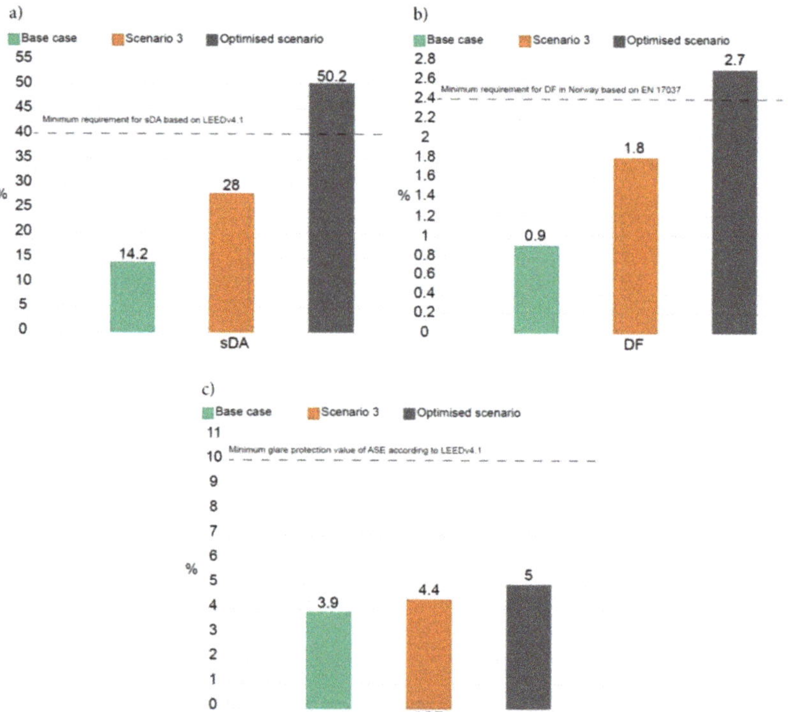

Figure 12. Results of simulated $sDA_{300,50\%}$, $ASE_{1000,250}$, and DF for Scenario 3 and optimized Scenario: (**a**) $sDA_{300,50\%}$; (**b**) DF; (**c**) $ASE_{1000,250}$.

Regarding DF, it increased from an initial value of 0.9% in the base-case to a value of 1.8% in Scenario 3, and further to a value of 2.7% after optimization. On the other hand, it is worth stressing that three large atria were needed to optimize daylighting, which resulted in less space available inside the building. Indeed, the optimized configuration will ensure effective usage with acceptable daylighting of the heritage building.

It is worth pointing out that both sDA and ASE are annual metrics that rely on threshold values. In more detail, sDA quantifies the fraction of space where the daylight autonomy

DA is over 50%, where DA quantifies the annual frequency of illuminances > 300 lx. It is a synthetic metric, which does not provide any information on the absolute illuminances. For this reason, a set of point-in-time simulations were run in ClimateStudio for some representative days (winter and summer solstice, and spring equinox), for an overcast sky and a clear sky, with the goal of visualizing the distribution of absolute illuminance inside the building for the base-case (Figure 13) and optimized atrium configuration (Figure 14).

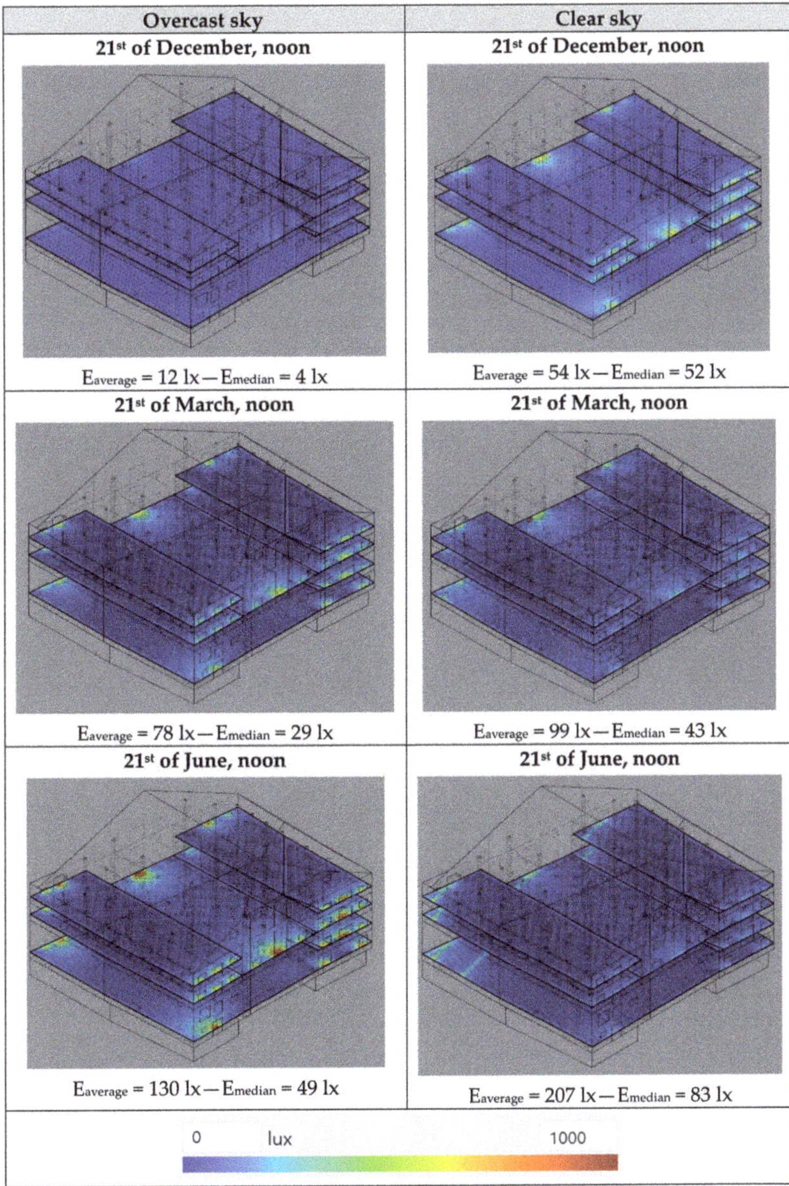

Figure 13. Point-in-time illuminances inside the building for the existing building (base-case).

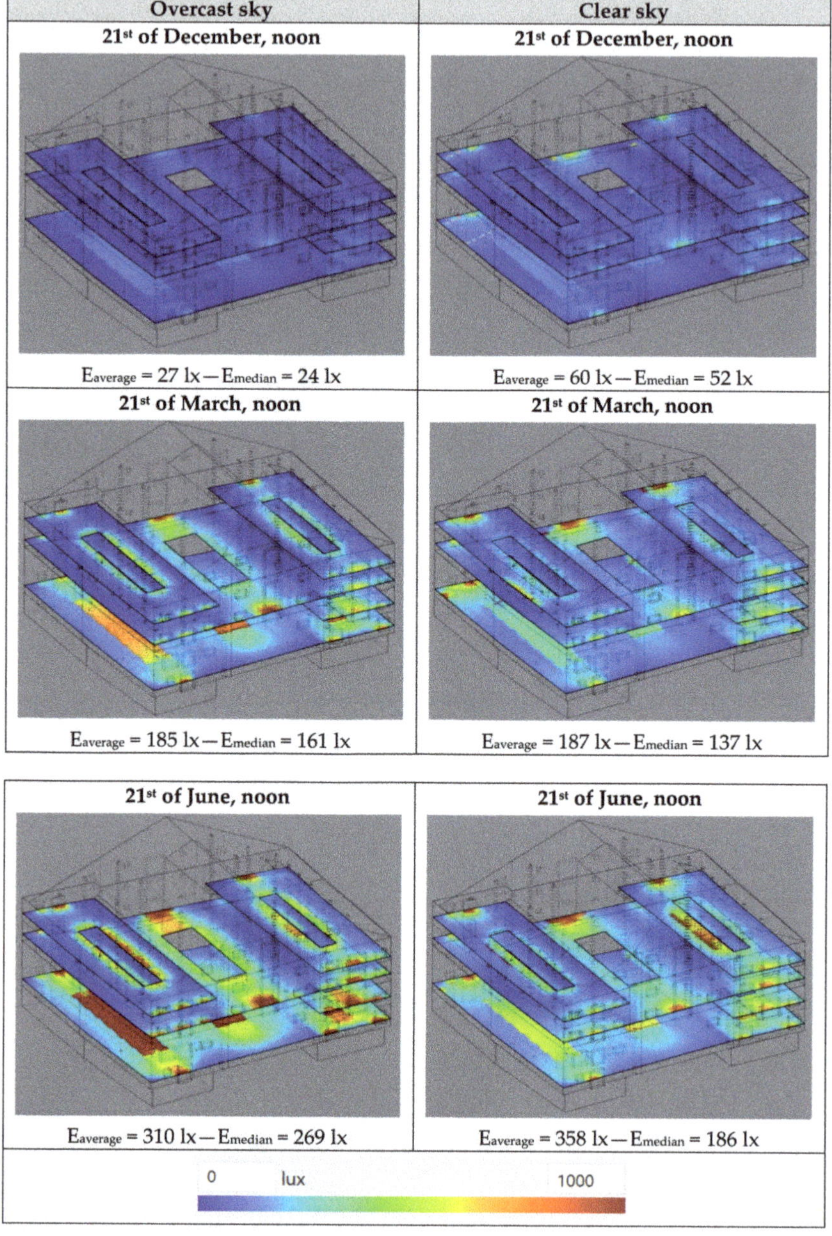

Figure 14. Point-in-time illuminances inside the building for the optimized atrium configuration.

Figure 13 shows the results obtained at noon for each day considered in base-case Scenario. According to the results, the highest average illuminance was achieved in June 21 under the clear sky with a value of 207 lx. However, for December 21, the average illuminance was 27 lx and 60 lx under the overcast and clear sky, respectively. The obtained results of point-in-time illuminance for the base-case Scenarios are as follow:

1. 21st of December, noon $E_{average}$ = 12 lx overcast sky $E_{average}$ = 54 lx clear sky
2. 21st of June, noon $E_{average}$ = 130 lx overcast sky $E_{average}$ = 207 lx clear sky
3. 21st of March, noon $E_{average}$ = 78 lx overcast sky $E_{average}$ = 99 lx clear sky.

As shown in Figure 14, the illuminance values remain low at the winter solstice, with an average illuminance of 27 lx for an overcast sky and 60 lx for a clear sky. Illuminance levels get increased at the spring equinox (similar average illuminance of 185 lx and 187 lx for an overcast and a clear sky, respectively) and results over 300 lx at the summer solstice, with an average value of 310 lx for an overcast sky and of 358 lx in the presence of a clear sky.

The point-in-time illuminances confirm that the poor daylighting in the baseline case was significantly improved through the optimized configuration.

4. Discussion

The study presented in this paper focused on the possibility of increasing daylighting levels in a real heritage building located in Trondheim (Norway) through rooflighting systems, which rely on a combination of skylights and corelighting systems (light atria). As is, the building has small windows and dark surfaces indoor, which determine scarce daylighting conditions in the internal spaces. Besides, as the building is listed, it is actually not possible to modify the layout of the façade by increasing the room area and therefore, the daylighting indoors through sidelighting systems. Therefore, rooflighting systems can represent the only way to admit daylight into the building, which is currently undergoing a deep renovation that will transform it from a warehouse (original usage) into a work area.

Daylighting inside the case-study building was analyzed through a combination of daylighting metrics:

1. daylight factor DF (point, average and median): although this is an obsolete metrics, it was included in the study as it is still assumed as a reference in many national regulations and legislations [61], including the latest European standard EN 17037 on 'Daylight in buildings'; besides, the calculation method included in the European standard EN 15193–1:2017 [62] to determine the energy demand for lighting relies on the daylight factor concept to determine the daylighting level in an indoor space [63] finally, as DF refers to an overcast sky condition, it seems particularly suitable for Nordic countries such as Norway, where the case-study building is located, where sunny skies occur less than 25% of the time on an annual basis and overcast skies are therefore predominant;
2. spatial Daylight Autonomy sDA and Annual Sunlight Exposure: unlike DF, which refers to an overcast sky and thus it does not include the dynamic variation of sunlight and skylight during a year, sDA and ASE are dynamic daylighting metrics that account for the presence of the sun on annual basis. Therefore, unlike DF, they account for all the factors that influence indoor daylighting and are more suitable to give preliminary information on the potential energy demand for lighting; besides, they are the metrics assumed in the LEED protocol for the credit on daylighting
3. point-in-time illuminances: as sDA and ASE are metrics with a threshold, they quantify the occurrence of time when illuminance values are over 300 lx (sDA) or over 1000 lx in direct sunlight only (ASE); no information is given on the absolute illuminance values during a year; for this reason, absolute illuminances were calculated for some reference days and sky conditions.

The research was conducted through a parametric approach to identify and optimize the shape and volume of different configurations of corelighting systems, using ClimateStudio and three Grasshopper tools: Ladybug, Honeybee and Galapagos. In order to answer the research questions that guided this investigation, the parametric study addressed six different Scenarios, which were defined by changing the atrium geometry (shape and volume). Therefore, after simulations, the sDA results showed that none of the Scenarios complied with LEED v4.1 and EN17037 requirements, as all the values were less than 50%.

Only the optimized configuration allowed minimum values of DF and sDA to be achieved in accordance with the standard EN 17037 and the LEED protocol.

As mentioned in the literature review, the shape and location of the atrium were the most influential parameters affecting the atria studied, along with the glazed area and the light reflectance properties of the atrium surfaces. Furthermore, genetic algorithm optimization has only recently been employed in designing new buildings and is rarely used in designing historical buildings [64,65]. Optimizing the building atrium has not been examined as thoroughly as optimizing other architectural variables such as form, side-lit openings, building façade, etc. To the authors' knowledge, parametric and optimization tools are quite new in their application to the heritage building. In this regard, an interesting study was carried out by Marzouk et al. [19]: in their study, the authors studied how to optimize skylights for an ancient building in Egypt, through Radiance, Daysim for daylighting analyses, and the multi-objective 'Octopus' component. The research presented in this paper addressed a similar problem, i.e., to admit daylight through rooflighting systems by relying on atria (corelighting).

In this regard, the Scenarios to enhance daylighting (starting from the base-case configuration with quite poor daylighting) and, even more, the optimized configuration result in relatively large volumes occupied by the corelighting systems. This resulted in a significant reduction of the available inner space (27% reduction in volume), which limits the exploitation of the internal layout in terms of usable space but guarantees better daylighting conditions in the exploited space; however, all the stages of the research were discussed and developed in accordance with the design team, which accepted the optimized configuration: such configuration will be built. Daylighting was conceived as one of the driving factors of the renovation, and the final, optimized solutions were considered satisfactory. Following adjustment, ASE was also tested to ensure visual comfort. The ASE increased to 5% after optimization, and the indoor daylight condition was greatly improved, and the risk of glare was averted.

Beside merits, the study also has some limits: for instance, the atrium shapes defined were quite simplistic from a geometrical viewpoint. Another issue was concerned with the fact that daylighting only was addressed in the research; other aspects of building physics (acoustics, thermal, energy demand) were not considered and should be the object of the following stages of the analysis. Increasing the window area in the roof (skylights + atria) may result in higher thermal losses in winter, with a potentially increased energy demand for heating, and in higher solar gains in summer, with a potential increase in the energy demand for cooling. As a solution to mitigate this latter problem, the upper surface of the atrium, as well as the skylights, were conceived as openable systems, so as to enhance natural ventilation in the summer period and to contrast the potential overheating problems due to direct sunlight. It would be interesting to analyze the overall performance of a building with rooflighting and corelighting solutions by integrating daylighting and thermal issues, as well as their related energy demands. At this stage of the research, though, the scope of the study was on daylighting only, as this was set as a primary design goal by the design team that is defining the retrofitting solutions: daylighting was given a priority over thermal and acoustical issues due to the high quality it brings in inside spaces, quality that is particularly important in Nordic climates.

Moreover, it is worth mentioning that the study is also limited by the extensive calculation time required to obtain high-quality results for large-scale building models with daylight simulation. Creating a model was a big challenge as the structure did not have single straight lines or angles. This affected both the size and complexity of the model, as well as the resolution of the results.

Moreover, the building was considered a heritage building, so finding the best way to minimize changes in the structure was so challenging that it took time to find a solution.

As far as future work is concerned, the present activity and its results only focused on the locations of the atriums; their sizes, the type of glasses, and their materials can be investigated in future work. Besides, another significant aspect of the atrium was that a considerable part of the atrium was transparent. Therefore, it results in heat losses or heat gain and overheating phenomena in the building. As a result, the thermal comfort and energy consumption when the atrium was considered can be investigated in future studies. More research is needed to develop strategies to improve daylight adequacy while achieving visual comfort utilizing diverse atrium arrangements.

Investigations regarding the daylight quality in the atrium could be further developed in future studies, as well as a proper sensitivity analysis of atrium design for the inputs and outputs daylight metrics that were exposed towards more daylight availability. It would also be interesting to assess the energy demand and verify how the new atrium configurations would affect the thermal conditions in the building.

5. Conclusions

This paper addressed a challenging task concerning heritage buildings: how to admit sufficient daylighting in a listed building, which means that no intervention in the façade can be considered. Particularly, an existing, listed wharf in Trondheim, Norway (L = 63° N) was selected as a case-study. Accordingly, the research question addressed in the study was: *"How to improve daylight conditions in heritage building locations in high latitudes?"*. Firstly, the base-case was a historical building with minimal modification to the original structure. Secondly, the base-case was deep and large, with small windows (WWR = 18.4%; WFR = 2.4%); for these reasons, this paper argued that an atrium should be used to increase daylight in construction.

To answer the research question, corelighting strategies were approached, and six configurations of skylights and atria were defined following the logic of the construction. Daylighting inside the six configurations was calculated using Climate Studio and Honeybee in Grasshopper simulations; surface reflectance measured in the building was used. Daylight metrics such as DF, sDA, and ASE were calculated and compared to discuss the daylight amount in the building using the target values required by the LEED v4.1 and EN 17037 regulations as benchmarks.

The simulations demonstrated that the ability to run dynamic daylight simulations for early-stage design was extremely viable. Therefore, supporting earlier studies and comparisons between computer-based simulations and physical models. After comparing the results, the optimal Scenario for supplying adequate daylighting was identified and this atrium configuration was optimized using Galapagos genetic algorithms.

The results showed that the sDA in the existing building (base-case) of 14.2% was increased to 50.2% after optimization, thus qualifying for the LEED credit. Accordingly, the initial average DF value of 0.9% was increased up to 2.7%, after the optimization, thus complying with EN 17037. Furthermore, ASE and views were also analyzed to guarantee visual comfort after the optimization. Results showed that the ASE value was 3.9% in the best case and increased to 5% after optimization. According to the standard, a value below 10% would provide a glare-free condition for building users. Based on the results, the interior daylight condition was improved adequately, and the risk of glare was prevented. On the other hand, the volume occupied by the atria was about 1290 m^3 which was equal to 13.7% compared to the whole building volume. Thus, implying a considerable reduction of the internal space available for activities.

Author Contributions: Conceptualization, B.M. and V.R.M.L.V.; methodology, F.P., B.M. and V.R.M.L.V.; software, F.P.; formal analysis, F.P. and V.R.M.L.V.; investigation, F.P. and V.R.M.L.V.; writing—original draft preparation, F.P.; writing—review and editing, V.R.M.L.V. and B.M.; visualization, F.P.; supervision, B.M. and V.R.M.L.V. All authors have read and agreed to the published version of the manuscript.

Funding: This research received no external funding.

Data Availability Statement: Not applicable.

Conflicts of Interest: The authors declare no conflict of interest.

References

1. Buda, A.; De Place Hansen, E.J.; Rieser, A.; Giancola, E.; Pracchi, V.N.; Mauri, S.; Marincioni, V.; Gori, V.; Fouseki, K.; López, C.S.P.; et al. Conservation-Compatible Retrofit Solutions in Historic Buildings: An Integrated Approach. *Sustainability* **2021**, *13*, 2927. [CrossRef]
2. Sibley, M. Let There Be Light! Investigating Vernacular Daylighting in Moroccan Heritage Hammams for Rehabilitation, Benchmarking and Energy Saving. *Sustainability* **2018**, *10*, 3984. [CrossRef]
3. *CIE 026/E:2018*; CIE System for Metrology of Optical Radiation for IpRGC-Influenced Responses to Light. International Commission on Illumination: Vienna, Austria, 2018.
4. MIsIrlIsoy, D.; Günçe, K. Adaptive Reuse Strategies for Heritage Buildings: A Holistic Approach. *Sustain. Cities Soc.* **2016**, *26*, 91–98. [CrossRef]
5. Günçe, K.; Misirlisoy, D. Adaptive Reuse of Military Establishments as Museums: Conservation vs. Museography. *WIT Trans. Built Environ.* **2014**, *143*, 125–136.
6. Langston, C.; Shen, L.Y. Application of the Adaptive Reuse Potential Model in Hong Kong: A Case Study of Lui Seng Chun. *Int. J. Strateg. Prop. Manag.* **2007**, *11*, 193–207. [CrossRef]
7. Balocco, C.; Calzolari, R. Natural Light Design for an Ancient Building: A Case Study. *J. Cult. Herit.* **2008**, *9*, 172–178. [CrossRef]
8. Khalil, A.M.R.; Hammouda, N.Y.; El-Deeb, K.F. Implementing Sustainability in Retrofitting Heritage Buildings. Case Study: Villa Antoniadis, Alexandria, Egypt. *Heritage* **2018**, *1*, 57–87. [CrossRef]
9. Singh, M.K.; Attia, S.; Mahapatra, S.; Teller, J. Assessment of Thermal Comfort in Existing Pre-1945 Residential Building Stock. *Energy* **2016**, *98*, 122–134. [CrossRef]
10. Marzouk, M.; ElSharkawy, M.; Mahmoud, A. Analysing User Daylight Preferences in Heritage Buildings Using Virtual Reality. *Build. Simul.* **2022**, *15*, 1561–1576. [CrossRef]
11. IEA SHC||Task 59||Renovating Historic Buildings Towards Zero Energy. Available online: https://task59.iea-shc.org/ (accessed on 30 September 2022).
12. Rose, J.; Thomsen, K.E. Comprehensive Energy Renovation of Two Danish Heritage Buildings within Iea Shc Task 59. *Heritage* **2021**, *4*, 2746–2762. [CrossRef]
13. Ahmad, S.S.; Ahmad, N.; Talib, A. Ceiling Geometry and Daylighting Performance of Side Lit Historical Museum Galleries Undertropical Overcast Sky Condition. *Pertanika J. Sci. Technol.* **2017**, *25*, 287–298.
14. Al-Maiyah, S.; Elkadi, H. The Role of Daylight in Preserving Identities in Heritage Context. *Renew. Sustain. Energy Rev.* **2007**, *11*, 1544–1557. [CrossRef]
15. Mehr, S.Y. Analysis of 19th and 20th Century Conservation Key Theories in Relation to Contemporary Adaptive Reuse of Heritage Buildings. *Heritage* **2019**, *2*, 920–937. [CrossRef]
16. Zaikina, V.; Moscoso, C.; Matusiak, B.; Chelli, A.; Balasingham, I. Users' Satisfaction of Indoor Environmental Quality Conditions in ZEB+ at High Latitudes. *IOP Conf. Ser. Earth Environ. Sci.* **2019**, *352*, 012001. [CrossRef]
17. Bastian, Z.; Schnieders, J.; Conner, W.; Kaufmann, B.; Lepp, L.; Norwood, Z.; Simmonds, A.; Theoboldt, I. Retrofit with Passive House Components. *Energy Effic.* **2022**, *15*, 10. [CrossRef]
18. Marzouk, M.; ElSharkawy, M.; Mahmoud, A. Optimizing Daylight Utilization of Flat Skylights in Heritage Buildings. *J. Adv. Res.* **2022**, *37*, 133–145. [CrossRef]
19. Marzouk, M.; ElSharkawy, M.; Eissa, A. Optimizing Thermal and Visual Efficiency Using Parametric Configuration of Skylights in Heritage Buildings. *J. Build. Eng.* **2020**, *31*, 101385. [CrossRef]
20. Medvedeva, N.; Kolesnikov, S. Specifics of Daylight in Atrium Spaces of Architectural Objects. *IOP Conf. Ser. Mater. Sci. Eng.* **2021**, *1079*, 022066. [CrossRef]
21. Matusiak, B.; Aschehoug, M.; Littlefair, P. Daylighting Strategies for an Infinitely Long Atrium: An Experimental Evaluation. *Int. J. Light. Res. Technol.* **1999**, *31*, 23–34. [CrossRef]
22. Lechner, N. *Heating, Cooling, Lighting: Sustainable Design Methods for Architects*; Wiley: Hoboken, NJ, USA, 2008.
23. Pilechiha, P.; Norouziasas, A.; Ghorbani Naeini, H.; Jolma, K. Evaluation of Occupant's Adaptive Thermal Comfort Behaviour in Naturally Ventilated Courtyard Houses. *Smart Sustain. Built Environ.* **2021**, *ahead of print*. [CrossRef]
24. Atrium | WBDG—Whole Building Design Guide. Available online: https://www.wbdg.org/space-types/atrium#spcatt (accessed on 12 February 2022).
25. Samant, S. A Critical Review of Articles Published on Atrium Geometry and Surface Reflectances on Daylighting in an Atrium and Its Adjoining Spaces. *Archit. Sci. Rev.* **2010**, *53*, 145–156. [CrossRef]
26. Dai, B.; Tong, Y.; Hu, Q.; Chen, Z. Characteristics of Thermal Stratification and Its Effects on HVAC Energy Consumption for an Atrium Building in South China. *Energy* **2022**, *249*, 123425. [CrossRef]
27. Calcagni, B.; Paroncini, M. Daylight Factor Prediction in Atria Building Designs. *Sol. Energy* **2004**, *76*, 669–682. [CrossRef]
28. Younis, G.M.; Abdulatif, F.S.; Mostafa, W.S. Impact of Design Characteristics of Daylight Elements to Creating Healthy Internal Environment for School Buildings Evaluation the Status of Schools in Mosul City. *Period. Eng. Nat. Sci.* **2019**, *7*, 1354–1372. [CrossRef]
29. Matusiak, B. Daylighting in Linear Atrium Buildings At High Latitudes. Ph.D. Thesis, Norwegian University of Science and Technology, Trondheim, Norway, 1998.

30. Ebrahimpour, A.; Maerefat, M. Application of Advanced Glazing and Overhangs in Residential Buildings. *Energy Convers. Manag.* **2011**, *52*, 212–219. [CrossRef]
31. Ochoa, C.E.; Aries, M.B.C.; van Loenen, E.J.; Hensen, J.L.M. Considerations on Design Optimization Criteria for Windows Providing Low Energy Consumption and High Visual Comfort. *Appl. Energy* **2012**, *95*, 238–245. [CrossRef]
32. Thalfeldt, M.; Pikas, E.; Kurnitski, J.; Voll, H. Facade Design Principles for Nearly Zero Energy Buildings in a Cold Climate. *Energy Build.* **2013**, *67*, 309–321. [CrossRef]
33. Yousefi, F.; Gholipour, Y.; Yan, W. A Study of the Impact of Occupant Behaviors on Energy Performance of Building Envelopes Using Occupants' Data. *Energy Build.* **2017**, *148*, 182–198. [CrossRef]
34. Obradovic, B.; Matusiak, B.S. Daylight Transport Systems for Buildings at High Latitudes. *J. Daylighting* **2019**, *6*, 60–79. [CrossRef]
35. Grynning, S.; Time, B.; Matusiak, B. Solar Shading Control Strategies in Cold Climates—Heating, Cooling Demand and Daylight Availability in Office Spaces. *Sol. Energy* **2014**, *107*, 182–194. [CrossRef]
36. Van Den Wymelenberg, K. Patterns of Occupant Interaction with Window Blinds: A Literature Review. *Energy Build.* **2012**, *51*, 165–176. [CrossRef]
37. Szybinska, B.S. Daylighting Is More than an Energy Saving Issue. In *Energy Efficient Buildings*; IntechOpen: London, UK, 2017.
38. Eltaweel, A.; Mandour, M.A.; Lv, Q.; Su, Y. Daylight Distribution Improvement Using Automated Prismatic Louvre. *J. Daylighting* **2020**, *7*, 84–92. [CrossRef]
39. Marzouk, M.; Eissa, A.; ElSharkawy, M. Influence of Light Redirecting Control Element on Daylight Performance: A Case of Egyptian Heritage Palace Skylight. *J. Build. Eng.* **2020**, *31*, 101309. [CrossRef]
40. Edwards, L.; Torcellini, P. *A Literature Review of the Effects of Natural Light on Building Occupants*; National Renewable Energy Laboratory: Golden, CO, USA, 2002.
41. Xue, P.; Mak, C.M.; Cheung, H.D. The Effects of Daylighting and Human Behavior on Luminous Comfort in Residential Buildings: A Questionnaire Survey. *Build. Environ.* **2014**, *81*, 51–59. [CrossRef]
42. Acosta, I.; Campano, M.Á.; Molina, J.F. Window Design in Architecture: Analysis of Energy Savings for Lighting and Visual Comfort in Residential Spaces. *Appl. Energy* **2016**, *168*, 493–506. [CrossRef]
43. Huitfeldtbrygga. Available online: https://www.huitfeldtbrygga.no/#sidewidgetarea (accessed on 12 February 2022).
44. Piraei, F. Evaluation and Optimization of Daylighting in Heritage Buildings: A Case Study at High Latitudes. Master Science's Thesis, Politecnico di Torino, Turin, Italy, 2022.
45. Ward Larson, G.; Shakespeare, R. *Rendering with Radiance: The Art and Science of Lighting Visualization*; Morgan Kaufman: Burlington, MA, USA, 1998.
46. Mardaljevic, J. Validation of a Lighting Simulation Program under Real Sky Conditions. *Light. Res. Technol.* **1995**, *27*, 181–188. [CrossRef]
47. CEN. *European Standard EN 17037*; Daylight in Buildings. Comité Européen de Normalisation: Brussels, Belgium, 2018.
48. IESNA Daylighting Metrics Committee. *Spatial Daylight Autonomy (SDA) and Annual Sunlight Exposure (ASE)*; Illuminating Engineering Society: New York, NY, USA, 2012.
49. U.S. Green Building Council. *LEED v4.1 Building Design and Construction*; U.S. Green Building Council: Washington, DC, USA, 2020.
50. Kharvari, F. An Empirical Validation of Daylighting Tools: Assessing Radiance Parameters and Simulation Settings in Ladybug and Honeybee against Field Measurements. *Sol. Energy* **2020**, *207*, 1021–1036. [CrossRef]
51. Roudsari, M.S.; Pak, M. Ladybug: A Parametric Environmental Plugin for Grasshopper to Help Designers Create an Environmentally-Conscious Design. In Proceedings of the BS 2013: 13th Conference of the International Building Performance Simulation Association, Chambéry, France, 26–28 August 2013.
52. Wong, I.L. A Review of Daylighting Design and Implementation in Buildings. *Renew. Sustain. Energy Rev.* **2017**, *74*, 959–968. [CrossRef]
53. Mahdavinejad, M.; Nazar, N.S. Daylightophil High-Performance Architecture: Multi-Objective Optimization of Energy Efficiency and Daylight Availability in BSk Climate. *Energy Procedia* **2017**, *115*, 92–101. [CrossRef]
54. Susa-Páez, A.; Piderit-Moreno, M.B. Geometric Optimization of Atriums with Natural Lighting Potential for Detached High-Rise Buildings. *Sustainability* **2020**, *12*, 6651. [CrossRef]
55. Rastegari, M.; Pournaseri, S.; Sanaieian, H. Daylight Optimization through Architectural Aspects in an Office Building Atrium in Tehran. *J. Build. Eng.* **2021**, *33*, 101718. [CrossRef]
56. Rastegari, M.; Pournaseri, S.; Sanaieian, H. Analysis of Daylight Metrics Based on the Daylight Autonomy (Dla) and Lux Illuminance in a Real Office Building Atrium in Tehran. *Energy* **2023**, *263*, 125707. [CrossRef]
57. Erlendsson, O. Daylight Optimization: A Parametric Study of Atrium Design. Master Science's Thesis, Royal Institute of Technology, Stockholm, Sweden, 2014.
58. Norouziasas, A. Active Transparent Facades: Experimental and Numerical Evaluation on Daylighting. Master Science's Thesis, Politecnico di Torino, Turin, Italy, 2021.
59. Rutten, D. Galapagos: On the Logic and Limitations of Generic Solvers. *Archit. Des.* **2013**, *83*, 132–135. [CrossRef]
60. Yi, Y.K.; Kim, H. Agent-Based Geometry Optimization with Genetic Algorithm (GA) for Tall Apartment's Solar Right. *Sol. Energy* **2015**, *113*, 236–250. [CrossRef]

61. Nigra, M.; Lo Verso, V.R.M.; Robiglio, M.; Pellegrino, A.; Martina, M. 'Re-Coding' Environmental Regulation–a New Simplified Metric for Daylighting Verification during the Window and Indoor Space Design Process. *Archit. Eng. Des. Manag.* **2022**, *18*, 521–544. [CrossRef]
62. CEN. *European Standard EN 15193-1*; Energy Performance of Buildings—Energy Requirements for Lighting—Part 1: Specifications, Module M9. European Committee for Standardisation (CEN): Brussels, Belgium, 2017; pp. 1–140.
63. Lo Verso, V.R.M.; Pellegrino, A. Energy Saving Generated Through Automatic Lighting Control Systems According to the Estimation Method of the Standard EN 15193-1. *J. Daylighting* **2019**, *6*, 131–147. [CrossRef]
64. Shahbazi, Y.; Heydari, M.; Haghparast, F. An Early-Stage Design Optimization for Office Buildings' Façade Providing High-Energy Performance and Daylight. *Indoor Built Environ.* **2019**, *28*, 1350–1367. [CrossRef]
65. Chen, X.; Yang, H.; Zhang, W. Simulation-Based Approach to Optimize Passively Designed Buildings: A Case Study on a Typical Architectural Form in Hot and Humid Climates. *Renew. Sustain. Energy Rev.* **2018**, *82*, 1712–1725. [CrossRef]

 *buildings*

Article

Evaluation of Passive Cooling and Thermal Comfort in Historical Residential Buildings in Zanzibar

Chang Liu [1,2], Hui Xie [1,2,*], Hartha Mohammed Ali [1] and Jing Liu [3]

[1] Faculty of Architecture and Urban Planning, Chongqing University, Chongqing 400030, China
[2] Key Laboratory of New Technology for Construction of Cities in Mountain Area, Ministry of Education, Chongqing University, Chongqing 400030, China
[3] Faculty of Engineering, Zunyi Normal University, Zunyi 563006, China
* Correspondence: xh@cqu.edu.cn

Abstract: Indoor thermal comfort is essential for occupants' well-being, productivity, and efficiency. Global climate change is leading to extremely high temperatures and more intense solar radiation, especially in hot, humid areas. Passive cooling is considered to be one of the environmental design strategies by which to create indoor thermal comfort conditions and minimize buildings' energy consumption. However, little evidence has been found regarding the effect of passive cooling on the thermal comfort of historical buildings in hot–dry or hot–humid areas. Therefore, we explored the passive cooling features (north-south orientation, natural ventilation, window shading, and light color painted walls) applied in historic residential buildings in Zanzibar and evaluated the residents' thermal responses and comfort perception based on questionnaires and field surveys. The results showed that the average predicted mean votes (PMVs) were 1.23 and 0.85 for the two historical case study buildings; the average predicted percentages of dissatisfaction (PPD) were 37.35% and 20.56%, respectively. These results indicate that the thermal conditions were not within the acceptable range of ASHRAE Standard 55. Further techniques, such as the use of lime plaster, wash lime, and appropriate organization, are suggested for the improvement of indoor thermal comfort in historical buildings in Zanzibar. This study provides guidelines to assist architects in designing energy-efficient residential buildings, taking into account cultural heritage and thermal comfort in buildings.

Keywords: historical residential buildings; passive cooling; thermal comfort

Citation: Liu, C.; Xie, H.; Ali, H.M.; Liu, J. Evaluation of Passive Cooling and Thermal Comfort in Historical Residential Buildings in Zanzibar. *Buildings* **2022**, *12*, 2149. https://doi.org/10.3390/buildings12122149

Academic Editors: Yue Wu, Zheming Liu and Zhe Kong

Received: 9 October 2022
Accepted: 30 November 2022
Published: 6 December 2022

Publisher's Note: MDPI stays neutral with regard to jurisdictional claims in published maps and institutional affiliations.

1. Introduction

Nowadays, global climate change is becoming one of the greatest environmental challenges [1]. It includes serious disruptions to the weather and climate patterns around the world, such as the impacts on rainfall, extreme weather events, and sea level rises, rather than just moderate temperature increases. One of the main reasons for the current environmental pollution problem is the excessive use of energy [2,3]. Across the entire world, buildings are major consumers of energy and major sources of greenhouse gas emissions. Around 30–40% of energy consumption and 30% of CO_2 emissions come from buildings [4–7]. Over the last 50 years, there has been a growing demand for houses and the energy necessary to run them due to a rapidly increasing world population [8,9]. According to the International Energy Agency (IEA), between 2012 and 2030, the building sector's total energy consumption will increase by 4.74 quadrillion Btu (QBtu).

IEA reports indicate that space heating and cooling are responsible for 30% of all energy consumption [7]. Passive cooling is considered part of an overall environmental design strategy that attempts to provide comfortable conditions in the interior of a building and to minimize buildings' energy consumption [10]. Previous research has reviewed passive cooling techniques for buildings [10], examined the influence of environmental and building factors on the performance of passive cooling, and focused on improving passive cooling techniques [11]. When incorporating passive cooling measures, thermal comfort should always be considered another important factor when evaluating the indoor environmental quality of buildings [12–15].

Over the last decade, the energy efficiency and thermal comfort of historical buildings have attracted increasing attention globally [16–18]. In addition to the balance between energy and societal climate, the improvement of the energy efficiency of historical buildings should also take the building conversation and cultural heritage into consideration [16]. Numerous previous studies have focused on the development of the numerical modeling of passive cooling techniques for historical buildings, the evaluation of air quality when using passive cooling techniques in historical buildings, and energy efficiency evaluations of heating systems in historical buildings; however, little research has systematically evaluated the passive cooling effect on thermal comfort [19–21]. Ricciardi et al. and Diler et al. evaluated the thermal comfort of historical theaters and mosques [12,22]. However, historical residential buildings, which are more numerous than theaters or mosques and are more closely related to daily life, are not taken into consideration. For historical buildings, passive cooling strategies in terms of orientation, construction materials, openings, and shadings are limited to the local culture. Therefore, there is still a need to evaluate the effect of the passive cooling effect on the thermal comfort of historical buildings in hot–dry or hot–humid areas.

The purposes of this research are to explore the passive cooling features used in historic residential buildings, as well as to evaluate the effects of north-south orientation, natural ventilation, pitched roof, light color finishing, and window shading on the thermal comfort of these historic buildings. This study is expected to provide guidelines to assist architects in designing energy-efficient residential buildings, taking into account both the cultural heritage and thermal comfort of the buildings.

2. Research Methods

The developing world faces greater challenges than the developed world in terms of both the impact of climate change and the capacity to respond to it. These factors are even crucial for a country, such as Zanzibar, with high population growth and increasing energy demand. The Stone Town of Zanzibar forms a unique urban settlement due to a combination of geographical and historical circumstances. Due to its unique architecture and culture, it was declared a UNESCO World Heritage Site in 2000 and is one of the best manifestations of trade and maritime history on the East African Coast [23,24]. Therefore, historical residential buildings in Zanzibar were selected as the case study areas for the evaluation of the passive cooling technique and its effect on thermal comfort.

The study area was Stone Town, located in the urban historical site of Zanzibar; it has a humid tropical monsoon climate and is in a tropical region of East Africa with an equator line passing through it. Stone Town is a city of prominent historical and artistic importance in East Africa. Its unique architectural style mostly dates back to the 19th century, which reflects not only its once powerful role as the capital of an empire spanning two continents, but also the cultural fusion that it has encouraged and nurtured over the centuries. Several historical buildings in Stone Town can be found on the seafront, such as former palaces of the sultans, fortifications, and mosques.

In order to evaluate the thermal comfort of the historical buildings of Stone Town in Zanzibar, questionnaires and field surveys were applied regarding two buildings. The questionnaire design, selection of case study buildings, and physical measurement details in the field survey are presented in this section.

2.1. Questionnaire Survey

The questionnaire surveys were conducted over a series of four weeks in February, the summer season, in Stone Town, Zanzibar. The questionnaire survey was conducted three times a day, between 7 a.m. and 10 a.m., 12 p.m. and 3 p.m., and 5 p.m. and 9 p.m.

The sample size of this study was 180, which accounts for 10% of the total population of Stone Town Historical City [25]. All participants lived in Stone Town, Zanzibar.

Subjective thermal comfort data were recorded using a questionnaire designed with considerations given to ASHRAE Standard 55 [26]. The questionnaire developed for this survey was divided into four main sections, as follows.

Part 1: Background and demography of the respondents. The first section consisted of general information about the participants, including their age, gender, and years of living in their current building.

Part 2: Building features. The second section covered several questions about the building's features, such as openings (door, window types, and materials), the height of the building, wall and roof materials, and the use of different thermal environment control means (windows and local fan/air condition/shading devices).

Part 3: Indoor and outdoor environment information. The indoor environment information included temperature, relative humidity, air velocity, and satisfaction evaluations regarding indoor thermal environment quality and personal environmental control. The outdoor environment information data during the field survey period were obtained from the Zanzibar Meteorological Department, including factors such as temperature, air velocity, and sun radiation.

Part 4: Thermal comfort evaluation. Thermal comfort was rated on an ASHRAE 7-point scale with the following aspects: thermal sensation vote (TSV, −3: cold, −2: cool, −1: slightly cool, 0: neutrality, 1: slightly warm, 2: warm and 3: hot), humidity predicted vote (HPV, −3 much too dry, −2 too dry, −1 slightly dry, 0 just right, +1 slightly humid, +2 too humid, +3 much too humid), and draft perception vote (DPV indicates that the air is perceived to be: −3 much too still, −2 too still, −1 slightly still, 0 just right, +1 slightly right, +2 too breezy, +3 much too breezy). The McIntyre scale was used to assess thermal preferences; it is a 3-point scale. The questions asked were, "Would you like to have it warmer (1), no change (0), cooler (−1)?" or "At this moment, would you prefer to feel warmer, cooler, or no change?" The thermal acceptability vote was used to determine the judgment regarding the perception of the thermal environment. The question was, "At this moment, do you consider the thermal environment acceptable (0) or unacceptable (1)? Detailed questionnaires can be found in Appendix A.

All the information collected by the questionnaire surveys was saved and analyzed in SPSS25.0.

2.2. Field Survey

To evaluate the thermal comfort level of the indoor environment in historical buildings, a field survey was carried out in two residential houses. The buildings were monitored for six consecutive days in February, as February is the hottest month and is the period in which the northeast monsoon wind blows relatively strongly. Each building was surveyed three times a day, between 7 a.m. and 10 a.m., 12 p.m. and 3 p.m., and 5 p.m. and 9 p.m., together with the questionnaire.

2.2.1. Selection of the Case Study Buildings

The criteria for selecting buildings for the field study were as follows:

- The buildings selected were grouped inside the Stone Town Historical Area.
- The buildings were built 200 to 270 years ago, around the middle of the 18th century.
- The buildings had almost the same structure and building envelope and were constructed with coral stones but with different building forms (typology).
- The buildings were easily accessible.
- Architectural drawings of the apartments were available.

Finally, two historical residential buildings, named HRB1 (Figure 1a) and HRB2 (Figure 1b), were chosen for the case study. The floor plans of HRB1 and HRB2 are presented in Figures 2 and 3.

HRB1 is a single-family residential house located in Shangani Ward and is 15 m from a main road. The building has a rectangular shape and a surface area of approximately 183.75 square meters. It is a three-story building; the ground floor is used as a shop, but it was not surveyed in this research, while the remaining upper two stories are used for residential purposes with a single family of four members. The first floor of this building is used as a living space, and the second floor is used for bedrooms (Figure 2).

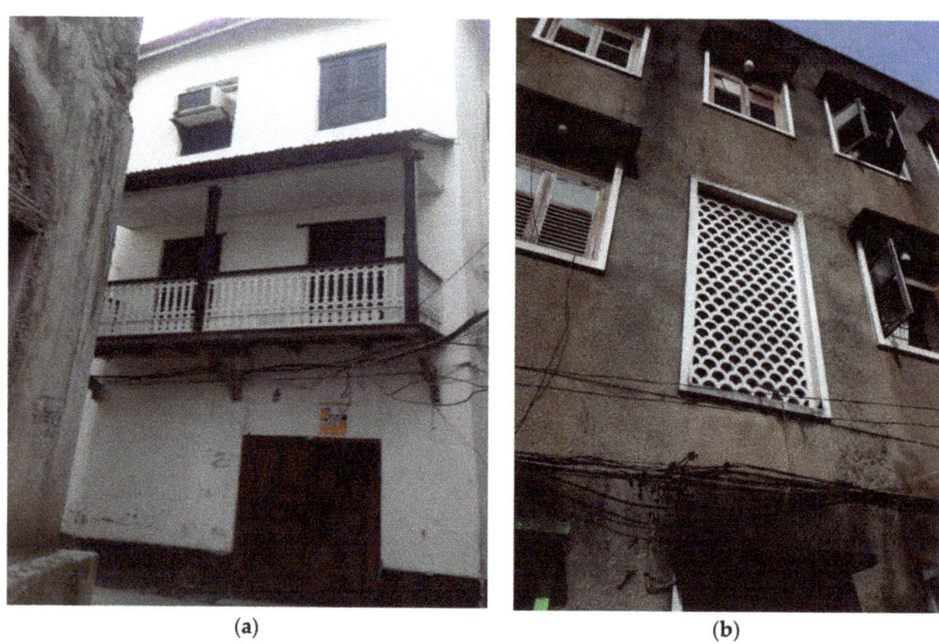

(**a**) (**b**)

Figure 1. Case study of historical residential buildings. (**a**) HRB1, (**b**) HRB2.

(**a**) (**b**)

Figure 2. Floor plan for HRB1. The positions of the thermohygrometers are marked with red dots (Points A and B). (**a**) First floor, (**b**) Second floor.

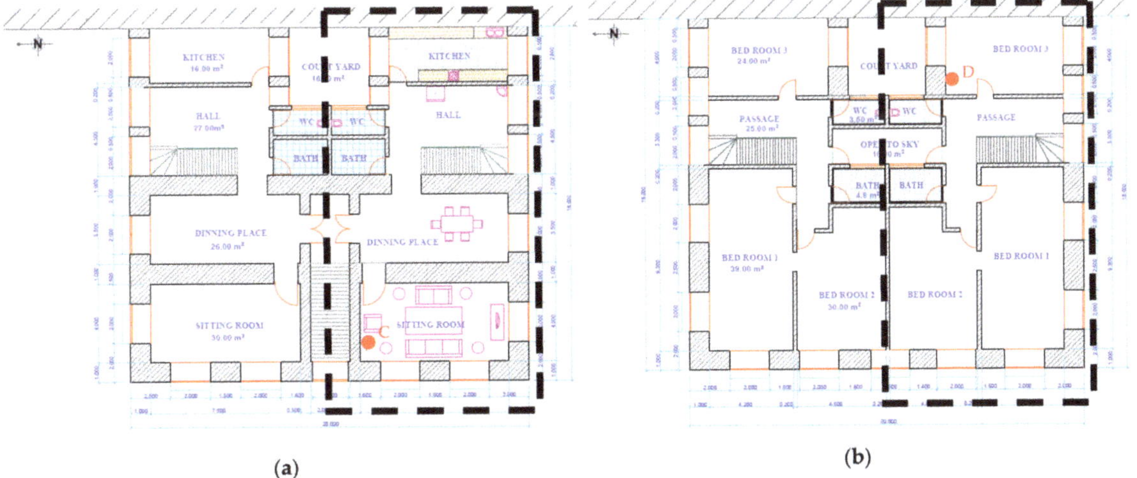

Figure 3. Floor plan for HRB2. The positions of the thermohygrometers are marked with red dots (Points C and D). Spaces enclosed with dashed lines were studied in this research. (**a**) First floor, (**b**) Second floor.

HRB2 is a multi-story building located in Malindi Ward. The building has a square shape and a surface area of approximately 360 square meters, which is two times the surface area of HRB1, with a 4 m-by-4 m courtyard accessible only by ground floor users. There are two dependent flats/units on the ground floor and another two flats/units on the first and second floors, which together give a total of four flats/units and four households in HRB2. Due to time and financial constraints, only one representative double-floor apartment was selected for the study. The first floor of this apartment is used as a living space, and the second floor is used for bedrooms (Figure 3).

2.2.2. Physical Measurements

The indoor environmental parameters, such as air temperature and relative humidity, were measured using a thermohygrometer device (TES1341) for a measurement period of approximately 10 min at an interval of 1 min. Three sampling points were selected, which resulted in 18 sampling points for this study in each surveyed building. For HRB1, the thermohygrometers were hung on the wall in the living room (Point A), bedroom 2 (Point B), and outside (Figure 2). For HRB2, the instruments were hung in the living room (Point C), bedroom 3 (Point D), and outside (Figure 3). The thermohygrometers were placed at a height of 1.1 m above the floor, which represents the height of the neck of a seated occupant, and 1.5 m away from the wall. Other environmental parameters, such as air velocity and direction, were assumed to be the same as those measured by meteorological stations.

3. Results

The results of the questionnaires and field studies from the study period are given in this section.

3.1. Questionnaire Survey Results

A total of 159 of 180 samples were valid for the questionnaire survey. The gender proportion of the samples was as follows: female, 96 (60.37%), and male, 63 (39.6%).

3.1.1. Building Envelope Evaluation

A total of 100% of the respondents' results showed that their buildings were built using coral stone, as shown in Table 1. Stone coral was actually beneficial to the thermal conditions of the building. Their heavy structure cools indoor spaces by delaying heat transmission from the outside, but in the evening, they release heat to the interior space (as well as to the outside). More air movement via the opening of windows is helpful for decreasing indoor temperature at night and in the early morning.

Table 1. Results of the building envelope evaluation.

Building Envelope		Percentage	
Wall materials		Coral Stone 100%	
Window openings	1–7 26.41%	7–14 63.52%	>14 10.06%
Window type	Louver/shutters 68.55%	Casement 31.44%	
Ceiling materials	Mangrove ceiling 68.55%	Cardboard ceiling 20.75%	Reinforced concrete ceiling 10.69%
Roof type	Hip roof 60.89%	Pitched roof 20.75%	Flat roof 16.36%
Sun shading device	CVB [1] 15.72%	VB [1] 10.69%	C [1] 73.58%

[1] CVB indicates curtains and verandah/balcony; VB indicates verandah/balcony; C indicates curtains.

Regarding openings, 26.41% of the occupants had between 1 and 7 openings, while 63.52% had between 7 and 14 openings, and the remaining 10.06% had >14 openings in their buildings. This shows the awareness of historical builders regarding natural ventilation. Nearly two-thirds (68.55%) of the respondents use louver/shutters; this kind of window can help to prevent sun rays from penetrating inside the building.

Regarding ceilings, the occupants' replies showed that 68.55% of their buildings have traditional ceilings, which are mangrove ceilings; this type of ceiling is made up of earth and mangrove poles. A total of 20.75% of the respondents had changed the ceiling to a hardboard ceiling, and 10.69% had reinforced concrete ceilings.

Regarding roof type, a total of 83.64% (hip and pitched roof) of the surveyed buildings had a roof structure that acts as a barrier against heat entering the living space and also creates a space under the roof, as it is the most common and economical heat insulator in every building.

A range of sun shading devices were listed in the survey, in which the respondents indicated which shading method they used and specified a reason for using the shading. A total of 15.72% of the survey participants indicated that they used curtains and a verandah/balcony (CVB) as their choice of shading, and 10.69% indicated that they used a verandah/balcony (VB). A total of 73.58% indicated that they used curtains (C).

3.1.2. Indoor Environment Parameter Assessment

According to ASHRAE Standard 55, the factors required for the evaluation of thermal comfort include metabolic rate, clothing insulation, air temperature, radiant temperature, air speed, and humidity. The two personal parameters, such as clothing insulation value and metabolic rate, were estimated in accordance with ASHRAE Standard 55 and ISO 7730 [26,27]. In this study, we set the metabolic rate as 1.2 met, which represents sedentary activities (office, dwelling school, and laboratory), and the clothing insulation value as 0.5 Clo.

The overall indoor air temperature taken in each dwelling during the questionnaire survey ranged between 26 °C and 32 °C, and the indoor relative humidity ranged from

58.90% to 75.05%, with a mean value of 63.44%. Based on these data, the operative temperature t_0 was calculated according to ASHRAE Standard 55 as:

$$t_0 = At_a + (1 - A)\overline{t_r},\tag{1}$$

where t_a is the averaged air temperature obtained from questionnaire, $\overline{t_r}$ is the mean radiant temperature, which is assumed to be equal to the air temperature, and A is a function of the average air speed. The air speeds were assumed to be the same as those measured by meteorological stations. The calculated operative temperature was 29.1 °C on average, which is outside the comfort zone limit of ASHRAE Standard 55.

The results of the evaluation of the indoor humidity perception vote (HPV) and the indoor draft perception vote (DPV) are presented in Figure 4. Only 63.51% of 159 respondents voted for HPV within the three central categories, showing that the occupants did not have acceptable thermal conditions. A total of 26.42% of the residents felt that the environment was much too humid, which is understandable in such a humid area, with a relative humidity between 58.90% and 75.05%.

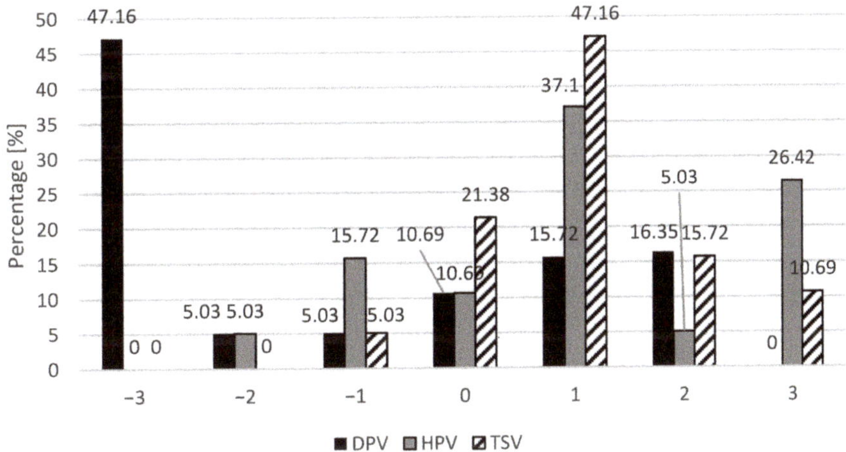

Figure 4. Evaluation of the indoor humidity perception vote (HPV), indoor draft perception vote (DPV), and thermal sensation vote (TSV).

Regarding the DPV, a large portion (52.19%) of the subjects perceived the air to be too steady (−3 to −2), while 31.44% of the respondents felt that the air velocity was just right (−1, 0, +1), and only 16.35% felt that the velocity should be less. These results show that air movement and thermal conditions in their buildings were not acceptable.

3.1.3. Thermal Sensation Vote (TSV)

The results of the TSV evaluation showed that the majority (73.57%) of the respondents voted between "−1 slightly cool" and "+1 slightly warm", while almost half (47.16%) of the respondents voted slightly warm. Linear regression was performed between thermal sensation and operative temperature to determine the strength of the relationship between them in residential buildings. The data are plotted in Figure 5. The fitted regression equation for the subjects' sensations versus the operative temperature was:

$$\text{TSV} = 0.650\, T_0 - 17.737.\tag{2}$$

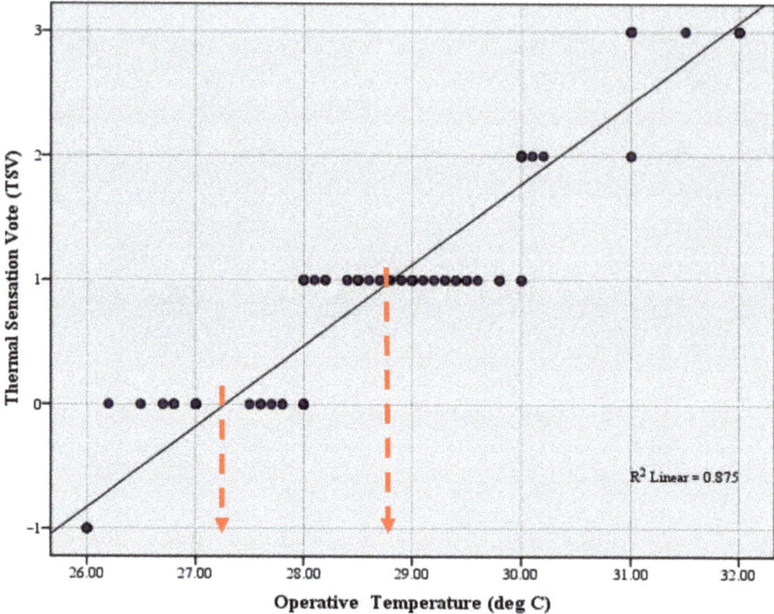

Figure 5. Linear regression of thermal sensation votes (TSV) versus operative temperature (OP).

The coefficient of determination (R^2) between TSV and the operative temperature was 0.875, which means TSV correlated strongly with the operative temperature. The slope of the regression line, which indicates the subjects' thermal sensitivity concerning the operative temperature, was high (0.65 unit/°C) compared with that in other warm places [28–30]. This demonstrates that the residents of Stone Town, Zanzibar, are less tolerant of a wider range of temperatures.

The neutrality condition derived by solving the regression equation for zero (neutral) yielded an estimated neutrality of 27.40 °C, as shown in Figure 5. A comfort band of 26.00–28.90 °C coincided with −1 and +1 sensation votes.

3.1.4. Thermal Preferences

Thermal preference was evaluated using the McIntyre scale (−1 (cooler), 0 (no change), +1 (warmer)). A total of 68.55% of the respondents preferred to be cooler in their buildings, and only 31.44% of the subjects indicated that no changes in the thermal environment were required. These results suggest that people prefer to feel cooler than neutral, which aligns well with previous results from studies regarding warm places [28,31]. The thermal acceptability results align well with those of thermal preference. A total of 68.55% of the participants judged that their thermal sensation was not acceptable, and 31.44% accepted the overall temperature of their building.

3.2. Field Survey Results

3.2.1. Passive Cooling Features

The passive cooling features of the two case study buildings (HRB1 and HRB2) regarding orientation, materials, and construction, and cooling and ventilation are presented in this subsection and are summarized in Table 2.

Table 2. Passive cooling features of case study buildings HRB1 and HRB2.

Case Study Buildings	Organization (Main Rooms)		Orientation [1]	Cooling and Ventilation	Materials and Construction	
HRB1	First floor	Living room and dining room	N, S, E	cross-ventilation balcony for shading	exterior wall	1 m thick coral rag painted in white color
		Kitchen	S, W	-	inner wall	0.5 m thick coral rag painted in white color
	Second floor	Bedroom 1	N, S, E	cross-ventilation AC (off)	floor	boriti poles
		Bedroom 2	S	-	roof	pitched roof with red tiles
HRB2	First floor	Living room	W, S	cross-ventilation	exterior wall	1 m thick coral rag with yellow cement–sand plastering
		Dining room	S	-		
		Kitchen	S	-		
	Second floor	Bedroom 1	W, S	cross-ventilation with shading	inner wall	0.5 m thick coral rag with yellow cement–sand plastering
		Bedroom 2	W	with shading		
		Bedroom 3	S	cross-ventilation (air shaft) with shading	roof	flat roof

[1] N indicates north; S indicates south; E indicates east; and W indicates west.

Orientation

HRB1 is surrounded by four streets of 1.5 m to 2 m in width and is approximately 15 m from the Kenyatta main road. Being surrounded by higher buildings, HRB1 receives shade. The long axis of the building runs from west to east, which means the façade on the north–south axis is bigger than that on the west–east elevation axis. It has a good orientation toward the wind direction and deflects the dominant orientation of the sun in the east–west direction. None of the bedrooms are located on the west side, which is a disadvantaged side. Even though the living room is located with an eastern orientation, it does not receive much heating during the day because of the shading element (balcony on the first floor) along the whole length of the façade. Traditionally, balconies are used to protect walls from direct solar radiation and act as transitional spaces to inner living areas. No air conditioning (AC) was in operation during the investigation.

HRB2 is bounded by three streets about 1.5 m to 2 m wide on the northern, southern, and western sides of the plot. The eastern side has shared a wall with another building, but not the whole length of the building because in the middle there is a courtyard/air well, which prevents the building from being totally attached. The long axis of the building runs north to south, which is contrary to the recommendation of the best orientation for buildings in tropical climates. The living room, bedroom 1, and bedroom 2 of HRB2 are located on the west side; therefore, they heat up in the afternoon, although there is shade protection on the window to break up the sun rays and curtains for protection from solar heat gain in the living room and bedrooms. Bedroom 3 is well-oriented at the southern side of the building; this room also benefits from air movement and natural light through the courtyard/air shaft.

Materials and Construction

Natural, local building materials made from stone, wood, and soil were found to be used in this case study. Both HRB1 and HRB2 are constructed with coral rags laid in random gravel. The exterior walls of these buildings are very thick, with a 1 m thickness, and act as load bearing walls; the interior walls are 500 mm thick, with the exception of bathroom/kitchen walls, which are 200 mm thick; the thick stone walls act as thermal mass against heat penetration.

The exterior and interior walls of HRB1 are painted in white, which saves the cooling energy of the building [32–34]. Similarly, the outside wall finish of HRB2 is rough with pale yellow paint, while the interior walls are painted white.

There is no insulation on the walls, windows, or roofs of HRB1 and HRB2 because they are expensive and difficult to maintain. HRB1 has a pitched roof of 32°, and the roof overhangs (600 mm), providing a sun shading effect to protect the walls. The existence of a ceiling also creates an attic space under the roof, as it is the most common and economical heat insulator that must be installed in every building. The roof overhangs can provide a sun shading effect to protect the walls. The floor slabs of this case study were found to be traditionally made of boriti poles (mangrove poles) with a lime–concrete cover laid over to form the floor above. The boritis or mangroves support the slab and help the walls remain upright. HRB2 has a flat roof; all rooms are prone to heat gain through the floor slab, even though the ceiling is painted with a whitewash.

Cooling and Ventilation

The living room, dining room, and bedroom 1 in HRB1 have provisions for cross-ventilation via four pairs of windows for bedroom 1, three pairs of windows for the living room with a jalousie door that opens up to provide maximum airflow inside the spaces, and lastly, two pairs of windows for the dining room; thus, it seems that the spaces inside the room are affected by the air flow.

Regarding HRB2, the living room and bedroom 1 have provisions for cross-ventilation via three pairs of windows for bedroom 1 (two openings on opposite and adjacent walls) and three pairs of windows for the living room (two openings on opposite and adjacent walls); thus, it seems that the spaces inside the room are affected by the air flow. The windows in HRB2 are all double-winged shuttered windows. Although the windows in most of the rooms have been designed to enhance cross-ventilation, in this case, there are some windows that are not open every day, e.g., the living room window with a southern orientation is not opened during the whole year; this is because of the furniture (showcase) kept in front of the window. Lattice windows are used on the west façades of this building to enhance cross-ventilation and daylight. The shading devices and shuttering of the windows in the case study building are 400 mm by 2000 mm in dimension on all sides of the building. The orientation for wind flow is quite challenging only on the northeastern side (blockage of air flow) because, on its eastern side, the building shares a common wall with another building, but the historical builders demonstrated common sense by introducing a 4 m-by-4 m courtyard/air well that allows air flow and natural daylight to enter the building.

3.2.2. Air Temperature and Humidity

In general, the indoor air temperature in the occupied zone for HRB1 varied between 28 °C and 31 °C, with an average value of 29.2 °C, and the indoor relative humidity was between 57.0% and 70.3%, with a mean value of 62.41% (Figure 6). For HRB2, the indoor air temperature as measured during the field survey varied between 26 °C and 29 °C, with a mean value of 28.16 °C (Figure 7). The indoor relative humidity in the building ranged between 49.4% and 76%, with a mean value of 63.89%. As presented in Section 3.2.1, the living room of HRB1 has six openings in three walls, which creates cross-ventilation. In bedroom 2, there is only one opening in the southern wall, and it is expected to have higher temperatures and humidity than in the living room. However, no significant differences between the temperature and relative humidity of the living room and bedroom 2 were found, as shown in Figure 6. A possible reason for this could be the diverse surface areas of the two rooms. For HRB2, both the living room and bedroom 3 have openings in different directions, which creates cross-ventilation. However, the temperature and humidity in the bedroom were found to be slightly lower than in the living room. This is because the openings in the living room are covered by furniture that cannot be opened frequently. Another reason is that the living room faces west and absorbs more sun radiation than bedroom 3, which has a southern orientation. Therefore, orientation, surface area, and air flow are all important factors in indoor environments.

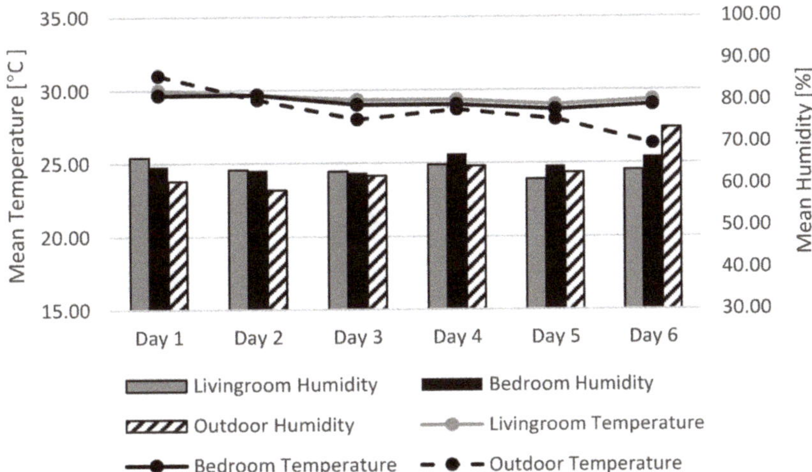

Figure 6. Indoor/outdoor temperatures and humidity averaged from 07:00–10:00, 12:00–15:00, and 17:00–21:00 (HRB1).

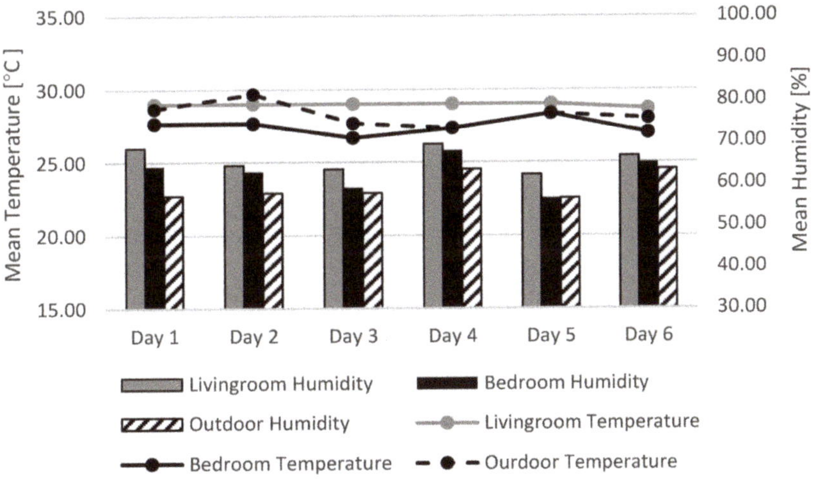

Figure 7. Indoor/outdoor temperatures and humidity averaged from 7.00–10:00, 12:00–15:00, and 17:00–21:00 (HRB2).

Figures 8 and 9 show the performance of both the living rooms and bedrooms in relation to the outdoor air temperature for HRB1 and HRB2. For HRB1, the outdoor air temperature value rose from 28.5 °C in the morning (07:00–10:00) to a peak of about 30.5 °C during the afternoon (12:00–15:00). The average temperature in the living room and bedroom was slightly different from that in the outdoors. This is because the indoor temperature rises as solar radiation increases. However, in the evening, the indoor temperature did not decrease as dramatically as the outdoor temperature, which means that, at night, the outdoor space is better than the indoor space. For HRB1, no significant change was found in the living room and bedroom temperatures, in contrast to the variation in the outdoor temperatures. However, for HRB2, the living room temperature was always higher than the outdoor temperature, while the bedroom (south) temperature was between 27.0 and 27.5 °C, which is much lower than the outdoor temperature.

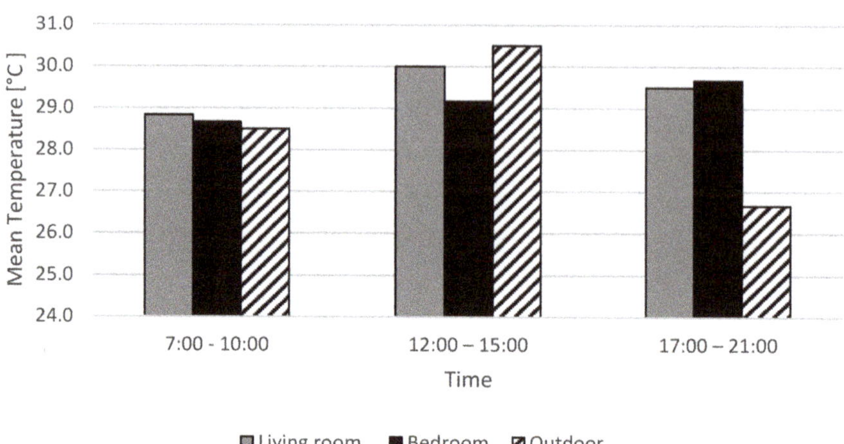

Figure 8. Temperature in the living room, bedroom, and outdoors as averaged over six days of measurements (HRB1).

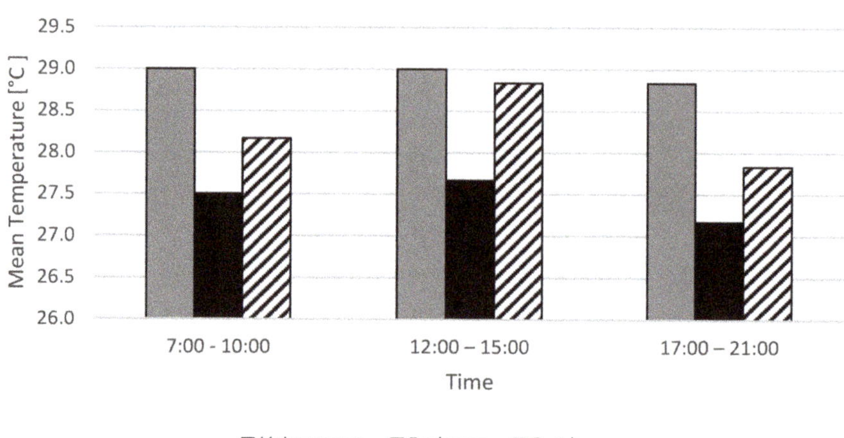

Figure 9. Temperature in the living room, bedroom, and outdoors as averaged over six days of measurements (HRB2).

3.2.3. Thermal Comfort Indices

The thermal-related input values for the calculation were determined via field measurements and are shown in Table 3 for HRB1 and Table 4 for HRB2. The calculations for the PMV and PPD were performed using a software comfort program that has been proposed by ISO 7730 [27].

The PMV of HRB1 was found to vary between 1 to 1.6, with a mean value of 1.23. A total of 80% satisfaction falls within the range of [−1.23 < PMV < +1.23]. The PPD varied from 26.1% to 56.3%, with a mean value of 37.35%. This indicates that 37.35% of the occupants in this survey building were expected to be dissatisfied (discomfort) with the thermal environment.

Table 3. Results of field measurements and thermal comfort indices (input variables for PMV and PPD) for HRB1.

| | | Input Variables | | | | | Thermal Comfort Indices | |
Date	Air Temperature (°C)	Relative Humidity (%)	Air Velocity (m/s²)	Mean Radiant (°C)	Clothing Insulation (clo)	Metabolic Rate (met)	PMV	PPD (%)
Day 1	29.8	65.36	0.19	29.8	0.5	1.2	1.6	56.3
Day 2	29.6	63.35	0.18	29.6	0.5	1.2	1.4	45.5
Day 3	29.1	62.85	0.18	29.1	0.5	1.2	1.1	30.5
Day 4	29.1	55.35	0.37	29.1	0.5	1.2	1.0	26.1
Day 5	28.8	62.68	0.33	28.8	0.5	1.2	1.1	30.5
Day 6	29.1	64.88	0.32	29.1	0.5	1.2	1.2	35.2

Table 4. Results of field measurements and thermal comfort indices (input variables for PMV and PPD) for HRB2.

| | | Input Variables | | | | | Thermal Comfort Indices | |
Date	Air Temperature (°C)	Relative Humidity (%)	Air Velocity (m/s²)	Mean Radiant (°C)	Clothing Insulation (clo)	Metabolic Rate (met)	PMV	PPD (%)
Day 1	28.3	66.18	0.2	28.3	0.5	1.2	1.0	26.1
Day 2	28.3	63.48	0.18	28.3	0.5	1.2	0.9	22.1
Day 3	27.8	61.11	0.19	27.8	0.5	1.2	0.7	15.3
Day 4	28.1	67.91	0.41	28.1	0.5	1.2	0.7	15.3
Day 5	28.6	59.15	0.35	28.6	0.5	1.2	1.0	26.1
Day 6	27.8	65.51	0.34	27.8	0.5	1.2	0.8	18.5

For HRB2, the PMV was found to range between 0.7 and 1, with a mean value of 0.85. A total of 80% satisfaction falls within the range of [−0.85 < PMV < +0.85]. The PPD varied from 15.3% to 26.1%, with a mean value of 20.56%. This indicates that 20.56% of the occupants in this survey building were expected to be slightly dissatisfied (discomfort); the required PPD value should be less than 10%. Additionally, in this case, we learned that the value of the PMV fell within the range of (−1, 0, +1), which represents 80% satisfaction, but, unfortunately, the values obtained are not in the ideal range of an acceptable thermal environment for general comfort, which is between [−0.5 < PMV < +0.5] (neutral).

4. Discussion

4.1. Thermal Comfort

The thermal comfort evaluation of the historical residential buildings of Stone Town based on subjective and objective surveys is summarized in Table 5.

Table 5. Summary of subjective and objective survey results.

| Factors | Subjective Survey | Objective Survey | |
		HRB1	HRB2
Air temperature (°C)	26–32	28–31	26–29
Relative humidity (%)	63.44	62.41	63.89
TSV(%)		73.57	
PMV	-	[−1.23, +1.23]	[−0.85, +0.85]
PPD (%)	-	37.35	20.56

The air temperature obtained from both the subjective and objective surveys was above 26 °C, which is higher than the comfort zone for summer conditions (23–26 °C).

Similarly, the relative humidity, which was around 63% as obtained from the questionnaires and measurements, is also higher than the value in the comfort zone (50–60%). Referring to the results obtained for the PMV and PPD values in the field measurements, neither of the case study buildings are controlled within the specified acceptable thermal comfort range based on the ASHRAE Standard 55, which is $[-0.5 < PMV < +0.5]$ with a PPD < 10%. Indeed, numerous thermal comfort studies carried out worldwide have revealed that the PMV-PPD index cannot evaluate thermal comfort under non-air-conditioned space without modification due to the ignorance of subjects' adaptations. Although at least 20% of occupants are not satisfied with the current thermal environment, the TSV reflecting real thermal sensations indicates that the majority (73.57%) of the respondents' thermal sensation votes are within the central three categories of the 7-point scale. It once again proves that the PMV-PPD index is not applicable to assess thermal comfort in naturally ventilated buildings. Therefore, in Table 5, we add TSV with the purpose of reflecting the actual thermal satisfaction of occupants.

4.2. Passive Cooling Strategies

Based on the questionnaire and field survey results, passive cooling strategies implemented in the historical residential buildings in Stone Town include the north-south orientation (HRB1), natural ventilation (HRB1 and HRB2), pitched roof (HRB1), light color finishing (HRB1 and HRB2), and various types of shading. The majority (73.57%) of respondents' TSV results are within the central three categories, which indicates that these passive techniques are beneficial for indoor thermal comfort in hot areas.

Similar results have been found in previous research in recent decades. Lapisa et al., and Ozarisoy indicated that the main façade the buildings with large area windows should avoid exposure to direct solar radiation [35,36]. In this study, bedroom 3 in HRB2 is facing south, while the living room is exposed to the worst orientation (west) to solar radiation. This is one of the possible reasons why the indoor temperature in bedroom 3 is much lower than that in the living room. Window shading is an optimal way to compensate for the non-ideal orientation in retrofitting historical buildings. Lapisa et al. confirmed that slope roofs can improve the thermal comfort effect of buildings, while high-slope roofs increase costs and cause potential safety problems in earthquakes [35]. Therefore, the application of pitched roofs to historical buildings in developing countries such as Zanzibar in this study should take the cost, potential safety risk, and thermal comfort into consideration. Previous research found that solar radiation can be effectively reflected by using cool roofs and walls, thereby decreasing the heat gain in the interior [35,37]. The exterior walls of the two case study buildings in this research are painted in white or light yellow, which contributes to thermal comfort inside the buildings. This technique can be further implemented on roofs to improve thermal performance. Natural ventilation is the most efficient way to reduce energy demand and increase thermal comfort in a hot climate by providing a large amount of airflow [36,38,39]. This technique is also implemented in the historical residential buildings in Stone Town, where a large number of openings and cross-ventilation can be found. However, large areas of openings will increase the potential energy demand for heating, so the decision making of the window areas should depend on the heating/cooling demand.

In addition, numerous studies have focused on the energy demand of residential buildings by using passive cooling strategies. Fernandez-Antolin et al. found that north-south orientation and light color finishing contribute to low cooling energy demand for residential buildings in Spain. This is consistent with the thermal comfort evaluation results obtained in this research, which implemented the same strategies [40]. Pero et al. Proposed four main processes for passive building design, including site planning, building's shape and envelope, window design, and renewable energy sources [41]. These strategies are effective for new buildings. However, many are not applicable to the renovation of historical buildings, such as the orientation and building shape. In some cases, window openings and shadings are also restricted due to the local culture. Therefore, design strategies for improving the thermal comfort of historical residential buildings are proposed in the following subsections.

4.3. Design Strategies for Historical Residential Buildings

Three factors should be considered for the improvement of thermal comfort in historical residential buildings: (1) improve the thermal comfort conditions; (2) respect the local culture and history; and (3) reduce energy consumption.

Sun radiation is the main source of heat gain inside buildings; therefore, one of the best means by which to improve the thermal comfort of buildings is to create a quality building envelope using construction materials that help reduce the sun's heat. For historical buildings, traditional building materials, such as lime plaster and lime wash, are suitable for solving the problem.

4.3.1. Lime Plaster

Lime plaster is generally made by heating lime rock (rock made of calcium carbonate, such as limestone or marble). Pure lime is white, lighter in weight than the original rock, and reacts violently with water. The lime plaster has the benefits of protecting masonry against weathering conditions, improving the thermal and acoustical performance.

Historical buildings in Stone Town were built hundreds of years ago, and damage is commonly detected nowadays. Lime plaster can improve the appearance of walls by hiding the imperfections of rough work, giving it an attractive texture compatible with the local environment.

The climate in Stone Town is hot and humid. Excessive humidity may enter the masonry in different ways and forms, such as the penetration of rainwater, fog, rising damp, and condensation, thereby leading to damage to the building's structure. Thus, lime plaster has an important role in allowing the walls of the property to breathe by letting out excess humidity.

4.3.2. Wash Lime

White and light colors are beneficial in reducing cooling energy [35,37]. Although the two case study buildings already have light-colored paint on the walls, as presented in Section 3.2.1, the indoor temperatures and thermal comfort are still under the requirements indicated by the ASHRAE; thus, wash lime is proposed. Wash lime is mostly applied to stone and brick walls and adobe. It is widely used due to its availability and low cost, making it a good choice for developing countries. In addition, wash lime can brighten up the exterior walls and form calcite crystals, which reflect the sun rays striking on the walls. A light color is preferred for wash lime, as indicated earlier.

5. Conclusions

A thermal comfort study of historical residential buildings in a tropical hot–humid climate was conducted in Zanzibar. A subjective survey was carried out among 159 occupants living in the historical site of Stone Town to investigate the passive cooling techniques used in historical residential buildings and the occupants' subjective evaluations of indoor environmental parameters. The objective indoor environmental parameters were collected by field measurements from two case study buildings over six continuous days to evaluate the thermal comfort in the historical buildings.

Even with the use of passive cooling technologies, such as the use of sun shading devices and more openings to create better ventilation, thermal comfort was still not acceptable, according to international ASHRAE Standard 55. In the questionnaire survey, 73.57% of the participants voted within the three central categories of the thermal sensation vote, and 68.55% of the occupants stated that they would prefer to be cooler. This indicates that their thermal environment is not acceptable. In the field survey, the indoor temperatures and relative humidity, as well as the PMV–PPD values, of the two case study buildings were found to be higher than the acceptable limits for thermal comfort zones, which indicates the necessity of improving the thermal environment in these historical residential buildings that already take passive cooling into consideration.

Based on the analysis of passive cooling technologies (north-south orientation, natural ventilation, pitched roof, light color finishing, and window shading) and their positive and negative effects on the thermal environment in historical residential buildings, design strategies are proposed to improve the local thermal environment. The design strategies include the application of lime plaster and wash lime, which have the potential to reduce energy consumption, improve thermal comfort, and respect the local culture and history.

This research provides guidelines to assist architects in designing energy-efficient residential buildings while also taking into account cultural heritage and thermal comfort. The retrofitting of historical buildings should avoid the large destruction of historical buildings, construct as much as possible with local materials, and preserve coordination with the local environment. Further studies concerning thermal comfort should be performed in suburban areas over a longer period, or more historical residential buildings, in order to obtain a broader picture of thermal comfort phenomena in areas with different cultural and behavioral patterns.

Author Contributions: Conceptualization, H.X. and H.M.A.; methodology, H.X. and H.M.A.; writing—original draft preparation, C.L. and H.M.A.; writing—review and editing, C.L., H.X. and J.L.; supervision, H.X. All authors have read and agreed to the published version of the manuscript.

Funding: This research was funded by the Science and Technology Research Program of Chongqing Municipal Education Commission, grant number KJQN201900113.

Data Availability Statement: Not applicable.

Conflicts of Interest: The authors declare no conflict of interest.

Appendix A. Questionnaire

FACULTY OF ARCHITECTURE AND URBAN PLANNING, CHONGQING UNIVERSITY
HISTORIC RESIDENTIAL BUILDING IN ZANZIBAR THERMAL COMFORT QUESTIONNAIRE SURVEY

The purpose of this survey is to evaluate the thermal conditions in the historic residential building in Zanzibar. The attached questionnaire is to collect information from the end-user of this building. Please participate in this survey by completing this questionnaire sheets.

Thanks.

Date: ___________ Time: ___________ Building Location: ___________ Floor of Building: ___________
Temperature: ___________ Humidity: ___________

1. Gender: ☐ Male ☐ Female
2. Age: ☐ less than 20, ☐ 20–39, ☐ 40–59, ☐ above 60
3. Height (m): ___________ Weight (kg): ___________
4. Education background: ___________ Occupation: ___________

5. Building envelope type: ☐ Coral Stone, ☐ Concrete, ☐ Timber, ☐ Others, please specify ___________
6. Which floor do you stay? ☐ 1st, ☐ 2nd, ☐ 3rd, ☐ Other, please specify ___________
7. What is the type of your house's ceiling? ☐ Mangrove ceiling, ☐ Hardboard ceiling, ☐ Other, please specify ___________
8. How many openings do you have in your house?
Windows: ☐ One, ☐ Two, ☐ Three, ☐ Other, please specify ___________
Door: ☐ One, ☐ Two, ☐ Three, ☐ Other, please specify ___________
9. Please indicate the style of your window in your building:
☐ Casement window, ☐ Louver/Shutters, ☐ Aluminum sliding window, ☐ Other, please specify ___________
10. Which side of your house doesn't have window? ☐ North, ☐ South, ☐ East, ☐ West
11. What is the roof type? ☐ Pitched roof, ☐ Hipped roof, ☐ Flat roof
12. What is your roofing material? ☐ Aluminum sheet, ☐ Concrete, ☐ Other, please specify ___________
13. What method of sun shading do you use? ☐ Curtains, ☐ Veranda/Balcony, ☐ Other, please specify ___________

14. How do you feel about the airflow at this moment?
☐ Much too still, ☐ Too still, ☐ Slightly still, ☐ Just right, ☐ Slightly breezy, ☐ Too breezy, ☐ Much too breezy
15. How do you feel in terms of humidity?
☐ Much too dry, ☐ Too dry, ☐ Slightly dry, ☐ Just right, ☐ Slightly humid, ☐ Too humid, ☐ Much too humid
16. How satisfied are you with the temperature in your house? ☐ Satisfied, ☐ Dissatisfied
17. If you are dissatisfied with the temperature in your house, which of the following contribute to your dissatisfaction:
☐ Always too hot, ☐ Often too hot, ☐ Occasionally too hot, ☐ Occasionally too cold, ☐ Often too cold, ☐ Always too cold
18. When is this most often a problem? ☐ Morning (before 11am), ☐ Mid-day (11 a.m.–2 p.m.), ☐ Afternoon (2 p.m.–5 p.m.),
☐ No particular time, ☐ Other, please specify ____________
19. Clothing situation of yours at this moment (please tick at appropriate):
☐ Short Sleeve Shirt/Blouse ☐ Dress ☐ Socks ☐ Shorts
☐ Long Sleeve Shirt/Blouse ☐ Shoes ☐ T-Shirt ☐ Undershirt
☐ Trousers/Long Skirt ☐ Vest ☐ Sandals ☐ Vest
☐ Sweater ☐ Other, please specify ____________
20. What is your activity level in the last 15 min? (Please tick at appropriate):
☐ Medium Activity Standing ☐ Reclining ☐ Seated ☐ Undershirt
☐ Standing Relaxedt ☐ Light Activity Standing ☐ High Activity ☐ Other, please specify ____________
21. Which of the following do you personally adjust or control in your house? (You can tick more than one option)
☐ Open internal door ☐ Open window ☐ Go outside ☐ Air condition on
☐ Fan on ☐ Portable Fan ☐ Other, please specify ____________
22. How do you feel at this moment?
☐ Cold, ☐ Cool, ☐ Slightly cool, ☐ Neutral, ☐ Slightly warm, ☐ Warm, ☐ Hot
23. Would you prefer to be: ☐ cooler, ☐ No change, ☐ warmer
24. How would you rate the overall acceptability or the temperature at this moment?
☐ Acceptable, ☐ Not acceptable

References

1. Izadyar, N.; Miller, W.; Rismanchi, B.; Garcia-Hansen, V. Impacts of facade openings' geometry on natural ventilation and occupants' perception: A review. *Build. Environ.* **2020**, *170*, 106613. [CrossRef]
2. Yadav, S.K. Environmental Pollution Problems Associated with Energy Production for High Demand. *Int. J. Instrum. Sci. Eng.* **2013**, *3*, 223–229.
3. Bose, B.K. Global Warming: Energy, Environmental Pollution, and the Impact of Power Electronics. *IEEE Ind. Electron. Mag.* **2010**, *4*, 6–17. [CrossRef]
4. Ming, R.; Yu, W.; Zhao, X.; Liu, Y.; Li, B.; Essah, E.; Yao, R. Assessing energy saving potentials of office buildings based on adaptive thermal comfort using a tracking-based method. *Energy Build.* **2020**, *208*, 109611. [CrossRef]
5. Calautit, J.K.; O'Connor, D.; Tien, P.W.; Wei, S.; Pantua, C.A.J.; Hughes, B. Development of a natural ventilation windcatcher with passive heat recovery wheel for mild-cold climates: CFD and experimental analysis. *Renew. Energy* **2020**, *160*, 465–482. [CrossRef]
6. Mostafavi, S.; Cox, R.; Futrell, B.; Ashafari, R. Calibration of white-box whole building energy models using a systems-identification approach. In *IECON 2018—44th Annual Conference of the IEEE Industrial Electronics Society, Proceedings of the 44th Annual Conference of the IEEE Industrial Electronics Society, Washington, DC, USA, 21–23 October 2018*; IEEE: New York, NY, USA, 2018; pp. 795–800.
7. IEA. *Energy Efficiency 2019*; IEA Publications: Paris, France, 2019.
8. Abdo, P.; Huynh, B.P.; Braytee, A.; Taghipour, R. An experimental investigation of the thermal effect due to discharging of phase change material in a room fitted with a windcatcher. *Sustain. Cities Soc.* **2020**, *61*, 102277. [CrossRef]
9. Soltani, M.; Dehghani-Sanij, A.; Sayadnia, A.; Kashkooli, F.; Gharali, K.; Mahbaz, S.; Dusseault, M. Investigation of Airflow Patterns in a New Design of Wind Tower with a Wetted Surface. *Energies* **2018**, *11*, 1100. [CrossRef]
10. Bhamare, D.K.; Rathod, M.K.; Banerjee, J. Passive cooling techniques for building and their applicability in different climatic zones—The state of art. *Energy Build.* **2019**, *198*, 467–490. [CrossRef]
11. Nejat, P.; Ferwati, M.S.; Calautit, J.; Ghahramani, A.; Sheikhshahrokhdehkordi, M. Passive cooling and natural ventilation by the windcatcher (Badgir): An experimental and simulation study of indoor air quality, thermal comfort and passive cooling power. *J. Build. Eng.* **2021**, *41*, 102436. [CrossRef]
12. Diler, Y.; Turhan, C.; Arsan, Z.D.; Akkurt, G.G. Thermal comfort analysis of historical mosques. Case study: The Ulu mosque, Manisa, Turkey. *Energy Build.* **2021**, *252*, 111441. [CrossRef]
13. Cheung, T.; Schiavon, S.; Parkinson, T.; Li, P.; Brager, G. Analysis of the accuracy on PMV—PPD model using the ASHRAE Global Thermal Comfort Database II. *Build. Environ.* **2019**, *153*, 205–217. [CrossRef]
14. Grygierek, K.; Sarna, I. Impact of Passive Cooling on Thermal Comfort in a Single-Family Building for Current and Future Climate Conditions. *Energies* **2020**, *13*, 5332. [CrossRef]

15. Netam, N.; Sanyal, S.; Bhowmick, S. Assessing the impact of passive cooling on thermal comfort in LIG house using CFD. *J. Therm. Eng.* **2019**, *5*, 414–421.

16. Berg, F.; Flyen, A.-C.; Godbolt, Å.L.; Broström, T. User-driven energy efficiency in historic buildings: A review. *J. Cult. Herit.* **2017**, *28*, 188–195. [CrossRef]

17. Magrini, A.; Franco, G. The energy performance improvement of historic buildings and their environmental sustainability assessment. *J. Cult. Herit.* **2016**, *21*, 834–841. [CrossRef]

18. Martínez-Molina, A.; Tort-Ausina, I.; Cho, S.; Vivancos, J.-L. Energy efficiency and thermal comfort in historic buildings: A review. *Renew. Sustain. Energy Rev.* **2016**, *61*, 70–85. [CrossRef]

19. Vitale, V.; Salerno, G. A numerical prediction of the passive cooling effects on thermal comfort for a historical building in Rome. *Energy Build.* **2017**, *157*, 1–10. [CrossRef]

20. Laurini, E.; de Vita, M.; de Berardinis, P. Monitoring the Indoor Air Quality: A Case Study of Passive Cooling from Historical Hypogeal Rooms. *Energies* **2021**, *14*, 2513. [CrossRef]

21. Varas-Muriel, M.J.; Fort, R.; Gómez-Heras, M. Assessment of an underfloor heating system in a restored chapel: Balancing thermal comfort and historic heritage conservation. *Energy Build.* **2021**, *251*, 111361. [CrossRef]

22. Ricciardi, P.; Ziletti, A.; Buratti, C. Evaluation of thermal comfort in an historical Italian opera theatre by the calculation of the neutral comfort temperature. *Build. Environ.* **2016**, *102*, 116–127. [CrossRef]

23. Selin, H. *Encyclopaedia of the History of Science, Technology, and Medicine in Non-Western Cultures*; Springer: Berlin/Heidelberg, Germany, 2016.

24. Available online: https://whc.unesco.org/en/list/173/ (accessed on 2 March 2022).

25. Zare, S.; Hasheminezhad, N.; Sarebanzadeh, K.; Zolala, F.; Hemmatjo, R.; Hassanvand, D. Assessing thermal comfort in tourist attractions through objective and subjective procedures based on ISO 7730 standard: A field study. *Urban Clim.* **2018**, *26*, 1–9. [CrossRef]

26. *ANSI/ASHRAE Standard 55-2017*; Thermal Environmental Conditions for Human Occupancy. ASHRAE: Atlanta, GA, USA, 2017.

27. *ISO 7730*; Ergonomics of the Thermal Environment: Analytical Determination and Interpretation of Thermal Comfort Using Calculation of the PMV and PPD Indices and Local Thermal Comfort Criteria. ISO: Geneva, Switzerland, 2005.

28. Indraganti, M.; Boussaa, D. Comfort temperature and occupant adaptive behavior in offices in Qatar during summer. *Energy Build.* **2017**, *150*, 23–36. [CrossRef]

29. Indraganti, M. Thermal comfort in naturally ventilated apartments in summer: Findings from a field study in Hyderabad, India. *Appl. Energy* **2010**, *87*, 866–883. [CrossRef]

30. Sharma, A.; Kumar, A.; Kulkarni, K.S. Thermal comfort studies for the naturally ventilated built environments in Indian subcontinent: A review. *J. Build. Eng.* **2021**, *44*, 103242. [CrossRef]

31. Damiati, S.A.; Zaki, S.A.; Rijal, H.B.; Wonorahardjo, S. Field study on adaptive thermal comfort in office buildings in Malaysia, Indonesia, Singapore, and Japan during hot and humid season. *Build. Environ.* **2016**, *109*, 208–223. [CrossRef]

32. Akbari, H.; Konopacki, S.; Pomerantz, M. Cooling energy savings potential of reflective roofs for residential and commercial buildings in the United States. *Energy* **1999**, *24*, 391–407. [CrossRef]

33. Eskin, N.; Türkmen, H. Analysis of annual heating and cooling energy requirements for office buildings in different climates in Turkey. *Energy Build.* **2008**, *40*, 763–773. [CrossRef]

34. Xi, W.; Liu, Y.; Zhao, W.; Hu, R.; Luo, X. Colored radiative cooling: How to balance color display and radiative cooling performance. *Int. J. Therm. Sci.* **2021**, *170*, 107172. [CrossRef]

35. Lapisa, R.; Arwizet; Kurniawan, A.; Krismadinata; Rahman, H.; Romani, Z. Optimized Design of Residential Building Envelope in Tropical Climate Region: Thermal Comfort and Cost Efficiency in an Indonesian Case Study. *J. Archit. Eng.* **2022**, *28*, 05022002. [CrossRef]

36. Ozarisoy, B. Energy effectiveness of passive cooling design strategies to reduce the impact of long-term heatwaves on occupants' thermal comfort in Europe: Climate change and mitigation. *J. Clean. Prod.* **2022**, *330*, 129675. [CrossRef]

37. Bradley, R.A. Assessing the effectiveness of several passive design strategies using the CIBSE overheating criteria: Case study of an Earth Brick Shell House in Johannesburg, South Africa. *Archit. Sci. Rev.* **2022**, *65*, 232–246. [CrossRef]

38. Mehmood, S.; Lizana, J.; Núñez-Peiró, M.; Maximov, S.A.; Friedrich, D. Resilient cooling pathway for extremely hot climates in southern Asia. *Appl. Energy* **2022**, *325*, 119811. [CrossRef]

39. Rahman, N.M.A.; Haw, L.C.; Fazlizan, A.; Hussin, A.; Imran, M.S. Thermal comfort assessment of naturally ventilated public hospital wards in the tropics. *Build. Environ.* **2022**, *207*, 108480. [CrossRef]

40. Fernandez-Antolin, M.M.; del Río, J.M.; Costanzo, V.; Nocera, F.; Gonzalez-Lezcano, R.A. Passive design strategies for residential buildings in different Spanish climate zones. *Sustainability* **2019**, *11*, 4816. [CrossRef]

41. Pero, C.D.; Bellini, O.E.; Martire, M.; Di Summa, D. Sustainable solutions for mass-housing design in Africa: Energy and cost assessment for the somali context. *Sustainability* **2021**, *13*, 4787. [CrossRef]

 *buildings*

Article

Lightweight Composite Partitions with High Sound Insulation in Hotel Interior Spaces: Design and Application

Ting Qu, Bo Wang and Hequn Min *

Key Laboratory of Urban and Architectural Heritage Conservation, Ministry of Education, School of Architecture, Southeast University, 2 Sipailou, Nanjing 210096, China
* Correspondence: hqmin@seu.edu.cn

Abstract: Sound insulation performance of partitions is one of the key factors contributing to the comfort of the hotel interior spaces. Based on the theory of constrained layer damping, this study proposed the light-weight composite partition structure with high sound insulation, which was composed of gypsum boards of different thicknesses and an isobutylene isoprene rubber board. The normal incidence sound transmission loss of the structure was evaluated through finite element simulations as well as experiments, which were conducted in a standing wave tube. The results show that the simulation and experimental results of two kinds of lightweight high sound insulation multi-layer composite partition walls are closely aligned; the surface density of the optimized partition wall was less than 42 kg/m^2, although the normal incidence STL exceeded 51.8 dB at 200 Hz and at 1/3 octave of 1000 Hz with the maximum value of 58.5 dB. The lightweight composite partition wall with high sound insulation has a huge application potential in enhancing the sound environment quality of hotels.

Keywords: hotel interior spaces; high sound insulation; lightweight composite structure partition; sound transmission loss

Citation: Qu, T.; Wang, B.; Min, H. Lightweight Composite Partitions with High Sound Insulation in Hotel Interior Spaces: Design and Application. *Buildings* **2022**, *12*, 2184. https://doi.org/10.3390/buildings12122184

Academic Editor: Andrea Petrella

Received: 9 November 2022
Accepted: 5 December 2022
Published: 9 December 2022

Publisher's Note: MDPI stays neutral with regard to jurisdictional claims in published maps and institutional affiliations.

1. Introduction

Sound insulation performance of the partition wall is one of the important indicators to measure the comfort of the hotel room environment, since high-quality sound environment can bring a pleasant experience for the residents [1]. Meanwhile, the sound insulation performance of the partition wall is a key factor of indoor environment quality (IEQ) [2–4]. It is worth mentioning that the lightweight partition wall has become one of the important considerations in building designing, due to the increasingly high requirements in lightweight design for green and energy-saving buildings. However, lightweight partitions usually have poor sound insulation performance according to the mass law.

Therefore, to address the contradiction between the low sound insulation performance of partition walls and higher IEQ requirements, many researchers have studied the sound insulation performance of lightweight partition walls from various perspectives [5]: the material types, surface density, and thickness [6]. Hagberg et al. used artificial neural networks, evaluating the acoustic performance of the sound insulation curves of different lightweight wood flooring materials, and acoustic data and subjective evaluation, examining the acoustic comfort of residential buildings [7–10]. Since the sound insulation is closely related to the material parameters of the composite structure layered panel, it can suppress the resonance effect of low frequency and the coincidence effect of high frequency.

Adding the viscoelastic material (VEM) core to the lightweight partition wall can effectively improve the sound insulation performance, because the VEM core is directly affected by the relative in-plane displacement of the constrained layer and the base layer, and the constrained layered damping (CLD) composite structure has a stronger energy dissipation ability than the free damping composite structure [11,12]. Quite a number of studies focused on the application of VEM in reducing the vibration of mechanical structures, however, few were on the sound insulation. Jung et al. [13,14] obtained the vibration isolation effect in a wide frequency range by periodically inserting VEM structures into the engine mounting system, using the finite element method and transfer matrix method in the theoretical analysis model. Gupta et al. [15] studied the free vibration of a damped stiffened sheet and periodically damped plates through the finite element numerical simulation. Zhou et al. [16] designed a new structure in which the VEM was inserted between two stiffeners, and found that the VEM had a significant impact on the vibration of the deflection and the propagation characteristics of low-frequency vibration energy.

To improve the sound transmission loss (STL) of composite structures, Moore et al. [17–21] proposed a new wall design with three interlayers, and found that when the Young's modulus of the panel was greater than that of the core material, the shear wave velocity increased with the thickness of the core material; when the damping loss factor was equal to that of the core material, the sound insulation performance of the plate was the worst. Thamburaj et al. [20] studied the acoustic and vibration transmission of sandwich beams with special materials, and found that the STL can be significantly increased through optimizing different damping, thickness, mass density and isotropic or anisotropic surface layer materials, and changing Young's modulus and shear modulus parameters [20]. Varanasi et al. [22–25] proposed that the orthogonal anisotropic materials were embedded in the composite structure as core layers. In addition, some researchers analyzed the STL and the three-dimensional vibration of composite structural plates [26–28], showing that the sensitivity of thin plate to material properties was slightly higher than that of thick plate, and the difference in high frequency region was caused by material damping.

Different core material parameters and sound insulation rules also play important roles in the STL of composite structure partition [29]. Kurtze and Watters [30–32] analyzed the sound insulation rules of sandwich plates with different composite materials and proposed that moving the coincidence frequency to a frequency higher than a given value could maximize the STL of sandwich plates with fixed mass. Through analyzing various characteristics of the plate, Nilsson A.C. [33] proposed that the acoustic characteristics of the sandwich plate could be self-optimized within the frequency range set. Niu et al. [34] proposed the discrete material optimization method for the design of fiber angle layer and the sequence selection of composite materials. Bouzouane et al. [35–37] studied the layered material parameters of sandwich plate and sandwich beam, and found that the thickness of VEM core greatly influenced the acoustic and vibration characteristics, and the values of STL increased with the core material thickness; for sandwich beams, the apparent bending stiffness is related not only to material parameters, but also frequency, boundary conditions and beam length parameters.

Some researchers have studied the numerical value and prediction method of the sound insulation performance of CLD sandwich structures. Assaf S. et al. [38] proposed an optimized damping sandwich structure using a set of high STL prediction methods. Santoni et al. [39] proposed a prediction model for the radiation effect of orthotropic plates using different methods, which considered the resonant and non-resonant contributions of each mode and the fluid loading effect on the sound radiation property of the plate. Guyader et al. [40,41] established the motion equation and natural boundary conditions of multilayer plates by using the dynamic equations, verified the continuity of displacement and shear stress at the interface, proposed the expression of transfer loss of viscoelastic orthogonal anisotropic finite multilayer plates, and proposed the influence on its main parameters. Huang et al. [42] developed an analytical theory of STL and reflected the unbounded plates composed of functionally gradient materials. Kang et al. [43] proposed a method to determine the critical apparent bending stiffness of sandwich plates. Howson et al. [44] proposed a method to determine the single and sandwich beam converging to any fixed frequency by a high-precision method to replace the traditional finite element technique. Backstrom et al. [45] deduced a sixth-order model that was in good agreement with the measured data within the frequency band of interest.

A large number of researchers have studied the application of high sound insulation composite plates from mechanical engineering in the electronics industry. However, few took the application perspective in studying the building space partition. The current study aimed to combine VEM and CLD in researching and designing the internal partition walls, so as to effectively improve the sound environment of the building space.

The current study examined the effects of four parameters (i.e., the material types, the surface density, the number of the VEM core, and the Young's modulus) on the sound insulation performance. This paper reporting the study is structured as follows: Section 2 gives the theoretical method of designing a single-layer plate and a double-layer composite structure; Section 3 presents the finite element method used for the numerical simulation and the analysis of the parameters of layers and core layers with different thicknesses, and discusses the influence of parameter setting on sound insulation performance and proposes an optimized structure; Section 4 presents the simulation and experimental results of the single-layer plate, which verified the feasibility of the method, and discusses the influence of the parameter-setting of surface layer and core layer on STL through comparing the simulation and experimental results of optimized configuration; Section 5 draws the conclusion.

2. Theory Analysis

2.1. Simulation of Single Layer Partition

Since the STL would increase with the increase in frequency, when the surface density of a single wall is constant, and the sound insulation performance of a single-layer homogeneous partition wall is affected by resonance and coincidence effect [46], as well as damping and stiffness, the sound insulation characteristic curve of a single-layer homogeneous partition wall is divided into three control zones according to frequency: stiffness control zone, mass law control zone, and (coincidence region [47], as shown in Figure 1.

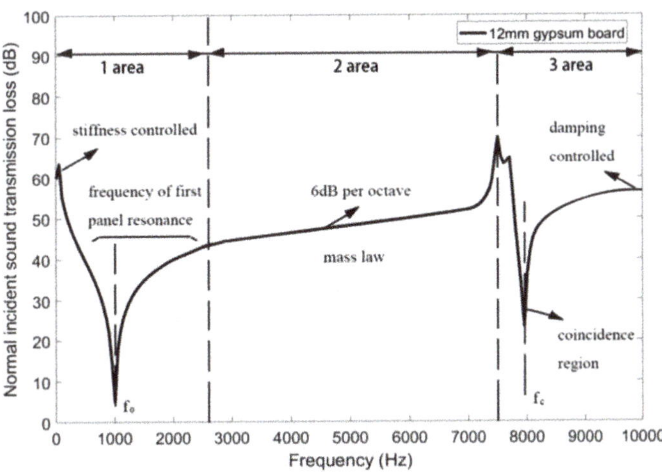

Figure 1. Spectrum of normal incident STL of one typical single layer partition (12 mm gypsum board).

Stiffness controlled zone (Area I): since each wall has its inherent natural resonance frequency range, the wall resonates when the external sound frequency is equal to the inherent frequency of the wall. In the area I, the sound insulation capacity of the single-layer partition is inversely proportional to the growth rate of the frequency, that is, the sound insulation capacity of the partition decreases by 6 dB with the increase in octave band frequency. With the increase in frequency, the wall enters the resonance region, and the natural frequency of the partition resonates with the frequency of the incident sound wave. In addition, the most influential is the first resonant frequency (f_0) within the range of 700 Hz~1500 Hz as in Figure 1, though there are many resonant frequencies in the resonance region. In the design process, therefore, the resonant frequency region should be narrowed as much as possible to exclude the first resonant frequency from the interesting frequency range.

Mass-law controlled zone (Area II): Figure 1 shows that the damping is positively related to the inhibition effect on the amplitude of resonance. As the frequency of incident sound wave increases, the resonance phenomenon disappears in the mass control area (Area II), where the STL is controlled by the mass law.

Damping controlled zone (Area III): When the frequency continues to increase across the mass control area, the sound insulation trough, called "coincidence region", occurs, when the mass effect and the bending stiffness effect of the partition wall cancel out. The sound insulation performance of the partition wall is affected by the coincidence effect (Area III), where the structural vibration mode is increasingly intense. However, due to the damping and internal friction of the partition wall, the vibration of the partition wall may or may not increase. As Figure 1 shows, the damping of the partition is negatively related to the vibration amplitude, which causes the less obvious coincidence valley.

2.2. Sound Insulation Theory of Multilayer Composite Partition

Based on the Kelvin-Voigt model of the constrained damping composite structure, a constitutive model was established, describing the Young's modulus, loss modulus, loss factor, storage modules, surface density, and relaxation time of the structure. Combined with the acoustic equation, a theoretical calculation method of STL was obtained.

Assuming that the composite plate is infinite in the XY direction of the coordinate system shown in Figure 2, the space on both sides is air medium, the incident acoustic wave is plane simple harmonic, the perpendicular line of the wave front is in the XZ plane, and the Angle between the wave front and the Z axis is θ_{n+1}, the acoustic wave equation has the two-dimensional form. The circular frequency of the incident sound wave is ω, the speed of sound in the air is c_0, and the x and z components of the acoustics wave vector are

$$k_x = k_0 \sin \theta_{n+1} \tag{1}$$

$$k_z = k_0 \cos \theta_{n+1} \tag{2}$$

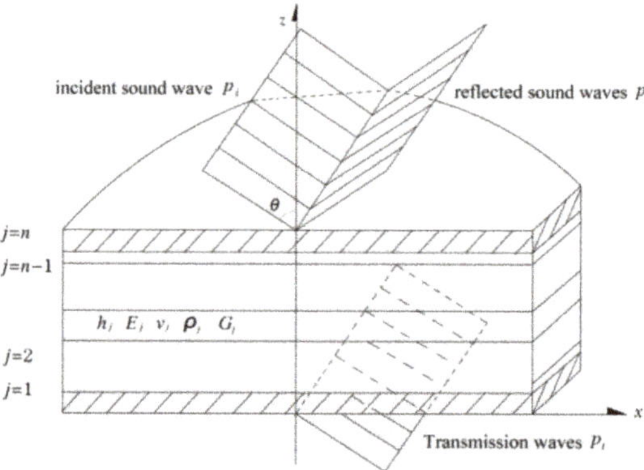

Figure 2. Schematic diagram of calculation model for infinitely damped composite plate.

The acoustic field in the air satisfies the Helmholtz equation:

$$\left(\frac{\partial^2}{\partial x^2} + \frac{\partial^2}{\partial z^2} \right) \varphi = \frac{1}{c_0^2} \frac{\partial^2 \varphi}{\partial t^2} \tag{3}$$

where φ is the velocity potential function of the sound field. This is referred to as $\dot{u}_z$ and the sound pressure as p, so the relationship between $\dot{u}_z$, p and φ is

$$\dot{u}_z = \frac{\partial \varphi}{\partial z} \tag{4}$$

$$p = -\rho_0 \frac{\partial \varphi}{\partial t} \tag{5}$$

The potential function in Equation (3) can be expressed as

$$\varphi(x,z,t) = \phi(z)e^{-ik_x x}e^{iwt} \tag{6}$$

Composite panels total thickness of $H\left(H = \sum_{j=1}^{j=n} h_j\right)$, integrating Equation (6) into Equation (3) can derive:

$$\varphi = \left(Ae^{ik_z(z-H)} + Be^{-ik_z(z-H)}\right)e^{-ik_x x}e^{iwt} \tag{7}$$

Fantasy Equation (7) into Equations (4) and (5), to obtain the incoming air pipe and transmission particle velocity and pressure, the incoming pipe subscript for in, transmission pipe subscript tr:

$$u_{zin} = ik_z\left(Ae^{ik_z(z-H)} - Be^{-ik_z(z-H)}\right)e^{-ik_x x}e^{iwt} \tag{8}$$

$$p_{in} = -\rho_0 iw\left(Ae^{ik_z(z-H)} + Be^{-ik_z(z-H)}\right)e^{-ik_x x}e^{iwt} \tag{9}$$

$$u_{ztr} = -ik_z Ce^{-ik_z z}e^{-ik_x x}e^{iwt} \tag{10}$$

$$p_{tr} = -\rho_0 iwCe^{-ik_z z}e^{-ik_x x}e^{iwt} \tag{11}$$

In the formula, A, B and C are the coefficients of sound pressure of incident sound, reflected sound and transmitted sound, respectively.

The relationship between the vibration dispositions of any elastic or viscoelastic layer of the composite plate under the action of the sound field and the potential functions of expansion and contraction waves and shear waves can be described by Navier equation:

$$\vec{u}_j = \text{grad}\,\varphi_j + crul\,\vec{\psi}_j \tag{12}$$

The potential functions φ and ψ satisfy the wave equation

$$\left(\frac{\partial^2}{\partial x^2} + \frac{\partial^2}{\partial z^2}\right)\varphi_j + \frac{1}{c_{Lj}^2}\varphi_j = 0 \tag{13}$$

The sound transmission coefficient of the composite plate is expressed as

$$\tau = |\Delta|^2 \tag{14}$$

Thus, the STL can be written as

$$\text{TL} = 10\lg\left(\frac{1}{\tau}\right) = -20\lg|\Delta| \tag{15}$$

3. Simulation and Experiment Setup

3.1. Configuration Design for Composite Partition

The sound insulation effect of a traditional single lightweight partition wall is weak, and could not reach the requirements of the standard value of 45 dB [48]. To overcome these shortcomings, the key parameters affecting sound insulation performance were studied. Thus, the thin and high sound insulation damping composite structure partition walls were proposed. Considering the general specification of the elastic panel and indoor partition in the market, gypsum boards were adopted for designing the partition configuration. The setting range of Young's modulus, thickness and surface density were 1600–3000 Mpa, 9.5–15 mm, and 600–1200 kg/m^3, respectively. The total thickness of CLD composite structure was within 80 mm, with the total surface density less than 50 kg/m^2. The VEM core was set to be 1–5 mm thick.

The configuration design process adopted in this study is as follows. Firstly, one typical configuration of the damping composite structure partition walls was established. The panel was composed of two gypsum boards as the surface layer and one VEM core layer with the same thickness. The Young's modulus, density, thickness of surface layer and VEM core were 1600 Mpa, 600 kg/m^3, 9.5 mm and 1 mm, respectively. Secondly, the sound insulation performance of the panel was compared with that of single layer gypsum board. Thirdly, the rules of improving sound insulation performance were established and the structure was further optimized. Table 1 shows the specifications and types in simulation.

Table 1. Simulated composite partition specifications and types.

Name	Type	Material Specification (mm)	Thickness (mm)	Surface Density (kg/m^2)
Case 1	Three-layer composite structure	12 gypsum board + 1 IIR + 12 gypsum board	25	15.4
Case 2	Five-layer composite structure	12 gypsum board + 2 IIR + 9.5 gypsum board + 1 IIR + 9.5 gypsum board	34	23.8
Case 3	Seven-layer composite structure	12 gypsum board + 3 IIR + 12 gypsum board + 2 IIR + 9.5 gypsum board + 1 IIR + 9.5 gypsum board	49	36.3

3.2. Simulation Models

The COMSOL Multiphysics software was used for the finite element modeling, in which a single physical field or a couple of physical fields can be selected according to the actual needs [49]. The acoustic module of the software was used to analyze the propagation process of acoustic waves, based on the signals of frequency domain and time domain. Since different acoustic interfaces need different physical fields, the model in this study mainly used the pressure acoustics and solid mechanics physical fields to model the configuration, and simulated the variation of STL under the normal sound waves incidence. As shown in Table 1, plasterboards and butyl rubber inserts with different thicknesses were combined in different forms to establish a 1:1 model in COMSOL. The STL simulation of configuration was carried out through the processes of model designing, geometric modeling, parameter setting, solution loading and simulation result processing.

3.3. Experimental Settings

The normal incidence STL experiments were conducted in a semi-anechoic chamber. Based on the Chinese standard of Acoustics-Determination of sound transmission loss in impendence tubes Transfer matrix method (GB/Z 22764-2011) and the American standard Test Method for Measurement of Normal Incidence Sound Transmission of Acoustical Materials Based on the Transfer Matrix Method (ASTM E2611-09) [50,51], the four-microphone transfer matrix method was utilized with a rectangular standing wave tube. Figure 3 shows the block diagram of the experimental measurement system, in which the parameters met the measurement standards, the wall body was an acrylic plate with a thickness of 15 mm, the cross-section size was 100 mm × 100 mm, and the cut-off frequency was 1700 Hz. Thus, the 1/3 octave frequency range between 50 and 1600 Hz was selected. Four G.R.A.S 40AP 1/2-inch sound pressure field microphones were used to measure the sound pressure signals on the inner wall of the tube at a distance of 5 cm from the center points. The specimen was tightened in the middle of the experiment device, and between the four microphones. The HIVI-M3S's 3.5-inch speakers were powered by an amplifier from the B&K 2716, which was mounted on top of the standing-wave tube and effectively sealed to the inner wall of the tube. In the experiment, the B&K Pulse 3560D system was used to send out sweep signals and carry out the real-time signal acquisition and analysis. Figure 3b is the field measurement diagram.

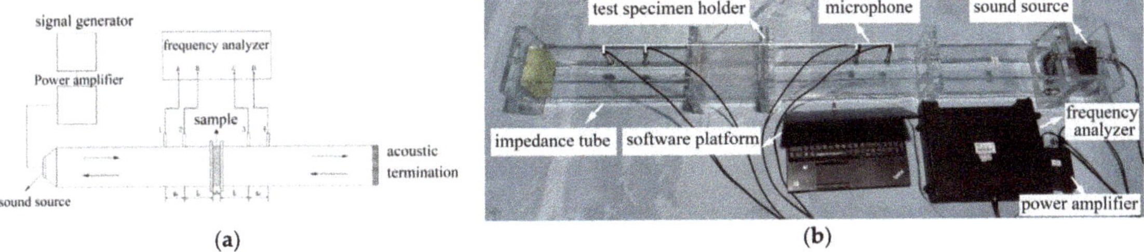

Figure 3. (**a**) Schematic diagram of experimental measurement system for transmission loss of composite partition structure; (**b**) Field drawing for standing wave tube test measurement field drawing.

4. Results and Discussion

4.1. Insulation Performance of the Base Case

This study used the 12 mm single-layer gypsum board as an example to carry out the simulation and compared the simulation results with the experimental results. It was found that due to the single-layer gypsum board effect, the simulation curve of sound insulation frequency had resonance phenomenon and resonance valley between 1200 Hz and 1300 Hz, resulting in a sharp decline in sound insulation performance. However, the simulation results fit well with the experimental results, which verifies the reliability of the theoretical method, as shown in Figure 4.

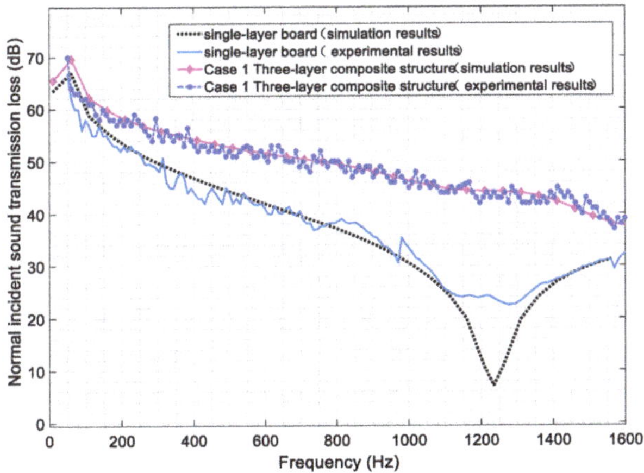

Figure 4. Comparison of simulation and experimental results.

Compared with the single layer board, the thickness of the single board in Case 1 was thicker, doubled to 25 mm. The sound insulation curve of Case 1 flattened the resonance valley of the single board and increased the single board by more than 10 dB, which greatly improved the STL. It can be seen that the improvement of sound insulation was not only related to the mass law, but also closely related to the VEM core layer, and the constrained layer damping formed a sandwich structure, which changed the frequency of the resonant mode and effectively improved the sound insulation performance. It can also be found that the sound insulation of sandwich partitions was better than that of the single-layer plate. Based on these results, the simulation of Young's modulus, thickness, density and VEM core thickness of the elastic panel and the combination modes of these parameters were expanded for revealing the influence of parameters more clearly.

4.2. Elastic Panel Parameters Discussed

This part mainly discusses the influence of the variation of Young's modulus, thickness and density parameters on the sound insulation performance of sandwich structure. Meanwhile, the parameters of Young's modulus at 2000 Mpa, thickness of 20 mm and density of 600 kg/m^3 were selected for experimental verification, which proved the validity of the simulation results.

4.2.1. Influence of Young's Modulus

Based on Case 1 (basic configuration), the influence of Young's modulus of sandwich plate was analyzed. When the thickness and density are unchanged, the ranges of Young's modulus were set at 1600 Mpa, 2000 Mpa and 3000 Mpa, and the thickness of elastic panels and core layer on both sides remained unchanged. Figure 5 shows that in the stiffness control region, with the increase in Young's modulus, the first resonance frequency moved to high frequency, and the STL significantly increased. In the mass low control area, the STL curve decreased slightly with the increase in Young's modulus, but the overall sound insulation performance increased steadily.

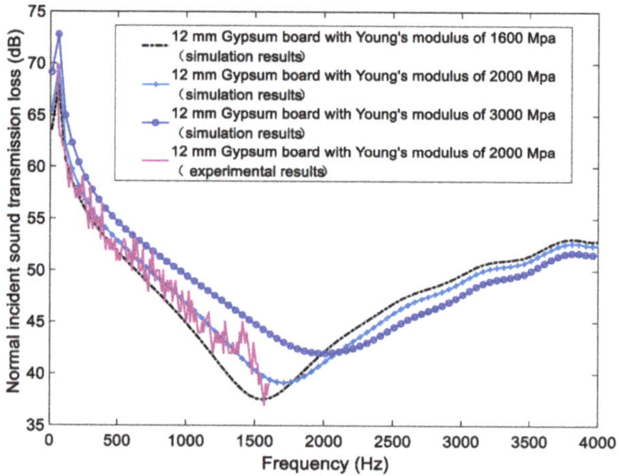

Figure 5. The STL spectrum of the change of Young's modulus with the set range.

4.2.2. Influence of Gypsum Board's Thickness

Based on Case 1 (basic configuration), the influence of sandwich plate thickness was analyzed. When Young's modulus and density remained unchanged, the thickness ranges of elastic surface layer were set at 9.5 mm, 12 mm and 15 mm. Figure 6 shows that in the stiffness control region, the first-order resonance frequency moved to the high frequency with the increase in the elastic surface layer's thickness, and the STL significantly increased. In the mass law control area, the STL of sandwich structure increased with the increase in thickness.

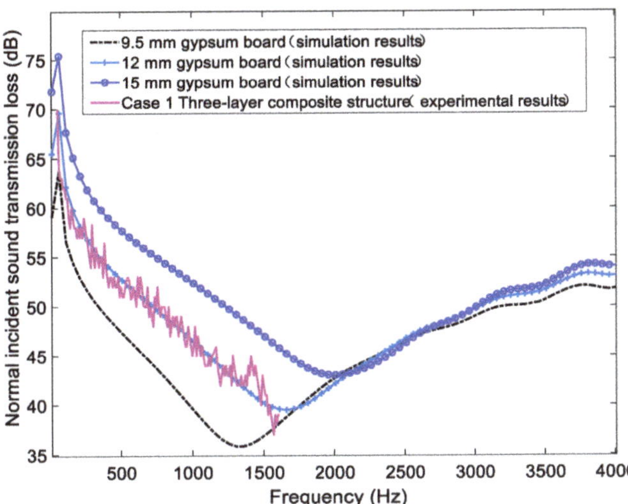

Figure 6. The STL spectrum of the composite partitions with the change of gypsum boards' thickness.

4.2.3. Influence of Density

Based on Case 1 (basic configuration), the influence of sandwich plate's density was analyzed. When Young's modulus and thickness remained unchanged, the density ranges of elastic surface layer were set at 600 kg/m^3, 800 kg/m^3, 1000 kg/m^3 and 1200 kg/m^3. Figure 7 shows that in the first-order stiffness region, the resonance valley moved to high frequency with the increase in density, and the overall STL increased to a certain extent. In the mass law control area, the sound insulation performance was significantly improved with the increase in density value.

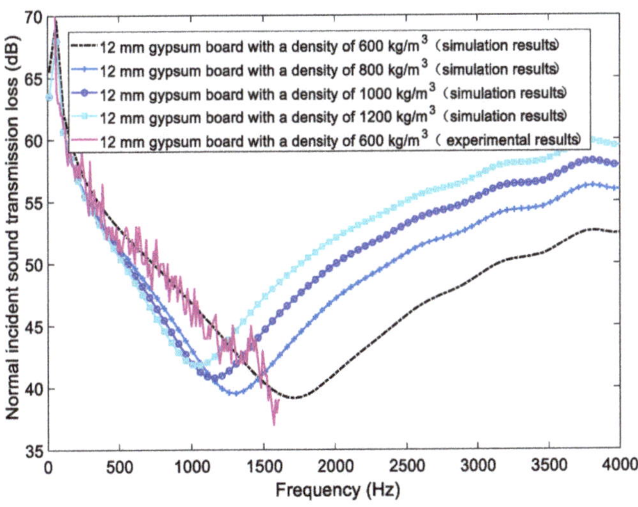

Figure 7. The STL spectrum of the composite partitions with the change of gypsum board's density.

4.3. Influence of the VEM Core's Thickness

Based on Case 1 (basic configuration), the influence of VEM core thickness of sandwich structure was analyzed. When Young's modulus and density were unchanged, the ranges of VEM core thickness were set at 1 mm, 3 mm and 5 mm. Meanwhile, a splint configuration with VEM core layer thickness of 1 mm was selected to verify the validity of the simulation results. Figure 8 shows that in the first-order stiffness control area, with the increase in VEM core layer's thickness, the resonance valley tended to be smooth and moved to high frequency, and the sound insulation performance was significantly improved. When the thickness of the core layer was 5 mm, the sound insulation of the first-order resonance valley can be improved by about 6 dB. In the mass law control area, the sound insulation performance of STL curve was also improved with the increase in thickness.

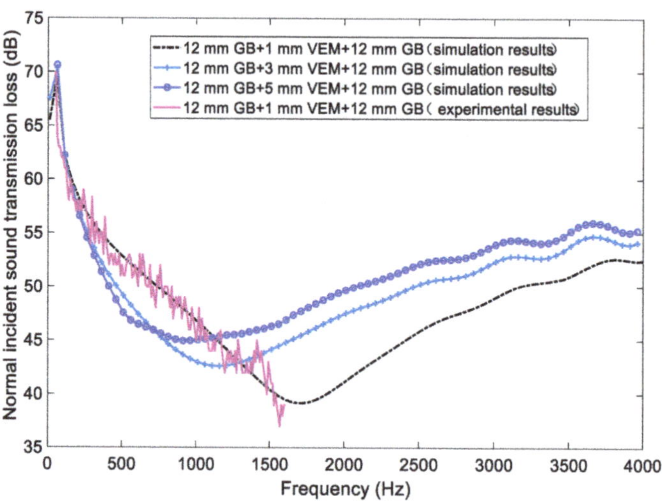

Figure 8. The STL spectrum of the composite partitions with the change of VEM core's thickness.

4.4. Influence of Composite Partition Walls' Configuration

To verify the influence of different combinations of configurations with the same thickness on STL, the overall thickness of the configuration was set at 43 mm. The configuration combination and specifications were set as the following three kinds: Composite structure 1: 19 mm gypsum board, 5 mm VEM, 19 mm gypsum board; Composite structure 2: 9.5 mm, 2 mm VEM 9.5 mm, 3 mm VEM, 19 mm gypsum board; Composite structure 3: 9.5 mm gypsum board, 2 mm VEM, 9.5 mm gypsum board, 1 mm VEM, 9.5 mm gypsum board, 2 mm VEM, 9.5 mm gypsum board. Meanwhile, composite structure 1 in the structure simulation was selected to verify the validity of the simulation results.

As is shown in Figure 9, composite structure 1 (two surface layers and one VEM core), there is a large trough between 1200 Hz and 1800 Hz, and the resonance phenomenon is obvious; the several valleys and peaks of different sizes appeared between 3000 Hz and 4000 Hz. The sound insulation performance is degraded. The sound insulation performance of composite structure 2 (three surface layers and two VEM cores) is obviously better than that of composite structure 1, due to the change of the combination mode from one splint to two splints. The low point between 1200 Hz and 1800 Hz was improved to about 4 dB. As for composite structure 3 (four surface layers and three VEM cores), the combination method also changed to some extent: the splint was recombined with three splints. The sound insulation performance of composite structure 3 was the best, and the troughs in composite structure 1 and composite structure 2 became smooth. Additionally, the STL of composite structure 3 had a stable performance between 2000 Hz and 4000 Hz, and the STL was significantly better than those of composite structure 1 and composite structure 2. According to the sound insulation law of the above configuration, it can be seen that composite structure 3 has obvious advantages in the sound insulation of the middle and high frequency band compared with the composite structure 2, but it decreases in the low frequency band.

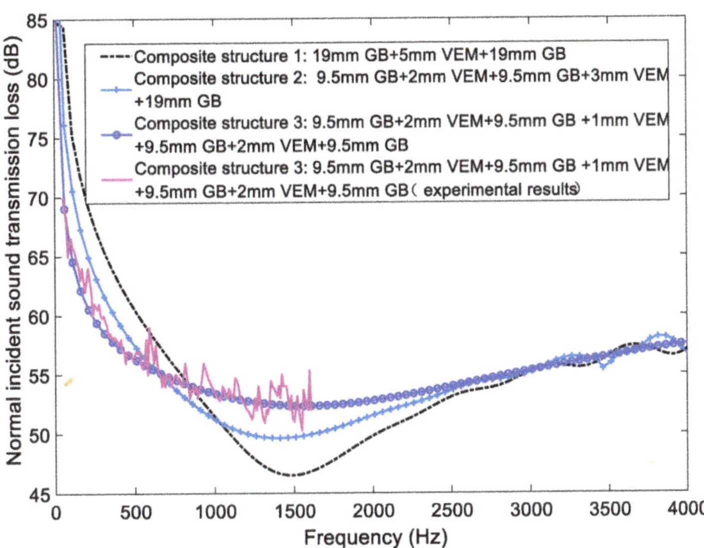

Figure 9. The STL spectrum of the composite partitions with various inter-layers for VEM core and gypsum board.

4.5. Optimal Configuration

The analysis revealed that Young's modulus, thickness, density of elastic panel, thickness of VEM core and combination of laminates were the key factors affecting STL. The thickness and density of the elastic panel were also important factors affecting the configuration. Controlled by the mass law, the thickness and density were positively correlated with the sound insulation performance of the partition. In the stiffness control area, with the increase in density and thickness, the bending stiffness of the structure increased, making the resonance valley gradually move to the high frequency. At the same time, the sound insulation performance was improved with the increase in Young's modulus, showing that the stiffness was positively correlated with the sound insulation performance. Additionally, the influence of VEM core layer thickness on the configuration cannot be ignored either. The VEM core layer directly affected by the displacement in the plane was related to the constraint layer and the base layer, resulting in shear deformation; VEM core layer had a stronger energy dissipation capacity and could effectively improve the valley of resonance caused by structural resonance. As a result, with a VEM core incorporated into the structure, the sound insulation performance was improved as the thickness of the composite sandwich increased. In addition, the effective combination of laminates is also one of the important factors to improve the sound insulation performance of partition wall. The simulation results show that when the configuration thickness remained unchanged, with the increase in the number of layers, the sound insulation performance of the configuration was significantly improved in the middle and high frequency bands, but slightly decreased in the low frequency region.

Based on the analysis of elastic panel and core layer parameters and the combination mode of partition structure, a better partition configuration of Case 3 was proposed according to the law affecting the sound insulation performance, and verified by experiments (Figure 10). The results show that the over-agreement between simulation and experimental junction was high, verifying the rationality of the configuration.

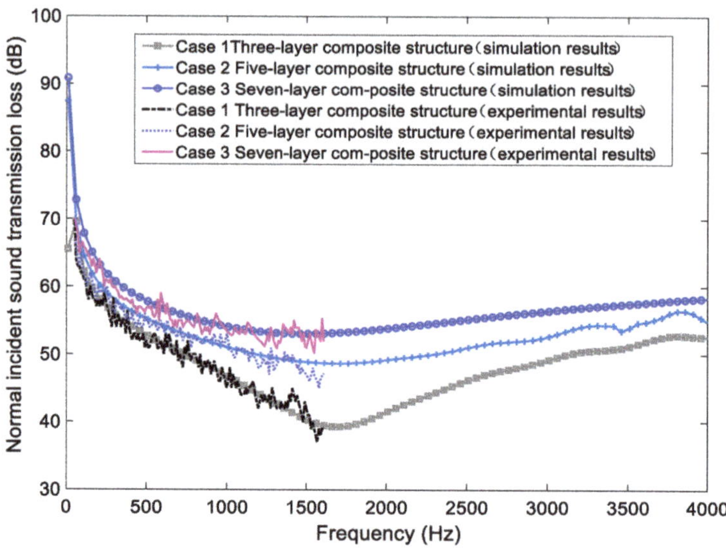

Figure 10. Comparison of simulation and experimental results; STL trends represent the simulation and experimental results of Case 1, Case 2, Case 3.

Case 1 was the basic reference configuration in this study, on the basis of which the parameter configuration was further optimized. Compared with Case 1, the sound insulation performance in Case 2 was improved by about 5 dB, and the average STL was 49 dB. However, further improvement was proved possible. In Case 3, based on the rules of parameter setting, the optimized partition configuration was composed of layered materials with different parameters, which effectively rectified the defects caused by resonance and coupling effects of single-layer wall and lightweight partition. The average STL value of the optimized configuration reached about 55 dB, forming a stable STL curve between the middle and low frequencies, indicating a significantly improved the sound insulation performance.

5. Conclusions

Based on the design process of CLD composite partitions, this study analyzed and discussed four parameters of the layered materials that influence the sound insulation performance, and proposed an optimized structure suitable for the space partition in the hotel. The proposed structure had a surface density of 36.3 kg/m^2 and a thickness of 49 mm, which could make up for the relatively poor sound insulation performance of light partition currently used in the market and may achieve the goal of lightweight and high sound insulation. The major contributions of the study are summarized as follows:

(1) The parameters of Young's modulus, thickness, surface density and VEM damping core layer thickness of elastic panel are important factors influencing the sound insulation performance of CLD partitions.

(2) As for lightweight and high sound insulation composite partition, the sound insulation performance can be improved by using different inter-layer combinations.

(3) Adding VEM layer to the configuration can effectively suppress and reduce the resonance valley in the first stiffness control area, which thus can effectively improve the sound insulation performance of the whole structure.

Author Contributions: T.Q.: Conceptualization, Methodology, Software, Investigation, Writing—original draft. B.W.: Writing—review & editing, Software, Validation Investigation. H.M.: Conceptualization, Methodology, Writing—review & editing, Supervision, Funding acquisition, Project administration. All authors have read and agreed to the published version of the manuscript.

Funding: This work is supported by Opening Foundation of Key Laboratory of New Technology for Construction of Cities in Mountain Area, Ministry of Education, China (LNTCCMA-20210104). This work was also supported by the Natural Science Foundation of China (Grant No. 51408113) and the Natural Science Foundation of Jiangsu Province, China (Grant No. BK20140632).

Institutional Review Board Statement: Not applicable.

Informed Consent Statement: Not applicable.

Data Availability Statement: Data can be obtained from the corresponding author upon reasonable request.

Acknowledgments: The authors sincerely thank the editor for her meticulous and professional work.

Conflicts of Interest: The authors declare that they have no known competing financial interest or personal relationships that could have appeared to influence the work reported in this paper.

References

1. Alves, J.A.; Silva, L.T.; Remoaldo, P.C.C. The Influence of Low-Frequency Noise Pollution on the Quality of Life and Place in Sustainable Cities: A Case Study from Northern Portugal. *Sustainability* **2015**, *7*, 13920–13946. [CrossRef]
2. Kong, Z.; Jakubiec, J.A. Evaluations of long-term lighting qualities for computer labs in Singapore. *Build. Environ.* **2021**, *194*, 107689. [CrossRef]
3. Kong, Z.; Utzinger, D.M.; Freihoefer, K.; Steege, T. The impact of interior design on visual discomfort reduction: A field study integrating lighting environments with POE survey. *Build. Environ.* **2018**, *138*, 135–148. [CrossRef]
4. Kong, Z.; Zhang, R.; Ni, J.; Ning, P.; Kong, X.; Wang, J. Towards an integration of visual comfort and lighting impression: A field study within higher educational buildings. *Build. Environ.* **2022**, *216*, 108989. [CrossRef]
5. Lin, J.; Tan, H.; Xu, C.; Shi, H.R. Research Progress of Acoustic Technology in Buildings. *Build. Sci.* **2013**, *29*, 10.
6. Mansilla, J.; Masson, F.; Palma, I.D.; Pepino, L.; Bender, L. Sound Insulation of Homogeneous Single Panels: A Comparison Between Real Construction Materials and Several Prediction Models. In Proceedings of the 24th International Congress on Sound and Vibration, London, UK, 23–27 July 2017; pp. 1–8.
7. Bader Eddin, M.; Ménard, S.; Bard Hagberg, D.; Kouyoumji, J.L.; Vardaxis, N.G. Prediction of Sound Insulation Using Artificial Neural Networks—Part I: Lightweight Wooden Floor Structures. *Acoustics* **2022**, *4*, 13. [CrossRef]
8. Bader Eddin, M.; Vardaxis, N.G.; Ménard, S.; Bard Hagberg, D.; Kouyoumji, J.L. Prediction of Sound Insulation Using Artificial Neural Networks—Part II: Lightweight Wooden Façade Structures. *Appl. Sci.* **2022**, *12*, 6983. [CrossRef]
9. Vardaxis, N.G.; Hagberg, D.B. Review of acoustic comfort evaluation in dwellings: Part III—airborne sound data associated with subjective responses in laboratory tests. *Build. Acoust.* **2018**, *25*, 289–305. [CrossRef]
10. Vardaxis, N.G.; Hagberg, D.B.; Dahlström, J. Evaluating Laboratory Measurements for Sound Insulation of Cross-Laminated Timber (CLT) Floors: Configurations in Lightweight Buildings. *Appl. Sci.* **2022**, *12*, 7642. [CrossRef]
11. Larbi, W.; Deü, J.F.; Ohayon, R. Vibroacoustic analysis of double-wall sandwich panels with viscoelastic core. *Comput. Struct.* **2016**, *174*, 92–103. [CrossRef]
12. Hwang, S.; Kim, J.; Lee, S.; Kwun, H. Prediction of sound reduction index of double sandwich panel. *Appl. Acoust.* **2015**, *93*, 44–50. [CrossRef]
13. Jung, W.; Gu, Z.; Baz, A. Mechanical filtering characteristics of passive periodic engine mount. *Finite Elem. Anal. Des.* **2010**, *46*, 685–697. [CrossRef]
14. Pellicier, A.; Trompette, N. A review of analytical methods, based on the wave approach, to compute partitions transmission loss. *Appl. Acoust.* **2007**, *68*, 1192–1212. [CrossRef]
15. Guptat, B.V.R.; Ganesan, N.; Narayanan, S. Finite element free vibration analysis of damped stiffened panels. *Comput. Struct.* **1986**, *24*, 485–489. [CrossRef]
16. Zhou, X.Q.; Yu, D.Y.; Shao, X.; Wang, S.; Zhang, S.Q. Simplified-super-element-method for analyzing free flexural vibration characteristics of periodically stiffened-thin-plate filled with viscoelastic damping material. *Thin-Walled Struct.* **2015**, *94*, 234–252. [CrossRef]
17. Moore, J.A.; Lyon, R.H. Sound transmission loss characteristics of sandwich panel constructions. *J. Acoust. Soc. Am.* **1991**, *89*, 777–791. [CrossRef]
18. Aloufi, B.; Behdinan, K.; Zu, J. Vibro-acoustic model of an active aircraft cabin window. *J. Sound Vib.* **2017**, *398*, 1–27. [CrossRef]
19. Tadeu, A.J.B.; Mateus, D.M.R. Sound transmission through single, double and triple glazing. *Exp. Evaluat. Appl. Acoust.* **2001**, *62*, 307–325. [CrossRef]
20. Thamburaj, P.; Sun, J.Q. Effect of Material and Geometry on the Sound and Vibration Transmission across a Sandwich Beam. *J. Vib. Acoust.* **2001**, *123*, 205–212. [CrossRef]

21. Yang, Y.; Fenemore, C.; Kingan, M.J.; Mace, B.R. Analysis of the vibroacoustic characteristics of cross laminated timber panels using a wave and finite element method. *J. Sound Vib.* **2021**, *494*, 115842. [CrossRef]

22. Varanasi, S.; Bolton, J.S.; Siegmund, T.H.; Cipra, R.J. The low frequency performance of metamaterial barriers based on cellular structures. *Appl. Acoust.* **2013**, *74*, 485–495. [CrossRef]

23. Li, J.; Li, S. Sound Transmission Through Metamaterial-Based Double-Panel Structures With Poroelastic Cores. *Acta Acust. United Acust.* **2017**, *103*, 869–884. [CrossRef]

24. Kim, Y.; ASME, M. Identification of Acoustic Characteristics of Honeycomb Sandwich Composite Panels Using Hybrid Analytical/Finite Element Method1. *J. Vib. Acoust.* **2013**, *135*, 011006-1–011006-11. [CrossRef]

25. Guerich, M. Optimization of Noise Transmission Through Sandwich Structures. *J. Vib. Acoust.* **2013**, *135*, 051010-1–051010-13. [CrossRef]

26. Zhou, X.Q.; Zhang, S.Q.; Lin, W.W. Sound radiation characteristics analysis for the honeycomb reinforced laminated structures with viscoelastic material fillers through the asymptotic homogenous method. *Compos. Struct.* **2020**, *245*, 112266. [CrossRef]

27. Li, Q.; Iu, V.P.; Kou, K.P. Three-dimensional vibration analysis of functionally graded material sandwich plates. *J. Sound Vib.* **2008**, *311*, 498–515. [CrossRef]

28. Huang, C.; Nutt, S. Sound transmission prediction by 3-D elasticity theory. *Appl. Acoust.* **2009**, *70*, 730–736. [CrossRef]

29. Chandra, N.; Raja, S.; Nagendra Gopal, K.V. Vibro-acoustic response and sound transmission loss analysis of functionally graded plates. *J. Sound Vib.* **2014**, *333*, 5786–5802. [CrossRef]

30. Kurtze, G.; Watters, B.G. New Wall Design for High Transmission Loss or High Damping. *J. Acoust. Soc. Am.* **1959**, *31*, 739–748. [CrossRef]

31. Li, J.; Li, S. Sound transmission through a damped sandwich panel. *Acta Acust. United Acust.* **2017**, *80*, 315–327.

32. Wang, J.; Lu, T.J.; Woodhouse, J.; Langley, R.S.; Evans, J. Sound transmission through lightweight double-leaf partitions: Theoretical modelling. *J. Sound Vib.* **2005**, *286*, 817–847. [CrossRef]

33. Nilsson, A.C. Wave propagation in and sound transmission through sandwich plates. *J. Sound Vib.* **1990**, *138*, 73–94. [CrossRef]

34. Niu, B.; Olhoff, N.; Lund, E.; Cheng, G. Discrete material optimization of vibrating laminated composite plates for minimum sound radiation. *Int. J. Solids Struct.* **2010**, *47*, 2097–2114. [CrossRef]

35. Bouzouane, B.; Ghorbel, A.; Akrout, A.; Abdennadher, M.; Boukharouba, T.; Haddar, M. Ultra-thin films effects on vibro-acoustic behaviour of laminated plate including a viscoelastic core. *Appl. Acoust.* **2019**, *147*, 121–132. [CrossRef]

36. Chandra, N.; Gopal, K.V.N.; Raja, S. Vibro-acoustic response of sandwich plates with functionally graded core. *Acta Mech.* **2017**, *228*, 2775–2789. [CrossRef]

37. Nilsson, A.; Baro, S.; Piana, E.A. Vibro-acoustic properties of sandwich structures. *Appl. Acoust.* **2018**, *139*, 259–266. [CrossRef]

38. Assaf, S.; Guerich, M. Numerical Prediction of Noise Transmission Loss through Viscoelastically Damped Sandwich Plates. *J. Sandw. Struct. Mater.* **2008**, *10*, 359–384. [CrossRef]

39. Santoni, A.; Schoenwald, S.; Fausti, P.; Tröbs, H.M. Modelling the radiation efficiency of orthotropic cross-laminated timber plates with simply-supported boundaries. *Appl. Acoust.* **2019**, *143*, 112–124. [CrossRef]

40. Guyader, J.L.; Lesueur, C. Acoustic transmission through orthotropic multilayered plates, part I: Plate vibration modes. *J. Sound Vib.* **1978**, *58*, 51–68. [CrossRef]

41. Guyader, J.L.; Lesueur, C. Acoustic transmission through orthotropic multilayered plates, part II: Transmission loss. *J. Sound Vib.* **1978**, *58*, 69–86. [CrossRef]

42. Huang, C.; Nutt, S. An analytical study of sound transmission through unbounded panels of functionally gradedmaterials. *J. Sound Vib.* **2011**, *330*, 1153–1165. [CrossRef]

43. Kang, L.; Sun, C.; An, F.; Liu, B. A bending stiffness criterion for sandwich panels with high sound insulation and its realization through low specific modulus layers. *J. Sound Vib.* **2022**, *536*, 117149. [CrossRef]

44. Howson, W.P.; Zare, A. Exact dynamic stiffness matrix for flexural vibration of three-layered sandwich beams. *J. Sound Vib.* **2005**, *282*, 753–767. [CrossRef]

45. Backström, D.; Nilsson, A.C. Modelling the vibration of sandwich beams using frequency-dependent parameters. *J. Sound Vib.* **2007**, *300*, 589–611. [CrossRef]

46. Wang, C. Modal sound transmission loss of a single leaf panel: Effects of intermodal coupling. *J. Acoust. Soc. Am.* **2015**, *6*, 3514–3522. [CrossRef] [PubMed]

47. Wang, C. Modal sound transmission loss of a single leaf panel: Asymptotic solutions. *J. Acoust. Soc. Am.* **2015**, *138*, 3964–13575. [CrossRef] [PubMed]

48. *GB 50118-2010*; Code for Design of Sound Insulation of Civil Buildings. Ministry of Housing and Urban-Rural Development of the People's Republic of China. China Architecture & Building Press: Beijing, China, 2010.

49. Comsol Multiphysics. *COMSOL Multiphysics User's Guide*; COMSOL, Inc.: Burlington, VT, USA, 2020.

50. *GB/Z 22764-2011*; Acoustics-Determination of Sound Transmission Loss in Impendence Tubes-Transfer Matrix Method. Standardization Administration of China: Beijing, China, 2012.

51. *ASTM E2611-09*; Standard Test Method for Measurement of Normal Incidence Sound Transmission of Acoustical Materials Based on the Transfer Matrix Method. ASTMInternational: West Conshohocken, PA, USA, 2009.

 *buildings*

Article

Optimizing Annual Daylighting Performance for Atrium-Based Classrooms of Primary and Secondary Schools in Nanjing, China

Jin Ma and Qingxin Yang *

School of Architecture, Southeast University, Nanjing 210096, China
* Correspondence: 220200118@seu.edu.cn

Abstract: Influenced by educational policies and newly emerging educational philosophies, the proportion of public space is expanding in primary and secondary schools in China. Consequently, the atrium in school design is increasingly drawing attention due to the consideration of space efficiency and its accommodability for diverse activities. Although many studies have already explored the daylighting performance of atriums, the particularities of primary and secondary schools are rarely noticed, which leads to the lack of a reliable basis for a quick judgment in the early design stage. This study used the annual daylight metrics of Spatial Daylight Autonomy ($sDA_{300,50\%}$) and Annual Sunlight Exposure ($ASE_{1000,250\,h}$) as the indicators, built a parametric model in Grasshopper, conducted the simulation using the Ladybug–Honeybee plug-in, and separately performed the linear regression analysis on the three groups of data from the different types of atriums. The results show that in Nanjing's climate, the north and east sides of atriums are the most suitable orientations for classrooms, and a corridor width of 3 m ensures high-quality daylight for the bottom floors. The optimal design equations for atrium width and length are provided for the three types of atriums, respectively, hopefully, to ensure that classrooms surrounding the atrium can reach the requirement of $sDA_{300,50\%} \geq 0.75$, and the design recommendations are offered based on the results.

Keywords: annual daylighting performance; atrium design; parametric simulation; primary and secondary school

Citation: Ma, J.; Yang, Q. Optimizing Annual Daylighting Performance for Atrium-Based Classrooms of Primary and Secondary Schools in Nanjing, China. *Buildings* **2023**, *13*, 11. https://doi.org/10.3390/buildings13010011

Academic Editors: Yue Wu, Zheming Liu and Zhe Kong

Received: 23 October 2022
Revised: 12 December 2022
Accepted: 15 December 2022
Published: 21 December 2022

1. Introduction

The school design patterns in China are changing from "single corridor with fixed classrooms" to "multiple public spaces connecting class groups" [1]. Previously, the "long and single corridors" characterized the spatial paradigm of most schools in China due to the efficiency-first belief [2]. However, the long traffic path and simple space create difficulties for teaching efficiency and teaching method adjustment [1,2]. With the implementation of "Quality Education" [3] and "Class Selection System" [4] policies, and the influence of new educational methods such as the "STEAM" method [5], Multi-age Classroom [6] and Informal Learning [7], compounding and diverse variations have appeared in school designs in recent years [8]; public spaces with high flexibility and adaptability which draw public attention are increasingly used in China [1,9].

The atrium is an important component in teaching spaces [1,8,10]. On the one hand, an atrium could promote physical activity and social interaction [10,11], shape cultural scenes and enhance the sense of space [12,13]. On the other hand, an atrium brings daylight into the center of the building and connects adjacent spaces to the outside world [14,15], which is significant because daylight from the atrium can replace or reduce artificial lighting, as well as lower energy consumption [16].

Daylight is considered a prime factor in school design because of its comprehensive effects on students [17–20]. Research has shown that natural daylight has both physiological and psychological effects on students' visual capabilities, productivity and comfort [21,22].

Good daylight is linked to better emotions and higher motivation, and can enhance the immune system, lower eyestrain and even improve achievements [23,24]. Additionally, daylight is beneficial in regulating students' circadian rhythms, as well as minimizing the physiological, cognitive, and health effects of circadian disruption caused by an electrical lighting environment [25].

Many studies have explored the daylighting of atriums and revealed that the daylighting performance of an atrium largely depends on its geometric characteristics [26–36] (Table 1). The main characteristics can be categorized into three types: (1) the skylight system (skylight height, shape, scale); (2) the atrium form (atrium height, shapes, scale); (3) the surrounding interface (corridors, windows, etc.).

Table 1. Review of the atrium daylight studies about the building types and parameter targets.

Year	Citation	Building Types	Parameter Targets
2012	[26]	Unspecified	Atrium scale (SAR, PAR) [1] Height and orientation Surface reflectance
2015	[27]	Unspecified	Atrium scale (WI) [2] Skylight form
2016	[28]	Offices	Atrium scale (SAR, PAR) Skylight height, Floor number
2017	[29]	Unspecified	Atrium scale (WI, WID) [3] Surface reflectance
2019	[30]	Commercial building	Atrium scale (shape, PAR) Building height, Skylight size
2021	[31]	Office	Atrium scale (WI, width to height ratio)
2021	[32]	Library	Atrium scale (SAR, PAR) Height
2022	[33]	Commercial building	Atrium shape and numbers Profile inclination Skylight height ratio
2022	[34]	Commercial building	Atrium shape and height
2022	[35]	School building	Unspecified
2022	[36]	Heritage building	Atrium shape and numbers

[1] SAR: the Section Aspect Ratio; PAR: the Plan Aspect Ratio. [2] WI: the Well Index [26]. [3] WID: the Well-Indexed Depth [26].

These studies greatly contribute to our understanding of atriums. However, few studies particularly focused on the atrium design for the classrooms of primary and secondary schools. The design of school buildings is different from that of other buildings, often due to the requirements of national standards. For example, the Code for Design of *School* GB50099-2011 issued in China [37] specifies the design modules and the number of floors for primary and secondary schools, which can cause the specificity of the classroom daylight and the inapplicability of the geometric indicators of the WI index and SAR; the vacations of primary and secondary schools also specify the occupancy profile for simulation.

Therefore, this study aims to evaluate the impact of atrium parameters on the daylighting environment of surrounding classrooms in primary and secondary schools. Specifically, this study:

(1) Analyzed the relationship between Spatial Daylight Autonomy ($sDA_{300,50\%}$), Annual Sunlight Exposure ($ASE_{1000,250\,h}$) and the parameters (orientation, floor and scale) of the atrium;

(2) Obtained design equations and provided suggestions for atrium design to optimize daylighting performance.

The final conclusions are expected to provide a reference for the designers, enable them to control daylighting performance in the early design stage through a convenient process, and hopefully improve the daylight environment of projects in the context of fast-paced construction in China.

2. Research Methodology

2.1. Research Process

The research process mainly consisted of four steps: (1) parametric simulation, (2) descriptive analysis, (3) regression analysis of the simulation results, and (4) derivation of design equations and suggestions (Figure 1).

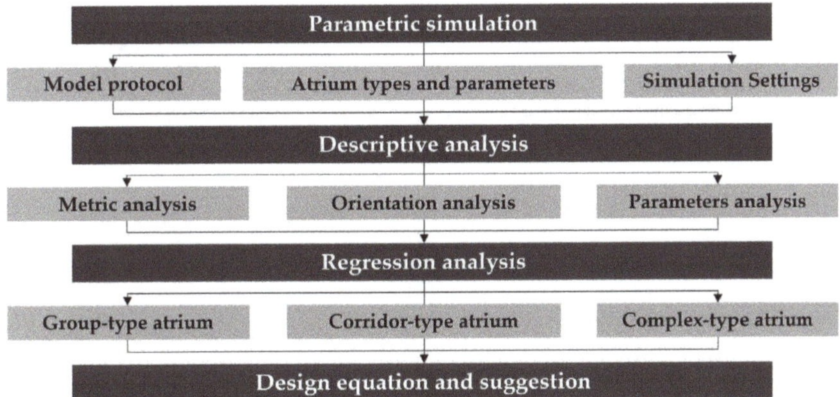

Figure 1. Research flow chart.

2.2. Model Configurations

The atrium-based teaching space was divided into three areas: classrooms and supporting areas on the outside, corridors and traffic spaces in the middle, and the atrium light box in the center. The classroom is the most important teaching space where students spend the most time, therefore the classroom units were set in four orientations as the simulation objects, and the surrounding spaces' design indexes (i.e., the atrium length L, atrium width W and corridor width x) were set as the parameters. Based on design conventions and the national standard [37], the setting of the classroom unit was defined, the interior and exterior windows were set to 2.1 m height, 2.4 m wide with 1.0 m of sill height and 0.9 m of shade depth, and the entire model was set to 4 floors with 3.9 m floor height; the skylight was set to flat form. The final model configurations are shown in Figure 2.

2.3. Atrium Types and Parameters

Based on practical experience, the main atrium forms of a school building in China can be categorized into 4 types: retreat type, group type, corridor type and complex type [38]. Among them, the retreat type reflecting health concerns is usually for the classrooms which always receive enough daylight because of the 1-floor daylight depth. The group type is mainly for mixed-age teaching and inter-class communication. The corridor type is always for informal learning spaces and the complex type of atrium is generally used for the STEAM center. The general scales of the 4 types are summarized in Figure 3.

Accordingly, a parametric model was built with Grasshopper and the parameter ranges were set up; the atrium lengths and widths were set in a range from 9 m to 54 m (as shown in Table 2, the adjacent two columns or rows are distanced by 9, as 9 m is the conventional length of the classroom), and the corridor widths were set to 3 m, 4.5 m and 6 m. Therefore, the following 36 atriums under the 3 different corridor widths were modeled for simulation.

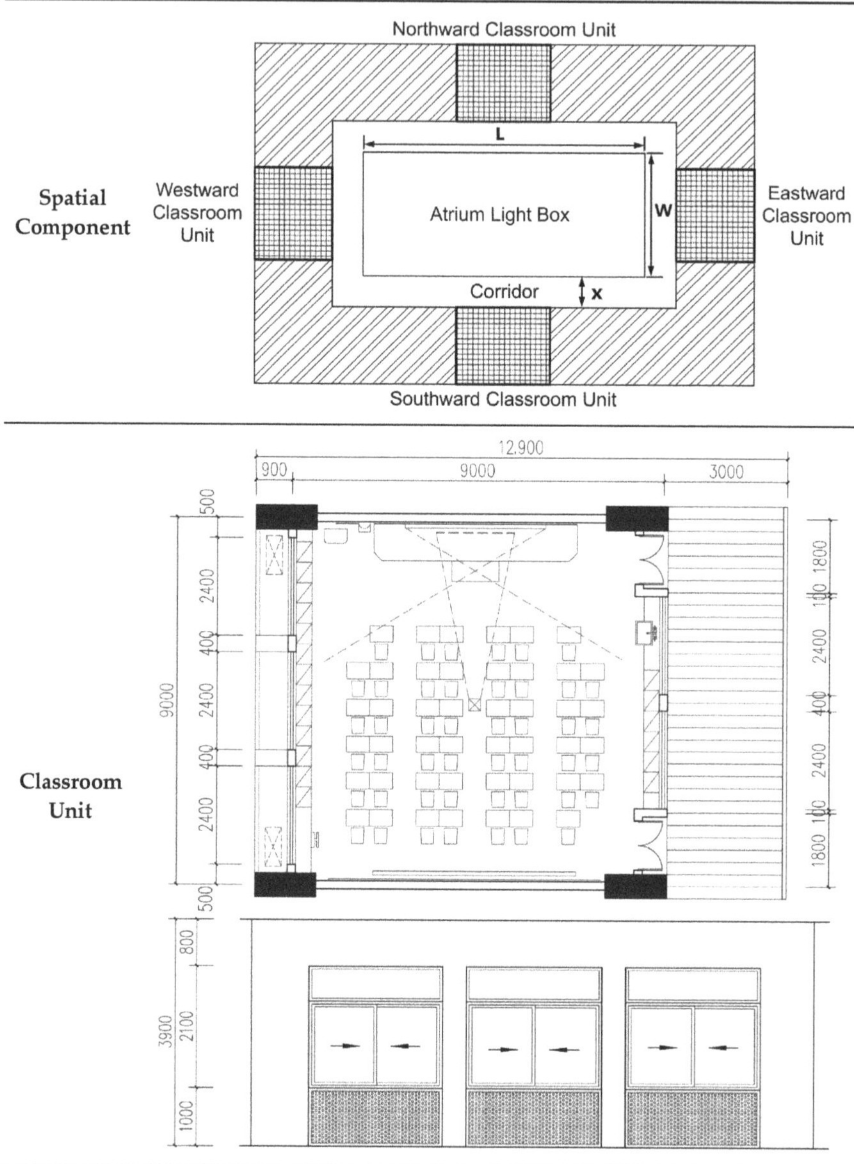

Figure 2. *Cont.*

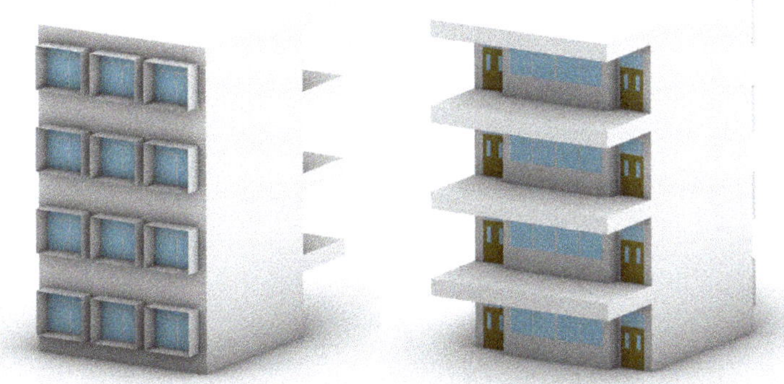

Combined
Model

Figure 2. The model configurations.

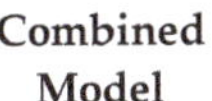

Spatial types

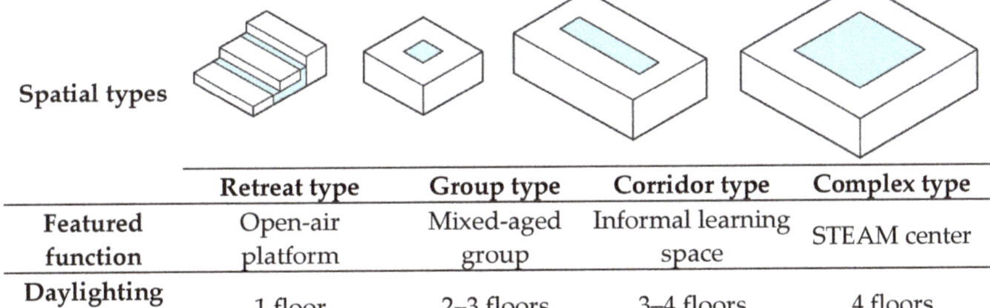

	Retreat type	Group type	Corridor type	Complex type
Featured function	Open-air platform	Mixed-aged group	Informal learning space	STEAM center
Daylighting depth	1 floor	2–3 floors	3–4 floors	4 floors
Atrium length range	18–36 m	9–18 m	27–54 m	27–54 m
Atrium width range	3–6 m	9–18 m	9–18 m	27–54 m

Figure 3. Atrium types and scales.

Table 2. Parameters (m) of simulation models.

W \ L	9	18	27	36	45	54
9	9 * 9	9 * 18	9 * 27	9 * 36	9 * 45	9 * 54
18	18 * 9	18 * 18	18 * 27	18 * 36	18 * 45	18 * 54
27	27 * 9	27 * 18	27 * 27	27 * 36	27 * 45	27 * 54
36	36 * 9	36 * 18	36 * 27	36 * 36	36 * 45	36 * 54
45	45 * 9	45 * 18	45 * 27	45 * 36	45 * 45	45 * 54
54	54 * 9	54 * 18	54 * 27	54 * 36	54 * 45	54 * 54

2.4. Simulation Settings

2.4.1. Software and Metrics

This study adopted the parametric simulation method. Taking advantage of the easy parametric control in Grasshopper, the simulation was conducted using the Ladybug–Honeybee plug-in, which uses Radiance as its calculation core, and whose accuracy was proven by quite a number of studies [31,34,39,40]. The simulation results were imported into SPSS for regression analysis [41,42].

The central daylight standard adopted in China [43] uses Daylight Factor (DF) as the evaluation indicator. However, DF lacks the concern of orientation and local climate and was gradually replaced by annual daylight factors such as Useful Daylight Illuminance (UDI) and Daylight Autonomy (DA) internationally. Spatial Daylight Autonomy ($sDA_{300,50\%}$) and Annual Sunlight Exposure ($ASE_{1000,250\,h}$), which were published by IESNA (Illuminating Engineering Society of North America) as the daylight evaluation methods in 2012 [44], were adopted by the latest LEED and WELL standards [45,46] as primary evaluation methods and used in quite a few studies [27,47]. Therefore, $sDA_{300,50\%}$ and $ASE_{1000,250\,h}$ were used as the major metrics for the simulation in this study.

2.4.2. Climate Data

The weather file of Nanjing (118.8° E, 31.0° N) from EnergyPlus [48] was adopted. Nanjing has a subtropical monsoon climate with distinct seasons and is assigned to the IV daylight zone in China (based on the annual average illuminance calculated with the climate data collected in 30 years) [43], which is an important zone that covers most of the southeast coast of China (Figure 4).

2.4.3. Occupancy Profile

The simulation was scheduled from 01 March to 30 June and from 01 September to 31 January, 8:00 am to 18:00 pm on weekdays, as the time from July to August, generally, is the summer vacation and February is the winter vacation in China.

2.4.4. Material Attributes

Based on the reflectance range described by the standard [37] in China, the reflectance of floor, ceiling and walls was set to 0.4, 0.8 and 0.8, respectively. Single-layer glass was used for interior windows with a visible transmittance of 0.8, while double-layer glass was used for exterior windows and skylights with a light transmittance of 0.6. The simulation platform was set to a desk height of 0.75 m.

2.4.5. Blinds System

A dynamic shading system that consisted of interior blinds with a diffuse visible transmittance of 20% and diffuse visible reflectance of 80% was taken into consideration for the classrooms to avoid sun glare. According to the requirement of IES [44], the blinds were set to be pulled down whenever more than 2% of the analysis points received direct sunlight.

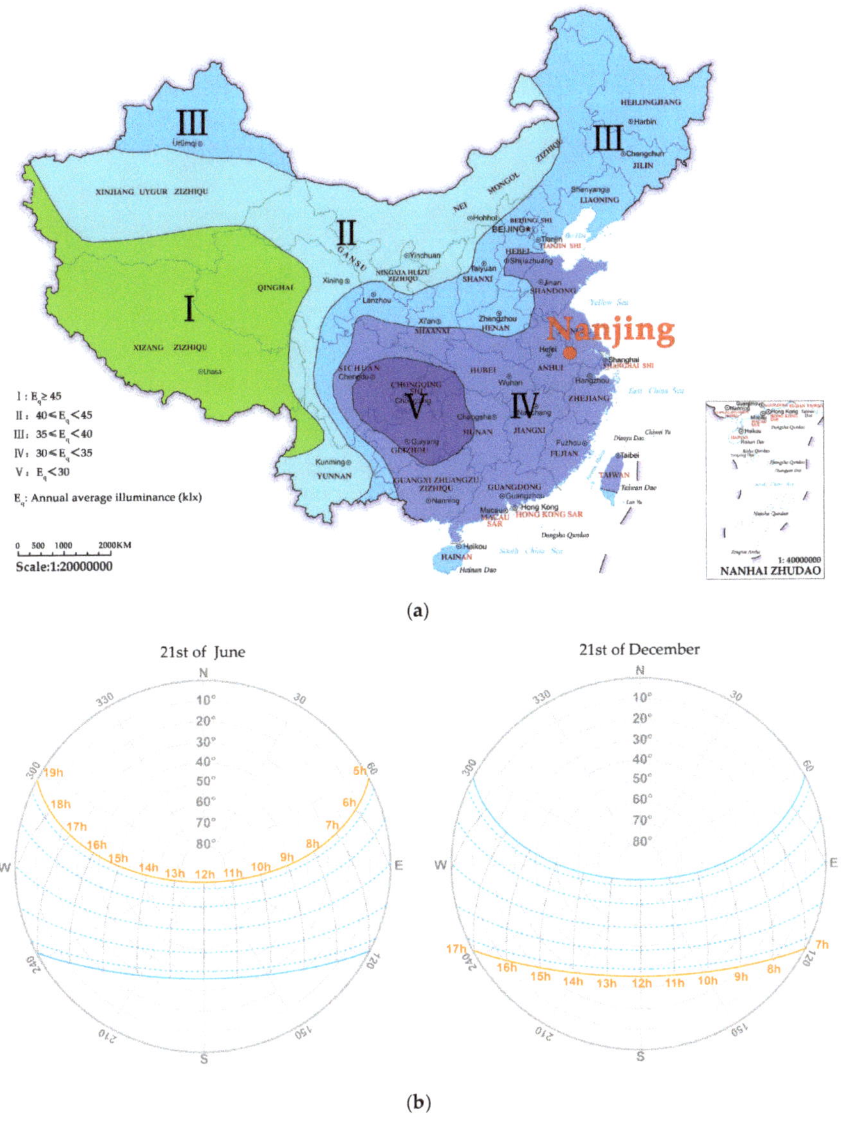

(a)

(b)

Figure 4. Geographic information on Nanjing. (**a**) Daylight Zone of China [1]. (**b**) Solar path of Nanjing. [1] the map is based on the standard map with the review number GS(2019)1673 downloaded from the National Platform for Common Geospatial Information Services of China [49], and the base map is not modified.

3. Results

3.1. Descriptive Analysis

3.1.1. Metric Analysis

From the simulation results of a fixed size model (atrium scale 27 m ∗ 27 m, corridor width 3 m, the top floor), the initial presentation of the metrics and their comparison were derived (Figure 5).

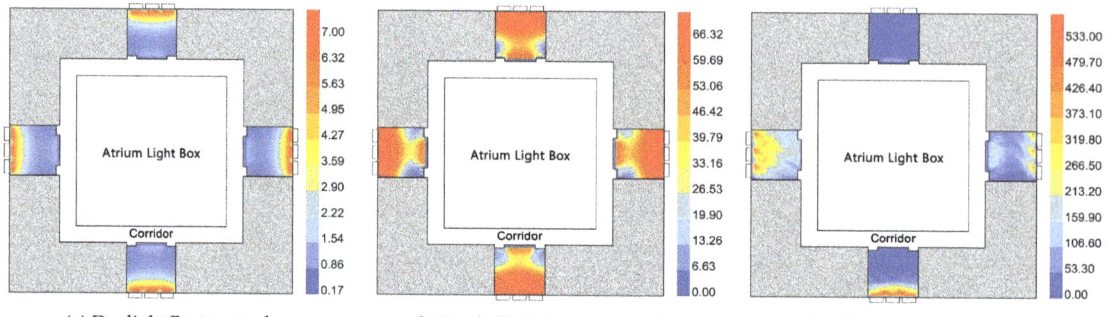

(**a**) Daylight Factor results
ADF = 1.75, 1.75, 1.75, 1.75
(from north to west)

(**b**) Daylight Autonomy results
sDA$_{300,50\%}$ = 0.62, 0.61, 0.51, 0.50
(from north to west)

(**c**) Direct sun hour results
ASE$_{1000,250\,h}$ = 0.00, 0.03, 0.13, 0.21
(from north to west)

Figure 5. Simulation results of the metrics in a fixed-size model.

The value of ADF did not meet the requirement of 3.0, while the sDA$_{300,50\%}$ values of the northward and eastward classrooms both exceeded the minimum requirement of 0.55. Simultaneously, the classrooms were protected from the risk of sun glare, which was proven to exist by the exceeded ASE$_{1000,250\,h}$. The consideration of sun glare, orientation and annual climate, which was not available in the overcast condition of DF, was proven to be significant in atrium-based classrooms of Nanjing by the comparison.

Specifically, classrooms of the four orientations presented various daylighting performances in an order of northward > eastward > southward > westward. Among them, the classrooms of the last two orientations were equipped with long-term blinds because of an ASE$_{1000,250\,h}$ higher than 0.10.

In addition, regardless of the orientations, the atrium scale showed a slight influence on ASE$_{1000,250\,h}$ and a stronger influence on sDA$_{300,50\%}$.

3.1.2. Orientation Analysis

Since it is unnecessary to analyze 108 models covering all four orientations and four floors, to select one orientation as the representative for subsequent analysis, the ASE$_{1000,250\,h}$ and sDA$_{300,50\%}$ results of the four orientations are gathered from a series of scaled models (Figure 6).

It was proved that the ASE$_{1000,250\,h}$ remained stable with a variation in atrium scales and floors, except for the eastward classrooms in which ASE$_{1000,250\,h}$ varied slightly. Therefore, the sDA$_{300,50\%}$ was the only metric considered in the following analysis.

With the dynamic blinds and exterior shades equipped, the sDA$_{300,50\%}$ of the four classroom orientations presented the order of northward ≈ eastward > southward > westward. This means that on account of daylight quality, the north and east sides of the atrium are the best orientations for a classroom.

To reduce the effect of the flexibility of the blind setting, the north orientation was selected for subsequent analysis because the northward classrooms rarely received glare and the blinds therefore rarely used. The shade depth was adjusted from 0.9 m to a structural thickness of 0.3 m because of the minor ASE$_{1000,250\,h}$ to fit the actual design.

3.1.3. Design Parameter Analysis

With the north orientation selected, the 108 models with varied parameters were simulated and the results of sDA$_{300,50\%}$ were compiled (Figure 7).

It can be concluded that:

(1) Under different atrium scales, the changing trends of sDA$_{300,50\%}$ varied significantly; the variation was gentle in the atrium width range of about 9–18 m and steep in the atrium width range of about 27–54 m;

(2) As the corridor width increased, the value of sDA$_{300,50\%}$ significantly decreased, especially using a corridor width of 6 m, and remained at a low level of less than 0.65.

To clarify the results of the complex charts with multiple variables, the complex relationship between the parameters and sDA$_{300,50\%}$ was simplified using the method of quantitative analysis. The data were imported into SPSS to conduct multivariable linear analysis and the relationship between sDA$_{300,50\%}$ and the design parameters were converted to equations, which thus provided clear and explicit guidance for atrium design.

3.2. Regression Analysis

The parameters for quantitative analysis were reclassified into the three atrium scales and some new scales were added to ensure consistency in the amount of data for each type. Large corridor widths were excluded to ensure the accuracy of the results (Table 3).

Table 3. Parameters of atrium scale for linear regression analysis.

Group Type					Corridor Type					Complex Type				
W \ L	9 m	12 m	15 m	18 m	W \ L	9 m	12 m	15 m	18 m	W \ L	27 m	36 m	45 m	54 m
9 m	9 * 9	9 * 12	9 * 15	9 * 18	27 m	27 * 9	27 * 12	27 * 15	27 * 18	27 m	27 * 27	27 * 36	27 * 45	27 * 54
12 m	12 * 9	12 * 12	12 * 15	12 * 18	36 m	36 * 9	36 * 12	36 * 15	36 * 18	36 m	36 * 27	36 * 36	36 * 45	36 * 54
15 m	15 * 9	15 * 12	15 * 15	15 * 18	45 m	45 * 9	45 * 12	45 * 15	45 * 18	45 m	45 * 27	45 * 36	45 * 45	45 * 54
18 m	18 * 9	18 * 12	18 * 15	18 * 18	54 m	54 * 9	54 * 12	54 * 15	54 * 18	54 m	54 * 27	54 * 36	54 * 45	54 * 54

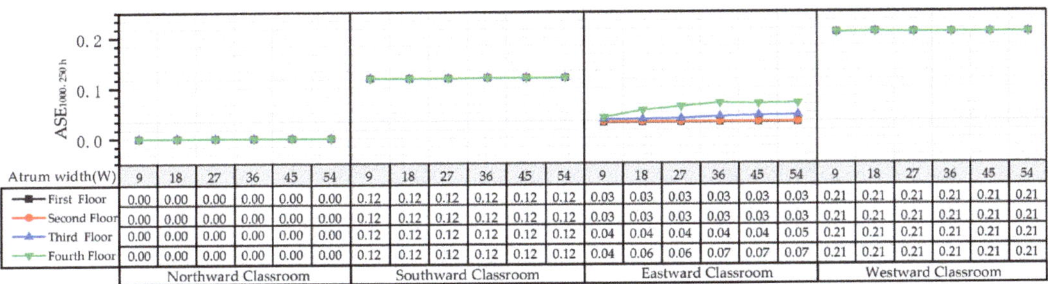

Atrium width(W)	9	18	27	36	45	54	9	18	27	36	45	54	9	18	27	36	45	54	9	18	27	36	45	54
First Floor	0.00	0.00	0.00	0.00	0.00	0.00	0.12	0.12	0.12	0.12	0.12	0.12	0.03	0.03	0.03	0.03	0.03	0.03	0.21	0.21	0.21	0.21	0.21	0.21
Second Floor	0.00	0.00	0.00	0.00	0.00	0.00	0.12	0.12	0.12	0.12	0.12	0.12	0.03	0.03	0.03	0.03	0.03	0.03	0.21	0.21	0.21	0.21	0.21	0.21
Third Floor	0.00	0.00	0.00	0.00	0.00	0.00	0.12	0.12	0.12	0.12	0.12	0.12	0.04	0.04	0.04	0.04	0.04	0.05	0.21	0.21	0.21	0.21	0.21	0.21
Fourth Floor	0.00	0.00	0.00	0.00	0.00	0.00	0.12	0.12	0.12	0.12	0.12	0.12	0.04	0.06	0.06	0.07	0.07	0.07	0.21	0.21	0.21	0.21	0.21	0.21
	Northward Classroom						Southward Classroom						Eastward Classroom						Westward Classroom					

Atrium Length(L)=45m, Corridor Width(x)=3m

(a)

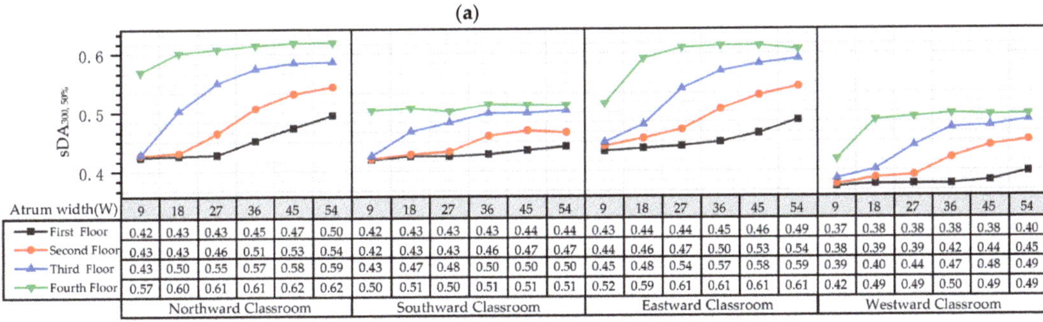

Atrium width(W)	9	18	27	36	45	54	9	18	27	36	45	54	9	18	27	36	45	54	9	18	27	36	45	54
First Floor	0.42	0.43	0.43	0.45	0.47	0.50	0.42	0.43	0.43	0.43	0.44	0.44	0.43	0.44	0.44	0.45	0.46	0.49	0.37	0.38	0.38	0.38	0.38	0.40
Second Floor	0.43	0.43	0.46	0.51	0.53	0.54	0.42	0.43	0.43	0.46	0.47	0.47	0.44	0.46	0.47	0.50	0.53	0.54	0.38	0.39	0.39	0.42	0.44	0.45
Third Floor	0.43	0.50	0.55	0.57	0.58	0.59	0.43	0.47	0.48	0.50	0.50	0.50	0.45	0.48	0.54	0.57	0.58	0.59	0.39	0.40	0.44	0.47	0.48	0.49
Fourth Floor	0.57	0.60	0.61	0.61	0.62	0.62	0.50	0.51	0.50	0.51	0.51	0.51	0.52	0.59	0.61	0.61	0.61	0.61	0.42	0.49	0.49	0.50	0.49	0.49
	Northward Classroom						Southward Classroom						Eastward Classroom						Westward Classroom					

Atrium Length(L)=45m, Corridor Width(x)=3m

(b)

Figure 6. Simulation results of the models in 4 orientations. (**a**) Annual Sunlight Exposure (ASE$_{1000,250\,h}$) results. (**b**) Spatial Daylight Autonomy (sDA$_{300,50\%}$) results.

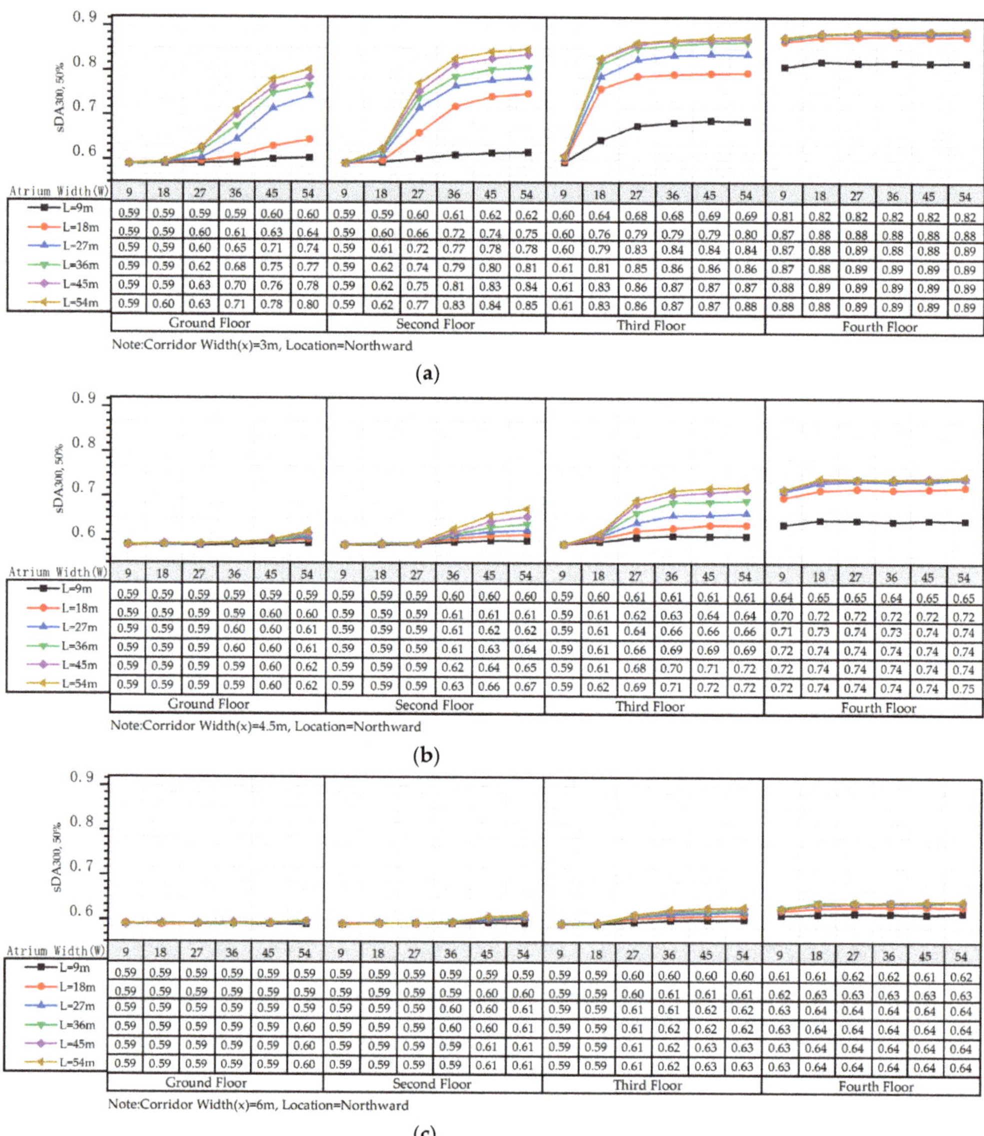

Note: Corridor Width(x)=3m, Location=Northward

Atrium Width (W)	Ground Floor						Second Floor						Third Floor						Fourth Floor					
	9	18	27	36	45	54	9	18	27	36	45	54	9	18	27	36	45	54	9	18	27	36	45	54
L=9m	0.59	0.59	0.59	0.59	0.60	0.60	0.59	0.59	0.60	0.61	0.62	0.62	0.60	0.64	0.68	0.68	0.69	0.69	0.81	0.82	0.82	0.82	0.82	0.82
L=18m	0.59	0.59	0.60	0.61	0.63	0.64	0.59	0.60	0.66	0.72	0.74	0.75	0.60	0.76	0.79	0.79	0.79	0.80	0.87	0.88	0.88	0.88	0.88	0.88
L=27m	0.59	0.59	0.60	0.65	0.71	0.74	0.59	0.61	0.72	0.77	0.78	0.78	0.60	0.79	0.83	0.84	0.84	0.84	0.87	0.88	0.89	0.88	0.88	0.89
L=36m	0.59	0.59	0.62	0.68	0.75	0.77	0.59	0.62	0.74	0.79	0.80	0.81	0.61	0.81	0.85	0.86	0.86	0.86	0.87	0.88	0.89	0.89	0.89	0.89
L=45m	0.59	0.59	0.63	0.70	0.76	0.78	0.59	0.62	0.75	0.81	0.83	0.84	0.61	0.83	0.86	0.87	0.87	0.87	0.88	0.89	0.89	0.89	0.89	0.89
L=54m	0.59	0.60	0.63	0.71	0.78	0.80	0.59	0.62	0.77	0.83	0.84	0.85	0.61	0.83	0.86	0.87	0.87	0.88	0.88	0.88	0.89	0.89	0.89	0.89

(**a**)

Note: Corridor Width(x)=4.5m, Location=Northward

Atrium Width (W)	Ground Floor						Second Floor						Third Floor						Fourth Floor					
	9	18	27	36	45	54	9	18	27	36	45	54	9	18	27	36	45	54	9	18	27	36	45	54
L=9m	0.59	0.59	0.59	0.59	0.59	0.59	0.59	0.59	0.59	0.60	0.60	0.60	0.59	0.60	0.61	0.61	0.61	0.61	0.64	0.65	0.65	0.64	0.65	0.65
L=18m	0.59	0.59	0.59	0.59	0.60	0.60	0.59	0.59	0.59	0.61	0.61	0.61	0.59	0.61	0.62	0.63	0.64	0.64	0.70	0.72	0.72	0.72	0.72	0.72
L=27m	0.59	0.59	0.59	0.60	0.60	0.61	0.59	0.59	0.59	0.61	0.62	0.62	0.59	0.61	0.64	0.66	0.66	0.66	0.71	0.73	0.74	0.73	0.74	0.74
L=36m	0.59	0.59	0.59	0.60	0.60	0.61	0.59	0.59	0.59	0.61	0.63	0.64	0.59	0.61	0.66	0.69	0.69	0.69	0.72	0.74	0.74	0.74	0.74	0.74
L=45m	0.59	0.59	0.59	0.59	0.60	0.62	0.59	0.59	0.59	0.62	0.64	0.65	0.59	0.61	0.68	0.70	0.71	0.72	0.72	0.74	0.74	0.74	0.74	0.74
L=54m	0.59	0.59	0.59	0.59	0.60	0.62	0.59	0.59	0.59	0.63	0.66	0.67	0.59	0.62	0.69	0.71	0.72	0.72	0.72	0.74	0.74	0.74	0.74	0.75

(**b**)

Note: Corridor Width(x)=6m, Location=Northward

Atrium Width (W)	Ground Floor						Second Floor						Third Floor						Fourth Floor					
	9	18	27	36	45	54	9	18	27	36	45	54	9	18	27	36	45	54	9	18	27	36	45	54
L=9m	0.59	0.59	0.59	0.59	0.59	0.59	0.59	0.59	0.59	0.59	0.59	0.59	0.59	0.59	0.60	0.60	0.60	0.60	0.61	0.61	0.62	0.62	0.61	0.62
L=18m	0.59	0.59	0.59	0.59	0.59	0.59	0.59	0.59	0.59	0.59	0.60	0.60	0.59	0.59	0.60	0.61	0.61	0.61	0.62	0.63	0.63	0.63	0.63	0.63
L=27m	0.59	0.59	0.59	0.59	0.59	0.59	0.59	0.59	0.59	0.60	0.60	0.61	0.59	0.59	0.61	0.61	0.62	0.62	0.63	0.64	0.64	0.64	0.64	0.64
L=36m	0.59	0.59	0.59	0.59	0.59	0.60	0.59	0.59	0.59	0.60	0.60	0.61	0.59	0.59	0.61	0.62	0.62	0.62	0.63	0.64	0.64	0.64	0.64	0.64
L=45m	0.59	0.59	0.59	0.59	0.59	0.60	0.59	0.59	0.59	0.59	0.61	0.61	0.59	0.59	0.61	0.62	0.63	0.63	0.63	0.64	0.64	0.64	0.64	0.64
L=54m	0.59	0.59	0.59	0.59	0.59	0.60	0.59	0.59	0.59	0.59	0.61	0.61	0.59	0.59	0.61	0.62	0.63	0.63	0.63	0.64	0.64	0.64	0.64	0.64

(**c**)

Figure 7. sDA$_{300,50\%}$ results under various atrium widths, atrium lengths and floors. (**a**) Results of 3 m corridor width. (**b**) Results of 4.5 m corridor width. (**c**) Results of 6 m corridor width.

3.2.1. Group-Type Atriums

With the atrium scale of group type, the atrium length (L) and width (W) ranged from 9 to 18 m; daylight depth was normally two or three floors. The simulation results are as follows: (Table 4).

The data of each floor were separately imported into SPSS and the multiple linear regression analysis was performed with the dependent variable of sDA$_{300,50\%}$ and the independent variables of atrium length and width. The results are as follows: (Table 5).

Table 4. Simulation results ($sDA_{300,50\%}$) of group-type atriums.

		The Bottom Floor				The Second Floor				The Third Floor			
L \ **W**		**9 m**	**12 m**	**15 m**	**18 m**	**9 m**	**12 m**	**15 m**	**18 m**	**9 m**	**12 m**	**15 m**	**18 m**
9 m		0.592	0.593	0.594	0.594	0.595	0.598	0.620	0.644	0.809	0.816	0.821	0.821
12 m		0.591	0.593	0.597	0.596	0.596	0.614	0.660	0.705	0.837	0.844	0.852	0.851
15 m		0.592	0.594	0.596	0.597	0.595	0.629	0.694	0.740	0.860	0.863	0.867	0.869
18 m		0.591	0.595	0.596	0.597	0.598	0.638	0.713	0.760	0.866	0.869	0.872	0.876

Table 5. Results of linear regression analysis.

	Model	Coefficients [a]			t	Sig.
		Unstandardized Coefficients		**Standardized Coefficients**		
		B	**Std. Error**	**Beta**		
1	(Constant)	0.586	0.002	—	380.754	<0.001
	Atrium length	0.0002	0.000	0.271	2.022	0.064
	Atrium width	0.0005	0.000	0.833	6.222	<0.001

[1] Notation: $F = 21.401$, $\rho = 0.000 < 0.01$, $R^2 = 0.767$. [a]. Dependent Variable: $sDA_{300,50\%}$ (the bottom floor).

The results ($F = 21.401$, $\rho = 0.000 < 0.05$, $R^2 = 0.767$) indicated the validity of the regression equations. The significance of atrium length is 0.064 (>0.05), indicating no significant change in $sDA_{300,50\%}$ due to variation in the atrium length. The significance of atrium width is 0.000 (<0.05), indicating that there is a significant change in $sDA_{300,50\%}$ due to a variation in the atrium width.

Similarly, the regression equations for the floors are as follows:

$$sDA_{300,50\%} \text{ (the bottom floor)} = 0.586 + 0.0002 \, L + 0.0005 \, W \tag{1}$$

$$sDA_{300,50\%} \text{ (the second floor)} = 0.375 + 0.007 \, L + 0.013 \, W \tag{2}$$

$$sDA_{300,50\%} \text{ (the third floor)} = 0.751 + 0.006 \, L + 0.001 \, W \tag{3}$$

As for the bottom floor, both the atrium length and width had a slight influence on the $sDA_{300,50\%}$ of the bottom floor, which was maintained at about 0.59 with small variations but meant that the minimum requirement of 0.55 can be reached.

As for the second floor, the $sDA_{300,50\%}$ was affected by the atrium scales and the related parameters were ranked as atrium width (coefficient of 0.013) > atrium length (coefficient of 0.007). By deforming the equation, the condition that $sDA_{300,50\%}$ can meet the requirement of 0.75 on the second floor was $L + 2\,W \geq 54$ m.

And the $sDA_{300,50\%}$ of the third floor can reach 0.75 no matter how the parameter changed.

In conclusion, on account of the small lighting box, the daylight performance of the bottom classrooms in the group-type atrium was at a low level but the minimum requirement of $sDA \geq 0.55$ can be reached. When the daylight depth was two floors and the combination of the atrium length and width meet $L + 2\,W \geq 54$ m, the $sDA_{300,50\%}$ of the bottom classrooms can reach the higher requirement of 0.75.

3.2.2. Corridor-Type Atriums

With the atrium scale of corridor type, the range of atrium width (W) was 9–18 m, while the range of atrium length (L) was 27–54 m; the daylight depth was normally three or four floors. The simulation results are as follows: (Table 6).

Table 6. Simulation results ($sDA_{300,50\%}$) of corridor-type atriums.

L \ W	The Bottom Floor				The Second Floor				The Third Floor				The Fourth Floor			
	9 m	12 m	15 m	18 m	9 m	12 m	15 m	18 m	9 m	12 m	15 m	18 m	9 m	12 m	15 m	18 m
27 m	0.60	0.65	0.75	0.79	0.59	0.59	0.60	0.61	0.59	0.59	0.59	0.59	0.87	0.88	0.88	0.88
36 m	0.61	0.66	0.77	0.81	0.59	0.59	0.59	0.61	0.59	0.59	0.59	0.59	0.87	0.88	0.88	0.88
45 m	0.61	0.66	0.78	0.83	0.59	0.59	0.59	0.61	0.59	0.59	0.59	0.59	0.88	0.88	0.88	0.89
54 m	0.61	0.67	0.79	0.83	0.59	0.59	0.60	0.62	0.59	0.59	0.59	0.59	0.88	0.88	0.88	0.88

The linear regression results showed that the inputs all passed the F-test (the bottom floor: $F = 28.365$, $\rho = 0.000 < 0.05$, the second floor: $F = 14.475$, $\rho = 0.000 < 0.05$, the third floor: $F = 23.267$, $\rho = 0.000 < 0.05$, the fourth floor: $F = 48.841$, $\rho = 0.000 < 0.05$), and the values of R^2 for the four floors were 0.778, 0.743, 0.967 and 0.883, respectively. The regression equations are as follows:

$$sDA_{300,50\%} \text{ (the bottom floor)} = 0.585 + 0.00004\,L + 0.0005\,W \tag{4}$$

$$sDA_{300,50\%} \text{ (the second floor)} = 0.554 + 0.0002\,L + 0.003\,W \tag{5}$$

$$sDA_{300,50\%} \text{ (the third floor)} = 0.349 + 0.001\,L + 0.024\,W \tag{6}$$

$$sDA_{300,50\%} \text{ (the fourth floor)} = 0.860 + 0.0001\,L + 0.001\,W \tag{7}$$

It was found that, except for the top floor, the effect of the atrium scale increases significantly with the number of floors (the coefficient ratio of 1:6:48), and the influence of the atrium width was obviously stronger than that of the atrium length with the average coefficient ratio of 1:16, which means that the increase in atrium width could be much more helpful to enhance the daylight in the bottom classrooms, compared with atrium length.

Similar to the group type, the $sDA_{300,50\%}$ of the bottom classrooms was basically not affected by atrium scales, which was maintained at about 0.59 with small variations and could reach the minimum requirement of 0.55.

As for the second floor, the $sDA_{300,50\%}$ could also not meet the higher requirement of 0.75 within the scale (0.625 at the maximum scale).

As for the third floor, when the scale satisfied the condition of $L + 24\,W \geq 400$ m, or more intuitively when the atrium width reached 15 m with a large atrium length of 36–54 m, the $sDA_{300,50\%}$ can reach 0.75.

As for the fourth floor, the $sDA_{300,50\%}$ was also above 0.75, no matter how the parameters changed.

In a word, the corridor-type atriums always had a narrow lighting box, thus resulting in poor daylight situation in the bottom classrooms. The increased atrium length performed poorly with an increase in daylight. Meanwhile, the daylight depths of corridor-type atriums may increase to four floors. On this condition, it was difficult for the classrooms on the bottom two floors to reach the $sDA_{300,50\%}$ of 0.75. On the higher floor, when the atrium scale met $L \geq 36$ m and $W \geq 15$ m, the values of classrooms' $sDA_{300,50\%}$ could meet the requirement.

3.2.3. Complex-Type Atriums

With the atrium scale of the complex type, the atrium lengths (L) and widths (W) all ranged from 27 to 54 m and the daylight depth was normally four floors. The simulation results are as follows: (Table 7).

Table 7. Simulation results ($sDA_{300,50\%}$) of complex-type atriums.

L \ W	The Bottom Floor				The Second Floor				The Third Floor				The Fourth Floor			
	27 m	36 m	45 m	54 m	27 m	36 m	45 m	54 m	27 m	36 m	45 m	54 m	27 m	36 m	45 m	54 m
27 m	0.60	0.65	0.71	0.74	0.72	0.77	0.78	0.78	0.83	0.84	0.84	0.84	0.89	0.88	0.88	0.89
36 m	0.62	0.68	0.75	0.77	0.74	0.79	0.80	0.81	0.85	0.86	0.86	0.86	0.89	0.89	0.89	0.89
45 m	0.63	0.70	0.76	0.78	0.75	0.81	0.83	0.84	0.86	0.87	0.87	0.87	0.89	0.89	0.89	0.89
54 m	0.63	0.71	0.78	0.80	0.77	0.83	0.84	0.85	0.86	0.87	0.87	0.88	0.89	0.89	0.89	0.89

The linear regression results in SPSS showed that the input data all passed the F-test (the bottom floor: $F = 127.559$, $\rho = 0.000 < 0.05$, the second floor: $F = 49.863$, $\rho = 0.000 < 0.05$, the third floor: $F = 38.754$, $\rho = 0.000 < 0.05$, the fourth floor: $F = 13.545$, $\rho = 0.000 < 0.05$), and the values of R^2 of the four floors were 0.952, 0.885, 0.856 and 0.676, respectively. The regression equations were as follows:

$$sDA_{300,50\%} \text{ (the bottom floor)} = 0.388 + 0.002\,L + 0.006\,W \tag{8}$$

$$sDA_{300,50\%} \text{ (the second floor)} = 0.594 + 0.002\,L + 0.003\,W \tag{9}$$

$$sDA_{300,50\%} \text{ (the third floor)} = 0.786 + 0.001\,L + 0.0005\,W \tag{10}$$

$$sDA_{300,50\%} \text{ (the fourth floor)} = 0.878 + 0.0001\,L + 0.0001\,W \tag{11}$$

It can be inferred that the atrium scales of the complex type had an obvious influence on the bottom classrooms compared with those of the other types, and the impact decreased as the number of floors increased (the ratio of atrium width coefficients as 2:1). The degrees of influence of the parameters differed as atrium width > atrium length, but the average coefficient ratio of 2:1 was much more balanced than that of the other types, indicating that both the length and the width of the atrium could effectively enhance the daylight performance of the atrium.

Deformation of the equations showed that the value of $sDA_{300,50\%}$ in the bottom classrooms can reach 0.75 when $L + 3\,W \geq 180$ m, or $L = W \geq 45$ m. On the second floor, the condition changed to $2\,L + 3\,W \geq 150$ m, or $L = W \geq 30$ m.

The classrooms on both the third and fourth floors can reach the $sDA_{300,50\%}$ of 0.75.

In short, complex-type atriums were large in scale and thus had better daylight performance. The classrooms on the top two floors can reach the $sDA_{300,50\%}$ requirement of 0.75 and the classrooms on lower floors can also easily reach the requirement under uncritical scale conditions. Therefore, it is possible to moderately increase the corridor width, construct informal learning spaces or add other space details in these atrium designs.

4. Design Recommendation

Accordingly, the recommendations for primary and secondary school design are summarized as follows:

(1) Classrooms are best placed on the north and east sides of the atrium and should not be placed on the west side;

(2) The atrium width can be maximized to efficiently improve natural daylighting;

(3) Corridor width should be kept to a minimum such as 3 m for the high-quality daylighting of the lower floor;

(4) It is better to use the bottom rooms as an office or for activities instead of classrooms, for a daylight depth of two floors is the best in the atrium scales of the group type and the corridor type;

(5) Considering the atrium scale of the complex type, it is possible to construct, in the atrium, both a high-quality daylighting environment and unconventional teaching spaces such as a learning corner, especially on the top two floors.

If the project requirements deviate from these recommendations, for example, an increase in corridor width or a deep daylight depth for conducting various teaching activities, the artificial lighting in classrooms should be qualified, and thus the increase in energy consumption should be taken into consideration.

5. Discussion

This study examined the daylighting performance for atrium-based classrooms in primary and secondary schools, connecting the two research topics of daylighting in the atrium and innovative school design; using the operable dimensions of the atrium as parameters, the study provided convenient optimization methods of a series of equations and related design suggestions for innovative school design.

Unlike the studies that focused on the space within the atrium or the surrounding open space [26–34], this study took the customized unit of the classroom as the object, and the variations of $sDA_{300,50\%}$ with the atrium scale were similarly regular as the previous studies [30,34]. Furthermore, previous studies generally used ADF as the only metric [27,28]. Compared to the smooth and ordered variation of ADF, $sDA_{300,50\%}$ requires more qualification and classification to make the variation clear and readable but was proved to be necessary on account of the consideration of local climate, orientation and glare control.

Nevertheless, the adoption of the annual daylighting metric also caused limitations. The simulation results may change with each of the input items, such as the site, boundary settings, occupancy schedule and blinds, which makes it difficult to apply the results in sites that have different climates or latitudes from Nanjing. Additionally, for the same reason, it is difficult to compare the results with the studies that have different settings, especially when the setting is not standard, for example, the specific occupancy of the school in this study. Therefore, the results are merely applicable to a quick judgment in the early design stage or serve as a reference, not for professional evaluation and detailed design.

Furthermore, daylighting is only one aspect of Indoor Environmental Quality (IEQ) that should be taken into account in the design process of a school [50]. Similarly, the atrium layout is also just one of the forms that the school buildings are developing [51]. Therefore, extended research on thermal comfort, indoor air quality and acoustic comfort in the atriums and the expansion of design features are planned to be gradually fulfilled in further work.

6. Conclusions

In this study, the model configurations of atrium-based teaching spaces were extracted, four types of atrium scales were summarized, and afterward, the investigation focused on the daylight performances of group-type, corridor-type and complex-type atriums; 132 operating conditions were arranged, through variations in atrium length, atrium width and corridor width, and were applied to classrooms on four floors in four orientations. The simulation was conducted using Ladybug–Honeybee tools with the weather file of Nanjing, and the traditional evaluation metric of DF was replaced by annual daylight metrics $sDA_{300,50\%}$ and $ASE_{1000,250\,h}$. Finally, practical suggestions on the atrium design for primary and secondary schools were derived.

(1) On account of the daylight and glare, classroom orientations to the atrium can be ordered as: northward, eastward, southward and westward, represented by the ratio of about 9:9:8:7 (ratio of the mean values of $sDA_{300,50\%}$ in four orientations) and the last southward and westward classrooms should be equipped with shading measures;

(2) The parameters affecting the daylight performance of the atrium can be arranged in descending order as corridor width, atrium width and atrium length (coefficient ratios of width and length vary from 16:1 to 2:1 as scales up);

(3) The design equations for the various atrium types had the common structure of $sDA_{300,50\%}\_floor = a + b * L + c * W$ (Table 8). In group-type atriums with three floors of daylight depth, the $sDA_{300,50\%}$ can reach 0.55 in the bottom floor classrooms and 0.75 in

the second or higher floors classrooms when L + 2 W $\geq$ 54 m. In corridor-type atriums with four floors of daylight depth, the sDA could still achieve 0.55 in the bottom two floors' classrooms and 0.75 in the third or higher floor classrooms when L + 24 W $\geq$ 400 m. In complex-type atriums with four floors of daylight depth, the $\text{sDA}_{300,50\%}$ of the bottom classrooms can reach 0.75 when L + 3 W $\geq$ 180 m, and 2 L + 3 W $\geq$ 150 m in the second and higher floor classrooms.

Table 8. Aggregation of Equation coefficients.

Atrium Type	Floor Number	Equation Coefficients			$\text{sDA}_{300,50\%} \geq 0.75$ Conditions
		a	b * 100	c * 100	
Group type	1	0.586	0.02	0.05	$\times$
	2	0.375	0.7	1.3	L + 2 W $\geq$ 54
	3	0.751	0.6	0.1	$\checkmark$
Corridor type	1	0.585	0.004	0.05	$\times$
	2	0.554	0.02	0.3	$\times$
	3	0.349	0.1	2.4	L + 24 W $\geq$ 400
	4	0.860	0.01	0.1	$\checkmark$
Complex type	1	0.388	0.2	0.6	L + 3 W $\geq$ 180
	2	0.594	0.2	0.3	2 L + 3 W $\geq$ 150
	3	0.786	0.1	0.05	$\checkmark$
	4	0.878	0.01	0.01	$\checkmark$

Author Contributions: Conceptualization, J.M. and Q.Y.; methodology, J.M. and Q.Y.; software, Q.Y.; validation, J.M. and Q.Y.; formal analysis, Q.Y.; investigation, Q.Y.; resources, J.M.; data curation, Q.Y.; writing—original draft preparation, Q.Y.; writing—review and editing, J.M.; visualization, Q.Y. All authors have read and agreed to the published version of the manuscript.

Funding: This research received no external funding.

Institutional Review Board Statement: Not applicable.

Informed Consent Statement: Not applicable.

Data Availability Statement: The data presented in this study are available on request from the authors.

Conflicts of Interest: The authors declare no conflict of interest.

References

1. Zhou, K.; Li, S.; Lin, Q.; Fang, K.; Wang, F.; Xu, J. Research on Secondary School Teaching Space Mode under the Teaching Organization Form of "Class Selection System". *Archit. J.* **2022**, *S1*, 109–116.
2. Li, S.; Li, Z.; Zhou, K.; Zhang, J. Research on the spatial pattern of primary and secondary school buildings adapted to the development of quality education. *Archit. J.* **2008**, *8*, 76–80.
3. Dello-Iacovo, B. Curriculum Reform and 'Quality Education' in China: An Overview. *Int. J. Educ. Dev.* **2009**, *29*, 241–249. [CrossRef]
4. Ministry of Education of the People's Republic of China. Opinions of the Ministry of Education on the Implementation of the General High School Academic Level Examination. 10 December 2014. Available online: http://www.moe.gov.cn/srcsite/A06/s3732/201808/t20180807_344610.html (accessed on 16 October 2022).
5. Boy, G.A. From STEM to STEAM: Toward a Human-Centred Education, Creativity & Learning Thinking. In Proceedings of the Proceedings of the 31st European Conference on Cognitive Ergonomics, Toulouse, France, 26–28 August; Association for Computing Machinery: New York, NY, USA, 2013; pp. 1–7.
6. Montessori, M.; Hunt, J.M.; Valsiner, J. *The Montessori Method*; Routledge: New York, NY, USA, 2017; ISBN 9781315133225.
7. Anderson, D.; Lucas, K.B.; Ginns, I.S. Theoretical Perspectives on Learning in an Informal Setting. *J. Res. Sci. Teach.* **2003**, *40*, 177–199. [CrossRef]
8. Ding, J.; Zheng, S.; Huang, Y. Discussion on the Design of Adaptive Learning Space under the Demand of Contemporary Education in China: Take Graphic Analysis of Cases in Recent 10 Years as an Example. *Archit. J.* **2021**, *S1*, 151–157.
9. Ding, J.; Zheng, S.; Huang, Y. Research on the Learning Space Design of Primary and Middle Schools in the New Era under the Concept of Flexibility. *New Archit.* **2022**, *1*, 77–83.

10. Wu, X.; Oldfield, P.; Heath, T. Spatial Openness and Student Activities in an Atrium: A Parametric Evaluation of a Social Informal Learning Environment. *Build. Environ.* **2020**, *182*, 107141. [CrossRef]

11. Weaver, M. Exploring Conceptions of Learning and Teaching Through the Creation of Flexible Learning Spaces: The Learning Gateway—A Case Study. *New Rev. Acad. Librariansh.* **2006**, *12*, 109–125. [CrossRef]

12. Kazemzadeh, M.; Asadi, F.S. The New Attention to Atrium for Creating Sustainable Townscape. *J. Civil Eng. Urban* **2014**, *4*, 98–102.

13. Adams, A.; Theodore, D.; Goldenberg, E.; McLaren, C.; McKeever, P. Kids in the Atrium: Comparing Architectural Intentions and Children's Experiences in a Pediatric Hospital Lobby. *Soc. Sci. Med.* **2010**, *70*, 658–667. [CrossRef]

14. Calcagni, B.; Paroncini, M. Daylight Factor Prediction in Atria Building Designs. *Sol. Energy* **2004**, *76*, 669–682. [CrossRef]

15. Ghasemi, M.; Noroozi, M.; Kazemzadeh, M.; Roshan, M. The Influence of Well Geometry on the Daylight Performance of Atrium Adjoining Spaces: A Parametric Study. *J. Build. Eng.* **2015**, *3*, 39–47. [CrossRef]

16. Sharples, S.; Lash, D. Daylight in Atrium Buildings: A Critical Review. *Archit. Sci. Rev.* **2007**, *50*, 301–312. [CrossRef]

17. Küller, R.; Lindsten, C. Health and Behavior of Children in Classrooms with and without Windows. *J. Environ. Psychol.* **1992**, *12*, 305–317. [CrossRef]

18. Hathaway, W.E. Effects of School Lighting on Physical Development and School Performance. *J. Educ. Res.* **1995**, *88*, 228–242. [CrossRef]

19. Schneider, M. Do School Facilities Affect Academic Outcomes? 2002. Available online: https://eric.ed.gov/?id=ED470979 (accessed on 28 November 2022).

20. Kong, Z.; Utzinger, D.M.; Freihoefer, K.; Steege, T. The Impact of Interior Design on Visual Discomfort Reduction: A Field Study Integrating Lighting Environments with POE Survey. *Build. Environ.* **2018**, *138*, 135–148. [CrossRef]

21. Boyce, P.R. *Human Factors in Lighting*, 2nd ed.; CRC Press: London, UK, 2003; ISBN 9780429221132.

22. Kong, Z.; Zhang, R.; Ni, J.; Ning, P.; Kong, X.; Wang, J. Towards an Integration of Visual Comfort and Lighting Impression: A Field Study within Higher Educational Buildings. *Build. Environ.* **2022**, *216*, 108989. [CrossRef]

23. Edwards, L.; Torcellini, P.; Laboratory, N.R.E. *A Literature Review of the Effects of Natural Light on Building Occupants*; U.S. Department of Energy: Washington, DC, USA, 2002; p. 58.

24. Kong, Z.; Liu, Q.; Li, X.; Hou, K.; Xing, Q. Indoor Lighting Effects on Subjective Impressions and Mood States: A Critical Review. *Build. Environ.* **2022**, *224*, 109591. [CrossRef]

25. Wright, K.P.; McHill, A.W.; Birks, B.R.; Griffin, B.R.; Rusterholz, T.; Chinoy, E.D. Entrainment of the Human Circadian Clock to the Natural Light-Dark Cycle. *Curr. Biol.* **2013**, *23*, 1554–1558. [CrossRef]

26. Du, J.; Sharples, S. The Assessment of Vertical Daylight Factors across the Walls of Atrium Buildings, Part 2: Rectangular Atria. *Light. Res. Technol.* **2012**, *44*, 124–138. [CrossRef]

27. Mohsenin, M.; Hu, J. Assessing Daylight Performance in Atrium Buildings by Using Climate Based Daylight Modeling. *Sol. Energy* **2015**, *119*, 553–560. [CrossRef]

28. Ghasemi, M.; Kandar, M.Z.; Noroozi, M. Investigating the Effect of Well Geometry on the Daylight Performance in the Adjoining Spaces of Vertical Top-Lit Atrium Buildings. *Indoor Built Environ.* **2016**, *25*, 934–948. [CrossRef]

29. Sudan, M.; Mistrick, R.G.; Tiwari, G.N. Climate-Based Daylight Modeling (CBDM) for an Atrium: An Experimentally Validated Novel Daylight Performance. *Sol. Energy* **2017**, *158*, 559–571. [CrossRef]

30. Li, J.; Ban, Q.; Chen, X.; Yao, J. Glazing Sizing in Large Atrium Buildings: A Perspective of Balancing Daylight Quantity and Visual Comfort. *Energies* **2019**, *12*, 701. [CrossRef]

31. Rastegari, M.; Pournaseri, S.; Sanaieian, H. Daylight Optimization through Architectural Aspects in an Office Building Atrium in Tehran. *J. Build. Eng.* **2021**, *33*, 101718. [CrossRef]

32. Wu, P.; Zhou, J.; Li, N. Influences of Atrium Geometry on the Lighting and Thermal Environments in Summer: CFD Simulation Based on-Site Measurements for Validation. *Build. Environ.* **2021**, *197*, 107853. [CrossRef]

33. Dong, L.; He, Y.; Qi, Q.; Wang, W. Optimization of Daylight in Atrium in Underground Commercial Spaces: A Case Study in Chongqing, China. *Energy Build.* **2022**, *256*, 111739. [CrossRef]

34. Xue, Y.; Liu, W. A Study on Parametric Design Method for Optimization of Daylight in Commercial Building's Atrium in Cold Regions. *Sustainability* **2022**, *14*, 7667. [CrossRef]

35. Ibrahim, I.; Al Badri, N.; Mushtaha, E.; Omar, O. Evaluating the Impacts of Courtyards on Educational Buildings, Case Study in the University of Sharjah. *Sustainability* **2022**, *14*, 141. [CrossRef]

36. Piraei, F.; Matusiak, B.; Lo Verso, V.R.M. Evaluation and Optimization of Daylighting in Heritage Buildings: A Case-Study at High Latitudes. *Buildings* **2022**, *12*, 2045. [CrossRef]

37. *GB 50099-2011*; Code for Design of School. Ministry of Housing and Urban-Rural Development (PRC): Beijing, China, 2012.

38. Fan, Q. Organization of Teaching Space in Primary and Secondary School based on Educational Sociality. Master's Thesis, Southeast University, Nanjing, China, 2021.

39. Kharvari, F. An Empirical Validation of Daylighting Tools: Assessing Radiance Parameters and Simulation Settings in Ladybug and Honeybee against Field Measurements. *Sol. Energy* **2020**, *207*, 1021–1036. [CrossRef]

40. Qingsong, M.; Fukuda, H. Parametric Office Building for Daylight and Energy Analysis in the Early Design Stages. *Procedia Soc. Behav. Sci.* **2016**, *216*, 818–828. [CrossRef]

41. Lo Verso, V.R.M.; Mihaylov, G.; Pellegrino, A.; Pellerey, F. Estimation of the Daylight Amount and the Energy Demand for Lighting for the Early Design Stages: Definition of a Set of Mathematical Models. *Energy Build.* **2017**, *155*, 151–165. [CrossRef]
42. Sepúlveda, A.; De Luca, F.; Kurnitski, J. Daylight and Overheating Prediction Formulas for Building Design in a Cold Climate. *J. Build. Eng.* **2022**, *45*, 103532. [CrossRef]
43. *GB 50033-2013*; Standard for Daylighting Design of Buildings. Ministry of Housing and Urban-Rural Development (PRC): Beijing, China, 2013.
44. *IES LM-83-12*; Approved Method: IES Spatial Daylight Autonomy (SDA) and Annual Sunlight Exposure (ASE). Illuminating Engineering Society (IES): New York, NY, USA, 2012.
45. *The Well Building Standard, V2*; International Well Building Institute (IWBI): New York, NY, USA, 2020.
46. *LEED v4.1 Building Design and Construction*; U.S. Green Building Council (GBC): Washington, DC, USA, 2019.
47. Kong, Z.; Jakubiec, J.A. Evaluations of Long-Term Lighting Qualities for Computer Labs in Singapore. *Build. Environ.* **2021**, *194*, 107689. [CrossRef]
48. EnergyPlus. Available online: https://energyplus.net/weather (accessed on 28 November 2022).
49. National Platform for Common Geospatial Information Services. Available online: https://www.tianditu.gov.cn/ (accessed on 28 November 2022).
50. Korsavi, S.S.; Montazami, A.; Mumovic, D. The Impact of Indoor Environment Quality (IEQ) on School Children's Overall Comfort in the UK; a Regression Approach. *Build. Environ.* **2020**, *185*, 107309. [CrossRef]
51. Gislason, N. Architectural Design and the Learning Environment: A Framework for School Design Research. *Learn. Environ. Res.* **2010**, *13*, 127–145. [CrossRef]

 *buildings*

Article

Profiling Students Based on the Overlap between IEQ and Psychosocial Preferences of Study Places

Amneh Hamida *, AnneMarie Eijkelenboom and Philomena M. Bluyssen

Chair Indoor Environment, Faculty of Architecture and the Built Environment, Delft University of Technology, 2628 Delft, The Netherlands
* Correspondence: a.b.hamida@tudelft.nl

Abstract: Research has shown that students differ in their preferences of indoor environmental quality (IEQ) and psychosocial aspects of their study places. Since previous studies have mainly focused on identifying these preferences rather than investigating the different profiles of students, this study aimed at profiling students based on their IEQ and psychosocial preferences of their study places. A questionnaire was completed by 451 bachelor students of the faculty of Architecture and the Built Environment. A TwoStep cluster analysis was performed twice separately. First, to cluster the students based on their IEQ preferences, and second based on their psychosocial preferences. This resulted in three clusters under each cluster model. Then, the overlap between these two models was determined and produced nine unique profiles of students, which are: (1) the concerned perfectionist, (2) the concerned extrovert, (3) the concerned non-perfectionist, (4) the visual concerned perfectionist, (5) the visual concerned extrovert, (6) visual concerned non-perfectionist, (7) the unconcerned introvert, (8) the unconcerned extrovert, and (9) the unconcerned non-perfectionist. A number of variables was found to be significantly different among these profiles. This study's outcome indicates that studying the overlap between IEQ and psychosocial preferences is required to understand the different possible profiles of students.

Keywords: IEQ preferences; psychosocial preferences; twostep cluster analysis; study place; students' profiles

Citation: Hamida, A.; Eijkelenboom, A.; Bluyssen, P.M. Profiling Students Based on the Overlap between IEQ and Psychosocial Preferences of Study Places. *Buildings* **2023**, *13*, 231. https://doi.org/10.3390/buildings13010231

Academic Editors: Yue Wu, Zheming Liu and Zhe Kong

Received: 15 December 2022
Revised: 6 January 2023
Accepted: 12 January 2023
Published: 13 January 2023

1. Introduction

Students in higher education spend their time carrying out study-related activities (e.g., individual studying) in indoor environments other than standard classrooms, such as informal learning/study places [1,2]. These places refer to spaces that are mainly used by students for performing such study-related activities [3]. Previous research has found that students generally conduct their study-related activities at home or in an educational building [4], and spend substantial time inside these places [5]. Therefore, understanding students' preferences of these places can help to provide indoor environments that support their academic performance and well-being [6,7]. These preferences can be related to indoor environmental aspects and psychosocial aspects [8].

Research on students' (primary, secondary, and university education) preferences is usually performed within the context of teaching-related activities (e.g., classroom setting) [9–18]. Few studies have examined university students' preferences of study places in informal learning settings (e.g., individual learning, collaborative learning outside the classroom) [19]. For example, Ramu et al. [1] explored students' preferences of informal academic study places on campus, and concluded that students were generally concerned about the layout and amenities (e.g., furniture) in these places. While both indoor environmental (e.g., lighting, temperature) and psychosocial (e.g., privacy, layout) preferences were included, the study was limited to study places used for collaborative study-related activities and located in educational buildings. Beckers et al. [4] investigated the reasons behind students' choices to use a certain place (at home or educational building) for studying. These reasons were significantly correlated with students' preferences, their personal characteristics, and study-related activities. Most of the study-related activities were conducted at home, and students were found to prefer studying at home because they had the ability to control the indoor environmental quality (IEQ)-factors (e.g., indoor air, thermal, sound and lighting quality). Another study conducted by Cunningham and Walton [20] found that 52 percent of university students chose the university library as a study place because it provided a quiet environment. Furthermore, Roetzel et al. [21] revealed that students' preferences of their study places can change with the study-related activities they perform. For instance, Braat-Eggen et al. [22] indicated that university students did not prefer background sounds, such as speech, in an open-plan study environment while they were performing cognitive tasks (e.g., studying for an exam).

So far, these previous studies generalized the preferences that were identified among the student sample. However, different students have different preferences that may change over time [2,8]. For example, in a study performed by Liu and Luther [23], students showed differences in their psychosocial preferences, such as privacy and interactions. Additionally, university students from different faculties can have distinct preferences of study places, found by Wilson and Cotgrave [7]. Students of the art and design discipline scored higher important scores for room layout, the ability to adjust furniture, and controlling the environmental factors than students within the built environment and engineering faculties. This was linked to the personality traits among the students from various faculties. Therefore, it is important to understand how university students' preferences of their study places vary.

An integrated analysis approach, which takes into account the differences in preferences and needs of occupants (profiles) and the different stressors at the environmental level (pattern of stressors), was recently introduced in the field of IEQ [8,24]. The approach claims that in order to provide a good IEQ for all occupants, determining profiles of clusters at the human level and matching those profiles with patterns of environmental stressors (positive and negative) in a certain indoor environment could be the right way to go. In other words, to be able to determine the pattern of stressors at the environmental level, clustering occupants based on their preferences is required to first identify the profiles of clusters to better understand how they interact in an indoor environment [25]. So far, a number of studies in which groups of occupants were clustered according to their preferences and needs have shown differences among the profiles of these clusters [26,27].

Profiles of clusters have been determined for various scenarios and situations, such as home occupants [28,29], primary school children [12], office workers [26,30], and outpatient staff of hospitals [27]. In two of those studies, TwoStep cluster analysis was performed to produce profiles of clusters with regards to (1) IEQ comfort and preferences, and (2) psychosocial comfort and preferences [26,27]. The study on the outpatient staff [27] resulted in six profiles of clusters based on IEQ comfort and preferences, and three profiles of clusters based on psychosocial comfort and preferences. Similar to that, the study on office workers during COVID-19 [26] resulted in two separated models; IEQ preferences model (including four profiles of clusters) and the psychosocial preferences model (including six profiles of clusters). In the latter study, Eijkelenboom and Bluyssen [27] stated that as the overlap between IEQ preferences and psychosocial preferences models was limited, it is essential to study both in future studies.

Profiles of students based on their preferences of both IEQ and psychosocial aspects of their study places are still to be explored. Thus, in this study the question was raised whether profiles of clusters for university students based on both their IEQ, and psychosocial preferences of their study places can be determined. If so, what are the distinctive preferences and characteristics of each student's profile? Accordingly, in this study an attempt was made to cluster simultaneously students' profiles based on both IEQ and psychosocial preferences of their study places.

2. Materials and Methods

Bachelor students of the Faculty of Architecture and the Built Environment at TU Delft were recruited for a survey in March 2021, October 2021 and March 2022. They were asked in this survey about their IEQ and psychosocial preferences of their study place. Students' names and emails were provided by the course coordinators. A brief introduction to the questionnaire was given to the students by the coordinators on the same day of sending the questionnaire. Then, each student received a unique link to the questionnaire via an invitation email. In addition, the students were informed that they had ten days to answer the questionnaire. Five days after sending the questionnaire, a reminder was sent to those students who had not submitted the questionnaire yet. Furthermore, the expected time (approximately 30 min) for answering the questionnaire was stated in the consent form (the first page of the questionnaire).

2.1. Questionnaire

The questionnaire, entitled "My Study Place", is based on previously validated questionnaires that were used for office workers such as the OFFICAR questionnaire [31], the preferences of office workers questionnaire [26], and the outpatient questionnaire staff [27]. The "My Study Place" questionnaire, built in the Qualtrics XM platform in both English and Dutch, consists of seven sections: personal information, psycho-social aspects, most used study place, preferences, comfort perception, lifestyle, and health. Appendix A includes details of the sections and sub-sections of the questionnaire. For example, the preferences section includes an IEQ preferences sub-section that comprises eight variables. This question is stated as "Please rate on a scale from 1 to 10, the importance of each of the following aspects for your study performance at your study place1: not important at all; 10: extremely important-e.g., temperature".

2.2. Participants

The questionnaire was completed by bachelor students of the faculty of Architecture and the Built Environment in March 2021, October 2021, and March 2022. In March 2021, 409 first-year bachelor students completed the questionnaire, in which two sections—the mostly used study place and the preferences—were not included, but the questions related to time spent at home during weekdays and weekend were included. In October 2021, the questionnaire (including these two sections, but excluding the questions related to time spent at home) was sent again to these students, of which 127 completed it. Nonetheless, 127 students were not sufficient to conduct the TwoStep cluster analysis. Accordingly, the "My Study Place" questionnaire including all seven sections was sent to another 472 bachelor students in March 2022, of which 347 students completed the questionnaire. Then, all the results were combined in one dataset with 474 (347 + 127) students. Subsequently, 22 students were excluded because they did not answer the preferences questions. Additionally, one student aged 49 years was excluded from the data set. Hence, the final dataset that was used for the analysis included 451 students.

2.3. Ethical Aspects

The Human Research Ethics Committee (HREC) at the Delft University of Technology approved the application to conduct this study on the 31st of January 2022. A consent form was included at the beginning of the questionnaire, stating all data will be treated anonymously. This form also mentioned that students could skip any part of the questionnaire if they felt uncomfortable answering it.

2.4. Data Management and Analysis

The data were exported from the Qualtrics XM platform to SPSS version 26.0 software (SPSS Inc, Chicago, IL, USA) for data analysis. Descriptive statistics were performed to calculate the frequencies, percentages, maximum, minimum and standard deviation (SD), and mean of the variables related to demographics, emotional state, IEQ comfort perception, psychosocial perception, IEQ preferences and the importance of IEQ-related items to study better, and psychosocial preferences.

TwoStep cluster analysis is a segmentation method that enables the creation of profiles of clusters based on any form of data, including categorical data [32]. This method was also used in previous studies within the domain of IEQ to determine profiles of clusters [12,24–28]. Accordingly, TwoStep cluster analysis was performed and validated twice and separately to create two distinct cluster models. The first TwoStep cluster analysis was performed to cluster the students based on their IEQ preferences, while the second one clustered them based on their psychosocial preferences. The input variables for the IEQ preferences model comprised eight variables: ventilation and fresh air, temperature, view to the outside, sounds from the outside, sounds from the inside, smells, artificial light, and daylight. The input variables for the psychosocial preferences model comprised nine variables: storage, cleanliness, amenities, chair type, presence and company of others, size of the room, bonding or identifying with the place, ability to adapt or control the place, and privacy. The settings of the TwoStep cluster analysis were based on selecting log-likelihood, determination of the number of clusters automatically, and Akaike's information criterion (AIC). Once the cluster model was generated, four validation steps were conducted : (1) silhouette measure of the cluster model is larger than 0.2 (fair and above); (2) Chi-square tests were performed to examine the relationship between the input variables of the cluster analysis and the final cluster model, with p-value less than 0.05 considered as statistically significant; (3) the predictor importance scores of the input variables were larger than 0.02; and (4) the dataset was randomly split half (50%) to re-run the final solution model on each half to ensure that both solutions were similar to the final solution.

After the TwoStep cluster analysis, descriptive analysis was conducted to calculate the frequencies, percentages, and SD for different variables of each cluster (e.g., health, IEQ perception, IEQ preferences). To compare differences between the clusters, Chi-square and ANOVA tests were used (for nominal and continuous variables, respectively). Each student belongs to two clusters, a cluster of IEQ preferences, and a cluster of psychosocial preferences, resulting in clusters of students with the same IEQ preferences but different psychosocial preferences, and vice versa. Hence, it is important to investigate the overlap between the two models to better understand in detail the profile of students within these two models. The overlap between the two cluster models was identified using cross-tabulation. In addition, frequencies, percentages, and SD for different variables of each profile within the overlap between the two models were calculated. The significant differences between the variables among the different profiles were tested using Chi-square and ANOVA tests. Chi-square calculations with less than 5 in one cell were excluded from the analysis.

3. Results

3.1. Students Characteristics

Table 1 presents several characteristics (e.g., age, gender, time spent at home, study place, and lifestyle) of the respondents in 2021 and 2022. Since only the mean time spent at home was significantly different between the two groups; students in 2021 spent more time at home than students in 2022, this study mainly focused on questions related to study places and excluded the questions related to students' homes, such as building-related symptoms. The mean age of the 451 students was 20 years old. The ratio of female to male students was 1.6. Students within this study spent their studying time mostly at their homes (74%), while 26% of them stayed in educational buildings for studying. The students stayed at their homes around 17 h per day during weekdays, and 16 h per day during the weekend.

Table 1. Students characteristics in 2021 and 2022.

	All Students	Students in 2021	Students in 2022	*p*-Value
Invited	878	409	472	-
Respondents	474	127	374	-
Response rate (%)	54.0	31.1	79.2	-
Age-mean (SD)	19.8 (1.6)	19.6 (1.1)	19.8 (1.8)	0.61
Gender-n (%)				0.70
Male	175 (39.0)	43 (40.6)	132 (38.5)	-
Female	274 (61.0)	63 (59.4)	211 (61.5)	-
Time spent at home during weekdays-mean (SD)				-
Weekdays	16.9 (3.6)	20.4 (2.8)	15.8 (3.1)	$p < 0.001$
Weekend	15.8 (4.2)	17.5 (4.2)	15.2 (4.0)	$p < 0.001$
Study place-n (%)				0.26
Home	333 (73.8)	85 (79.4)	248 (72.1)	-
Educational building	116 (25.7)	22 (20.6)	94 (27.3)	-
Lifestyle-n (%)				
Smoking	134 (29.7)	22 (20.6)	112 (32.5)	0.12
Alcohol	384 (85.1)	92 (86.0)	292 (84.9)	0.74
Physical activity	407 (90.2)	98 (91.6)	309 (89.8)	0.59

3.2. Students' Preferences of Their Study Places

Figure 1 presents the mean and SD values of the eight IEQ preferences aspects. Daylight (8.4 ± 1.5) was the most important aspect of the whole study sample. This is followed by both view to the outside (8.2 ± 1.8) and temperature (8.2 ± 1.3). In contrast, smells (6.2 ± 2.3), artificial light (6.2 ± 2.0), and sounds from the outside (6.3 ± 2.2) were the least important IEQ aspects. Figure 2 illustrates the mean and SD values of the nine psychosocial preference aspects. Amenities (8.0 ± 1.5) and cleanliness (7.6 ± 1.7) were the most important psychosocial aspects of the study place. On the other hand, students in this study reported the lowest scores on three psychosocial aspects: presence and company of others (5.3 ± 2.5), bounding or identifying with the place (5.4 ± 2.5), and size of the room (5.5 ± 2.0).

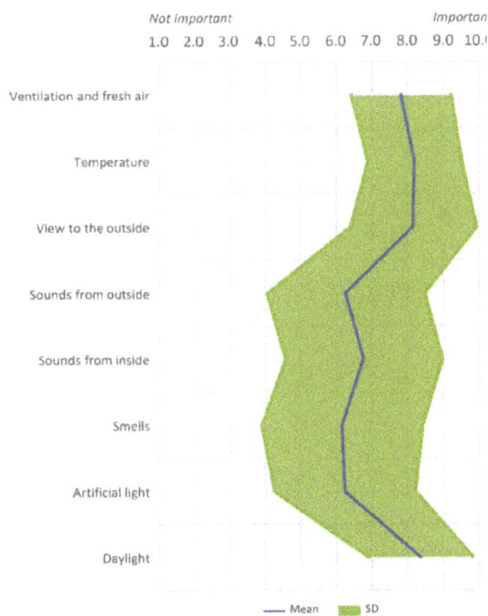

Figure 1. IEQ preferences of study places.

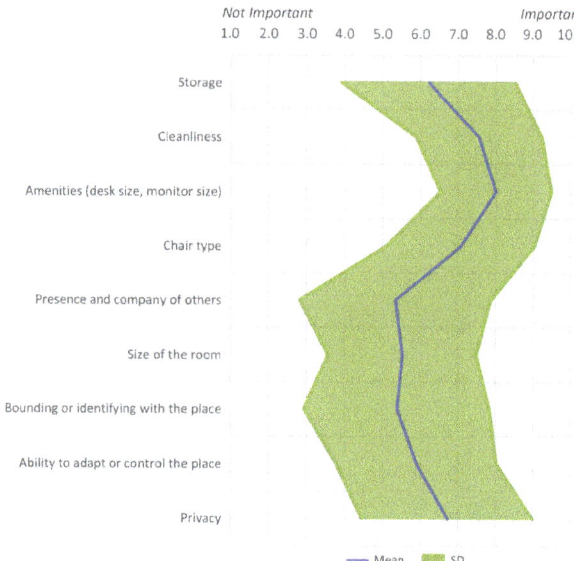

Figure 2. Psychosocial preferences of study places.

3.3. TwoStep Cluster Analysis

TwoStep cluster analysis was carried out to categorize profiles of students based on their IEQ preferences and separate psychosocial preferences in their study places. This was carried out by using the original variables that consists of eight variables of the IEQ preferences and nine variables for the psychosocial preferences. The results of the TwoStep cluster analysis resulted in two models: the IEQ preferences model, and the psychosocial preferences model. Each of these two models comprised three distinct clusters. The Silhouette coefficient was fair for both models; 0.3 for the IEQ preferences model, and 0.2 for the psychosocial preferences model.

The predictor importance of the eight input variables for the IEQ preferences model, as well as the nine input variables for the psychosocial preferences model, was found to be strong and larger than 0.02. Additionally, after randomly splitting the dataset into two halves, only a few changes were found between the two halves and the final solution (Table 2). Furthermore, all eight IEQ preference variables were found to be statistically significant in relation to the IEQ preferences model ($p < 0.05$). Similarly, the nine psychosocial preference variables were found to be statistically significant in relation to the psychosocial preferences model.

Table 2. Predictor importance of the input variables for both models.

	Predictor Importance	Final Solution	First Half Solution	Second Half Solution
IEQ preferences model	0.60–1.00	Daylight (1.00) Sounds from inside (0.80) View to the outside (0.75) Smells (0.68)	Sounds from the inside (1.00) View to the outside (0.84) Daylight (0.62)	Sounds from the inside (1.00)
	0.30–0.59	Sounds from the outside (0.57) Ventilation and fresh air (0.30)	Smells (0.52) Sounds from the outside (0.42)	Daylight (0.58) View to the outside (0.44) Smells (0.40) Sounds from the outside (0.40)
	0.02–0.29	Artificial light (0.21) Temperature (0.20)	Ventilation and fresh air (0.21) Temperature (0.07) Artificial light (0.03)	Ventilation and fresh air (0.20) Artificial light (0.20) Temperature (0.06)
Psychosocial preferences model	0.60–1.00	Bonding or identifying with the place (1.00) Ability to adapt or control the place (0.91) Size of the room (0.71) Cleanliness (0.63)	Presence and company of others (1.00) Ability to adapt or control the place (0.78) Privacy (0.71)	Ability to adapt or control the place (1.00) Bonding or identifying with the place (0.82)
	0.30–0.59	Storage (0.54) Presence and company of others (0.51) Chair type (0.36) Amenities (0.33) Privacy (0.30)	Size of the room (0.56) Bonding or identifying with the place (0.54) Storage (0.53) Chair type (0.43) Amenities (0.38) Cleanliness (0.30)	Storage (0.53) Amenities (0.52) Size of the room (0.52) Presence and company of others (0.52) Chair type (0.47) Privacy (0.35)
	0.02–0.29	-	-	Cleanliness (0.22)

3.3.1. IEQ Preferences Model

The clusters of the IEQ preference clusters are described in Table 3 and Appendix B. Table 3 only includes the variables that were statistically different among the clusters within the IEQ preferences model ($p < 0.05$). The IEQ preferences model resulted in three clusters: IEQC1 (concerned with all IEQ aspects), IEQC2 (concerned with daylight and view to the outside), and IEQC3 (concerned with only temperature). These three clusters scored a high importance level for daylight (ranged from 7.0 to 9.0), view to the outside (ranged from 6.7 to 9.0), and temperature (ranged from 7.8 to 8.7).

Table 3. Descriptives of IEQ clusters.

	IEQC1	IEQC2	IEQC3	*p*-Value
N (%within the total sample)	159 (35.5)	149 (33.3)	140 (31.3)	-
Gender -N (%within cluster level)				$p < 0.001$
Male	42 (26.4)	63 (42.6)	68 (48.9)	-
Female	117 (73.6)	85 (57.4)	71 (51.1)	-
Study place-N (%within cluster level)				0.007
Home	103 (64.8)	117 (78.5)	110 (78.6)	-
Educational building	55 (34.6)	31 (20.8)	30 (21.4)	-
IEQ preferences-mean (SD)				
Ventilation and fresh air	8.5 (1.1)	7.7 (1.3)	7.2 (1.6)	$p < 0.001$
Temperature	8.7 (1.1)	7.8 (1.3)	8.0 (1.3)	$p < 0.001$
View to the outside	8.7 (1.3)	9.0 (1.1)	6.7 (1.9)	$p < 0.001$
Sounds from the outside	7.6 (1.7)	4.8 (1.9)	6.3 (2.1)	$p < 0.001$
Sounds from the inside	8.1 (1.4)	5.0 (2.1)	7.2 (1.8)	$p < 0.001$
Smells	7.8 (1.4)	4.9 (2.0)	5.7 (2.2)	$p < 0.001$
Artificial light	7.1 (1.7)	6.0 (2.0)	5.5 (1.9)	$p < 0.001$
Daylight	9.0 (0.9)	9.0 (0.9)	7.0 (1.4)	$p < 0.001$
Importance of IEQ-related aspects-mean (SD)				
Lamp on my desk	6.6 (2.3)	5.9 (2.4)	6.2 (2.2)	0.026
Personal desk ventilation and fresh air	7.6 (2.2)	7.1 (2.2)	6.4 (2.0)	$p < 0.001$
Control of surrounding sounds	7.7 (1.6)	5.6 (2.2)	6.8 (1.9)	$p < 0.001$
Control of shading	7.8 (1.7)	6.5 (2.2)	7.2 (1.7)	$p < 0.001$
Control of room ventilation	7.8 (1.6)	6.2 (2.0)	6.5 (2.0)	$p < 0.001$
Control of room temperature	7.7 (1.5)	6.7 (1.9)	6.8 (2.0)	$p < 0.001$
Headphones	7.7 (2.4)	7.3 (2.6)	6.6 (2.4)	0.004

3.3.2. Psychosocial Preferences Model

Descriptions of the psychosocial preference clusters are presented in Table 4 and Appendix C. Table 4 only illustrates the variables that were found to be statistically different among the three clusters within the psychosocial preferences model ($p < 0.05$). This model consists of three distinct clusters: PSC1 (Preference for most of psychosocial aspects), PSC2 (preference for presence and company of others), and PSC3 (preference only for amenities and cleanliness). Generally, the students within these clusters reported a high importance for two aspects, which are cleanliness (ranged from 7.1 to 9.0) and amenities (ranged from 7.5 to 8.9).

Table 4. Descriptive of psychosocial clusters.

	PSC1	PSC2	PSC3	*p*-Value
N (%within the total sample)	110 (25.0)	186 (42.3)	144 (32.7)	-
Lifestyle-N (%within cluster level)				
Smoking	21 (19.0)	56 (26.9)	52 (36.1)	0.025
Alcohol	85 (77.3)	161 (86.6)	128 (88.9)	0.021
Study place-N (%within cluster level)				$p < 0.001$
Home	98 (89.1)	117 (62.9)	110 (76.4)	-
Educational building	12 (10.9)	68 (36.6)	33 (22.9)	-
Psychosocial preferences-mean (SD)				
Storage	8.1 (1.3)	5.6 (2.1)	5.6 (2.5)	$p < 0.001$
Cleanliness	9.0 (1.0)	7.1 (1.4)	7.1 (1.8)	$p < 0.001$
Amenities	8.9 (1.1)	7.9 (1.2)	7.5 (1.8)	$p < 0.001$
Chair type	8.0 (1.7)	7.4 (1.5)	6.0 (2.3)	$p < 0.001$
Presence and company of others	5.1 (2.4)	6.6 (2.0)	4.0 (2.4)	$p < 0.001$
Size of the room	6.4 (1.8)	6.2 (1.5)	4.1 (1.8)	$p < 0.001$
Bonding or identifying with the place	6.6 (1.8)	6.3 (2.0)	3.3 (2.0)	$p < 0.001$
Ability to adapt or control the place	7.2 (1.6)	6.4 (1.7)	4.2 (1.9)	$p < 0.001$
Privacy	8.1 (1.4)	6.2 (2.2)	6.3 (2.4)	$p < 0.001$

3.4. Overlap between the IEQ and the Psychosocial Preferences Model

The overlap between the IEQ and psychosocial preferences model resulted in nine distinct profiles that are illustrated in Figure 3. Descriptions of these profiles, presented in Table 5, are statistically significantly different between the profiles. A comprehensive description for these nine groups is illustrated in Appendix D.

In general, all nine profiles are concerned with three IEQ preferences, which are daylight (ranged from 6.6 to 9.3), view to the outside (ranged from 6.4 to 9.1), and temperature (ranged from 7.6 to 8.9). Pertaining to the psychosocial preferences, most of the profiles scored high importance levels for two aspects: amenities (ranged from 7.4 to 8.9) and cleanliness (ranged from 6.8 to 9.0). Therefore, the description for each profile is based on highlighting which profile scored the highest and/or lowest importance level for both IEQ and psychosocial preferences among all profiles.

Each name of the nine profiles consists of two parts: the first part is related to IEQ preferences, and the second part is related to psychosocial preferences. The IEQ preferences part consists of one of three names that are; (1) concerned, which means all IEQ preferences are important, (2) visual concerned, which implies that daylight and view to the outside are important, and (3) unconcerned, which indicates that almost all IEQ preferences are not very important except for temperature. The psychosocial preferences part includes one of the four categories; (1) perfectionist, which implies high importance levels for most of the psychosocial aspects, (2) extrovert, which reflects the high importance level for the presence and company of others, (3) introvert, which means that privacy is highly important, (4) non-perfectionist, which indicates that most of the psychosocial aspects are not highly important, except amenities.

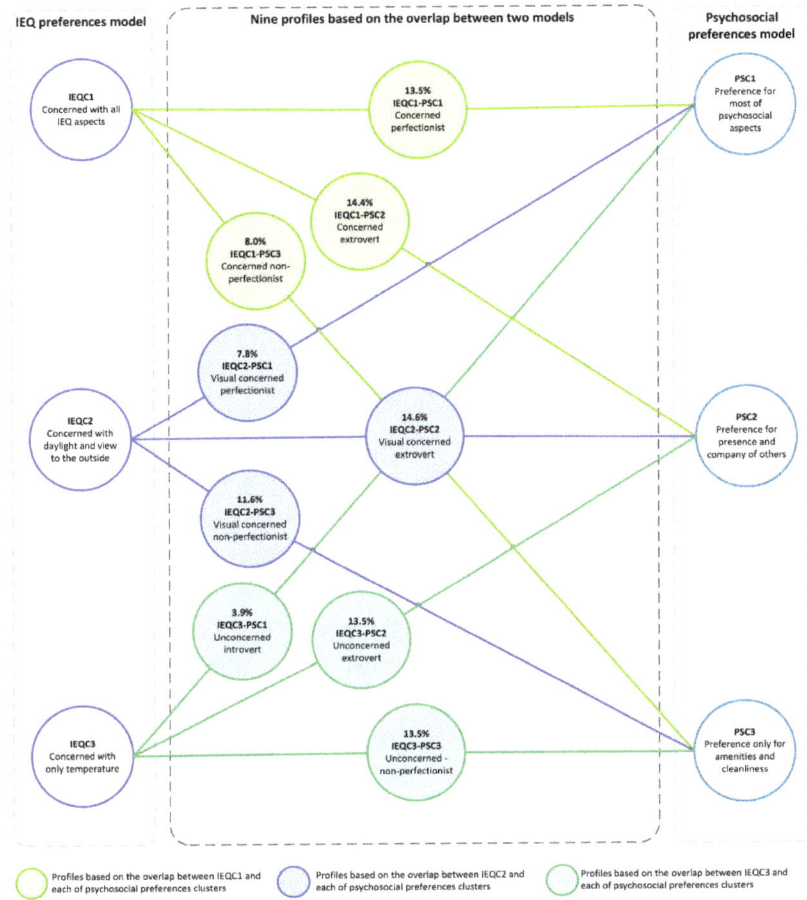

Figure 3. The nine profiles of students based on the overlap between the IEQ preferences model and psychosocial preferences model.

Table 5. Description of the overlap profiles between the two clusters models.

	IEQC1-PSC1	IEQC1-PSC2	IEQC1-PSC3	IEQC2-PSC1	IEQC2-PSC2	IEQC2-PSC3	IEQC3-PSC1	IEQC3-PSC2	IEQC3-PSC3	*p*-Value
N (%within the total sample)	59 (13.5)	63 (14.4)	35 (8.0)	34 (7.8)	64 (14.6)	48 (11.0)	17 (3.9)	59 (13.5)	59 (13.5)	-
Age										$p < 0.001$
Mean (SD)	19.6 (1.8)	19.7 (1.9)	19.6 (1.1)	19.9 (1.2)	19.7 (1.2)	19.7 (1.5)	20.1 (0.9)	20.0 (2.1)	19.8 (1.4)	-
Maximum	29	31	24	23	24	26	22	23	23	-
Minimum	17	17	18	18	18	18	19	18	18	-
Gender–N (%within profile level)										$p < 0.001$
Male	17 (28.8)	18 (28.6)	6 (17.1)	16 (47.1)	26 (40.6)	19 (39.6)	10 (58.8)	24 (40.7)	32 (54.2)	-
Female	42 (71.2)	45 (71.4)	29 (82.9)	18 (52.9)	38 (59.4)	28 (58.3)	7 (41.2)	34 (57.6)	27 (45.8)	-

Table 5. *Cont.*

	IEQC1-PSC1	IEQC1-PSC2	IEQC1-PSC3	IEQC2-PSC1	IEQC2-PSC2	IEQC2-PSC3	IEQC3-PSC1	IEQC3-PSC2	IEQC3-PSC3	p-Value
Recently experienced events-N (%within profile level)										
Positive events	22 (37.3)	22 (34.9)	9 (25.7)	8 (23.5)	18 (28.1)	12 (25.0)	5 (29.4)	14 (23.7)	16 (27.1)	0.012
Lifestyle-n (%within profile level)										
Alcohol	44 (74.6)	53 (87.3)	32 (91.4)	28 (82.4)	56 (87.5)	43 (89.6)	13 (76.5)	50 (84.7)	51 (86.4)	$p < 0.001$
Physical activity	53 (89.8)	58 (92.1)	33 (94.3)	32 (94.1)	60 (93.8)	43 (89.6)	12 (70.6)	52 (88.1)	51 (86.4)	$p < 0.001$
PANAS-Mean (SD)										
Positive affect	17.8 (2.7)	17.1 (2.6)	17.6 (2.2)	18.3 (2.4)	17.4 (2.5)	17.4 (2.6)	17.9 (2.4)	16.9 (2.3)	17.1 (2.9)	$p < 0.001$
Negative affect	11.9 (3.0)	11.6 (3.1)	12.1 (3.0)	11.7 (2.7)	11.4 (2.8)	11.0 (3.1)	11.7 (3.1)	11.8 (3.0)	10.9 (3.1)	$p < 0.001$
Health-n (%within profile level)										
Depression	12 (20.3)	12 (19.0)	7 (20.0)	6 (17.6)	12 (17.7)	11 (22.9)	5 (29.4)	9 (15.3)	17 (28.8)	$p < 0.001$
Anxiety	17 (28.8)	19 (30.2)	7 (20.0)	8 (23.5)	12 (18.8)	15 (31.3)	5 (29.4)	11 (18.6)	13 (22.0)	$p < 0.001$
IEQ perception-n (%within profile level)										
Dissatisfied with air freshness	31 (52.5)	41 (65.1)	21 (60.0)	20 (58.8)	35 (54.7)	32 (66.7)	9 (52.9)	30 (50.8)	44 (74.6)	0.011
Dissatisfied with air smell	27 (45.8)	28 (44.4)	17 (48.6)	12 (35.3)	40 (62.5)	19 (39.6)	6 (35.3)	26 (44.1)	26 (44.1)	0.003
IEQ preferences mean (SD)										
Ventilation and fresh air	8.6 (1.1)	8.3 (1.1)	9.0 (1.1)	7.8 (1.3)	7.7 (1.4)	7.6 (1.2)	7.5 (1.5)	7.3 (1.2)	7.1 (1.9)	$p < 0.001$
Temperature	8.7 (1.1)	8.6 (1.0)	8.9 (1.1)	8.1 (1.5)	7.8 (1.3)	7.6 (1.3)	8.4 (1.7)	7.9 (1.2)	8.1 (1.4)	$p < 0.001$
View to the outside	8.6 (1.3)	8.8 (1.2)	8.7 (1.4)	9.1 (0.9)	8.9 (1.0)	9.2 (1.1)	7.1 (1.7)	6.9 (1.9)	6.4 (1.9)	$p < 0.001$
Sounds from the outside	7.8 (1.4)	7.5 (1.8)	7.4 (2.1)	5.6 (1.9)	4.8 (1.7)	4.3 (1.9)	6.9 (2.0)	6.1 (2.0)	6.4 (2.1)	$p < 0.001$
Sounds from the inside	8.2 (1.3)	7.9 (1.4)	8.2 (1.5)	4.9 (2.2)	5.1 (1.9)	5.0 (2.2)	7.4 (1.7)	7.1 (1.6)	7.2 (2.0)	$p < 0.001$
Smells	7.9 (1.4)	7.8 (1.3)	7.5 (1.7)	5.3 (1.7)	5.1 (1.8)	4.3 (2.1)	6.5 (2.4)	5.7 (1.8)	5.6 (2.4)	$p < 0.001$
Artificial light	7.2 (1.9)	7.0 (1.5)	6.8 (1.6)	6.6 (1.9)	5.9 (1.7)	5.5 (2.1)	6.1 (2.1)	5.5 (1.5)	5.5 (2.2)	$p < 0.001$
Daylight	9.0 (0.9)	9.0 (0.9)	8.8 (1.0)	9.3 (0.7)	8.8 (0.9)	9.0 (1.0)	7.2 (1.3)	7.1 (1.2)	6.6 (1.6)	$p < 0.001$
Psychosocial preferences-mean (SD)										
Storage	7.8 (1.3)	5.3 (2.2)	6.3 (2.6)	8.4 (1.1)	5.6 (2.1)	5.4 (2.3)	8.2 (1.3)	5.9 (1.9)	5.4 (2.5)	$p < 0.001$
Cleanliness	9.0 (1.1)	7.4 (1.2)	7.7 (1.8)	9.0 (1.0)	7.0 (1.5)	7.0 (1.7)	9.0 (0.9)	6.8 (1.3)	6.8 (1.9)	$p < 0.001$
Amenities	8.9 (1.2)	8.0 (1.1)	7.8 (2.0)	8.9 (0.9)	7.9 (1.3)	7.4 (1.8)	8.9 (1.0)	7.8 (1.1)	7.4 (1.5)	$p < 0.001$
Chair type	8.1 (1.7)	7.6 (1.5)	6.5 (2.6)	7.8 (1.7)	7.3 (1.5)	5.7 (2.2)	7.9 (1.9)	7.2 (1.4)	6.1 (2.1)	$p < 0.001$
Presence and company of others	5.3 (2.4)	6.9 (2.0)	4.8 (2.8)	4.5 (2.4)	6.5 (2.2)	3.8 (2.3)	5.4 (2.5)	6.3 (1.8)	3.6 (2.2)	$p < 0.001$
Size of the room	6.4 (1.9)	6.4 (1.3)	4.0 (1.8)	6.1 (1.6)	6.2 (1.7)	4.1 (1.8)	6.6 (1.9)	6.0 (1.5)	4.0 (1.8)	$p < 0.001$
Bonding or identifying with the place	6.8 (1.6)	6.5 (1.9)	3.0 (1.8)	6.3 (1.9)	6.3 (2.3)	3.2 (2.2)	6.2 (2.1)	6.3 (1.9)	3.5 (2.0)	$p < 0.001$
Ability to adapt or control the place	7.4 (1.6)	6.4 (1.7)	3.9 (1.6)	7.0 (1.6)	6.6 (1.6)	4.3 (1.9)	7.2 (1.6)	6.3 (1.7)	4.2 (2.0)	$p < 0.001$
Privacy	8.2 (1.4)	6.5 (2.0)	6.5 (2.7)	7.9 (1.3)	5.9 (2.5)	6.2 (2.4)	8.5 (1.6)	6.4 (2.1)	6.4 (2.4)	$p < 0.001$
Importance of IEQ-related aspects-mean (SD)										
Chair seat heating	5.0 (2.8)	3.8 (2.3)	4.0 (3.4)	3.8 (2.7)	4.5 (2.8)	3.3 (2.6)	4.5 (2.6)	3.6 (2.3)	2.8 (2.4)	$p < 0.001$
Chair backrest eating	4.9 (2.8)	4.0 (2.5)	3.9 (3.4)	3.7 (2.8)	4.6 (3.1)	3.1 (2.7)	4.5 (3.1)	3.7 (2.7)	2.8 (2.6)	$p < 0.001$
Heating on my desk	4.4 (2.7)	3.6 (2.5)	3.1 (2.6)	3.8 (2.9)	4.2 (2.8)	2.7 (2.3)	3.9 (2.7)	3.6 (2.3)	2.6 (2.6)	$p < 0.001$
Lamp on my desk	6.9 (2.2)	6.4 (2.2)	6.6 (2.8)	7.1 (2.0)	5.5 (2.1)	5.8 (2.7)	6.9 (2.2)	6.2 (2.0)	6.3 (2.3)	$p < 0.001$
Personal desk ventilation and fresh air	8.1 (1.7)	7.0 (2.4)	7.7 (2.3)	7.3 (1.9)	7.1 (2.3)	7.1 (2.3)	6.4 (2.5)	6.2 (2.0)	6.5 (1.9)	$p < 0.001$
Control of surrounding sounds	8.1 (1.5)	7.3 (1.5)	7.6 (1.8)	6.7 (2.2)	5.4 (1.8)	5.0 (2.2)	6.7 (2.2)	6.6 (1.7)	7.1 (1.8)	$p < 0.001$
Control of shading	8.2 (1.6)	7.4 (1.6)	7.9 (1.9)	7.2 (2.0)	6.2 (2.1)	6.4 (2.4)	7.4 (1.8)	7.2 (1.6)	7.3 (1.7)	$p < 0.001$
Control of room ventilation	8.2 (1.2)	7.6 (1.6)	7.4 (1.9)	7.0 (2.0)	6.1 (1.8)	6.0 (2.1)	7.3 (1.7)	7.0 (1.5)	6.0 (2.2)	$p < 0.001$
Control of room temperature	8.3 (1.1)	7.4 (1.7)	7.3 (1.4)	7.5 (1.7)	6.3 (1.9)	6.7 (1.7)	8.3 (2.0)	6.8 (2.0)	6.7 (1.7)	$p < 0.001$
Headphones	7.4 (2.6)	7.6 (2.3)	7.9 (2.4)	7.5 (2.4)	7.7 (1.3)	6.6 (2.9)	7.2 (2.0)	6.6 (2.2)	6.6 (2.6)	$p < 0.001$
Presence of plants	7.0 (2.3)	5.9 (2.6)	4.8 (2.3)	7.0 (2.0)	5.7 (2.4)	5.5 (2.6)	6.9 (2.5)	5.6 (2.3)	4.0 (2.8)	$p < 0.001$
Personal control over the most used study place-mean (SD)										
Temperature	4.6 (1.7)	3.8 (1.9)	4.3 (2.1)	5.3 (1.3)	4.5 (1.8)	4.4 (1.7)	4.7 (1.3)	4.3 (1.9)	4.5 (1.7)	$p < 0.001$
Ventilation	5.1 (1.7)	3.9 (2.2)	4.9 (1.9)	5.5 (1.5)	4.6 (1.9)	4.6 (1.9)	5.8 (1.5)	4.4 (1.8)	4.8 (1.8)	$p < 0.001$
Shading from the sun	5.2 (2.0)	4.2 (2.1)	4.6 (2.3)	5.1 (2.0)	4.3 (2.2)	4.4 (2.1)	5.4 (1.7)	5.0 (2.0)	4.5 (2.1)	$p < 0.001$
Lighting	5.4 (2.0)	4.3 (2.2)	4.8 (2.2)	5.9 (1.4)	4.8 (2.1)	5.0 (1.9)	6.0 (1.0)	5.0 (2.0)	5.0 (2.1)	$p < 0.001$
Noise	3.3 (1.6)	2.4 (1.1)	3.0 (1.4)	3.8 (1.8)	3.1 (1.5)	2.9 (1.4)	3.1 (1.4)	2.9 (1.5)	2.8 (1.5)	$p < 0.001$

3.4.1. Overlap between IEQC1 with Psychosocial Clusters

The overlap between the IEQC1 and the three psychosocial clusters resulted in three profiles: IEQC1-PSC1: the concerned perfectionist; IEQC1-PSC2: the concerned extrovert; and IEQC1-PSC3: the concerned the non-perfectionist.

- IEQC1-PSC1: the concerned perfectionist.

The concerned perfectionist profile comprises 59 students (14%), of which 29% are male and 71% are female students. These students are the largest group that experienced positive events (37%). In terms of lifestyle, this group has the lowest number of students that consume alcohol (75%). Regarding the IEQ preferences, the concerned perfectionist students rated the highest importance for sounds from the inside (8.2), smells (7.9), sounds from the outside (7.8), and artificial light (7.2). Furthermore, they rated the highest importance (as compared to the other groups) for six IEQ-related items, which are control of room temperature (8.3), control of room ventilation (8.2), control of shading (8.2), control of surrounding sounds (8.1), personal desk ventilation and fresh air (8.1) and presence of plants (7.0). As this profile overlaps with PSC1 who are concerned with all psychosocial preferences (except presence and company of others), it is the most concerned with cleanliness (9.0), amenities (8.9), chair type (8.1), ability to adapt or control the place (7.4), and bonding or identifying with the place (6.8).

- IEQC1-PSC2: the concerned extrovert.

The concerned extrovert profile consists of 63 students (14%), of which the percentages of male and female students are similar as the concerned perfectionist profile (29% and 71%, respectively). This profile is the second highest group that experienced recently positive events (35%). Regarding health, students within this profile are the second highest group that suffered from anxiety (30%). It can be noted that these students are the group to have the least control over all IEQ factors in their most used study place. Pertaining to IEQ preferences, the concerned extrovert students are concerned about all IEQ aspects, especially daylight (9.0) and view to the outside (8.8). As this profile overlaps with PSC2, it is the profile that is most concerned with the presence and company of others in their study places (6.9), while they rated the lowest importance for storage (5.3).

- IEQC1-PSC3: the concerned non-perfectionist

The concerned non-perfectionist profile comprises 35 students (8%), which includes the lowest percentage of male students (17%) and the highest percentage of female students (83%). Nearly half of them (45%) were feeling relaxed when they were completing the questionnaire. It can be noted that this profile rated the highest for negative affect (12.1). With regards to their lifestyle, students within this profile are the highest in terms of alcohol consumption (91%), as well as doing physical activity (94%). In terms of IEQ preferences, the concerned non-perfectionist students rated the highest importance for ventilation and fresh air (9.0), temperature (8.9) and sounds from the inside (8.2). Regarding IEQ-related items, these students rated the highest importance for headphones (7.9). As this profile overlaps with PSC3, which rated the least importance scores for most of the psychosocial preferences, it is the least concerned with bonding or identifying with the place (3.0), ability to adapt or control the place (3.9) and size of the room (4.0).

3.4.2. Overlap between IEQC2 with Psychosocial Clusters

The overlap between the IEQC2 and the three psychosocial clusters resulted in three profiles: IEQC2-PSC1: the visual concerned perfectionist; IEQC2-PSC2: the visual concerned extrovert; and IEQC2-PSC3: the visual concerned non-perfectionist.

- IEQC2-PSC1: the visual concerned perfectionist

The visual concerned perfectionist profile is the second smallest profile size that comprises 34 students (8%), of which 47% are male students and 53% are female students. It is the profile that least experienced recently positive events (23%), while it rated the highest positive affect (18.3). Most of the students (97%) within this profile spent their studying time at their homes. Pertaining to IEQ perception, these students comprise the profile that is least dissatisfied with air smell (35%). With regards to the IEQ preferences, the visual concerned perfectionist students rated the highest importance for daylight (9.3) and view to the outside (9.1). Regarding the IEQ-related items, these students rated the highest importance for lamp on my desk (7.1), and the presence of plants (7.0). With regards to psychosocial preferences, they rated the highest importance for cleanliness (9.0), amenities (8.9) and storage (8.4). In addition, the visual concerned perfectionist students scored a high importance level for privacy (7.9).

- IEQC2-PSC2: the visual concerned extrovert

The visual concerned extrovert profile is the largest profile size consisting of 64 students (15%), of which 41% of them are male students and 59% are female students. They tended to feel relaxed while they were answering the questionnaire (35%). Regarding IEQ perception, the visual concerned extrovert students are the most dissatisfied with air smell (63%) in their most used study places. However, view to the outside (8.9) and daylight (8.8) are the highest important IEQ preference aspects for these students. Pertaining to the IEQ-related items, these students are the least concerned with personal desk ventilation and fresh air (5.5). Furthermore, they rated the lowest importance scores for the IEQ-related items control of shading (6.2) and control of room temperature (6.3). On the other hand, they are the second profile that scored a high importance score for headphones (7.7). In terms of psychosocial preferences, this profile is the least concerned with privacy (5.9). However, it is the second highest profile that is concerned with the presence and company of others (6.5).

- IEQC2-PSC3: the visual concerned non-perfectionist

The visual concerned non-perfectionist profile comprises 48 students (11%), in which the reported percentages of male and female students (40% and 58%, respectively) are similar to the visual concerned extravert profile. The visual concerned non-perfectionist students are the highest group that suffered from anxiety (31%). With regards to IEQ preferences, they rated the lowest importance for sounds from the outside (4.3), smells (4.3), sounds from the inside (5.0), artificial light (5.5), and temperature (7.6) among the other profiles. However, these students scored high importance levels for the view to the outside (9.2) and daylight (9.0) Furthermore, they rated the lowest importance for two IEQ-related items which are control of room sounds (5.0) and control of room ventilation (6.0). However, personal desk ventilation and fresh air (7.1) is the highest important IEQ-related item for this profile. Pertaining to psychosocial preferences, this profile rated the lowest importance for chair type (5.7). Although the amenities (7.4) aspect was scored the lowest importance level by the visual concerned non-perfectionist students, it was considered the highest important psychosocial aspect.

3.4.3. Overlap between IEQC3 with Psychosocial Clusters

The overlap between the IEQC3 and the three psychosocial clusters resulted in three profiles: IEQC3-PSC1: the unconcerned introvert; the unconcerned extrovert; and the unconcerned non-perfectionist.

- IEQC3-PSC1: the unconcerned introvert

The unconcerned introvert profile is the smallest profile size that comprises 17 students (4%), of which the percentage of male students (59%) is higher than the percentage of female students (41%). The study places for the majority of these students (94%) were located at their homes. Regarding lifestyle, the unconcerned introvert students are the second lowest group that consumes alcohol (77%), and the profile that takes part in the least physical activities (71%). On the contrary, they are the group that suffered most from depression among the other profiles (29%), as well as one of the profiles that suffered most from anxiety (29%). Pertaining to the IEQ perception, this profile reported the least dissatisfaction percentage with the air smell of their study places (35%). In terms of the IEQ-related items, this profile is the most concerned with control of temperature (8.3). In contrast, the unconcerned introvert students are the second profile that is not concerned with personal desk ventilation and fresh air (6.4). However, they do have the highest control over lighting (6.0), ventilation (5.8), and shading from the sun (5.4) in their study places. With regards to the psychosocial preferences, of the four aspects, these students rated these aspects as the highest importance: cleanliness (9.0), amenities (8.9), privacy (8.5), and size of the room (6.6).

- IEQC3-PSC2: the unconcerned extrovert

The unconcerned extrovert profile is considered as a large profile size that consists of 59 students (14%), of which the female students' percentage (58%) is higher than the male students' percentage (41%). These students recorded the highest percentage of feeling neutral while they were completing the questionnaire (28%). They are the second lowest profile to experience recently positive events (24%). Furthermore, they rated the lowest positive affect (16.9). It can be noted that this profile has the least students that suffered from both depression (15%) and anxiety (19%). In terms of IEQ preferences, the unconcerned extrovert students are the least concerned with artificial light in their study places (5.5). Nonetheless, temperature (7.9) is the most important IEQ preference. With regards to IEQ perception, the unconcerned extrovert students reported the least dissatisfaction percentage with air freshness (50%). Pertaining to the IEQ-related items, these students rated both personal ventilation and fresh air the least important (6.2) and headphones (6.6). Nevertheless, control of shading (7.2) and control of room ventilation (7.0) are the most important items for them. Regarding the psychosocial preferences, this profile rated of lowest importance cleanliness (6.8) in their study places. However, the unconcerned extrovert profile is one of the profiles that rated the highest importance for the presence and company of others (6.3).

- IEQC3-PSC3: the unconcerned non-perfectionist

The unconcerned non-perfectionist profile has the same profile size as the unconcerned extrovert profile, with 59 students (14%), of which the percentage of male students (54%) is higher than the percentage of female students (46%). Students within this profile rated the lowest negative affect score among other profiles (10.9). Regarding health, this profile is the second highest group that suffered from depression (29%). Pertaining to IEQ perception, this profile that is dissatisfied the most with air freshness (75%). In terms of IEQ preferences, the unconcerned non-perfectionist students are the least concerned with artificial light (5.5), view to the outside (6.4), daylight (6.6), and ventilation and fresh air (7.1). On the other hand, they are only concerned about temperature (8.1). With regards to IEQ-related items, they rated the least importance for the presence of plants (4.0), control of room ventilation (6.0) and headphones (6.6). Nonetheless, they are concerned about the control of shading (7.3) and surrounding sounds (7.0). Regarding the psychosocial preferences, this profile of students is the least concerned with the presence and company of others (3.6) and the size of the room (4.0). While amenities (7.4) and cleanliness (6.8) are the most important for these students, they are rated the least important among other profiles.

4. Discussion

4.1. Comparison with Previous Studies

The majority of students (74%) within this study spent most of their studying time at their homes. A previous study indicated that a home can be considered as an off-campus informal study place, and that most students studied at home as well before the COVID-19 outbreak [33].

Students in this study were generally concerned with three IEQ preferences: daylight, view of the outside, and temperature in their study places. Furthermore, they rated high importance levels for two psychosocial aspects: amenities and cleanliness. Previous studies found similar findings with regards to these preferences. For example, temperature [34] and daylight [3,25] (which is also known as natural lighting) were found to be important criteria by university students in informal study places. Due to the development of the information and communication technologies (ICT), amenities including PCs and laptops were considered important aspects by students in informal study places [3]. In addition, the presence of windows, which also refers to the view to the outside, was also preferred by university students for their study places at the library [35]. Cleanliness has also been affirmed to be an important aspect for students in informal study places such as university libraries [36] and university campus facilities [37]. While students in the current study rated a high importance score for the view to the outside, university students in another study rated a low importance score for the window view in the university library [36]. Yet, in another two studies [38,39], university students tended to choose their study places in the campus library that is close to the window. The latter outcome is similar to the findings of the current study: students generally preferred to have a view to the outside in their study places, whether at home or on campus. A previous study concluded that window views of the natural environment outside (e.g., green spaces) have a positive psychological impact on university students in terms of recovery from attentional fatigue [40]. During COVID-19 lockdown, the poor view to the outside negatively affected the mental health of university students while they were staying at home [41]. Hence, these preferences have a significant role in fulfilling students' preferences, as well as promoting their health.

In a study conducted by Zhang et al. [12], six profiles of primary school children based on their IEQ preferences and needs in classrooms were determined. While the most important three environmental aspects for these children were "hearing the teacher", "fresh air", and "air temperature", university students from the faculty of architecture were mainly concerned with visual aspects including "daylight" and "view to the outside". In both studies, one IEQ profile was concerned with light or visual aspects, although the primary school children were mainly concerned with artificial light and the university students with natural light. Furthermore, in both studies, one profile was concerned with all IEQ aspects and one profile was not concerned with any of the IEQ-aspects. The difference can be seen in the additional profiles concerned with sound, thermal and air quality aspects. These differences could be associated with the population. In other words, the respondents in this study were all bachelor students of the faculty of Architecture studying to become an architect, a profession in which visual aspects are important. On the contrary, primary school children comprise pupils that are yet to choose their profession or field of study.

4.2. Students' Profiles Based on the Overlap between the Two Cluster Models

While previous studies on office workers [26] and outpatient staff [27] conducted the cluster analysis separately based on IEQ preferences/perception and psychosocial preferences, the present study explored the overlap among the IEQ preferences and psychosocial preferences clusters. This resulted in several advantages. For instance, the number of variables that were significantly different among the profiles was higher than in the separated cluster models. In this study, health (e.g., depression) was not significantly different among both cluster models (IEQ and psychosocial preferences). However, this variable was found to be significantly different among the nine profiles resulting from the overlap. According to the study of office workers [26], the health variables such as anxiety were only significantly different among the IEQ clusters, while not found to be significantly different among the psychosocial preferences clusters. Similarly, in the study on outpatient staff [27], some variables only varied significantly among the IEQ clusters (e.g., preference for control of temperature), while it was not significantly different among the psychosocial clusters. Therefore, the overlap facilitates a more detailed understanding of the distinct characteristics among the profiles. IEQ, as well as psychosocial preferences, is also important to support comfort; combined profiles contribute to more realistic insights.

Students that had similar IEQ preferences within IEQC1, who were mainly concerned with all IEQ aspects, showed differences in various psychosocial aspects. The results showed that concerned perfectionist students were concerned with all IEQ preferences (specifically sounds and smells), as well as all the psychosocial preferences, except the presence and company of others. On the contrary, the concerned extroverts rated similar importance scores for IEQ preferences as the concerned perfectionists, but they were the most concerned cluster in terms of the presence and company of others in their study places. Additionally, the concerned extrovert students belonged to the profile that had the least personal control over IEQ aspects in their most used study places. Furthermore, both the concerned perfectionists and the concerned extroverts experienced the most recent positive events. In contrast, the concerned non-perfectionist students who were concerned with all IEQ aspects (specifically ventilation and fresh air) scored the highest negative affect compared to the other clusters. In addition, this profile scored the least importance for bonding or identifying with the place, ability to adapt or control the place, and size of the room. However, this profile rated the highest importance score for headphones as a significant IEQ-related item that helps them to study better.

Students within cluster IEQC2 that were generally concerned with daylight scored different in their psychosocial preferences. The overlap between IEQC2 and the three psychosocial preference clusters showed a significant difference in several characteristics. Visual concerned perfectionist students were the most concerned with daylight and the view to the outside. However, the students in this profile experienced the least positive events, while they scored the highest positive affect. Additionally, they were more concerned with a lamp on their desks to study better than the visual concerned-extrovert students, who were the least concerned with privacy in their study places. While the visual concerned perfectionists were the most concerned with amenities, the visual concerned non-perfectionists were the least concerned with amenities. In terms of health, the visual concerned non-perfectionists suffered the most from anxiety.

IEQC3 students were the least concerned with both artificial light as well as ventilation and fresh air. They showed different characteristics in the overlap between IEQC3 and the three psychosocial preferences profiles. While both the unconcerned extroverts and unconcerned non-perfectionists suffered the most from depression, the unconcerned extroverts suffered the least from depression. In addition, the unconcerned extroverts suffered the least from anxiety. It is interesting to note that all three profiles were the least concerned with having personal desk ventilation and fresh air in their study places. However, only the unconcerned non-perfectionists reported the highest dissatisfaction with air freshness in their most used study places, while both the unconcerned introverts and the unconcerned extroverts reported the least dissatisfaction with air freshness. Additionally, the unconcerned introverts were the least dissatisfied with the smell in their most used study places. While the unconcerned introverts rated the highest importance score for control of room temperature, both the unconcerned extroverts and the unconcerned non-perfectionists rated low scores for this IEQ-related item. Additionally, the unconcerned introverts had the highest control level over IEQ aspects in their study places (specifically, ventilation, shading from the sun, and lighting). Furthermore, while the unconcerned introverts rated the highest importance for cleanliness, both the unconcerned extroverts and the unconcerned non-perfectionists were the least concerned with cleanliness. Furthermore, although the unconcerned introverts were the most concerned with privacy in their study places, both the unconcerned extroverts and the unconcerned non-perfectionists were not very concerned with privacy.

4.3. Differences in Preferences of Profiles in Relation to Design Implementations

In general, there were significant differences among the profiles in terms of IEQ and psychosocial preferences, which means that generalizing the preferences of the whole study sample is not appropriate. For example, while privacy was indicated as highly preferred by students in their study places [34], this study reveals that there are two opposite profiles in terms of the importance of privacy in the study places: one of them (the unconcerned introvert) was highly concerned with privacy, while the other one (the visual concerned extrovert) rated privacy in their study places as less important. This result is similar to the findings of a previous study [23] in which the outcome showed that students' characteristics (birthplace and current educational level) have an influence on students' preferences, such as privacy and interaction.

While all the nine profiles were found to be statistically different in all IEQ preference aspects, their mean importance scores were higher than the mid-scale point (5.0). Nonetheless, there were profiles for which their mean importance scores for sounds from the outside (visual concerned extroverts and visual concerned non-perfectionists), sounds from the inside (visual concerned perfectionists), and smells (visual concerned non-perfectionists) were less than the mid-point scale (5.0). This means that there are profiles of students which are not highly concerned about sounds in their study places. The current study found that three profiles are not concerned about the sounds (from the outside or the inside) at their study places. This is in line with another study, conducted by Cunningham and Walton [20], which indicated that the preferences of university students to study in a quiet environment (e.g., university library) vary. In contrast, Beckers et al. [2] found that most university students prefer studying in quiet learning spaces.

The overlap among the IEQ and psychosocial models contributed to understanding in-depth students' profiles based on their different preferences in their study places. Different approaches can be applied to fulfil the different preferences of each profile. For instance, soundscape approach considers the individual's sound preferences in a certain environment. According to ISO 12913-1 [42], the soundscape is defined as: "acoustic environment as perceived or experienced and/or understood by a person or people in a context". This approach can understand the sound preference of each profile at study places. Additionally, the soundscape is mainly focused on using the sound as a resource that fulfils the sound preference rather than focusing on quiet spaces [43]. For instance, Shu and Ma [9] concluded that natural sound sources, such as birdsongs and stream sounds, had restorative effects on classroom children after performing a cognitive task, while the quiet condition did not show an effect. In addition, a study conducted by Topak and Yılmazer [44] found that students' sound preferences differ based on the context of the space, classroom or computer laboratory. They also found that natural sounds (e.g., birdsongs) were preferred by students to hear in their learning environments. Moreover, Xiao and Aletta [45] concluded that the soundscape approach could facilitate architects and interior designers to understand the students' experiences to provide high-quality sound environments or study places, such as libraries, by identifying different types of users. Accordingly, soundscape can be accounted for during the design process to understand the sound preferences of each profile of students at their study places. Another approach that can be applied to fulfil the different preferences is the application of customized (i.e., personalized) designs. These applications can match the preferences of each profile and could provide comfort for them, such as customized and personalized shading [46], ceiling fans [47], and heating [48], which allow users to have control over the surrounding environment based on their preferences.

4.4. Limitations

The sample of this study is limited to bachelor university students (specifically of the faculty of Architecture and the Built environment), whose mean age was 20 years old. The questionnaire was also completed at the time of the COVID-19 outbreak, which may have influenced students' preferences during this situation. It was sent to students during the fall and spring (October and March) seasons in the Netherlands, which could have had an impact on students' responses such as whether they scored high importance for both daylight and temperature. Furthermore, the IEQ and psychosocial preferences were asked within the context of studying at study places in general, while the learning activities/styles (e.g., individual, collaborative) were not investigated in the present study. The nine profiles in this study were identified based on the preferences (IEQ and psychosocial preferences) of bachelor students at the faculty of Architecture and the Built Environment in the Netherlands. Hence, further studies could validate these nine profiles with students from other faculties, as well as other universities with a different cultural background. As this study is based on a survey (questionnaire) with 451 students who were studying either at their homes or in educational buildings, space geometry and physical measurements of IEQ factors were not included in this study.

5. Conclusions

In conclusion, students with similar IEQ preferences have different psychosocial preferences, and vice versa. This was affirmed by determining nine profiles of university students based on the overlap between the IEQ and psychosocial preferences. These profiles showed significant differences among them in terms of various variables, including perception, lifestyle, health, and gender. It is worthwhile to note that the number of variables that were significantly different between the profiles is higher within the overlap between the IEQ and psychosocial preferences than clustering the students based on these preferences separately. The outcome of this study provides insight into different profiles of university students, each with their own preferences of study places. For instance, the concerned perfectionists are highly concerned with sounds (from the outside and inside) of their study places, while the visual concerned non-perfectionists are not highly concerned with sounds. These findings show the need for designing study places for more than one profile and not just for the "average" student.

The novelty of this study lies in the overlap of the IEQ and psychosocial preferences models that resulted in nine profiles, which showed significant differences among a number of variables. Therefore, it is recommended for future studies to determine the profiles of occupants (e.g., students, office workers, home occupants) within different scenarios (e.g., classrooms, study places, offices, homes) by the analysis of the overlap between the two sets of clusters.

Since this study is based on a survey in which physical measurements were not considered, it is suggested for future research to investigate these study places in-depth. For instance, field studies such as exploring the soundscapes of these study places can be investigated by measuring the sound pressure level (SPL), identifying sound sources as well as space geometry, and conducting in-depth interviews with the students from different profiles.

Author Contributions: Conceptualization, A.H., A.E. and P.M.B.; methodology, A.H., A.E., and P.M.B.; formal analysis, A.H.; investigation, A.H. and P.M.B.; data curation, A.H.; writing—original draft preparation, A.H., A.E. and P.M.B.; writing—review and editing, A.H., A.E. and P.M.B.; supervision, A.E. and P.M.B. All authors have read and agreed to the published version of the manuscript.

Funding: This research received no external funding.

Institutional Review Board Statement: The study was conducted according to the guidelines of the Declaration of Helsinki, and approved by the Institutional Review Board (HREC) of DELFT UNIVERSITY OF TECHNOLOGY (31 January 2022).

Informed Consent Statement: Informed consent was obtained from all students involved in the study.

Data Availability Statement: The data are not publicly available due to restrictions regarding the privacy of the participants of this study.

Acknowledgments: The authors would like to thank the first-year bachelors students (2021 and 2022) of the Faculty of Architecture and the Built Environment at TU Delft for completing this questionnaire.

Conflicts of Interest: The authors declare no conflict of interest.

Appendix A. Sections and Sub-Sections of the Questionnaire

Section	Sub-Section	Instrument
Personal information	Age Gender	-
Psycho-social aspects	Mood	OFFICAR, select one out of nine moods (e.g., cheerful) [31,49,50].
	Recently experienced positive events (e.g., wedding) and negative events (e.g., funeral).	OFFICAR, select either yes or no [31,49,50].
	Positive and Negative Affect Schedule (PANAS)	I-PANAS-SF, including five positive effects and five negative effects, on a scale 1 to 5 (1: never, 5: always) [51].
Mostly used study place	Study place type	Select one of the three options: home, educational building, or other.
Preferences	IEQ preferences *Please rate on a scale from 1 to 10, the importance of each of the following aspects for your study performance at your study place1: Not important at all; 10: Extremely important*-e.g., temperature".	Eight aspects on a scale 1 to 10 (1: not important at all, 10: extremely important) [26].
	Psychosocial preferences: *"Please rate on a scale from 1 to 10, the importance of each of the following aspects for your study performance at your study place1: Not important at all; 10: Extremely important*-e.g., privacy".	Nine aspects on a scale 1 to 10 (1: not important at all, 10: extremely important) [26].
	Importance of IEQ-related items: *"Please rate on a scale from 1 to 10, the importance of each of following the items that would help you to study better; 1: Not important at all; 10: Extremely important*-e.g., lamp on my desk".	Eleven aspects on a scale 1 to 10 (1: not important at all, 10: extremely important) [26].
Comfort	IEQ perception: *"On a scale of 1 to 7, how would you describe the general indoor comfort of your MOST used study place in the past 3 months? e.g., temperature satisfaction".*	Eighteen aspects on a scale 1 to 7 (1: dissatisfied, 7: satisfied [26,27,31].
	Control over IEQ factors: *"How much control do you personally have over the following aspects of your MOST used study place?-e.g., ventilation".*	Five aspects on a scale 1 to 7 (1: not at all, 7: full control) [26].
	Psychosocial perception: *How satisfied are you with the following in your MOST used study place-e.g., amount of privacy".*	Five aspects on a scale 1 to 7 (1: unsatisfactory, 7: satisfactory) [26].
Lifestyle	Physical activity	OFFICAR, select either yes or no [31].
	Smoking	OFFICAR, select one out of four options (e.g., no never, yes former, yes incidentally, yes daily) [31].
	Alcohol	OFFICAR, select one out of three options (e.g., yes daily, yes occasionally, no) [31].
Health and medical history	Suffering from diseases: *"Have you ever been told by your doctor that you are suffering from: e.g., asthma"*	OFFICAR, includes eighteen diseases, each disease is rated one out of three options: never, yes in the last 12 months, yes but not in the last 12 months [31].

Appendix B. IEQ Preferences Clusters

	IEQC1	IEQC2	IEQC3	*p*-Value
n (%within the total sample)	159 (35.5)	149 (33.3)	140 (31.3)	-
Age				0.325
Mean (SD)	19.6 (1.7)	19.7 (1.3)	19.9 (1.7)	-
Maximum	31	26	29	-
Minimum	17	18	18	-
Mood-n (%within cluster level)				0.375
Cheerful	12 (7.5)	14 (9.4)	10 (7.1)	-
Relaxed	42 (26.4)	43 (28.9)	39 (27.9)	-
Calm	31 (19.5)	20 (13.4)	15 (10.7)	-
Neutral	31 (19.5)	28 (18.8)	33 (23.6)	-
Sad	10 (6.3)	8 (5.4)	17 (12.1)	-
Bored	21 (13.2)	23 (15.4)	17 (12.1)	-
Recently experienced events-n (%within cluster level)				
Positive events	45 (34.0)	40 (26.8)	36 (25.7)	0.226
Negative events	56 (35.2)	44 (29.5)	50 (35.7)	0.455
Lifestyle-n (%)				
Smoking	42 (26.4)	46 (30.9)	46 (32.8)	0.380
Alcohol	133 (83.7)	129 (86.6)	119 (85.0)	0.502
Physical activity	146 (91.8)	138 (92.6)	120 (85.7)	0.098
PANAS-Mean (SD)				
Positive affect	17.5 (2.6)	17.6 (2.5)	17.1 (2.7)	0.122
Negative affect	11.8 (3.0)	11.3 (2.9)	11.4 (3.0)	0.617
Health-n (%within cluster level)				
Hay fever	35 (22.2)	30 (20.1)	33 (23.6)	0.205
Rhinitis	52 (32.9)	51 (34.2)	43 (30.2)	0.074
Eczema	18 (6.3)	25 (16.8)	22 (15.7)	0.517
Other skin conditions	12 (7.6)	15 (10.1)	18 (12.9)	0.590
Migraine	24 (15.2)	23 (15.5)	21 (15.0)	0.314
Depression	31 (19.5)	29 (19.5)	32 (22.9)	0.477
Anxiety	44 (27.8)	31 (20.8)	30 (21.4)	0.126
Mental health problems	32 (20.3)	23 (15.5)	23 (16.4)	0.677
IEQ perception of study-n (%within cluster level)				
Temperature in general dissatisfaction	31 (19.5)	27 (18.1)	38 (28.4)	0.084
Temperature not stable	42 (26.4)	36 (24.2)	43 (30.7)	0.588
Dissatisfied with air freshness	93 (58.5)	88 (59.1)	84 (60.0)	0.730
Dissatisfied with air smell	72 (45.3)	71 (47.7)	58 (41.4)	0.393
Air quality in general dissatisfaction	23 (14.5)	18 (12.1)	18 (12.9)	0.521
Daylight dissatisfaction	22 (13.8)	15 (10.1)	15 (10.7)	0.928
Reflection from the sun dissatisfaction	27 (17.0)	25 (16.8)	12 (8.6)	0.167
Artificial light dissatisfaction	24 (15.1)	33 (22.1)	25 (17.9)	0.182
Lighting in general dissatisfaction	13 (8.2)	16 (10.7)	9 (6.4)	0.867
Noise from outside dissatisfaction	36 (22.6)	37 (24.8)	36 (25.7)	0.391
Noise from installations dissatisfaction	19 (11.9)	23 (15.4)	20 (14.3)	0.907
Noise other than installations dissatisfaction	38 (23.9)	29 (19.5)	29 (20.7)	0.745
Noise in general dissatisfaction	28 (17.6)	30 (20.1)	20 (14.3)	0.921
Vibration dissatisfaction	21 (13.2)	16 (10.7)	16 (11.4)	0.836
Psychosocial perception of study place- n (%within cluster level)				
Amount of privacy dissatisfaction	14 (8.8)	10 (6.7)	12 (8.6)	0.754
Layout dissatisfaction	12 (7.5)	10 (6.7)	6 (4.7)	0.498
Decoration dissatisfaction	8 (5.0)	10 (6.7)	13 (9.3)	0.337
Cleanliness dissatisfaction	22 (13.8)	21 (14.1)	19 (13.6)	0.993
View to the outside dissatisfaction	16 (10.1)	20 (13.4)	18 (12.9)	0.611
Psychosocial preferences-mean (SD)				
Storage	6.5 (2.3)	6.2 (2.3)	6.0 (2.3)	0.503
Amenities	8.3 (1.5)	8.0 (1.5)	7.8 (1.4)	0.064
Presence and company of others	5.8 (2.5)	5.1 (2.6)	5.0 (2.6)	0.308
Size of the room	5.9 (1.9)	5.4 (2.0)	5.2 (2.0)	0.133
Bonding or identifying with the place	5.8 (2.3)	5.3 (2.6)	5.1 (2.4)	0.090
Ability to adapt or control the place	6.2 (2.1)	5.9 (2.1)	5.1 (2.6)	0.249
Importance of IEQ-related aspects-mean (SD)				
Chair seat heating	4.3 (2.8)	4.0 (2.8)	3.4 (2.5)	0.106
Chair backrest eating	4.4 (2.9)	3.9 (3.0)	3.5 (2.8)	0.095
Heating on my desk	3.9 (2.7)	3.6 (2.7)	3.2 (2.5)	0.141
Presence of plants	6.1 (2.5)	5.9 (2.5)	5.1 (2.7)	0.62
Personal control over the most used study place-mean (SD)				
Temperature	4.2 (1.9)	4.7 (1.7)	4.4 (1.8)	0.206
Ventilation	4.6 (1.8)	4.8 (1.8)	4.6 (1.8)	0.311
Shading from the sun	4.7 (2.2)	4.5 (2.1)	4.8 (2.0)	0.772
Lighting	4.8 (2.2)	5.1 (1.9)	5.1 (1.9)	0.377
Noise	2.9 (1.4)	3.2 (1.6)	2.9 (1.5)	0.168

Appendix C. Psychosocial Preferences Clusters

	PSC1	PSC2	PSC3	*p*-Value
n (%within total sample)	110 (25.0)	186 (42.3)	144 (32.7)	-
Age				0.084
Mean (SD)	19.7 (1.5)	19.8 (1.8)	19.7 (1.3)	-
Maximum	29	31	26	-
Minimum	17	17	18	-
Gender -n (%within cluster level)				0.776
Male	43 (39.1)	68 (36.6)	58 (40.3)	-
Female	67 (60.9)	117 (62.9)	85 (59.0)	-
Mood-n (%)				0.262
Cheerful	9 (8.1)	20 (10.8)	7 (4.9)	-
Relaxed	26 (23.7)	48 (25.8)	50 (34.7)	-
Calm	18 (16.4)	29 (15.6)	19 (13.2)	-
Neutral	23 (20.9)	42 (22.6)	23 (16.0)	-
Sad	10 (9.1)	14 (7.5)	9 (6.3)	-
Bored	20 (18.2)	18 (9.7)	21 (14.6)	-
Recently experienced events-n (%within cluster level)				
Positive events	35 (31.8)	54 (29.0)	37 (25.7)	0.557
Negative events	34 (30.9)	64 (34.4)	47 (32.6)	0.822
Lifestyle-n (%)				
Physical activity	97 (88.2)	170 (91.4)	129 (89.6)	0.658
PANAS-Mean (SD)				
Positive affect	18.0 (2.5)	17.1 (2.5)	17.3 (2.6)	0.168
Negative affect	11.8 (2.9)	11.6 (2.9)	11.2 (3.1)	0.301
Health-n (%within cluster level)				
Asthma	4 (3.6)	9 (8.0)	4 (2.8)	0.204
Hay fever	21 (19.1)	45 (24.2)	31 (21.5)	0.796
Rhinitis	26 (23.6)	70 (37.7)	48 (33.3)	0.194
Eczema	16 (7.3)	26 (14.0)	23 (16.0)	0.984
Other skin conditions	6 (5.4)	22 (11.8)	17 (11.8)	0.262
Migraine	16 (14.5)	30 (16.2)	22 (15.3)	0.697
Depression	23 (20.9)	33 (17.8)	35 (24.3)	0.923
Anxiety	30 (27.3)	42 (22.6)	35 (24.3)	0.181
Mental health problems	21 (19.1)	23 (16.7)	24 (16.7)	0.701
IEQ perception of study-n (%within cluster level)				
Temperature in general dissatisfaction	23 (20.9)	43 (23.1)	29 (20.1)	0.832
Temperature not stable	32 (29.1)	48 (25.8)	41 (28.5)	0.744
Dissatisfied with air smell	45 (40.9)	94 (50.5)	63 (43.8)	0.261
Air quality in general dissatisfaction	18 (16.4)	20 (10.8)	21 (14.6)	0.324
Daylight dissatisfaction	9 (8.2)	24 (12.9)	18 (12.5)	0.434
Reflection from the sun dissatisfaction	11 (10.0)	28 (15.1)	26 (18.1)	0.188
Artificial light dissatisfaction	14 (12.7)	37 (19.9)	29 (20.1)	0.227
Lighting in general dissatisfaction	7 (6.4)	18 (9.7)	12 (8.3)	0.623
Noise from outside dissatisfaction	30 (27.3)	50 (26.9)	30 (20.8)	0.399
Noise from installations dissatisfaction	16 (14.5)	24 (12.9)	20 (13.9)	0.902
Noise other than installations dissatisfaction	26 (23.6)	39 (21.0)	32 (22.2)	0.842
Noise in general dissatisfaction	20 (18.2)	32 (17.2)	26 (18.1)	0.953
Vibration dissatisfaction	16 (14.5)	24 (12.9)	13 (9.0)	0.385
Psychosocial perception of study place- n (%within cluster level)				
Cleanliness dissatisfaction	10 (9.1)	28 (15.1)	24 (16.7)	0.193
View to the outside dissatisfaction	11 (10.0)	25 (13.4)	18 (12.5)	0.676
IEQ preferences-mean (SD)				
Ventilation and fresh air	8.2 (1.3)	7.8 (1.3)	7.7 (1.7)	0.065
View to the outside	8.5 (1.4)	8.2 (1.7)	7.9 (2.0)	0.075
Sounds from the inside	7.1 (2.2)	6.7 (2.0)	6.7 (2.4)	0.154
Importance of IEQ-related aspects-mean (SD)				
Personal desk ventilation and fresh air	7.6 (2.0)	6.7 (2.3)	7.0 (2.2)	0.138
Headphones	7.4 (2.4)	7.3 (2.3)	6.9 (2.7)	0.734
Personal control over the most used study place-mean (SD)				
Temperature	4.8 (1.6)	4.2 (1.9)	4.4 (1.8)	0.087
Shading from the sun	5.2 (2.0)	4.5 (2.1)	4.5 (2.1)	0.051
Lighting	5.7 (1.7)	4.7 (2.1)	5.0 (2.0)	0.065
Noise	3.4 (1.6)	2.8 (1.4)	2.9 (1.4)	0.069

Appendix D. Descriptive of the Overlap Nine Profiles

	IEQC1-PSC1	IEQC1-PSC2	IEQC1-PSC3	IEQC2-PSC1	IEQC2-PSC2	IEQC2-PSC3	IEQC3-PSC1	IEQC3-PSC2	IEQC3-PSC3	p-Value
n (%within the total sample)	59 (13.5)	63 (14.4)	35 (8.0)	34 (7.8)	64 (14.6)	48 (11.0)	17 (3.9)	59 (13.5)	59 (13.5)	-
Mood-N (%)										
Cheerful *	5 (8.5)	5 (7.9)	2 (5.7)	3 (8.8)	9 (14.1)	2 (4.2)	1 (5.9)	6 (10.2)	3 (5.1)	-
Relaxed *	14 (23.7)	12 (19.0)	16 (45.7)	4 (11.8)	23 (35.9)	8 (16.7)	4 (23.5)	13 (22.0)	9 (15.3)	-
Calm *	11 (18.6)	14 (22.2)	6 (17.1)	6 (17.6)	7 (10.9)	7 (14.6)	1 (5.9)	8 (13.6)	6 (10.2)	-
Neutral *	13 (22.0)	14 (22.2)	3 (8.6)	7 (20.6)	11 (17.2)	8 (16.7)	3 (17.6)	17 (28.8)	12 (20.3)	-
Sad *	3 (5.1)	4 (6.3)	1 (8.6)	1 (2.9)	2 (3.1)	4 (8.3)	2 (11.8)	8 (13.6)	5 (8.5)	-
Bored *	9 (15.3)	8 (12.7)	3 (8.6)	6 (17.6)	7 (10.9)	10 (20.8)	5 (29.4)	3 (5.1)	8 (13.6)	-
Recently experienced events-n (%within profile level)										
Negative events	20 (33.9)	22 (34.9)	13 (37.1)	8 (23.5)	20 (31.3)	15 (31.3)	6 (35.3)	22 (37.3)	19 (32.2)	0.054
Lifestyle-n (%within profile level)										
Smoking *	8 (13.6)	22 (34.9)	12 (34.3)	10 (29.4)	19 (29.7)	15 (31.2)	3 (17.7)	15 (25.4)	21 (35.6)	-
Study place-N (%within profile level)										
Home	49 (83.1)	31 (49.2)	22 (62.9)	33 (97.1)	43 (67.2)	40 (83.3)	16 (94.1)	43 (72.9)	46 (78.0)	-
Educational building *	10 (16.9)	31 (49.2)	13 (37.1)	1 (2.9)	21 (32.8)	7 (14.6)	1 (5.9)	16 (27.1)	13 (22.0)	-
Health-n (%within profile level)										
Hay fever *	9 (15.3)	18 (28.6)	8 (22.9)	8 (23.5)	11 (17.2)	10 (20.9)	4 (23.5)	16 (27.1)	13 (22.0)	-
Rhinitis *	14 (23.7)	28 (44.5)	10 (28.6)	8 (23.5)	24 (37.5)	17 (35.4)	4 (23.5)	18 (30.5)	21 (35.6)	-
Eczema *	6 (10.2)	7 (11.1)	5 (14.3)	7 (20.5)	11 (17.2)	7 (14.6)	3 (17.6)	8 (13.6)	11 (18.6)	-
Other skin conditions *	2 (3.4)	8 (12.7)	2 (5.7)	3 (8.8)	6 (9.4)	5 (10.4)	1 (5.9)	7 (11.9)	10 (17.0)	-
Migraine *	6 (10.2)	11 (17.5)	7 (20.0)	8 (23.5)	10 (15.6)	5 (10.4)	2 (11.8)	9 (15.3)	10 (16.9)	-
Mental health problems *	11 (18.6)	14 (22.2)	6 (17.1)	8 (23.5)	8 (12.5)	7 (14.6)	2 (11.8)	9 (15.3)	11 (18.6)	-
IEQ perception of study place-n (% level)										
Temperature in general dissatisfaction *	13 (22.0)	13 (20.6)	5 (14.3)	4 (11.8)	14 (21.9)	8 (16.7)	6 (35.3)	16 (27.1)	16 (27.1)	-
Temperature not stable	14 (23.7)	20 (31.7)	8 (22.9)	11 (32.4)	12 (18.8)	13 (27.1)	7 (41.2)	16 (27.1)	19 (32.2)	0.093
Air quality in general dissatisfaction *	8 (13.6)	10 (15.9)	5 (14.3)	5 (14.7)	5 (7.8)	8 (16.7)	5 (29.4)	5 (8.5)	8 (13.6)	-
Daylight dissatisfaction *	6 (10.2)	11 (17.5)	5 (14.3)	2 (5.9)	8 (12.5)	4 (8.3)	1 (5.9)	5 (8.5)	9 (15.3)	-
Reflection from the sun dissatisfaction *	6 (10.2)	15 (23.8)	6 (17.1)	4 (11.8)	11 (17.2)	10 (20.8)	1 (5.9)	2 (3.4)	9 (15.3)	-
Artificial light dissatisfaction *	8 (13.6)	13 (20.6)	3 (8.6)	6 (17.6)	12 (18.8)	14 (29.2)	-	12 (20.3)	12 (20.3)	-
Lighting in general dissatisfaction *	3 (5.1)	8 (12.7)	2 (5.7)	4 (11.8)	9 (14.1)	2 (4.2)	-	1 (1.7)	8 (13.6)	-
Noise from outside dissatisfaction *	17 (28.8)	15 (23.8)	4 (11.4)	9 (26.5)	18 (28.1)	10 (20.8)	4 (23.5)	17 (28.8)	15 (25.4)	-
Noise from installations dissatisfaction *	8 (13.6)	7 (11.1)	4 (11.4)	4 (11.8)	8 (12.5)	10 (20.8)	4 (23.5)	9 (15.3)	6 (10.2)	-
Noise other than installations dissatisfaction *	14 (23.7)	14 (22.2)	10 (28.6)	8 (23.5)	11 (17.2)	10 (20.8)	4 (23.5)	14 (23.7)	11 (18.6)	-
Noise in general dissatisfaction *	10 (16.9)	10 (15.9)	8 (22.9)	7 (20.6)	14 (21.9)	9 (18.8)	3 (17.6)	8 (13.6)	9 (15.3)	-
Vibration dissatisfaction *	7 (11.9)	10 (15.9)	4 (11.3)	6 (17.6)	7 (10.9)	3 (6.3)	3 (17.6)	7 (11.9)	6 (10.2)	-
Psychosocial perception of study place- n (%within profile level)										
Amount of privacy dissatisfaction *	1 (1.7)	12 (19.0)	1 (2.9)	1 (2.9)	2 (3.1)	7 (14.6)	-	6 (10.2)	6 (10.2)	-
Layout dissatisfaction *	3 (5.1)	7 (11.1)	2 (5.7)	-	5 (7.8)	5 (10.4)	-	2 (3.4)	4 (6.8)	-
Decoration dissatisfaction *	2 (3.4)	6 (9.5)	-	-	6 (9.4)	4 (8.3)	-	3 (5.1)	10 (16.9)	-
Cleanliness dissatisfaction *	6 (10.2)	12 (19.0)	4 (11.4)	3 (8.8)	9 (14.1)	9 (18.8)	1 (5.9)	7 (11.9)	11 (18.6)	-
View to the outside dissatisfaction *	4 (6.8)	10 (15.9)	2 (5.7)	3 (8.8)	11 (17.2)	6 (12.5)	4 (23.5)	4 (6.8)	10 (16.9)	-

* N < 5, thus chi-squared test not performed.

References

1. Ramu, V.; Taib, N.; Massoomeh, H.M. Informal academic learning space preferences of tertiary education learners. *J. Facil. Manag.* **2021**. [CrossRef]
2. Beckers, R.; Van der Voordt, T.; Dewulf, G. Learning space preferences of higher education students. *Build. Environ.* **2016**, *104*, 243–252. [CrossRef]
3. Harrop, D.; Turpin, B. A Study Exploring Learners' Informal Learning Space Behaviors, Attitudes, and Preferences. *New Rev. Acad. Librariansh.* **2013**, *19*, 58–77. [CrossRef]
4. Beckers, R.; Van der Voordt, T.; Dewulf, G. Why do they study there? Diary research into students' learning space choices in higher education. *High Educ. Res. Dev.* **2016**, *35*, 142–157. [CrossRef]
5. Cox, A.M. Space and embodiment in informal learning. *High Educ.* **2018**, *75*, 1077–1090. [CrossRef]
6. Wang, S.; Han, C. The Influence of Learning Styles on Perception and Preference of Learning Spaces in the University Campus. *Buildings* **2021**, *11*, 572. [CrossRef]
7. Wilson, H.K.; Cotgrave, A. Factors that influence students' satisfaction with their physical learning environments. *Struct. Surv.* **2016**, *34*, 256–275. [CrossRef]
8. Bluyssen, P.M. Patterns and Profiles for understanding the indoor environment and its occupants. In Proceedings of the 2022: CLIMA 2022 The 14th REHVA HVAC World Congress, Rotterdam, The Netherlands, 22–25 May 2022; pp. 1–7. [CrossRef]
9. Shu, S.; Ma, H. Restorative Effects of Classroom Soundscapes on Children's Cognitive Performance. *Int. J. Environ. Res. Public Health* **2019**, *16*, 293. [CrossRef] [PubMed]
10. Ricciardi, P.; Buratti, C. Environmental quality of university classrooms: Subjective and objective evaluation of the thermal, acoustic, and lighting comfort conditions. *Build. Environ.* **2018**, *127*, 23–36. [CrossRef]
11. Corgnati, S.P.; Filippi, M.; Viazzo, S. Perception of the thermal environment in high school and university classrooms: Subjective preferences and thermal comfort. *Build. Environ.* **2007**, *42*, 951–959. [CrossRef]
12. Zhang, D.; Ortiz, M.A.; Bluyssen, P.M. Clustering of Dutch school children based on their preferences and needs of the IEQ in classrooms. *Build. Environ.* **2019**, *147*, 258–266. [CrossRef]
13. Nico, M.A.; Liuzzi, S.; Stefanizzi, P. Evaluation of thermal comfort in university classrooms through objective approach and subjective preference analysis. *Appl. Ergon.* **2015**, *48*, 111–120. [CrossRef] [PubMed]
14. Liu, Q.; Huang, Z.; Li, Z.; Pointer, M.R.; Zhang, G.; Liu, Z. A Field Study of the Impact of Indoor Lighting on Visual Perception and Cognitive Performance in Classroom. *Appl. Sci.* **2020**, *10*, 7436. [CrossRef]
15. Teli, D.; Jentsch, M.F.; James, P.A.B. Naturally ventilated classrooms: An assessment of existing comfort models for predicting the thermal sensation and preference of primary school children. *Energy Build.* **2012**, *53*, 166–182. [CrossRef]
16. Cankaya, S.; Yilmazer, S. The effect of soundscape on the students' perception in the high school environment. In Proceedings of the Inter-Noise 2016 45th International Congress and Exposition on Noise Control Engineering: Towards a Quiter Future, Hamburg, Germany, 21–24 August 2016; pp. 4809–4816.
17. Mishra, A.K.; Derks, M.T.H.; Kooi, L.; Loomans, M.G.L.C.; Kort, H.S.M. Analysing thermal comfort perception of students through the class hour, during heating season, in a university classroom. *Build. Environ.* **2017**, *125*, 464–474. [CrossRef]
18. De Giuli, V.; Da Pos, O.; De Carli, M. Indoor environmental quality and pupil perception in Italian primary schools. *Build. Environ.* **2012**, *56*, 335–345. [CrossRef]
19. Peng, L.; Jin, S.; Deng, Y.; Gong, Y. Students ' Perceptions of Active Learning Classrooms from an Informal Learning Perspective: Building a Full-Time Sustainable Learning Environment in Higher Education. *Sustainability* **2022**, *14*, 8578. [CrossRef]
20. Cunningham, M.; Walton, G. Informal learning spaces (ILS) in university libraries and their campuses: A Loughborough University case study. *New Libr. World* **2016**, *117*, 49–62. [CrossRef]
21. Roetzel, A.; DeKay, M.; Nakai Kidd, A.; Klas, A.; Sadick, A.M.; Whittem, V.; Zinkiewicz, L. Architectural, indoor environmental, personal and cultural influences on students' selection of a preferred place to study. *Archit. Sci. Rev.* **2020**, *63*, 275–291. [CrossRef]
22. Braat-Eggen, E.; Reinten, J.; Hornikx, M.; Kohlrausch, A. The Effect of Background Noise on a "Studying for an Exam" Task in an Open-Plan Study Environment: A Laboratory Study. *Front. Built. Environ.* **2021**, *7*, 1–12. [CrossRef]
23. Liu, C.; Luther, M. Privacy and interaction preferences of students in informal learning spaces on university campus. *Facilities* **2022**. *ahead-of-print*. [CrossRef]
24. Bluyssen, P.M. Towards an integrated analysis of the indoor environmental factors and its effects on occupants. *Intell. Build. Int.* **2020**, *12*, 199–207. [CrossRef]
25. Altomonte, S.; Allen, J.; Bluyssen, P.M.; Brager, G.; Heschong, L.; Loder, A.; Schiavon, S.; Veitch, J.A.; Wang, L.; Wargocki, P. Ten questions concerning well-being in the built environment. *Build. Environ.* **2020**, *180*, 106949. [CrossRef]
26. Ortiz, M.A.; Bluyssen, P.M. Profiling office workers based on their self-reported preferences of indoor environmental quality and psychosocial comfort at their workplace during COVID-19. *Build. Environ.* **2022**, *211*, 108742. [CrossRef] [PubMed]
27. Eijkelenboom, A.; Bluyssen, P.M. Profiling outpatient staff based on their self-reported comfort and preferences of indoor environmental quality and social comfort in six hospitals. *Build. Environ.* **2020**, *184*, 107220. [CrossRef]
28. Ortiz, M.A.; Bluyssen, P.M. Proof-of-concept of a questionnaire to understand occupants ' comfort and energy behaviours: First results on home occupant archetypes. *Build. Environ.* **2018**, *134*, 47–58. [CrossRef]
29. Ortiz, M.A.; Bluyssen, P.M. Developing home occupant archetypes: First results of mixed-methods study to understand occupant comfort behaviours and energy use in homes. *Build. Environ.* **2019**, *163*, 106331. [CrossRef]

30. Kim, D.H.; Bluyssen, P.M. Clustering of office workers from the OFFICAIR study in The Netherlands based on their self-reported health and comfort. *Build. Environ.* **2020**, *176*, 106860. [CrossRef]
31. Bluyssen, P.M.; Roda, C.; Mandin, C.; Fossati, S.; Carrer, P.; de Kluizenaar, Y.; Mihucz, V.G.; de Oliveira Fernandes, E.; Bartzis, J. Self-reported health and comfort in ' modern ' office buildings: First results from the European OFFICAIR study. *Indoor Air* **2016**, *26*, 298–317. [CrossRef]
32. Dietrich, T.; Rundle-Thiele, S.; Kubacki, K. *Segmentation in Social Marketing*; Springer Nature: Singapore, 2017.
33. Vanichvatana, S.; Rd, R.; Mak, H. Who uses home as informal learning spaces: A Bangkok private university case study. *World J. Educ. Technol. Curr. Issues* **2020**, *12*, 37–47. [CrossRef]
34. Wu, X.; Kou, Z.; Oldfield, P.; Heath, T.; Borsi, K. Informal Learning Spaces in Higher Education: Student Preferences and Activities. *Buildings* **2021**, *11*, 252. [CrossRef]
35. Kim, Y.; Yang, E. Academic library spaces and student activities during the COVID-19 pandemic. *J. Acad. Librariansh.* **2022**, *48*, 102529. [CrossRef] [PubMed]
36. Hyun, S.; Wan, T. What Matters for Students' Use of Physical Library Space? *J. Acad. Librariansh.* **2015**, *41*, 274–279. [CrossRef]
37. Kärnä, S.; Julin, P.; Nenonen, S. User satisfaction on a university campus by students and staff. *Intell. Build. Int.* **2013**, *5*, 69–82. [CrossRef]
38. Declercq, C.P.; Cranz, G. Moving Beyond Seating-centered Learning Environments: Opportunities and Challenges Identified in a POE of a Campus Library. *J. Acad. Librariansh.* **2014**, *40*, 574–584. [CrossRef]
39. Webb, K.; Schaller, M.A.; Hunley, S.A. Measuring Library Space Use and Preferences: Charting a Path Toward Increased Engagement. *Portal. Libr. Acad.* **2008**, *8*, 407–422. [CrossRef]
40. Felsten, G. Where to take a study break on the college campus: An attention restoration theory perspective. *J. Environ. Psychol.* **2009**, *29*, 160–167. [CrossRef]
41. Amerio, A.; Brambilla, A.; Morganti, A.; Aguglia, A.; Bianchi, D.; Santi, F.; Costantini, L.; Odone, A.; Costanza, A.; Signorelli, C.; et al. COVID-19 Lockdown: Housing Built Environment's Effects on Mental Health. *Int. J. Environ. Res. Public Health* **2020**, *17*, 5973. [CrossRef]
42. *ISO 12913-1*; Acoustics—Soundscape—Part 1: Definition and Conceptual Framework. ISO: Geneva, Switzerland, 2014.
43. Kang, J.; Aletta, F.; Gjestland, T.T.; Brown, L.A.; Botteldooren, D.; Schulte-Fortkamp, B.; Lercher, P.; van Kamp, I.; Genuit, K.; Fiebig, A.; et al. Ten questions on the soundscapes of the built environment. *Build. Environ.* **2016**, *108*, 284–294. [CrossRef]
44. Topak, S.C.; Yılmazer, S. A comparative study on indoor soundscape assessment via a mixed method: A case of the high school environment. *Appl. Acoust.* **2022**, *189*, 108554. [CrossRef]
45. Xiao, J.; Aletta, F. A soundscape approach to exploring design strategies for acoustic comfort in modern public libraries: A case study of the Library of Birmingham. *Noise Mapp.* **2016**, *3*, 264–273. [CrossRef]
46. Li, Z.; Zhu, H.; Dong, B.; Xu, X. Development of a systematic procedure to establish customized shading behavior identification model. *Energy Build.* **2021**, *239*, 110793. [CrossRef]
47. Rissetto, R.; Schweiker, M.; Wagner, A. Personalized ceiling fans: Effects of air motion, air direction and personal control on thermal comfort. *Energy Build.* **2021**, *235*, 110721. [CrossRef]
48. Veselý, M.; Molenaar, P.; Vos, M.; Li, R.; Zeiler, W. Personalized heating e Comparison of heaters and control modes. *Build. Environ.* **2017**, *112*, 223–232. [CrossRef]
49. Eijkelenboom, A.M.; Kim, D.H.; Bluyssen, P.M. First results of self-reported health and comfort of staff in outpatient areas of hospitals in the Netherlands. *Build. Environ.* **2020**, *177*, 106871. [CrossRef]
50. Bluyssen, P.M.; Ortiz-sanchez, M.; Roda, C. Self-reported rhinitis of students from different universities in the Netherlands and its association with their home environment. *Build. Environ.* **2016**, *110*, 36–45. [CrossRef]
51. Thompson, E.R. Development and validation of an internationally reliable short-form of the positive and negative affect schedule (PANAS). *J. Cross Cult. Psychol.* **2007**, *38*, 227–242. [CrossRef]

 *buildings*

Article

Effect of Interior Space and Window Geometry on Daylighting Performance for Terrace Classrooms of Universities in Severe Cold Regions: A Case Study of Shenyang, China

Yingjie Jia, Zheming Liu *, Yaoxuan Fang, Huiying Zhang, Caiyi Zhao and Xuqiang Cai

Jangho Architecture College, Northeastern University, Shenyang 110819, China
* Correspondence: liuzheming@mail.neu.edu.cn

Abstract: Good daylighting performance positively affects students' physical and mental health, learning efficiency, and the building's energy-saving capability. Due to the terrace classroom having ample space, large capacity, the ability to avoid obstructing sight, and the ability to meet various use needs, it is the most important place in university buildings. However, research on the daylighting performance of university terrace classrooms is limited, leading to a lack of quantitative guidance in early design stages. This study aims to explore the effects of interior space and window geometry of terrace classrooms in universities in severe cold regions on daylighting performance. This research took Shenyang as an example; spatial daylight autonomy ($sDA_{300,50\%}$) and useful daylight illuminance ($UDI_{100-2000}$) were selected as daylighting performance evaluation indices. Based on the Grasshopper parametric platform, the simulation was carried out using Ladybug and Honeybee plugins. Correlation and regression analyses revealed the relationship between interior space and window geometry parameters and the evaluation indices. The results showed the following: window-to-floor ratio (WFR), classroom height (H_{tc}), window height (H_w), window-to-wall ratio (WWR), classroom width (W_{tc}), and window width (W_w) have positive effects on improving the daylight sufficiency of the terrace classrooms facing each orientation, and the degree of the effect decreases in order. To ensure the overall daylighting performance, the W_{tc} can be maximized. The width of walls between windows for south-facing and west-facing classrooms should be 0.9 m. The WWR and WFR for south-facing classrooms should be 0.3–0.5 and 0.11–0.14, respectively. The WWR and WFR for north-facing classrooms should be 0.6–0.7 and 0.14–0.20, respectively. Prediction models are established for the $sDA_{300,50\%}$ and $UDI_{100-2000}$ of the terrace classrooms facing each orientation.

Keywords: daylighting performance; terrace classrooms; interior space geometry; window geometry; severe cold regions

Citation: Jia, Y.; Liu, Z.; Fang, Y.; Zhang, H.; Zhao, C.; Cai, X. Effect of Interior Space and Window Geometry on Daylighting Performance for Terrace Classrooms of Universities in Severe Cold Regions: A Case Study of Shenyang, China. *Buildings* **2023**, *13*, 603. https://doi.org/10.3390/buildings13030603

Academic Editor: Alessandro Cannavale

Received: 27 January 2023
Revised: 22 February 2023
Accepted: 23 February 2023
Published: 24 February 2023

1. Introduction

Daylighting for buildings is natural and does not exhibit stroboscopic effects while simultaneously being rich in changes, and it exhibits unique and innate advantages that are incomparable to other artificial light sources. Appropriate interior space and window geometry can effectively play the positive role of natural light and provide a comfortable light environment. This is beneficial to the physical and mental health of users. Furthermore, it can improve learning and work efficiency and can also reduce energy consumption [1–3].

Currently, scholars have performed substantial research on the influence of the interior space and window geometry of multi-functional buildings in different areas on daylighting performance [4–6]. Ghisi and Tinker [7] and Susorova et al. [8] studied the impact of window size, orientation, and room geometry on the daylighting performance of office buildings in different climatic zones, such as desert climate, subtropical climate, Mediterranean climate, and temperate broad-leaved forest climate, which aimed to optimize daylighting quality and energy-saving levels. Lee et al.'s research explored the best window designs of office buildings in five typical climatic zones in Asia (tropical monsoon climate, subtropical monsoon climate, subtropical marine monsoon climate, temperate monsoon climate, and temperate marine monsoon climate), by simulating the influence of window-to-wall ratios and window orientations on natural light. By conducting an analysis on solar radiation and building energy consumption, design guidelines for energy-saving lighting windows were proposed [9]. With the help of field measurements and numerical simulation methods, Omar et al. explored the promotion effect of window geometry, room depth, and other factors on natural light in a university library in the Mediterranean climate zone to enhance daylighting performance [10]. Fan et al. selected a gymnasium in Xiong'an, Beijing, located in the warm temperate continental monsoon climate zone as the research object. They proposed an optimization method for the window design of the stadium facade based on the average illumination, average solar radiation accumulation per hour, and glare index when the stadium was opened in summer [11]. Huang et al. carried out research on the daylighting performance of painting and calligraphy exhibition halls in museums and took daylight factor (DF), DF uniformity, and discomfort glare index (DGI) as evaluation indicators, by comparing and analyzing the daylighting performance of four lighting modes—low side window, high side window, flat skylight, and sawtooth skylight. Huang et al. proposed the window geometry optimization design method [12]. In addition, many studies evaluated the impact of different interior spaces and window geometries on the natural daylighting performance and visual comfort of residential buildings [13–18].

As the central place to carry out educational activities, educational buildings' indoor physical environment quality is closely related to students' physical and mental health, growth, and living quality [19–21]. In addition, a high-quality indoor environment can effectively improve students' learning efficiency and performance, as well as the interaction and social state between teachers and students, which is conducive to students' subsequent professional development and future social development [15,22]. Daylighting environments are essential parts of indoor physical environments. Providing good daylighting conditions can create a good development environment for students. Therefore, daylighting performance in educational buildings has attracted significant attention. Currently, many research studies on the light environment of educational buildings emphasize improving environment quality and energy-saving levels. Only a few research studies focus on evaluating and improving the natural light environment [23], mainly concentrating on ordinary classrooms, professional classrooms, and laboratories [19,24]. Pellegrino et al. evaluated the daylighting performance of ordinary classrooms in an Italian school with the hope of providing reasonable solutions to improving daylight availability [25]. Rubies et al. took an academic classroom in L'Aquila University as a case to explore the influence of interior space and window geometry on daylighting performance and building lighting energy consumption, by simulating and analyzing parameters such as room geometry, window-to-floor ratio, window shape, window orientation, etc. [26]. Ashrafian and Moazzen selected west-facing and east-facing classrooms in a temperate continental climate zone for simulation to explore whether the average illuminance provided under different window-to-wall ratios met the needs of users; they obtained a window geometry design method that is beneficial for lighting energy saving [22]. Moreover, most regions selected for existing research studies on educational buildings are in middle and low latitudes, such as tropical or warm and humid climates, which makes it difficult to apply the relevant research results to areas with different climates or latitudes [27–30].

With the expansion in enrollment observed in Chinese universities, teaching locations that are spacious and have large capacities are needed to meet the increasing number of students and diversified use needs. Terrace classrooms have the following advantages: A large area, a large number of people, and a stepped floor, which can meet the requirement for more indoor seats and avoid the problem of vision blocking. As the most important teaching location in universities, a terrace classroom's use functions are flexible and changeable and can be used as the principal space for students' self-study, examination, meetings, and community activities. Therefore, the terrace classroom has become the most frequently used classroom type for university students. The quality of its daylighting performance is of great significance for meeting the needs of students' study life, maintaining physical and mental health, and reducing building energy consumption.

However, currently, the research on the natural light environment of terrace classrooms needs to be improved, and most existing research studies are aimed at artificial lighting and energy consumption [31,32]. In addition, China's related standards and design datasets lack specific guidance strategies for daylighting design in terrace classrooms. Therefore, in order to improve the daylighting performance of university terrace classrooms so that they can play a positive role in students' learning and life and saving architectural lighting energy consumption, it is urgent to study the daylighting performance of university terrace classrooms and propose corresponding design suggestions [33].

China's severe cold regions have higher latitudes and long cold winters. The total illuminance and sunshine hours are fewer than in other areas, so it is necessary to improve the utilization rate of sunlight and play an active role in natural daylighting. Thus, improving the daylighting performance of terrace classrooms in severe cold regions has become an urgent problem. This paper aims to study the relationship between interior space and window geometry and the daylighting performance of university terrace classrooms in severely cold areas. Based on the Grasshopper parameterization platform, the daylighting performance simulation used the Ladybug and Honeybee plug-in in Shenyang, China. Via correlation analysis, box diagram analysis, and multiple linear regression analysis, quantitative analyses were conducted on the relationship between all terrace classroom orientations: (1) Interior space geometry parameters include the following: classroom width (W_{tc}), classroom depth (D_{tc}), and classroom height (H_{tc}). (2) Window geometry parameters include the following: window height (H_w), window width (W_w), number of windows (N_w), the width of wall between windows (W_{wbw}), windowsill height (H_{ws}), window-to-wall ratio (WWR), window-to-floor ratio (WFR), and selected daylighting performance evaluation indices ($sDA_{300,50\%}$ and $UDI_{100-2000}$). Then, the prediction model of the daylighting performance evaluation indices of terrace classrooms facing each orientation was established. The research results provide a reference and evaluation basis for designing university terrace classrooms in severe cold regions.

2. Methodology

2.1. Workflow

The research process mainly includes four steps: (1) problem formulation and objectives; (2) simulation model building, initial parameter setting, and simulation running; (3) parameter analysis and prediction model establishment; (4) description and interpretation (Figure 1).

2.2. Location and Climate

Shenyang is located at $41°48'$ northern latitude and $123°25'$ eastern longitude, in the south of northeast China and the middle of Liaoning Province. Shenyang has four distinct seasons, with long winters and short summers throughout the year. It belongs to the temperate subhumid continental climate and the severe cold area in China's thermal division [34]. Meteorological data released by the National Meteorological Center (2009–2018) show that the average monthly temperature is $-12 \sim 25\ °C$; the average monthly relative humidity is 49%~79%; and the average temperature in spring, summer, autumn, and winter is $3 \sim 16\ °C$, $18 \sim 29\ °C$, $3 \sim 15\ °C$, and $-15 \sim -3\ °C$, respectively [35].

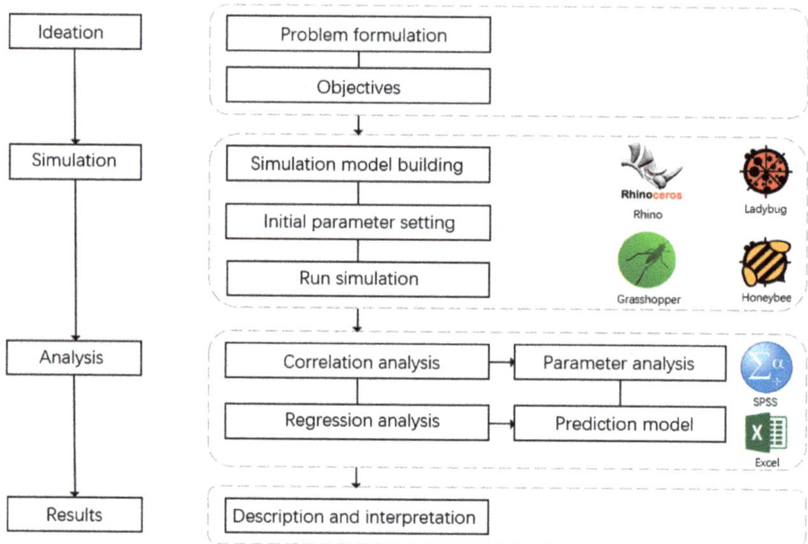

Figure 1. Research workflow.

According to GB 50033-2013 [34] and the annual mean total illuminance obtained by different solar elevation angles, cloud types, cloud cover, and sunshine duration in other regions of China, Shenyang belongs to class III (Figure 2), and the mean total illuminance of natural light years is between 35 and 40 klx. According to the National Meteorological Data Center statistics, the average annual peak sunshine duration was 1406–1582 h in Shenyang from 2000 to 2022. The average annual global horizontal irradiance can reach 5063–5096 MJ/m^2, and the average annual direct horizontal irradiance can reach 4804–4812 MJ/m^2 (Figure 3a). The average monthly peak sunshine duration is 64–184 h, the average monthly global horizontal irradiance is 230–663 MJ/m^2, and the average monthly direct horizontal irradiance is 178–561 MJ/m^2 (Figure 3b).

2.3. Daylighting Performance Evaluation Criteria

The daylighting performance evaluation index is divided into static and dynamic, forming two types. Among the static daylighting evaluation indices, the daylight factor (DF) is the most widely used index for evaluating light environments [34]. DF lacks orientation and local climate concerns and has been gradually replaced internationally by the more useful daylight illuminance (UDI) and daylight autonomy (DA) indices. Compared with DF, DA reflects the influence of regional photo-climate on the natural lighting performance of buildings and can evaluate their natural lighting level more comprehensively and accurately [36]. Still, DA cannot reflect the proportion of the area reaching the given illuminance in the space. To comprehensively evaluate the DA values of all calculation points, the Illuminating Engineering Society (IES) proposed the sDA index in 2013. If only $sDA_{300\,lx,50\%}$ is used as the evaluation index, the glare caused by excessive daylighting will be ignored. If only UDI is used as the daylighting performance evaluation index, the adequacy of illuminance cannot be judged [37]. Therefore, this paper selects sDA and UDI as the evaluation indices of daylighting performance.

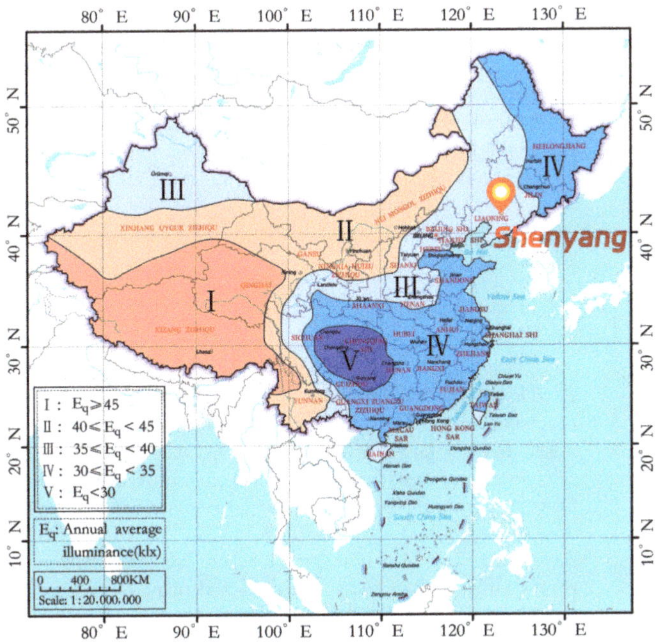

Figure 2. Daylight zone of China.

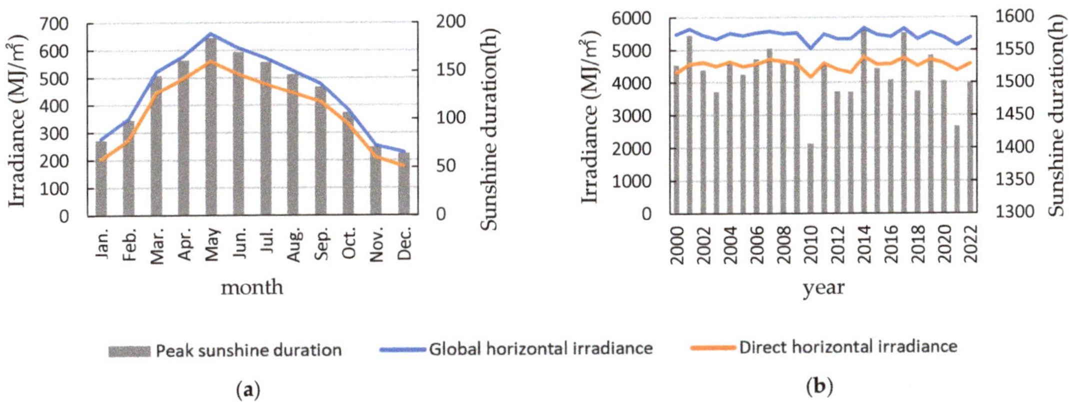

Figure 3. Statistics of solar radiation intensity and peak sunshine duration in Shenyang from 2000 to 2022: (**a**) mean monthly meteorological data statistics; (**b**) mean annual meteorological data statistics.

Spatial daylight autonomy (sDA) measures the sufficiency of daylight illuminance in a given area. It is defined as the percentage of the floor area of a building that receives natural light above a specified minimum illuminance value (300 lx for educational buildings) during a specified working period throughout the year (for example, 50% of the time from 8 a.m. to 6 p.m.) [38]. Based on natural illuminance, this dynamic lighting evaluation index specifies the range or percentage of illuminance, which can accurately determine and evaluate the natural illuminance of building spaces [39]. Therefore, $sDA_{300\,lx,\,50\%}$ is selected as the evaluation index in this paper. $sDA_{300\,lx,\,50\%}$ is the measure recommended by the Illuminating Engineering Society of North America (IES) and LEED. In addition, IES LM-83-12 rated the sufficiency of the ambient daylight available according to $sDA_{300\,lx,\,50\%}$. When $sDA_{300\,lx,\,50\%}$ is greater than or equal to 75%, the daylight sufficiency of the analysis area is rated as "preferred". When $sDA_{300\,lx,\,50\%}$ is in the range of 55–75%, the daylight sufficiency of the analysis area is rated as "nominally acceptable".

Useful daylight illuminance (UDI) represents the frequency at which natural lighting illuminance reaches a certain range throughout the year. The distribution of natural lighting illuminance is divided into three fields: UDI is defined as the available illuminance within the scope of 100–2000 lx, <100 lx is the range of too-low illuminance, and >2000 lx is the range of too-high illuminance [40]. Natural lighting illuminance that is too low leads to low daylight availability, and natural lighting illuminance that is too high easily causes glare and visual discomfort. Therefore, $UDI_{100-2000}$ is selected as the evaluation index in this paper.

2.4. Simulation Settings

2.4.1. Simulation Tools

Based on the Rhino and Grasshopper platforms, Ladybug and Honeybee plug-ins were used to set the simulation parameters, and the Radiance software was used to simulate the optical environment. This daylighting analysis platform is widely used in building daylighting research, and many studies have verified that this platform has good accuracy [41–44].

As a parameter modeling programming tool of the modeling software Rhino [43], Grasshopper runs on the Rhino platform and is one of the mainstream software used in data design. Radiance, a relatively advanced illuminance forecasting tool, was used as the daylight engine in this study. Ladybug is a microclimate analysis plugin software based on the Grasshopper parameterized platform. The plugin can import EnergyPlus Weather (EPW) data on demand for data analysis and visualization [45]. The Honeybee plug-in can invoke various building performance analysis software for the simulation analysis of natural lighting, thermal comfort, building energy consumption, and output visualization results. This paper used Honeybee to connect Grasshopper to Radiance for the building lighting simulation experiments.

2.4.2. Establishment of Simulation Models

Figure 4 shows the geometric parameter information of the terrace classroom simulation model. This paper determined simulation models by using six parameters, including W_{tc}, D_{tc}, H_w, H_{ws}, W_{wbw}, and the number of walls between windows. According to the relevant provisions in the Architectural Design Data Set [46], the horizontal distance (a) between the front edge of the first step and the front wall of the classroom was set as 5.5 m, the depth of the platform at the back of the classroom was set as 3 m, the slope of the step was set as 9°, and the distance (c) between the upper edge of the window and the ceiling was set as 0.2 m. The width (d) of the window end wall was 1 m [47]. W_{tc}, H_{ws}, and H_w determined H_{tc}. In addition, the windows in the model were evenly distributed horizontally. W_{tc}, W_{wbw}, and the number of walls between the windows determined W_w.

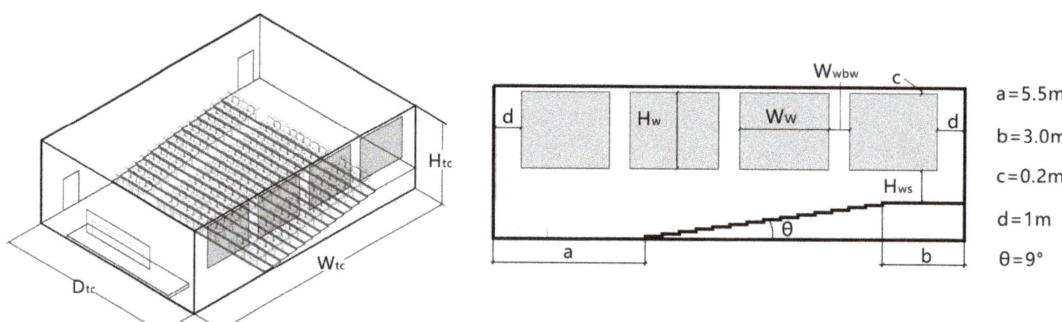

Figure 4. The geometric parameter information of simulation models.

This paper adopted the orthogonal experimental method [48] for the experimental design. Orthogonal experiment design is a scientific method for studying and dealing with multi-factor experiments. This method uses an orthogonal table to scientifically select the test conditions and reasonably carry out the multi-factor test by analyzing part of the test's results to master the overall test situation, and then optimize the optimal horizontal combination. According to the relevant regulations of terrace classrooms in the public teaching buildings of colleges and universities [46], the size of terrace classrooms was set as the common 150–360 students, and the corresponding room space varied from 16 to 23 m. The classroom depth varied from 12 to 15.5 m. According to the GB 50352–2019 standard [49], the windowsill height range was 0.6–1.3 m and the window height was 1.4–2.8 m. It is required in the GB 50352-2019 standard [47] that the width of the walls between windows should not be greater than 1.2 m, and when the width of the walls between windows is less than 1 m, constructional columns should be set. Therefore, the width range of the walls between windows was set to be 0–1.2 m, varying at unequal spacing. The number of walls between windows ranges from 1 to 8. The values of each factor and level are shown in Table 1. SPSS was used to generate orthogonal experiments [50], and 64 terrace classroom simulation models were formed. The experiment combined east, south, west, and north (four orientations) and finally determined 256 simulation models. The direction of the classroom in this paper refers to the orientation of the window wall.

Table 1. Design factors and levels of terrace classrooms.

Level	Classroom Width	Classroom Depth	Window Height	Windowsill Height	Width of Wall Between Windows	Number of Walls Between Windows
1	16 m	12 m	1.4 m	0.6 m	0 m	1
2	17 m	12.5 m	1.6 m	0.7 m	0.3 m	2
3	18 m	13 m	1.8 m	0.8 m	0.45 m	3
4	19 m	13.5 m	2 m	0.9 m	0.6 m	4
5	20 m	14 m	2.2 m	1 m	0.75 m	5
6	21 m	14.5 m	2.4 m	1.1 m	0.9 m	6
7	22 m	15 m	2.6 m	1.2 m	1.05 m	7
8	23 m	15.5 m	2.8 m	1.3 m	1.2 m	8

2.4.3. Initial Parameter Settings

1. Weather Data File and Material Properties

The US Department of Energy website's weather data file retrieved via Ladybug uses the Shenyang (123.25° E, 41.48°N) weather file by EnergyPlus [45].

According to the reflectance range described in GB50033-2013 [34], combined with Shenyang's geographical location and the terrace classroom's actual situation, the reflectance of the floor, wall, and ceiling was set as 0.30, 0.55, and 0.75, respectively. The interior windows comprised laminated glass with a visible light transmission ratio of 0.88.

2. Measuring Point Setup and Simulation Process

According to the GB/T 5699-2017 standard [51], the horizontal plane with a vertical height of 0.75 m from the ground was set as the illuminance calculation plane. According to the size of the interior space, the calculation plane was divided into analysis grids of 1.0 m × 1.0 m. The measuring points were located in the center of each grid, and the spacing between the measuring points was 1.0 m. The simulation time was set from 8:00 to 18:00 every day.

The meteorological data files, material parameters, analysis grid, and simulation time determined above were inputted into the designated positions in the program. The built-in daylight simulation engines, Ladybug and Honeybee, were run. The 256 models determined above were simulated and sorted into groups according to their orientation; finally, the data were obtained.

2.5. Statistical Analysis

This paper used statistical methods such as correlation analysis and multiple linear regression analysis to explore the influence of interior space and window geometry parameters on daylighting evaluation indicators in a terrace classroom. Firstly, a correlation analysis was conducted between independent variables (classroom interior space and window geometry) and dependent variables ($sDA_{300,50\%}$ and $UDI_{100-2000}$) to determine the degree and linear trend of linear correlations between each parameter and the evaluation index of the natural lighting environment. Pearson's correlation coefficient in SPSS was used to determine the existence of correlation and the closeness of the relationship between variables [52]. Next, multiple linear regression analysis [42,53], which has been used in several relevant studies, was selected to establish the prediction models of the terrace classrooms' interior space and window geometry parameters, and the $sDA_{300,50\%}$ and $UDI_{100-2000}$ in the terrace classroom facing each orientation, by using SPSS software.

3. Results

3.1. Correlation Analysis

This study conducted a linear correlation analysis between the interior space and window geometry of terrace classrooms facing east, south, west, and north using $sDA_{300,50\%}$ and $UDI_{100-2000}$. Table 2 shows the results of the correlation analysis. According to the definition of a significant correlation, when the significance value (Sig.) is less than 0.05, the independent and dependent variables are significantly correlated. In addition, the correlation coefficient (r) indicates the degree of closeness between the independent and dependent variables. The greater the absolute value of the correlation coefficient, the closer the relationship.

Table 2. Correlation analysis results of interior space and window geometry parameters with $sDA_{300,50\%}$ and $UDI_{100-2000}$.

Orientation	Index	W_{tc}	D_{tc}	H_{tc}	W_w	H_w	W_{wbw}	H_{ws}	N_w	WWR	WFR
East	SDA	0.331 **	−0.405 **	0.663 **	0.309 *	0.664 **	−0.232	0.156	−0.246	0.591 **	0.883 **
	UDI	0.415 **	−0.097	0.322 **	0.060	0.215	0.057	0.291 *	−0.057	0.159	0.230
South	SDA	0.309 *	−0.315 *	0.702 **	0.267 *	0.693 **	−0.208	0.183	−0.211	0.575 **	0.844 **
	UDI	0.273 *	0.022	−0.246 *	−0.119	−0.367 **	0.313 *	0.184	0.244	−0.381 **	−0.454 **
West	SDA	0.294 *	−0.328 **	0.693 **	0.254 *	0.694 **	−0.230	0.162	−0.203	0.583 **	0.858 **
	UDI	0.376 **	−0.084	0.042	−0.059	−0.098	0.266 *	0.292 *	0.067	−0.212	−0.163
North	SDA	0.329 **	−0.399 **	0.646 **	0.321 **	0.642 **	−0.245	0.159	−0.260 *	0.595 **	0.877 **
	UDI	0.412 **	−0.190	0.492 **	0.135	0.399 **	−0.031	0.301 *	−0.118	0.311 *	0.463 **

Notes: * and ** represent significant test results at 5% and 1% levels, respectively.

W_{tc} is positively correlated with $sDA_{300,50\%}$ and $UDI_{100-2000}$ for each classroom orientation, while other parameters have different linear correlations with $sDA_{300,50\%}$ and $UDI_{100-2000}$ for each classroom orientation. WFR, H_{tc}, H_w, WWR, W_{tc}, and W_w had a significant linear positive correlation with $sDA_{300,50\%}$ in the terrace classrooms facing each orientation, and the correlation degree decreased in turn. This indicates that the above parameters positively affect $sDA_{300,50\%}$, and the effect degree decreases successively. In addition, D_{tc} and $sDA_{300,50\%}$ for each classroom orientation had significant linear negative correlations, with an $|r|$ value between 0.315 and 0.405, which shows that the D_{tc} for the $sDA_{300,50\%}$ for each classroom orientation had certain inhibitions. In addition, N_W was only negatively correlated with $sDA_{300,50\%}$ in north-facing classrooms but to a low degree.

In addition, for east-facing classrooms, W_{tc}, H_{tc}, and H_{ws} were significantly positively correlated with $UDI_{100-2000}$, and the degree of correlation decreased successively. W_{tc}, H_{ws}, and W_{wbw} were positively correlated with $UDI_{100-2000}$, and their effects decreased successively for west-facing classrooms. WFR, WWR, H_w, and H_{tc} were negatively correlated with $UDI_{100-2000}$ in south-facing classrooms but positively correlated with $UDI_{100-2000}$ in north-facing classrooms. In addition, the correlation coefficients between $UDI_{100-2000}$

and each parameter for each orientation of the classroom were relatively small, with an | r | value between 0.25 and 0.50.

3.2. Exploratory Analysis

The relationships between terrace classrooms' $sDA_{300,50\%}$ and $UDI_{100-2000}$ and geometry parameters that have significant influences on them were further analyzed by a boxplot. The lines in the box are the medians; the top and bottom edges of the box indicate the 75th and 25th percentile of all values sorted from the smallest to the largest, and the top and bottom T-bars indicate maximum and minimum values, respectively.

3.2.1. Interior Space Geometry Parameters and Daylighting Performance Evaluation Indices

In terms of the terrace classroom width, when W_{tc} was within the range of 16–19 m, with the increase in W_{tc}, the median value of $sDA_{300,50\%}$ of the terrace classrooms facing each orientation showed an obvious upward trend (Figure 5a), among which the south-facing classroom's $sDA_{300,50\%}$ increased the most. When W_{tc} increased from 19 m to 20 m, the median value of $sDA_{300,50\%}$ changed a little in terrace classrooms facing each orientation. When W_{tc} ranged from 20 to 23 m, the value of the terrace classrooms facing each orientation resumed an upward trend, among which the increase in the value of south-facing classrooms was the lowest. In addition, as W_{tc} gradually increased, the median $sDA_{300,50\%}$ was always the highest in south-facing classrooms, followed by west-facing classrooms, and it was similar and the lowest in east-facing and north-facing classrooms.

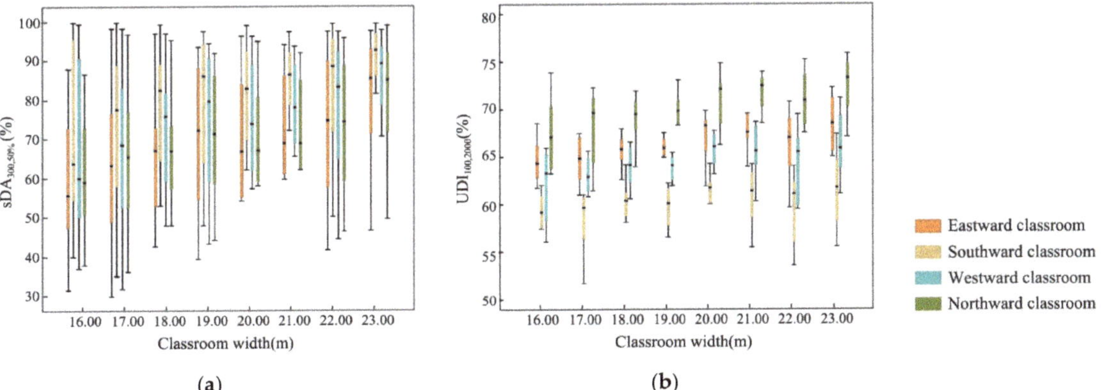

Figure 5. The analysis results of the distribution interval between the classroom width and the evaluation indices: (**a**) the analysis results of $sDA_{300,50\%}$; (**b**) the analysis results of $UDI_{100-2000}$.

When W_{tc} was 16–20 m, with the increase in W_{tc}, the median values of $UDI_{100-2000}$ in the terrace classrooms facing each orientation showed an upward trend (Figure 5b). When the W_{tc} was between 20 and 23 m, with the increase in W_{tc}, the median values of south and west $UDI_{100-2000}$ tended to be stable, while the median values of $UDI_{100-2000}$ in the east- and north-facing classrooms showed an upward trend. In addition, the median values of $UDI_{100-2000}$ in the north-, east-, west-, and south-facing classrooms decreased in turn.

In terms of the terrace classroom depth, with the increase in D_{tc}, the $sDA_{300,50\%}$ of the terrace classrooms facing each orientation showed an overall downward trend (Figure 6). When D_{tc} increased from 12.5 m to 13 m and from 14 m to 14.5 m, the median value of $sDA_{300,50\%}$ of the terrace classrooms facing each orientation decreased significantly. In addition, the median value of $sDA_{300,50\%}$ was the highest in south-facing classrooms, followed by west-facing classrooms. East-facing and north-facing classrooms exhibit similar sufficiency and the lowest daylight sufficiency.

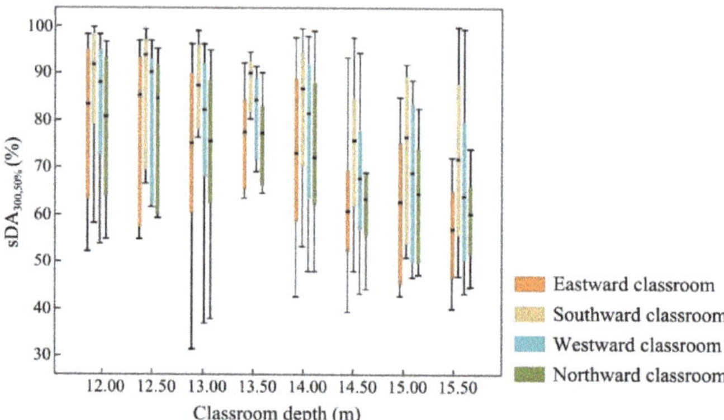

Figure 6. The analysis results of the distribution interval between the classroom depth and the evaluation index.

In terms of the terrace classroom height, when H_{tc} was within the range of 2.5–3.4 m, with the increase in H_{tc}, the median and interquartile values of $sDA_{300,50\%}$ of the terrace classrooms facing each orientation showed an obvious upward trend (Figure 7a). When H_{tc} increased from 3.4 m to 3.7 m, the median value of $sDA_{300,50\%}$ of the terrace classroom facing each orientation changed a little. When the H_{tc} was 3.7–4.3 m, the median value of $sDA_{300,50\%}$ showed a slight upward trend. In addition, when the H_{tc} was 2.5–2.8 m, the median $sDA_{300,50\%}$ of west-facing and north-facing classrooms was similar and about 5% higher than that of east-facing classrooms. When the H_{tc} was 3.1–4.3 m, the median $sDA_{300,50\%}$ of the east- and north-facing classrooms was similar and lower than that of the east-facing classroom. The median $sDA_{300,50\%}$ was always the highest in the south-facing classroom.

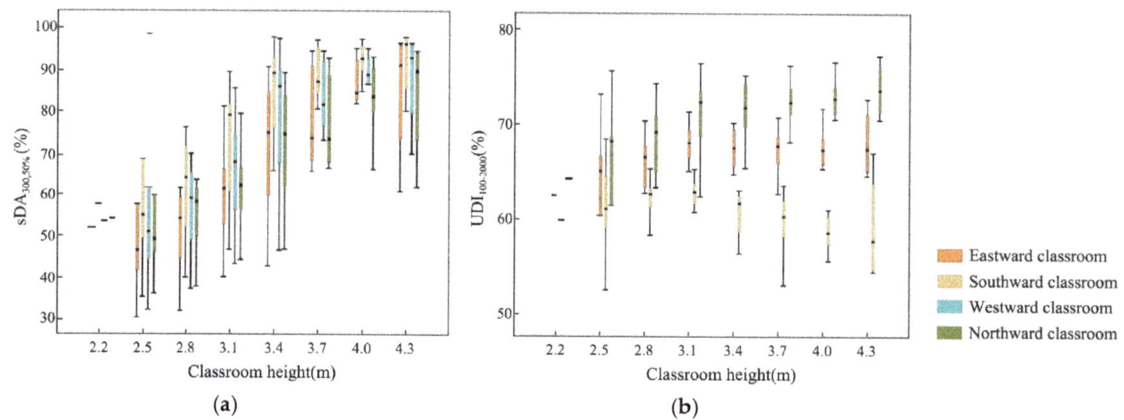

Figure 7. The analysis results of the distribution interval between the classroom height and the evaluation indices: (a) the analysis results of $sDA_{300,50\%}$; (b) the analysis results of $UDI_{100–2000}$.

When H_{tc} was within the range of 2.2–3.1 m, with the increase in H_{tc}, the median value of $UDI_{100–2000}$ in east-, south-, and north-facing terrace classrooms showed an obvious upward trend (Figure 7b). With the increase in H_{tc} in the range of 3.1–4.3 m, the changing trend of $UDI_{100–2000}$ was different for the terrace classrooms facing each orientation, among which the median values of $UDI_{100–2000}$ in south-facing, east-facing, and north-facing classrooms decreased, stabilized, and slightly increased, respectively.

3.2.2. Window Geometry Parameters and Daylighting Performance Evaluation Indices

In terms of the window width, when W_w was 0.5–4.1 m, the median value of the $sDA_{300,50\%}$ of the terrace classrooms facing each orientation showed a slight upward trend with the increase in W_w (Figure 8). When W_w increased from 4.1 m to 5.3 m, the median value of the $sDA_{300,50\%}$ of the terrace classrooms facing each orientation decreased slightly. When W_w was 5.3–8.9 m, the $sDA_{300,50\%}$ of the terrace classrooms facing each orientation resumed a slight upward trend. In addition, the median value of $sDA_{300,50\%}$ in south-facing classrooms was the highest, followed by west-facing classrooms. East-facing and north-facing classrooms exhibited similar sufficiency and the lowest daylight sufficiency.

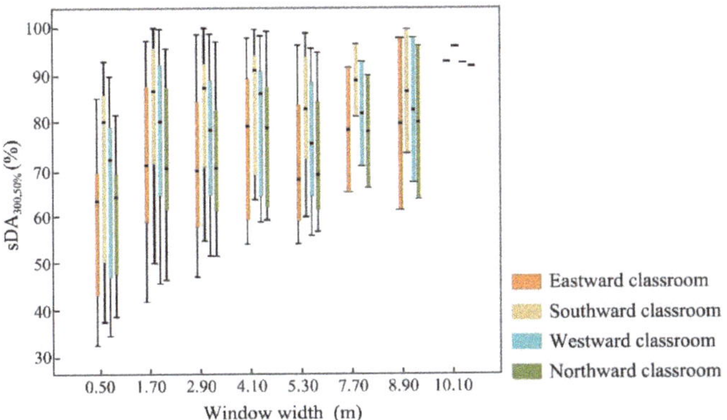

Figure 8. The analysis results of the distribution interval between the window width and the evaluation index.

In terms of the window height, as H_w increased from 1.4 m to 2.4 m, the $sDA_{300,50\%}$ of the terrace classrooms facing each orientation showed a significant increasing trend (Figure 9a). When H_w was within 1.4–2.4 m, the median $sDA_{300,50\%}$ of east-facing and north-facing classrooms was close. The difference between them and the median $sDA_{300,50\%}$ of west-facing and south-facing classrooms gradually increased with an increase in H_w. In addition, the median value of $sDA_{300,50\%}$ in south-facing classrooms was always the highest. When H_w increased within 2.4–2.8 m, the $sDA_{300,50\%}$ tended to be stable in the terrace classrooms facing each orientation, with the median and interquartile range values of $sDA_{300,50\%}$ being higher in south-facing classrooms, moderate in west-facing classrooms, and similar and lowest in east-facing and north-facing classrooms, respectively.

Additionally, for south-facing terrace classrooms, the median values of $UDI_{100-2000}$ tended to be stable when H_w was within the range of 1.4–1.8 m (Figure 9b). Moreover, the median values of $UDI_{100-2000}$ showed a slightly decreasing trend as H_w gradually increased from 1.8 m to 2.8 m. For north-facing terrace classrooms, the median $UDI_{100-2000}$ showed an apparent upward trend when H_w increased in 1.4–2.0 m, after which the $UDI_{100-2000}$ median values leveled off as H_w continued increasing. Furthermore, the median value of the $UDI_{100-2000}$ of north-facing classrooms was always larger than those of south-facing classrooms, with a maximum difference of about 15%.

In terms of the width of the wall between windows, with the gradual increase in W_{wbw} within the range of 0–0.9 m, the $UDI_{100-2000}$ of south-facing and west-facing terrace classrooms showed an upward trend (Figure 10). After that, as W_{wbw} continued increasing, the median value of $UDI_{100-2000}$ in south-facing and west-facing classrooms decreased and changed significantly. In addition, the $UDI_{100-2000}$ of the west-facing classrooms was consistently larger than that of the south-facing classrooms, with a maximum difference of about 5%.

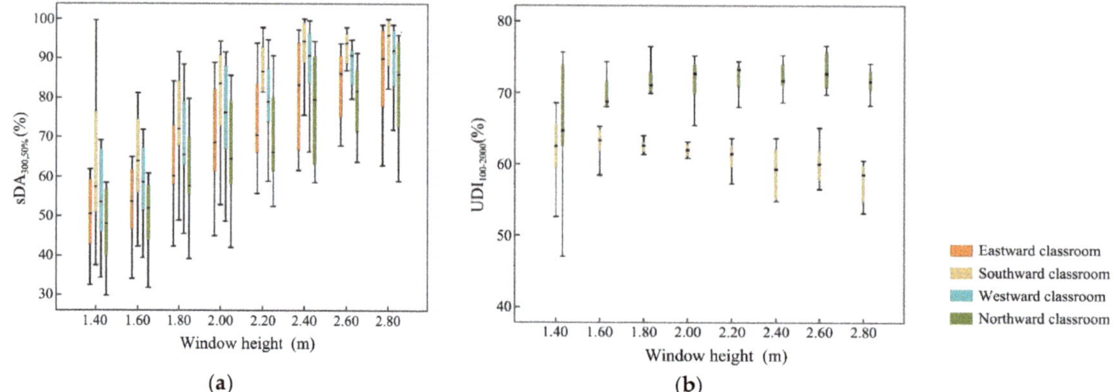

Figure 9. The analysis results of the distribution interval between the window height and the evaluation indices: (**a**) the analysis results of $sDA_{300,50\%}$; (**b**) the analysis results of $UDI_{100-2000}$.

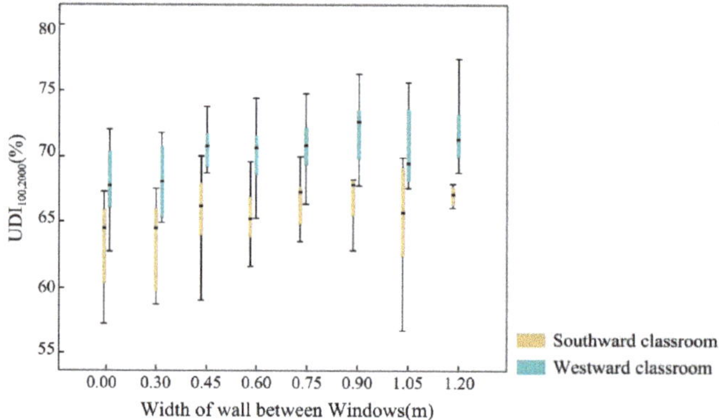

Figure 10. The analysis results of the distribution interval between the width of the wall between windows and the evaluation index.

In terms of the windowsill height, the median values of $UDI_{100-2000}$ for north-facing classrooms showed an increasing trend as H_{ws} increased (Figure 11). Moreover, as H_{ws} gradually increased in the range of 0.6–1.0 m, the median values of $UDI_{100-2000}$ for east-facing and west-facing classrooms showed a steady and slightly decreasing trend, respectively. With the gradual increase in H_{ws} in 1.0–1.3 m, the median $UDI_{100-2000}$ value in east-facing and west-facing classrooms showed an upward trend. In addition, the median value of $UDI_{100-2000}$ in north-facing, east-facing, and west-facing classrooms decreased sequentially.

In terms of the number of windows, only N_w in north-facing classrooms had a significant impact on $sDA_{300,50\%}$ (Figure 12). When N_w was 4, the median value of $sDA_{300,50\%}$ was about 15% higher than other Nw. When N_w was 5–9, with the increase in N_w, the median value of $sDA_{300,50\%}$ changed a little, while the maximum and minimum values showed a clear downward trend.

In terms of the window–wall ratio, with the increase in WWR, the $sDA_{300,50\%}$ of the terrace classrooms facing each orientation showed an upward trend, and the median value of $sDA_{300,50\%}$ in south-facing classrooms was always the highest (Figure 13a). Among them, when the WWR was 0.20–0.30, the $sDA_{300,50\%}$ of west-facing and north-facing classrooms was similar, and the median value was greater than that of east-facing classrooms by about 5%. When the WWR was in the range of 0.30–0.60, the $sDA_{300,50\%}$ of east-facing and north-facing classrooms was similar, with median values that were about 10% and 17% smaller than those in west-facing and south-facing classrooms, respectively. When the WWR was

0.60–0.70, the median $sDA_{300,50\%}$ in east-, west-, and north-facing classrooms was similar and about 5% smaller than in the south.

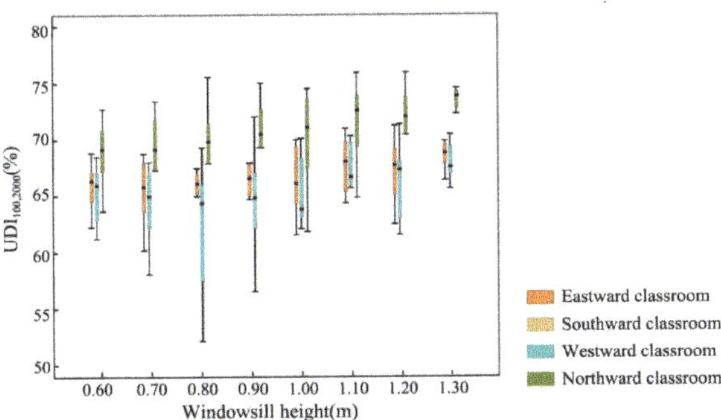

Figure 11. The analysis results of the distribution interval between the windowsill height and the evaluation index.

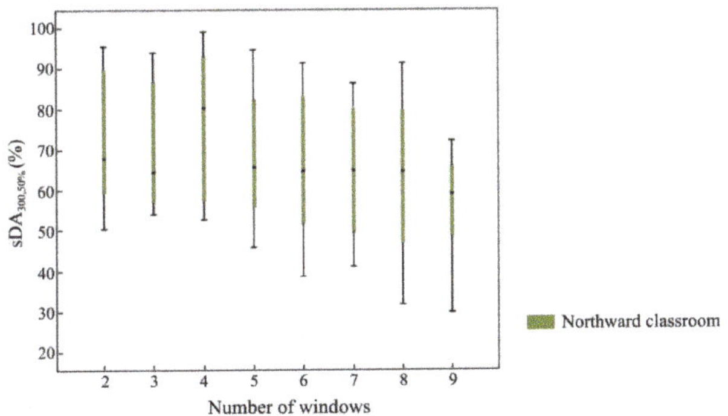

Figure 12. The analysis results of the distribution interval between the number of windows and the evaluation index.

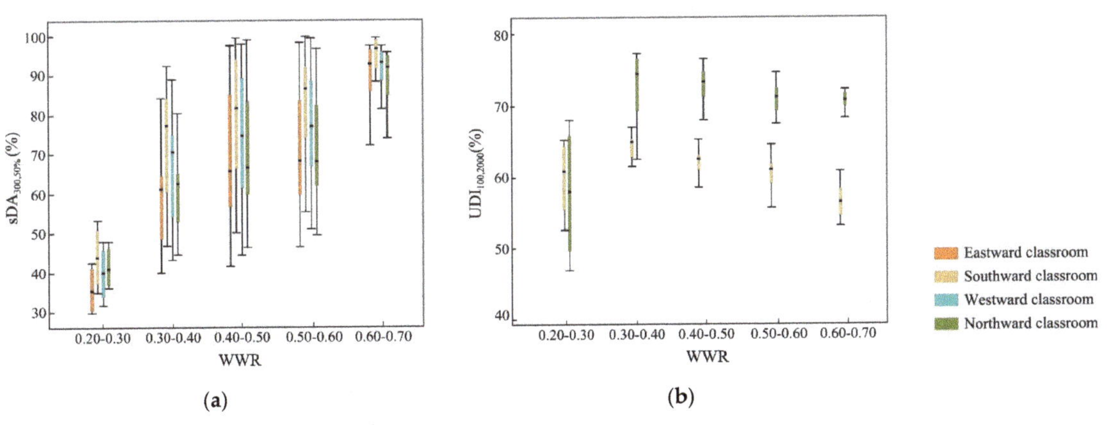

Figure 13. The analysis results of the distribution interval between WWR and the evaluation indices: (**a**) the analysis results of $sDA_{300,50\%}$; (**b**) the analysis results of $UDI_{100–2000}$.

For south-facing and north-facing terrace classrooms, when the WWR increased from 0.2–0.3 to 0.3–0.4, $UDI_{100–2000}$ showed a clear upward trend, and the increase in north-facing classrooms was significantly larger (Figure 13b). As the WWR gradually increased in the range of 0.3–0.7, the $UDI_{100–2000}$ of the two classrooms showed a downward trend, and the median value of $UDI_{100–2000}$ in the north-facing classroom was about 10% greater than that of the south.

In terms of the window–floor ratio, with the increase in WFR, the $sDA_{300,50\%}$ of the terrace classrooms facing each orientation showed a clear upward trend, and the median value of $sDA_{300,50\%}$ in the south-facing classroom was always the highest (Figure 14a). When the WFR was within the range of 0.04–0.11, the median $sDA_{300,50\%}$ value of west- and north-facing classrooms was similar and about 4% greater than that of east-facing classrooms. When the WFR was in the range of 0.11–0.17, the $sDA_{300,50\%}$ of east-facing and north-facing classrooms was similar, and the median value was lower than that of west-facing classrooms by about 5–10%. When the WFR was 0.17–0.20, the median $sDA_{300,50\%}$ value difference between classrooms was small.

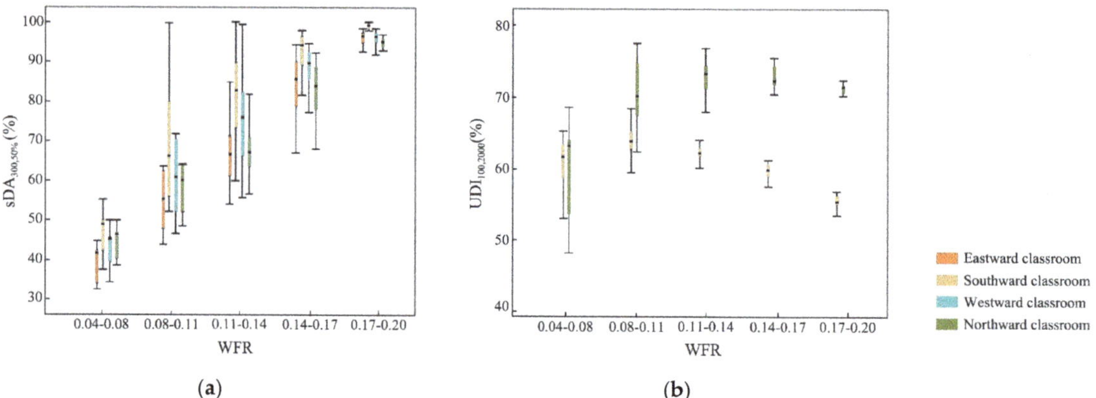

Figure 14. The analysis results of the distribution interval between WFR and the evaluation indices: (**a**) the analysis results of $sDA_{300,50\%}$; (**b**) the analysis results of $UDI_{100–2000}$.

As the WFR gradually increased within the range of 0.04–0.11, the median value of $UDI_{100–2000}$ in the south-facing terrace classrooms showed an upward trend. Later, as the WFR continued to increase, $UDI_{100–2000}$ showed a clear downward trend (Figure 14b). When the WFR gradually increased in the range of 0.04–0.14, the median $UDI_{100–2000}$ value in north-facing classrooms showed an upward trend; then, $UDI_{100–2000}$ decreased slightly as the WFR continued to increase. In addition, the $UDI_{100–2000}$ in the north-facing classroom was always larger than in the south-facing classroom, and the maximum difference between the two-facing $UDI_{100–2000}$ was about 15%.

3.3. Prediction Model

According to the correlation analysis results of the interior space geometry parameters and window geometry parameters of terrace classrooms with $sDA_{300,50\%}$ and $UDI_{100–2000}$, the geometry parameters with a significant influence on $sDA_{300,50\%}$ and $UDI_{100–2000}$ were taken as independent variables, and $sDA_{300,50\%}$ and $UDI_{100–2000}$ were taken as dependent variables for the multiple linear regression analysis. Prediction models for the $sDA_{300,50\%}$ and $UDI_{100–2000}$ of terrace classrooms facing each orientation were constructed.

Firstly, to describe the quantitative relationship between variables more accurately, the curve estimation method was used to fit the selected geometrical parameters (independent variables) with $sDA_{300,50\%}$ and $UDI_{100–2000}$ (dependent variables) with various curve types, and based on the significance index (Sig.) values, to determine the curve relationship between variables. This study focuses on several common curve models, including linear, conic, cubic, and logarithmic curve models.

To comprehensively consider the interactive effects of geometry parameters in the regression model and seek the optimal combination of geometry parameters that can explain the variation law of differently oriented terrace classrooms' $sDA_{300,50\%}$ and $UDI_{100-2000}$, in this study, the curve fitting model exhibiting statistical significance between each geometrical parameter, and terrace classroom's $sDA_{300,50\%}$ and $UDI_{100-2000}$, were selected as independent variables and comprehensively applied to the regression analysis. Using multiple linear regression step-by-step methods, multiple iterative regression analyses for different variable combinations were carried out. Finally, the optimal variable combination was selected, and the prediction model was built.

The regression equation for the $sDA_{300,50\%}$ (%) of the terrace classrooms facing each orientation is as follows.

$$sDA_{300,50\%}(\text{eastward}) = 0.002 \cdot W_{tc}^3 - 103.920 \cdot WWR^3 + 613.203 \cdot WFR + 0.256, \quad (1)$$

$$sDA_{300,50\%}(\text{southward}) = 0.001 \cdot W_{tc}^3 - 78.776 \cdot WWR^2 + 60.087 \cdot \ln(WFR) + 223.122, \quad (2)$$

$$sDA_{300,50\%}(\text{westward}) = 0.001 \cdot W_{tc}^3 - 75.377 \cdot WWR + 63.075 \cdot \ln(WFR) + 242.057, \quad (3)$$

$$sDA_{300,50\%}(\text{northward}) = 0.002 \cdot W_{tc}^3 - 114.202 \cdot WWR^3 + 693.732 \cdot WFR - 12.771 \cdot \ln(H_w) - 0.945, \quad (4)$$

The adjusted R^2 of the multiple linear regression model for east-, south-, west-, and north-facing classrooms' $sDA_{300,50\%}$ is 0.885, 0.861, 0.844, and 0.882, respectively, indicating that the interpretation degree of the model for the east-, south-, west-, and north-facing classrooms' $UDI_{100-2000}$ is 88.5%, 86.1%, 84.4%, and 88.2%, respectively. The model has high goodness of fit. The significant values (Sig.) of the regression models in the variance analyses were all lower than 0.05, indicating that the established regression models had significant statistical significance. The Sig. values in the t-test were all less than 0.05, indicating that the regression coefficients were significant, and there was a significant correlation between the independent and dependent variables. In addition, the variance inflation factors (VIF) of the independent variables ranged from 1.029 to 5.622, all of which were lower than 10, indicating no multicollinearity in the models (Table 3).

Table 3. Comprehensive analysis results of multiple regression models of $sDA_{300,50\%}$ in differently orientated terrace classrooms.

Orientation	Adjusted R^2	Sig.	Independent Variable	Standardized Coefficients	t	Sig.	VIF
East	0.885	0.000	Constant	——	0.073	0.042	——
			W_{tc}^3	0.260	6.000	0.000	2.442
			WWR^3	−0.383	−5.687	0.000	1.029
			WFR	1.147	17.158	0.000	2.479
South	0.861	0.000	Constant	——	19.317	0.000	——
			W_{tc}^3	0.207	4.338	0.000	1.032
			WWR^2	−0.427	−5.624	0.000	2.619
			$\ln(WFR)$	1.180	15.611	0.000	2.594
West	0.844	0.000	Constant	——	14.630	0.000	——
			W_{tc}^3	0.200	3.947	0.000	1.037
			WWR	−0.434	−4.996	0.000	3.054
			$\ln(WFR)$	1.198	13.895	0.000	3.010
North	0.882	0.000	Constant	——	−0.254	0.008	——
			W_{tc}^3	0.249	5.467	0.000	1.038
			WWR^3	−0.420	−5.496	0.000	2.921
			WFR	1.294	12.214	0.000	5.622
			$\ln(H_w)$	−0.162	−2.177	0.034	2.791

In addition, according to the standardized regression coefficient, the influence of WFR, WWR^2, and W_{tc}^3 on the $sDA_{300,50\%}$ in the east-facing and north-facing classrooms decreased successively. For the north-facing classrooms, $\ln(H_w)$ was only more influential than W_{tc}^3. For the $sDA_{300,50\%}$ prediction model for south- and west-facing classrooms, WWR^2 and WWR had the greatest influence, followed by Wtc^3, and $\ln(WFR)$ had the least influence.

The regression equation for the $UDI_{100-2000}$ (%) of the terrace classrooms facing each orientation is as follows.

$$UDI_{100\text{-}2000} \text{ (eastward)} = 0.68{\cdot}W_{tc} - 0.96{\cdot}Hw^2 + 18.13{\cdot}\ln(H_{tc}) + 37.01, \tag{5}$$

$$UDI_{100\text{-}2000} \text{ (southward)} = -1670.36{\cdot}WFR^3 + 12.45{\cdot}\ln(W_{tc}) + 26.83, \tag{6}$$

$$UDI_{100\text{-}2000} \text{ (westward)} = 13.98{\cdot}\ln(W_{tc}) + 2.13{\cdot}H_{ws}^3 + 3.08{\cdot}W_{wbw} + 19.40, \tag{7}$$

$$UDI_{100\text{-}2000}(\text{northward}) = 15.00{\cdot}\ln(W_{tc}) - 65.73{\cdot}WWR^3 + 21.41{\cdot}\ln(WFR) - 0.21{\cdot}H_w^3 + 84.07, \tag{8}$$

The adjusted R^2 of the multiple linear regression model for east-, south-, west-, and north-facing classrooms' $UDI_{100-2000}$ is 0.310, 0.385, 0.287, and 0.774, respectively, indicating that the interpretation degree of the model for the east-, south-, west-, and north-facing classrooms' $UDI_{100-2000}$ is 31%, 38.5%, 28.7%, and 77.4%, respectively. The model has high goodness of fit. The significant values (Sig.) of the regression models in variance analyses were all less than 0.05, indicating that the established regression models had significant statistical significance. The Sig. values in the t-test were all less than 0.05, indicating that the regression coefficients were significant and there was a significant correlation between the independent and dependent variables. In addition, the variance inflation factor (VIF) of the independent variable ranged from 1.000 to 4.451, all of which are less than 10, indicating no multicollinearity in the models (Table 4).

Table 4. Comprehensive analysis results of multiple regression models of $UDI_{100-2000}$ in differently-orientated terrace classrooms.

Orientation	Adjusted R^2	Sig.	Independent Variable	Standardized Coefficients	t	Sig.	VIF
East	0.310	0.000	Constant	——	6.683	0.000	——
			W_{tc}	0.415	3.963	0.000	1.000
			H_w^2	−0.490	−2.221	0.030	4.451
			$\ln(H_{tc})$	0.773	3.501	0.001	4.451
South	0.385	0.000	Constant	——	2.475	0.016	——
			WFR^3	−0.574	−5.779	0.000	1.010
			$\ln(W_{tc})$	0.337	3.393	0.001	1.010
West	0.287	0.015	Constant	——	1.675	0.009	——
			$\ln(W_{tc})$	0.382	3.595	0.001	1.000
			H_{ws}^3	0.322	3.024	0.004	1.000
			W_{wbw}	0.266	2.504	0.015	1.000
North	0.774	0.000	Constant	——	8.215	0.000	——
			$\ln(W_{tc})$	0.349	5.496	0.000	1.056
			WWR^3	−0.842	−8.784	0.000	2.400
			$\ln(WFR)$	1.360	11.083	0.000	3.929
			H_w^3	−0.260	−2.916	0.005	2.081

In addition, according to the standardized regression coefficient, $\ln(H_{tc})$, H_w^2, and W_{tc} successively reduced the influence on the $UDI_{100-2000}$ of the east-facing classrooms. For south-facing classrooms, WFR^3 had a greater influence on $UDI_{100-2000}$ than $\ln(W_{tc})$. For west-facing classrooms' $UDI_{100-2000}$, the influence of $\ln(W_{tc})$, H_{ws}^3, and W_{wbw} gradually decreased. $\ln(WFR)$ had the greatest influence on the north-facing classrooms' $UDI_{100-2000}$, followed by WWR^3 and $\ln(W_{tc})$, and H_w^3 exhibited the least influence.

4. Discussion

4.1. Impact of Interior Space Geometry

H_{tc}, D_{tc}, and W_{tc} are three essential parameters affecting interior space. The correlation analysis showed a significant positive correlation between the H_{tc} and the $sDA_{300,50\%}$ of the terrace classrooms facing each orientation. With the increase in H_{tc}, the increasing range of $sDA_{300,50\%}$ of the classrooms facing each orientation gradually decreased. This is because daylight illuminance increased with H_{tc}, which led to the improvement of overall illuminance levels and an increase in the effective natural light area. When H_{tc} increased to 3.7 m, most interior areas reached the standard of $sDA_{300,50\%}$. After that, the increasing speed of the available lighting area in the terrace classrooms slowed down. Moreover, the results showed that the $UDI_{100-2000}$ of the east-facing, south-facing, and north-facing terrace classrooms had an apparent upward trend when H_{tc} increased gradually within the 2.2–3.1 m range. Later, with the increase in H_{tc}, the $UDI_{100-2000}$ of south-, east-, and north-facing classrooms showed a noticeable decline, a steady trend, and a slight increase, respectively. This is because the direct sunlight received by classrooms in the south, east, and north orientations decreased in turn. Therefore, when H_{tc} was low, with the increase in H_{tc}, the increasing gains in the amount of light entering mainly extended the time period of when illuminance exceeded the lower limit of $UDI_{100-2000}$, thus increasing the cumulative time of available illuminance for natural lighting. However, when H_{tc} continued increasing, the increase in the amount of light entering gradually extended the time period within which the south-facing classroom exceeded the upper limit illuminance of $UDI_{100-2000}$, resulting in a decrease in the cumulative time of available illuminance for natural lighting. In the east-facing terrace classrooms, the time exceeding the lower and upper limits of $UDI_{100-2000}$ illuminance increased with the continuous gain of daylighting, so the cumulative time of available illuminance for natural lighting tended to be stable. The increase in the amount of light entering for the north-facing classroom mainly extended the time period within which the lower limit illuminance of $UDI_{100-2000}$ was exceeded, thus continuing to increase the cumulative time of available illuminance for natural lighting.

With the increase in W_{tc}, the $sDA_{300,50\%}$ of the classrooms facing each orientation showed an upward trend. $UDI_{100-2000}$ in east- and north-facing classrooms continued to increase, while $UDI_{100-2000}$ in west- and south-facing classrooms increased and tended to be stable. During the construction of the simulation model, W_{tc} largely determined the range of W_w. When W_w increased, the increase in incoming daylight would increase the overall illuminance level and the available lighting area. Additionally, with the increase in W_{tc} within 16–20 m, the illuminance time below the lower limit of $UDI_{100-2000}$ gradually decreased, so $UDI_{100-2000}$ showed an upward trend. Later, with the further increase in W_{tc}, the cumulative time of available illuminance for natural lighting in east- and north-facing classrooms continued to increase. However, the amount of direct sunlight received in south- and west-facing classrooms amounts to more than that in east- and north-facing classrooms. With the further increase in W_{tc}, the time period in which the lower limit of $UDI_{100-2000}$ at the far window was exceeded increased continuously. In contrast, the time period in which the upper limit of $UDI_{100-2000}$ at the close window was exceeded increased gradually. Thus, the cumulative time of available illuminance for natural lighting tended to be stable.

Compared with W_{tc}, D_{tc} had a more significant impact on $sDA_{300,50\%}$ and negatively influenced the $sDA_{300,50\%}$ of the terrace classrooms facing each orientation. Voll et al. concluded that when the room was shallower and windows were larger, daylighting was better, which indicated that classroom depth was negatively correlated with $sDA_{300,50\%}$. This is the same as the result obtained in this paper [54]. Susorova et al. chose the ratio of room width to depth as one of the design parameters when studying the daylighting performance and energy consumption level of commercial office buildings. It was observed that classroom width determines the amount of received daylight, and classroom depth determines the distance of daylight penetration. Wide and shallow rooms have better daylighting performance. The increase in classroom depth will decrease the available lighting area, resulting in a significant increase in energy consumption [8]. Although the use function and spatial form of research objects in this paper have differences, the research results are consistent with those of our predecessors. Therefore, attention should be focused upon the proportional relationship between D_{tc} and W_{tc} under different orientations and the choice of H_{tc} when designing the interior space of terrace classrooms.

Different orientations of the terrace classrooms have a great impact on daylighting performance. The results of the correlation analysis showed that in most cases, the interior space and window geometry parameters except D_{tc} were positively correlated with $sDA_{300,50\%}$ and $UDI_{100-2000}$ in east-, west-, and north-facing classrooms. However, due to the large amount of direct daylight received by south-facing terrace classrooms, H_w, H_{tc}, WWR, and WFR were significantly negatively correlated with the $UDI_{100-2000}$ of south-facing classrooms. In addition, due to the difference in the amount of light entering relative to differently oriented terrace classrooms, the south-facing classrooms performed best in $sDA_{300,50\%}$, followed by west-facing, and east- and north-facing classrooms performed worst. Ignacio Acosta et al. found that when the windows' geometries remained unchanged, the $DA_{250\,lx}$ of the south-facing room was three times that of the north-facing room, and the $DA_{250\,lx}$ of the east-facing or west-facing room was nearly twice that of the north-facing room [18]. Due to the differences between the study's site and object, the results of their analysis are different from this paper's numerical values. Nevertheless, the research results of the overall natural illuminance level of classrooms with different orientations in this paper again confirmed the above conclusions.

4.2. Impact of Window Geometry

H_w, W_w, H_{ws}, W_{wbw}, and N_w are the five important parameters that determine the geometry of windows. Among them, W_w and H_w determine the area of the individual windows. H_{ws}, W_{wbw}, and N_w determine the distribution of windows on the exterior wall.

Different H_w and W_w will have a great impact on daylighting performance. As H_w increased, the $sDA_{300,50\%}$ of the terrace classrooms facing each orientation first increased and then stabilized. $UDI_{100-2000}$ in south-facing classrooms first stabilized and then decreased, while the north-facing $UDI_{100-2000}$ and the south-facing $UDI_{100-2000}$ changed in reverse. Direct sunlight is the main source of indoor daylighting [55]. The higher the H_w, the less direct sunlight is blocked, and the higher the natural light illuminance level. When the amount of light entering reaches a certain level, almost all measurement points receive natural light above 300 lx all year round, and the time reaches 50%. At this moment, increasing H_w will not help the natural light illuminance levels much, and it will also increase the energy consumption of heating and cooling. Since the south-facing classroom can receive significantly more direct sunlight than the north-facing classroom, as the amount of light entering increased, the time for exceeding the upper limit illuminance of $UDI_{100-2000}$ in the south-facing classroom increased gradually. In contrast, the time below the lower limit illuminance of $UDI_{100-2000}$ in the north-facing classroom gradually decreased. Therefore, the north-facing classroom and the south-facing classroom changed in reverse. In addition, W_w is only positively correlated with $sDA_{300,50\%}$, and compared with H_w, W_w has less influence on daylighting performances. Deng et al. found that the increase in window height caused $sDA_{150,50\%}$ or $sDA_{450,50\%}$ to increase significantly more than the effect of a window's width on it in the daylighting performance of the reading room [56]. Although their research objects and study areas differed from this paper's, the results were consistent. In order to improve the quality of natural light, H_w and W_w should not be too large. Rupp et al. stated in their paper that while an increase in window area can increase lighting, controlling a certain window area can improve daylight availability [57]. Zheng simulated a total of seven groups of classrooms facing 0–30° southeast and concluded that $UDI_{100-2000}$ decreased with an increase in window height, which is consistent with the local curve trend of south-facing classrooms in this paper [58]. After simulating the influence of north–south facing classroom window height on the natural light environment, Zhang pointed out that for south-facing classrooms, choosing a larger window height size can improve the daylighting quality of the classroom, which is inconsistent with the results of this paper [59];

this is because her research subjects were ordinary classrooms with interior spaces and window geometry that were quite different from those in this study.

The results of the correlation analysis showed that the south- and west-facing terrace classrooms' $UDI_{100-2000}$ are positively correlated with W_{wbw}. Deng et al. confirmed that the enlargement of the width of the wall between windows could improve the illuminance uniformity of the deeper areas of the room while introducing a certain shading effect [56]. In addition, this paper's results showed a significant positive correlation between $UDI_{100-2000}$ and H_{ws} in east-, west-, and north-facing classrooms. With the increase in H_{ws}, the north-facing classrooms' $UDI_{100-2000}$ always reflected an upward trend, and the east- and west-facing classrooms' $UDI_{100-2000}$ reflected a downward and then upward trend. For east- and west-facing classrooms, when H_{ws} was less than 0.75 m of the working plane's height, as H_{ws} increased, direct sunlight near the window increased, and the $UDI_{100-2000}$ in the classroom decreased; when H_{ws} was greater than 0.75 m, as H_{ws} increased, the light reaching the depth of the classroom increased, while direct light near the window decreased, and the $UDI_{100-2000}$ in the classroom increased. Since there was no direct light in the north-facing classroom, the $UDI_{100-2000}$ always increased as H_{ws} increased. Allam et al. mentioned in their analysis of the relationship between windowsill height and lighting energy that when the windowsill's height is higher, the amount of blocked direct sunlight increases. That is, the natural light illuminance level will decrease with the increase in windowsill height, and sDA will decrease [55]. Therefore, in architectural designs, the relationship between the time period between the available illuminance for natural lighting and the illuminance adequacy of natural light should be balanced, and an appropriate H_{ws} should be used.

For WWR and WFR, as WWR or WFR increased, the $sDA_{300,50\%}$ of the terrace classrooms facing each orientation showed a clear upward trend. Both the south-facing and north-facing classrooms' $UDI_{100-2000}$ showed increasing and then decreasing trends, but the decreasing trend of north-facing classrooms was smaller than that of south-facing classrooms. Since the south-facing classroom can receive more direct sunlight than the north-facing classroom, and when the WWR or WFR is small, with the increase in WWR or WFR, the increase in the amount of light entering mainly increases the time period in which the south- and north-facing classrooms exceed the lower limit illuminance of $UDI_{100-2000}$, thereby increasing the cumulative time of available illuminance for natural lighting. However, when WWR and WFR increased to a certain extent, as the amount of light entering continued to grow, the excess direct sunlight in the south-facing classroom led to a significant increase in the time period within which illuminance exceeded the upper limit of $UDI_{100-2000}$; thus, the cumulative time of available illuminance for natural lighting decreased significantly. The intensity of sunlight incidence in the north-facing classroom is relatively low, so the decrease in $UDI_{100-2000}$ is relatively small. Fila et al. pointed out in their study that when WWR reached 10%, the overall illuminance level significantly improved. When WWR increases to 30% or even 50%, the room will contain too much sunlight, which will increase the time period within which the upper limit illuminance of $UDI_{100-2000}$ is exceeded, which is not conducive for the improvement of daylighting performance [60]. The change trend of $UDI_{100-2000}$ obtained from the above studies is basically consistent with that in this paper, but because the research site is located in a low-latitude tropical region and the research object is located in mid-latitude temperate zones, there are great differences in the variation intervals of WWR.

4.3. Limitation

The study site of this paper was in Shenyang, so applying results to regions with different climates or latitudes than Shenyang is difficult. In addition, this paper only analyzes the most common unilateral natural lighting conditions in the terrace classrooms of universities in severe cold regions, and the classroom's size is set to the most common 150~360 people. Moreover, considering the need for the overall effect of the facade in architectural designs, the windows are only evenly arranged horizontally. Therefore, comparing the study's results with studies in different settings is difficult.

Various scales of terrace classrooms and window arrangements in different climate zones will be studied in the future. In addition, glare and reflection problems in the classroom and other indoor physical environment variables will be combined to conduct a comprehensive study to effectively improve the quality of the indoor environment.

5. Conclusions

This paper used Shenyang as an example. The relationship between interior spaces and window geometries and the daylighting performance of university terrace classrooms in severe cold areas was studied by combining numerical simulations and statistical analyses, which provided a quantitative reference and evaluation basis for designing university terrace classrooms. The findings and design recommendations are as follows:

1. WFR, H_{tc}, H_w, WWR, W_{tc}, and W_w positively affected the $sDA_{300,50\%}$ of the terrace classrooms facing each orientation, and the degree of effect decreased sequentially. D_{tc} has an inhibitory effect on $sDA_{300,50\%}$ of the terrace classrooms facing each orientation, and the effects on the east-facing and north-facing classrooms are more obvious. In addition, when the classrooms' geometry is the same, the south-facing classrooms have the highest daylight sufficiency, followed by west-facing classrooms, and the east- and north-facing classrooms are relatively low. For $UDI_{100-2000}$, the geometry parameters that significantly impact classrooms facing different orientations vary greatly. H_w, H_{tc}, WWR, and WFR showed significant negative and positive correlations with $UDI_{100-2000}$ in south-facing and north-facing classrooms, respectively.

2. For east- and north-facing terrace classrooms, enlarging W_{tc} and H_{tc} can effectively improve the daylighting performance, so it can be set to the maximum. In addition, when W_{tc} and H_{tc} are greater than 22 m and 3.4 m, respectively, and D_{tc} is less than 14 m, the sufficiency of the ambient daylight available is preferred ($sDA_{300,50\%} \geq 0.75$). For south-facing terrace classrooms, W_{tc} can be maximized to enhance the overall natural lighting. When H_{tc} is greater than 3.1 m, the daylight sufficiency increases with the increase in H_{tc}, but the cumulative time of available illuminance for natural lighting decreases. In addition, when W_{tc} and H_{tc} are greater than 17 m and 3.1 m, respectively, and D_{tc} is less than 15 m, daylight sufficiency is preferred. For west-facing terrace classrooms, W_{tc} can be maximized. In addition, when W_{tc} and H_{tc} are greater than 19 m and 3.4 m, respectively, and D_{tc} is less than 14 m, daylight sufficiency is preferred. Therefore, when designing the interior space geometry of the terrace classroom, attention should be focused upon grasping the appropriate size relationship between D_{tc}, W_{tc}, and H_{tc} in different orientations.

3. The addition of H_w helps improve daylight sufficiency in the terrace classrooms facing each orientation. When the H_w of south- and west-facing classrooms is greater than or equal to 2 m and the H_w of east- and north-facing classrooms is greater than or equal to 2.4 m, daylight sufficiency is preferred. In addition, for north-facing classrooms, adding H_w can increase the cumulative time of the available illuminance for natural lighting, so H_w can be maximized to enhance the overall daylighting performance. For south-facing classrooms, when H_w is greater than 1.8 m, $UDI_{100-2000}$ decreases with the increase in H_w, which is not conducive to the improvement of the overall daylighting performance. Compared with H_w, W_w introduces fewer effects on daylighting quality in terrace classrooms. When the east-facing and north-facing classrooms' W_w is greater than or equal to 4.1 m and the west-facing W_w is greater than or equal to 1.7 m, the sufficiency of the available ambient daylight is preferred. For south-facing and west-facing terrace classrooms, appropriately increasing W_{wbw} is conducive to improving the cumulative time of the available illuminance for natural lighting, and it is best when W_{wbw} is 0.9 m. There is a significant positive correlation between H_{ws} and $UDI_{100-2000}$ in east-, west-, and north-facing terrace classrooms. For east- and west-facing classrooms, H_{ws} should be larger than the indoor desktop height.

With the increase in WWR and WFR, the $sDA_{300,50\%}$ of the terrace classrooms facing each orientation always reflected an upward trend, and south-facing and north-facing terrace classrooms' $UDI_{100-2000}$ reflected a trend that first increased and then decreased. To improve the overall daylighting performance, the WWR and WFR of the south-facing classroom should be 0.3–0.5 and 0.11–0.14, respectively, and the WWR and WFR of the north-facing classroom should be 0.6–0.7 and 0.14–0.20, respectively.

4. Taking the interior space and window geometry parameters of the terrace classrooms as the independent variables, the prediction models of the $sDA_{300,50\%}$ and $UDI_{100-2000}$ of the terrace classrooms facing each orientation were constructed, and the model's fit was high and had significant statistical significance. The prediction model can provide an accurate quantitative evaluation of the daylighting performance of the terrace classrooms of universities in severe cold regions.

Author Contributions: Z.L. and Y.J. contributed to the article equally and should be regarded as co-first authors. Z.L. conceived the paper and designed the numerical study; Y.J., Y.F. and H.Z. performed the numerical simulation; Y.J., Y.F., H.Z., C.Z. and X.C. analyzed the data and created the figures.; Y.J., Y.F. and H.Z. drafted the paper; Z.L. revised the paper. All authors have read and agreed to the published version of the manuscript.

Funding: This work was financially supported by the National Training Program of Innovation and Entrepreneurship for Undergraduates (Grant Number: 220244), the Fundamental Research Program of the Education Department of Liaoning Province (Grant Number: LJKQZ2021006), the Liaoning Social Science Planning Fund Project (Grant Number: L22CGL012), the Fundamental Research Funds for the Central Universities (Grant Number: N2111001), and the National Training Program of Innovation and Entrepreneurship for Undergraduates (Grant Number: 230244).

Data Availability Statement: The data presented in this study are available on request from the authors.

Conflicts of Interest: The authors declare no conflict of interest.

References

1. Haddad, S.; Synnefa, A.; Marcos, M.Á.P.; Paolini, R.; Delrue, S.; Prasad, D.; Santamouris, M. On the potential of demand-controlled ventilation system to enhance indoor air quality and thermal condition in Australian school classrooms. *Energy Build.* **2021**, *238*, 110838. [CrossRef]
2. Heschong, L.; Manglani, P.; Wright, R.; Peet, R.; Howlett, O. Windows and Classrooms: Student Performance and the Indoor Environment. In Proceedings of the ACEEE Summer Study, Pacific Grove, CA, USA, 22–27 August 2004.
3. Roche, L.; Dewey, E.; Littlefair, P. Occupant reactions to daylight in offices. *Light. Res. Technol.* **2000**, *32*, 119–126. [CrossRef]
4. Galasiu, A.D.; Veitch, J.A. Occupant preferences and satisfaction with the luminous environment and control systems in daylit offices: A literature review. *Energy Build.* **2006**, *38*, 728–742. [CrossRef]
5. Cheung, H.; Chung, T. A study on subjective preference to daylit residential indoor environment using conjoint analysis. *Build. Environ.* **2008**, *43*, 2101–2111. [CrossRef]
6. Lim, Y.-W.; Kandar, M.Z.; Ahmad, M.H.; Ossen, D.R.; Abdullah, A.M. Building façade design for daylighting quality in typical government office building. *Build. Environ.* **2012**, *57*, 194–204. [CrossRef]
7. Ghisi, E.; Tinker, J.A. An Ideal Window Area concept for energy efficient integration of daylight and artificial light in buildings. *Build. Environ.* **2005**, *40*, 51–61. [CrossRef]
8. Susorova, I.; Tabibzadeh, M.; Rahman, A.; Clack, H.L.; Elnimeiri, M. The effect of geometry factors on fenestration energy performance and energy savings in office buildings. *Energy Build.* **2013**, *57*, 6–13. [CrossRef]
9. ALee, J.W.; Jung, H.J.; Park, J.Y.; Lee, J.B.; Yoon, Y. Optimization of building window system in Asian regions by analyzing solar heat gain and daylighting elements. *Renew. Energy* **2013**, *50*, 522–531.
10. Omar, O.; García-Fernández, B.; Fernández-Balbuena, A.; Vázquez-Moliní, D. Optimization of daylight utilization in energy saving application on the library in faculty of architecture, design and built environment, Beirut Arab University. *Alex. Eng. J.* **2018**, *57*, 3921–3930. [CrossRef]
11. Fan, Z.; Liu, M.; Tang, S. A multi-objective optimization design method for gymnasium facade shading ratio integrating energy load and daylight comfort. *Build. Environ.* **2022**, *207*, 108527. [CrossRef]
12. Huang, X.; Wei, S.; Zhu, S. Study on Daylighting Optimization in the Exhibition Halls of Museums for Chinese Calligraphy and Painting Works. *Energies* **2020**, *13*, 240. [CrossRef]

13. Li, D.H.W.; Lo, S.M.; Lam, J.C.; Yuen, R.K.K. Daylighting Performance in Residential Buildings. *Arch. Sci. Rev.* **1999**, *42*, 213–219. [CrossRef]

14. Sun, C.Y.Y.; Han, Y.S. Parametric Daylight Performance Simulation Research of Mezzanine Space in Office Buildings in Severe Cold Regions. *J. Hum. Settl. West China* **2021**, *36*, 24–30.

15. Krüger, E.L.; Fonseca, S.D. Evaluating daylighting potential and energy efficiency in a classroom building. *J. Renew. Sustain. Energy* **2011**, *3*, 063112. [CrossRef]

16. Al-Dossary, A.M.; Kim, D.D. A Study of Design Variables in Daylight and Energy Performance in Residential Buildings under Hot Climates. *Energies* **2020**, *13*, 5836. [CrossRef]

17. Dabe, T.J.; Adane, V.S. The impact of building profiles on the performance of daylight and indoor temperatures in low-rise residential building for the hot and dry climatic zones. *Build. Environ.* **2018**, *140*, 173–183. [CrossRef]

18. Acosta, I.; Campano, M.; Molina, J.F. Window design in architecture: Analysis of energy savings for lighting and visual comfort in residential spaces. *Appl. Energy* **2016**, *168*, 493–506. [CrossRef]

19. Michael, A.; Heracleous, C. Assessment of natural lighting performance and visual comfort of educational architecture in Southern Europe: The case of typical educational school premises in Cyprus. *Energy Build.* **2017**, *140*, 443–457. [CrossRef]

20. Junaibi, A.A.A.; Zaabi, E.J.A.; Nassif, R.; Mushtaha, E. Daylighting in Educational Buildings: Its Effects on Students and How to Maximize Its Performance in the Architectural Engineering Department of the University of Sharjah. In Proceedings of the 3rd International Sustainable Buildings Symposium (ISBS 2017), Dubai, United Arab Emirates, 15–17 March 2017.

21. Peters, T.; D'Penna, K. Biophilic Design for Restorative University Learning Environments: A Critical Review of Literature and Design Recommendations. *Sustainability* **2020**, *12*, 7064. [CrossRef]

22. Ashrafian, T.; Moazzen, N. The impact of glazing ratio and window configuration on occupants' comfort and energy demand: The case study of a school building in Eskisehir, Turkey. *Sustain. Cities Soc.* **2019**, *47*, 101483. [CrossRef]

23. Goia, F.; Haase, M.; Perino, M. Optimizing the configuration of a facade module for office buildings by means of integrated thermal and lighting simulations in a total energy perspective. *Appl. Energy* **2013**, *108*, 515–527. [CrossRef]

24. Wong, N.H.; Istiadji, A.D. Effect of external shading devices on daylighting penetration in residential buildings. *Light. Res. Technol.* **2004**, *36*, 317–333. [CrossRef]

25. Pellegrino, A.; Cammarano, S.; Savio, V. Daylighting for Green Schools: A Resource for Indoor Quality and Energy Efficiency in Educational Environments. *Energy Procedia* **2015**, *78*, 3162–3167. [CrossRef]

26. de Rubeis, T.; Nardi, I.; Muttillo, M.; Ranieri, S.; Ambrosini, D. Room and window geometry influence for daylight harvesting maximization—Effects on energy savings in an academic classroom. *Energy Proc.* **2018**, *148*, 1090–1097. [CrossRef]

27. Mangkuto, R.A.; Rohmah, M.; Asri, A.D. Design optimisation for window size, orientation, and wall reflectance with regard to various daylight metrics and lighting energy demand: A case study of buildings in the tropics. *Appl. Energy* **2016**, *164*, 211–219. [CrossRef]

28. Galal, K.S. The impact of classroom orientation on daylight and heat-gain performance in the Lebanese Coastal zone. *Alex. Eng. J.* **2019**, *58*, 827–839. [CrossRef]

29. Lakhdari, K.; Sriti, L.; Painter, B. Parametric optimization of daylight, thermal and energy performance of middle school classrooms, case of hot and dry regions. *Build. Environ.* **2021**, *204*, 108173. [CrossRef]

30. Ekasiwi, S.N.N.; Antaryama, I.G.N.; Krisdianto, J.; Ulum, M.S. Correlation of classroom typologies to lighting energy performance of academic building in warm-humid climate (case study: ITS Campus Sukolilo Surabaya). *IOP Conf. Series Earth Environ. Sci.* **2018**, *126*, 012049. [CrossRef]

31. Freewan, A.A.; Al Dalala, J.A. Assessment of daylight performance of Advanced Daylighting Strategies in Large University Classrooms; Case Study Classrooms at JUST. *Alex. Eng. J.* **2020**, *59*, 791–802. [CrossRef]

32. Milone, D.; Pitruzzella, S.; Franzitta, V.; Viola, A.; Trapanese, M. Energy savings through integration of the illuminance natural and artificial, using a system of automatic dimming: Case study. In Proceedings of the 2nd International Conference on Advanced Materials Design and Mechanics (ICAMDM 2013), Kuala Lumpur, Malaysia, 17–18 May 2013; p. 253.

33. Davidsson, H.; Perers, B.; Karlsson, B. Performance of a multifunctional PV/T hybrid solar window. *Sol. Energy* **2010**, *84*, 365–372. [CrossRef]

34. *GB50033-2013*; Standard for Daylighting Design of Buildings. China Architecture & Building Press: Beijing, China, 2013. (In Chinese)

35. Weather China. Available online: http://www.weather.com.cn/forecast/history.shtml?areaid=101070101&month=11 (accessed on 25 November 2022).

36. Mohsenin, M.; Hu, J. Assessing daylight performance in atrium buildings by using Climate Based Daylight Modeling. *Sol. Energy* **2015**, *119*, 553–560. [CrossRef]

37. ElBatran, R.M.; Ismaeel, W.S. Applying a parametric design approach for optimizing daylighting and visual comfort in office buildings. *Ain Shams Eng. J.* **2021**, *12*, 3275–3284. [CrossRef]

38. *IES LM-83-12*; Approved Method: IES Spatial Daylight Autonomy (SDA) and Annual Sunlight Exposure (ASE). Illuminating Engineering Society (IES): New York, NY, USA, 2012.

39. Wang, M. *Research on Top Daylighting Design of Classrooms in Primary and Secondary Schools in Guangdong Based on Visual Health*; South China University of Technology: Guangzhou, China, 2018. (In Chinese)

40. Nabil, A.; Mardaljevic, J. Useful daylight illuminances: A replacement for daylight factors. *Energy Build.* **2006**, *38*, 905–913. [CrossRef]
41. Rastegari, M.; Pournaseri, S.; Sanaieian, H. Daylight optimization through architectural aspects in an office building atrium in Tehran. *J. Build. Eng.* **2021**, *33*, 101718. [CrossRef]
42. Xue, Y.; Liu, W. A Study on Parametric Design Method for Optimization of Daylight in Commercial Building's Atrium in Cold Regions. *Sustainability* **2022**, *14*, 7667. [CrossRef]
43. Kharvari, F. An empirical validation of daylighting tools: Assessing radiance parameters and simulation settings in Ladybug and Honeybee against field measurements. *Sol. Energy* **2020**, *207*, 1021–1036. [CrossRef]
44. Ma, Q.S.; Fukuda, H. Parametric office building for daylight and energy analysis in the early design stages. In Proceedings of the Conference on Urban Planning and Architectural Design for Sustainable Development (UPADSD), Lecce, Italy, 14–16 October 2015; pp. 818–828.
45. Energyplus, Weather Data. Available online: https://energyplus.net/weather (accessed on 17 December 2022).
46. *Architectural Design Data Set*; China Architecture & Building Press: Beijing, China, 2019.
47. *GB 50099-2011*; Code for Design of School. China Architecture & Building Press: Beijing, China, 2011. (In Chinese)
48. Xu, Z.; Wang, T.; Li, C.; Bao, L.; Ma, Q.; Miao, Y. Brief Introduction to the Orthogonal Test Design. *J. Libr. Inf. Sci.* **2002**, *12*, 148–150. (In Chinese)
49. *GB 50352-2019*; Uniform Standard for Design of Civil Buildings. China Architecture & Building Press: Beijing, China, 2019. (In Chinese)
50. Liu, R.; Zhang, Y.; Wen, C.; Tang, J. Study on the design and analysis methods of orthogonal experiment. *Exp. Technol. Manag.* **2010**, *27*, 52–59. (In Chinese) [CrossRef]
51. *GB/T 5699-2017*; Method of Daylighting Measurements. Standardization Administration: Beijing, China, 2017. (In Chinese)
52. Leng, H.; Chen, X.; Ma, Y.; Wong, N.H.; Ming, T. Urban morphology and building heating energy consumption: Evidence from Harbin, a severe cold region city. *Energy Build.* **2020**, *224*, 110143. [CrossRef]
53. Nault, E.; Moonen, P.; Rey, E.; Andersen, M. Predictive models for assessing the passive solar and daylight potential of neighborhood designs: A comparative proof-of-concept study. *Build. Environ.* **2017**, *116*, 11–16. [CrossRef]
54. Voll, H.; Seinre, E. A method of optimizing fenestration design for daylighting to reduce heating and cooling loads in offices. *J. Civ. Eng. Manag.* **2014**, *20*, 714–723. [CrossRef]
55. Allam, A.S.; Bassioni, H.A.; Kamel, W.; Ayoub, M. Estimating the standardized regression coefficients of design variables in daylighting and energy performance of buildings in the face of multicollinearity. *Sol. Energy* **2020**, *211*, 1184–1193. [CrossRef]
56. Deng, X.; Wang, M.; Fan, Z.; Liu, J. Dynamic daylight performance oriented design optimizations for contemporary reading room represented deep open-plan spaces. *J. Build. Eng.* **2022**, *62*, 105145. [CrossRef]
57. Rupp, R.F.; Ghisi, E. Assessing window area and potential for electricity savings by using daylighting and hybrid ventilation in office buildings in southern Brazil. *Simulation* **2017**, *93*, 935–949. [CrossRef]
58. Zheng, C. *Study on Parametric Design of Natural Lighting Performance of University Building in Hubei Province*; Yangtze University: Jingzhou, China, 2021. [CrossRef]
59. Zhang, Q. *Research on Daylighting Design of Teaching Buildings in Hanzhong Area*; Xi'an University of Architecture and Technology: Xi'an, China, 2021. [CrossRef]
60. Fela, R.F.; Utami, S.S.; Mangkuto, R.A.; Suroso, D.J. The Effects of Orientation, Window Size, and Lighting Control to Climate-Based Daylight Performance and Lighting Energy Demand on Buildings in Tropical Area. In Proceedings of the International Building Performance Simulation Association International Conference and Exbition, Rome, Italy, 2–4 September 2019. [CrossRef]

MDPI

St. Alban-Anlage 66

4052 Basel

Switzerland

Tel. +41 61 683 77 34

Fax +41 61 302 89 18

www.mdpi.com

Buildings Editorial Office

E-mail: buildings@mdpi.com

www.mdpi.com/journal/buildings